OXFORD LOGIC GUIDES

Series Editors

A.J. MACINTYRE D. SCOTT

Emeritus Editors

D. M. Gabbay
John Shepherdson

OXFORD LOGIC GUIDES

For a full list of titles please visit
http://www.oup.co.uk/academic/science/maths/series/OLG/

21. C. McLarty: *Elementary categories, elementary toposes*
22. R.M. Smullyan: *Recursion theory for metamathematics*
23. Peter Clote and Jan Krajícek: *Arithmetic, proof theory, and computational complexity*
24. A. Tarski: *Introduction to logic and to the methodology of deductive sciences*
25. G. Malinowski: *Many valued logics*
26. Alexandre Borovik and Ali Nesin: *Groups of finite Morley rank*
27. R.M. Smullyan: *Diagonalization and self-reference*
28. Dov M. Gabbay, Ian Hodkinson, and Mark Reynolds: *Temporal logic: Mathematical foundations and computational aspects, volume 1*
29. Saharon Shelah: *Cardinal arithmetic*
30. Erik Sandewall: *Features and fluents: volume I: A systematic approach to the representation of knowledge about dynamical systems*
31. T.E. Forster: *Set theory with a universal set: Exploring an untyped universe, second edition*
32. Anand Pillay: *Geometric stability theory*
33. Dov M. Gabbay: *Labelled deductive systems*
35. Alexander Chagrov and Michael Zakharyaschev: *Modal logic*
36. G. Sambin and J. Smith: *Twenty-five years of Martin-Löf constructive type theory*
37. María Manzano: *Model theory*
38. Dov M. Gabbay: *Fibring logics*
39. Michael Dummett: *Elements of intuitionism, second edition*
40. D.M. Gabbay, M.A. Reynolds and M. Finger: *Temporal logic: Mathematical foundations and computational aspects, volume 2*
41. J.M. Dunn and G. Hardegree: *Algebraic methods in philosophical logic*
42. H. Rott: *Change, choice and inference: A study of belief revision and nonmonotoic reasoning*
43. Johnstone: *Sketches of an elephant: A topos theory compendium, volume 1*
44. Johnstone: *Sketches of an elephant: A topos theory compendium, volume 2*
45. David J. Pym and Eike Ritter: *Reductive Logic and Proof Search: Proof theory, semantics and control*
46. D.M. Gabbay and L. Maksimova: *Interpolation and definability: Modal and intuitionistic logics*
47. John L. Bell: *Set Theory: Boolean-valued models and independence proofs, third edition*
48. Laura Crosilla and Peter Schuster: *From sets and types to topology and analysis: Towards practicable foundations for constructive mathematics*
49. Steve Awodey: *Category theory*
50. Roman Kossak and James Schmerl: *The structure of models of Peano arithmetic*
51. André Nies: *Computability and Randomness*
52. Steve Awodey: *Category theory, second edition*

Computability and Randomness

André Nies
The University of Auckland

OXFORD
UNIVERSITY PRESS

Great Clarendon Street, Oxford OX2 6DP

Oxford University Press is a department of the University of Oxford. It furthers the University's objective of excellence in research, scholarship, and education by publishing worldwide in

Oxford New York

Auckland Cape Town Dar es Salaam Hong Kong Karachi Kuala Lumpur Madrid Melbourne Mexico City Nairobi New Delhi Shanghai Taipei Toronto

With offices in

Argentina Austria Brazil Chile Czech Republic France Greece Guatemala Hungary Italy Japan Poland Portugal Singapore South Korea Switzerland Thailand Turkey Ukraine Vietnam

Oxford is a registered trade mark of Oxford University Press in the UK and in certain other countries

Published in the United States by Oxford University Press Inc., New York

First Published 2009
First published in paperback 2012

British Library Cataloguing in Publication Data
Data available

Library of Congress Cataloging in Publication Data
Data available

Typeset by Newgen Imaging Systems (P) Ltd., Chennai, India

ISBN 978–0–19–923076–1 (Hbk.)
ISBN 978–0–19–965260–0 (Pbk.)

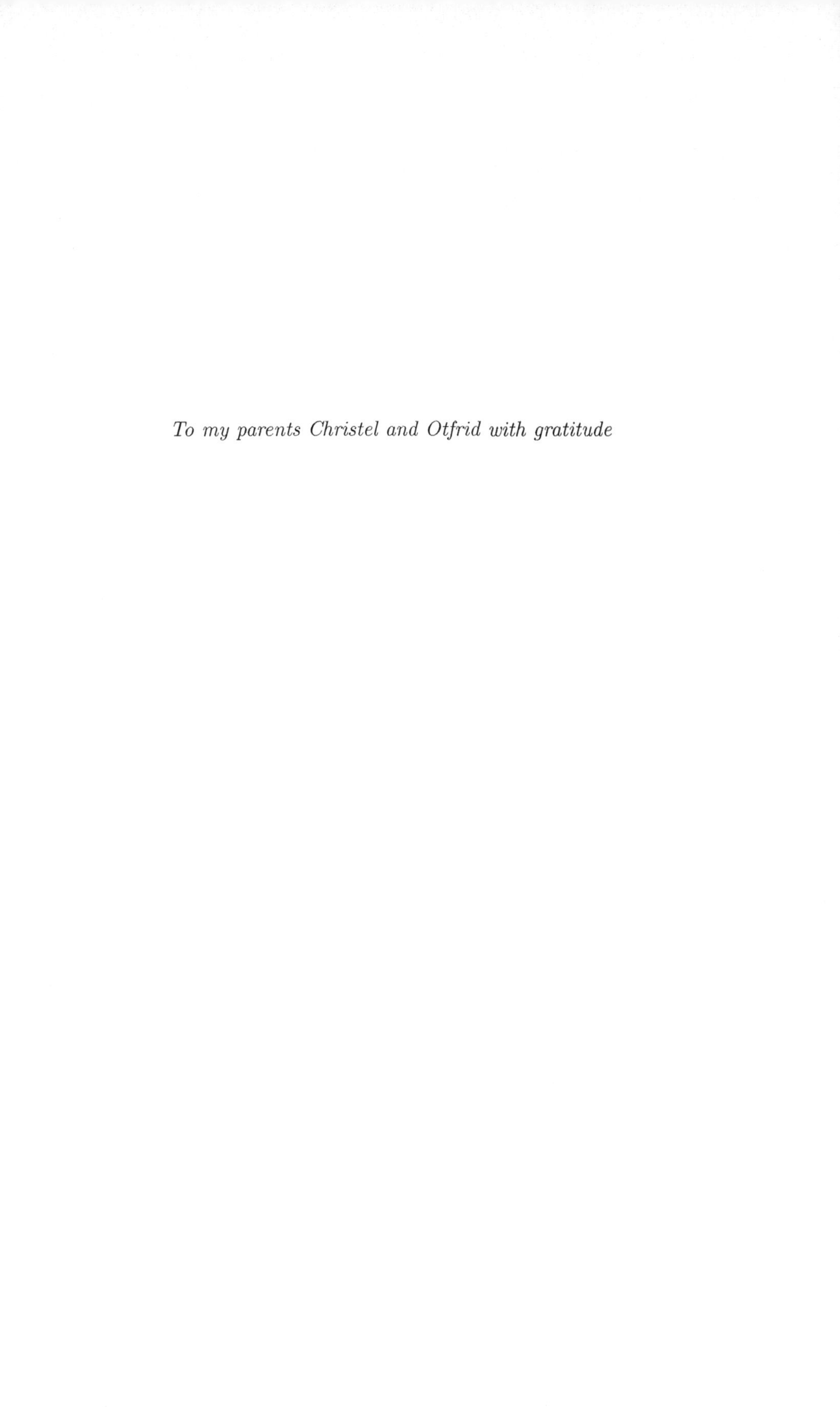

To my parents Christel and Otfrid with gratitude

PREFACE

The complexity and randomness aspects of sets of natural numbers are closely related. Traditionally, computability theory is concerned with the complexity aspect. However, computability theoretic tools can also be used to introduce mathematical counterparts for the intuitive notion of randomness of a set. Recent research shows that, conversely, concepts and methods originating from randomness enrich computability theory.

This book is about these two aspects of sets of natural numbers and about their interplay. Sets of natural numbers are identified with infinite sequences of zeros and ones, and simply called sets.

Chapters 1 and 6 are mostly about the complexity aspect. We introduce lowness and highness properties of sets.

Chapters 2, 3, and 7 are mostly about the randomness aspect. Firstly we study randomness of finite objects. Then we proceed to sets. We establish a hierarchy of mathematical randomness notions. Each notion matches our intuition of randomness to some extent.

In Chapters 4, 5, and 8 we mainly study the interplay of the computability and randomness aspects. Section 6.3 also touches upon this interplay. Chapter 9 looks at analogs of results from the preceding chapters in higher computability theory.

In the area or research connecting complexity and randomness, several times, properties of sets were studied independently for a while, only to be shown to coincide later. Some important results in this book show such coincidences. Other results separate properties that are conceptually close. Even if properties introduced in different ways coincide, we still think of them as conceptually distinct.

This book can be used in various ways: (1) as a reference by researchers; (2) for self-study by students; and (3) in courses at the graduate level.

Such a course can lean towards computability (Chapter 1, some of Chapters 4 and 6), randomness (Chapters 2, 3, 7, and 1 to the extent needed), or the interplay between the two (Chapters 4, 5, 8, and as much as needed from other chapters).

Figure 1 displays major and minor dependencies between chapters. The latter are given by dashed lines; the labels indicate the section which depends on the preceding chapter.

The book contains many exercises and a number of problems. Often the exercises extend the material given in the main text in interesting ways. They should be attempted seriously by the student before looking at the solutions at the back of the book. The problems are currently open, possibly only because no one has tried.

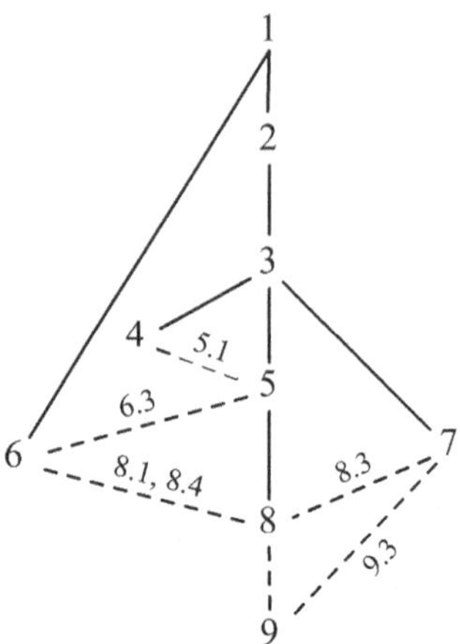

FIG. 1. Major and minor dependencies between chapters.

Notation is listed from page 416 on for reference. The absolute value of a number $r \in \mathbb{R}$ is denoted by abs(r). The cardinality of a set X is denoted by $\#X$. We use the bracket notation in sums as explained in Knuth (1992). For instance, $\sum_n n^{-2} [\![n \text{ is odd}]\!]$ denotes $1 + 1/9 + 1/25 + \ldots = \pi^2/8$.

The following conventions on variables apply.

n, m, k, l	natural numbers
x, y, z, v, w	binary strings (often identified with numbers)
σ, ρ, τ	binary strings when seen as descriptions or oracle strings
$A, \ldots E, V, \ldots, Z$	subsets of $\mathbb{N}$
f, g, h	functions $\mathbb{N} \to \mathbb{N}$
$\mathcal{A}, \mathcal{B}, \ldots$	classes.

This book would not exist without the help of my colleagues and friends. Special thanks to Santiago Figueira, Noam Greenberg, Bjørn Kjos-Hanssen, Antonín Kučera, Antonio Montalbán, Joseph Miller, Selwyn Ng, Alex Raichev, and Jan Reimann. Substantial help was also provided by George Barmpalias, David Belanger, Laurent Bienvenue, Helen Broome, Peter Cholak, Barbara Csima, David Diamondstone, Nick Hay, Greg Hjorth, Bart Kastermans, Steffen Lempp, Ken Harris, Chris Porter, Richard Shore, Stephen Simpson, Sebastiaan Terwijn, Paul Vitanyi, and Liang Yu. I am grateful to the University of Auckland, and especially to the department of computer science. I gratefully acknowledge support by the Marsden fund of the Royal Society of New Zealand. I thank Oxford University Press, and in particular Dewi Jackson, for their support and patience.

The book originally published in February 2009. In the current version, some minor errors have been corrected, and the references and status of open problems have been updated.

Auckland, August 2011

CONTENTS

1

THE COMPLEXITY OF SETS

We study the complexity of sets of natural numbers. There are two interrelated types of complexity.

Computational. Informally, we ask how much (or little) the set knows.
Descriptive. We ask how well the set can be described.

In both cases, to understand the complexity of sets, we introduce *classes of similar complexity*, namely classes of sets sharing a certain complexity property. For both types of complexity, the smallest class we will consider is the class of computable sets.

Classes of computational complexity. To give a mathematical definition for the intuitive notion of a computable function $f\colon \mathbb{N} \to \mathbb{N}$, a formal model of computation is used, for instance Turing machines. The model can be extended to allow queries to an "oracle set" Z during the computations, thereby defining what it means for a function to be computable with oracle Z. In most cases, a class of computational complexity is given by a condition indicating the strength of Z as an oracle. For instance, Z is *low* if deciding whether a computation using Z converges is no harder than deciding whether a computation without an oracle converges. Z is *computably dominated* if each function computed by Z is bounded by a computable function, and Z is *high* if some function computed by Z grows faster than each computable function. These classes will be studied in Section 1.5.

Classes of descriptive complexity. One introduces *description systems.* The descriptions are finite objects, such as first-order formulas or Turing programs, which can be encoded by natural numbers in an effective way. Formally, a description system is simply a function $F\colon I \to \mathcal{P}(\mathbb{N})$ where $I \subseteq \mathbb{N}$. If $F(e) = Z$, then e is a description of Z in that system, and the class of descriptive complexity given by F is the range of F. Examples include the computably enumerable sets, where $F(e) = W_e$ is the set of inputs on which the e-th Turing program halts, the arithmetical sets, and the Π^1_1 sets, a high-level analog of the c.e. sets where the enumeration takes place at stages that are computable ordinals. (C.e. sets are introduced in Definition 1.1.8, arithmetical sets in 1.4.10, and Π^1_1 sets in 9.1.1.) Since descriptions can be encoded by natural numbers, all classes of descriptive complexity are countable. (In other areas of logic it can also be the case that descriptions are infinite, though they should be simpler than the objects they describe. In descriptive set theory, say, certain functions from $\mathbb{N}$ to $\mathbb{N}$, called Borel codes, describe Borel sets of real numbers. In model theory, for some complete first-order theories T, sequences of ordinals can describe countable models of T up to isomorphism.)

The classes of descriptive complexity we consider usually form an almost linear hierarchy given by the inclusion of classes. (The class of Π^0_2 singletons, Definition 1.8.61, is one of the few exceptions to this rule.) This contrasts with the case of computational complexity. For instance, the only sets that are low and computably dominated are the computable sets. Another difference is that being in a class of descriptive complexity means the set is well-behaved in a particular sense. In contrast, classes of computational complexity will be of two types: the ones consisting of sets that know little, and the ones consisting of sets that know a lot. Classes of the first type are given by lowness properties, such as being low, or computably dominated, while classes of the second type are given by highness properties, such as being high.

The counterpart of knowing a lot in descriptive complexity might be being hard to describe. We will see that this is one aspect of the intuitive notion of randomness for sets. The other, related, aspect is not satisfying any exceptional properties (in the sense of the uniform measure on Cantor space $2^{\mathbb{N}}$). In Chapters 3, 7 and 9 we will introduce various classes capturing the degree of randomness of a set. A central one is the class of Martin-Löf random sets.

So far we have only discussed the *absolute* complexity of a set Z, by looking at its membership in certain classes. The *relative computational complexity* is measured by comparing Z to other sets. To do so, one introduces preorderings $\leq_r$ on sets, called *reducibilities*: $X \leq_r Y$ means that X is no more complex than Y in the sense of $\leq_r$. Traditionally, a reducibility specifies a way to determine whether $n \in X$ with the help of queries of the form "$k \in Y$?" Such a method to compute X from Y is called a reduction procedure. We also study weak reducibilities, which can be used to compare the computational complexity of sets even if there is no reduction procedure; see Section 5.6 and page 339.

There are a few examples of preorderings $\leq_r$ used to compare the relative descriptive complexity of sets, such as enumeration reducibility (Odifreddi, 1999, Ch. XIV) and the reducibility $\leq_K$ (5.6.1). Furthermore, some preorderings $\leq_r$ have been introduced where $A \leq_r B$ expresses in some way that B is at least as random as A, for instance $\leq_K$ again, $\leq_S$ on the left-c.e. sets (3.2.28), and $\leq_{vL}$ on the Martin-Löf random sets (5.6.2). No general theory has emerged so far.

One of the aims of this chapter is to give a brief, but self-contained introduction to computability theory, focussing on the material that is needed later on. Topics left out here can often be found in Soare (1987) or Odifreddi (1989, 1999).

We will rely on important meta-concepts: uniformity, relativization, and universality. Uniformity is discussed in Remark 1.1.4, and relativization before Proposition 1.2.8 on page 10.

We will return to the complexity of sets in Sections 1.2, 1.4, and 1.5 of this chapter, as well as in Chapters 5, 8, and 9.

1.1 The basic concepts

We review fundamental concepts of computability theory: partial computable functions, computably enumerable sets, and computable sets. We provide an important tool, the Recursion Theorem.

Partial computable functions

One main achievement of mathematical logic is a formal definition for the intuitive concept of a computable function. We mostly consider functions $f\colon \mathbb{N}^k \mapsto \mathbb{N}$ (where $k \geq 1$). This is an inessential restriction since other finite objects that could be considered as inputs, say finite graphs, can be encoded by natural numbers in some efficient way. One obtains a mathematical definition of computable functions by introducing Turing machines (Turing, 1936). Such a machine has k tapes holding the inputs (say in binary), one output tape, and several internal work tapes. A Turing machine reads and writes symbols from a finite alphabet (which includes the symbols 0 and 1) on these tapes. The input tapes are read-only, while the output tape is write-only. The behavior of the machine is described by a finite sequence of instructions, called a *Turing program*, which is carried out in a step-wise fashion one instruction at a time. See Odifreddi (1989) for details. The function f is computable if there is a Turing program P for the machine model with k input tapes which, for all inputs $x_0, \ldots, x_{k-1}$, halts with $f(x_0, \ldots, x_{k-1})$ on the output tape.

Of course, for certain inputs, a Turing program may run forever. There is no algorithm to decide whether a program halts even on a single input, let alone on all. Thus it is natural to include partial functions in our mathematical definition of computability.

1.1.1 Definition. Let ψ be a function with domain a subset of $\mathbb{N}^k$ and range a subset of $\mathbb{N}$. We say that ψ is *partial computable* if there is a Turing program P with k input tapes such that $\psi(x_0, \ldots, x_{k-1}) = y$ iff P on inputs $x_0, \ldots, x_{k-1}$ outputs y. We write $\psi(x_0, \ldots, x_{k-1})\downarrow$ if P halts on inputs $x_0, \ldots, x_{k-1}$. We say that ψ is *computable* if ψ is partial computable and the domain of ψ is $\mathbb{N}^k$.

Many other formal definitions for the intuitive notion of a computable function were proposed. All turned out to be equivalent. This lends evidence to the Church–Turing thesis which states that any intuitively computable function is computable in the sense of Definition 1.1.1. More generally, each informally given algorithmic procedure can be implemented by a Turing program. We freely use this thesis in our proofs: we give a procedure informally and then take it for granted that a Turing program implementing it exists.

Fix k and an effective listing of the Turing programs for k inputs. Let P^k_e be the program for k inputs given by the e-th program. Let Φ^k_e denote the partial computable function with k arguments given by P^k_e. If $\Phi = \Phi^k_e$ then e is called an *index* for Φ. Often there is only one argument, and instead of Φ^1_e we write

$$\Phi_e. \tag{1.1}$$

The following notation is frequently used when dealing with partial functions. Given expressions α, β,

$$\alpha \simeq \beta$$

means that either both expressions are undefined, or they are defined with the same value. For instance, $\sqrt{(r+1)(r-1)} \simeq \sqrt{r^2-1}$ for $r \in \mathbb{R}$.

A universal Turing program. The function $\Xi(e, x) \simeq \Phi_e(x)$ is partial computable in the intuitive sense. The informal procedure is: on inputs e, x, fetch the e-th Turing program P_e^1 and run it on input x. If its computation halts with output y then give y as an output. By the Church–Turing thesis, Ξ is partial computable. A Turing program computing Ξ is called a *universal Turing program*. The following theorem states that Ξ can emulate partial computable functions in two arguments whenever the Turing programs are listed in an appropriate effective way. For details see Odifreddi (1989).

1.1.2 Theorem. (Parameter Theorem) *For each partial computable function Θ in two variables there is a computable strictly increasing function q such that*

$$\forall e \, \forall x \, \Phi_{q(e)}(x) \simeq \Theta(e, x).$$

An index for q can be obtained effectively from an index for Θ.

Proof idea: Given a Turing program P for Θ, we obtain the program $P_{q(e)}^1$ by making the first input e part of the program code. □

For a formal proof, one would need to be more specific about the effective listing of Turing programs. The same applies to the next result.

1.1.3 Lemma. (Padding Lemma) *For each e and each m, one may effectively obtain $e' > m$ such that the Turing program $P_{e'}$ behaves exactly like P_e.*

Proof idea: We obtain the program $P_{e'}$ from P_e by adding sufficiently much code which is never executed. □

1.1.4 Remark. (Uniformity) In subsequent results we will often make statements like the one in the last line of Theorem 1.1.2: we do not merely assert the existence of an object, but actually that its description (within some specified description system) can be computed from descriptions of the given objects. The formal version of the last line in Theorem 1.1.2 is: there is a computable function h such that if $\Theta = \Phi_e^2$, then $\forall i \, \forall x \, \Phi_{q(i)}(x) \simeq \Theta(i, x)$ holds for $q = \Phi_{h(e)}^1$. One says that the construction is *uniform*, namely, there is a single procedure to obtain the desired object from the ingredients which works for each collection of given objects. Proofs of basic results are usually uniform. More complex proofs can be nonuniform. We will at times be able to show this is necessarily so, for instance in Proposition 5.5.5.

The Recursion Theorem is an important technical tool. It was proved by Kleene (1938) in a paper on ordinal notations. Informally, it asserts that one cannot change in an effective way the input/output behavior of all Turing programs.

1.1.5 Recursion Theorem. *Let $g\colon \mathbb{N} \mapsto \mathbb{N}$ be computable. Then there is an e such that $\Phi_{g(e)} = \Phi_e$. We say that e is a* fixed point *for g.*

Proof. By the Parameter Theorem 1.1.2 there is a computable function q such that $\Phi_{q(e)}(x) \simeq \Phi_{g(\Phi_e(e))}(x)$ for all e, x. Choose an i such that $q = \Phi_i$, then

$$\Phi_{q(i)} = \Phi_{\Phi_i(i)} = \Phi_{g(\Phi_i(i))}. \tag{1.2}$$

So $e = \Phi_i(i) = q(i)$ is a fixed point. □

We obtained the index i for q effectively from an index for g, by the uniformity of the Parameter Theorem. Thus, if g is a computable function of two arguments, we can compute a fixed point $e = f(n)$ for each function g_n given by $g_n(e) = g(e, n)$. Taking the uniformity one step further, note that an index for f can be obtained effectively from an index for g. This yields an extended version:

1.1.6 Recursion Theorem with Parameters. *Let $g\colon \mathbb{N}^2 \mapsto \mathbb{N}$ be computable. Then there is a computable function f, which can be obtained effectively from g, such that $\Phi_{g(f(n),n)} = \Phi_{f(n)}$ for each n.* □

The incompleteness theorem of Gödel (1931) states that for each effectively axiomatizable sufficiently strong consistent theory T in the language of arithmetic one can find a sentence ϵ which holds in $\mathbb{N}$ but is not provable in T. Peano arithmetic is an example of such a theory. The incompleteness theorem relies on a fixed point lemma proved in a way analogous to the proof of the Recursion Theorem. One represents a formula σ in the language of arithmetic by a natural number $\underline{\sigma}$. This is the analog of representing a partial computable function Ψ by an index e, in the sense that $\Psi = \Phi_e$. Notice the "mixing of levels" that is taking place in both cases: a partial computable function of one argument is applied to a number, which can be viewed as an index for a function. A formula in one free variable is evaluated on a number, which may be a code for a further formula. The fixed point lemma says that, for each formula $\Gamma(x)$ in one free variable, one can determine a sentence ϵ such that

$$T \vdash \epsilon \leftrightarrow \Gamma(\underline{\epsilon}).$$

Informally, ϵ asserts that it satisfies Γ itself. Roughly speaking, if $\Gamma(x)$ expresses that the sentence x is not provable from T, then ϵ asserts of itself that it is not provable, hence ϵ holds in $\mathbb{N}$ but $T \nvdash \epsilon$.

In the analogy between Gödel's and Kleene's fixed point theorems, Γ plays the role of the function g. Equivalence of sentences under T corresponds to equality of partial computable functions. One obtains the fixed point ϵ as follows: the map $F(\sigma) = \sigma(\underline{\sigma})$, where σ is a formula in one free variable, is computable, and hence can be represented in T by a formula ψ in two free variables (here one uses that T is sufficiently strong; we skip technical details). Hence there is a formula α expressing "$\Gamma(F(\sigma))$", or more precisely $\exists y[\psi(\underline{\sigma}, y)\ \&\ \Gamma(y)]$. Thus, for each formula σ

$$T \vdash \alpha(\underline{\sigma}) \leftrightarrow \Gamma(\underline{\sigma(\underline{\sigma})}).$$

Forming the sentence $\sigma(\underline{\sigma})$ is the analog of evaluating $\Phi_e(e)$, and α is the analog of the function q. Now let ϵ be $\alpha(\underline{\alpha})$. Since $\underline{\alpha}$ is the analog of the index i for q, ϵ (that is, the result of evaluating α on its own code number) is the analog of $\Phi_i(i)$ (the result of applying q to its own index). As in the last line (1.2) of the proof of the Recursion Theorem, one obtains that $T \vdash \epsilon \leftrightarrow \alpha(\underline{\alpha}) \leftrightarrow \Gamma(\underline{\alpha(\underline{\alpha})})$.

1.1.7 Exercise. Extend the Recursion Theorem by showing that each computable function g has infinitely many fixed points. Conclude that the function f in Theorem 1.1.6 can be chosen one-one.

Computably enumerable sets

1.1.8 Definition. We say that a set $A \subseteq \mathbb{N}$ is *computably enumerable* (*c.e.*, for short) if A is the domain of some partial computable function.

The reason for choosing this term will become apparent in 1.1.15. Let

$$W_e = \mathrm{dom}(\Phi_e). \tag{1.3}$$

Then $(W_e)_{e \in \mathbb{N}}$ is an effective listing of all c.e. sets. A sequence of sets $(S_e)_{e \in \mathbb{N}}$ such that $\{\langle e, x\rangle : x \in S_e\}$ is c.e. is called *uniformly computably enumerable*. An example of such a sequence is $(W_e)_{e \in \mathbb{N}}$.

The characteristic function f of a set A is given by $f(x) = 1$ if $x \in A$ and $f(x) = 0$ otherwise; A and f are usually identified. A is called *computable* if its characteristic function is computable; otherwise A is called *incomputable*.

1.1.9 Proposition. *A is computable $\Leftrightarrow$ A and $\mathbb{N} - A$ are c.e.*

Proof. $\Rightarrow$: If A is computable, there is a program Q_0 that halts on input x iff $x \in A$, and a program Q_1 that halts on input x iff $x \notin A$.

$\Leftarrow$: By the Church–Turing thesis it suffices to give an informal procedure for computing A. Fix programs Q_0, Q_1 such that Q_0 halts on input x iff $x \in A$, and Q_1 halts on input x iff $x \notin A$. To decide whether $x \in A$, run the computations of Q_0 on x and of Q_1 on x in parallel until one of them halts. If Q_0 halts first, output 1, otherwise output 0. □

We may obtain a c.e. incomputable set denoted $\emptyset'$ by a direct diagonalization. We define $\emptyset'$ in such a way that $\mathbb{N} - \emptyset'$ differs from W_e at e: let

$$\emptyset' = \{e\colon\ e \in W_e\}.$$

The reason for choosing this notation becomes apparent in 1.2.9. The set $\emptyset'$ is called the *halting problem*, since $e \in \emptyset'$ iff program P^1_e halts on input e. (It is often denoted by K, but we reserve this letter for prefix-free Kolmogorov complexity.)

1.1.10 Proposition. *The set $\emptyset'$ is c.e. but not computable.*

Proof. $\emptyset'$ is c.e. since $\emptyset' = \mathrm{dom}(J)$, where J is the partial computable function given by $J(e) \simeq \Phi_e(e)$. If $\emptyset'$ is computable then there is e such that $\mathbb{N} - \emptyset' = W_e$. Then $e \in \emptyset' \leftrightarrow e \in W_e \leftrightarrow e \notin \emptyset'$, contradiction. (This is similar to Russell's paradox in set theory.) □

The sequence $(W_e)_{e \in \mathbb{N}}$ is universal for uniformly c.e. sequences.

1.1.11 Corollary. *For each uniformly c.e. sequence $(A_e)_{e \in \mathbb{N}}$ there is a computable function q such that $A_e = W_{q(e)}$ for each e.*

Proof. Define the partial computable function Θ by $\Theta(e, x) \simeq 0$ iff $x \in A_e$, and $\Theta(e, x)$ is undefined otherwise. Then the function q obtained by the Parameter Theorem is as required. □

1.1.12 Exercise. Suppose $(\widehat{W}_e)_{e\in\mathbb{N}}$ is a further universal uniformly c.e. sequence. Assume that $(\widehat{W}_e)_{e\in\mathbb{N}}$ also has the padding property, namely, for each e and each m, one may effectively obtain $e' > m$ such that $\widehat{W}_{e'} = \widehat{W}_e$. Show that there is a computable permutation π of $\mathbb{N}$ such that $\widehat{W}_e = W_{\pi(e)}$ for each e.

Indices and approximations

A construction of an object, say a c.e. set, is usually carried out by giving an informal procedure that runs at *stages* s. We need effective approximations at stages of the objects that are *given* (if there are such objects). These approximations can often be derived from the descriptions of the objects via a Turing program. If the e-th Turing program describes an object in some specified way, then we say e is an index for that object. This terminology has already been used, for instance, after (1.1) on page 3.

1.1.13 Definition. We write

$$\Phi_{e,s}(x) = y$$

if $e, x, y < s$ and the computation of program P_e on input x yields y in at most s computation steps. We write $\Phi_{e,s}(x)\downarrow$ if there is some y such that $\Phi_{e,s}(x) = y$, and $\Phi_{e,s}(x)\uparrow$ otherwise. Further, we let $W_{e,s} = \mathrm{dom}(\Phi_{e,s})$.

At stage s we have complete information about $\Phi_{e,s}$ and $W_{e,s}$ (which is precisely what we need in a construction). To state this more formally, we need to specify an effective listing $D_0, D_1, \ldots$ of the finite subsets of $\mathbb{N}$.

1.1.14 Definition. Let $D_0 = \emptyset$. If $n > 0$ has the form $2^{x_1} + 2^{x_2} + \ldots + 2^{x_r}$, where $x_1 < \ldots < x_r$, then let $D_n = \{x_1, \ldots, x_r\}$. We say that n is a *strong index* for D_n. For instance, $D_5 = \{0, 2\}$ since $5 = 2^0 + 2^2$.

There is a computable function f such that $f(e, s)$ is a strong index for $W_{e,s}$. We think of a computable enumeration of a set A as an effective listing $a_0, a_1, \ldots$ of the elements of A in some order. To include the case that A is finite, we rather formalize this via an effective union of finite sets (A_s). We view A_s as the set of elements enumerated by the end of stage s. At certain stages we may decide not to enumerate any element.

1.1.15 Definition. A *computable enumeration* of a set A is an effective sequence $(A_s)_{s\in\mathbb{N}}$ of (strong indices for) finite sets such that $A_s \subseteq A_{s+1}$ for each s, and $A = \bigcup_s A_s$.

Each c.e. set W_e has the computable enumeration $(W_{e,s})_{s\in\mathbb{N}}$. Conversely, if A has a computable enumeration then A is c.e., for $A = \mathrm{dom}(\Phi)$ where Φ is the partial computable function given by the following informal procedure: at stage s we let $\Phi(x) = 0$ if $x \in A_s$. An *index for a c.e. set* A is a number e such that

$A = W_e$. When a c.e. set A is described in such a way, then we automatically have a computable enumeration $(A_s)_{s \in \mathbb{N}}$ of A given by $A_s = W_{e,s}$.

Here is an easy application of computable enumerations.

1.1.16 Proposition. *For each partial computable function* Φ, $\text{ran}(\Phi)$ *is c.e.*

Proof. The given object is $\Phi = \Phi_e$, and we enumerate $A = \text{ran}(\Phi)$. Since we have complete information about Φ_s at stage s, we can compute from s a strong index for $A_s = \text{ran}(\Phi_s)$. Then $(A_s)_{s \in \mathbb{N}}$ is the required computable enumeration of A. □

Exercises. Use computable enumerations and the Church–Turing thesis.

1.1.17. Given a c.e. set A, one can uniformly obtain a partial computable function ψ with domain an initial segment of $\mathbb{N}$ such that the range of ψ is A.

1.1.18. A function Φ is partial computable iff its graph $\{\langle x, y\rangle : \Phi(x) = y\}$ is c.e.

1.1.19. Each infinite c.e. set has an infinite computable subset.

1.1.20. (Reduction Principle) For each pair of c.e. sets A, B one can effectively determine disjoint c.e. sets $\widetilde{A} \subseteq A$ and $\widetilde{B} \subseteq B$ such that $A \cup B = \widetilde{A} \cup \widetilde{B}$.

1.2 Relative computational complexity of sets

Recall from the beginning of this chapter that the *relative computational complexity* of a set A is measured by comparing A to other sets via preorderings called reducibilities. To introduce a reducibility $\leq_r$ one specifies a particular type of procedure. It determines whether $n \in X$ with the help of queries of the form "is k in Y?" Each procedure of this type is called an *r-reduction procedure*. There is a hierarchy of reducibilities. The most restricted one we consider is usually many-one reducibility $\leq_m$. An important more general one is Turing reducibility $\leq_T$.

Given a reducibility $\leq_r$ on sets we will write $X \equiv_r Y$ for the corresponding equivalence relation $X \leq_r Y \leq_r X$. The equivalence classes are called *r-degrees*. The r-degree of X consists of the sets having the same complexity as X with respect to $\leq_r$. The r-degrees form a partial order denoted $\mathcal{D}_r$. Some properties of such structures $\mathcal{D}_r$ are sketched on page 16.

Many-one reducibility

One of the simplest examples of a reducibility is the following.

1.2.1 Definition. X is *many-one reducible* to Y, denoted $X \leq_m Y$, if there is a computable function f such that $n \in X \leftrightarrow f(n) \in Y$ for all n.

Thus, the many-one reduction procedures are given by computable functions. Such reductions occur in various areas of mathematics. For instance, interpretations of theories are many-one reductions. For a further example, if G is a finitely generated subgroup of the finitely generated group H then the word problem of G is many-one reducible to the word problem of H. (In both examples we have assumed an effective encoding of the objects in question by natural numbers.)

If X is computable, $Y \neq \emptyset$, and $Y \neq \mathbb{N}$, then $X \leq_m Y$: choose $y_0 \in Y$ and $y_1 \notin Y$. Let $f(n) = y_0$ if $n \in X$, and $f(n) = y_1$ otherwise. Then $X \leq_m Y$ via f. Thus, disregarding $\emptyset$ and $\mathbb{N}$, the computable sets form the least many-one degree.

For each set Y the class $\{X : X \leq_m Y\}$ is countable. In particular, there is no greatest many-one degree. However, $\emptyset'$ is the most complex among the c.e. sets in the sense of $\leq_m$:

1.2.2 Proposition. *A is c.e. $\Leftrightarrow$ $A \leq_m \emptyset'$.*
An index for the many-one reduction as a computable function can be obtained effectively from a c.e. index for A, and conversely.

Proof. $\Leftarrow$: If $A \leq_m \emptyset'$ via h, then $A = \mathrm{dom}(\Psi)$ where $\Psi(x) \simeq J(h(x))$ (recall that $J(e) \simeq \Phi_e(e)$). So A is computably enumerable.

$\Rightarrow$: We claim that there is a computable function g such that

$$W_{g(e,n)} = \begin{cases} \{e\} & \text{if } n \in A, \\ \emptyset & \text{else.} \end{cases}$$

For let $\Theta(e, n, x)$ converge if $x = e$ and $n \in A$. By a three-variable version of the Parameter Theorem 1.1.2, there is a computable function g such that $\forall e, n, x\, [\Theta(e, n, x) \simeq \Phi_{g(e,n)}(x)]$. By Theorem 1.1.6, there is a computable function h such that $W_{g(h(n),n)} = W_{h(n)}$ for each n. Then

$$\begin{aligned} n \in A &\Rightarrow W_{h(n)} = \{h(n)\} \Rightarrow h(n) \in \emptyset', \text{ and} \\ n \notin A &\Rightarrow W_{h(n)} = \emptyset \qquad \Rightarrow h(n) \notin \emptyset'. \end{aligned}$$

The uniformity statements follow from the uniformity of Theorem 1.1.6. □

1.2.3 Definition. A c.e. set C is called *r-complete* if $A \leq_r C$ for each c.e. set A.

Usually $\leq_m$ implies the reducibility $\leq_r$ under consideration. Then, since $\emptyset'$ is m-complete, a c.e. set C is r-complete iff $\emptyset' \leq_r C$. An exception is 1-reducibility, which is more restricted than $\leq_m$: we say that $X \leq_1 Y$ if $X \leq_m Y$ via a one-one function f.

Exercises.

1.2.4. The set $\emptyset'$ is 1-complete. (This will be strengthened in Theorem 1.7.18.)

1.2.5. (Myhill) $X \equiv_1 Y \Leftrightarrow$ there is a computable permutation p of $\mathbb{N}$ such that $Y = p(X)$. (For a solution see Soare 1987, Thm. I.5.4.)

Turing reducibility

Many-one reducibility is too restricted to serve as an appropriate measure for the relative computational complexity of sets. Our intuitive understanding of "Y is at least as complex as X" is: X can be computed with the help of Y (or, "X can be computed relative to Y"). If $X \leq_m Y$ via h, then this holds via a very particular type of relative computation procedure: on input x, compute $k = h(x)$ and output 1 ("yes") if $k \in Y$, and 0 otherwise. To formalize more general ways of relative computation, we extend the machine model by a one-way infinite "oracle" tape which holds all the answers to oracle questions of the form "is k in Y?". The tape has a 1 in position k if $k \in Y$, otherwise it has a 0. To make the query, the machine moves the head on the oracle tape to position k and checks whether the entry at that position is 1.

Extending the definitions at (1.1) to oracle Turing machines, from now on we will view the effective listing $(\Phi_e)_{e\in\mathbb{N}}$ as a listing of partial functions depending on two arguments, the oracle set and the input. We write $\Phi_e^Y(n)\downarrow$ if the program P_e halts when the oracle is Y and the input is n; we write $\Phi_e(Y;n)$, or $\Phi_e^Y(n)$ for this output. We also use the notation $\Phi_e^Y(n)\uparrow$ for the negation of $\Phi_e^Y(n)\downarrow$. The Φ_e are called *Turing functionals*. Extending (1.3), we let

$$W_e^Y = \mathrm{dom}(\Phi_e^Y). \tag{1.4}$$

In this context we call W_e a *c.e. operator*. Turing functionals will be studied in more detail in Section 6.1, and c.e. operators in Section 6.3.

1.2.6 Definition. A total function $f\colon \mathbb{N} \mapsto \mathbb{N}$ is called *Turing reducible* to Y, or *computable relative to* Y, or *computable in* Y, if there is an e such that $f = \Phi_e^Y$. We denote this by $f \le_T Y$. We also say that Y *computes* f. For a set A, we write $A \le_T Y$ if the characteristic function of A is Turing reducible to Y.

Sometimes we also consider Turing reductions to total functions g. Then $f \le_T g$ means that f is Turing reducible to the *graph* of g, that is, to $\{\langle n, g(n)\rangle \colon n \in \mathbb{N}\}$.

1.2.7 Exercise. Verify that $\le_m$ and $\le_T$ are preorderings of the subsets of $\mathbb{N}$.

Relativization and the jump operator

The process of extending definitions, facts, and even proofs from the case involving plain computations to the case of computations relative to an oracle is called *relativization*. For instance, in Definition 1.2.6, we relativized the notion of a computable function to obtain the notion of a function computable in Y. Recall that $W_e^Y = \mathrm{dom}(\Phi_e^Y)$. A set A is *c.e. relative to* Y (or c.e. in Y) if $A = W_e^Y$ for some e. Any notation introduced for the unrelativized case will from now on be viewed as a shorthand for the oracle version where the oracle is $\emptyset$. For instance, we view Φ_e as a shorthand for $\Phi_e^{\emptyset}$.

The relativization of Proposition 1.1.9 is as follows.

1.2.8 Proposition. *A is computable in Y $\Leftrightarrow$ A and $\mathbb{N} - A$ are c.e. in Y.*

It is proved by viewing the proof of Proposition 1.1.9 relative to an oracle. (Note that we now assume a version of the Church–Turing thesis with oracles.)

Relativizing the halting problem to Y yields its Turing jump Y'. This important operation was introduced by Kleene and Post (1954).

1.2.9 Definition. We write $J^Y(e) \simeq \Phi_e^Y(e)$. The set $Y' = \mathrm{dom}(J^Y)$ is the *Turing jump* of Y. The map $Y \to Y'$ is called the *jump operator*.

By the oracle version of the Church–Turing thesis, J is a Turing functional.

When relativizing, special care should be applied that every computation involved becomes a computation with oracle Y. However, in some lucky cases a proof designed for the computable case and yielding some computable object actually works for all oracles, and the resulting object is always computable. For instance, in the relativized proof of the Parameter Theorem 1.1.2, the function q is computable, and independent of the oracle. Thus, for each functional Θ there is a computable function q such that, for each oracle Y and each pair of arguments e, x, we have $\Phi^Y_{q(e)}(x) \simeq \Theta^Y(e, x)$. As a consequence one obtains a version of the Recursion Theorem 1.1.6 for Turing functionals.

1.2.10 Theorem. *For each computable binary function g there is a computable function f such that $\Phi^Y_{g(f(n),n)} = \Phi^Y_{f(n)}$ for each set Y and each number n.* □

The proof of Proposition 1.2.2 (that the halting problem is m-complete) uses the Recursion Theorem. So we obtain a version relative to an oracle, but still with unrelativized m-reducibility:

1.2.11 Proposition. *A is c.e. in Y iff $A \leq_m Y'$.* □

Relativizing Proposition 1.1.10 (the halting problem is c.e. but incomputable), we obtain that the jump produces a set that is c.e. relative to the given set and not Turing below it.

1.2.12 Proposition. *For each Y, the set Y' is c.e. relative to Y. Also, $Y \leq_m Y'$ and $Y' \not\leq_T Y$, and therefore $Y <_T Y'$.*

Proof. Y' is c.e. in Y since $Y' = \text{dom}(J^Y)$. As Y is c.e. relative to itself, by Proposition 1.2.11 $Y \leq_m Y'$. If $Y' \leq_T Y$ then there is e such that $\mathbb{N} - Y' = W^Y_e$. Then $e \in Y' \leftrightarrow e \in W^Y_e \leftrightarrow e \notin Y'$, contradiction. □

1.2.13 Definition. We define $Y^{(n)}$ inductively by $Y^{(0)} = Y$ and $Y^{(n+1)} = (Y^{(n)})'$. Thus $Y <_T Y^{(1)} <_T Y^{(2)} <_T \ldots$ by Proposition 1.2.12.

The following relates the reducibilities $\leq_m$ and $\leq_T$ via the jump operator.

1.2.14 Proposition. *For each Y, Z, we have $Y \leq_T Z \Leftrightarrow Y' \leq_m Z'$.*

Proof. $\Rightarrow$: The set Y' is c.e. in Y and hence c.e. in Z. Therefore $Y' \leq_m Z'$ by Proposition 1.2.11.

$\Leftarrow$: By Proposition 1.2.8, Y and $\mathbb{N} - Y$ are c.e. in Y. So $Y, \mathbb{N} - Y \leq_m Y' \leq_m Z'$, whence both Y and $\mathbb{N} - Y$ are c.e. in Z by Proposition 1.2.8 again. Hence $Y \leq_T Z$. □

We will frequently use the fact that the jump is a universal Turing functional:

1.2.15 Fact. *From a Turing functional $\Phi = \Phi_e$ one can effectively obtain a computable strictly increasing function p, called a* reduction function *for Φ, such that $\forall Y\, \forall x\, \Phi^Y(x) \simeq J^Y(p(x))$.*

Proof. Let $\Theta^Y(x,y) \simeq \Phi^Y(x)$ (an index for Θ is obtained effectively). By the oracle version of the Parameter Theorem, there is a computable strictly increasing function p such $\forall Y \forall y\, \Phi^Y_{p(x)}(y) \simeq \Theta^Y(x,y) \simeq \Phi^Y(x)$. Letting $y = p(x)$ we obtain $J^Y(p(x)) = \Phi^Y_{p(x)}(p(x)) = \Phi^Y(x)$. $\square$

We often apply this fact without an oracle. Thus, from a partial computable function $\alpha = \Phi_e^\emptyset$ one can effectively obtain a reduction function p such that $\forall x\, \alpha(x) \simeq J(p(x))$.

Fact 1.2.15 yields a machine independent characterization of the jump: if $\widehat{J}$ is a further universal Turing functional, there is a computable permutation π of $\mathbb{N}$ such that $\widehat{J}^Y(x) \simeq J^Y(\pi(x))$ for each Y, x. This is proved in the same way as Myhill's Theorem (Exercise 1.2.5). An example of such an alternative jump operator is $\widehat{J}^X(y) \simeq \Phi_e^X(n)$ where $y = \langle e, n\rangle$.

We provide a further useful variant of the Recursion Theorem due to Smullyan. Given computable functions g and h one may obtain a *pair* of fixed points.

1.2.16 Double Recursion Theorem. *Given computable binary functions g, h, one can effectively obtain numbers a, b such that $\Phi^Y_{g(a,b)} = \Phi^Y_a$ & $\Phi^Y_{h(a,b)} = \Phi^Y_b$ for each Y.*

Proof. By the Recursion Theorem with Parameters 1.2.10, there is a computable function f such that for each Y and each n we have $\Phi^Y_{g(f(n),n)} = \Phi^Y_{f(n)}$. Now apply the Recursion Theorem to the function $\lambda n.h(f(n), n)$ in order to obtain a fixed point b, and let $a = f(b)$. Then a, b is a pair as required. $\square$

Strings over $\{0, 1\}$

To proceed we need some more terminology and notation. An element of $\{0, 1\}$ is called a *bit*. Finite sequences of bits are called *strings*.

String notation

The set of all strings is denoted by $\{0, 1\}^*$. The letters $\sigma, \rho, \tau, x, y, z$ will usually denote strings. The following notation is fairly standard.

$\sigma\tau$	concatenation of σ and τ
σa	σ followed by the symbol a
$\sigma \preceq \tau$	σ is a prefix of τ, that is, $\exists \rho\, [\sigma\rho = \tau]$
$\sigma \mid \tau$	σ, τ are *incompatible* i.e. neither of σ, τ is a prefix of the other
$\sigma <_L \tau$	σ is to the left of τ, that is, $\exists \rho\, [\rho 0 \preceq \sigma$ & $\rho 1 \preceq \tau]$
$\lvert\sigma\rvert$	the length of σ
$\varnothing$	the empty string, that is, the string of length 0.

We picture $\{0, 1\}^*$ as a tree growing upwards, with $\sigma 0$ to the left of $\sigma 1$. The relation $<_L$ is a linear order called the lexicographical order. For a set Z,

$$\boxed{Z \upharpoonright_n \text{ denotes the string } Z(0)Z(1)\ldots Z(n-1).}$$

The notations $\sigma \preceq Z$, $\sigma <_L Z$, etc., are used in the obvious senses. For sets Y, Z we write $Y <_L Z$ if $\exists \rho\, [\rho 0 \prec Y$ & $\rho 1 \prec Z]$.

Identifying strings with natural numbers
Frequently strings, not merely natural numbers, are the bottom objects. In this case we want to apply notation developed for natural numbers in the setting of strings, so it will be useful to identify strings with natural numbers. We could identify σ with a strong index for the nonempty finite set $\{|\sigma|\} \cup \{i \colon \sigma(i) = 1\}$, in other words, with the number that has binary representation 1σ. The only problem with this is that 0 does not correspond to a string. To avoid this,

we identify $\sigma \in \{0,1\}^*$ *with* $n \in \mathbb{N}$ *s.t. the binary representation of* $n+1$ *is* 1σ.

For instance, the string 000 is identified with the number 7, since the binary representation of 8 is 1000. Also, the string 10 is identified with the number 5. The empty string $\varnothing$ is identified with 0. If we want to make the identification explicit, we write

$$n = \mathsf{number}(\sigma) \text{ and } \sigma = \mathsf{string}(n). \tag{1.5}$$

Note that $\mathsf{number}(0^i) = 2^i - 1$ and $\mathsf{number}(1^i) = 2^{i+1} - 2$. Thus, the interval $[2^i - 1, 2^{i+1} - 1)$ is identified with the strings of length i. The *length-lexicographical order* where the more significant bits are on the left is the linear ordering on binary strings given by $\{\langle x, y\rangle \colon \mathsf{number}(x) < \mathsf{number}(y)\}$.

Logarithm
For $n \in \mathbb{N}^+$, we let $\log n = \max\{k \in \mathbb{N} \colon 2^k \le n\}$. Then $n \ge 2^{\log n} > n/2$. By convention $\log 0 = 0$.

If $\sigma = \mathsf{string}(n)$, then $|\sigma| = \log(n+1)$. For instance, if $n = 2^i - 1$, then $\sigma = \mathsf{string}(n) = 0^i$, and $|\sigma| = \log 2^i = i$; if $n = 2^{i+1} - 2$, then $\sigma = \mathsf{string}(n) = 1^i$, and $|\sigma| = \log(2^{i+1} - 1) = i$. (The usual real-valued logarithm is denoted by $\log_2 x$.)

Approximating the functionals Φ_e, and the use principle

A general convention when we are dealing with an approximation at a stage s is that *all numbers that matter to a computation at stage s should be less than s.* For instance, when approximating a functional Φ_e, at stage s we only allow oracle questions less than s. We extend Definition 1.1.13 to oracle computations.

1.2.17 Definition. We write $\Phi^Y_{e,s}(x) = y$ if $e, x, y < s$ and the computation of program P_e on input x yields y in at most s computation steps, with all oracle queries less than s. We write $\Phi^Y_{e,s}(x)\downarrow$ if there is y such that $\Phi_{e,s}(x) = y$, and $\Phi^Y_{e,s}(x)\uparrow$ otherwise. Further, we let $W^Y_{e,s} = \mathrm{dom}(\Phi^Y_{e,s})$.

The *use principle* is the fact that a terminating oracle computation only asks finitely many oracle questions. Hence $(\Phi^Y_{e,s})_{s\in\mathbb{N}}$ approximates Φ^Y_e, namely,

$$\Phi^Y_e(x) = y \ \leftrightarrow\ \exists s\, \Phi^Y_{e,s}(x) = y.$$

1.2.18 Definition. The *use* of $\Phi^Y_e(x)$, denoted $\mathsf{use}\,\Phi^Y_e(x)$, is defined if $\Phi^Y_e(x)\downarrow$, in which case its value is 1+the largest oracle query asked during this computation (and 1 if no question is asked at all). Similarly, $\mathsf{use}\,\Phi^Y_{e,s}(x)$ is 1+the largest oracle question asked up to stage s.

We write

$$\Phi_e^\sigma(x) = y$$

if $\Phi_e^F(x)$ yields the output y, where $F = \{i < |\sigma| : \sigma(i) = 1\}$, and the use is at most $|\sigma|$. We write $\Phi_e^\sigma(x)\uparrow$ if there is no such y. Then, for each set Y,

$$\Phi_e^Y(x) = y \leftrightarrow \Phi_e^{Y \upharpoonright_u}(x) = y,$$

where $u = \mathsf{use}\, \Phi_e^Y(x)$. We write $\Phi_{e,s}^\sigma(x) = y$ if $\Phi_e^\sigma(x) = y$ in at most s steps, and $\Phi_{e,s}^\sigma(x)\uparrow$ if there is no such y.

Weak truth-table reducibility and truth-table reducibility

If a Turing functional Φ_e is given then $\lambda Y, x.\, \mathsf{use}\, \Phi_e^Y(x)$ is also a Turing functional (namely, there is i such that $\Phi_i^Y(x) \simeq \mathsf{use}\, \Phi_e^Y(x)$ for each Y and x). Thus, if Y is an oracle such that $f = \Phi_e^Y$ is total, the function $\mathsf{use}\, \Phi_e^Y$ is computable in Y. This function may grow very quickly. A reducibility stronger than $\leq_T$ is obtained when we require that $f = \Phi_e^Y$ for some e such that $\mathsf{use}\, \Phi_e^Y$ is bounded by a computable function.

1.2.19 Definition. A function $f \colon \mathbb{N} \to \mathbb{N}$ is *weak truth-table* reducible to Y, denoted $f \leq_{wtt} Y$, if there is a Turing functional Φ_e and a computable bound r such that $f = \Phi_e^Y$ and $\forall n\, \mathsf{use}\, \Phi_e^Y(n) \leq r(n)$. For a set A, we write $A \leq_{wtt} Y$ if the characteristic function of A is weak truth-table reducible to Y.

It may happen that $f \leq_{wtt} Y$ via Φ_e and r such that Φ_e^Z is not a total function for some oracle $Z \neq Y$. We obtain an even stronger reducibility when requiring that Φ_e^Z is total for each Z; this implies a computable bound on the use.

1.2.20 Definition. A function $f \colon \mathbb{N} \to \mathbb{N}$ is *truth-table* reducible to Y, denoted $f \leq_{tt} Y$, if there is a Turing functional Φ_e such that $f = \Phi_e^Y$ and Φ_e^Z is total for each oracle Z (we call such a Φ_e a truth-table reduction). For a set A, we write $A \leq_{tt} Y$ if the characteristic function of A is truth-table reducible to Y.

The reducibility between sets is called truth-table reducibility because $\Phi_e^Z(n)$ can be obtained by first computing a Boolean expression from n, and then evaluating it on the answers to oracle questions. Recall strong indices for finite sets from Definition 1.1.14, and note that the expression on the right in (i) below corresponds to a Boolean formula in disjunctive normal form.

1.2.21 Proposition.

(i) $X \leq_{tt} Y \Leftrightarrow$ *there is a computable function g such that, for each n,*

$$n \in X \leftrightarrow \bigvee_{\sigma \in D_{g(n)}} [\sigma \preceq Y].$$

(ii) $X \leq_{tt} Y$ *implies* $X \leq_{wtt} Y$.

Proof. (i) $\Rightarrow$: Suppose $X \leq_{tt} Y$ via a truth-table reduction $\Phi = \Phi_e$. The tree $T_n = \{\sigma \colon \Phi_{|\sigma|}^\sigma(n)\uparrow\}$ is finite for each n, for otherwise it has an infinite path Z by König's Lemma (1.8.2 below), and $\Phi^Z(n)\uparrow$. Given n one can compute a strong index $\widetilde{g}(n)$ for the finite set of minimal strings σ (under the prefix relation) such that $\Phi_{|\sigma|}^\sigma(n)\downarrow$. Hence one can also compute a strong index $g(n)$ for the set of all minimal strings σ such that $\Phi_{|\sigma|}^\sigma(n)\downarrow = 1$. Then $D_{g(n)}$ is as required.

$\Leftarrow$: Consider the following procedure relative to an oracle Z: on input n, first compute $D_{g(n)}$. If $\sigma \preceq Z$ for some $\sigma \in D_{g(n)}$, output 1, otherwise output 0. By the oracle version of the Church–Turing thesis, a Turing program P_e^1 formalizing the procedure exists, and Φ_e is the corresponding functional. Clearly Φ_e^Z is total for each oracle Z.

(ii) Let Φ_e be the Turing functional of the previous paragraph. Then for each Z the function use $\Phi_e^Z(n)$ is bounded by $\max\{|\sigma|: \sigma \in D_{g(n)}\}$. □

By (i), the truth-table reduction procedures correspond to computable functions g. Thus, our effective listing of the partial computable functions yields an effective listing of reduction procedures which includes all the tt-reductions.

1.2.22 Proposition. *$f \leq_{tt} A \Leftrightarrow$ there is a Turing functional Φ and a computable function t such that $f = \Phi^A$ and the number of steps needed to compute $\Phi^A(n)$ is bounded by $t(n)$.*

Proof. $\Rightarrow$: Suppose $f = \Phi^A$ and Φ^Z is total for each oracle Z. Let g be as in the proof of implication "$\Rightarrow$" of Proposition 1.2.21(i). Then $t(n) = \max\{|\sigma|: \sigma \in D_{g(n)}\}$ bounds the running time of the computation $\Phi^Z(n)$ for each oracle Z.

$\Leftarrow$: Let $\widetilde{\Phi}$ be the Turing functional such that $\widetilde{\Phi}^Z(n) = \Phi^Z_{t(n)}(n)$ if the latter is defined, and $\widetilde{\Phi}^Z(n) = 0$ otherwise. Then $\widetilde{\Phi}^Z$ is total for each Z and $\widetilde{\Phi}^A = \Phi^A$. Hence $f \leq_{tt} A$. □

Note that, by the proof of "$\Rightarrow$", every function $f \leq_{tt} A$ is bounded from above by a computable function: for each n, $f(n) \leq \max\{\Phi^\sigma_{|\sigma|}(n): \sigma \in D_{\widetilde{g}(n)}\}$.

Clearly $X \leq_m Y$ implies $X \leq_{tt} Y$ (as an exercise, specify a computable function g as in Proposition 1.2.21). To summarize, the implications between our reducibilities are

$$\leq_m \Rightarrow \leq_{tt} \Rightarrow \leq_{wtt} \Rightarrow \leq_T.$$

None of the converse implications hold. In fact, the classes of complete sets differ. (Recall from Definition 1.2.3 that a c.e. set C is r-complete if $A \leq_r C$ for each c.e. set A.) A hypersimple set (1.7.5 below) can be Turing-complete, but is never wtt-complete by 4.1.15. A simple set (1.6.2 below) can be tt-complete, but is never m-complete (see Odifreddi 1989, Ch. III). A natural example of a set which is wtt- but not tt-complete comes from algorithmic randomness: the set of (binary) rationals in $[0,1]$ that are less than Chaitin's halting probability Ω. See Section 3.2 for definitions, Proposition 3.2.30 for wtt-completeness of this set and Theorem 4.3.9 for its tt-incompleteness. A direct construction of such a set is also possible, but cumbersome (Odifreddi 1989, III.9).

Exercises. The effective disjoint union of sets A and B is

$$A \oplus B = \{2n: n \in A\} \cup \{2n+1: n \in B\}.$$

1.2.23. (i) Show that $A, B \leq_m A \oplus B$.
(ii) Let $\leq_r$ be one of the reducibilities above. Then, for any set X,

$$A, B \leq_r X \leftrightarrow A \oplus B \leq_r X.$$

1.2.24. Let $C = A_0 \cup A_1$ where A_0, A_1 are c.e. and $A_0 \cap A_1 = \emptyset$. Then $C \equiv_{wtt} A_0 \oplus A_1$.

1.2.25. Show that $\exists Z\, f \leq_{tt} Z \Leftrightarrow$ there is a computable h such that $\forall n\, f(n) \leq h(n)$.

Degree structures

One can abstract from the particularities of a set and only consider its relative computational complexity, measured by a reducibility $\le_r$. The equivalence classes of the equivalence relation given by

$$X \equiv_r Y \;\leftrightarrow\; X \le_r Y \le_r X$$

are called r-degrees. The r-degree of a set X is denoted by $\deg_r(X)$. The r-degrees form a partial order denoted by $\mathcal{D}_r$. We state some basic facts about $\mathcal{D}_r$ for a reducibility $\le_r$ between $\le_m$ and $\le_T$. In the case of many-one reducibility we disregard the sets $\emptyset$ and $\mathbb{N}$. For proofs see for instance Odifreddi (1989).

A structure $(U, \le, \vee)$ is an *uppersemilattice* if $(U, \le)$ is a partial order and, for each $x, y \in U$, $x \vee y$ is the least upper bound of x and y.

1.2.26 Fact.

(i) The least element of $\mathbf{D}_r$ *is* $\mathbf{0} = deg_r(\{0\})$, *the degree consisting of the computable sets.*
(ii) $\mathbf{D}_r$ *is an* uppersemilattice, *where the least upper bound of the degrees of sets* A *and* B *is given by the degree of* $A \oplus B$.
(iii) For each $\mathbf{a} \in \mathbf{D}_r$ *the set* $\{\mathbf{b}\colon\ \mathbf{b} \le \mathbf{a}\}$ *is countable.*
(iv) $\mathbf{D}_r$ *has cardinality* $2^{\aleph_0}$. □

By 1.2.12, $\mathbf{D}_r$ has no maximal elements: for each A, we have $A' >_r A$. By 1.2.14, the jump operator induces a map $'\colon \mathbf{D}_T \mapsto \mathbf{D}_T$, given by $\deg_T(X) \mapsto \deg_T(X')$, called the *Turing jump*. This map is monotonic: for each pair $\mathbf{x}, \mathbf{y} \in \mathbf{D}_T$ we have $\mathbf{x} \le \mathbf{y} \to \mathbf{x}' \le \mathbf{y}'$. Note that $\mathbf{0} < \mathbf{0}' < \mathbf{0}'' < \dots$ is an infinite ascending sequence of Turing degrees. In Corollary 1.6.6 we will see that the jump is not one-one. In fact, for each $\mathbf{x} \in \mathbf{D}_T$ there is $\mathbf{y} > \mathbf{x}$ such that $\mathbf{x}' = \mathbf{y}'$.

1.2.27 Definition. Let $(U, \le, \vee)$ be an uppersemilattice, and let $I \subseteq U$ be nonempty. We say I is an *ideal* of U if I is closed downward, and $x, y \in I$ implies $x \vee y \in I$. For instance $I = \{\mathbf{x}\colon\ \exists n\, \mathbf{x} \le \mathbf{0}^{(n)}\}$ is an ideal in $\mathbf{D}_T$, called the ideal of arithmetical degrees. The ideal I is called *principal* if $I = \{b\colon\ b \le a\}$ for some $a \in U$.

1.3 Sets of natural numbers

Sets of natural numbers are important objects of study in computability theory. They can be identified with infinite sequences of bits, and also with the real numbers r such that in $0 \le r < 1$ via the representation in base 2 (here one disregards the cofinite sets). For instance, the set of even numbers is identified with $101010\dots$, and also with the real number $0.101010\dots = 2/3$. The term *set* will refer to sets of natural numbers unless otherwise stated.

There are two extremes as to how to view a set Z: the *local view*, where the set is understood by looking at its initial segments $Z\restriction_n$, and the *global view*, where the set is appreciated all at once. In the local view, the set is revealed bit by bit, similar to the outcomes of an experiment that proceeds in time (see page 73 for more on this). Strings are important for us because they represent the finite initial segments of sets. The global view is to think of the set as a single entity. In the ideal case we would like to give a description of the set.

Let us step back and consider some alternatives to sets of natural numbers. How essential is the use of the natural numbers for the indexing of bits? One can also use other effectively given domains D, such as the tree $\{0,1\}^*$ of finite strings over $\{0,1\}$, or the rationals. Instead of subsets of $\mathbb{N}$, one now studies subsets of $\{0,1\}^*$, or subsets of $\mathbb{Q}$.

1. The local view changes, since we have a different perception of what the finite "parts" of a subset of D are. For instance, if $D = \{0,1\}^*$, a finite part might consist of the labeled finite tree of the bits up to level n. If $D = \mathbb{Q}$ it is unclear how one would define a notion of finite part taking the order structure of D into account. The following definition provides a reasonable notion of finite part for a subset of an arbitrary domain D. A *finite assignment* for D is a sequence $\alpha = (\langle d_0, r_0\rangle, \ldots, \langle d_{k-1}, r_{k-1}\rangle)$ where all $d_i \in D$ are distinct, $k \in \mathbb{N}$, and $r_i \in \{0,1\}$. If $Z\colon\ D \to \{0,1\}$, we think of α as a part of Z if $Z(d_i) = r_i$ for each $i < k$. For more on finite assignments see page 297.

2. In contrast, for the global view the choice of the domain D matters little as long as the elements of D can be effectively encoded by natural numbers. The reason is that, with very few exceptions, classes $\mathcal{C}$ of sets introduced in computability (and randomness) theory are closed under computable permutations π, namely, $Z \in \mathcal{C} \;\leftrightarrow\; Z \circ \pi \in \mathcal{C}$ for each set Z. Thus, it does not matter that the choice of D, and its encoding by natural numbers, is arbitrary.

The indexing of bits by natural numbers is a convenience rather than a necessity. It is convenient for us because we may use our intuition based on discrete physical processes that proceed in time, and we have a clear idea of what the finite parts of the set are, namely strings that are initial segments of the set. Also, using $\mathbb{N}$ for indexing, we have the rich structure of arithmetic $(\mathbb{N}, +, \times)$ at our disposal: most descriptions of sets will use some extension of the language of arithmetic (see Section 1.4). At the same time, the indexing of bits by natural numbers is not essential for the global view, since we may use a domain other than the natural numbers and still study the same properties of sets.

Sets of natural numbers are identified with functions $\mathbb{N} \to \{0,1\}$, so why not study directly functions $f\colon\ \mathbb{N} \to \mathbb{N}$? After all, they are equally fundamental. In fact, most of the theory of descriptive and computational complexity can be developed in the same way for functions. Both sets and functions can be put into topological context, using product topologies. Sets are the elements of Cantor space $2^{\mathbb{N}}$ (Section 1.8), while functions are the elements of Baire space $\mathbb{N}^{\mathbb{N}}$. Cantor space has the nicer properties, since on the one hand it is compact, on the other hand it allows us to define the uniform measure λ, which assigns the quantity $2^{-|\sigma|}$ to each basic open cylinder $[\sigma] = \{Z\colon\ \sigma \prec Z\}$ (see 1.9.7 on page 70). The failure of these properties in the case of Baire space is the reason why we give preference to sets. In particular, the theory of algorithmic randomness relies on the uniform measure and therefore only works for sets.

We could be slightly more general and work with functions $\mathbb{N} \to \{0, \ldots, b-1\}$ for some fixed $b \in \mathbb{N} - \{0, 1\}$. They form a compact space on which the uniform measure can be defined, and can be identified with the real numbers r such that in $0 \leq r < 1$ via the representation in base b (here one disregards the functions that eventually have the value $b - 1$). Also see Remark 3.2.34.

1.4 Descriptive complexity of sets

In this section we develop some more theory on the descriptive complexity of sets. In the next section we do the same for the absolute computational complexity. In the beginning of this chapter we discussed the method of introducing classes of sets sharing a certain complexity property. The class of computable sets is contained in each such class, for both types of complexity, because a Turing program P_e^1 that computes a set Z is also a very simple way of describing Z. In the terminology introduced earlier, we use the description system F_{comp} given by $F_{comp}(e) = Z$ if P_e^1 computes Z. A larger class of descriptive complexity, the class of c.e. sets, is given by the description system $F_{c.e.}(e) = W_e = \{x\colon\ P_e^1(x) \text{ halts}\}$. Note that every number is a description with respect to $F_{c.e.}$, while the domain of F_{comp} is incomputable by Exercise 1.4.20(iii).

Δ_2^0 sets and the Shoenfield Limit Lemma

In a computable enumeration $(Z_s)_{s\in\mathbb{N}}$ of a set Z, for each x, $Z_s(x)$ can change at most once, namely from 0 to 1. Which sets Z are described if we allow an arbitrary finite number of changes? These sets are called Δ_2^0 sets and form an important class of descriptive complexity. The reason for choosing this terminology will become apparent in Definition 1.4.15.

1.4.1 Definition. We say that a set Z is Δ_2^0 if there is a computable sequence of strong indices $(Z_s)_{s\in\mathbb{N}}$ such that $Z_s \subseteq [0, s)$ and $Z(x) = \lim_s Z_s(x)$. We say that $(Z_s)_{s\in\mathbb{N}}$ is a *computable approximation* of Z.

A computable enumeration is a special case of a computable approximation.

The following notation is very useful. Given an expression E that is approximated during stages s,

$$E[s]$$

denotes its value at the *end of* stage s. For instance, given a Δ_2^0 set Z with a computable approximation, instead of $\Phi_{e,s}^{Z_s}(x)$ we simply write $\Phi_e^Z(x)[s]$. $E[s]$ can usually be evaluated in an effective way. We say that the expression E is *stable at* s if $E[t] = E[s]$ for all $t \geq s$.

Shoenfield (1959) proved that the Δ_2^0 sets coincide with the sets that are Turing reducible to the halting problem. For this reason, the Δ_2^0 sets can also be viewed as a class of computational complexity, namely the class of oracles that are at most as powerful as the halting problem. In the proof of Shoenfield's result we introduce the notion of a change set which will be important later on.

1.4.2 Lemma. (Shoenfield Limit Lemma)
Z is Δ^0_2 $\Leftrightarrow$ $Z \le_T \emptyset'$. The equivalence is uniform.

Proof. $\Leftarrow$: Fix a Turing functional Φ_e such that $Z = \Phi_e^{\emptyset'}$. Then the required computable approximation is given by $Z_s = \{x < s\colon \Phi_e^{\emptyset'}(x)[s] = 1\}$. This approximation was obtained from Φ_e in an effective way.

$\Rightarrow$: We define a c.e. set C such that $Z \le_T C$. This is sufficient because $C \le_m \emptyset'$ by Proposition 1.2.2. The set C is called the *change set* because it records the changes of the computable approximation. If $Z_s(x) \neq Z_{s+1}(x)$ we put $\langle x, i\rangle$ into C_{s+1}, where i is least such that $\langle x, i\rangle \notin C_s$. To show that $Z \le_T C$, on input x, using the oracle C compute the least i such that $\langle x, i\rangle \notin C$. If i is even then $Z(y) = Z_0(y)$, otherwise $Z(y) = 1 - Z_0(y)$.

We have obtained C and the Turing reduction of Z to C effectively from the computable approximation of Z. Proposition 1.2.2 is also effective. □

If $Z = \Phi_e^{\emptyset'}$ we say that e is a Δ^0_2 *index* for Z. A number e is a Δ^0_2 index only if $\Phi_e^{\emptyset'}$ is total (and also 0, 1-valued). In contrast, each number i describes a c.e. set W_i. The set of Δ^0_2 indices is far from computable; see Exercise 1.4.21.

Sets and functions that are n-c.e. or ω-c.e.

One obtains classes of descriptive complexity between the classes of c.e. sets and Δ^0_2 sets by restricting the number changes in a computable approximation.

1.4.3 Definition. (i) We say that a set Z is *ω-c.e.* if there is a computable approximation $(Z_s)_{s\in\mathbb{N}}$ of Z and a computable function b such that

$$b(x) \ge \#\{s > x\colon Z_s(x) \neq Z_{s-1}(x)\} \text{ for each } x.$$

(ii) If $Z_s(s-1) = 0$ for each $s > 0$ and $b(x)$ can be chosen constant of value n, then we say Z is *n-c.e.*

Thus, Z is 1-c.e. iff Z is c.e., and Z is 2-c.e. iff $Z = A - B$ for c.e. sets A, B.

1.4.4 Proposition. *Z is ω-c.e. $\Leftrightarrow$ $Z \le_{wtt} \emptyset'$ $\Leftrightarrow$ $Z \le_{tt} \emptyset'$.*
The equivalences are effective.

Proof. First suppose that $Z \le_{wtt} \emptyset'$ via a functional Φ_e with computable use bound f. To show that Z is ω-c.e., as before let $Z_s = \{x < s\colon \Phi_e^{\emptyset'}(x)[s] = 1\}$. Since $\Phi_e^{\emptyset'}(x)[s]$ only becomes undefined when a number less than $f(x)$ enters $\emptyset'$, the number of changes of $Z_s(x)$ is bounded by $2f(x)$.

Now suppose that Z is ω-c.e. via the computable approximation $(Z_s)_{s\in\mathbb{N}}$ and the function b bounding the number of changes. We show that $Z \le_{tt} \emptyset'$. Let C be the change set introduced in the proof of the implication "$\Rightarrow$" of the Shoenfield Limit Lemma. Since $b(x) \ge \min\{i\colon \langle x, i\rangle \notin C\}$, the reduction of Z to C given there can be carried out by computing a truth-table from the input x and evaluating it on the answers to oracle questions to C. Hence $Z \le_{tt} C \le_m \emptyset'$. □

By Proposition 1.2.21, truth-table reduction procedures may be viewed as partial computable functions mapping inputs to truth tables. Let $(\Theta_e)_{e\in\mathbb{N}}$ be an effective listing of all such (possibly partial) truth-table reduction procedures defined on initial segments of $\mathbb{N}$. Then Proposition 1.4.4 yields an indexing of the ω-c.e. sets that includes computable approximations:

1.4.5 Definition. The ω-c.e. set with index e is $V_e = \{x : \Theta_e^{\emptyset'}(x) = 1\}$. A computable approximation of V_e is given by $V_{e,s} = \{x : \Theta_e^{\emptyset'}(x)[s] = 1\}$.

Thus, if $\Theta_e(x)$ is undefined then $V_{e,s}(x) = 0$ for each s. By 1.2.2, for each e we uniformly have a many-one reduction of W_e to $\emptyset'$. Hence there is a computable function g such that $W_e = V_{g(e)}$ for each e.

The hierarchy of descriptive complexity classes introduced so far is

$$\text{computable} \subset \text{c.e.} \subset \text{2-c.e.} \subset \text{3-c.e.} \subset \ldots \subset \omega\text{-c.e.} \subset \Delta_2^0. \tag{1.6}$$

It is proper even when one considers the Turing degrees of sets at the various levels (see Odifreddi 1989). For instance, there is a 2-c.e. set Z such that no set $Y \equiv_T Z$ is c.e., and there is a Δ_2^0 set Z such that no set $Y \equiv_T Z$ is ω-c.e.

The definitions of Δ_2^0 sets and ω-c.e. sets can be extended to functions $g\colon \mathbb{N} \to \mathbb{N}$.

1.4.6 Definition. A function g is Δ_2^0 if there is a binary computable function $\lambda x, s. g_s(x)$ such that $\forall x\, g(x) = \lim_s g_s(x)$. Moreover, g is ω-c.e. if if there is, in addition, a computable bound b such that $b(x) \geq \#\{s > x\colon g_s(x) \neq g_{s-1}(x)\}$ for each x.

One can extend the Limit Lemma and the first equivalence of Proposition 1.4.4 to functions. The second equivalence in 1.4.4 fails for functions by Exercise 1.2.25 and since an ω-c.e. function need not be bounded by a computable function.

1.4.7 Exercise. Let $g\colon \mathbb{N} \to \mathbb{N}$ be a function.
(i) g is $\Delta_2^0 \Leftrightarrow g \leq_T \emptyset'$.
(ii) g is ω-c.e. $\Leftrightarrow g \leq_{wtt} \emptyset'$.

1.4.8 Exercise. (Mohrherr, 1984) Let $E \geq_{tt} \emptyset'$. Then $Z \leq_{wtt} E$ implies $Z \leq_{tt} E$.

The following will be needed later.

1.4.9 Fact. *There is a binary function $q \leq_T \emptyset'$ with the following property: for each ω-c.e. function g there is an e such that $\forall n\, g(n) = q(e, n)$.*

Proof. Define a computable function $\widetilde{q}$ as follows. For $e = \langle e_0, e_1 \rangle$, let $\widetilde{q}(e, n, s) = \Phi_{e_0}^{\emptyset'}(n)[s]$ if this converges with use $\leq \Phi_{e_1,s}(n)$, and $\widetilde{q}(e, n, s) = 0$ otherwise.

Let $q(e, n) = \lim_s \widetilde{q}(e, n, s)$. Then q is as required by Exercise 1.4.7(ii). □

Degree structures on particular classes ⋆

Recall from the beginning of this chapter that the three approaches to measure the complexity of sets are via the descriptive complexity, the absolute computational, and the relative computational complexity. These approaches are related in several ways. One connection is that classes of computational complexity can actually comprise the least degree of a reducibility; see Section 5.6. Here we consider another type of connection. For a reducibility $\leq_r$, one can study the r-degrees of the sets in a particular class of similar complexity $\mathcal{C}$. In this way, one arrives at interesting degree structures, for instance $\mathcal{R}_T$, the Turing degrees of c.e. sets, and $\mathcal{D}_T(\leq \mathbf{0}')$, the Turing degrees of the Δ^0_2 sets. This is a bit more natural when $\mathcal{C}$ is closed downward under $\leq_r$, say, when $\mathcal{C}$ is the class of Δ^0_2 sets and $\leq_r$ is Turing reducibility. Given n, the class of n-c.e. sets is merely closed downward under $\leq_m$. Recall from Fact 1.2.26 that $\mathbf{0}$ is the degree consisting of the computable sets. By the results in Section 1.2 and the Limit Lemma, all the degree structures discussed here have a greatest degree, the degree of the halting problem, denoted by $\mathbf{1}$. All degree structures are uppersemilattices because the relevant classes are closed under $\oplus$.

In the following we list some properties of the Turing degree structures on classes in the hierarchy (1.6). Like most theorems on degree structures, they can be expressed in the first-order language of partial orders. For details on these often difficult results see Odifreddi (1989, 1999) or Soare (1987).

In $\mathcal{D}_T(\leq \mathbf{0}')$, and even in the Turing degrees of ω-c.e. sets, there is a minimal element, that is, there is a degree $\mathbf{a} > \mathbf{0}$ such that $\mathbf{x} < \mathbf{a}$ implies $\mathbf{x} = \mathbf{0}$ (Sacks, 1963*c*). The n-c.e. degrees do not have minimal elements (Lachlan; see Odifreddi, 1999, XI.5.9b).

$\mathcal{R}_T$ is dense, i.e., for each $\mathbf{a} < \mathbf{b}$ there is $\mathbf{c}$ such that $\mathbf{a} < \mathbf{c} < \mathbf{b}$ (Sacks, 1964).

The structures of n-c.e. and of ω-c.e. degrees have a maximal incomplete element, i.e., there is $\mathbf{a} < \mathbf{1}$ such that $\mathbf{a} < \mathbf{x}$ implies $\mathbf{x} = \mathbf{1}$ (Cooper, Harrington, Lachlan, Lempp and Soare, 1991). In contrast, $\mathcal{D}_T(\leq \mathbf{0}')$ has no maximal incomplete element.

The arithmetical hierarchy

Up to now we have defined classes of descriptive complexity via computable approximations, possibly with extra conditions. This led to the hierarchy (1.6) on page 20. To obtain more powerful description systems, we will replace this dynamic way of describing a set by descriptions using the first-order language of arithmetic (with signature containing the symbols $+, \times$). Computable relations are first-order definable in the language of arithmetic (Kaye, 1991, Thm. 3.3), so we may as well suppose that, for $k \geq 1$ and each computable relation on $\mathbb{N}^k$, the signature contains a k-place relation symbol. For each description system, we use as descriptions the first-order formulas in this extended language satisfying certain syntactic conditions. In this way we will also obtain alternative, equivalent description systems for the computable, the c.e., and the Δ^0_2 sets.

1.4.10 Definition. Let $A \subseteq \mathbb{N}$ and $n \geq 1$.

(i) A is Σ^0_n if $x \in A \leftrightarrow \exists y_1 \forall y_2 \ldots Q y_n\, R(x, y_1, \ldots, y_n)$, where R is a symbol for a computable relation, Q is "$\exists$" if n is odd and Q is "$\forall$" if n is even.

(ii) A is Π^0_n if $\mathbb{N} - A$ is Σ^0_n, that is $x \in A \leftrightarrow \forall y_1 \exists y_2 \ldots Q y_n\, S(x, y_1, \ldots, y_n)$, where S is a symbol for a computable relation, Q is "$\forall$" if n is odd and Q is "$\exists$" if n is even.

(iii) A is *arithmetical* if A is Σ^0_n for some n.

One can show that these classes are closed under finite unions and intersections. A *bounded quantifier* is one of the form "$\exists x < n$" or "$\forall x < n$". We still obtain the same classes if we intersperse bounded quantifiers of any type in the quantifier part of expressions above, or replace single quantifiers Q by whole blocks of quantifiers of the same type as Q. For instance, the expression

$\exists y \forall z < y \exists u \forall v\, R(x, y, z, u, v)$ yields a Σ_2^0 set. See Odifreddi (1989, Prop. IV.1.4) for details.

Definition 1.4.10 can be viewed relative to an oracle C.

1.4.11 Definition. For $C \subseteq \mathbb{N}$ and $n \in \mathbb{N}$, we define $\Sigma_n^0(C)$ classes and $\Pi_n^0(C)$ classes as in Definition 1.4.10, but with relations $R, S \leq_T C$.

Note that we now interpret the formulas in the structure $(\mathbb{N}, +, \times)$ extended by a unary predicate for C, in which R and S are first-order definable (see 1.4.24).

1.4.12 Fact. *A is $\Sigma_1^0 \Leftrightarrow A$ is c.e. The equivalence is uniform.*

Proof. $\Rightarrow$: Suppose $x \in A \leftrightarrow \exists y\, R(x, y)$ for computable R. Let Φ be the partial computable function given by the Turing program that on input x looks for a witness y such that $R(x, y)$, and halts when such a witness is found. Then $A = \mathrm{dom}(\Phi)$, so A is c.e. according to Definition 1.1.8.

$\Leftarrow$: Suppose $A = \mathrm{dom}(\Phi)$ for a partial computable function Φ. Let R be the computable relation given by $R(x, s) \leftrightarrow \Phi(x)[s]\downarrow$. Then $x \in A \leftrightarrow \exists s\, R(x, s)$, so A is Σ_1^0. □

The next result is due to Post. Statement (i) generalizes Fact 1.4.12. Recall from 1.2.2 that $\emptyset'$ is many-one complete for the c.e. sets. In (ii) we generalize this to the Σ_n^0 sets. A Σ_n^0 set C is called *Σ_n^0-complete* if $A \leq_m C$ for each Σ_n^0 set A. In a similar way one defines Π_n^0-completeness. $\emptyset^{(n)}$ is defined in 1.2.13.

1.4.13 Theorem. *Let $n \geq 1$.*

(i) A is $\Sigma_n^0 \Leftrightarrow A$ is c.e. relative to $\emptyset^{(n-1)}$.

(ii) $\emptyset^{(n)}$ is Σ_n^0-complete.

Proof. We use induction on n. For $n = 1$, (i) is Fact 1.4.12 and (ii) is Proposition 1.2.2. Now let $n > 1$.

(i) First suppose A is Σ_n^0, namely $x \in A \Leftrightarrow \exists y_1 \forall y_2 \ldots Q y_n\, R(x, y_1, \ldots, y_n)$ for some computable relation R. Then the set

$$B = \{\langle x, y_1 \rangle \colon \forall y_2 \ldots Q y_n\, R(x, y_1, \ldots, y_n)\}$$

is Π_{n-1}^0, and A is c.e. relative to B. By (ii) for $n-1$ we have $B \leq_m \mathbb{N} - \emptyset^{(n-1)}$. So A is c.e. relative to $\emptyset^{(n-1)}$.

Now suppose A is c.e. relative to $\emptyset^{(n-1)}$. Then there is a Turing functional Φ such that $A = \mathrm{dom}(\Phi^{\emptyset^{(n-1)}})$. By the use principle,

$$x \in A \Leftrightarrow \exists \eta, s \left[\Phi_s^\eta(x)\downarrow \ \&\ \forall i < |\eta| \begin{array}{c} (\eta(i) = 1 \rightarrow i \in \emptyset^{(n-1)}\ \& \\ \eta(i) = 0 \rightarrow i \notin \emptyset^{(n-1)}) \end{array} \right].$$

The innermost part can be put into Σ_n^0-form, so A is Σ_n^0 because the quantifier $\forall i$ is bounded; (ii) now follows by Proposition 1.2.11 where $Y = \emptyset^{(n-1)}$. □

Recall that in 1.4.1 we introduced Δ_2^0 sets via computable approximations. They can also be characterized using the language of arithmetic.

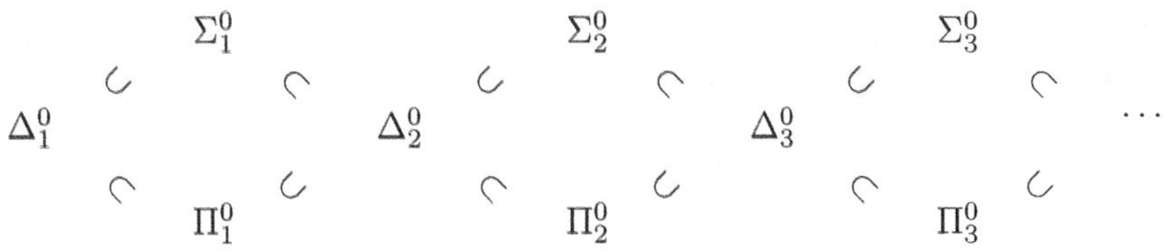

$$\begin{array}{ccccccccccccc} & & \Sigma^0_1 & & & & \Sigma^0_2 & & & & \Sigma^0_3 & & \\ & \subseteq & & \supseteq & & \subseteq & & \supseteq & & \subseteq & & \supseteq & \\ \Delta^0_1 & & & & \Delta^0_2 & & & & \Delta^0_3 & & & & \cdots \\ & \supseteq & & \subseteq & & \supseteq & & \subseteq & & \supseteq & & \subseteq & \\ & & \Pi^0_1 & & & & \Pi^0_2 & & & & \Pi^0_3 & & \end{array}$$

FIG. 1.1. The arithmetical hierarchy.

1.4.14 Proposition. *A is* $\Delta^0_2 \Leftrightarrow$ *A is both* Σ^0_2 *and* Π^0_2.

Proof.

$A \in \Delta^0_2$	$\Leftrightarrow$	$A \le_T \emptyset'$	by the Limit Lemma 1.4.2
	$\Leftrightarrow$	A and $\mathbb{N} - A$ are c.e. in $\emptyset'$	by Proposition 1.2.8
	$\Leftrightarrow$	$A \in \Sigma^0_2 \cap \Pi^0_2$	by Theorem 1.4.13. □

The following is therefore consistent with Definition 1.4.1.

1.4.15 Definition. We say that A is Δ^0_n if A is both Σ^0_n and Π^0_n.

The computable sets coincide with the Δ^0_1 sets by Fact 1.4.12. The hierarchy of classes introduced in Definitions 1.4.10 and 1.4.15 is called the *arithmetical hierarchy* (Figure 1.1).

1.4.16 Proposition. *Let* $n \ge 1$. *Then A is* $\Delta^0_n \Leftrightarrow A \le_T \emptyset^{(n-1)}$.

Proof. By Theorem 1.4.13, A is $\Delta^0_n \Leftrightarrow A$ and $\mathbb{N} - A$ are c.e. in $\emptyset^{(n-1)}$. By Proposition 1.2.8, this condition is equivalent to $A \le_T \emptyset^{(n-1)}$. □

The Σ^0_2-sets Z can still be reasonably described by a suitable computable sequence of finite sets $(Z_s)_{s\in\mathbb{N}}$.

1.4.17 Proposition. *Z is* $\Sigma^0_2 \Leftrightarrow$ *there is a computable sequence of strong indices* $(Z_s)_{s\in\mathbb{N}}$ *such that* $Z_s \subseteq [0, s)$ *and* $x \in Z \leftrightarrow \exists s \forall t \ge s\, Z_t(x) = 1$.
The equivalence is uniform.

Proof. $\Rightarrow$: By Theorem 1.4.13(i) there is a Turing functional Φ such that $Z = \mathrm{dom}(\Phi^{\emptyset'})$. Now let $Z_s = \{x < s \colon \Phi^{\emptyset'}(x)[s]\downarrow\}$.
$\Leftarrow$: The expression "$\exists s \forall t \ge s\, [Z_t(x) = 1]$" is in Σ^0_2 form. □

In the exercises we give further examples of complete sets at the lower levels of the arithmetical hierarchy. They are somewhat more natural than the sets $\emptyset^{(n)}$ because they are not obtained by relativization. Rather, they describe properties of c.e. sets.

1.4.18 Definition. The *index set of* a class S of c.e. sets is the set $\{i \colon W_i \in S\}$. In a similar way, using the listing $(V_e)_{e\in\mathbb{N}}$ from 1.4.5, we define the index set of a class of ω-c.e. sets.

Exercises. Show the following.

1.4.19. $\emptyset'$ is not an index set.

1.4.20. (i) The set $\{e\colon W_e \neq \emptyset\}$ is Σ^0_1 complete.
(ii) The set $\{e\colon W_e \text{ finite}\}$ is Σ^0_2-complete. In fact, for each Σ^0_2 set S there is a uniformly c.e. sequence $(X_n)_{n\in\mathbb{N}}$ of initial segments of $\mathbb{N}$ such that $\forall n\,[n \in S \leftrightarrow X_n \text{ finite}]$, and this sequence itself is obtained effectively from a description of S.
(iii) The set $\mathsf{Tot} = \{e\colon \mathrm{dom}(\Phi_e) = \mathbb{N}\} = \{e\colon W_e = \mathbb{N}\}$ is Π^0_2-complete.
(iv) Both $\{e\colon W_e \text{ cofinite}\}$ and $\{e\colon W_e \text{ computable}\}$ are Σ^0_3-complete.

1.4.21. The set $\{e\colon \mathrm{dom}\,\Phi^{\emptyset'}_e = \mathbb{N}\}$ is Π^0_3-complete.

1.4.22. Let $\mathcal{S}$ be a class of c.e. sets [ω-c.e. sets] containing all the finite sets. Suppose the index set of S is Σ^0_3. Then $\mathcal{S}$ is uniformly c.e. [uniformly ω-c.e.]

1.4.23.$^\diamond$ Let $\mathcal{S}$ be a class of c.e. sets closed under finite variants that contains the computable sets but not all the c.e. sets. If the index set of $\mathcal{S}$ is Σ^0_3 then it is Σ^0_3-complete.

1.4.24. Let $X \subseteq \mathbb{N}$. (i) Each relation $R \leq_T X$ is first-order definable in the structure $(\mathbb{N}, +, \cdot, X)$. (ii) The index set $\{e\colon W_e \leq_T X\}$ is $\Sigma^0_3(X)$.

1.4.25. A is $\Delta^0_n \Leftrightarrow \forall x\, A(x) = \lim_{k_1}\lim_{k_2} \ldots \lim_{k_{n-1}} g(x, k_1, \ldots, k_{n-1})$ for some computable $\{0,1\}$-valued function g.

1.5 Absolute computational complexity of sets

At the beginning of this chapter we discussed classes of similar complexity. A *lowness property* of a set specifies a sense in which the set is computationally weak. Usually this means that it is not very useful as an oracle. Naturally, we require that such a property be closed downward under Turing reducibility; in particular it only depends on the Turing degree of the set. If a set is computable then it satisfies any lowness property. A set that satisfies a lowness property can be thought of as almost computable in a specific sense.

Highness properties say that the set is computationally strong. They are closed upward under Turing reducibility. If a set satisfies a highness property it is almost Turing above $\emptyset'$ in a specific sense.

Classes of computational complexity are frequently defined in terms of how fast the functions computed by the set grow. To compare the growth rate of functions one can use the domination preordering on functions.

1.5.1 Definition. Let $f, g\colon \mathbb{N} \to \mathbb{R}$. We say that f *dominates* g if $f(n) \geq g(n)$ for almost every n.

In this section we introduce two lowness properties and one highness property of a set A. We also consider some of their variants.

(a) A is *low* if $A' \leq_T \emptyset'$.
(b) A is *computably dominated* if each function $g \leq_T A$ is dominated by a computable function.
(c) A is *high* if $\emptyset'' \leq_T A'$.

The classes given by (a) and (c) can be characterized by domination properties. For (c), by Theorem 1.5.19, A is high iff there is a function $g \leq_T A$ dominating each computable function, which shows that highness is opposite to being computably dominated. For (a) see Exercise 1.5.6.

A general framework for lowness properties and highness properties will be given in Section 5.6. We consider weak reducibilities $\leq_W$. Such a reducibility determines a lowness property $C \leq_W \emptyset$ and a dual highness property $C \geq_W \emptyset'$. For instance, we could define $A \leq_W B$ iff $A' \leq_T B'$. Then the associated lowness and highness properties are (a) and (c) above. Table 8.3 on page 363 contains further examples of such dual properties given by a weak reducibility.

It can be difficult to determine whether a lowness property is satisfied by more than the computable sets, and whether a highness property applies to sets other than the sets Turing above $\emptyset'$. We will need to introduce new methods to do so for the properties (a)–(c): the priority method (page 32) or basis theorems (page 56) for (a) and (b), and, for instance, pseudojump inversion (page 249) for (c). So far, we actually have not seen any example of a set that is neither computable nor Turing above $\emptyset'$.

A pair of lowness properties can be "orthogonal" in the sense that the only sets that satisfy them both are the computable sets. For instance, the properties in (a) and (b) are orthogonal. In contrast, classes of descriptive complexity form an almost linear hierarchy, disregarding cases like the Σ^0_1 versus the Π^0_1 sets where one class is simply obtained by taking the complements of the sets in the other class. For a further difference between computational and descriptive complexity, the downward closed class given by (b), say, is uncountable, while classes of descriptive complexity are countable.

Given a lowness property $\mathcal{L}$, we will be interested in the question whether it is null or conull (Definition 1.9.8), and whether there are sets $A, B \in \mathcal{L}$ such that $\emptyset' \leq_T A \oplus B$. Similarly, we will be interested in whether a highness property $\mathcal{H}$ is null or conull, and whether there are sets $A, B \in \mathcal{H}$ that form a minimal pair (only the computable sets are Turing below both A and B). Being conull means that the property is not very restrictive. Most of the properties we study will be null, including the ones in (a)–(c). If sets A, B as above exist then the property is not that close to being computable (for lowness properties), or being Turing above $\emptyset'$ (for highness properties). We will return to the topic of minimal pairs satisfying a highness property on page 258.

In this section we concentrate on the properties (a)–(c) and their variants. We introduce further properties in subsequent chapters, often using concepts related to randomness. In Chapter 4 we study the conull highness property of having diagonally noncomputable degree, and the stronger highness property of having the same degree as a completion of Peano arithmetic (which is null by Exercise 5.1.15). In Chapter 5 we consider lowness for Martin-Löf randomness. Lowness for other randomness notions is studied in Chapter 8. Figure 8.1 on page 361 gives an overview of all the downward closed properties.

Sets that are low_n

Recall from 1.2.13 that $C^{(n)}$ is the result of n applications of the jump operator, beginning with the set C. A hierarchy of absolute computational complexity is obtained by considering $C^{(n)}$ within the Turing degrees, for $n \geq 0$. Note that $C^{(n)} \geq_T \emptyset^{(n)}$ by Proposition 1.2.14.

1.5.2 Definition. Let $n \geq 0$. We say that C is *low_n* if $C^{(n)} \equiv_T \emptyset^{(n)}$.

The most important among these classes is low_1, the class of sets C such that $C' \equiv_T \emptyset'$. If $C \in low_1$ we simply say that C is *low*. Each low set is Δ^0_2. Thus, such a set is computationally weak in the sense that $\emptyset'$ can determine whether $\Phi^C_e(x)$ converges for each e, x, and in case it does find the output. We will see in Theorem 1.6.4 that an incomputable low c.e. set exists. In particular, the jump is not a one-one map on the Turing degrees.

Each class low_n is closed downward under Turing reducibility, and contained in Δ^0_{n+1}. The hierarchy

$$\text{computable} \subset low_1 \subset low_2 \subset \ldots \subset \{Z \colon Z \not\geq_T \emptyset'\} \qquad (1.7)$$

is proper by Theorem 6.3.6 below.

The following property due to Mohrherr (1986) will be important later on.

1.5.3 Definition. We say that a set C is *superlow* if $C' \equiv_{tt} \emptyset'$.

It suffices to require that $C' \leq_{tt} \emptyset'$, because $\emptyset' \leq_m C'$ for any C by 1.2.14. By 1.4.4, it is also equivalent to ask that C' be ω-c.e., namely, C' can be computably approximated with a computable bound on the number of changes.

The class of superlow sets is closed downwards under Turing reducibility. It lies strictly between the classes of computable and of low sets. In Theorem 1.6.5 we build a c.e. incomputable superlow set, and Exercise 1.6.7 asks for a low but not superlow c.e. set. Also see Remark 6.1.5.

The following states that A' is as simple as possible compared to A.

1.5.4 Definition. A is *generalized low_1*, or in GL_1 for short, if $A' \equiv_T A \oplus \emptyset'$.

Equivalently, $\emptyset'$ is Turing complete relative to A. Clearly the class GL_1 coincides with low_1 on the Δ^0_2 sets, and no set $A \geq_T \emptyset'$ is in GL_1. However, a set in GL_1 is not necessarily computationally weak. In fact, if $B \not\geq_T \emptyset'$ then there is $A \geq_T B$ such that A is in GL_1 by a result of Jockusch (1977); also see Odifreddi (1999, Ex. XI.3.11).

We extend Definition 1.5.1. For functions $f, \psi \colon \mathbb{N} \to \mathbb{N}$, where possibly ψ is partial, we say that f *dominates* ψ if $\forall^\infty n\, [\psi(n)\downarrow \ \rightarrow \ f(n) \geq \psi(n)]$.

Exercises. Show the following.

1.5.5. If C is superlow, there is a computable function h such that $Y \leq_T C$ implies $Y \leq_{tt} \emptyset'$ with use function bounded by h for each Y. (Here we view truth table reductions as functions mapping inputs to truth tables; see before 1.4.5.)

1.5.6. A is in GL_1 $\Leftrightarrow$ some function $f \leq_T A \oplus \emptyset'$ dominates each function that is partial computable in A.

1.5.7. If B is *low*$_2$ then the index set $\{e\colon W_e \le_T B\}$ is Σ^0_3.

1.5.8. B is $\mathrm{Low}_2 \Leftrightarrow \mathrm{Tot}^B = \{e\colon \Phi^B_e \text{ total}\}$ is Σ^0_3.

Computably dominated sets

We study a lowness property of a set A stating that the functions computed by A do not grow too quickly.

1.5.9 Definition. A is called *computably dominated* if each function $g \le_T A$ is dominated by a computable function.

Exercise 1.5.17 shows that we cannot effectively determine the dominating function from the Turing reduction of g to A, unless A is computable.

We say that $E \subseteq \mathbb{N}$ is *hyperimmune* if E is infinite and p_E is not dominated by a computable function, where p_E is the listing of E in order of magnitude (also see Definition 1.7.1 below). The intuition is that E is a very sparse set.

1.5.10 Proposition. *A is* not *computably dominated* $\Leftrightarrow$ *there is a hyperimmune set $E \equiv_T A$.*

Proof. $\Leftarrow$: Immediate since $p_E \le_T E$.

$\Rightarrow$: Suppose $g \le_T A$ is not dominated by a computable function. Let $E = \mathrm{ran}(h)$, where the function h is defined as follows: $h(0) = 0$, and for each $n \in \mathbb{N}$, $h(2n+1) = h(2n) + g(n) + 1$ and $h(2n+2) = h(2n+1) + p_A(n) + 1$. Clearly $E \equiv_T h \equiv_T A$. Moreover $g(n) \le h(2n+1)$, so that $h = p_E$ is not dominated by a computable function. □

For this reason, a set that is not computably dominated is also called a set of *hyperimmune degree*. In the literature, a computably dominated set is usually called a set of hyperimmune-free degree. The study of hyperimmune-free degrees was initiated by Martin and Miller (1968).

1.5.11 Proposition. *A is computably dominated $\Leftrightarrow$ for each function f,*

$$f \le_T A \rightarrow f \le_{tt} A.$$

Proof. $\Rightarrow$: Suppose $f = \Phi^A$. Let $g(x) = \mu s\, \Phi^A_s(x)\downarrow$. Then $g \le_T A$, so there is a computable function t such that $t(x) \ge g(x)$ for each x. Thus t bounds the running time of Φ^A, whence $f \le_{tt} A$ by Proposition 1.2.22.

$\Leftarrow$: By the remark after Proposition 1.2.22, each function $f \le_{tt} A$ is dominated by a computable function. □

Each computable set A is computably dominated. Are there others? We will answer this question in the affirmative in Theorem 1.8.42. Here we observe that there are none among the incomputable Δ^0_2 sets.

1.5.12 Proposition. *If A is Δ^0_2 and incomputable, then A is not computably dominated.*

Proof. Let $(A_s)_{s\in\mathbb{N}}$ be a computable approximation of A. Then the following function g is total:

$$g(s) \simeq \mu t \geq s.\, A_t\restriction_s = A\restriction_s.$$

Note that $g \leq_T A$. Assume that there is a computable function f such that $g(s) \leq f(s)$ for each s. Then A is computable: for each n and each $s > n$ we have $A_t(n) = A(n)$ for some $t \in [s, f(s))$, namely $t = g(s)$. On the other hand, if s is sufficiently large then $A_u(n) = A_s(n)$ for *all* $u \geq s$. Thus, to compute A, on input n determine the least $s > n$ such that $A_u(n) = A_s(n)$ for all $u \in [s, f(s))$. Then $A_s(n) = A(n)$, so the output $A_s(n)$ is correct. □

Exercises.

1.5.13. (a) Strengthen Proposition 1.5.12 as follows: if $C <_T A \leq_T C'$ for some set C, then A is of hyperimmune degree. (b) Conclude that, if A is computably dominated and $C <_T A$, then $C' <_T A'$.

1.5.14. Strengthen Proposition 1.5.12 in yet another way: if A is Σ^0_2 and incomputable then A is of hyperimmune degree.

1.5.15. Show that if A is computably dominated then $A'' \leq_T A' \oplus \emptyset''$. In particular, each computably dominated set is in GL_2 (Definition 1.5.20 below).

1.5.16.$^\diamond$ (Jockusch, 1969) Show that if $X \leq_T A \;\rightarrow\; X \leq_{tt} A$ for each set X, then A is already computably dominated.

1.5.17.$^\diamond$ Let us call a set A uniformly computably dominated if there is a computable function r such that for each e, if Φ^A_e is total then $\Phi_{r(e)}$ is total and dominates Φ^A_e. Show that the only uniformly computably dominated sets are the computable ones.

Sets that are $high_n$

Recall from Definition 1.5.2 that a set C is low_n if $C^{(n)} \equiv_T \emptyset^{(n)}$, namely $C^{(n)}$ is as low as possible. How about having a *complex* n-th jump?

1.5.18 Definition. Let $n \geq 0$. A set C is $high_n$ if $\emptyset^{(n+1)} \leq_T C^{(n)}$.

Of course, $C^{(n)}$ could be even more complex. However, if C is Δ^0_2 then $C^{(n)} \leq_T \emptyset^{(n+1)}$. So in that case, to be $high_n$ means that the n-th jump is as complex as possible.

All the classes $high_n$ are closed upward under Turing reducibility, so the complementary classes non-$high_n = 2^{\mathbb{N}} - high_n$ are closed downward. We have refined the hierarchy (1.7):

$$comp. \subset low_1 \subset low_2 \subset \ldots \subset \text{non-}high_2 \subset \text{non-}high_1 \subset \{Z\colon\ Z \not\geq_T \emptyset'\}. \quad (1.8)$$

This hierarchy of downward closed classes is a proper one as we will see in 6.3.6. Also, there is a c.e. set that is not in low_n or $high_n$ for any n by 6.3.8.

Of particular interest is the class $high_1 = \{C\colon\ \emptyset'' \leq_T C'\}$ (simply called the *high* sets), because such sets occur naturally in various contexts. For instance, a theorem of Martin (1966*a*) states that a c.e. set C is high iff $C \equiv_T A$ for some maximal set A; also see Soare 1987, Thm. XI.2.3. Here a co-infinite c.e. set A is called *maximal* if for each c.e. set $W \supseteq A$, either W is cofinite or $W - A$ is finite. For another example, C is high iff there is a computably random but not Martin-Löf random set $Z \equiv_T C$ by Theorem 7.5.9 below.

We are not yet in the position to show the existence of a high set $C \not\geq_T \emptyset'$. Our first example will be Chaitin's halting probability Ω relative to $\emptyset'$ (3.4.17). In Corollary 6.3.4 we prove that there is a high c.e. set $C <_T \emptyset'$.

It is easy to define a function $f \leq_T \emptyset'$ that dominates all computable functions: the set $\{\langle e, x\rangle \colon \Phi_e(x)\downarrow\}$ is c.e., and hence many-one reducible to $\emptyset'$ via a computable function h. Let $f(x) = \max\{\Phi_e(x) \colon e \leq x \ \& \ h(\langle e, x\rangle) \in \emptyset'\}$. If Φ_e is total then $f(x) \geq \Phi_e(x)$ for all $x \geq e$. This property of $\emptyset'$ in fact characterizes the high sets. Thus, being high is opposite to being computably dominated.

1.5.19 Theorem. (Martin, 1966b) *C is high $\Leftrightarrow$ some function $f \leq_T C$ dominates all computable functions.*

Proof. $\Rightarrow$: We define a function $f \leq_T C$ that dominates each total Φ_e, extending the argument for the case $C = \emptyset'$ just given. Note that $\{e \colon \Phi_e \text{ total}\}$ is Π^0_2, and hence $\{e \colon \Phi_e \text{ total}\} \leq_m \mathbb{N} - \emptyset'' \leq_T C'$. By the Limit Lemma 1.4.2 there is a binary function $p \leq_T C$ such that for each e, $\lim_s p(e, s)$ exists, and $\lim_s p(e, s) = 1$ iff Φ_e is total. To compute $f(x)$ with oracle C, let $s \geq x$ be least such that for each $e \leq x$, either $\Phi_{e,s}(x)\downarrow$ or $p(e, s) = 0$, and let $f(x) = \max\{\Phi_{e,s}(x) \colon e \leq x \ \& \ \Phi_{e,s}(x)\downarrow\}$.

If Φ_e is total then there is $s_0 \geq e$ such that $p(e, s) = 1$ for all $s \geq s_0$, so that $f(x) \geq \Phi_e(x)$ for all $x \geq s_0$.

$\Leftarrow$: Suppose that $f \leq_T C$ dominates all computable functions. We show that $\mathbb{N} - \emptyset'' = \{e \colon \Phi^{\emptyset'}_e(e)\uparrow\}$ is Turing reducible to C'. Note that $\Phi^{\emptyset'}_e(e)\uparrow$ iff the computation is undefined at infinitely many stages, that is, no computation $\Phi^{\emptyset'}_e(e)[s]$ is stable. Thus $\Phi^{\emptyset'}_e(e)\uparrow$ iff the partial computable function

$$g(s) \simeq \mu t > s\, [\Phi^{\emptyset'}_e(e)[t]\uparrow]$$

is total. In that case g is dominated by f, and therefore

$$e \notin \emptyset'' \Leftrightarrow \exists n_0 \forall n \geq n_0 \exists t\, [n \leq t \leq f(n) \ \& \ \Phi^{\emptyset'}_e(e)[t]\uparrow].$$

Since t is bounded by $f(n)$ and $f \leq_T C$, this shows that $\mathbb{N} - \emptyset''$ is $\Sigma^0_2(C)$. Also $\emptyset'' \in \Sigma^0_2 \subseteq \Sigma^0_2(C)$. Therefore $\emptyset'' \leq_T C'$ by Prop. 1.4.14 relative to C. □

The following class contains the class GL_1 of Definition 1.5.4.

1.5.20 Definition. A is *generalized low*$_2$, or GL_2 for short, if $A'' \equiv_T (A \oplus \emptyset')'$. Equivalently, $\emptyset'$ is high relative to A.

1.5.21 Exercise. Show that A is $\mathrm{GL}_2 \Leftrightarrow$ some function $f \leq_T \emptyset' \oplus A$ dominates each (total) function $g \leq_T A$. (Compare this with Exercise 1.5.6 characterizing GL_1.)

1.6 Post's problem

Post (1944) asked whether a c.e. set can be incomputable and Turing incomplete, that is, whether there is a c.e. set A such that $\emptyset <_T A <_T \emptyset'$. It took 12 years to answer his question. Kleene and Post (1954) made a first step by building a pair of Turing incomparable Δ^0_2 sets. To do so they introduced the *method of finite extensions*. Post's question was finally answered in the affirmative by Friedberg (1957*b*) and Muchnik (1956) independently. They built a pair of Turing incomparable sets that are also computably enumerable, strengthening the Kleene–Post result. Their proof technique is nowadays called the priority method with finite injury. For more background on Post's problem see Chapter III of Odifreddi (1989) and our discussion on page 34.

Turing incomparable Δ^0_2-sets

For sets Y, Z we write $Y \mid_T Z$ if $Y \not\leq_T Z$ & $Z \not\leq_T Y$. The following result is due to Kleene and Post (1954).

1.6.1 Theorem. *There are sets $Y, Z \leq_T \emptyset'$ such that $Y \mid_T Z$.*

Proof idea. Note that $Y \mid_T Z$ is equivalent to the conjunction of the statements R_i for all i, where

$$\begin{aligned} R_{2e} &: \exists n \neg Y(n) = \Phi^Z_e(n) \\ R_{2e+1} &: \exists n \neg Z(n) = \Phi^Y_e(n). \end{aligned}$$

Thus we may divide the overall task that $Y \mid_T Z$ into subtasks, called *requirements*. To *meet* a requirement means to make its statement true.

The construction of Y and Z is relative to $\emptyset'$. We meet the requirements one by one in the given order. We define sequences $\sigma_0 \prec \sigma_1 \prec \ldots$ and $\tau_0 \prec \tau_1 \prec \ldots$, and let $Y = \bigcup_i \sigma_i$ and $Z = \bigcup_i \tau_i$. At stage $i+1$ we meet R_i by defining σ_{i+1} and τ_{i+1} appropriately. Since we have $\emptyset'$ at our disposal as an oracle, we may ask whether $\Phi^Z_e(n)$ can be made defined for a particular number n. Then we may define $Y(n)$ in such a way that it differs from $\Phi^Z_e(n)$. This method of providing a counterexample to an equality of sets is called *diagonalization*, and a number such as n above is called a diagonalization witness.

Construction. Let $\sigma_0 = \tau_0 = \emptyset$.

Stage $i+1$, $i = 2e$. Let $n = |\sigma_i|$. Using $\emptyset'$ as an oracle, check whether there is $\tau \succ \tau_i$ such that $y = \Phi^\tau_e(n)\downarrow$. (Note that this is a Σ^0_1 question, so it can be answered by $\emptyset'$.) If so, let $\tau_{i+1} = \tau$ and $\sigma_{i+1} = \sigma_i x$, where $x = \max(1-y, 0)$. Otherwise, let $\sigma_{i+1} = \sigma_i 0$ and $\tau_{i+1} = \tau_i 0$ (merely to ensure the strings are extended at every stage).

Stage $i+1$, $i = 2e+1$. Similar, with the sides interchanged: let $n = |\tau_i|$. Using $\emptyset'$ as an oracle, see if there is $\sigma \succ \sigma_i$ such that $y = \Phi^\sigma_e(n)\downarrow$. If so, let $\sigma_{i+1} = \sigma$ and $\tau_{i+1} = \tau_i x$, where $x = \max(1-y, 0)$. Otherwise, let $\sigma_{i+1} = \sigma_i 0$ and $\tau_{i+1} = \tau_i 0$.

Verification. Clearly $Y, Z \leq_T \emptyset'$. Each requirement R_{2e} is met due to the actions at stage $i+1$ where $i = 2e$: if we cannot find an extension $\tau \succ \tau_i$ such that $\Phi^\tau_e(n)\downarrow$ then by the use principle (see after 1.2.17) $\Phi^Z_e(n)\uparrow$, because Z extends τ_i. Otherwise we ensure that $Y(n) \neq \Phi^Z_e(n)$. Either way R_{2e} is met. The case of a requirement R_{2e+1} is similar. □

Simple sets

We now turn our attention to the c.e. sets and their Turing degrees. The halting problem is a rather special example of an incomputable c.e. set. Here we investigate a whole class of incomputable sets, the co-infinite c.e. sets that meet each infinite c.e. set nontrivally.

1.6.2 Definition. A c.e. set A is *simple* if $\mathbb{N} - A$ is infinite and $A \cap W \neq \emptyset$ for each infinite c.e. set W.

In particular, $\mathbb{N} - A$ is not c.e., so A is not computable. The intuition is that A is so large that it meets each infinite c.e. set. To call such a set "simple" is misleading, but it has been done so for decades and no one intends to change the term. At least the halting problem $\emptyset'$ is not simple: by the Padding Lemma 1.1.3 one can obtain an infinite c.e. set W of indices for the empty set. Then $\emptyset' \cap W = \emptyset$.

1.6.3 Theorem. *There is a simple set.*

Proof, version 1. This argument is due to Post (1944). Let $A = \text{ran}(\psi)$ where

$$\psi(i) \simeq \text{the first element} \geq 2i \text{ enumerated into } W_i.$$

Since ψ is partial computable, A is c.e. If $x < 2i$ is in A then $x = \psi(k)$ for some $k < i$. Hence $\#A \cap [0, 2i) \leq i$, so A is co-infinite. By definition A is simple. □

Proof, version 2. We present the foregoing proof in a different language in order to introduce some terminology which will be used frequently in later constructions of c.e. sets. More terminology will be developed in the proof of Theorem 1.6.4 which strengthens the present result.

We build A by a computable enumeration (Definition 1.1.15). As in the proof of Theorem 1.6.1, we divide the overall task to make A simple into requirements

$$S_i : \#W_i = \infty \;\Rightarrow\; W_i \cap A \neq \emptyset, \tag{1.9}$$

while keeping A co-infinite. (The S_i will be called *simplicity requirements.*) The construction of A is in stages. We let $A_0 = \emptyset$. At each stage $s > 0$ we have a finite set A_{s-1} of elements that have been enumerated so far. A_{s-1} is given by a strong index. During stage s we determine a finite set $F \subseteq \mathbb{N}$ and let $A_s = A_{s-1} \cup F$. We say that the elements of F are *enumerated* into A. Note that we think of A_s as the value by the *end* of stage s. We let A be the c.e. set $\bigcup_s A_s$.

Construction of A. Let $A_0 = \emptyset$.
Stage $s > 0$. For each $i < s$, if the requirement S_i is not satisfied at stage s, namely $A_{s-1} \cap W_{i,s-1} = \emptyset$, and there is an $x \in W_{i,s}$ such that $x \geq 2i$, then enumerate the least such x into A_s. We say that S_i acts.

Unlike the proof of Theorem 1.6.1, the stage number is no longer directly tied to a requirement to be met. Rather, a requirement S_i may act at any stage $s > i$. In the present construction, each S_i acts at most once.

After giving the formal construction one has to verify that it actually builds objects as desired:

Verification. If W_i has an element $\geq 2i$ then by construction $W_i \cap A \neq \emptyset$. Thus S_i is met. A number $< 2e$ can only be enumerated by a requirement S_i, $i < e$, and each requirement enumerates at most one number. Thus $\#A \cap [0, 2e) \leq e$ for each e, whence A is co-infinite. □

A c.e. set that is neither computable nor Turing complete

The easiest way to solve Post's problem is by building a low simple set A. To do so, we use the priority method with finite injury introduced by Friedberg (1957*b*) and Muchnik (1956).

Let us begin by noting that the proof of Theorem 1.6.3 above necessarily makes the set A Turing complete: we say a co-infinite c.e. set A is *effectively simple* if there is a computable function g such that $\#W_e \geq g(e) \rightarrow W_e \cap A \neq \emptyset$. The proof of 1.6.3 actually yields an effectively simple set where $g(e) = 2e$. In Proposition 4.1.13 we will show that each effectively simple set is Turing complete.

To ensure that A is low we will add a further type of requirements to the proof of Theorem 1.6.3 (in its second version). While the simplicity requirements want to enumerate elements into A, the new requirements restrict A.

1.6.4 Theorem. *There is a low simple set A.*

Proof idea. Unlike the Kleene–Post Theorem, we cannot use a construction relative to $\emptyset'$ because we want to build a computable enumeration. In the construction we meet the requirements S_i in (1.9) and lowness requirements

$$L_e : \exists^\infty s \, J^A(e)[s-1]\downarrow \;\Rightarrow\; J^A(e)\downarrow \tag{1.10}$$

that restrict A. If L_e is met then $A'(e) = \lim_s f(e, s)$, where

$$f(e,s) = \begin{cases} 1 & \text{if } J^A(e)[s]\downarrow, \\ 0 & \text{otherwise.} \end{cases}$$

So $A' \leq_T \emptyset'$ by the Limit Lemma 1.4.2. That is, A is low.

The strategy for L_e is as follows. When $J^A(e)[s-1]$ newly converges then L_e *restrains* A up to s; in other words, L_e attempts to prevent numbers $< s$ from entering A.

The conflict between the A-positive requirements S_i and the A-restricting requirements L_e is resolved by imposing an effective *priority ordering*, for instance

$$S_0 > L_0 > S_1 > L_1 > \ldots$$

Requirements further to the left are said to have stronger (or higher) priority. A requirement can only restrain requirements of weaker (or lower) priority during the construction. In the verification one shows by induction on descending priority that the action of each requirement is *finitary*, and hence each single requirement is not restrained from some stage on. The strategies for the requirements have to be designed in such a way that they can live with finitely many disturbances. For instance, S_i has to cope with the restraints of finitely many stronger priority lowness requirements. So it needs an element of W_i that is larger than the eventual values of these restraints. There is no computable upper bound for the maximum of these eventual values, so the construction does not any longer make A effectively simple.

An undesirable situation for L_e is the following: it thought it had already secured a computation $J^A(e)\downarrow$, but then it is *injured* because a number $x <$ use $J^A(e)$ is enumerated into A, destroying that computation. To get around this, L_e needs sufficiently many chances, provided by new convergences of $J^A(e)$.

L_e imposes its restraint by *initializing* weaker priority simplicity requirements: when L_e sees a new convergence of $J^A(e)[s-1]$, it tells these requirements to start from the beginning. If a requirement is initialized at stage s, it can afterwards only put numbers $\geq s$ into A. Since oracle questions occurring in the computation $J^A(e)[s-1]$ are less than s, an enumeration of such numbers cannot injure L_e.

Unlike previous constructions, it may now take many attempts to make a requirement permanently satisfied.

Construction. Let $A_0 = \emptyset$.

Stage $s > 0$.

(1) For each $e < s$, if $J^A(e)[s-1]\downarrow$ but $J^A(e)[s-2]\uparrow$ in case $s > 1$, then initialize the requirements S_i for $i > e$. We say that L_e *acts.*

(2) For each $e < s$, if $A_{s-1} \cap W_{e,s-1} = \emptyset$ and there is $x \in W_{e,s}$ such that $x \geq 2e$ and x is no less than the last stage when S_e was initialized, then enumerate the least such x into A. We say that S_e *acts.*

Claim. *Each requirement acts only finitely often, and is met.*

We suppose inductively that the claim holds for all requirements of stronger priority. So we can choose t such that no requirement of stronger priority acts from stage t on. If the requirement in question is S_e, in fact it acts at most once, in which case it is met; if it does not act after stage t, then $W_e \subseteq [0, t)$ because S_e is not initialized after t, so it is met as well.

If the requirement is L_e then by the choice of t it is not injured from stage t on. If $J^A(e)[s-1]\downarrow$ for no $s \geq t$ then it never acts after t, and, since its hypothesis fails it is met. If $s \geq t$ is least such that $J^A(e)[s-1]\downarrow$ then it acts at stage s. By the initialization it carries out at stage s, it never acts again as the computation $J^A(e)[s-1]$ is preserved forever; again L_e is met. □

The construction actually makes A is superlow (see Definition 1.5.3):

1.6.5 Theorem. (Extends 1.6.4) *There is a superlow simple set A.*

Proof. The number of injuries to L_e is computably bounded, for L_e is only injured when some requirement S_i acts such that $i \leq e$. Since each S_i acts at most once, this can happen at most $e+1$ times. If f is as above then

$$\#\{s > e\colon\ f(e,s) \neq f(e,s-1)\} \leq 2e+2.$$

Hence A' is ω-c.e., whence $A' \leq_{tt} \emptyset'$. □

Relativizing the proof of Theorem 1.6.4 yields for each oracle C a set $A \not\leq_T C$ such that A is c.e. relative to C and $(A \oplus C)' = C'$. Thus the jump fails to be one-one on the structure $\mathcal{D}_T$ of all Turing degrees in the following strong sense.

1.6.6 Corollary. *For each* $\mathbf{c} \in \mathcal{D}_T$ *there is* $\mathbf{a} \in \mathcal{D}_T$ *such that* $\mathbf{a} > \mathbf{c}$ *and* $\mathbf{a}' = \mathbf{c}'$.

1.6.7.$^\diamond$ **Exercise.** Prove a further variant of Theorem 1.6.4: there is a low c.e. set A that is not superlow. (A detailed solution requires some of the terminology on building Turing functionals introduced in Section 6.1 below.)

Is there a natural solution to Post's problem?

Post may have hoped for a different kind of solution to the problem he posed, one that is more natural. The meaning of the word "natural" in real life might be: something that exists independent of us humans. In mathematics, to be natural an *object* must be more than a mere artifact of arbitrary human-made definitions (for instance, the particular way we defined a universal Turing program). Natural *properties* should be conceptually easy. Being a simple set is such a property, satisfying the requirements in the proof of Theorem 1.6.4 is not. In computability theory a natural class of sets should be closed under computable permutations. With very few exceptions, classes we study satisfy this criterion; see also Section 1.3. On the other hand, the class of sets satisfying the lowness requirements in the proof of Theorem 1.6.4 may fail this criterion (depending on the particular choice of a universal Turing program). Also, what we put into the c.e. set A constructed depends on the particular way the sets W_e are defined, even in which order they are enumerated. So the set A is an artifact of the way we specified the universal Turing program (defined before 1.1.2). Neither the property (satisfying the requirements) is natural, nor the set A for which the property holds.

Let us say a *Post property* is a property of c.e. sets which is satisfied by some incomputable set and implies Turing incompleteness. Post was not able to define such a property; the closest he came was to show that each hypersimple set (Definition 1.7.5) is truth-table incomplete (also see 4.1.15). The first result in the direction of a natural Post property was by Marchenkov (1976), who introduced a "structural" Post property (to be maximal relative to some c.e. equivalence relation, and also a left cut in a computable linear order). Harrington and Soare (1991) found a Post property that is even first-order definable in $\mathcal{E}$, the lattice of c.e. sets under inclusion. While Marchenkov's property relies on other effective notions, the Harrington–Soare property is based purely on the interaction of c.e. sets given by the inclusion relation. However, in both cases, the construction showing that an incomputable set with the Post property exists is more complex than the one in the proof of Theorem 1.6.4.

In Section 5.2 we will encounter a further Post property, being *K-trivial.* The construction of a c.e. incomputable K-trivial set only takes a few lines and has no injury to requirements. However, it is harder to show that each K-trivial set is Turing-incomplete.

Kučera (1986) gave an injury-free solution to Post's problem where the construction to show existence is somewhat more difficult than in the case of a K-trivial set, but the verification that the set is Turing incomplete is easier. The Post property of Kučera is to be below a Turing incomplete Martin-Löf random set. No injury is needed to show that there is a Turing incomplete (or even low) Martin-Löf random set, and Kučera's construction produces a c.e. incomputable set Turing below it (see Section 4.2).

All these examples of c.e. incomputable sets with a Post property are still far from natural, because they depend, for instance, on our particular version of the universal Turing program. Perhaps the injury-free solutions are more natural, but the choice of a universal program still matters since we have to diagonalize to make the set incomputable. Any reasonable solution W to Post's problem should be relativizable to an oracle set X, so one would expect that $X <_T W^X <_T X'$ for each X. If the solution does not depend on the choice of the universal program, it should also be degree invariant: if $X \equiv_T Y$, then $W^X \equiv_T W^Y$. The existence of such a degree invariant solution to Post's problem is a long-standing open question posed by Sacks (1963*b*).

Turing incomparable c.e. sets

The priority method was introduced by Friedberg (1957*b*) and Muchnik (1956) when they extended the Kleene–Post Theorem 1.6.1 to the c.e. case: there are Turing incomparable c.e. sets A and B. There is only one type of requirement now, but in two symmetric forms, one for $A \not\leq_T B$, and the other for $B \not\leq_T A$. The strategy combines elements from the strategies for simplicity and lowness requirements in the proof of Theorem 1.6.4. For instance, a requirement to ensure $A \neq \Phi_e^B$ enumerates into A and restricts the enumeration of B.

1.6.8 Theorem. (Extends 1.6.1) *There are c.e. sets A and B such that $A \mid_T B$.*

Proof idea. We meet the same requirements as in Theorem 1.6.1, namely

$$\begin{aligned} R_{2e} &: \exists n \neg A(n) = \Phi_e^B(n) \\ R_{2e+1} &: \exists n \neg B(n) = \Phi_e^A(n). \end{aligned}$$

Here $\neg A(n) = \Phi_e^B(n)$ means that either $\Phi_e^B(n)\uparrow$ or $\Phi_e^B(n)\downarrow \neq A(n)$. Again, we cannot use a construction relative to $\emptyset'$, because we want to build *computable* enumerations of A and B. The strategy for R_{2e} is somewhat similar to the one in Theorem 1.6.1: it looks for an unused candidate n such that currently $\Phi_e^B(n) = 0$, and puts n into A; it also attempts to protect this computation by initializing the requirements of weaker priority. If later the B-enumeration of some requirement (necessarily of stronger priority) destroys the computation, then R_{2e} is initialized (in particular, declared unsatisfied), and has to start anew. In the verification, one shows that R_{2e} acts only finitely often. Once it stops acting it is met: otherwise, if actually $A = \Phi_e^B$, there would be yet another candidate available, so it would act another time. To ensure the candidate is not put into A by some other strategy, a requirement R_i chooses its candidates from $\mathbb{N}^{[i]} = \{\langle x, i\rangle : x \in \mathbb{N}\}$.

Construction of A and B. Let $A_0 = B_0 = \emptyset$. All requirements are initialized.

Stage $s > 0$. Let i be least such that R_i is currently unsatisfied and, where $r < s$ is the greatest stage such that R_i was initialized at r,

if $i = 2e$ then $\Phi_e^B(n)[s-1] = 0$ for the least $n \in \mathbb{N}^{[i]} - A_{s-1}$ such that $n > r$,
if $i = 2e+1$ then $\Phi_e^A(n)[s-1] = 0$ for the least $n \in \mathbb{N}^{[i]} - B_{s-1}$ such that $n > r$.

In the first case put n into A, in the second case n into B. Declare R_i satisfied and initialize all requirements of weaker priority. We say that R_i *acts*. Note that $n < s$.

Claim. *Each requirement R_i is initialized only finitely often, acts only finitely often, and is met.*
By induction, we may choose a stage t such that no requirement R_k, $k < i$, acts at a stage $\geq t$. Then R_i is not initialized from t on. Say $i = 2e$ for some e.

Case 1. R_i acts at a stage $s \geq t$ by putting a number n into A. Then R_i is declared satisfied at s and remains so. Moreover, $\Phi^B(n) = 0$ from stage s on, because, by choice of t and its initialization of weaker priority requirements,

$$\Phi^B(n)[s-1] = \Phi^{B\restriction_u}(n)[s-1] = \Phi^{B\restriction_u}(n) = \Phi^B(n),$$

where $u = \text{use}\, \Phi^B(n)[s-1]$. So $1 = A(n) \neq \Phi^B(n) = 0$, and R_i is met.

Case 2. Otherwise, i.e., R_i never acts at a stage $\geq t$. Then $A(n) = 0$ for any $n \geq t$ in $\mathbb{N}^{[i]}$, because before stage t only numbers less than t can be put into A. If $\Phi^B(n) = 0$ then $\Phi^B(n)[s-1] = 0$ for some $s \geq t$, so R_i acts after all, contradiction. Therefore $\neg A(n) = \Phi_e^B(n)$ and again R_i is met. □

1.6.9 Remark. It is useful to view the foregoing construction as a game between us and an *opponent* (whom I like to call Otto). We build the c.e. sets A and B, trying to meet the requirements R_i. Otto controls the functionals Φ_e. He defines them in a way to make our life as hard as possible. For instance, suppose R_{2e} has not been satisfied and consider $n \in \mathbb{N}^{[i]}$ as in the construction. Otto may wait as long as he wants before he lets $\Phi_e^B(n) = 0$. In this case R_{2e} wakes up and enumerates n into A, thereby injuring the weaker priority requirements that want to preserve A. If he never lets $\Phi_e^B(n) = 0$ after the last stage when R_{2e} is initialized then we win R_{2e} because Φ_e^B is not total. Our strategy must be designed in such a way that we win no matter what Otto does.

The following result, known as the *Sacks Splitting Theorem*, is due to Sacks (1963*c*).

1.6.10 Exercise. Given a c.e. set C, one may uniformly determine low c.e. sets A_0 and A_1 such that $C = A_0 \cup A_1$ and $A_0 \cap A_1 = \emptyset$.

This leads to an alternative proof of Theorem 1.6.8: by 1.2.24 we have $C \equiv_T A_0 \oplus A_1$. If we let $C = \emptyset'$ then $A_0 \,|_T\, A_1$, otherwise $\emptyset'$ would be low.

Hint. For each $e \in \mathbb{N}$ and each $i \in \{0,1\}$, meet the lowness requirements for A_i

$$G_{2e+i} : \exists^\infty s\, J^{A_i}(e)[s-1]\downarrow \;\Rightarrow\; J^{A_i}(e)\downarrow.$$

When x enters C at stage s, you have to decide which side to put it in (of course, no one wants it). Choose the strongest priority requirement G_{2e+i} that would be injured by the enumeration of x into A_i (namely, $x < \text{use}\, J^{A_i}(e)[s-1]$) and put x into the other side A_{1-i}. After spelling out the formal construction, verify by induction on n that each G_n is met. □

Consider, for instance, the requirement G_2, which tries to preserve a computation $J^{A_0}(1)$. The number of injuries to G_2 depends on the use $u = \text{use}\, J^{A_1}(0)$. Each time an $x < u$ enters C, we might injure G_2 because G_1 has stronger priority. Thus, unlike Theorem 1.6.5, the number of injuries to a requirement is not computably bounded, and we cannot guarantee that the A_i are superlow. Indeed, Bickford and Mills (1982) showed that there are no superlow c.e. sets A_0, A_1 such that $A_0 \oplus A_1$ is weak truth-table complete. On the other hand, one can achieve that $A_0 \oplus A_1$ is Turing complete by Theorem 6.1.4.

1.7 Properties of c.e. sets

We introduce three properties of a co-infinite c.e. set A and discuss how they relate to its relative computational complexity. Recall from 1.6.2 that such a set A is simple if $A \cap W_e \neq \emptyset$ for each e such that W_e is infinite. We show that each incomputable c.e. weak truth-table degree contains a simple set. Thus, simplicity has no implication on the complexity beyond being incomputable, unless our measure of relative complexity is finer than weak truth-table reducibility (see Remark 1.7.4). The first two of the properties we introduce strengthen simplicity.

(1) A is *hypersimple* if $\mathbb{N} - A$ is hyperimmune, namely, the function mapping n to the n-th element of $\mathbb{N} - A$ is not dominated by a computable function (see after 1.5.9). Such sets exist in each incomputable c.e. Turing degree, but no longer in each c.e. weak truth-table degree.

(2) A is *promptly simple* if for some computable enumeration, for each infinite W_e some element of W_e enters A with only a computable delay from the stage on when it entered W_e. The Turing degrees of promptly simple sets form a proper subclass of the c.e. incomputable Turing degrees. In fact they coincide with the c.e. degrees $\mathbf{d}$ that are *non-cappable*, namely $\forall \mathbf{y} \neq \mathbf{0}\, \exists \mathbf{b}\, [\mathbf{0} < \mathbf{b} \leq \mathbf{d}, \mathbf{y}]$ holds in the partial order of c.e. Turing degrees. (Non-cappable Turing degrees are complex in the sense that they share incomputable knowledge with each c.e. degree $\mathbf{y} > \mathbf{0}$.) This coincidence, due to Ambos-Spies, Jockusch, Shore and Soare (1984), implies that the promptly simple degrees are first-order definable in $\mathcal{R}_T$, the partial order of c.e. Turing degrees. We will only prove the easier implication that each promptly simple degree is non-cappable.

(3) A is *creative* if it is incomputable in a uniform way, namely, there is a computable function p such that for each e, the number $p(e)$ shows that W_e is not the complement of A. The intuition is that such sets are far from computable; in fact we show that A is creative iff A is m-complete. By a result of Harrington, the class of creative sets can be characterized using only the Boolean operations on c.e. sets. Thus, being creative is first-order definable in the lattice $\mathcal{E}$ of c.e. sets under inclusion.

Both the first-order definability of the promptly simple degrees in the c.e. Turing degrees and the first-order definability of the creative sets in $\mathcal{E}$ are rather unexpected because these properties were defined in terms of concepts that appear to be external to the structure. The definability results show that the properties are in fact intrinsic. Another example of this is the first-order definability in $\mathcal{R}_T$ of the *low*$_n$ degrees for $n \geq 2$, and of the *high*$_n$ degrees for $n \geq 1$. These results are due to Nies, Shore and Slaman (1998).

The properties (1) and (3) will only play a marginal role for us, but (2) will be important later on, because several strong lowness properties, such a being low for Martin-Löf randomness (Definition 5.1.7), hold for some promptly simple set.

We will need the following notation throughout this section.

1.7.1 Definition. (i) For a set $B \subseteq \mathbb{N}$ let $\overline{B}$ denote its complement $\mathbb{N} - B$.
(ii) For a set $S \subseteq \mathbb{N}$, the function p_S lists S in the order of magnitude. That is, $p_S(0) < p_S(1) < \ldots$, and $S = \operatorname{ran}(p_S)$. We say that $p_S(i)$ is the i-th element of S. The domain of p_S is $\{i\colon i < \#S\}$.
If $S = \overline{B}$ for some c.e. set B we build, we write $p_{\overline{B},s}(i)$ for $p_{\overline{B}_s}(i)$.

Each incomputable c.e. wtt***-degree contains a simple set***

To obtain a simple set Turing below a given incomputable c.e. set C, we use the *permitting method*.

1.7.2 Theorem. (Extends 1.6.4) *For each c.e. incomputable set C, there is a simple set A such that $A \leq_{wtt} C$.*

Proof. Once again we meet the simplicity requirements S_i in (1.9). To ensure that $A \leq_{wtt} C$, we only to put x into A at stage s if C *permits* it, in the sense that $C_{s-1} \restriction_x \neq C_s \restriction_x$. Then $C \restriction_{x+1} = C_s \restriction_{x+1}$ implies $A(x) = A_s(x)$, so to determine $A(x)$, we compute the least s such that $C \restriction_{x+1} = C_s \restriction_{x+1}$, using C as an oracle, and output $A_s(x)$. Thus $A \leq_{wtt} C$.

Why does C permit any number for S_i? If W_i is infinite, then there will be infinitely many x that S_i wishes to put into A. If C never permitted any x then C would be computable, contrary to our hypothesis.

Construction of A. Let $A_0 = \emptyset$.

Stage $s > 0$. For each $i < s$, if $A_{s-1} \cap W_{i,s-1} = \emptyset$ and there is an $x \in W_{i,s}$ such that $x \geq 2i$ and $C_{s-1} \restriction_x \neq C_s \restriction_x$, then enumerate the least such x into A_s.

Verification. As before, A is co-infinite. To show that each requirement S_i is met, suppose that W_i is infinite but $W_i \cap A = \emptyset$. Then C is computable: on input y, find the least t such that $x \in W_{i,t}$ for some $x > y$. Then $C(y) = C_t(y)$, else we would put x into A at the least stage $s \geq t$ such that $C_{s-1}(y) \neq C_s(y)$. This contradiction shows that S_i is met. □

To obtain the full result, we refine this proof by also coding C into A.

1.7.3 Theorem. (Extends 1.6.4) *For each incomputable c.e. set C, there is a simple set A such that $A \equiv_{wtt} C$.*

Proof. For $C \leq_{wtt} A$, we put $p_{\overline{A},s}(3y)$ into A when y enters C. We need to make sure that A is still co-infinite.

Construction of A. Let $A_0 = \emptyset$.
Stage $s > 0$. For each $i < s$, if $A_{s-1} \cap W_{i,s-1} = \emptyset$, and there is an $x \in W_{i,s}$ such that $x \geq 3i$ and $C_{s-1} \restriction_x \neq C_s \restriction_x$, then enumerate the least such x into A_s.

If $y \in C_s - C_{s-1}$, put $p_{\overline{A},s-1}(3y)$ into A.

Verification. Note that $\#\overline{A} \cap [0, 3e) \geq 2e$ for each e, since at most e elements less than $3e$ enter A due to the requirements S_i and at most e elements for the coding of C into A. Hence $p_{\overline{A}}(y) \leq 3y$ for each y. Then $A_s\upharpoonright_{3y+1} = A\upharpoonright_{3y+1}$ implies $C_s(y) = C(y)$, so $C \leq_{wtt} A$. The rest is as before. □

1.7.4 Remark. The result cannot be strengthened to truth-table degrees: Jockusch (1980) proved that there is a nonzero c.e. truth-table degree which contains no simple set. In fact, there is a c.e. incomputable set C such that no simple set A satisfies $A \leq_{tt} C$ and $C \leq_{wtt} A$.

Hypersimple sets

Informally, A is simple if the complement of A is thin in the sense that it does not contain any infinite c.e. set. Stronger properties have been studied (see Odifreddi 1999, pg. 393). They all state in some way that the complement of A is thin (while still being infinite). The strongest is being maximal: for each c.e. set $W \supseteq A$, either W is cofinite or $W - A$ is finite. A moderate strengthening of being simple is hypersimplicity, namely, the complement of A is hyperimmune (1.5.10).

1.7.5 Definition. A is *hypersimple* if A is co-infinite and $\exists^\infty n\, f(n) < p_{\overline{A}}(n)$ for each computable function f.

It would be sufficient to require $\exists n\, f(n) < p_{\overline{A}}(n)$, for if $f(n) \geq p_{\overline{A}}(n)$ for almost all n then a finite modification $\widetilde{f}$ of f satisfies $\forall n\, \widetilde{f}(n) \geq p_{\overline{A}}(n)$. Each hypersimple set A is simple, for if A is not simple then by 1.1.19 $\overline{A}$ has an infinite computable subset S. Then p_S is computable and $p_{\overline{A}}(n) \leq p_S(n)$ for each n.

To show that each incomputable c.e. Turing degree contains a hypersimple set one could modify Theorem 1.7.3 and its proof. We give an alternative method to turn an incomputable c.e. set C into a hypersimple set $A \equiv_T C$.

1.7.6 Proposition. *For each incomputable c.e. set C there is a hypersimple set A such that $A \leq_{tt} C \leq_T A$.*

Proof. By Proposition 1.1.17, we can effectively obtain a partial computable function ψ defined on an initial segment of $\mathbb{N}$ such that $C = \mathrm{ran}(\psi)$: at stage s we enumerate $\psi(s)$. Let A be the set of stages s such that some element less than $\psi(s)$ enters C later than stage s, namely, $A = \{s\colon \exists t > s\, [\psi(t) < \psi(s)]\}$. These are called the deficiency stages of ψ. The description of A is in Σ^0_1 form since "$\psi(t) < \psi(s)$" is Σ^0_1. So A is c.e. by Proposition 1.4.12. Since C is incomputable, $\mathrm{dom}(\psi) = \mathbb{N}$. Let $C_s = \{\psi(0), \ldots, \psi(s-1)\}$. Then

$$s \in A \leftrightarrow \bigvee\{y \in C\colon y < \psi(s) \;\&\; y \notin C_s\}.$$

By Proposition 1.2.21 this effective assignment of a Boolean expression to s shows that $A \leq_{tt} C$. For $C \leq_T A$, note that for each m, where $x_m = \psi(p_{\overline{A}}(m))$, at a stage greater than $p_{\overline{A}}(m)$ only a number greater than x_m can be enumerated into C. In particular, $x_0 < x_1 < \ldots < x_m$ and hence $m \in C \leftrightarrow m \in C_{f(m)}$ for every function f such that $\forall m\, f(m) \geq p_{\overline{A}}(m)$. Letting $f = p_{\overline{A}}$ this shows that $C \leq_T A$. Also, no such function f is computable, so A is hypersimple. □

We say that Y is *introreducible* if Y is infinite and $Y \leq_T X$ for each infinite $X \subseteq Y$. The proof of 1.7.6 shows that $C \leq_T f$ for every function f such that $\forall m\, f(m) \geq p_{\overline{A}}(m)$.

If $X \subseteq \overline{A}$ is infinite, then $f(m) = p_X(m) \geq p_{\overline{A}}(m)$ for each m. Therefore $\overline{A} \leq_T X$. Thus $\overline{A}$ is introreducible.

1.7.7 Exercise. Show that each truth table-degree contains an introreducible set.

1.7.8 Remark. We have seen two ways of obtaining a c.e. set. At first sight they seem to be very different.

(a) The set can be given by a *definition*, "in one piece". Examples are the first proof of Theorem 1.6.3 to obtain a simple set, and the proof of Proposition 1.7.6 (where the definition of a simple set is based on a computable enumeration of the given set C). An example of a simple weak truth-table complete set obtained by a definition is in 2.1.28.

(b) We can build the set using a stage-by-stage *construction*. Examples are the second proof of Theorem 1.6.3, and the proof of Theorem 1.7.3.

Distinguishing these two ways is helpful for our understanding, even though formally, a construction of a c.e. set A is just a definition of a computable enumeration of A via an effective recursion on stages. (In more complex constructions, like the ones encountered in Chapter 6, one also builds auxiliary objects such as Turing functionals.) For a construction, at stage s we need to know what happened at the previous stages, which is not the case for direct definitions (such as the first proof of 1.6.3).

Constructions have the advantage of being flexible. For instance, we extended the construction in the second proof of Theorem 1.6.3 in order to obtain a low simple set. However, constructions also introduce more artifacts and thereby tend to make the set less natural. Definitions of c.e. sets seem to be more natural, but often they are not available.

Let us briefly skip ahead to Kučera's injury-free solution to Post's problem (Section 4.2) and the construction of a c.e. K-trivial set (Section 5.2), both already mentioned on page 34. These are constructions in that the action at stage s depends on the past: we have to keep track of whether a requirement has already been satisfied. On the other hand, they are very close to direct definitions because the requirements do not interact. Both sets are obtained by applying a simple operation to a given object, similar to the proof of Proposition 1.7.6. This given object is a low Martin-Löf random set for Kučera's construction, and the standard cost function for the construction of a K-trivial set.

Promptly simple sets

The definition of simplicity is static: we are not interested in the stage when an element of an infinite set W_e appears in A, only in that it appears at all. However, since the sets W_e are equipped with a computable enumeration $(W_{e,s})_{s \in \mathbb{N}}$, one can also consider a stronger, dynamic version of the concept, where an element appearing in an infinite set W_e at stage s is enumerated into A at the same stage or earlier. Given a computable enumeration $(B_s)_{s \in \mathbb{N}}$, for $s > 0$ we let $B_{\mathsf{at}\ s} = B_s - B_{s-1}$.

1.7.9 Definition. A c.e. set A is *promptly simple* (Maass, 1982) if A is co-infinite and, for some computable enumeration $(A_s)_{s \in \mathbb{N}}$ of A, for each e,

$$\#W_e = \infty \;\rightarrow\; \exists s > 0\, \exists x\, [x \in W_{e,\mathsf{at}\ s} \cap A_s]. \tag{1.11}$$

We would rather wish to say that x enters W_e and A at the *same* stage, but this is impossible: for each effective enumeration $(A_s)_{s\in\mathbb{N}}$ of a c.e. set A, there is an e such that $W_e = A$ but $x \in A_{\text{at } s} \to x \notin W_{e,s}$, that is, every element enters W_e later than A. This is immediate if we assume a reasonable implementation of the universal Turing program. It has to simulate the enumeration of A, and each simulated step takes at least as long as a step of the given enumeration of A.

The following seemingly more general variant of Definition 1.7.9 can be found in the literature: there is a computable enumeration $(A_s)_{s\in\mathbb{N}}$ of A and a computable function p such that $\#W_e = \infty \to \exists s \exists x\ [x \in W_{e,\text{at } s} \cap A_{p(s)}]$. However, this formulation is equivalent via the computable enumeration $(A_{p(s)} \cap [0,s))_{s\in\mathbb{N}}$ of A.

With a minor modification, the construction of a low simple set A in Theorem 1.6.4 produces a promptly simple set. This is so because, when a simplicity requirement wants to enumerate a number into A that is greater than the restraint is has to obey, this wish can be granted without delay.

1.7.10 Theorem. (Extends 1.6.4) *There is a low promptly simple set.*

Proof. We modify the proof of Theorem 1.6.4. Instead of the simplicity requirements (S_e) in (1.9) we now meet the *prompt* simplicity requirements

$$PS_e\colon \#W_e = \infty \Rightarrow \exists s\, \exists x\, [x \in W_{e,\text{at } s}\ \&\ x \in A_s].$$

In the construction on page 33 we replace (2) by the following:
For each $e < s$, if PS_e has not been met yet and there is $x \geq 2e$ such that $x \in W_{e,\text{at } s}$ and x is no less than the last stage when PS_e was initialized, then enumerate the least such x into A and declare PS_e met. □

Exercises. Prompt simplicity of a set is formulated in terms of the existence of a particular computable enumeration. Maass (1982) gave a condition equivalent to prompt simplicity which does not involve any enumeration:

1.7.11. Let A be a c.e. co-infinite set. Show that A is promptly simple $\Leftrightarrow$ there is a computable function q such that for each e (1) $W_{q(e)} \subseteq W_e$ and $W_e - A = W_{q(e)} - A$ and (2) $\#W_e = \infty \to W_e - W_{q(e)} \neq \emptyset$.

1.7.12. In 1.7.11 replace (2) by the seemingly stronger condition $\#W_e = \infty \to \#(W_e - W_{q(e)}) = \infty$. Show that this also characterizes prompt simplicity.

Minimal pairs and promptly simple sets

1.7.13 Definition. We say that incomputable sets A and B form a *minimal pair* if every set $Z \leq_T A, B$ is computable.

Lachlan (1966) and Yates (1966) independently proved that minimal pairs of c.e. sets exist (see Soare 1987, Thm. IX.1.2). Lachlan also showed that for c.e. sets A and B, it suffices to require in 1.7.13 that each *c.e.* set $Z \leq_T A, B$ be computable (see Odifreddi 1999, X.6.12).

We show that a promptly simple set E cannot be part of a minimal pair of c.e. sets, namely, its Turing degree is non-cappable. We already mentioned at the beginning of this section that this property characterizes the Turing degrees of promptly simple sets, a result of Ambos-Spies, Jockusch, Shore and Soare (1984). A further characterization of this class of c.e. Turing degrees from the same paper will be provided in Theorem 6.2.2 below: A has promptly simple degree iff A is low cuppable, namely, there is a low c.e. set Z such that $\emptyset' \leq_T A \oplus Z$.

1.7.14 Theorem. *Let the set E be promptly simple. From an incomputable c.e. set C one can effectively obtain a simple set A such that $A \leq_{wtt} C, E$.*

Proof. Choose a computable enumeration $(E_s)_{s\in\mathbb{N}}$ via which E is promptly simple. We ensure that $A \leq_{wtt} C$ by direct permitting (see the proof of Theorem 1.7.2). For $A \leq_{wtt} E$ we use a more general type of permitting called *delayed permitting*: if x enters A at stage s then $E_{s-1}\restriction_x \neq E\restriction_x$, that is, E changes below x at some stage $t \geq s$. Since E is c.e. this implies $A \leq_{wtt} E$.

We make A simple by meeting the requirements S_i in (1.9). We uniformly enumerate auxiliary sets G_i to achieve the E-changes. By the Recursion Theorem we have in advance a computable function g such that $G_i = W_{g(i)}$ for each i. In more detail, given a parameter $r \in \mathbb{N}$, we let g be a computable function effectively obtained from r such that $W_{g(i)} = W_r^{[i]}$ for each i. Based on g, we enumerate a uniformly c.e. sequence $(G_i)_{i\in\mathbb{N}}$. Hence there is a computable function f such that $W_{f(r)} = \bigcup_i G_i \times \{i\}$. Let r^* be such that $W_{f(r^*)} = W_{r^*}$, then the function g obtained for parameter r^* is as required, because $G_i = W_{f(r^*)}^{[i]} = W_{r^*}^{[i]} = W_{g(i)}$ for each i.

Suppose at stage s we are in the situation that we want to put some x into A in order to meet S_i, namely, $x \in W_{i,s}$ and x is permitted by C. We first put a number $y < x$ not yet in $E \cup G_i$ into $G_{i,s}$, which therefore later appears in $W_{g(i)}$. If we try this sufficiently often, then eventually some such y must enter E after a computable delay, for otherwise the infinite set $W_{g(i)}$ would show that E is not promptly simple. We can test whether y enters E within the allowed delay. If so, we put x into A at stage s, thereby meeting S_i.

Construction of the c.e. sets A and G_i, $i \in \mathbb{N}$. Let $A_0 = \emptyset$ and $G_{i,0} = \emptyset$ for each i. All the requirements are declared unsatisfied.

Stage $s > 0$. If there is $i < s$ such that S_i is not satisfied, $W_{g(i),s-1} = G_{i,s-1}$ and there are numbers $y < x$ such that $x \geq 2i$, $x \in W_{i,s}$, $C_s\restriction_x \neq C_{s-1}\restriction_x$, and $y \notin E_{s-1} \cup G_{i,s-1}$, then let i be least, let $\langle x, y\rangle$ be least for i, and put y into $G_{i,s}$. We say that S_i acts. Search for the least $t \geq s$ such that $y \in W_{g(i),t}$ (in the fixed point case t exists). If $y \in E_t$ then put x into A_s and declare S_i satisfied.

Verification. By the Recursion Theorem with Parameters 1.1.6, the fixed point r^* discussed above, and hence the function g, can be obtained effectively from C. Hence A is obtained effectively from C. Clearly $A \leq_{wtt} C$ by direct permitting. Further, $A \leq_{wtt} E$ by delayed permitting. The following shows that A is simple.

Claim. *Each requirement S_i acts only finitely often, and is met.*

We suppose inductively that the claim holds for all requirements S_j, $j < i$. So we can choose s_0 such that no requirement of stronger priority than S_i acts at a stage $s \geq s_0$. The claim holds if W_i is finite, so suppose that W_i is infinite. Assume for a contradiction that S_i is not declared satisfied. Since C is incomputable and E is co-infinite, there are arbitrarily large stages $s \geq s_0$ at which we attempt to meet S_i via numbers $y < x$. Thus $G_i = W_{g(i)}$ is infinite. Note that for each such attempt, y enters $W_{g(i)}$ at a stage $t \geq s$ because $W_{g(i),s-1} = G_{i,s-1}$. Then, since E is promptly simple via the given enumeration, there is some attempt where $y \in E_t$. Thus S_i is declared satisfied, contradiction. □

Creative sets ⋆

We study c.e. sets that are incomputable in a uniform way:

1.7.15 Definition. A c.e. set C is *creative* if there is a computable one-one function p such that $\forall e\, [W_e \cap C = \emptyset \;\rightarrow\; p(e) \notin W_e \cup C]$.

In 1.1.10 we used a diagonalization argument to prove that the c.e. set $\emptyset' = \{e \colon e \in W_e\}$ is incomputable. This argument actually does more:

1.7.16 Fact. *$\emptyset'$ is creative via the function $p(e) = e$.*

Proof. For each e we have $e \in \emptyset' \leftrightarrow e \in W_e$. If W_e and $\emptyset'$ are disjoint then this implies $e \notin W_e \cup \emptyset'$. □

1.7.17 Lemma. *If B is c.e. and $C \leq_m B$ for some creative set C then B is creative.*

Proof. Suppose that C is creative via the function p, and $C \leq_m B$ via h. Let f be a computable function such that $W_{f(e)} = h^{-1}(W_e)$ for each e. If $W_e \cap B = \emptyset$ then $W_{f(e)} \cap C = \emptyset$, so $p(f(e)) \notin W_{f(e)} \cup C$ and hence $h(p(f(e))) \notin W_e \cup B$. Thus B is creative via the function $h \circ p \circ f$. □

1-reducibility $\leq_1$ was introduced after Definition 1.2.3. The following is due to Myhill (1955).

1.7.18 Theorem. *The following are equivalent for a c.e. set C.*

(i) C is creative.
(ii) C is m-complete.
(iii) C is 1-complete.

Proof. (iii)⇒(ii) is trivial, and (ii)⇒(i) follows from the foregoing facts since $\emptyset' \leq_m C$.
(i)⇒(iii): Suppose C is creative via p. Given a c.e. set B, we show that $B \leq_1 C$. Let g be a computable binary function such that

$$W_{g(x,y)} = \begin{cases} \{p(x)\} & \text{if } y \in B \\ \emptyset & \text{otherwise.} \end{cases}$$

By the Recursion Theorem with Parameters 1.1.6 and Exercise 1.1.7, there is a one-one function f such that $W_{g(f(y),y)} = W_{f(y)}$ for each y. Let $h = p \circ f$. If $y \in B$ then $W_{f(y)} = \{p(f(y))\}$, so $h(y) \in C$, otherwise this would contradict the hypothesis that C is creative via p. If $y \notin B$ then $W_{f(y)} = \emptyset$, so $h(y) \notin C$. Thus $B \leq_1 C$ via h. □

From Myhill's Theorem 1.2.5 we conclude the following.

1.7.19 Corollary. *For every pair of creative sets C_1, C_2 there is a computable permutation p of $\mathbb{N}$ such that $p(C_1) = C_2$.* □

Harrington proved that creativity is first-order definable in the lattice of c.e. sets under inclusion, a result that was first published in Soare (1987). The first-order property defining creativity of C states that there is an auxiliary set F such that, for each Z, there is a piece R which is incomputable as shown by F, and on which C coincides with Z.

For c.e. sets X and R such that $X \subseteq R$ we write $X \sqsubset R$ if $R - X$ is c.e. Note that if R is computable, then so is X. If $X \subseteq R$ but $X \not\sqsubset R$ then there are infinitely many elements that first appear in R and later enter X (that is, $\#\{x\colon \exists s\, [x \in R_{s-1} \cap X_{\mathrm{at}\ s}]\} = \infty$), for otherwise $R - X$ is c.e. as $R - X$ is almost equal to the set $\{x\colon \exists t\, [x \in R_t - X_t]\}$.

1.7.20 Theorem. *Let C be c.e. Then C is creative $\Leftrightarrow$*

$$\exists F\, \forall Z\, \exists R\, [R \cap F \not\sqsubset R \ \&\ R \cap C = R \cap Z], \tag{1.12}$$

where the quantifiers range over c.e. sets.

Proof idea. For the implication "$\Rightarrow$", as all creative sets are computably isomorphic by 1.7.19, it suffices to provide a particular creative set satisfying (1.12). We can take

$$\widehat{C} = \{\langle x, e\rangle\colon \langle x, e\rangle \in W_e\},$$

as we will verify below. It is harder to show the implication "$\Leftarrow$". If C satisfies (1.12) then we want to define a computable function p via which C is creative. The condition (1.12) is designed to make the construction of such a p work. We are given the c.e. set F and enumerate a c.e. set Z. Assume first that we actually know a witness R for Z in (1.12). Note that R is infinite because $R \cap F \not\sqsubset R$. We have an infinite list of elements x for which we can dictate membership in C because we control $Z(x)$. We let $p(e) = x$ be the e-th element in this list. At first $x \notin Z$ and hence $x \notin C$; if x enters W_e we put x into Z, so that $x \in W_e \cap C$. Thus C is creative via p.

Actually we do not know such a witness R. So we play the strategy above simultaneously for all c.e. sets W_i as possible witnesses. We will write R_i instead of W_i to improve the readability. We define a partial computable function p_i based on the assumption that R_i is the witness for Z in (1.12). At each stage we extend p_i for the least i such that the condition $R_i \cap C = R_i \cap Z$ looks correct so far. When at stage s we define $p_i(e)$ for the next e, we need a value $x \in R_i$ such that $x \notin Z_{s-1}$, so we have to be sure that x is not already taken at a previous stage as a value $p_{i'}(e')$ for some i', e'. This is where the condition $R_i \cap F \not\sqsubset R_i$ comes in: as explained above, there are infinitely many numbers that enter F at a stage s when they are already in $R_{i,s-1}$. We can only use elements x of this kind as values $p_i(e) = x$. There is no conflict with the previous values because they are in F_{s-1}, nor with later values as they are not in F_s.

Proof details. $\Rightarrow$: To see that the set $\widehat{C}$ defined above is 1-complete, let e^* be a c.e. index such that $W_{e^*} = \emptyset' \times \mathbb{N}$. Then $x \in \emptyset' \leftrightarrow \langle x, e^* \rangle \in \widehat{C}$ for each x. Next, (1.12) is satisfied via $F = \emptyset' \times \mathbb{N}$: given a c.e. set $Z = W_k$, let R be the computable set $\mathbb{N} \times \{k\}$. Then $R \cap F = \emptyset' \times \{k\} \not\subseteq^* R$, otherwise $\emptyset'$ is computable. For each element $y = \langle x, k \rangle$ of R, we have $y \in \widehat{C} \leftrightarrow y \in W_k$, so that $R \cap \widehat{C} = R \cap Z$.

$\Leftarrow$: Suppose C satisfies (1.12) via F.
Construction of Z and partial computable functions p_i ($i \in \mathbb{N}$).
Let $Z_0 = \emptyset$, and declare $p_{i,0}(e)$ undefined for each i, e.

Stage $s > 0$. Let i be least such that

(a) if $t < s$ is greatest such that $t = 0$ or we defined a new value of p_i at stage t, then $(R_i \cap C) \upharpoonright_t [s] = (R_i \cap Z) \upharpoonright_t [s]$,
(b) there is an $x \in R_{i,s-1} \cap F_{\text{at } s}$.

If i exists, we say that we *choose* i at stage s. Define $p_{i,s}(e) = x$ where e is least such that $p_{i,s-1}(e)$ is undefined. If $x \in W_{e,u}$ for some (possibly later) stage $u \geq s$, then put x into Z at that stage.

Claim 1. *There is i such that p_i is total.*
Let i be least such that R_i witnesses (1.12) for Z. If $p_{i'}$ is total for some $i' < i$ we are done. Otherwise there is s_0 such that we do not choose any $i' < i$ at a stage $s \geq s_0$. Since $R_i \cap F \not\subseteq^* R_i$, there is an infinite stream of numbers from R_i into F, thus for infinitely many s we choose i at stage s. Hence p_i is total.
Claim 2. *C is creative via p_i.*
Since p_i is total and we always check (a) before we define a further value of p_i, we have $R_i \cap C = R_i \cap Z$. Given e, we define $p_i(e) = x$ at some stage s, and x is not a value $p_{i'}(e')$ at stage $s - 1$ since $x \in F_{\text{at } s}$ while those values are in F_{s-1}. Also x is not taken as a value $p_{i'}(e')$ at any later stage as these are in $\mathbb{N} - F_s$. Since $x \in R_i$, this implies that $x \in W_e \leftrightarrow x \in Z \leftrightarrow x \in C$. Clearly p_i is one-one. This establishes Claim 2 and the Theorem. □

1.8 Cantor space

So far, the natural numbers have been the atomic objects. We studied sets of natural numbers and functions mapping numbers to numbers. From now on we will often view *sets* of natural numbers as the atomic objects. We study sets of sets and functions mapping sets to sets. Sets of natural numbers are identified with infinite sequences over $\{0, 1\}$. These sequences are the elements of Cantor space $\{0, 1\}^{\mathbb{N}}$ (usually denoted by $2^{\mathbb{N}}$). It is equipped with the product topology where the topology on $\{0, 1\}$ is discrete. Subsets of $2^{\mathbb{N}}$ will be called *classes* to distinguish them from sets of numbers. (The open and the closed *sets* are exceptions to this rule; here we use the terms "set" and "class" interchangeably.)

A Stone space is a compact topological space such that the sets that are simultaneously closed and open (called clopen sets) form a basis. The space $2^{\mathbb{N}}$ is an example of a Stone space. In this section we develop a bit of Stone duality, a correspondence between topological concepts in Stone spaces and concepts related to Boolean algebras. The dual algebra of a Stone space S is its Boolean algebra B of clopen sets. The dual space of a Boolean algebra B is the space where the points are the ultrafilters U, and the basic open sets are the sets of the form $\{U\colon x \in U\}$ for $x \in B$ (which are also closed). Open sets correspond to ideals, and closed sets to filters. The dual algebra of $2^{\mathbb{N}}$ is a countable dense Boolean algebra. Such a Boolean algebra is unique up to isomorphism.

We only develop Stone duality for $2^{\mathbb{N}}$, to the extent relevant to us, namely, for representing open or closed sets. Instead of working with the countable dense Boolean algebra, we will restrict ourselves to $\{0,1\}^*$. Filters then become binary trees without dead branches. Thus each closed set is represented by such a tree. The advantage of this representation of closed sets is that we have "descended down one level": we are looking at sets of strings rather than at classes. As a result, we may apply the usual algorithmic notions developed for numbers (or strings) as the atomic objects. For instance, we will study closed sets where the representing tree is a Π^0_1 set, called Π^0_1 classes. We prove some important existence theorems, such as the Low Basis Theorem 1.8.37 that each nonempty Π^0_1 class P contains a low set. The Π^0_1 classes are at the first level of the arithmetical hierarchy for classes defined in 1.8.55.

Given a set Z and a number n, we may regard $Z \upharpoonright_n$ as a partial description of Z since it specifies the first n bits. Classes will be used as a tool to study sets, because they constitute a more general type of partial descriptions: we may think of a class $\mathcal{C}$ as a partial description of any set $Z \in \mathcal{C}$. This enables us to switch to the global view of a set, where the set is appreciated all at once (Section 1.3). We will be interested in partial descriptions given by classes that have special topological properties, or are easy to describe themselves, or both (like Π^0_1 classes).

If the class $\mathcal{C}$ is small in a particular sense, it provides a close description of Z in that sense. Usually we will take smallness in the sense of the uniform measure on Cantor space (also called the product measure, and defined in 1.9.7). For instance, the class corresponding to the partial description $z = Z \upharpoonright_n$ of Z is the basic open cylinder $\{Y\colon\ z \prec Y\}$. It has uniform measure 2^{-n} and therefore gets smaller as n increases.

In Section 1.4 we defined computable approximations $(Z_r)_{r\in\mathbb{N}}$ of a Δ^0_2 set Z, where Z_r is contained in $[0,r)$ (equivalently, one can let Z_r be a string of length r). We think of Z_r as a different type of partial description of Z. It also improves as r increases. The idea to approximate a set in stages can be generalized: we may weaken the effectiveness condition on the approximation, or we can replace each Z_r by a class as a partial description. In the proof of the Low Basis Theorem 1.8.37 we even do both: we define a $\emptyset'$-computable sequence $(P^r)_{r\in\mathbb{N}}$ of Π^0_1 classes, where P^0 is the given class P. The low set $Z \in P$ is determined by $\{Z\} = \bigcap_r P^r$.

In the introduction to Chapter 3 we will consider the idea that a random set is a set which does not even admit a close description. For instance, a Martin-Löf test (see 3.2.1) is a further way to approximate a set in stages r by classes G_r. Now the r-th partial description G_r is an open set which has a uniform measure of at most 2^{-r} and is uniformly c.e. in r. A set is Martin-Löf random if it is not in $\bigcap_r G_r$ for any such test. Note that $\bigcap_r G_r$ has uniform measure 0 but is usually not a singleton.

Open sets

For a string y, the class of infinite binary sequences extending y is denoted by

$$[y] = \{Z\colon\ y \preceq Z\}.$$

These classes are called basic open cylinders, or *cylinders* for short. Clearly $[x] \supseteq [y] \leftrightarrow x \preceq y$. The cylinders form a basis of a topology: $R \subseteq 2^{\mathbb{N}}$ is *open* if R is the union of the cylinders contained in R, or, in other words, if for each $Z \in 2^{\mathbb{N}}$ we have $Z \in R \leftrightarrow \exists n\, [Z\upharpoonright_n] \subseteq R$.

A Turing functional Φ can be viewed as a partial map $\Phi : 2^{\mathbb{N}} \to 2^{\mathbb{N}}$ where $\Phi(Y)$ is defined iff Φ^Y is total. Unless the domain of this map is empty, it can be considered as a subspace of $2^{\mathbb{N}}$. One reason why we use the product topology on $2^{\mathbb{N}}$ is that we want this partial map to be continuous. This is the case because of the use principle (see after 1.2.17) which states that a converging oracle computation only depends on a finite initial segment of the oracle (Exercise 1.8.8).

Binary trees and closed sets

A subset P of a topological space is called *closed* if its complement is open. We represent closed sets in Cantor space by subtrees of $\{0,1\}^*$.

1.8.1 Definition. (i) A *binary tree* is a subset B of $\{0,1\}^*$ closed under taking prefixes. That is, $x \in B$ and $y \preceq x$ implies $y \in B$. (ii) Z is a *path* of B if $Z\upharpoonright_n \in B$ for each n. The set of paths of B is denoted by *Paths*(B).

For instance, $T = \{0^i\colon\ i \in \mathbb{N}\} \cup \{0^i1\colon\ i \in \mathbb{N}\}$ is a binary tree such that *Paths*$(T) = \{0^\infty\}$. Since binary trees are subtrees of $\{0,1\}^*$, we may apply the visual terminology of "above/left/right" introduced on page 12. The following is known as König's Lemma.

1.8.2 Lemma. *If B is an infinite binary tree then Paths$(B) \neq \emptyset$.*

Proof. For each n let x_n be the leftmost string x of length n such that $B \cap \{y\colon\ y \succeq x\}$ is infinite. Then $x_n \prec x_{n+1}$ for each n, and $\bigcup_n x_n$ is a path of B. □

We say that $x \in B$ is a *dead branch* of a binary tree B if $B \cap \{y\colon\ y \succeq x\}$ is finite. By König's Lemma, this is equivalent to the condition that no path of B extend x.

Paths(B) is a closed set: if $Z \notin$ *Paths*(B) then there is n such that $Z\upharpoonright_n \notin B$, so $[Z\upharpoonright_n] \cap$ *Paths*$(B) = \emptyset$. However, there are "more" binary trees than closed sets. For instance, if we take the tree T above and cut off the dead branches, we obtain $T' = \{0^i\colon\ i \in \mathbb{N}\}$. The trees T and T' have the same paths. Closed sets correspond to trees B without dead branches, that is, $x \in B$ implies $x0 \in B$ or $x1 \in B$ for each $x \in \{0,1\}^*$.

1.8.3 Fact. (Stone duality for closed sets)

(i) *If P is closed then*

$$T_P = \{x\colon\ [x] \cap P \neq \emptyset\}$$

is a tree without dead branches such that Paths$(T_P) = P$.

(ii) If B is a tree without dead branches then $B = T_{Paths(B)}$.

Proof. (i) Clearly T_P is closed under prefixes. Moreover, T_P has no dead branches because $[x] = [x0] \cup [x1]$. Since $2^{\mathbb{N}} - P$ is open,

$Z \in P \leftrightarrow \forall n\, [Z\restriction_n] \cap P \neq \emptyset \leftrightarrow Z \in Paths(T_P)$.

(ii) Clearly $T_{Paths(B)} \subseteq B$. For the converse inclusion, suppose that $x \in B$. Since B has no dead branches, $\{y \succeq x\colon y \in B\}$ is infinite, so by König's Lemma 1.8.2 there is a set $Z \in Paths(B)$ extending x. □

We will sometimes identify P and T_P. For instance, when we say that x *is on* P we mean that $x \in T_P$, or equivalently that $[x] \cap P \neq \emptyset$.

Representing open sets

If R is open we let $P = 2^{\mathbb{N}} - R$ and represent R by $A_R = \{0,1\}^* - T_P$. Thus,

$$A_R = \{x\colon [x] \subseteq R\}.$$

This set is closed under extensions of strings, namely $x \in A_R$ and $x \prec y$ implies $y \in A_R$. Further, for each $x \in \{0,1\}^*$, $x0 \in A_R$ and $x1 \in A_R$ implies $x \in A_R$. A subset of $\{0,1\}^*$ with these two properties is sometimes called an ideal (of strings). Ideals are the complements in $\{0,1\}^*$ of trees without dead branches. We occasionally identify R and A_R and write $x \in R$ for $[x] \subseteq R$.

The open set generated by a set $S \subseteq \{0,1\}^*$ is

$$[S]^{\prec} = \{X \in 2^{\mathbb{N}}\colon \exists y \in S\; y \preceq X\}.$$

There is a correspondence between ideals and open sets analogous to Fact 1.8.3. It is immediate from 1.8.3 by complementing:

1.8.4 Fact. (Stone duality for open sets)

(i) If R is open, then $A_R = \{x\colon [x] \subseteq R\}$ is an ideal and $R = [A_R]^{\prec}$.

(ii) If C is an ideal of strings then $C = A_{[C]^{\prec}}$. □

The strings x in A_R that are minimal under the prefix ordering form an antichain $D = (x_i)_{i<N}$, $N \in \mathbb{N} \cup \{\infty\}$, such that $[D]^{\prec} = R$. For instance, consider $R = 2^{\mathbb{N}} - \{0^\infty\}$. The corresponding ideal is $A_R = \{x\colon \forall i \in \mathbb{N}\, [x \neq 0^i]\}$, and the antichain is $D = \{0^i 1\colon i \in \mathbb{N}\}$.

Compactness and clopen sets

A topological space $\mathcal{X}$ is called *compact* if, whenever $\mathcal{X}$ is the union of a collection of open sets, then $\mathcal{X}$ is already the union of a finite subcollection. Equivalently, whenever the intersection of a collection of closed sets is empty, then already the intersection of a finite subcollection is empty. If the space has a countable basis then we may assume that the given collection is countable. We will prove that Cantor space $2^{\mathbb{N}}$ is compact. It has a countable basis consisting of the cylinders $[x]$, so it suffices to show that each countable descending sequence $(P^i)_{i\in\mathbb{N}}$ of nonempty closed sets has a nonempty intersection. This is how we usually apply the compactness of $2^{\mathbb{N}}$: $(P^i)_{i\in\mathbb{N}}$ describes a list of desirable conditions on sets, and compactness tells us that there is a set satisfying all the conditions. (We will use this method for the first time in the proof of Theorem 1.8.37.)

1.8.5 Proposition. *If $(P^i)_{i\in\mathbb{N}}$ is a descending sequence of nonempty closed sets, then $\bigcap_i P^i \neq \emptyset$.*

Proof. Let v_n be the leftmost string of length n on P^n. For each e, as n grows $v_n \upharpoonright_e$ only moves to the right on $\{0,1\}^*$, so $z_e = \lim_{n\geq e}(v_n \upharpoonright_e)$ exists.

Let $Z = \bigcup_e z_e$. Fix n. For all e we have $[z_e] \cap P^n \neq \emptyset$. Since P^n is closed, this implies that $Z \in P^n$. □

A subset of a topological space is called *clopen* if it is both open and closed. The clopen sets form a Boolean algebra with the usual union and intersection operations. The clopen sets in Cantor space play a role similar to the finite subsets of $\mathbb{N}$. For instance, we will provide computable approximations of c.e. open sets as effective unions of clopen sets. By the following, the strong indices for finite sets $F \subseteq \{0,1\}^*$ given by Definition 1.1.14 can also be used as indices for the clopen sets. They will be called *strong indices for clopen sets.*

1.8.6 Proposition. *$\mathcal{C} \subseteq 2^{\mathbb{N}}$ is clopen $\Leftrightarrow$ $\mathcal{C} = [F]^{\prec}$ for some finite set $F \subseteq \{0,1\}^*$.*

Proof. $\Rightarrow$: Since the cylinders form a basis of $2^{\mathbb{N}}$, there are sets $D, E \subseteq \{0,1\}^*$ such that $\mathcal{C} = [D]^{\prec} = \bigcup_{\sigma\in D}[\sigma]$ and $2^{\mathbb{N}} - \mathcal{C} = [E]^{\prec} = \bigcup_{\rho\in E}[\rho]$. By the compactness of $2^{\mathbb{N}}$ there are finite sets F, G such that $F \subseteq D$, $G \subseteq E$ and $[F]^{\prec} \cup [G]^{\prec} = 2^{\mathbb{N}}$. Then $\mathcal{C} = [F]^{\prec}$, so F is as required.

$\Leftarrow$: Each cylinder $[\sigma]$ is clopen since its complement is $\bigcup\{[\rho] \colon \rho \neq \sigma \;\&\; |\rho| = |\sigma|\}$. Thus each set of the form $[F]^{\prec}$ is clopen. □

Exercises.

1.8.7. Let $X, Y \in 2^{\mathbb{N}}$. If $X \neq Y$ let $d(X,Y) = 2^{-n}$ where n is least such that $X(n) \neq Y(n)$. Let $d(X,X) = 0$. Show that $(2^{\mathbb{N}}, d)$ is a *metric space* which induces the usual product topology on $2^{\mathbb{N}}$.

A function F between topological spaces is *continuous* if the preimage under F of each open set is open. Consider a map $F\colon\; D \to 2^{\mathbb{N}}$ where D is a subspace of $2^{\mathbb{N}}$. Since the basic open cylinders form a basis, F is continuous iff for each $\rho \in \{0,1\}^*$ such that $[\rho] \cap \mathrm{ran}(F) \neq \emptyset$ there is $\sigma \in \{0,1\}^*$ such that $F([\sigma] \cap D) \subseteq [\rho]$.

1.8.8 Exercise. We view a Turing functional Φ such that $\mathrm{dom}(\Phi) \neq \emptyset$ as a map from the subspace $D = \mathrm{dom}(\Phi)$ to $2^{\mathbb{N}}$. Show that Φ is continuous.

Recall that $A \oplus B = \{2n\colon\; n \in A\} \cup \{2n+1\colon\; n \in B\}$.

1.8.9 Exercise. Show that $L\colon\; 2^{\mathbb{N}} \to 2^{\mathbb{N}}$ is continuous $\Leftrightarrow$ there is a Turing functional Φ and a set A such that $L(Z) = \Phi(Z \oplus A)$ for each $Z \in 2^{\mathbb{N}}$.

The correspondence between subsets of $\mathbb{N}$ *and real numbers*

Co-infinite subsets of $\mathbb{N}$ and real numbers in $[0,1)_{\mathbb{R}}$ are often identified. When we make this identification we usually indicate it. We give the details.

1.8.10 Definition. The map

$$F\colon \{Z \in 2^{\mathbb{N}}\colon Z \text{ is co-infinite}\} \to [0,1)_{\mathbb{R}} \tag{1.13}$$

is defined by $F(Z) = 0.Z = \sum_{i \in Z} 2^{-i-1}$.

We will determine the inverse G of F, thereby showing that F is a bijection. Each real number $r \in [0,1)_{\mathbb{R}}$ can be written in the form $r = \sum_{i \geq 0} r_i 2^{-i-1}$ where $r_i \in \{0,1\}$. We say $0.r_0 r_1 \ldots$ is a binary expansion of r. The set of *dyadic rationals* is

$$\mathbb{Q}_2 = \{z 2^{-n} : z \in \mathbb{Z}, n \in \mathbb{N}\}.$$

A binary expansion of r is unique unless $r \in \mathbb{Q}_2$, in which case we give preference to the finite binary expansion. Thus we view $1/4$ as 0.01 rather than as $0.00111\ldots$ Via this binary expansion a real number $r \in [0,1)_{\mathbb{R}}$ can be identified with a sequence $Z = G(r) = r_0 r_1 \ldots$ that has infinitely many zeros, that is, with a co-infinite set Z. Clearly F and G are inverses.

For a finite string y, we sometimes use the notation $0.y$ as a shorthand for $0.y000\ldots$. Usually we identify a dyadic rational in $(0,1)$ with a finite string ending in 1.

1.8.11 Remark. The bijection F maps the basic open cylinder $[\sigma]$ (restricted to $\{Z \in 2^{\mathbb{N}}\colon Z \text{ is co-infinite}\}$) to the interval $I(\sigma) = [0.\sigma, 0.\sigma + 2^{-|\sigma|})$. For a linear order L with least but no greatest element let *Intalg* L denote the Boolean algebra generated by the intervals $[a,b)$ where $a,b \in L$. Thus, *Intalg* L consists of the sets of the form $\bigcup_{0 \leq i < n} [a_i, b_i)$ where $a_0 < b_0 < a_1 < \ldots < b_{n-1}$, and possibly $b_{n-1} = 1$, where 1 is a greatest element adjoined to L. The Boolean algebra of clopen sets in Cantor space corresponds to the subalgebra of *Intalg* $[0,1)_{\mathbb{R}}$ generated by the intervals of the form $I(\sigma)$. For instance, if $p,q \in \mathbb{Q}_2$, $p < q$, then the corresponding clopen set is $\{Z\colon p \leq 0.Z < q\}$.

Exercises.

1.8.12. To be able to work with the usual topology on $\mathbb{R}$, we also have to remove the finite sets from Cantor space, and the dyadic rationals from $[0,1)_{\mathbb{R}}$. Let $\mathcal{X} = \{Z \in 2^{\mathbb{N}}\colon Z \text{ is co-infinite and infinite}\}$. By restricting the bijection F we obtain a bijection

$$\widetilde{F}\colon \mathcal{X} \to [0,1)_{\mathbb{R}} - \mathbb{Q}_2.$$

Show that $\widetilde{F}$ is a homoeomorphism of the subspace topologies inherited from Cantor space on the left, and the usual topology of $\mathbb{R}$ on the right.

1.8.13. Show that $\{Z \in 2^{\mathbb{N}}\colon Z \text{ is co-infinite}\}$ with the subspace topology is homeomorphic to Baire space $\mathbb{N}^{\mathbb{N}}$ with the product topology.

Effectivity notions for real numbers

We fix some effective encoding of $\mathbb{Q}_2$ by natural numbers. Then a notion $\mathcal{C}$ defined for sets can be applied to a real number r via the left cut $\{q \in \mathbb{Q}_2\colon q < r\}$ (we say that r is *left-*$\mathcal{C}$), or the right cut $\{q \in \mathbb{Q}_2\colon q > r\}$ (we say that r is *right-*$\mathcal{C}$).

1.8.14 Definition. Let r be a real number.

(i) r is *computable* if $\{q \in \mathbb{Q}_2\colon q < r\}$ is computable, and r is Δ^0_2 if this set is Δ^0_2.

(ii) r is *left-c.e.* if $\{q \in \mathbb{Q}_2 \colon q < r\}$ is c.e., and r is *right-c.e.* if $\{q \in \mathbb{Q}_2 \colon q > r\}$ is c.e.

(iii) $Z \subseteq \mathbb{N}$ is *left-c.e.* if the real number $0.Z$ is left-c.e., or, equivalently, if $\{\sigma \colon \sigma <_L Z\}$ is c.e. Similarly we define right-c.e. sets.

(iv) r is *difference left-c.e.* if there are left-c.e. reals $\alpha, \beta \in \mathbb{R}$ such that $r = \alpha - \beta$.

These classes of reals can be characterized via effective approximations by rationals. One may in fact require that the rationals be dyadic. The characterization (v) below is due to Ambos-Spies, Weihrauch and Zheng (2000).

1.8.15 Fact. Let $r \in \mathbb{R}$. The following equivalences hold uniformly.

(i) r is $\Delta^0_2 \Leftrightarrow r = \lim_n q_n$ for a computable sequence $(q_n)_{n\in\mathbb{N}}$ of rationals.

(ii) r is left-c.e. $\Leftrightarrow r = \lim_n q_n$ for a non-descending computable sequence $(q_n)_{n\in\mathbb{N}}$ of rationals.

(iii) r is right-c.e. $\Leftrightarrow r = \lim_n q_n$ for a non-ascending computable sequence $(q_n)_{n\in\mathbb{N}}$ of rationals.

(iv) r is computable $\Leftrightarrow r = \lim_n q_n$ for a computable sequence $(q_n)_{n\in\mathbb{N}}$ of rationals such that $\mathrm{abs}(r - q_n) \le 2^{-n}$ for each n
$\Leftrightarrow$ given n one can compute $q \in \mathbb{Q}$ such that $\mathrm{abs}(r - q) \le 2^{-n}$.

(v) r is difference left-c.e. $\Leftrightarrow r = \lim_n q_n$ for a computable sequence $(q_n)_{n\in\mathbb{N}}$ of rationals such that $\sum_n \mathrm{abs}(q_{n+1} - q_n) < \infty$.

Proof. We leave (i), (iii) and (iv) as exercises.
(ii) $\Rightarrow$: If $W_e = \{q \in \mathbb{Q}_2 \colon q < r\}$ (via our identification), then let $q_n = \max(W_{e,n})$.
$\Leftarrow$: The left cut of r is c.e. because for $q \in \mathbb{Q}_2$ we have $q < r \leftrightarrow \exists n\, q < q_n$.

(v) $\Leftarrow$: We have $\lim_n(a_n - b_n) = \lim_n a_n - \lim_n b_n$ for every pair of converging sequences $(a_n)_{n\in\mathbb{N}}$ and $(b_n)_{n\in\mathbb{N}}$ of reals. Therefore

$$\begin{aligned} r &= q_0 + \sum_n (q_{n+1} - q_n) \\ &= q_0 + \sum_n (q_{n+1} - q_n)\, [\![q_{n+1} \ge q_n]\!] - \sum_n (q_n - q_{n+1})\, [\![q_{n+1} \le q_n]\!]. \end{aligned}$$

$\Rightarrow$: Let $r = \alpha - \beta$ for left-c.e. real numbers α and β. Let $(\alpha_n)_{n\in\mathbb{N}}$ and $(\beta_n)_{n\in\mathbb{N}}$ be nondescending approximations of α, β by dyadic rationals as in (ii). Let $q_n = \alpha_n - \beta_n$, then $\lim_n q_n = r$ and, since $q_{n+1} - q_n = (\alpha_{n+1} - \alpha_n) - (\beta_{n+1} - \beta_n)$, we have $\sum_n \mathrm{abs}(q_{n+1} - q_n) \le \alpha + \beta < \infty$. □

We will usually show that a real number is computable by proving the last condition in (iv): on input n, we compute an approximation $q \in \mathbb{Q}$ of r such that $\mathrm{abs}(r - q) \le 2^{-n}$. Note that r is computable iff r is both left-c.e. and right-c.e. by Proposition 1.1.9.

A further way to apply a set notion to a real number r is the following: r can be uniquely written in the form $r = z + 0.B$ for $z \in \mathbb{Z}$ and co-infinite $B \subseteq \mathbb{N}$;

now one requires the relevant property for B. This leads to the same class if the property is being computable, or being Δ^0_2. In contrast, if A is a c.e. set, then $0.A$ is left-c.e., but not conversely. A counterexample is, for instance, Chaitin's number Ω; see page 108. Each left-c.e. set Z is truth-table equivalent to the c.e. set $\{\sigma\colon \sigma <_L Z\}$.

The computable real numbers form a field. More generally, for any ideal L in the Turing degrees, the real numbers with Turing degree in L form a field, because a real number t obtained from real numbers r, s by a field operation is computable in $r \oplus s$. The left-c.e. real numbers are closed under addition but not under the operation $x \to -x$. In contrast, Ambos-Spies, Weihrauch and Zheng (2000) proved the following surprising fact.

1.8.16 Proposition. *The set D of difference left-c.e. real numbers is a subfield of $\mathbb{R}$.*

Proof. Throughout, we use Fact 1.8.15(v). Clearly D is closed under the operation $x \to -x$ and under addition. For the closure under product, suppose $r, s \in D$. There are $M \in \mathbb{N}$ and effective sequences $(x_n)_{n\in\mathbb{N}}$ and $(y_n)_{n\in\mathbb{N}}$ of rationals converging to r, s, respectively, such that $\forall n\, \mathrm{abs}(x_n) \le M$, $\forall n\, \mathrm{abs}(y_n) \le M$,

$$\textstyle\sum_n \mathrm{abs}(x_{n+1} - x_n) \le M, \text{ and } \sum_n \mathrm{abs}(y_{n+1} - y_n) \le M.$$

Then $\lim_n x_n y_n = rs$ and $\sum_n \mathrm{abs}(x_{n+1}y_{n+1} - x_n y_n) \le \sum_n \mathrm{abs}(x_{n+1}y_{n+1} - x_n y_{n+1}) + \sum_n \mathrm{abs}(x_n y_{n+1} - x_n y_n) \le 2M^2$, so $rs \in D$.

It remains to show that $1/r \in D$ in case that $r \neq 0$. We may assume $M > \mathrm{abs}(1/r)$, so there is $n_0 \in \mathbb{N}$ such that $x_n \neq 0$ and $\mathrm{abs}(1/x_n) \le M$ for all $n \ge n_0$. Then

$$\sum_{n \ge n_0} \mathrm{abs}\Big(\frac{1}{x_{n+1}} - \frac{1}{x_n}\Big) = \sum_{n \ge n_0} \frac{\mathrm{abs}(x_n - x_{n+1})}{\mathrm{abs}(x_n x_{n+1})} \le M^3.$$

Note that the closure under the field operations is effective. □

Exercises.

1.8.17. If r is left-c.e. then e^r is left-c.e. as well.

1.8.18. Suppose r_i is a computable real number uniformly in $i \in \mathbb{N}$ and $0 \le r_i \le 2^{-i}$ for each i. Show that $r = \sum_i r_i$ is computable.

Effectivity notions for classes of sets

This subsection introduces two essential concepts: co-c.e. closed sets (also called Π^0_1 classes) and c.e. open sets (also called Σ^0_1 classes). We show how to index such classes, and how to obtain effective approximations for them.

By Stone duality (Facts 1.8.3 and 1.8.4), any complexity notion for sets of strings can be applied to closed sets and to open sets. To be in a closed set P is a property related to universal quantification: Z is in P if *each* initial segment of Z lies on the representing tree. Similarly, to be in an open set is an existential property. This is why the most fruitful effectiveness notion for closed sets is to be Π^0_1, and the most fruitful one for open sets is to be Σ^0_1. We may call these objects either sets (when viewing them as sets of strings) or classes (when viewing them as sets of subsets of $\mathbb{N}$).

1.8.19 Definition.

(i) A closed set P is *co-c.e.* if the corresponding binary tree $T_P = \{x\colon [x] \cap P \neq \emptyset\}$ has a c.e. complement in $\{0,1\}^*$. A co-c.e. closed set is usually called a Π^0_1 *class*.

(ii) An open set R is *c.e.* if the corresponding set of strings $A_R = \{x\colon [x] \subseteq R\}$ is c.e.; such an open set is also called a Σ^0_1 *class*.

Representing Π^0_1 *classes*

Usually we show that a class P is Π^0_1 by defining *some* Π^0_1 tree B such that $P = \mathit{Paths}(B)$, and using the following fact.

1.8.20 Fact. *Let* $B \subseteq \{0,1\}^*$ *be a* Π^0_1 *tree. Then* $P = \mathit{Paths}(B)$ *is a* Π^0_1 *class.*

Proof. The subtree of B given by

$$B^* = \{\sigma\colon \forall n \geq |\sigma|\, \exists \rho \in B\, [|\rho| = n \,\&\, \rho \succeq \sigma]\} \tag{1.14}$$

is a Π^0_1 tree since the quantifier $\exists\rho$ is bounded. Moreover, B^* has no dead branches and $P = \mathit{Paths}(B^*)$. Hence $B^* = T_P$ by the correspondence in Fact 1.8.3, and P is a Π^0_1 class. □

On the other hand, if we are willing to admit dead branches we can find a representing tree that is computable.

1.8.21 Fact. *Each* Π^0_1 *class is of the form* $\mathit{Paths}(B)$ *for some computable tree* B.

Proof. T_P is a Π^0_1 tree, so $A = \{0,1\}^* - T_P$ is c.e., and therefore has a computable enumeration $(A_s)_{s\in\mathbb{N}}$. The tree $B = \{\sigma\colon \forall \rho \preceq \sigma\, [\rho \notin A_{|\sigma|}]\}$ is computable. Clearly $T_P \subseteq B$ and hence $P \subseteq \mathit{Paths}(B)$. For the inclusion $\mathit{Paths}(B) \subseteq P$, if $Z \notin P$, then we can choose n such that $Z\restriction_n \in A_s$ for some s. Then $Z\restriction_n$ is a dead branch of B because none of its extensions of length s are in B. Hence $Z \notin \mathit{Paths}(B)$. □

A Π^0_1 class P can be viewed as a problem. Each $Z \in P$ is a solution of this problem. A Π^0_1 tree B such that $P = \mathit{Paths}(B)$ is a description of the problem. By Fact 1.8.21, each problem given by a Π^0_1 class has a computable description. Fact 1.8.31 below shows that sometimes there exists a solution but not a computable one. However, there is always a low solution by the Low Basis Theorem 1.8.37. Later on we will encounter more examples of nonempty Π^0_1 classes without computable members, for instance the sets that are Martin-Löf-random for a constant b (see 3.2.9).

If T_P is computable then its leftmost path is computable. Thus, if P has no computable member, the tree B in the proof of 1.8.21 necessarily contains dead branches.

Representing c.e. open sets

1.8.22 Fact. $R \subseteq 2^{\mathbb{N}}$ *is c.e. open* $\Leftrightarrow$ $R = [W_e]^{\prec}$ *for some* e.

Proof. $\Rightarrow$: By definition A_R is c.e. Then $R = [A_R]^{\prec}$ by Fact 1.8.4(i).

$\Leftarrow$: Let
$$\widehat{W}_e = \{\sigma : \exists \rho \in W_e\,[\rho \preceq \sigma]\}. \tag{1.15}$$

Then $R = [\widehat{W}_e]^{\prec}$. Since $\widehat{W}_e$ is closed under extensions, $B = \{0,1\}^* - \widehat{W}_e$ is a Π^0_1 tree. So $P = \mathit{Paths}(B)$ is a Π^0_1 class by Fact 1.8.20, and $R = 2^{\mathbb{N}} - P$. Thus $A_R = \{0,1\}^* - T_P$. Since T_P is Π^0_1, A_R is c.e. So R is a c.e. open set. □

1.8.23 Remark. We say the number e is an *index* for a c.e. open set R if $R = [W_e]^{\prec}$. An index for an object usually provides us with an effective approximation of the object (see page 7). In the case of c.e. open sets, the approximation for index e is denoted by $(\widehat{W}_{e,s})_{s\in\mathbb{N}}$. It is chosen in such a way that $\widehat{W}_{e,s}$ is closed under extensions within the strings of length up to s: let

$$\widehat{W}_{e,s} = \{x\colon\ |x| \le s \ \&\ \exists y \in W_{e,s}\,[y \preceq x]\}. \tag{1.16}$$

We suppress the index e for R and simply write R_s for $\widehat{W}_{e,s}$. Thus $R = \bigcup_s [R_s]^{\prec}$.

1.8.24 Example. Let Φ be a partial computable functional. For each string x, the class
$$S_x = \{Z\colon\ \forall i < |x|\,[\Phi^Z(i) = x(i)]\}$$
is c.e. open uniformly in x, because $S_x = [W]^{\prec}$ for the c.e. set $W = \{\sigma\colon\ \Phi^\sigma = x\}$.

1.8.25 Definition. A set $C \subseteq \{0,1\}^*$ such that $x \mid y$ for every pair of distinct strings $x, y \in C$ is called an *antichain* of $\{0,1\}^*$, or a *prefix-free set.*

Every c.e. open set is generated by a computable antichain:

1.8.26 Fact. *One may uniformly in an index e for a c.e. open set R obtain a computable antichain B such that $[B]^{\prec} = R$. Moreover, B is effectively given in the form $B = \{x_i\}_{i<N}$, $N \in \mathbb{N} \cup \{\infty\}$, where $x_i \ne x_j$ for $i \ne j$ and $|x_i| \le |x_{i+1}|$.*

Proof. Let $(\widehat{W}_{e,s})_{s\in\mathbb{N}}$ be as in (1.16). Begin with an empty antichain. At stage s add to the antichain all strings of length s which are in $\widehat{W}_{e,s}$ and do not extend a string in $\widehat{W}_{e,s-1}$. □

1.8.27 Remark. Note that $\widehat{W}_e = \bigcup \widehat{W}_{e,s}$ is usually not an ideal of strings. As a remedy we could modify the definition to $\widehat{W}_{e,s} = \{\sigma\colon\ |\sigma| \le s \ \&\ [\sigma] \subseteq [W_{e,s}]^{\prec}\}$. Then $\widehat{W}_{e,s}$ would be closed under extensions within the strings of length up to s, and satisfy the condition that $\sigma 0, \sigma 1 \in \widehat{W}_{e,s} \rightarrow \sigma \in \widehat{W}_{e,s}$. Thus, in fact $\widehat{W}_e$ would be an ideal and $[\widehat{W}_e]^{\prec} = R$, so $\widehat{W}_e = A_R$ by Fact 1.8.4. However, this is not worth the additional effort.

An effective listing of the Π^0_1 classes

Suppose that e is an index for a c.e. open set R, and $\widehat{W}_e$ is as in (1.15). Let $P = 2^{\mathbb{N}} - R$, then $B_e = \{0,1\}^* - \widehat{W}_e$ is a Π^0_1 tree such that $\mathit{Paths}(B_e) = P$. In this way an index e for R can also be used as an index for the Π^0_1 class $P = 2^{\mathbb{N}} - R$. The approximation of P derived from this index is an effective sequence

$$(P_s)_{s\in\mathbb{N}} \tag{1.17}$$

of (strong indices for) clopen sets, where $P_s \supseteq P_{s+1}$ and $P = \bigcap_s P_s$. To obtain this sequence let $P_s = [\{\sigma\colon |\sigma| = s \ \&\ \sigma \not\in \widehat{W_{e,s}}\}]^{\prec}$. Thus P_s is generated by the strings of length s that are still on the tree B_e at stage s. (This differs from the case of an approximation $(R_s)_{s\in\mathbb{N}}$ of a c.e. open set, where R_s is a set of strings of length at most s.)

1.8.28 Fact. *Let G^e be the Π^0_1 class given by index e. Then the set $\{e\colon G^e = \emptyset\}$ is Σ^0_1 .*

Proof. Clearly $G^e = \emptyset \leftrightarrow \exists s\, G^e_s = \emptyset$, and the latter condition is Σ^0_1. □

1.8.29 Exercise. Show that the operations $\cup$ and $\cap$ are effective on indices for c.e. open sets, as well as on indices for Π^0_1 classes.

Examples of Π^0_1 *classes*

We are now in the position to give some interesting examples of Π^0_1 classes. The first one will be important later on, for instance in Section 4.3. Recall that $J(e)$ denotes $\Phi_e(e)$. A $\{0,1\}$-valued function f is called two-valued diagonally noncomputable, or *two-valued d.n.c.* for short, if it avoids being equal to $J(e)$ whenever $J(e)$ is defined, namely, $\neg f(e) = J(e)$ for each e.

1.8.30 Remark. A simple example of a two-valued d.n.c. function f is obtained as follows. Since $\{e\colon J(e) = 1\}$ is c.e., there is a computable function q such that $J(e) = 1 \Leftrightarrow q(e) \in \emptyset'$. Now let $f(e) = 1 - \emptyset'(q(e))$. If $J(e) = 1$ then $f(e) = 0$, otherwise $f(e) = 1$. Notice that $f \leq_{tt} \emptyset'$.

1.8.31 Fact. *The two-valued d.n.c. functions form a nonempty Π^0_1 class P without computable members. In particular, the tree T_P is not computable.*

Proof. The set of strings $B = \{\sigma\colon \forall s\, \forall e < s\, \neg\sigma(e) = J_s(e)\}$ is a Π^0_1 tree and $P = \mathit{Paths}(B)$. Hence P is a Π^0_1 class by Fact 1.8.20. We have already seen in Remark 1.8.30 that $P \neq \emptyset$. Suppose $f = \Phi_e$ is a two-valued function. Then $f(e) = J(e)$, so $f \not\in P$. □

The following examples of Π^0_1 classes will be reconsidered later on.

1.8.32 Examples.

(i) Let Φ be a partial computable functional. Then for each n, $\{Z\colon \Phi^Z(n)\uparrow\}$ is a Π^0_1 class.

(ii) Let ψ be a partial computable function with values in $\{0,1\}$. Then the total $\{0,1\}$-valued functions extending ψ form a Π^0_1 class.

(iii) Let T be an effectively axiomatized theory. Then the completions of T form a Π^0_1 class. (Here we fix some effective encoding of the sentences in the language of T by numbers.)

Proof. (i) Recall the definition of $\Phi^\sigma_s(n)$ from page 14. Let $B = \{\sigma\colon \forall s\, \Phi^\sigma_s(n)\uparrow\}$, then B is a Π^0_1 tree and $\mathit{Paths}(B)$ is the class under consideration. We leave (ii) and (iii) as exercises. □

Consider the class in (iii) when T is Peano arithmetic. Then T has no computable completion, so once again there is no computable solution for the problem described by the Π^0_1 class of completions of T.

Isolated points and perfect sets

A point Z in a topological space $\mathcal{X}$ is called *isolated* if the singleton $\{Z\}$ is open. For instance, if $\mathcal{X}$ satisfies the separation axiom T_1 (each singleton is closed) and $\mathcal{X}$ is finite then *each* point is isolated. Consider the case that $\mathcal{X}$ is the subspace *Paths*(B) of Cantor space for a binary tree B. Then a path Z of B is isolated iff there is a number n_0 such that Z is the only path extending $Z\restriction_{n_0}$. For example, consider the tree

$$B = \{0^i : i \in \mathbb{N}\} \cup \{0^i10^k : i, k \in \mathbb{N}\}.$$

Then *Paths*(B) is infinite, and every path except for 0^∞ is isolated in $\mathcal{X}$.

A nonempty closed set P in a topological space is called *perfect* if it has no isolated points. If $P \subseteq 2^{\mathbb{N}}$ this amounts to saying that the binary tree T_P has no isolated path. Thus each perfect class in Cantor space has size $2^{\aleph_0}$.

The following fact is usually applied to computable trees, and often in relativized form, but it actually holds for Π^0_1 trees.

1.8.33 Fact. *Let $B \subseteq \{0,1\}^*$ be a Π^0_1 tree. Then each isolated path Z of B is computable.*

Proof. Let B^* be the Π^0_1 subtree of extendable nodes of B defined in (1.14). Then Z is an isolated path of B^* as well. So choose a number n_0 such that Z is the unique path of B^* extending $Z\restriction_{n_0}$. To compute $Z(n)$ for $n \geq n_0$, enumerate $\{0,1\}^* - B^*$ until there remains a unique $\sigma \succeq Z\restriction_{n_0}$ such that $|\sigma| = n+1$ and σ is on B^*. This must happen, for otherwise some path of B^* other than Z also extends $Z\restriction_{n_0}$. Thus $\sigma \prec Z$, so output $\sigma(n)$. □

In particular, every nonempty Π^0_1 class without computable members is perfect.

1.8.34 Corollary. *Let B be a binary tree.*

(i) If Z is an isolated path of B then $Z \leq_T B$.
(ii) If Paths(B) is finite, then $Z \leq_T B$ for each path Z.

Proof. (i). Relativize Fact 1.8.33 (the case where the binary tree is computable).
(ii). It suffices to observe that each path of B is isolated. □

1.8.35 Exercise. Suppose the partial computable function ψ in 1.8.32(ii) has a co-infinite domain. Then the class of total $\{0,1\}$-valued extensions of ψ is perfect.

The Low Basis Theorem

A basis theorem (for Π^0_1 classes) states that each nonempty Π^0_1 class has a member with a particular property. We begin with an example of such a theorem, the Kreisel Basis Theorem, that the left-c.e. sets form a basis for the Π^0_1 classes.

1.8.36 Fact. *Every nonempty Π^0_1 class P has a left-c.e. member, namely its leftmost path.*

Proof. By Fact 1.8.3 we may identify a closed $P \subseteq 2^{\mathbb{N}}$ with the tree $T_P = \{x \colon [x] \cap P \neq \emptyset\}$, the set of strings that are on P. We show that the leftmost path Y of P is left-c.e. Note that

$$\tau <_L Y \leftrightarrow \forall \sigma \leq_L \tau \left[|\sigma| = |\tau| \rightarrow \sigma \text{ is not on } P\right].$$

This is in Σ^0_1 form because the universal quantifier is bounded. Thus the set $\{\tau \colon \tau <_L Y\}$ is c.e. □

The proof of a basis theorem is usually uniform: from an index for the given Π^0_1 class one can compute a description of a member with the desired property. Thus, a basis theorem provides an effective choice function, picking a member with a particular property from the class in case the class is nonempty.

To verify that the proof of Fact 1.8.36 is uniform we provide a little more detail: from an index for P (that is, an index for a Π^0_1 tree B such that $Paths(B) = P$) we can effectively obtain an index for the Π^0_1 tree B^* without dead branches in (1.14). Since $B^* = \{\sigma \colon \sigma \text{ is on } P\}$, we have $\tau <_L Y \leftrightarrow \forall \sigma \leq_L \tau \left[|\sigma| = |\tau| \rightarrow \sigma \notin B^*\right]$, which gives the required c.e. index for $\{\tau \colon \tau <_L Y\}$.

The desired property of member Y of the given Π^0_1 class is usually a lowness property. On the other hand, $Y \in P$ often means that Y is complex in some sense (computational or descriptive). For instance, P could be the class of two-valued d.n.c. functions in Fact 1.8.31. Then a basis theorem implies that a set can be computationally weak in a given sense, but complex in the sense of being in P. In particular, the Π^0_1 classes we are interested in here have no computable paths, otherwise the basis theorem is trivial.

The following is due to Jockusch and Soare (1972*b*).

1.8.37 Theorem. (Low Basis Theorem)
Every nonempty Π^0_1 class has a low member.

Proof. Let P be the given Π^0_1 class. We define a descending sequence of nonempty Π^0_1 classes $(P^e)_{e \in \mathbb{N}}$, where $P^0 = P$. Then, by the compactness of $2^{\mathbb{N}}$ (in the form of Proposition 1.8.5), there is a set Y in $\bigcap_e P^e$.

The class P^{e+1} determines whether $e \in Y'$: either this holds for no $Y \in P^{e+1}$, or for all such Y. The halting problem $\emptyset'$ can decide which case applies, so $Y' \leq_T \emptyset'$. (Note that Y is the only element of $\bigcap_e P^e$, since $Y \leq_m Y'$ via some fixed many-one reduction.)

Construction relative to $\emptyset'$ of Π^0_1 classes $(P^e)_{e \in \mathbb{N}}$. At stage e we define P^e. Let $P^0 = P$.

Stage $e + 1$. Suppose that P^e has been defined. If $P^e \cap \{Z \colon J^Z(e)\uparrow\} \neq \emptyset$ then let P^{e+1} be this class, otherwise let $P^{e+1} = P^e$.

Notice that by Example 1.8.32(i) one may effectively determine an index for the Π^0_1 class $P^e \cap \{Z \colon J^Z(e)\uparrow\}$. Thus by Fact 1.8.28, it is a Π^0_1 property of e that this class is nonempty. Therefore this can be decided by $\emptyset'$. Clearly, $e \notin Y'$ iff this first alternative applies at stage $e + 1$. Hence $Y' \leq_T \emptyset'$. □

In the foregoing proof we approximated a set Y not by specifying initial segments but rather via the classes P^e. In fact P^e determines $Y' \upharpoonright_e$. We discussed approximations by classes on page 46 at the beginning of this section.

The proof of Theorem 1.8.37 actually produces a set $Y \in P$ such that Y' is left-c.e. Note that this property of a set Y depends on the particular way the jump operator is defined. However, the property implies that $Y' \equiv_{tt} \emptyset'$ (namely, Y is superlow, Definition 1.5.3). To verify that Y' is left-c.e., we rewrite the proof, avoiding a construction relative to $\emptyset'$. This also stresses its uniformity.

1.8.38 Theorem. (Extends 1.8.37) *Every nonempty Π^0_1 class P has a member Y such that Y' is left-c.e.; a c.e. index for $\{\sigma\colon \sigma <_L Y'\}$ (and hence a reduction procedure for $Y' \leq_{tt} \emptyset'$) can be obtained effectively from an index for P.*

Proof idea: First consider $Y'(0)$. We maintain the guess that $J^Y(0)\uparrow$ as long as possible, namely, till it becomes apparent that $J^Y(0)\downarrow$ for *all* Y in P (note that this is a Σ^0_1 event). For $Y'(1)$ we do the same, but at first within the restricted Π^0_1 class $\widetilde{P} = P \cap \{Y\colon J^Y(0)\uparrow\}$. Once our guess at $Y'(0)$ changes to 1, we remove this restriction on $\widetilde{P}$. We may already have discovered that $J^Y(1)\downarrow$ for all Y in $\widetilde{P}$. This led us to make a guess that $Y'(1) = 1$. This guess may now be revised to $Y'(1) = 0$, in case there remains some Y in P such that $J^Y(1)\uparrow$. It should be clear how to continue this procedure in order to guess at $Y'(e)$ for each e.

Proof details. We apply the Kreisel Basis Theorem 1.8.36 to the Π^0_1 class of strings τ such that for some $Y \in P$, whenever $e < |\tau|$ and $\tau(e) = 0$ then $J^Y(e)\uparrow$ (without a prediction in the case that $\tau(e) = 1$). For a string τ, let

$$Q_\tau = \{Y \in P\colon \forall e < |\tau| \left[\tau(e) = 0 \ \rightarrow\ J^Y(e)\uparrow\right]\}.$$

By Fact 1.8.28, "$Q_\tau = \emptyset$" is a Σ^0_1 property of τ. Trivially, $Q_\varnothing = P \neq \emptyset$ and $Q_\tau \neq \emptyset \ \rightarrow\ Q_{\tau 1} \neq \emptyset$. Thus $B = \{\tau\colon Q_\tau \neq \emptyset\}$ is a Π^0_1 tree without dead branches such that $1^\infty \in \mathit{Paths}(B)$. The leftmost path V of B is left-c.e. by 1.8.36. By the compactness of $2^{\mathbb{N}}$ (1.8.5) there is a set Y in $\bigcap_e Q_{V\upharpoonright_e}$. To show that $Y' = V$, inductively assume that $\tau = Y'\upharpoonright_e = V\upharpoonright_e$. Then $J^Y(e)\uparrow \leftrightarrow Q_\tau \cap \{Y\colon J^Y(e)\uparrow\} \neq \emptyset \ \leftrightarrow\ V(e) = 0$. Thus $Y'(e) = V(e)$.

For the uniformity statement, note that from an index for the Π^0_1 class P we effectively obtained a c.e. index for $\{0,1\}^* - B$, which is an index for the Π^0_1 class $\mathit{Paths}(B)$. From this we obtain a c.e. index for $\{\sigma\colon \sigma <_L Y'\}$ by the uniformity of Fact 1.8.36. ☐

The following extension of Theorem 1.8.37 provides more information on the Turing degrees of members of a Π^0_1 class P: there is a set $Y \in P$ not Turing above a given incomputable set B. If B is Δ^0_2 then, in addition, we may choose Y low. The result is again due to Jockusch and Soare (1972*b*).

1.8.39 Theorem. (Extends 1.8.37) *Given a Π^0_1 class $P \neq \emptyset$ and an incomputable set B, there is a set $Y \in P$ such that $B \not\leq_T Y$. Moreover $Y' \leq_T B \oplus \emptyset'$.*

Proof. Recall that $(\Phi_e)_{e\in\mathbb{N}}$ is a listing of the Turing functionals. As in the proof of Theorem 1.8.37, we define a descending sequence of nonempty Π^0_1 classes

$(P^e)_{e\in\mathbb{N}}$, and let Y be an element of $\bigcap_e P^e$. However, now the class P^{2e+1} is used to determine whether $e \in Y'$, while P^{2e+2} prevents that $B = \Phi_e^Y$.

Construction relative to $\emptyset' \oplus B$ *of* Π^0_1 *classes* $(P^e)_{e\in\mathbb{N}}$. Let $P^0 = P$.

Stage $2e+1$. This is similar to stage $e+1$ in the proof of Theorem 1.8.37. If $P^{2e} \cap \{X\colon J^X(e)\uparrow\} \neq \emptyset$, then let P^{2e+1} be this class. Otherwise, let $P^{2e+1} = P^{2e}$.

Stage $2e+2$. Using $\emptyset' \oplus B$ as an oracle, search in parallel for the following:

(a) a string σ and a number k such that σ is on P^{2e+1} (i.e., $[\sigma] \cap P^{2e+1} \neq \emptyset$) and $B(k) \neq \Phi_e^\sigma(k)$. If such a pair is found let $P^{2e+2} = P^{2e+1} \cap [\sigma]$;

(b) a number n such that $P^{2e+1} \cap \{X\colon \Phi_e^X(n)\uparrow\} \neq \emptyset$. If such a number is found let P^{2e+2} be this class.

The parallel search carried out at an even stage terminates: otherwise, Φ_e^X is total for each $X \in P^{2e+1}$ since (b) does not terminate. In that case, B is computable, contrary to our assumption: given k, we may compute s such that for all $\sigma \in P_s^{2e+1}$, if $|\sigma| = s$ then $\Phi_e^\sigma(k)$ is defined, and in addition, since (a) does not terminate, all such computations $\Phi_e^\sigma(k)$ give the same output. Then this common output is $B(k)$. (Here we have used the approximation given by (1.17), which is automatically obtained from an index for the Π^0_1 class.)

To see that $Y' \leq_T B \oplus \emptyset'$, as before it suffices to let $B \oplus \emptyset'$ decide which case applies at each stage as this determines Y'. The halting problem $\emptyset'$ suffices for the odd stages, and $B \oplus \emptyset'$ can decide whether the parallel search carried out at each single even stage terminates first in (a), or first in (b). □

Even if B is c.e., one cannot in general achieve that $Y \leq_{wtt} \emptyset'$. A counterexample can be derived from Exercise 8.5.23.

Exercises.

1.8.40. Extend 1.8.39 as follows: given incomputable sets $B_1, \ldots, B_k$, there is a set $Y \in P$ such that $B_1, \ldots, B_k \not\leq_T Y$ and $Y' \leq_T B_1 \oplus \ldots \oplus B_k \oplus \emptyset'$.

1.8.41. Suppose that $P \neq \emptyset$ is a Π^0_1 class and $B \subseteq \mathbb{N}$. Show that there is a set $Y \in P$ such that $(Y \oplus B)' \leq_{tt} B'$. (This is more than a mere relatization of 1.8.37.)

The basis theorem for computably dominated sets

Martin and Miller (1968) proved that the computably dominated sets (Definition 1.5.9) form a basis for the Π^0_1 classes. Since there is a Π^0_1 class without a computable member (Fact 1.8.31), this shows that a computably dominated set can be incomputable. An extension of the result states that there are $2^{\aleph_0}$ computably dominated sets.

1.8.42 Theorem. *Every nonempty* Π^0_1 *class* P *has a computably dominated member* Y.

Proof. Once again we define a descending sequence of nonempty Π^0_1 classes $(P^e)_{e\in\mathbb{N}}$ and use Proposition 1.8.5 to conclude that there is $Y \in \bigcap_e P^e$. We begin with $P^0 = P$. The class P^{e+1} either shows that Φ_e^Y is partial, or that there is a computable function f dominating Φ_e^Y.

Construction relative to $\emptyset''$ of Π^0_1 classes $(P^e)_{e\in\mathbb{N}}$. Let $P^0 = P$.
Stage $e+1$. Suppose P^e has been determined.

(a) $Q_x = P^e \cap \{Z\colon \Phi^Z_e(x)\uparrow\}$ is a Π^0_1 class, uniformly in x. If there is x such that $Q_x \neq \emptyset$, let x be least such and let P^{e+1} be this class. This ensures that Φ^Z_e is partial for each set $Z \in P^{e+1}$.

(b) Otherwise let $P^{e+1} = P^e$. (We will show that there is a computable function f dominating Φ^Z_e for each $Z \in P^{e+1}$.)

Let $Y \in \bigcap_e P^e$. If $g = \Phi^Y_e$ is a total function then at stage $e+1$, we are in case (b) since $Y \in P^{e+1}$. To compute a function f that dominates g, we use the approximation $(P^e_s)_{s\in\mathbb{N}}$ obtained in (1.17) on page 55. Given an input x, compute $s = s(x)$ such that

$$\forall\sigma \in P^e_s\,[\,|\sigma| = s \;\rightarrow\; \Phi^\sigma_{e,s}(x)\downarrow].$$

Such an s exists, otherwise the class Q_x considered at stage $e+1$ is nonempty. Now let $f(x)$ be the maximum of all the values $\Phi^\sigma_e(x)$ where $|\sigma| = s(x)$ and $\sigma \in P^e_{s(x)}$. Then f dominates Φ^Z_e for each $Z \in P^e$. □

While an incomputable computably dominated set is not Δ^0_2 by 1.5.12, the set Y above automatically satisfies $Y'' \le_{tt} \emptyset''$. This fact requires some additional effort to verify. Recall from Exercise 1.4.20(iii) that $\mathsf{Tot} = \{e\colon \mathrm{dom}(\Phi_e) = \mathbb{N}\}$ is Π^0_2-complete. The solution of 1.4.20 actually shows that $\mathsf{Tot}^Y = \{e\colon \mathrm{dom}(\Phi^Y_e) = \mathbb{N}\} \equiv_m \mathbb{N} - Y''$ for each Y, and hence $\mathsf{Partial}^Y := \mathbb{N} - \mathsf{Tot}^Y \equiv_m Y''$. The following is analogous to Theorem 1.8.38.

1.8.43 Theorem. (Extends 1.8.42) *Every nonempty Π^0_1 class P has a computably dominated member Y such that $Y'' \le_{tt} \emptyset''$, and indeed $\{\tau\colon \tau <_L \mathsf{Partial}^Y\}$ is Σ^0_2. A truth-table reduction of Y'' to $\emptyset''$ can be obtained effectively from an index for P.*

Proof. Let $Y \in \bigcap_e P^e$ as above. We effectively obtain from P a Σ^0_2 index for the set $\{\tau\colon \tau <_L \mathsf{Partial}^Y\}$. This suffices because $Y'' \equiv_m \mathsf{Partial}^Y \equiv_{tt} \{\tau\colon \tau <_L \mathsf{Partial}^Y\}$ via fixed reduction procedures.

The following construction enumerates $\{\tau\colon \tau <_L \mathsf{Partial}^Y\}$ relative to $\emptyset'$. At each stage $s > 0$, for each $e < s$ we have a guess $\widetilde{P}^e$ at the class P^e from the foregoing proof which is (an index for) a Π^0_1 class, and a guess τ_s of length s at $\mathsf{Partial}^Y$ which only moves to the right on the tree $\{0,1\}^*$. At stage s, for each $e < s$ in ascending order, a procedure S_e carries out one instruction, thereby defining the current guesses $\widetilde{P}^e[s]$ and $\tau_s(e)$. At stage 1 we let $\widetilde{P}^0 \doteq P$, $\tau_1 = \varnothing$, and start procedure S_0.

Procedure S_e

(a) Define $\widetilde{P}^{e+1} = \widetilde{P}^e$ and $\tau_s(e) = 0$.

(b) If $e+1 < s$ start procedure S_{e+1}. From now on, using $\emptyset'$ as an oracle, for $x = 0, 1, \ldots$ ask whether $Q_x = \widetilde{P}^e \cap \{Z\colon \Phi^Z_e(x)\uparrow\} \neq \emptyset$. If so, initialize the procedures S_i, $e < i < s$; from now on let $\widetilde{P}^{e+1} = Q_x$, and $\tau_s(e) = 1$. Otherwise let $\tau_s(e) = 0$.

Claim. *We have* $\lim_s \widetilde{P}^e[s] = P^e$ *and* $\lim_s \tau_s(e) = \mathsf{Partial}^Y(e)$.
This implies $\tau <_L \mathsf{Partial}^Y \leftrightarrow \exists s\,[\tau <_L \tau_s]$.

To prove the claim we use induction on e. Clearly $\lim_s \widetilde{P}^0[s] = P^0$, and $\lim_s \tau_s(0) = 1$ iff some x is found by S_0 such that $P^0 \cap \{Z: \Phi_0^Z(x)\uparrow\} \neq \emptyset$ iff Φ_0^Y is partial. Now suppose that the claim holds for all $i \leq e$, and let s_0 be a stage by which all the limits for $i \leq e$ have been reached. Then S_e is not initialized at any stage $s \geq s_0$, so if S_e finds x at a stage $s \geq s_0$, it defines $\widetilde{P}^{e+1}[s]$ and $\tau_s(e)$ correctly. Otherwise, $\widetilde{P}^{e+1}[s]$ and $\tau_s(e)$ are already correct at stage $s = s_0$. □

Recall from page 56 that a nonempty closed set in a topological space is called perfect if it has no isolated points.

1.8.44 Theorem. (Extends 1.8.42) *Every nonempty Π_1^0 class P without computable members has a perfect subclass $\mathcal{S}$ of computably dominated sets.*

Proof. We modify the proof of Theorem 1.8.42. Instead of a descending sequence $(P^e)_{e\in\mathbb{N}}$, we build a tree $(P^\sigma)_{\sigma\in\{0,1\}^*}$ of Π_1^0 classes.

Stage 0. Let $P^\varnothing = P$.

Stage $e+1$. Suppose that $P^\sigma \neq \emptyset$ has been determined for each σ such that $|\sigma| = e$. Firstly, for every such σ let $\widehat{P}^{\sigma 0}$ and $\widehat{P}^{\sigma 1}$ be nonempty disjoint subclasses of P^σ. They exist since there are incompatible strings τ_0, τ_1 on P^σ, so we may let $\widehat{P}^{\sigma i} = P^\sigma \cap [\tau_i]$ $(i \in \{0,1\})$. Now proceed as before, but with both classes $\widehat{P}^{\sigma i}$ separately: if there is x such that $\widehat{P}^{\sigma i} \cap \{Z: \Phi_e^Z(x)\uparrow\} \neq \emptyset$ then let x be least such and let $P^{\sigma i}$ be this class. Otherwise let $P^{\sigma i} = \widehat{P}^{\sigma i}$.

Verification. For each set C, there is a set $Y_C \in \bigcap_e P^{C\restriction_e}$, and, by the same argument as before, each such Y_C is computably dominated. If $C \neq D$ then $Y_C \neq Y_D$. So the class $\mathcal{S} = \{Y_C : C \subseteq \mathbb{N}\}$ is as required. □

1.8.45 Corollary. *There are $2^{\aleph_0}$ many computably dominated sets.*

Proof. Apply Theorem 1.8.44 to any Π_1^0 class without computable members, for instance the class of two-valued d.n.c. functions from Fact 1.8.31. □

A further basis theorem is 4.3.2 below: if D computes a two-valued d.n.c. function, then each nonempty Π_1^0 class contains a set $Y \leq_T D$.

1.8.46.$^\diamond$ **Exercise.** (Kučera and Nies) Let P be a nonempty Π_1^0 class. Suppose that $B >_T \emptyset'$ is Σ_2^0. Then there is a computably dominated set $Y \in P$ such that $Y' \leq_T B$.
Hint. Combine the techniques of the Low Basis Theorem and Theorem 1.8.42 with permitting below B relative to $\emptyset'$. Fix an enumeration $(B_s)_{s\in\mathbb{N}}$ of B relative to $\emptyset'$, and use the function $c_B \leq_T B$ given by $c_B(i) = \mu t > i.\, B_t\restriction_i = B\restriction_i$ for the permitting.

Weakly 1-generic sets

1-genericity for sets is an effective version of the notion of Cohen genericity from set theory. We first introduce the simpler concept of weak 1-genericity. Each weakly 1-generic set is hyperimmune. Each hyperimmune set is Turing equivalent to a weakly 1-generic set.

A subset D of a topological space is called *dense* if $D \cap R \neq \emptyset$ for each nonempty open set R. A set $D \subseteq 2^{\mathbb{N}}$ is dense iff $D \cap [x] \neq \emptyset$ for each cylinder $[x]$ (namely, each string is extended by a set in D). For instance, for each n, the open set

$E_n = \{Z\colon \exists i \ge n\,[Z(i) = 1]\}$ is dense in $2^{\mathbb{N}}$. Note that E_n is c.e. open, being an effective union of clopen sets.

In Proposition 1.8.5 we interpreted a descending sequence $(P^i)_{i\in\mathbb{N}}$ of nonempty closed classes as desirable conditions. Compactness showed that there is a set $Z \in \bigcap_i P^i$. A desirable condition can also be given by a dense open set. Such a condition cannot be ruled out by any finite initial segment of a set $G \subseteq \mathbb{N}$. Baire's category theorem for $2^{\mathbb{N}}$ states that the intersection of a countable collection $(D_n)_{n\in\mathbb{N}}$ of dense open sets contains a set G. One builds G by the method of finite extensions. Let $[\sigma_0] \subseteq D_0$, and if σ_n has been defined choose $\sigma_{n+1} \succ \sigma_n$ such that $[\sigma_{n+1}] \subseteq D_{n+1}$. Then $G = \bigcup_n \sigma_n$ is as required. (This result holds in any uncountable Polish space, such as Baire space $\mathbb{N}^{\mathbb{N}}$.)

The weakly 1-generic sets are the ones that meet each condition of this kind where the dense set is also computably enumerable.

1.8.47 Definition. $G \subseteq \mathbb{N}$ is *weakly 1-generic* if G is in each dense c.e. open set $D \subseteq 2^{\mathbb{N}}$.

1.8.48 Proposition. *Each weakly 1-generic set G is hyperimmune. In particular, G is not computably enumerable.*

Proof. G is infinite since G is in the dense set $E_n = \{Z : \exists i \ge n\,[Z(i) = 1]\}$ for each n. Next, given a computable function f, we will show that there is n such that $p_G(n) \ge f(n)$. The c.e. open set

$$D_f = [\{\sigma 0^{f(|\sigma|)}\colon \sigma \neq \varnothing\}]^{\prec}$$

is dense. Thus $G \in D_f$, and we may choose σ such that $\sigma 0^{f(|\sigma|)} \prec G$. Let $n = |\sigma|$. In the worst case, σ consists only of ones, so that $p_G(n-1) = n-1$ and G has the string

$$\underbrace{1\ldots1}_{n}\overbrace{0\ \ldots\ 0}^{f(n)}$$

as an initial segment. Even then we have $p_G(n) \ge f(n)$. □

To build a weakly 1-generic set, we use the construction in the proof of Baire's category theorem.

1.8.49 Theorem. *There is a weakly 1-generic left-c.e. set G.*

Proof. (1) It is somewhat simpler to show that there is a weakly 1-generic Δ^0_2 set. We build an effective sequence $(\sigma_s)_{s\in\mathbb{N}}$ of strings such that $|\sigma_s| = s$. This sequence is a computable approximation of the set G, namely $G(n) = \lim_{s>n}\sigma_s(n)$ for each n.

We use the effective listing of c.e. open sets $([\widehat{W}_e]^{\prec})_{e\in\mathbb{N}}$, where, as in (1.15), $\widehat{W}_e = \{\sigma\colon \exists\rho \in W_e\,[\rho \preceq \sigma]\}$. Note that $[\widehat{W}_e]^{\prec} = [W_e]^{\prec}$. We meet the requirements

$$R_e:\ [\widehat{W}_e]^{\prec} \text{ is dense} \Rightarrow G \in [\widehat{W}_e]^{\prec}.$$

Construction. Let $\sigma_0 = \varnothing$. Initialize all the requirements.

Stage $s > 0$. For each e, let $t_{e,s}$ be the maximum of e and the last stage when R_e was initialized. R_e is called *satisfied* at stage s if $\sigma_{s-1} \in \widehat{W}_{e,s}$. If there is $e < s$ such that R_e is not satisfied and there is $\sigma \in \widehat{W}_{e,s}$ such that $|\sigma| = s$ and

$$\sigma_{s-1} \upharpoonright_{t_{e,s}} \preceq \sigma, \tag{1.18}$$

then choose e least. Let $\sigma_s = \sigma 0$ and initialize the requirements R_j for $j > e$. We say that R_e acts. Otherwise let $\sigma_s = \sigma_{s-1}0$.

Verification. First we show by induction that each requirement R_e acts finitely often: when all R_i, $i < e$, have stopped acting, R_e acts at most once at a stage s, since from then on $G \upharpoonright_s$ is preserved by the initialization, so that R_e is permanently satisfied. Since R_e can only change $G(n)$ for $n \geq e$, this implies that $G(n) = \lim_{s>n}\sigma_s(n)$ exists for each n. Moreover, $t_e = \lim_s t_{e,s}$ exists for each e. If $[\widehat{W}_e]^{\prec}$ is dense then $\widehat{W}_e$ contains some $\sigma \succeq G \upharpoonright_{t_e}$, so R_e is met. Thus G is weakly 1-generic.
(2) To make G left-c.e., in (1.18) we only allow extensions σ such that $\sigma_{s-1} <_L \sigma$. To see that R_e is met, note that G is co-infinite by construction, so $G \upharpoonright_{t_e} 1^r 0 \prec G$ for some r. Now R_e can be satisfied permanently via some string in $\widehat{W}_e$ extending $G \upharpoonright_{t_e} 1^{r+1}$. □

Kurtz (1981, 1983) proved a converse of Proposition 1.8.48 for Turing degrees.

1.8.50 Theorem. *A is hyperimmune* $\Rightarrow$ $A \equiv_T G$ *for some weakly 1-generic* G. □

Thus, one can characterize the hyperimmune degrees using "effective topology". For a recent proof of Kurtz's theorem see Downey and Hirschfeldt (2010).

1-generic sets

Let X be a topological space, $S \subseteq X$ and $A \in X$. We say that A is in the *closure* of S if $S \cap V \neq \emptyset$ for every open set V containing A. If X is Cantor space and S is open, this means that for each $\sigma \prec A$ there is $\rho \succeq \sigma$ such that $[\rho] \subseteq S$. In this case we say that S is *dense along* A.

1.8.51 Definition. A is *1-generic* if $A \in S$ for every c.e. open set S that is dense along A.

1-generic sets were introduced by Jockusch (1977). If S is a dense open set then S is dense along any A. Therefore each 1-generic set is weakly 1-generic (Definition 1.8.47). Note that A is 1-generic iff for each c.e. open set S, either $A \in S$ or there is $\sigma \prec A$ such that $[\sigma] \cap S = \emptyset$. Intuitively, if $A \notin S$ then an initial segment of A already precludes A from being in S. For instance, let $S = \{Z\colon \Phi^Z(n)\downarrow\}$ for a Turing functional Φ. If $\Phi^A(n)\uparrow$ then there is $\sigma \prec A$ such that $\Phi^Y(n)\uparrow$ for each $Y \succ \sigma$.

Part (1) of the proof of Theorem 1.8.49 actually builds a 1-generic set. It suffices to observe that our strategy meets the requirements

$$R_e: \ [\widehat{W_e}]^{\prec} \text{ is dense along } G \Rightarrow \ G \in [\widehat{W_e}]^{\prec}.$$

1.8.52 Theorem. *There is a 1-generic set* $G \in \Delta^0_2$. □

A 1-generic set is not left-c.e., and in fact it does not even Turing bound an incomputable c.e. set. For this result of Jockusch (1977) see Odifreddi (1999, p. 662). However, the construction of a 1-generic set can be combined with permitting:

Exercises.

1.8.53. For each incomputable c.e. set C there is a 1-generic set $A \le_T C$.

1.8.54. Each 1-generic set G is in GL_1, namely, $G' \le_T G \oplus \emptyset'$.

The arithmetical hierarchy of classes

Sets of natural numbers are our primary objects of study. In particular, we are interested in their computational complexity, their descriptive complexity, and their randomness properties. Recall from the beginning of this chapter that to understand these aspects of a set we study classes of sets sharing certain complexity or randomness properties. It will be useful to measure the descriptive complexity of classes as well. With the exception of Chapter 9, all classes of sets X studied in this book can be described by a formula in the language of arithmetic involving initial segments of X. Similar to the case of the arithmetical hierarchy of sets defined in 1.4.10, the descriptive complexity of a class is measured by the complexity of its description, in terms of how many alternations of (number) quantifiers it has. The variable y_n of the innermost quantifier is now used to determine the initial segment $X\upharpoonright_{y_n}$ of X.

It is useful to extend the definitions to relations of k numbers and one set. The case $k = 0$ refers to classes.

1.8.55 Definition. Let $\mathcal{A} \subseteq \mathbb{N}^k \times 2^{\mathbb{N}}$ and $n \ge 1$.

(i) $\mathcal{A}$ is Σ^0_n if

$$\langle e_1, \ldots, e_k, X\rangle \in \mathcal{A} \leftrightarrow \exists y_1 \forall y_2 \ldots Q y_n \, R(e_1, \ldots, e_k, y_1, \ldots, y_{n-1}, X\upharpoonright_{y_n}),$$

where R is a computable relation, and Q is "$\exists$" if n is odd and Q is "$\forall$" if n is even.

(ii) $\mathcal{A}$ is Π^0_n if the complement of $\mathcal{A}$ is Σ^0_n, that is,

$$\langle e_1, \ldots, e_k, X\rangle \in \mathcal{A} \leftrightarrow \forall y_1 \exists y_2 \ldots Q y_n \, S(e_1, \ldots, e_k, y_1, \ldots, y_{n-1}, X\upharpoonright_{y_n}),$$

where S is a computable relation, and Q is " $\forall$" if n is odd and Q is " $\exists$" if n is even.

A relation is *arithmetical* if it is Σ^0_n for some n.

For instance, a Π^0_2 class is of the form $\{X\colon \forall y_1 \exists y_2 \, S(y_1, X\upharpoonright_{y_2})\}$, and a Σ^0_3 class is of the form $\{X\colon \exists y_1 \forall y_2 \exists y_3 \, R(y_1, y_2, X\upharpoonright_{y_3})\}$, where S and R are computable relations. We have already introduced Π^0_1 and Σ^0_1 classes in 1.8.19. Let us verify that these two ways of introducing Π^0_1 and Σ^0_1 classes are equivalent.

1.8.56 Fact.

(i) $P \subseteq 2^{\mathbb{N}}$ is Π^0_1 in the sense of 1.8.55 $\Leftrightarrow$ P is a co-c.e. closed set, that is, P is Π^0_1 in the sense of 1.8.19.

(ii) $U \subseteq 2^{\mathbb{N}}$ is Σ^0_1 in the sense of 1.8.55 $\Leftrightarrow$ U is a c.e. open set, that is, U is Σ^0_1 in the sense of 1.8.19.

Proof. (i) $\Rightarrow$: Suppose that $P = \{X \colon \forall y\, S(X \upharpoonright_y)\}$ for a computable set S and let $B = \{z \in \{0,1\}^* \colon \forall z' \preceq z\, S(z')\}$. Then B is a computable tree and $P = \mathit{Paths}(B)$, so by Fact 1.8.20 P is Π^0_1 in the sense of 1.8.19.
$\Leftarrow$: By 1.8.21, $P = \mathit{Paths}(B)$ for a computable tree B. Therefore $X \in P \leftrightarrow \forall n\, X \upharpoonright_n \in B$, hence P is Π^0_1 in the sense of 1.8.55.
The statement (ii) is now immediate by taking complements. □

Most of the classes of sets we consider will be arithmetical (the only exceptions are the classes of Chapter 9). Often arithmetical classes are introduced by an appropriate property of sets or functions computed by a set X. This makes them arithmetical by the use principle. Sometimes they are defined by an arithmetical growth condition on the initial segment complexity of their members.

We verify that some classes or relations we have introduced are arithmetical:

1.8.57 Proposition.

(i) The relation $\{\langle e, X\rangle \colon \Phi^X_e$ is total$\}$ is Π^0_2.

(ii) The class $\{A \colon A$ is computably dominated$\}$ is Π^0_4.

Proof. (i) For each e, X we have Φ^X_e is total $\leftrightarrow$ $\forall x \exists s\, \Phi^{X\upharpoonright_s}_{e,s(x)} \downarrow$. The relation $\{\langle x, e, \sigma\rangle \colon \Phi^{\sigma}_{e,|\sigma|}(x)\downarrow\}$ is computable.
(ii) This class is given by the description in Π^0_4 form

$$\forall e \exists i \forall y, s, w \exists t [(\forall y' \leq y\, \Phi^{A\upharpoonright_s}_{e,s}(y') \downarrow \;\&\; \Phi^{A\upharpoonright_s}_{e,s}(y) = w) \rightarrow w \leq \Phi_{i,t}(y)].$$

For, if Φ^A_e is total then there is i such that $\Phi^A_e(y) \leq \Phi_i(y)$ for each y. On the other hand, if z is least such that $\Phi^A_e(z)\uparrow$, there are i, t such $\Phi^A_e(y) \leq \Phi_{i,t}(y)$ for each $y < z$. □

1.8.58 Remark. Let $n > 1$ and consider a Σ^0_n class

$$\mathcal{A} = \{X \colon \exists y_1 \forall y_2 \ldots Q y_n\, R(y_1, \ldots, y_{n-1}, X \upharpoonright_{y_n})\}.$$

Then $\mathcal{A} = \bigcup_{y_1} \mathcal{B}_{y_1}$ where $\mathcal{B}_{y_1} = \{X \colon \forall y_2 \ldots Q y_n\, R(y_1, \ldots, y_{n-1}, X \upharpoonright_{y_n})\}$ is a Π^0_{n-1} class uniformly in y_1. Thus the Σ^0_n classes are the unions of effective sequences of Π^0_{n-1} classes. Similarly, the Π^0_n classes are the intersections of effective sequences of Σ^0_{n-1} classes.

The notions of Π^0_n and of Σ^0_n relations can be relativized.

1.8.59 Definition. For $C \subseteq \mathbb{N}$ and $n \in \mathbb{N}$, we define $\Sigma^0_n(C)$ relations and $\Pi^0_n(C)$ relations $\mathcal{A} \subseteq \mathbb{N}^k \times 2^{\mathbb{N}}$ as in Definition 1.8.55, but for $R, S \leq_T C$.

In the exercises we show that each $\Pi^0_n(\emptyset')$ class is Π^0_{n+1} but not conversely.

Remark 1.8.58 allows us to simplify the presentation of Π^0_2 classes.

1.8.60 Proposition. *Let $\mathcal{A} \subseteq 2^{\mathbb{N}}$. Then $\mathcal{A}$ is $\Pi^0_2 \Leftrightarrow$ there is a computable $S \subseteq \{0,1\}^*$ such that $\mathcal{A} = \{X\colon \forall y_1 \exists y_2 > y_1 S(X\upharpoonright_{y_2})\} = \{X\colon \exists^\infty n\, S(X\upharpoonright_n)\}$.*

Proof. We only have to prove the implication "$\Rightarrow$". Let $\mathcal{A} = \bigcap_n G_n$ where $(G_n)_{n\in\mathbb{N}}$ is an effective sequence of Σ^0_1 classes. By Remark 1.8.23, given t we can compute an approximation $G_{n,t}$ of G_n consisting of strings of length up to t. Let

$$m(x) = \max\{n\colon x \in G_{n,|x|}\},$$

and let $S = \{ya\colon y \in \{0,1\}^* \;\&\; a \in \{0,1\} \;\&\; m(ya) > m(y)\}$. Then S is computable, and for each X we have $\exists^\infty n\, X\upharpoonright_n \in S \leftrightarrow \exists^\infty n\, X \in G_n \leftrightarrow Z \in \mathcal{A}$. □

The *Borel classes* are the subclasses of $2^{\mathbb{N}}$ that can be obtained from the basic open cylinders $[x]$ via applications of the operations of complementation and countable unions. Each arithmetical class is Borel by Remark 1.8.58. The arithmetical hierarchy of classes is an effective version of the finite levels of a hierarchy of Borel classes.

The description of classes in arithmetic also provides new ways of describing sets.

1.8.61 Definition. Let $n > 1$. A set Z is a Π^0_n*-singleton* if $\{Z\}$ is a Π^0_n class.

The Π^0_1-singletons coincide with the computable sets by Fact 1.8.33. The Π^0_2 singletons already takes us beyond the arithmetical sets. Let $\emptyset^{(\omega)} = \bigcup_n(\emptyset^{(n)} \times \{n\})$ be the effective union of all the jumps $\emptyset^{(n)}$.

1.8.62 Proposition. *$\emptyset^{(\omega)}$ is a Π^0_2 singleton that is not an arithmetical set.*

Proof. There is a computable g such that $(X^{[n]})' = W^X_{g(n)}$ for each X, n, so

$$\begin{aligned} X = \emptyset^{(\omega)} &\Leftrightarrow X^{[0]} = \emptyset \;\&\; \forall n\, X^{[n+1]} = (X^{[n]})' \\ &\Leftrightarrow \forall n\, \forall k, s \left[\langle k, 0\rangle \notin X \;\&\; \exists t \geq s[\langle k, n+1\rangle \in X \leftrightarrow k \in W^{X\upharpoonright_t}_{g(n),t}]\right]. \end{aligned}$$

Thus the class $\{\emptyset^{(\omega)}\}$ is Π^0_2. □

Recall the discussion of the local versus the global view of sets in Section 1.3. If $Y \in \Sigma^0_n$, the description showing this embodies the local view. This is even more apparent at the lower levels Σ^0_1 and Σ^0_2 of the arithmetical hierarchy, where we still have a reasonable effective approximation of Y. On the contrary, a description of a set Z as a Π^0_n singleton, such as the description of $\emptyset^{(\omega)}$ above, embodies the global view. The description only works because it involves the set as a whole.

In Proposition 3.6.2 we will see that each Δ^0_2 set is a Π^0_2 singleton. However, there is a Δ^0_3 (even left-Σ^0_2) set that is not a Π^0_2 singleton, for instance the 2-random set $\Omega^{\emptyset'}$ (page 136). Thus, there are two incomparable classes of descriptive complexity, the arithmetical sets and the Π^0_2 singletons. This is an exception to the rule that the classes of descriptive complexity we consider form a nearly linear hierarchy.

Exercises. Show the following.
1.8.63. The class of c.e. sets is Σ^0_3. The class of computable sets is Σ^0_3. (Also see 3.2.3.)
1.8.64. The class of c.e. [computable] sets is not Π^0_3.
1.8.65. Let $\mathcal{A}$ be a Π^0_2 class and let Ψ be a Turing functional. Then the class $\mathcal{C} = \{Z : \Psi^Z \text{ is total } \& \; \Psi^Z \in \mathcal{A}\}$ is Π^0_2.
1.8.66. For each Σ^0_3 class $\mathcal{B}$ the Turing upward closure $\mathcal{G} = \{Z : \exists A \in \mathcal{B}\,[A \leq_T Z]\}$ is Σ^0_3 as well.
1.8.67. (i) Each $\Pi^0_1(\emptyset')$ class is Π^0_2. (ii) Some Π^0_2 singleton class is not $\Pi^0_1(\emptyset')$.
1.8.68. Extend 1.8.67(i): each $\Pi^0_n(\emptyset')$ class is a Π^0_{n+1} class.

Comparing Cantor space with Baire space

Let $\mathbb{N}^*$ be the set of finite sequences of numbers. König's Lemma 1.8.2 relies on the fact that the given tree is finitely branching. It fails in $\mathbb{N}^*$: for instance, the infinite tree $\{n0^n : n \in \mathbb{N}\}$ has no infinite path.

For $\sigma \in \mathbb{N}^*$ let $[\sigma] = \{f \in \mathbb{N}^\mathbb{N} : \sigma \prec f\}$. The sets $[\sigma]$ form a base of a topology on $\mathbb{N}^\mathbb{N}$, and $\mathbb{N}^\mathbb{N}$ equipped with this topology is called the *Baire space*. This space is not compact because $\mathbb{N}^\mathbb{N} = \bigcup_n [n]$. Nonetheless, many notions we have discussed for Cantor space can also be studied in the setting of Baire space. This includes not only the notions from topology, such as open, closed, and compact sets and their representations (page 47), but also arithmetical definability of classes (1.8.55). As an example we consider closed sets. The class $\mathcal{C} \subseteq \mathbb{N}^\mathbb{N}$ is closed iff $\mathcal{C} = \mathit{Paths}(B)$ for a subtree B of $\mathbb{N}^*$, and $\mathcal{C} \subseteq \mathbb{N}^\mathbb{N}$ is a Π^0_1 class of functions iff there is a computable set $R \subseteq \mathbb{N}^*$ such that $\mathcal{C} = \{f : \forall n\, [f\restriction_n \in R]\}$ iff $\mathcal{C} = \mathit{Paths}(B)$ for a computable subtree of $\mathbb{N}^*$. The problem whether $\mathit{Paths}(B) = \emptyset$ for a computable tree $B \subseteq \mathbb{N}^*$ is very complex, namely Π^1_1-complete (see Chapter 9, page 368). For a computable binary tree the corresponding problem is merely Σ^0_1. In particular, the Low Basis Theorem 1.8.37 fails in Baire space.

A function $f \in \mathbb{N}^\mathbb{N}$ can be encoded by a set $X \in 2^\mathbb{N}$, its graph $\Gamma_f = \{\langle n, f(n)\rangle : n \in \mathbb{N}\}$. For functions totality is automatic. For sets X, we have to require that X codes a function (Π^0_1) that is total (Π^0_2). In the following we will see that descriptions of functions as Π^0_1 singletons have the same expressive power as descriptions of sets as Π^0_2 singletons. We first introduce some notation. Let $D\colon \mathbb{N}^* \to \mathbb{N}$ be the computable injection given by $D(n_0, \ldots, n_{k-1}) = \prod_{i=0}^{k-1} p_i^{n_i+1}$, where p_i is the i-th prime number. Let *Seq* denote the range of D, the computable set of *sequence numbers*. If $\alpha = \prod_{i=0}^{k-1} p_i^{n_i+1}$ as above then we write $\alpha(i) = n_i$.

1.8.69 Proposition. *Suppose that $f \in \mathbb{N}^\mathbb{N}$ and $\{f\}$ is Π^0_1. Then $\{\Gamma_f\}$ is Π^0_2.*

Proof. Let $T(X)$ be the Π^0_2 condition expressing that X is the graph of a function. Suppose f is the unique function satisfying the condition $\forall n\, R(f\restriction_n)$ where R is computable. Then Γ_f is the unique set X satisfying the Π^0_2 condition

$$T(X)\ \&\ \forall n\, \exists \alpha \in \mathit{Seq}\, [|D^{-1}(\alpha)| = n\ \&\ R(\alpha)\ \&\\ \forall y < \alpha\, \forall i < n\, (\alpha(i) = y \leftrightarrow \langle i, y\rangle \in X)]. \qquad \square$$

The converse holds up to Turing degree, because the witnesses for the existential statement in a Π^0_2 condition on a set A can be incorporated into the function:

1.8.70 Proposition. *For each Π^0_2 singleton A, there is a Π^0_1 function singleton f such that $f \equiv_T A$.*

Proof. Suppose A is the unique set satisfying the Π^0_2 condition $\forall n\, \exists m\, S(n, A\restriction_m)$. Let $f\colon \mathbb{N} \to \mathit{Seq}$ be the function given by $f(n) = \alpha$ if $\alpha = D(A\restriction_m)$ for the minimal m such that $S(n, A\restriction_m)$. Clearly $f \le_T A$. Also f is unbounded, for if $|D^{-1}(f(n))| \le d$ for each n, then each set $Z \succ A\restriction_d$ also satisfies the Π^0_2 condition describing A. Thus $A \le_T f$. Finally $\{f\}$ is Π^0_1, because f is the unique function satisfying the condition

$$\forall n\, [f(n) \in \mathrm{ran}(D)\ \&\ D^{-1}(f(n)) \in \{0,1\}^*\ \&\ \forall x \prec D^{-1}(f(n))[\neg S(n, x)]\ \&\\ S(n, D^{-1}(f(n)))\ \&\ \forall k\, [D^{-1}(f(n)) \preceq D^{-1}(f(k)) \vee D^{-1}(f(k)) \preceq D^{-1}(f(n))]]. \qquad \square$$

By 1.8.62, this shows that there is a function $f \equiv_T \emptyset^{(\omega)}$ such that $\{f\}$ is Π^0_1.

Exercises.

1.8.71. A tree $T \subseteq \mathbb{N}^*$ is *finitely branching* if $T \subseteq \{\sigma \in \mathbb{N}^*: \forall i < |\sigma| \, [\sigma(i) < g(i)]\}$ for some function g. Show that a closed set $P \subseteq \mathbb{N}^{\mathbb{N}}$ is compact $\Leftrightarrow$ the associated tree $T_P = \{\sigma \in \mathbb{N}^*: [\sigma] \cap P \neq \emptyset\}$ is finitely branching.

1.8.72. Suppose $P \subseteq \mathbb{N}^{\mathbb{N}}$ is a nonempty Π^0_1 class such that $T_P = \{\sigma: [\sigma] \cap P \neq \emptyset\}$ is finitely branching via some computable function g (we say that P is bounded). Show that P has a low member.

1.9 Measure and probability

We will introduce a notion of size for certain classes $\mathcal{C} \subseteq 2^{\mathbb{N}}$ by assigning to $\mathcal{C}$ a nonnegative real number $\mu\mathcal{C}$. All the Borel classes will be assigned a size. As an auxiliary notion we discuss outer measures μ, where $\mu\mathcal{C}$ is defined for all classes $\mathcal{C}$. Outer measures satisfy three conditions: $\mu(\emptyset) = 0$, monotonicity, and countable subadditivity. The restriction of an outer measure to the Borel classes yields a measure, a function with the stronger property of countable additivity. In fact, μ is countably additive on a larger domain, the μ-measurable sets.

We mostly use the uniform measure λ where each cylinder $[\sigma]$ is assigned the size $2^{-|\sigma|}$. The theory can be developed in a similar way for the unit interval $[0,1]_{\mathbb{R}}$. Taking into account the identifications in Definition 1.8.10, the uniform measure on Cantor space corresponds to the Lebesgue measure on $[0,1]_{\mathbb{R}}$.

Outer measures

Let $\mathbb{R}^+_0$ denote the set of nonnegative real numbers.

1.9.1 Definition. A function $\mu\colon \mathcal{P}(2^{\mathbb{N}}) \to \mathbb{R}^+_0$ is called an *outer measure* if it satisfies the following.

(i) $\mu(\emptyset) = 0$;
(ii) $\mathcal{C} \subseteq \mathcal{D} \subseteq 2^{\mathbb{N}} \;\rightarrow\; \mu(\mathcal{C}) \leq \mu(\mathcal{D})$ (*monotonicity*);
(iii) $\mu(\bigcup_i \mathcal{C}_i) \leq \sum_i \mu(\mathcal{C}_i)$ for each family $(\mathcal{C}_i)_{i \in \mathbb{N}}$ of classes (*countable subadditivity*).

To introduce an outer measure, one begins with an appropriate assigment of values to basic open cylinders and then extends it to all classes.

1.9.2 Definition. A *measure representation* is a function $r\colon \{0,1\}^* \to \mathbb{R}^+_0$ that satisfies for every $\sigma \in \{0,1\}^*$ the equality

$$r(\sigma 0) + r(\sigma 1) = r(\sigma). \tag{1.19}$$

The extension process is in two steps.

1. If $\mathcal{A} \subseteq 2^{\mathbb{N}}$ is open, choose a prefix-free set $E \subseteq \{0,1\}^*$ such that $[E]^{\prec} = \mathcal{A}$ (for instance let E be the set of minimal strings σ such that $[\sigma] \subseteq \mathcal{A}$ as on page 48). Let

$$\mu_r(\mathcal{A}) = \textstyle\sum_{\sigma \in E} r(\sigma).$$

The property (1.19) ensures that this sum does not exceed $r(\varnothing)$, and that any choice of a prefix-free set E yields the same value $\mu_r(\mathcal{A})$.

2. For a class $\mathcal{C} \subseteq 2^{\mathbb{N}}$ we let

$$\mu_r(\mathcal{C}) = \inf\{\mu_r(\mathcal{A}) \colon \mathcal{C} \subseteq \mathcal{A} \;\&\; \mathcal{A} \text{ is open}\}. \tag{1.20}$$

It is not hard to verify that μ_r is indeed an outer measure (Exercise 1.9.5).

In the following we provide a useful fact about outer measures. It is a special case of the Lebesgue Density Theorem. First a definition.

1.9.3 Definition. Suppose μ is an outer measure. If $\mathcal{C} \subseteq 2^{\mathbb{N}}$ and $\sigma \in \{0,1\}^*$ is such that $\mu[\sigma] \neq 0$, then the local outer measure of $\mathcal{C}$ in $[\sigma]$ is

$$\mu(\mathcal{C} \mid \sigma) = \mu(\mathcal{C} \cap [\sigma])/\mu[\sigma].$$

The theorem states that if $\mu\mathcal{C} > 0$ for an outer measure μ, then the local outer measure $\mu(\mathcal{C} \mid \sigma)$ can be arbitrarily close to 1.

1.9.4 Theorem. *Let μ be an outer measure, and let $\mathcal{C} \subseteq 2^{\omega}$ be such that $\mu\mathcal{C} > 0$. Then for any δ, $0 < \delta < 1$, there exists $\sigma \in \{0,1\}^*$ such that $\mu(\mathcal{C} \mid \sigma) \geq \delta$.*

Proof. Let $\epsilon = (\frac{1}{\delta} - 1)\mu\mathcal{C}$. By the definition of an outer measure, there is an open set $\mathcal{A} \supseteq \mathcal{C}$ such that $\mu\mathcal{A} - \mu\mathcal{C} \leq \epsilon$. Then $\mu\mathcal{A} \leq \mu\mathcal{C} + \epsilon = \frac{1}{\delta}\mu\mathcal{C}$, that is, $\delta \cdot \mu\mathcal{A} \leq \mu\mathcal{C}$. There is a prefix-free set $D \subseteq \{0,1\}^*$ such that $\mathcal{A} = [D]^{\prec} = \bigcup_{\sigma \in D}[\sigma]$. We claim that some $\sigma \in D$ satisfies the conclusion of the theorem. Otherwise, for each $\sigma \in D$, $\mu[\sigma] \neq 0$ implies $\mu(\mathcal{C} \mid \sigma) < \delta$. Since $\mu\mathcal{C} > 0$, we have $\mu[\sigma] > 0$ for some $\sigma \in D$. Then

$$\begin{aligned}
\mu\mathcal{C} &= \mu(\textstyle\bigcup_{\sigma \in D} \mathcal{C} \cap [\sigma]) && \text{since } \mathcal{C} \subseteq \mathcal{A} \\
&\leq \textstyle\sum_{\sigma \in D} \mu(\mathcal{C} \cap [\sigma]) && \text{by countable sub-additivity} \\
&< \delta \cdot \textstyle\sum_{\sigma \in D} \mu[\sigma] && \text{since } \mu(\mathcal{C} \mid \sigma) < \delta \text{ for each } \sigma \in D \\
& && \text{such that } \mu[\sigma] > 0 \text{, and there is such a } \sigma \\
&= \delta \cdot \mu\mathcal{A} \leq \mu\mathcal{C},
\end{aligned}$$

contradiction. □

1.9.5 Exercise. Show that μ_r is an outer measure for each measure representation r.

Measures

For details on the following see for instance Doob (1994).

1.9.6 Definition.

(i) A set $\mathcal{B} \subseteq \mathcal{P}(2^{\mathbb{N}})$ is called a *σ-algebra* if $\mathcal{B}$ is nonempty, closed under complements, and the operations of countable union and intersection.

(ii) A function $\mu \colon \mathcal{B} \to \mathbb{R}_0^+$ is called a *measure* if $\mu(\emptyset) = 0$ and μ is *countably additive*, namely, $\mu(\bigcup_i \mathcal{D}_i) = \sum_i \mu(\mathcal{D}_i)$ for each countable family $(\mathcal{D}_i)_{i \in \mathbb{N}}$ of pairwise disjoint sets. If $\mu(2^{\mathbb{N}}) = 1$ then μ is called a *probability measure*.

Clearly countable additivity implies monotonicity. It also implies countable subadditivity, since for each familiy $(\mathcal{C}_i)_{i\in\mathbb{N}}$, $\bigcup_i \mathcal{C}_i$ is the disjoint union of the sets $\mathcal{D}_i = \mathcal{C}_i - \bigcup_{k<i} \mathcal{C}_k$, and $\mu\mathcal{D}_i \leq \mu\mathcal{C}_i$ for each i.

For an outer measure μ, a set $\mathcal{G} \subseteq 2^{\mathbb{N}}$ is called μ-measurable if for each $\mathcal{C} \subseteq 2^{\mathbb{N}}$ we have $\mu(\mathcal{C}) = \mu(\mathcal{C} \cap \mathcal{G}) + \mu(\mathcal{C} - \mathcal{G})$. It is not hard to see that each clopen set is μ-measurable. A central result of measure theory due to Carathéodory (1968) states that the measurable classes form a σ-algebra, and the restriction of μ to this σ-algebra is a measure. In particular, a Borel class is μ-measurable for any outer measure μ, since the Borel classes form the smallest σ-algebra containing the clopen sets. Most of the subclasses of $2^{\mathbb{N}}$ we will encounter are arithmetical and hence Borel.

Uniform measure and null classes

In the following let r be the measure representation given by $r(x) = 2^{-|x|}$.

1.9.7 Definition. The *uniform (outer) measure*, denoted by λ, is the outer measure obtained from r via the extension process described after 1.9.1.

1.9.8 Definition. A class $\mathcal{A} \subseteq 2^{\mathbb{N}}$ is called *null* if $\lambda\mathcal{A} = 0$.
If $2^{\mathbb{N}} - \mathcal{A}$ is null we say that $\mathcal{A}$ is *conull.*

By step 2 of the extension process we have

1.9.9 Fact. *$\mathcal{A} \subseteq 2^{\mathbb{N}}$ is null $\Leftrightarrow$ there is a sequence $(G_m)_{m\in\mathbb{N}}$ of open sets such that $lim_m \lambda G_m = 0$ and $\mathcal{A} \subseteq \bigcap_m G_m$.*

Note that the class $\mathcal{B} = \bigcap_m G_m$ is Borel and $\lambda\mathcal{B} = 0$. Later on, we will introduce randomness notions by imposing effectiveness or definability restrictions on the condition in Fact 1.9.9 characterizing null sets. For instance, we do this to introduce Martin-Löf randomness in Definition 3.2.1.

For each pair of sets X, Y (not necessarily subsets of $\mathbb{N}$) we let

$$X \triangle Y = (X - Y) \cup (Y - X).$$

It can be shown that $\mathcal{C} \subseteq 2^{\mathbb{N}}$ is λ-measurable iff there is a Borel class $\mathcal{B}$ such that $\mathcal{C}\triangle\mathcal{B}$ is null, and in this case $\lambda\mathcal{C} = \lambda\mathcal{B}$. Also, the restriction of λ to the λ-measurable sets is the unique measure μ given by the measure representation $r(x) = 2^{-|x|}$. We will henceforth call λ-measurable classes simply *measurable*. For each class $\mathcal{C}$ and each set $F \subseteq \mathbb{N}$, the operation $\mathcal{C} \to \mathcal{C}_F = \{Z\triangle F \colon Z \in \mathcal{C}\}$ switches the values of the bits of all sets in $\mathcal{C}$ at the positions given by F. The uniform outer measure λ is invariant under this operation:

1.9.10 Fact. *For each measurable class $\mathcal{C} \subseteq 2^{\mathbb{N}}$ and each set $F \subseteq \mathbb{N}$, the class $\mathcal{C}_F$ is measurable, and $\lambda\mathcal{C}_F = \lambda\mathcal{C}$.*

Proof. Clearly $\lambda([\sigma]_F) = \lambda[\sigma]$ for each string σ. Since λ is countably additive this implies $\lambda\mathcal{A}_F = \lambda\mathcal{A}$ for each open set $\mathcal{A}$. Thus $\lambda\mathcal{C}_F = \lambda\mathcal{C}$ by (1.20). □

Using the axiom of choice one can define a non-measurable class. For $X, Y \subseteq \mathbb{N}$ we write $X =^* Y$ if $X\triangle Y$ is finite. We say that Y is a *finite variant* of X. Clearly $=^*$ is an equivalence relation.

1.9.11 Proposition. *Suppose a class $\mathcal{V} \subseteq 2^{\mathbb{N}}$ contains exactly one member of each equivalence class of $=^*$. Then $\mathcal{V}$ is not measurable.*

Proof. Assume $\mathcal{V}$ is measurable. Then $2^{\mathbb{N}}$ is the disjoint union of the measurable classes $\mathcal{V}_F$ where F is finite. By countable additivity, $1 = \lambda(2^{\mathbb{N}}) = \sum_F \lambda\mathcal{V}_F$ $[\![F \subseteq \mathbb{N}$ is finite$]\!]$. This is impossible since $\lambda\mathcal{V}_F = \lambda\mathcal{V}$ for each F. □

As a consequence, there is no Borel well-order of Cantor space, for otherwise one could take as $\mathcal{V}$ the Borel class of minimal elements in each equivalence class. Note that $\mathcal{V}$ is a version for Cantor space of the Vitali set, where instead of $2^{\mathbb{N}}$ and $=^*$ one has $[0,1]_{\mathbb{R}}$ with the equivalence relation $\{\langle r, s\rangle : r - s \in \mathbb{Q}\}$.

For a class $\mathcal{C}$ and a string σ, let $C \mid \sigma = \{X : \sigma X \in \mathcal{C}\}$. The conditional outer measure $\lambda(\mathcal{C} \mid \sigma)$ of Definition 1.9.3 equals the uniform outer measure of the class $C \mid \sigma$.

With few exceptions, classes $\mathcal{C}$ relevant to computability theory are Borel and closed under finite variants, that is, $X \in \mathcal{C}$ and $X =^* Y$ implies $Y \in \mathcal{C}$. By the following, the uniform measure can only distinguish between small and large classes of this kind.

1.9.12 Proposition. (Zero-one law) *If a measurable class $\mathcal{C}$ is closed under finite variants then $\lambda\mathcal{C} = 0$ or $\lambda\mathcal{C} = 1$.*

Proof. Suppose that $\lambda\mathcal{C} > 0$. We show that $\lambda\mathcal{C} > \delta$ for every δ, $0 \le \delta < 1$. If σ and ρ are strings of the same length, then by the invariance property of λ above, for each X we have $\sigma X \in \mathcal{C} \leftrightarrow \rho X \in \mathcal{C}$. Therefore,

$$\lambda(\mathcal{C} \cap [\sigma]) = \lambda(\{\sigma X : \sigma X \in \mathcal{C}\}) = \lambda(\{\rho X : \rho X \in \mathcal{C}\}) = \lambda(\mathcal{C} \cap [\rho]).$$

By Theorem 1.9.4 choose σ such that $\lambda(\mathcal{C} \mid \sigma) > \delta$ and let $n = |\sigma|$. Then $\lambda(\mathcal{C} \cap [\sigma]) > \delta 2^{-n}$, hence $\lambda\mathcal{C} = \sum_{|\rho|=n} \lambda(\mathcal{C} \cap [\rho]) > 2^n \delta 2^{-n} = \delta$. □

Exercises.

1.9.13. If S is a prefix-free set then $\lim_n \#(S \cap \{0,1\}^n)/2^n = 0$.

1.9.14. An ultrafilter $\mathcal{U} \subseteq \mathcal{P}(\mathbb{N})$ is called free if $\{n\} \notin \mathcal{U}$ for each n. Show that a free ultrafilter is not measurable when viewed as a subset of $2^{\mathbb{N}}$.

1.9.15.$^{\diamond}$ Suppose that $N \in \mathbb{N}$, $\epsilon > 0$, and for $1 \le i \le N$, the class $\mathcal{C}_i$ is measurable and $\lambda\mathcal{C}_i \ge \epsilon$. If $N\epsilon > k$ then there is a set $F \subseteq \{1, \ldots, N\}$ such that $\#F = k+1$ and $\lambda \bigcap_{i \in F} \mathcal{C}_i > 0$. For instance, if $N = 5$ sets of measure at least $\epsilon = 1/2$ are given, then three of them have an intersection that is not a null class. (This is applied in 8.5.18.) *Hint.* Think of integrating a suitable function $g \colon 2^{\mathbb{N}} \to \{0, \ldots, N\}$.

Uniform measure of arithmetical classes

By 1.8.58 each arithmetical class is Borel and hence measurable. The uniform measure of a Σ^0_n class is left-Σ^0_n. The uniform measure of a Π^0_n class is left-Π^0_n. We prove this for $n = 1$ and leave the general case as Exercise 1.9.22.

1.9.16 Fact. *If $R \subseteq 2^{\mathbb{N}}$ is c.e. open then λR is left-c.e. in a uniform way. If $P \subseteq 2^{\mathbb{N}}$ is a Π^0_1 class then λP is right-c.e. in a uniform way.*

Proof. According to 1.8.23, from an index for R we obtain the effective approximation $(R_s)_{s \in \mathbb{N}}$ such that $R = \bigcup_s [R_s]^{\prec}$. Then $\lambda R = \sup_s \lambda R_s$, so λR is left-c.e. The second statement follows by taking complements. □

We discuss c.e. open sets R such that λR is computable. Recall from 1.8.19 that $A_R = \{\sigma\colon [\sigma] \subseteq R\}$. First we note that none of the properties "λR computable" and "A_R computable" implies the other.

1.9.17 Example. There is a c.e. open set R such that $\lambda R = 1$ and A_R is not computable.

Proof. Let $R = 2^{\mathbb{N}} - P$ where P be the class of two-valued d.n.c. functions from Fact 1.8.31. Then $A_R = \{0,1\}^* - T_P$ is not computable. Since there are infinitely many x such that $J(x) = 0$, we have $\lambda P = 0$. □

On the other hand, the uniform measure of a c.e. open set R with computable A_R can be an arbitrary left-c.e. real by Exercise 1.9.20.

We provide two results needed later on when we study Schnorr randomness.

1.9.18 Fact. *Let R be a c.e. open set such that λR is computable. Then $\lambda(R \cap C)$ is computable uniformly in R, λR, and an index for the clopen set C.*
The same holds for Π^0_1 classes of computable measure.

Proof. We use Fact 1.8.15(iv). Given $n \in \mathbb{N}$, we can compute $t \in \mathbb{N}$ such that $\lambda R - \lambda[R_t]^{\prec} \leq 2^{-n}$. Then $\lambda(R \cap C) - \lambda([R_t]^{\prec} \cap C) \leq 2^{-n}$. Thus the rational $q_n = \lambda([R_t]^{\prec} \cap C)$ is within 2^{-n} of $\lambda(R \cap C)$.

For Π^0_1 classes, one relies on the first statement and takes complements. □

1.9.19 Lemma. *Let S be a c.e. open set such that λS is computable. From a rational q such that $1 \geq q \geq \lambda S$ we may in an effective way obtain a c.e. open set $\widetilde{S}$ such that $S \subseteq \widetilde{S}$ and $\lambda \widetilde{S} = q$.*

Proof. We identify sets and real numbers according to Definition 1.8.10. Let

$$\widetilde{S} = S \cup \bigcup\{[0,x)\colon\ x \in \mathbb{Q}_2\ \&\ \lambda(S \cup [0,x)) < q\}.$$

To check that $\widetilde{S}$ is as required, note that the function $f\colon [0,1)_{\mathbb{R}} \to [0,1)_{\mathbb{R}}$ given by $f(r) = \lambda(S \cup [0,r))$ is non-decreasing and satisfies $f(s) - f(r) \leq s - r$ for $s > r$. Thus f is continuous. Further, $f(0) \leq q$ and $f(r) \geq r$ for each r, so there is a least t such that $f(t) = q$. Then $\widetilde{S} = S \cup [0,t)$, and hence $\lambda \widetilde{S} = q$.

To see that $\widetilde{S}$ is c.e., note that $f(x) = x + \lambda(S \cap [x,1))$ for each $x \in \mathbb{Q}_2$, so by 1.9.18 $f(x)$ is a computable real uniformly in x. Since $x < t$ iff $f(x) < q$, this shows that t is a left-c.e. real, whence the open set $\widetilde{S} = S \cup [0,t)$ is c.e.

Note that f is obtained effectively from S and λS. Thus we uniformly obtain a c.e. index for $\{p\colon\ p < t\}$, and hence an index for $\widetilde{S}$. □

Exercises. Show the following.

1.9.20. For each left-c.e. real number $r \in [0,1)$ there is a c.e. open set R such that A_R is computable and $\lambda R = r$.

1.9.21. Each Π^0_1 class P of computable positive measure has a computable member.

1.9.22. The measure of a Σ^0_n class is left-Σ^0_n in a uniform way.
The measure of a Π^0_n class is left-Π^0_n in a uniform way.

Probability theory

We briefly discuss binary strings and sets of natural numbers from the viewpoint of probability theory. Unexplained terms in italics represent standard terminology found in text books such as Shiryayev (1984).

A *sample space* is given by a set $\mathcal{X}$ together with a σ-algebra and a probability measure on it. For instance, a string of length n is an element of the sample space $\{0,1\}^n$, and a set is an element of the sample space $2^{\mathbb{N}}$. The probability measure P is given by $P(\{x\}) = 2^{-|x|}$ in the case of strings and by the uniform measure λ in the case of sets.

For each appropriate i we have a *random variable* $\xi_i \colon \mathcal{X} \to \{0,1\}$ given by $\xi_i(x) = x(i)$. We think of $\xi_i(x)$ as the i-th outcome in the sequence of experiments described by x. If $\mathcal{X}$ is Cantor space, this is a dynamic version of the local view of sets (see Section 1.3): the set is revealed bit by bit. Note that

$$P(\xi_i = 0) = P(\xi_i = 1) = 1/2.$$

Moreover, the ξ_i are *independent.* Such a sequence of random variables is called a *Bernoulli scheme* for the probability 1/2.

For $n \in \mathbb{N}$ let $S_n = \sum_{i<n} \xi_i$. If the sample space is $\{0,1\}^n$ then $S_n(x)$ is the number of occurrences of ones in the string x. The *expectation* of S_n/n is 1/2. In probability theory one is interested in bounding the probability of the *event*

$$\mathrm{abs}(S_n/n - 1/2) \geq \epsilon \qquad (1.21)$$

for $\epsilon > 0$, namely that the number of ones differs by at least ϵn from the expected value $n/2$. The *Chebycheff inequality* shows that this probability is at most $1/(4n\epsilon^2)$. (For a numerical example, if $n = 1000$ and $\epsilon = 1/10$, then 1/40 bounds the probability that the number of ones is at least 600 or at most 400.) In Section 2.5 we will use the improved estimate $2e^{-2n\epsilon^2}$ for the probability of the event (1.21) in order to bound the number of strings of length n where the number of zeros and of ones is unbalanced.

2

THE DESCRIPTIVE COMPLEXITY OF STRINGS

In contrast to the remainder of the book, this chapter is on finite objects. We may restrict ourselves to (binary) strings, because other types of finite objects, such as strings over alphabets other than $\{0, 1\}$, natural numbers, or finite graphs, can be encoded by binary strings in an effective way. We are interested in giving a description σ of a string x, and if possible one that is shorter than x itself. Such descriptions are binary strings as well. To specify how σ describes a string x, we will introduce an optimal machine, which outputs the string x when the input is the description σ. The descriptive complexity $C(x)$ is the length of a shortest description of x. We first study the function $x \to C(x)$. Some of its drawbacks can be addressed by introducing a variant K, where the set of descriptions is prefix-free. The function K also has its drawbacks as a measure of descriptive string complexity, but is the more fruitful one where the interaction of computability and randomness is concerned. The Machine Existence Theorem 2.2.17 is an important tool for showing that the elements of a collection of strings have short descriptions in the sense of K. It will be used frequently in later chapters.

A string x is b-incompressible (in the sense of C, or K) if it has no description σ (in that sense) such that $|\sigma| \leq |x| - b$. In Section 2.5 we will see that incompressibility can serve as a mathematical counterpart for the informal concept of randomness for strings. Using some basic tools from probability theory, we show that an incompressible string x has properties one would intuitively expect from a random string. For instance, x has only short runs of zeros.

Optimal machines act as description systems for strings (analogous to the description systems for sets mentioned at the beginning of Chapter 1). However, the theory of describing strings differs from the theory of describing sets. Only a few reasonable description systems have been introduced. Every string can be described in each system. For a set, the question is whether it can be described at all, while for strings we are interested in the length of a shortest description. For strings, we can formalize randomness by being hard to describe (that is, being incompressible). It takes more effort to formalize randomness for sets (see the introduction to Chapter 3).

The prefix-free descriptive complexity K can be used to determine the degree of randomness of a set $Z \subseteq \mathbb{N}$, because to a certain extent it is measured by the growth rate of the function $n \to K(Z \upharpoonright_n)$. For instance, a central notion, Martin-Löf randomness, is equivalent to the condition that for some b, each initial segment $Z \upharpoonright_n$ is b-incompressible in the sense of K (Theorem 3.2.9).

Much of the material in this chapter goes back to work of Solomonoff (1964), Kolmogorov (1965), Levin and Zvonkin (1970), Levin (1973, 1976), and Chaitin (1975). A standard textbook is Li and Vitányi (1997).

Comparing the growth rate of functions

We frequently want to measure the growth of a function $g\colon \mathbb{N} \to \mathbb{R}$. One way is to compare g to the particularly well-behaved functions of the following type.

2.0.1 Definition. An *order function* is a computable nondecreasing unbounded function $f\colon \mathbb{N} \to \mathbb{N}$. Examples of order functions are $\lambda n. \log n$, $\lambda n.n^2$, and $\lambda n.2^n$.

In the following let $f, g\colon \mathbb{N} \to \mathbb{R} \cup \{\infty\}$. We will review three ways of saying that f grows at least as fast as g: by domination, up to an additive constant, and up to a multiplicative constant (if f and g only have nonnegative values).

(1) Recall the domination preordering on functions from Definition 1.5.1: f dominates g if $\forall^\infty n\, [f(n) \geq g(n)]$. An example of a growth condition on a function g saying that g grows very slowly is to require that g is dominated by each order function. We will study an unbounded function g of this type in 2.1.22.

(2) To compare f and g up to additive constants, let

$$g \leq^+ f : \leftrightarrow \exists c \in \mathbb{N}\, \forall n\, [g(n) \leq f(n) + c]$$

(where $\infty + c = \infty$). This preordering is a bit weaker than the domination preordering. We usually avoid explicitly mentioning the functions f, g. Instead, we write expressions defining them, such as in the statement $\log n + 16 \leq^+ n$. The preordering $\leq^+$ gives rise to the equivalence relation $g =^+ f : \leftrightarrow g \leq^+ f \leq^+ g$. We often try to characterize the growth rate of a function f by determining its equivalence class with respect to $=^+$. For instance, the class of bounded functions coincides with the equivalence class of the function that is constant 0.

(3) Recall that $\mathrm{abs}(r)$ denotes the absolute value of a number $r \in \mathbb{R}$. Let

$$g = O(f) : \leftrightarrow \exists c \geq 0\, \forall^\infty n\, [\mathrm{abs}(g(n)) \leq c\, \mathrm{abs}(f(n))].$$

Moreover, $g = h + O(f)$ means that $g - h = O(f)$. We think of $O(f)$ as an error term, namely, an unspecified function $\widetilde{f}$ such that $\mathrm{abs}(\widetilde{f}(n)) \leq c\, \mathrm{abs}(f(n))$ for almost all n. In particular, $O(1)$ is an unspecified function with absolute value bounded by a constant. Thus $g \leq^+ f \;\leftrightarrow\; g \leq f + O(1)$.

For $f, g\colon \mathbb{N} \to \mathbb{R}^+$ we let $f \sim g : \leftrightarrow \lim_n f(n)/g(n) = 1$. If $f \sim g$, one can replace f by g in error terms of the form $O(f)$.

2.1 The plain descriptive complexity C

Machines and descriptions

Recall that elements of $\{0,1\}^*$ are called strings. A partial computable function mapping strings to strings is called a *machine*. (As we identify a string σ with the natural number n such that the binary representation of $n+1$ is 1σ, formally speaking a machine is the same a partial computable function. It is useful, though, to have this term for the particular context of strings.) If M is a machine and $M(\sigma) = x$ then we say σ is an *M-description* of x. If M is the identity function then x is an M-description of itself. (We will call this machine the *copying machine*.) In general, an M-description of x may be shorter than x, in which case one can view σ as a compressed form of x. The machine M carries out the decompression necessary to re-obtain x from σ.

Our main interest is in measuring how well a string x can be compressed. That is, we are more interested in the length of a shortest description than in the description itself. We introduce a special notation for this length of a shortest M-description:

$$C_M(x) = \min\{|\sigma|\colon\ M(\sigma) = x\}. \tag{2.1}$$

Here $\min \emptyset = \infty$. For an example of a drastic compression, let M be the machine that takes an input σ, views it as a natural number and outputs

$$x = 2^{2^{2^\sigma}}.$$

Then $C_M(x) = |\sigma| =^+ \log^{(4)} x$. For instance, the string $\sigma = 10$ corresponds to the number 5, so $M(\sigma) = 2^{2^{32}} = 2^{4294967296} = x$. The string σ is an M-description of x, and $C_M(x) = 2$, while x has length 2^{32}.

2.1.1 Definition. The machine R is called *optimal* if for each machine M, there is a constant e_M such that

$$\forall \sigma, x\, [M(\sigma) = x \rightarrow \exists \tau\, (R(\tau) = x \ \&\ |\tau| \leq |\sigma| + e_M)], \tag{2.2}$$

or, equivalently, $\forall x\, [C_R(x) \leq C_M(x) + e_M]$.

Thus, for each M, the length of a shortest R-description of a string exceeds the length of a shortest M-description only by a constant e_M. Optimal machines are often called *universal* machines.

The constant e_M is called the *coding constant* for M (with respect to R). It bounds the amount the length of a description increases when passing from M to R. Although we are usually only interested in the length of an R-description, we can in fact obtain an appropriate R-description τ from the M-description σ in an effective way, by trying out in parallel all possible R-descriptions τ of length at most $|\sigma| + e_M$ till one is found such that $R(\tau) = M(\sigma)$.

An optimal machine exists since there is an effective listing of all the machines. The particular choice of an optimal machine is usually irrelevant; most of the inequalities in the theory of string complexity only hold up to an additive constant, and if R and S are optimal machines then $\forall x\, C_R(x) =^+ C_S(x)$. Nonetheless, we will specify a particularly well-behaved optimal machine. Recall the effective listing $(\Phi_e)_{e \in \mathbb{N}}$ of partial computable functions from (1.1) on page 3. We will from now on assume that Φ_1 is the identity.

2.1.2 Definition. The *standard optimal plain machine* $\mathbb{V}$ is given by letting

$$\mathbb{V}(0^{e-1}1\rho) \simeq \Phi_e(\rho)$$

for each $e > 0$ and $\rho \in \{0,1\}^*$.

This definition makes sense because each string σ can be written uniquely in the form $0^{e-1}1\rho$. The machine $\mathbb{V}$ is optimal because for each machine M there is an $e > 0$ such that $M = \Phi_e$, so $M(\sigma) = x$ implies $\mathbb{V}(0^{e-1}1\sigma) = x$. The coding constant with respect to $\mathbb{V}$ is simply an index $e > 0$ for M. Regarding (2.2), note that a $\mathbb{V}$-description τ of x is obtained in a particularly direct way from an index $e > 0$ for M and an M-description σ of x: by putting $0^{e-1}1$ in front of σ. From now on, we simply write $C(x)$ for $C_{\mathbb{V}}(x)$. For each optimal machine R we have $C_R(x) \leq^+ |x|$ since R emulates the copying machine. In particular, since this machine is Φ_1, for each x

$$C(x) \leq |x| + 1. \tag{2.3}$$

This upper bound cannot be improved in general by Exercise 2.1.19.

In Exercises 2.1.6 and 2.1.24 we study necessary conditions for a set to be the domain of an optimal machine. These conditions only depend on the number of strings in the set at each length.

Exercises. Show the following.

2.1.3. A shortest $\mathbb{V}$-description cannot be compressed by more than a constant: there is $b \in \mathbb{N}$ such that, if σ is a shortest $\mathbb{V}$-description of a string x, then $C(\sigma) \geq |\sigma| - b$.

2.1.4. There is an optimal machine R such that for each x, m, the string x has at most one R-description of length m.

2.1.5. Let $d \geq 1$. There is an optimal machine R such that d divides $|\rho|$ for each $\rho \in \operatorname{dom} R$.

2.1.6. Let $D = \operatorname{dom} R$ for an optimal machine R. Then there is $b \in \mathbb{N}$ such that for each n we have $2^n \leq s_{n,b} < 2^{n+b}$, where $s_{n,b} = \#\{\sigma \in D \colon n \leq |\sigma| < n+b\}$.

2.1.7. If x is a string such that $x(i) = 0$ for each even $i < |x|$ then $C(x) \leq^+ |x|/2$.

The counting condition, and incompressible strings

In Theorem 2.1.16 we will provide a machine-independent characterization of the function C. In the first step we characterize the class of functions of the form C_M up to the equivalence relation $=^+$ (see page 75). In the second step we single out C as the least function in this class (again up to $=^+$).

The functions of the form C_M are characterized by two conditions:

(1) being computably approximable from above (an effectivity property), and
(2) the counting condition in 2.1.9 below (which is related to incompressibility).

2.1.8 Definition. A function $D\colon \{0,1\}^* \to \mathbb{N} \cup \{\infty\}$ is *computably approximable from above* if $D(x) = \lim_s D_s(x)$ for a binary computable function $x, s \to D_s(x)$ with values in $\mathbb{N} \cup \{\infty\}$ such that $\forall x\, \forall s\, [D_{s+1}(x) \leq D_s(x)]$.

Each machine M equals some partial computable function Φ_e, so by Definition 1.1.13 we have an approximation $M_s(\sigma) = \Phi_{e,s}(\sigma)$. Therefore each function C_M is computably approximable from above via the approximation

$$C_{M,s}(x) := \min\{|\sigma| \colon M_s(\sigma) = x\}.$$

The counting condition for a function D says that not too many strings x yield small values $D(x)$.

2.1.9 Definition. A function $D\colon \{0,1\}^* \to \mathbb{N} \cup \{\infty\}$ satisfies the *counting condition* if $\#\{x\colon D(x) < k\} < 2^k$ for each k.

This implies that for each k, there is a string x of length k such that $D(x) \geq k$.

2.1.10 Fact. *For each machine M the function C_M satisfies the counting condition. That is, for each k, fewer than 2^k strings have an M-description that is shorter than k.*

Proof. At most $\sum_{0 \leq i < k} 2^i = 2^k - 1$ M-descriptions are shorter than k. □

Short descriptions can yield very long strings and can take a very long time to be decompressed. Thus, even though there are fewer than 2^k strings x such that $C(x) < k$, we cannot predict when they have all appeared. This is confirmed by Proposition 2.1.22.

Incompressible strings
An important property of strings is incompressibility, the formal counterpart of the intuitive concept of randomness for strings. See page 99 for more details.

2.1.11 Fact. *For each machine M, and for each $n \in \mathbb{N}$, there is a string x of length n such that $C_M(x) \geq |x|$.*

Proof. Immediate by Fact 2.1.10. □

By Exercise 2.1.19(ii), for almost every n there is a string x of length n such that $C(x) \geq |x| + 1$.

2.1.12 Definition. Let $d \in \mathbb{N}$. A string x is *d-compressible$_C$* if $C(x) \leq |x| - d$. Otherwise, x is called *d-incompressible$_C$*.
Let Cpr_d denote the set of d-compressible$_C$ strings.

By Fact 2.1.11 the set $\mathsf{Cpr}_1 = \{x\colon C(x) < |x|\}$ is co-infinite. Proposition 2.1.28 below shows that Cpr_1 is a simple set. In Exercise 2.1.3 we have seen that shortest $\mathbb{V}$-descriptions are d-incompressible$_C$ for some fixed d.

The proportion of d-compressible$_C$ strings of a fixed length is less than 2^{-d+1}:

2.1.13 Fact. *Let $d \in \mathbb{N}$. For each $n \in \mathbb{N}$, the number of d-compressible$_C$ strings of length n is less than 2^{n-d+1}.*

Proof. The function C satisfies the counting condition. For $k = n - d + 1$ this yields $\#\{x\colon C(x) \leq n - d\} < 2^{n-d+1}$. □

A method to build machines
The following method to build a machine M will be used to characterize C. A *request* is a pair $\langle n, x\rangle$ from $\mathbb{N} \times \{0,1\}^*$. Informally, issuing a request $\langle n, x\rangle$ means to ask for an M-description of x with length n.

2.1.14 Proposition. *Suppose W is a c.e. set of requests such that for each n, there are at most 2^n requests with first component n. Then there is a machine M such that $\langle n, x\rangle \in W \leftrightarrow \exists \sigma\, (|\sigma| = n \;\&\; M(\sigma) = x)$ for each n, x.*

Proof. We may assume that at most one request is enumerated into W at each stage. We define the machine M by giving a computable enumeration $(M_s)_{s\in\mathbb{N}}$ of its graph (see Exercise 1.1.18).

Let $M_0 = \emptyset$. For $s > 0$, if a request $\langle n, x\rangle$ is in $W_s - W_{s-1}$ then put $\langle \sigma, x\rangle$ into M_s, i.e., let $M_s(\sigma) = x$, for the leftmost string σ of length n that is not in $\mathrm{dom}(M_{s-1})$. By our hypothesis such a string can be found. Clearly M is a machine as required. □

A characterization of C

2.1.15 Proposition. *Suppose that $D\colon \{0,1\}^* \to \mathbb{N} \cup \{\infty\}$ is computably approximable from above and satisfies the counting condition. Then there is a machine M such that $\forall x\, C_M(x) = D(x) + 1$ (where $\infty + 1 = \infty$).*

Proof. Suppose the function $\lambda x, s.D_s(x)$ is a computable approximation of D from above. The c.e. set of requests W is given as follows: *whenever $s > 0$ and $n = D_s(x) < D_{s-1}(x)$, then enumerate $\langle n+1, x\rangle$ into W at stage s*. There is no request with first component 0, and for each n,

$$\#\{x\colon \langle n+1, x\rangle \in W\} \le \#\{x\colon D(x) \le n\} < 2^{n+1}.$$

Applying proposition 2.1.14 to W now yields a machine M as required. □

Recall that a machine R is called optimal iff $C_R \le^+ C_M$ for each machine M. Proposition 2.1.15 yields a machine-independent characterization of C up to $=^+$ (and thus of any function C_R for an optimal machine R).

2.1.16 Theorem. *C can be characterized as follows up to $=^+$: it is the least with respect to $\le^+$ among the functions D that are computably approximable from above and satisfy the counting condition $\forall k\, \#\{x\colon D(x) < k\} < 2^k$.* □

Exercises.

2.1.17. Fix $d \ge 2$, and let r_n be the number of d-incompressible$_C$ strings of length n. Show that $C(r_n) \ge^+ n$. That is, the binary representation of r_n is incompressible$_C$.

2.1.18. We cannot improve the conclusion in Proposition 2.1.15 to $\forall^\infty x\, C_M(x) = D(x)$: there is a function D as in Proposition 2.1.15 such that for each machine M, $D(x) = C_M(x)$ fails for infinitely many x.

2.1.19. (i) Show that, for any optimal machine R, there is a constant d such that for almost all n, there is x with $n \le |x| < n + d$ and $C_R(x) > |x|$.

(ii) (M. Cai) Show that if R is an optimal machine and C_R satisfies (2.3), then we can in fact let $d = 1$. In other words, for almost all lengths the bound $C_R(x) \le |x| + 1$ in (2.3) cannot be improved. *Hint.* Use that C_R is not computable, Cor. 2.1.29 below.

(iii) Show that for each $b \ge 2$, there is an optimal machine S that uses only strings of length divisible by b as descriptions. Hence, if R is a machine extending S such that $R(x) = x$ whenever b does not divide $|x|$, we have $C_R(x) \le |x|$ for any such x. Hence (i) cannot be strengthened for an arbitrary optimal machine R.

Invariance, continuity, and growth of C

We list some properties of the function C. They all hold in fact for C_R, where R is any optimal machine. Usually we view the described finite objects as numbers, not as strings. This subsection mostly follows Li and Vitányi (1997).

Invariance
A machine N on input x cannot increase $C(x)$ by more than a constant:

2.1.20 Fact. *For each machine N and each x such that $N(x)\downarrow$ we have $C(N(x)) \le^+ C(x)$. If N is a one-one function then $C(N(x)) =^+ C(x)$.*

Proof. Let M be a machine such that $M(\sigma) \simeq N(\mathbb{V}(\sigma))$ for each σ. Then $C_M(N(x)) \le C(x)$, so that $C(N(x)) \le C(x) + e_M$ where e_M is the coding constant for M. If N is one-one then the same argument applies to its inverse N^{-1}, so $C(x) \le^+ C(N(x))$ as well. □

In particular, if π is a computable permutation of $\mathbb{N}$, then $C(x) =^+ C(\pi(x))$ for each x. Also, $C(|x|) \le^+ C(x)$.

Continuity properties of C
Recall that $\mathrm{abs}(z)$ denotes the absolute value of $z \in \mathbb{Z}$.

2.1.21 Proposition.

(i) $\mathrm{abs}(C(x) - C(y)) \le^+ 2\log \mathrm{abs}(x-y)$ *for each pair of strings x, y.*

(ii) If the string y is obtained from the string x by changing the bit in one position, then $\mathrm{abs}(C(x) - C(y)) \le^+ 2\log |x|$.

Proof. (i) follows from the stronger Proposition 2.4.4 below, so we postpone the proof till then. We leave (ii) as an exercise. □

The growth of C
The function C is unbounded but slowly growing in the sense that large arguments can have small values. To make this more precise, we consider the fastest growing non-decreasing function that bounds C from below: let

$$\overline{C}(x) = \min\{C(y) \colon y \ge x\}.$$

Fact 2.1.10 states that C satisfies the counting condition, that is, for each k there are fewer than 2^k strings x such that $C(x) < k$. Thus $\overline{C}$ is unbounded. The function $\overline{C}$ is computably approximable from above, but it fails the counting condition. The following says that $\overline{C}$ grows indeed very slowly.

2.1.22 Proposition. *Each order function h dominates $\overline{C}$.*

Proof. We will compare the numbers r_n and y_n ($n \in \mathbb{N}$), where

$$r_n = \min\{z \colon h(z) \ge n\}$$
$$y_n = \min\{z \colon \overline{C}(z) \ge n\}.$$

The function $\lambda n.r_n$ is computable, so by Fact 2.1.20 and (2.3) on page 77 we have $C(r_n) \le^+ C(n) \le^+ \log n$. Now $y_n \le r_n$ implies $n \le \overline{C}(y_n) \le C(r_n) \le^+ \log n$. For each c, $n \le \log n + c$ can only hold for finitely many n. Thus $y_n > r_n$ for almost all n. Since h is non-decreasing, this implies that h dominates $\overline{C}$. □

Exercises.

2.1.23. (Stephan) Let z_m be the largest number such that $C(z_m) < m$. Show that there is a constant d such that $\forall n\, C(z_{n+d}) \geq n$.

2.1.24.$^\diamond$ (Stephan) Let b be the maximum of d in the previous exercise and the constant from Exercise 2.1.6. Let $D = \text{dom}\, R$ for an optimal machine R and

$$s_{n,b} = \#\{\sigma \in D\colon\ n \leq |\sigma| < n + b\}$$

as in 2.1.6. Show that $\forall n\, C(s_{n,b}) \geq^+ n$. (Since $s_{n,b} < 2^{n+b}$, this means that the binary representation of $s_{n,b}$ is incompressible$_C$. Calude, Nies, Staiger and Stephan (2008) have proved that, conversely, if D is a c.e. set such that $\forall n\, C(s_{n,b}) \geq^+ n$ for some b, then D is the domain of an optimal machine.)

2.1.25. The *deficiency set* for a machine R is $D_R = \{x\colon\ \exists y > x\, [C_R(y) \leq C_R(x)]\}$. This set is co-infinite. Build an optimal machine R with a c.e. deficiency set.

2.1.26. Let R be an optimal machine. Show that $p_{\overline{D_R}}$, the listing of $\mathbb{N} - D_R$ in order of magnitude, dominates each computable function. (A c.e. set D such that $p_{\overline{D}}$ dominates each computable function is called *dense simple.*)

Algorithmic properties of C

We show that the set $B = \{x\colon\ C(x) < |x|\}$ of 1-compressible$_C$ strings is simple and *wtt*-complete. Therefore C is incomputable, and in fact $\emptyset' \leq_T C$. This restricts the usefulness of C as a practical measure for the descriptive complexity of a string.

The function C is incomputable because we allowed an arbitrary number of steps for the decompression. One can also consider time-bounded versions of C. Recall from Definition 2.1.2 that $\mathbb{V}$ is the standard optimal machine. For any computable g such that $\forall n\, g(n) \geq n$, the function C^g is computable, where

$$C^g(x) = \min\{\,|\sigma|\colon\ \mathbb{V}(\sigma) = x \text{ in } g(|x|) \text{ steps}\}.$$

We bound the number of computation steps in terms of the length of the described string rather than the length of its description because the description may be much shorter than the described string. Time-bounded versions of C are used mostly in the theory of feasible computability. We will work with a time-bounded version of C on page 136.

2.1.27 Proposition. *$C_M \leq_{wtt} \emptyset'$ for each machine M.*

Proof. By 1.4.7 it suffices to show that the function C_M is ω-c.e. Fix c such that $\forall x\, [C_M(x) \leq |x| + c]$. Then the function $g_s(x) = \min(|x| + c, C_{M,s}(x))$ is a computable approximation of C_M. The number of changes for x is at most $|x| + c$. Thus C_M is ω-c.e. □

2.1.28 Proposition. *The set $B = \{x\colon\ C(x) < |x|\}$ is simple and wtt-complete.*

Proof. B is c.e. via the computable enumeration $B_s = \{x\colon\ C_s(x) < |x|\}$. By Fact 2.1.11, for each length m there is an $x \notin B$ such that $|x| = m$, so B is co-infinite. If B is not simple then there is an infinite computable set $\{r_0, r_1, \ldots\}$ contained in the complement of B such that $\forall i\, |r_i| < |r_{i+1}|$. Then $\forall i\, C(r_i) \leq^+ C(i) \leq^+ \log i$. For sufficiently large i this contradicts $C(r_i) \geq |r_i| \geq i$.

To show that B is weak truth-table complete we define a c.e. set of requests W. If $n \in \emptyset'_s - \emptyset'_{s-1}$ then we put the request $\langle n, x \rangle$ into W, where x is the leftmost string of length $2n$ such that $C_s(x) \geq 2n$. Note that x exists by Fact 2.1.13. Let M be the machine obtained from W via Proposition 2.1.14, and let d be the coding constant for M. We provide a weak truth-table reduction of a finite variant of $\emptyset'$ to B:

> *on input $n > d$, using B as an oracle, compute a stage s such that $B_s(x) = B(x)$ for all strings x of length $2n$ and output $\emptyset'_s(n)$.*

If $n \in \emptyset'_t - \emptyset'_{t-1}$ for some $t > s$, then for some string x of length $2n$ such that $C_s(x) \geq 2n$ this causes $C_M(x) \leq n$ and hence $C(x) \leq n + d < 2n$. Therefore $x \in B - B_s$ contrary to the choice of s. Thus $n \in \emptyset' \leftrightarrow n \in \emptyset'_s$ for each $n > d$. The use of this reduction procedure is computably bounded. □

2.1.29 Corollary. *The function C is not computable.* □

The argument proving Corollary 2.1.29 works in fact for any optimal machine. Proposition 2.1.28 can be strenghtened: Kummer (1996) proved that the set $\{x\colon C(x) < |x|\}$ is in fact truth-table complete.

Exercises.

2.1.30. Show that the set $A = \{\langle x, n\rangle\colon C(x) \leq n\}$ is c.e. and *wtt*-complete.

2.1.31. Prove that $\exists^\infty x\,[C(x) < C^g(x)]$ for each computable function g such that $\forall n\, g(n) \geq n$. Show that this cannot be improved to $\forall^\infty x\,[C(x) < C^g(x)]$.

2.2 The prefix-free complexity K

Using $C(x)$ as a measure for the descriptive complexity of a string x is conceptually simple, because we do not restrict the machines carrying out the decompression of the descriptions. However, this simplicity leads to certain drawbacks of C. In this section we will first discuss these drawbacks in more detail, and then introduce a variant of C which addresses these particular problems (but also has its drawbacks). A machine M is called prefix-free if the domain of M is a prefix-free set. We will develop a theory of string complexity that parallels the one for C but is based on prefix-free machines. The resulting descriptive complexity of a string x is called its prefix-free complexity and is denoted by $K(x)$. We will see in subsequent chapters that K, rather than C, is the appropriate measure of descriptive complexity for strings when one is concerned with the interplay of computational complexity and randomness. Occasionally there are, however, important applications of C, such as the initial segment characterization of 2-random sets in Theorem 3.6.10. We also use C to understand jump traceability in Theorem 8.4.10.

In the previous subsection we have discussed the fact that C is incomputable and how one could get around this by using a time-bounded version C^g. However, since we will be using descriptive complexity of strings for theoretical purposes, we do not consider this incomputability as a drawback – the prefix-free version K is not computable either.

Drawbacks of C

Complexity dips

Recall from 2.1.12 that a string w is d-incompressible$_C$ if $C(w) > |w| - d$. We proved in Fact 2.1.13 that the number of d-incompressible$_C$ strings of length n is at least $2^n - 2^{n-d+1}$. One may ask whether there is a string w of length n such that each prefix of w is d-incompressible$_C$ when n is large compared to d. The answer is negative, because a machine N can "cheat" in the decompression, by encoding some extra information into the length of a description.

2.2.1 Proposition. *There is a constant c with the following property. For each $d \in \mathbb{N}$ and each string w of length at least $2^{d+1} + d$ there is an $x \preceq w$ such that $C(x) \leq |x| - d + c$.*

Proof. We use the notation introduced in (1.5) on page 13. The machine N is given by $N(\sigma) = \mathsf{string}(|\sigma|)\sigma$. It is sufficient to obtain a prefix x of w with an N-description of length $|x| - d$. Let $k = \text{number}(w \upharpoonright_d)$ (so that $k < 2^{d+1}$), and let $x = w \upharpoonright_{d+k}$ be the prefix of w with length given by $d + m$ where m is the number represented by the first d bits of w. Let σ be the string of length k such that $x = x \upharpoonright_d \sigma$, then $N(\sigma) = x$. Thus $C_N(x) \leq |x| - d$. □

For example let $d = 3$ and $w = 010101110001011011$. Then $k = \mathsf{number}(010) = 9$, so $x = w \upharpoonright_{3+9} = \boxed{010|101110001}$. Thus $N(\sigma) = x$ where $\sigma = 101110001$.

Failure of subadditivity

A further desirable property of a complexity measure for strings would be the following: descriptions of strings x and y can be put together to obtain a description of $\langle x, y\rangle$. This would imply $C(\langle x, y\rangle) \leq^+ C(x) + C(y)$. Concatenating the descriptions of x and y does not work since we cannot tell where the description of x ends and the description of y begins. We prove that (1) $C(\langle x, y\rangle) \leq^+ C(x) + C(y)$ fails. Note that $C(xy) \leq^+ C(\langle x, y\rangle)$, so the failure of (2) $C(xy) \leq^+ C(x) + C(y)$ is even stronger (but see 2.2.3 for a direct proof that (1) fails). The following says that for each d there is a string w of length $O(2^d)$ such that for some decomposition $w = xy$ (2) fails for the constant d.

2.2.2 Corollary. *Let $d \in \mathbb{N}$. Suppose w is a string of length $2^{d+1} + d$ such that $C(w) \geq |w|$. If $x \preceq w$ is as in 2.2.1 and $w = xy$ then $C(w) \geq^+ C(x) + C(y) + d$.*

Proof. This follows from $C(x) \leq^+ |x| - d$ and $C(y) \leq |y| + 1$. □

2.2.3 Exercise. Show that subadditivity in the form (1) above fails badly: for each n there are strings x, y such that $|xy| = n$ and $C(\langle x, y\rangle) \geq^+ C(x) + C(y) + \log n$.

Prefix-free machines

2.2.4 Definition. A machine M is *prefix-free* if its domain is a prefix-free set, that is, $\forall \sigma, \rho \in \operatorname{dom} M[\sigma \preceq \rho \rightarrow \sigma = \rho]$. In order to indicate that a machine M is prefix-free, we write $K_M(x)$ instead of $C_M(x)$.

Consider the following experiment due to Chaitin: start the machine M, and whenever it requests a new input bit, toss a coin and feed the resulting bit to the machine. Since M is prefix-free the following sum converges and expresses the probability that M halts in this experiment:

$$\Omega_M = \sum_{\sigma} 2^{-|\sigma|} [\![M(\sigma)\downarrow]\!]. \tag{2.4}$$

Thus $\Omega_M = \lambda[\mathrm{dom}\, M]^{\prec}$ is the measure of the open set generated by the domain of M. Since this open set is c.e., Ω_M is a left-c.e. real number (see 1.9.16). Let $\Omega_{M,s} = \lambda[\mathrm{dom}(M_s)]^{\prec}$, then $(\Omega_{M,s})_{s\in\mathbb{N}}$ is a nondecreasing computable approximation of Ω_M.

2.2.5 Example. Suppose M is a machine which halts on input σ iff σ is of the form 0^i1 for even i. Then $\Omega_M = 0.101010\ldots = 2/3$.

Optimal prefix-free machines

2.2.6 Definition. We say that a machine R is an *optimal prefix-free machine* if R is prefix-free, and for each prefix-free machine M there is a constant d_M such that $\forall x\, K_R(x) \le K_M(x) + d_M$. As before, the constant d_M is called the *coding constant* of M (with respect to R).

2.2.7 Proposition. *An optimal prefix-free machine R exists.*

Proof. We first provide an effective listing $(M_d)_{d\in\mathbb{N}}$ of all the prefix-free machines. M_d is a modification of the partial computable function Φ_d in (1.1) on page 3. When the computation $\Phi_d(\sigma) = x$ converges at stage s, then declare $M_{d,s}(\sigma) = x$ unless $M_{d,t}(\tau)$ has already been defined at a stage $t < s$ for some τ such that $\tau \prec \sigma$ or $\sigma \prec \tau$. Clearly M_d is prefix-free and $M_d = \Phi_d$ if Φ_d is prefix-free. Now define R by $R(0^{d-1}1\sigma) \simeq M_d(\sigma)$ for $d > 0$. By the Padding Lemma 1.1.3 we can leave out M_0, so R is optimal. □

For the moment we will fix an arbitrary optimal prefix-free machine R and write $K(x)$ for $C_R(x) = \min\{|\sigma|\colon R(\sigma) = x\}$. Like any other machine, R is emulated by $\mathbb{V}$. Thus $C(x) \le^+ K(x)$.

The string complexity K amends the drawbacks of C discussed on page 83. Firstly, $K(\langle x, y\rangle) \le^+ K(x) + K(y)$, because given an R-description σ of x and an R-description τ of y, a new prefix-free machine can decompress the description $\rho = \sigma\tau$. It recovers σ and τ, then it runs R on both to compute x and y, and finally it outputs $\langle x, y\rangle$. (In Theorem 2.3.6 we give an explicit expression for $K(\langle x, y\rangle)$ using a conditional version of K.)

Secondly, by Proposition 2.5.4 there are arbitrarily long strings x such that each prefix of x is d-incompressible in the sense of K, for some fixed d.

Upper bounds for K

By (2.3) we have $C(x) \le |x|+1$. A drawback of K is the absence of a computable upper bound, in terms of the length, that is tight for almost all lengths. For C, the upper bound was based on the copying machine Φ_1, which is not prefix-free. In fact this upper bound fails for K (see Remark 2.5.3 below). To obtain a crude computable upper bound (to be improved subsequently) consider the machine M given by $M(0^{|x|}1x) = x$ for each string x. This machine is prefix-free since $0^{|x|}1x \preceq 0^{|y|}1y$ first implies $|x| = |y|$, and then $x = y$. Thus

$$K(x) \leq^+ 2|x|. \tag{2.5}$$

The idea in the foregoing argument was to precede x by an encoding of its length $n = |x|$ in such a way that the strings encoding lengths form a prefix-free set. Instead of 0^n1, we can take any other such encoding. So, why not take a shortest possible prefix-free encoding of n, using R itself for the decompression? Then we get as close to the copying machine as we possibly can within the prefix-free setting. This yields the following improved upper bounds.

2.2.8 Proposition. *$K(x) \leq^+ K(|x|) + |x| \leq^+ 2\log|x| + |x|$ for each string x.*

Proof. The second inequality follows by applying (2.5) to $|x|$. For the first inequality, define a prefix-free machine N as follows:

> *on input τ search for a decomposition $\tau = \boxed{\sigma|x}$ such that $n = R(\sigma)\downarrow$ and $|x| = n$. If one is found output x.*

Clearly N is prefix-free. Given x, let σ be a shortest R-description of $|x|$. Then $N(\sigma x) = x$. This shows that $K(x) \leq^+ K(|x|) + |x|$. □

Incorporating N into the machine R obtained in the proof of Proposition 2.2.7, one may achieve that the constant in the inequality $K(x) \leq^+ K(|x|)+|x|$ is 1. The idea is to emulate N with a loss in compression of only 1. Since N is already based on the optimal machine to be defined, we have to use the Recursion Theorem.

2.2.9 Theorem. (Extends 2.2.7) *There is an optimal prefix-free machine $\mathbb{U}$ such that $K(x) \leq K(|x|) + |x| + 1$ for each x, where $K(x) = K_{\mathbb{U}}(x)$.*

Proof. Let $(M_d)_{d\in\mathbb{N}}$ be the effective listing of all the prefix-free machines from the proof of 2.2.7. Given e, we define a prefix-free machine N_e by $N_e(\sigma x) = x$ iff $M_e(\sigma) = |x|$. By the Parameter Theorem 1.1.2 there is a computable function q such that

$$\Phi_{q(e)}(0^{d-1}1\rho) \simeq \begin{cases} N_e(\rho) & \text{if } d = 1 \\ M_d(\rho) & \text{if } d > 1. \end{cases}$$

By the Recursion Theorem 1.1.5 (and 1.1.7) there is an $i > 1$ such that $\Phi_{q(i)} = \Phi_i$. Since $\Phi_{q(e)}$ is prefix-free for each e, Φ_i is prefix-free and hence $\Phi_i = M_i$. By the Padding Lemma 1.1.3 it does not matter to leave out M_0 and M_1 in the emulation, so $\mathbb{U} := \Phi_i$ is an optimal prefix-free machine.

Given x, let σ be a shortest $\mathbb{U}$-description of $|x|$, then $\mathbb{U}(1\sigma x) = \Phi_{q(i)}(1\sigma x) = N_i(\sigma x) = x$. Therefore $K(x) \leq K(|x|) + |x| + 1$. □

From now on, we will simply write $K(x)$ for $K_{\mathbb{U}}(x) = \min\{|\sigma|\colon \mathbb{U}(\sigma) = x\}$. We also let

$$K_s(x) = \min\{|\sigma|: \mathbb{U}_s(\sigma) = x\} \tag{2.6}$$

(where $\min \emptyset = \infty$). Note that $C(x) \le^+ K(x)$ since $\mathbb{U}$ is emulated by $\mathbb{V}$.

Exercises.

2.2.10. Redo the Exercises 2.1.3–2.1.5, 2.1.17 and 2.1.23 for prefix-free machines and K.

2.2.11. Show that $K(x) \le^+ 2\log\log|x| + \log|x| + |x|$.

2.2.12. (Calude, Nies, Staiger and Stephan, 2008) Let U be an optimal prefix-free machine, and let $\mathcal{C} = \{S \subseteq \{0,1\}^*: S \text{ is prefix-free } \& \operatorname{dom} U \subseteq S\}$.
Show that (i) $\mathcal{C}$ is a Π^0_1 class, and therefore contains a low set; (ii) no Π^0_1 set S is in $\mathcal{C}$.

The Machine Existence Theorem and a characterization of K

We proceed as in the characterization of C on page 79: firstly, we characterize, up to $=^+$, the class of functions of the form K_M for a prefix-free machine M, by

(1) being computably approximable from above, and
(2) satisfying the weight condition defined in 2.2.13.

Secondly, K is the least function in this class with respect to $\le^+$.

2.2.13 Definition. A function $D\colon \mathbb{N} \to \mathbb{N} \cup \{\infty\}$ satisfies the *weight condition* if $\sum_x 2^{-D(x)} \le 1$. (Here $2^{-\infty} := 0$.)

The weight condition implies $\#\{x: D(x) = i\} \le 2^i$ for each i, which in turn implies the counting condition in 2.1.9. The weight condition holds for any function of the type K_M because $\sum_x 2^{-K_M(x)} \le \sum_\sigma 2^{-|\sigma|} [\![M(\sigma)\downarrow]\!] = \Omega_M \le 1$.

To proceed with the characterization of the functions K_M we need an analog of Proposition 2.1.14, which we will call the *Machine Existence Theorem* (it is also refered to in the literature as the Kraft-Chaitin Theorem). Unlike 2.1.14, it is nontrivial. It will be an important tool in future constructions. First we discuss Kraft's Theorem. Then we introduce our tool which is an effectivization of Kraft's Theorem. It appeared in Chaitin (1975). A similar result had already been obtained by Levin (1973); see Exercise 2.2.23. We prefer the Machine Existence Theorem to the setting of Levin because its notation is more convenient in our applications.

Let $N \in \mathbb{N} \cup \{\infty\}$ and suppose we want to encode strings x_i, $i < N$, by strings σ_i in such a way that the set of strings σ_i is prefix-free. In this case we call the (finite or infinite) list $\langle \sigma_0, x_0 \rangle, \langle \sigma_1, x_1 \rangle, \ldots$ a *prefix-free code.* For instance, if we let $\sigma_i = 0^{|x_i|}1x_i$ we obtain a prefix-free code. (For practical applications, such a code is useful when each x_i represents a letter in an alphabet. Then a text is a concatenation of the x_i's, and the text is encoded by the corresponding concatenation τ of the strings σ_i. Now τ can be decoded in a unique way by scanning from left to right, splitting off one string σ_i at a time.)

Recall from 2.1.14 that a *request* is a pair $\langle r, x\rangle$ from $\mathbb{N} \times \{0,1\}^*$, meaning that we want a description of x with length r. The following is due to Kraft (1949).

2.2.14 Proposition. *Suppose $(\langle r_i, x_i \rangle)_{i<N}$ is a list of requests, where $N \in \mathbb{N} \cup \{\infty\}$. A prefix-free code $\langle \sigma_0, x_0 \rangle, \langle \sigma_1, x_1 \rangle, \ldots$ such that $|\sigma_i| = r_i$ for each $i < N$ exists if and only if the function $\lambda i.r_i$ satisfies the weight condition.*

Proof. The necessity of the weight condition is clear since the basic cylinders $[\sigma_i]$ are pairwise disjoint, and therefore $\sum_{i<N} 2^{-r_i} = \lambda \bigcup_i [\sigma_i] \leq 1$. For its sufficiency, note that we may reorder the list of requests and achieve that $r_0 \leq r_1 \leq \ldots$; now let σ_i be the string of length r_i such that $0.\sigma_i = \sum_{j<i} 2^{-r_j}$. □

To illustrate this proof, suppose that $N = 4$ and the sequence $(r_i)_{i<N}$ is $1, 3, 3, 3$. Then $\sigma_0 = 0$, $\sigma_1 = 100$, $\sigma_2 = 101$, and $\sigma_3 = 110$. The following figure shows how each strings σ_i is associated with a subinterval of $[0, 1)$ of length 2^{-r_i}.

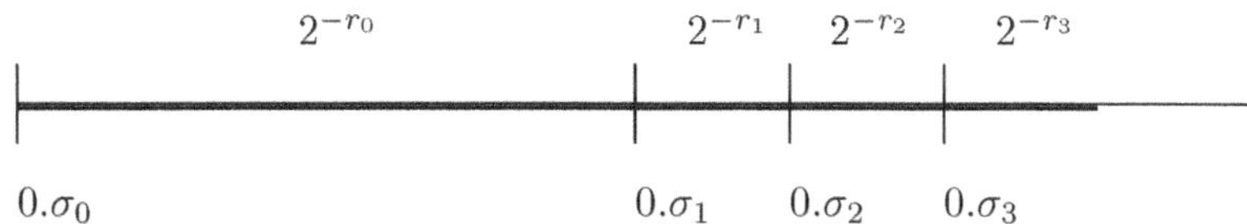

Example. Suppose we want to encode strings in such a way that, if 0 is twice as likely as 1, then the encoding string can be expected to be shorter than the given string. We will find a prefix code for the four possible blocks of two bits in such a way that the codeword for the most likely block 00 has length 1. The list of requests $\langle r_i, x_i \rangle$ $(0 \leq i < 4)$ is $\langle 1, 00 \rangle$, $\langle 2, 01 \rangle$, $\langle 3, 10 \rangle$, $\langle 3, 11 \rangle$. Note that $\sum_{i<4} 2^{-r_i} = 1/2+1/4+1/8+1/8 = 1$, so the method in the proof of 2.2.14 yields the prefix-free code $\langle 0, 00 \rangle$, $\langle 10, 01 \rangle$, $\langle 110, 10 \rangle$, $\langle 111, 11 \rangle$. For instance, the codeword for the string 000010 is 00110. If 0 is twice as likely as 1, then the expected length of a codeword for a string of length n is $17n/18$.

Suppose that the function $i \mapsto \langle r_i, x_i \rangle$ is computable. Then we would like to have a prefix-free code that is an effective list. In other words, we want the function $\sigma_i \mapsto x_i$ to be given by a prefix-free machine. The Machine Existence Theorem provides such a machine. Its proof is harder than the proof of Proposition 2.2.14 because in the effective setting we are no longer allowed to reorder the list to make the sequence (r_i) non-descending.

In our applications we usually deal with a c.e. set W of requests, rather than an effective list (which may have repetitions). For a request $\rho = \langle r, x \rangle$ we let $(\rho)_0$ denote the first component r, and $(\rho)_1$ the second component x.

2.2.15 Definition. A c.e. set $W \subseteq \mathbb{N} \times 2^{<\omega}$ is a *bounded request set* if

$$\sum_{\rho} 2^{-(\rho)_0} [\![\rho \in W]\!] \leq 1. \tag{2.7}$$

If $S \subseteq \{0,1\}^*$, the *weight* of S given by W is

$$\mathsf{wgt}_W(S) = \sum_{\rho,x} 2^{-(\rho)_0} [\![\rho \in W \ \& \ x \in S \ \& \ (\rho)_1 = x]\!]. \tag{2.8}$$

We say that $\mathsf{wgt}_W(\{0,1\}^*)$ is the *total weight* of W.

2.2.16 Example. For computable B the set $W = \{\langle n+1, B\upharpoonright_n \rangle : n \in \mathbb{N}\}$ is a bounded request set.

2.2.17 Theorem. (Machine Existence Theorem) *For each bounded request set W, one can effectively obtain a prefix-free machine $M = M_d$, $d > 1$, such that*

$$\forall r, y\,[\langle r, y\rangle \in W \;\leftrightarrow\; \exists w\,(|w| = r \;\&\; M(w) = y)].$$

Moreover, Ω_M equals the total weight of W.

We say that M_d is a *prefix-free machine for W*. Recall that d is the coding constant for M_d (with respect to the standard optimal prefix-free machine). In order to avoid explicit mention of M_d, we will also refer to d as the *coding constant* for W.

Bounded request sets are useful because they are easier to handle than prefix-free machines. The machine existence theorem provides the coding constant d for the corresponding machine. So we know that, if we enumerate the request $\langle r, y\rangle$, then $K(y) \le r + d$ (usually this is all we care about).

Proof of Theorem 2.2.17. Let $\langle r_n, y_n\rangle_{n<N}$ be an effective enumeration of W, where $N \in \mathbb{N}$ or $N = \infty$. As in the remark after Proposition 2.2.14 we associate with each string σ the half-open interval $I(\sigma) = [0.\sigma, 0.\sigma + 2^{-|\sigma|})$ of real numbers such that the binary representation (containing infinitely many zeros) extends σ. For instance $I(011) = [3/8, 1/2)$. In the construction of M, at stage $n \ge 0$ we will find a string w_n of length r_n and set $M(w_n) = y_n$. The idea is to let w_n be the leftmost string such that the associated interval is disjoint from the previous intervals. For instance, if $r_0 = r_1 = 3$ and $r_2 = 1$ we assign the intervals as follows:

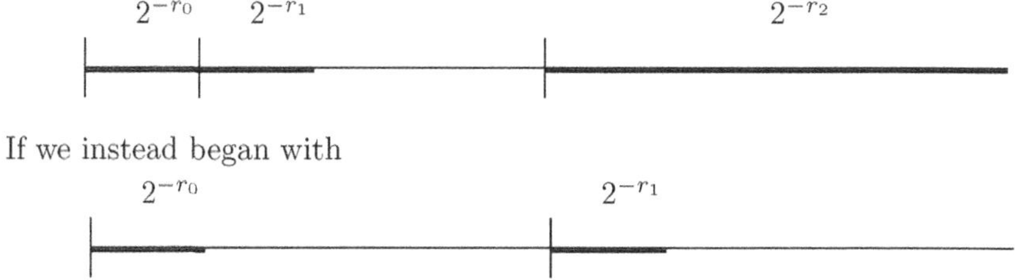

If we instead began with

then we would be stuck in the third step because no interval of length $1/2$ is available any longer. (However, see Exercise 2.2.24.)

We let $R_{-1} = \{\varnothing\}$. At the beginning of each stage $n \ge 0$ we have a finite prefix-free set R_{n-1} of strings where all extensions are unused.

Construction of strings w_n and finite sets of strings R_n.
Stage n.

(1) Let z_n be the longest string in R_{n-1} of length $\le r_n$ (below we will verify that z_n exists).

(2) Choose w_n so that $I(w_n)$ is the leftmost subinterval of $I(z_n)$ of length 2^{-r_n}, i.e., let $w_n = z_n 0^{r_n - |z_n|}$.

(3) To obtain R_n, first remove z_n from R_{n-1}. If $w_n \neq z_n$ then also add the strings $z_n 0^i 1$, $0 \leq i < r_n - |z_n|$, to R_n.

We verify inductively that for each $n \geq 0$ the following hold:

(a) The string z_n exists.

(b) All the strings in R_n have different lengths. (In fact, for $x, y \in R_n$ we have $|x| < |y| \leftrightarrow x <_L y$, that is, the intervals $I(x)$ get longer as one moves to the right.)

(c) $\{I(z)\colon z \in R_n\} \cup \{I(w_i)\colon i \leq n\}$ is a partition of $[0,1)$.

We prove (a) for $n \geq 0$, assuming (b) and (c) for $n-1$ (for $n = 0$ they are trivial statements). If z_n fails to exist, then r_n is less than the length of each string in R_{n-1}, so that $2^{-r_n} > \sum_z 2^{-|z|} [\![z \in R_{n-1}]\!]$ by (b) for $n-1$. Then $\sum_{i=0}^{n} 2^{-r_i} > 1$ since $\sum_z 2^{-|z|} [\![z \in R_{n-1}]\!] + \sum_{i=0}^{n-1} 2^{-r_i} = 1$ by (c) for $n-1$. This contradicts the assumption that W is a bounded request set.

Clearly (b) for n holds if $w_n = z_n$. If $w_n \neq z_n$ then $|z_n| < |w_n|$ but also $|w_n|$ is less than the length of the shortest string in R_{n-1} that is longer that z_n, so (b) holds by the definition of R_n. Finally, (c) is satisfied by the definition of R_n.

The machine M given by $M(w_n) = y_n$ is prefix-free by (c). As M was determined effectively from W we may obtain an index $d > 1$ for M. For each request $\rho = \langle r, y\rangle \in W$ we put a new description w of length r into $\mathrm{dom}(M)$, so Ω_M equals the total weight of W. □

To give an example how to apply Theorem 2.2.17, we reprove the inequality $K(x) \leq^+ |x| + 2\log|x|$ in 2.2.8. If c is chosen sufficiently large then the set $W = \{\langle |x| + 2\log|x| + c, x\rangle\colon x \in \{0,1\}^*\}$ is a bounded request set, since $\sum_\rho 2^{-(\rho)_0} [\![\rho \in W]\!] \leq 2^{-c}(\sum_{n>0} 2^n 2^{-n-2\log n}) \leq 1$. Let M be the prefix-free machine for W. Then $K_M(x) \leq |x| + 2\log|x| + c$ and thus $K(x) \leq^+ |x| + 2\log|x|$.

Theorem 2.2.17 can be used to characterize the functions of the type K_M for a prefix-free machine M. This is analogous to Proposition 2.1.15.

2.2.18 Proposition. *Suppose that $D\colon \mathbb{N} \to \mathbb{N} \cup \{\infty\}$ is computably approximable from above and satisfies the weight condition $\sum_x 2^{-D(x)} \leq 1$. Then there is a prefix-free machine M such that $\forall x\, K_M(x) = D(x) + 1$.*

Proof. Suppose that $\lambda x, s.\, D_s(x)$ is a computable approximation of D from above. The c.e. set W is defined exactly as in the proof of 2.1.15: whenever $r = D_s(x) < D_{s-1}(x)$ then enumerate the request $\langle r+1, x\rangle$ into W at stage s. For each x, at the least stage s such that $D_s(x) = D(x) < \infty$, the last request of the form $\langle u, x\rangle$ is enumerated into W, where $u = D(x) + 1$. This enumeration contributes $2^{-D(x)-1}$ to the sum in (2.7). The contribution of all the requests $\langle m, x\rangle$ enumerated at previous stages is less than $\sum_{m>u} 2^{-m} = 2^{-u}$ since $m > u$ for such a request. All the requests with second component x together contribute less than $2^{-D(x)}$. Since D satisfies the weight condition, W is a bounded request set. Applying the Machine Existence Theorem 2.2.17 to W yields a machine M as required. □

We now obtain a machine-independent characterization of K (and in fact of any function K_R where R is an optimal prefix-free machine as defined in 2.2.6).

2.2.19 Theorem. *K can be characterized as follows up to $=^+$: it is the least with respect to $\leq^+$ among the functions D that are computably approximable from above and satisfy the weight condition $\sum_x 2^{-D(x)} \leq 1$.* □

The following application of 2.2.19 will be needed for Theorem 8.1.9. It can be seen as an effective version of Exercise 1.9.13.

2.2.20 Proposition. *Let B be a c.e. prefix-free set and let $b_m = \#(B \cap \{0,1\}^m)$ for $m \in \mathbb{N}$. Then $\forall m\, [K(m) \leq^+ m - \log b_m]$.*

Proof. Clearly the function D given by $D(m) = m - \log b_m$ is computably approximable from above. For each m we have $\lambda[B \cap \{0,1\}^m]^{\prec} = 2^{-m} b_m$. Since $2^{\log b_m} \leq b_m$, we have $\sum_m 2^{-m+\log b_m} \leq \lambda[B]^{\prec} \leq 1$. Therefore D satisfies the weight condition. □

2.2.21 Remark. (Coding constants given in advance.) In constructing a bounded request set L, by the Recursion Theorem we often assume that a coding constant d for a machine M_d is given, and we think of M_d as a prefix machine for L. This is useful because d represents the "loss" that occurs when putting a request $\langle r, x\rangle$ into L; its enumeration merely ensures that $K(x) \leq r + d$.

It is paradoxical that we can assume that d is given even if we are building L. Let us justify this carefully.

(1) From an index for a c.e. set $G \subseteq \mathbb{N} \times 2^{<\omega}$ we may effectively obtain an index for a bounded request set $\widetilde{G}$ such that $\widetilde{G} = G$ in case G already is a bounded request set.
(2) Let M_d $(d > 1)$ be the machine effectively obtained from $\widetilde{G}$ via the Machine Existence Theorem.
(3) Our construction yields a bounded request set L uniformly in d.

We have to show that L is a bounded request set for *each* d. If $G = L$, which will happen for some effectively given G by the Recursion Theorem, then we may conclude that G is a bounded request set and M_d is a machine for L. (We need to show that L is always a bounded request set, for otherwise we might end up with a fixed point where $G = L$ but $\widetilde{G} \neq G$, which would be of no interest to us.) See Proposition 5.2.13 for our first application of this method.

Exercises.

2.2.22. Show that $\sum_x f(K(x))2^{-K(x)} = \infty$ for each order function f.

2.2.23. (Levin) A *discrete c.e. semimeasure* on $\{0,1\}^*$ is a function $m\colon \{0,1\}^* \to \mathbb{R}_0^+$ such that $m(x)$ is a left-c.e. real number uniformly in x and $\sum_x m(x) \leq 1$. Note that $S \to m(S) = \sum_{x \in S} m(x)$ defines a measure on all subsets S of $\{0,1\}^*$. For instance, the function $m(x) = 2^{-x-1}$ is a discrete c.e. semimeasure; a further example is $m(x) = 2^{-K(x)}$. Verify that a function $D\colon \mathbb{N} \to \mathbb{N} \cup \{\infty\}$ is computably approximable from above and satisfies the weight condition iff $m(x) = 2^{-D(x)}$ is a discrete c.e.

semimeasure. Next, show that there is a close correspondence between discrete c.e. semimeasures and bounded request sets:

(a) For each bounded request set W, the function $m(x) = \mathsf{wgt}_W(\{x\})$ is a discrete c.e. semimeasure on $\{0,1\}^*$ such that $m(S) = \mathsf{wgt}_W(S)$.
(b) If m is a discrete c.e. semimeasure, there is a bounded request set W such that $m(x) = \mathsf{wgt}_W(\{x\})$ for each x. Also, $L = \{\langle k+1, x\rangle : 2^{-k} \le m(x)\}$ is a bounded request set such that $4 \cdot \mathsf{wgt}_L(\{x\}) \ge m(x) \ge \mathsf{wgt}_L(\{x\})$ for each x.

For details see Li and Vitányi (1997).

2.2.24. (Gács) (i) Let $p, \delta \in [0,1)_{\mathbb{R}}$, $\delta > 0$, $p + \delta \le 1$. Show that there is a string w such that $\delta < 2^{-|w|+2}$ and $I(w) \subseteq [p, p+\delta)$.
(ii) Use (i) to prove the slightly weaker version of 2.2.17 where the conclusion is $\forall r, y\, [\langle r, y\rangle \in W \leftrightarrow \exists w\ (|w| \le r + 2\ \&\ M(w) = y)]$ (that is, the M-description corresponding to the request $\langle r, y\rangle$ may be chosen by 2 longer than r).

The Coding Theorem

The probability that a prefix-free machine M outputs a string x is

$$P_M(x) = \sum_\sigma 2^{-|\sigma|}\, [\![M(\sigma) = x]\!] = \lambda[\{\sigma : M(\sigma) = x\}]^{\prec}.$$

Thus $\Omega_M = \sum_x P_M(x)$.

2.2.25 Theorem. (Coding Theorem) *From a prefix-free machine M we may effectively obtain a constant c such that $\forall x\ 2^c 2^{-K(x)} > P_M(x)$.*

Proof. We show that the function $D(x) = \lceil -\log_2 P_M(x)\rceil$ (where $\lceil r\rceil$ denotes the least integer no less than r) is computably approximable from above and satisfies the weight condition $\sum_x 2^{-D(x)} \le 1$. Then we apply Theorem 2.2.19.

Let $P_{M,s}(x) = \lambda[\{\sigma : M_s(\sigma) = x\}]^{\prec}$, then $D_s(x) = \lceil -\log_2 P_{M,s}(x)\rceil$ is a computable approximation of D from above.

By definition $D(x) \ge -\log_2 P_M(x) > D(x) - 1$, so

$$2^{-D(x)} \le P_M(x) < 2^{-D(x)+1}.$$

This implies the weight condition for D, since $\sum_x 2^{-D(x)} \le \sum_x P_M(x) \le 1$.

By 2.2.19 we have $K \le^+ D$, so there is a constant c such that $\forall x\, K(x) - c \le D(x) - 1$. Then for each x we have $2^c 2^{-K(x)} \ge 2^{-D(x)+1} > P_M(x)$. □

Note that $2^{-K_M(x)} \le P_M(x)$ because a shortest M-description of x contributes $2^{-K_M(x)}$ to $P_M(x)$. If $M = \mathbb{U}$ then by the Coding Theorem we also have $2^c 2^{-K(x)} > P_{\mathbb{U}}(x)$ for each x, so $2^{-K(x)}$ and $P_{\mathbb{U}}(x)$ are proportional. By another application of the Coding Theorem, for each prefix-free machine M we have $P_M(x) = O(P_{\mathbb{U}}(x))$.

In Fact 2.1.13 we proved that if $d \in \mathbb{N}$ then for each $n \in \mathbb{N}$ the number of d-compressible$_C$ strings of length n is less than 2^{n-d+1}. We apply the Coding Theorem to obtain an analog of this for K. By Theorem 2.2.9, $K(x) \le |x| + K(|x|) + 1$. We will calculate an upper bound for the number of strings of length n such that $K(x) \le |x| + K(|x|) - d$.

2.2.26 Theorem. *There is a constant $\mathbf{c} \in \mathbb{N}$ such that the following hold.*

(i) $\forall d \in \mathbb{N}\ \ \forall n\ \ \#\{x : |x| = n\ \&\ K(x) \le n + K(n) - d\} < 2^{\mathbf{c}} 2^{n-d}$

(ii) $\forall b \in \mathbb{N}\ \ \forall n\ \ \#\{x : |x| = n\ \&\ K(x) \le K(n) + b\} < 2^{\mathbf{c}} 2^{b}$.

Proof. (i). Let M be the prefix-free free machine given by $M(\sigma) = |\mathbb{U}(\sigma)|$. By the Coding Theorem, let $\mathbf{c}$ be the constant such that $2^{\mathbf{c}}2^{-K(n)} > P_M(n)$ for each n. If $|x| = n$ and $K(x) \le n + K(n) - d$, then a shortest $\mathbb{U}$-description of x, being an M-description of n, contributes at least $2^{-n-K(n)+d}$ to $P_M(n)$. If there are at least $2^{n+\mathbf{c}-d}$ such x, then $P_M(n) \ge 2^{n+\mathbf{c}-d}2^{-n-K(n)+d} = 2^{\mathbf{c}}2^{-K(n)}$, a contradiction.
(ii) follows from (i) letting $d = n - b$. (Note that (i) is vacuously true for negative d.) □

The estimate in (ii) will be applied in Section 5.2 to show that each K-trivial set is Δ^0_2. Another application of the Coding Theorem is Theorem 2.3.6 where we give an expression for the descriptive complexity of an ordered pair $K(\langle x, y\rangle)$.

2.2.27 Exercise. Use Exercise 2.2.23 and the Coding Theorem to show that $m(x) = 2^{-K(x)}$ is an optimal discrete c.e. semimeasure on $\{0,1\}^*$, namely, for each discrete c.e. semimeasure v we have $\forall x\, [e \cdot m(x) \ge v(x)]$ for an appropriate constant $e > 0$.

2.3 Conditional descriptive complexity

We study the conditional descriptive complexity of a string x. An auxiliary string y is available to help with the decompression. We consider conditional versions of both C and K.

Basics

The conditional descriptive complexity $C(x \mid y)$ is the length of a shortest description of x using the string y as auxiliary information. For a simple example, if y is the first half of a string x of even length, we expect $C(x \mid y)$ to be quite a bit smaller than $C(x)$. Also $C(f(x) \mid x)$ should be bounded by a constant for each computable function f.

To develop the formal theory of conditional C-complexity, one simply allows all the machines involved to have two inputs, the second being the auxiliary information y. Such a machine is called *binary machine*. The standard optimal binary machine is given by

$$\mathbb{V}^2(0^{e-1}1\sigma, y) \simeq \Phi^2_e(\sigma, y)$$

where $e > 0$ and $\sigma \in \{0,1\}^*$. Now define

$$C(x \mid y) = \min\{|\sigma| \colon \mathbb{V}^2(\sigma, y) = x\}.$$

Next we introduce a conditional version of the prefix-free string complexity K. A binary machine M is called *prefix-free when fixing the second component* if for each string y the set $\{\sigma \colon M(\sigma, y)\downarrow\}$ is prefix-free. There is an effective listing $(M^2_d)_{d\in\mathbb{N}}$ of all such machines. Now we may adapt the proof of Theorem 2.2.9, using the Recursion Theorem with Parameters 1.1.6, in order to obtain an optimal binary machine $\mathbb{U}^2$ that is prefix-free when fixing the second component and such that $\mathbb{U}^2(0^{d-1}1\sigma, y) = M^2_d(\sigma, y)$ for $d > 1$. We let

$$K(x \mid y) = \min\{|\sigma| \colon \mathbb{U}^2(\sigma, y) = x\}.$$

Then $K(x \mid y) \leq K(n \mid y) + n + 1$ where $n = |x|$.

The empty string is of no use as an auxiliary information. Hence $C(x \mid \varnothing) =^+ C(x)$ and $K(x \mid \varnothing) =^+ K(x)$. (We could modify the definitions of $\mathbb{V}$ and of $\mathbb{U}$ in order to achieve an equality here.)

2.3.1 Fact. *Let N be a machine.*
(i) For each z and each y such that $N(y)\downarrow$ we have $C(N(y) \mid z) \leq^+ C(y \mid z)$.
(ii) For each y and each z such that $N(z)\downarrow$ we have $C(y \mid z) \leq^+ C(y \mid N(z))$.
The same holds with K in place of C, and still for an arbitrary machine N.

Proof. (i) is a version of Fact 2.1.20 for conditional complexity and proved in the same way; (ii) is left as an exercise. The case of K is similar. □

In the following we will need a notation for a particular shortest $\mathbb{U}$-description of a string x such that the decompression takes the least amount of time.

2.3.2 Definition. For a string x, if $t = \mu s.\, K_s(x) = K(x)$, let x^* be the leftmost string σ such that $|\sigma| = K(x)$ and $\mathbb{U}_t(\sigma) = x$.

The reason for this particular choice of a shortest description becomes apparent in the proof of the following fact.

2.3.3 Fact. $K(y \mid x^*) =^+ K(y \mid \langle x, K(x)\rangle)$.

Proof. We apply Fact 2.3.1(ii) for K for both inequalities.

For $\leq^+$, let N be a machine that on an input of the form x^* computes the pair $\langle x, K(x)\rangle = \langle \mathbb{U}(x^*), |x^*|\rangle$.

For $\geq^+$, let N' be a machine that on an input of the form $\langle x, K(x)\rangle$ first computes $t = \mu s.\, K_s(x) = K(x)$, and then uses t to compute the output x^*. (It doesn't matter what the machines do on other inputs.) □

Exercises. Show the following.

2.3.4. $C(x) =^+ C(\langle x, C(x)\rangle)$ and $K(x) =^+ K(\langle x, K(x)\rangle)$.

2.3.5. For all x, y, z we have $K(x \mid z) \leq^+ K(x \mid y) + K(y \mid z)$.

An expression for $K(x, y)$ ⋆

Using conditional K-complexity we will find an expression for the prefix-free complexity of an ordered pair $\langle x, y\rangle$. We write $K(x, y)$ as a shorthand for $K(\langle x, y\rangle)$.

(1) Clearly $K(x, y) \leq^+ K(x) + K(y)$ via the prefix-free machine which on input ρ tries to find a decomposition $\rho = \sigma\tau$ such that $\mathbb{U}(\sigma)\downarrow = x$ and $\mathbb{U}(\tau)\downarrow = y$, and in that case outputs $\langle x, y\rangle$.

(2) This can be improved to $K(x, y) \leq^+ K(x) + K(y \mid x)$ because the machine already has x when it decompresses τ to obtain y.

(3) A further improvement is the inequality $K(x, y) \leq^+ K(x) + K(y \mid x^*)$: the machine looks for a decomposition $\rho = \sigma\tau$ such that $\mathbb{U}(\sigma)\downarrow = x$ and $\mathbb{U}^2(\tau, \sigma)\downarrow = y$. Once it finds this, it outputs $\langle x, y\rangle$. This machine is prefix-free because $\mathbb{U}^2$ is prefix-free when fixing the second component. The last upper bound for $K(x, y)$ is our final one by the following result of Levin in Gács (1974).

2.3.6 Theorem. $K(x, y) =^+ K(x) + K(y \mid \langle x, K(x)\rangle) =^+ K(x) + K(y \mid x^*)$.

Proof. By the preceding discussion and Fact 2.3.1, it remains to show that $K(x) + K(y \mid x^*) \le^+ K(x, y)$. We will apply the Coding Theorem 2.2.25 to the machine $G(\sigma) \simeq (\mathbb{U}(\sigma))_0$ which outputs the first component of $\mathbb{U}(\sigma)$ viewed as an ordered pair in case $\mathbb{U}(\sigma)\downarrow$. Let c be a constant such that $\forall x\, 2^c 2^{-K(x)} > P_G(x)$.

Construction of a uniformly c.e. sequence of bounded request sets (L_σ).

Fix σ. Suppose $\mathbb{U}(\sigma)\downarrow = x$ at stage s. If $\mathbb{U}(\rho) = \langle x, y\rangle$ at a stage $t > s$, put the request $\langle |\rho| - |\sigma| + c, y\rangle$ into L_σ, but only as long as the total weight of L_σ defined after (2.8) does not exceed 1. (Putting the request into L_σ causes an increase of this weight by $2^{-|\rho|+|\sigma|-c}$.)

Let $M_{f(\sigma)}$ be the prefix-free machine effectively obtained from σ via the Machine Existence Theorem 2.2.17. The binary machine N on inputs ρ, σ first waits for $\mathbb{U}(\sigma)\downarrow$. If so, it simulates $M_{f(\sigma)}(\rho)$. If $\sigma = x^*$ then $2^c 2^{-|\sigma|} > P_G(x) = \sum_\rho 2^{-|\rho|} [\![\exists z\, \mathbb{U}(\rho) = \langle x, z\rangle]\!]$. Therefore the total weight put into L_σ is at most $P_G(x) 2^{|\sigma|-c} \le 1$, hence we never threaten to exceed the weight 1 in the construction of L_σ. Thus $N(\rho, \sigma) = y$ for some ρ such that $|\rho| = K(x, y) - |\sigma| + c$, which implies that $K(y \mid x^*) \le^+ |\rho| =^+ K(x, y) - K(x)$, as required. □

2.4 Relating C and K

Basic interactions

Proposition 2.2.8 states that $K(x) \le^+ K(|x|) + |x|$. In the proof we constructed a prefix-free machine N that tries to split the input τ in the form $\tau = \sigma\rho$, where σ is a prefix-free description. The point is that such a decomposition is unique if it exists. We will obtain some facts relating C and K by varying this idea. For instance, a modification of N, where we attempt to compute $\mathbb{V}(\rho)$ instead of copying ρ, yields the following.

2.4.1 Proposition. $K(x) \le^+ K(C(x)) + C(x)$.

Proof. On input τ, the machine $\widetilde{N}$ first searches for a decomposition $\tau = \sigma\rho$ such that $n = \mathbb{U}(\sigma)\downarrow$ and $|\rho| = n$. Once it is found $\widetilde{N}$ simulates $\mathbb{V}(\rho)$. (That is, $\widetilde{N}$ behaves exactly like $\mathbb{V}$ on input ρ, including its output.)

Clearly $\widetilde{N}$ is prefix-free. Given x, let ρ be a shortest $\mathbb{V}$-description of x and let σ be a shortest $\mathbb{U}$-description of $|\rho|$, then $\widetilde{N}(\sigma\rho) = x$. So $K(x) \le^+ |\sigma| + |\rho| = K(C(x)) + C(x)$. □

In the following we use this fact to show that $K(x)$ exceeds $C(x)$ by at most $2\log(C(x)) \le^+ 2\log(|x|)$, up to a small additive constant. Hence $C \sim K$, so in several results C can be replaced by K. An example is Proposition 2.1.22.

2.4.2 Corollary. $C(x) \le^+ K(x) \le^+ C(x) + 2\log(C(x)) \le^+ C(x) + 2\log(|x|)$.

Proof. We have already observed that $C(x) \le^+ K(x)$. For the second inequality, let $m = C(x)$. Then $K(m) \le^+ 2\log m$ by (2.5) on page 85 and since $|m| = \log(m+1)$ when m is viewed as a string (see page 13). Now apply 2.4.1. □

The term $2\log(C(x))$ in the middle could even be decreased to $\log(C(x))+2\log\log(C(x))$ by 2.2.8. If x has length 10000, say, then $K(x) - C(x)$ is bounded by about 20.

One more time, we apply the idea to split off a prefix-free description.

2.4.3 Proposition. *For all strings x and y we have $C(xy) \leq^+ K(x) + C(y)$.*

Proof. On input τ, the machine M first searches for a decomposition $\tau = \sigma\rho$ such that $\mathbb{U}(\sigma)\downarrow= x$. Once found, M simulates the computation of $\mathbb{V}$ on input ρ. If $\mathbb{V}(\rho) = y$ then M outputs xy. Clearly, $C_M(xy) \leq K(x) + C(y)$. □

Next, we prove a continuity property of C that strengthens Proposition 2.1.21.

2.4.4 Proposition. $\text{abs}(C(x) - C(y)) \leq^+ K(\text{abs}(x - y)) \leq^+ 2\log\text{abs}(x - y)$.

Proof. First suppose that $C(x) \geq C(y)$. Note that $x = y \pm \text{abs}(x - y)$. We may describe x based on (1) a $\mathbb{U}$-description σ of $z = \text{abs}(x - y)$, (2) a $\mathbb{V}$-description ρ of y, and (3) a further bit b telling us whether to add or to subtract z. If we put this information into the form $\tau = \sigma b \rho$, then a machine S on input τ can first find $\sigma \prec \tau$ such that $\mathbb{U}(\sigma) = z$, then read b, and then apply $\mathbb{V}$ to the rest ρ to obtain y. Now it has all the information needed in order to calculate x. Thus $C(x) \leq^+ K(\text{abs}(x-y))+C(y)$.

Note that $y = x \pm \text{abs}(x - y)$, so if $C(x) < C(y)$ then the same machine S shows that $C(y) \leq^+ K(\text{abs}(x - y)) + C(x)$. □

The following result of Levin (see Li and Vitányi 1997) relates plain descriptive complexity C with the conditional prefix-free complexity K. Like 2.4.2, is states that $K(x) - C(x)$ is small.

2.4.5 Proposition. $C(x) =^+ K(x \mid C(x))$.

Proof. "$\geq^+$:" Let M be the binary machine such that $M(\sigma, |\sigma|) \simeq \mathbb{V}(\sigma)$ and $M(\sigma, m)\uparrow$ for $m \neq |\sigma|$. Clearly M is prefix-free when fixing the second component. If σ is a $\mathbb{V}$-description of x then $M(\sigma, |\sigma|) = x$. Hence, if $|\sigma| = C(x)$, then $|\sigma| + d \geq K(x \mid C(x))$, where $d > 1$ is an index for M in the effective listing of binary machines that are prefix-free when fixing the second component.

"$\leq^+$:" Let N be the machine that, on input τ, searches for a decomposition $\tau = \sigma\rho$ such that $n = \mathbb{U}(\sigma)\downarrow$ and then simulates $\mathbb{U}^2(\rho, n + |\rho|)$. If $C(x) < K(x \mid C(x))$ we are done, so we may assume that $n = C(x) - K(x \mid C(x)) \geq 0$.

Let ρ be a shortest string such that $\mathbb{U}^2(\rho, C(x)) = x$, so that $|\rho| = K(x \mid C(x))$ and $|\rho| + n = C(x)$. Since $N(n^*\rho) = x$, we have $C_N(x) \leq K(n) + |\rho|$ and hence $C(x) \leq^+ K(n) + |\rho|$. Thus, since $K(w) \leq^+ 2|w|$ by (2.5), $n = C(x) - |\rho| \leq^+ K(n) \leq^+ 2\log n$. Therefore, $n = C(x) - K(x \mid C(x))$ is bounded from above by a constant. □

Solovay's equations ⋆

By Corollary 2.4.2 $K(x) - C(x) \leq^+ 2\log|x|$. Solovay's equations tell us how to express C in terms of K and conversely up to an error of only $O(\log\log|x|)$: within such an error we have $C(x) = K(x) - K(K(x))$ and $K(x) = C(x) + C(C(x))$. The present proof is due to Miller (2008) and builds on Theorem 2.3.6, Proposition 2.4.5, and ideas from the original proof of Solovay (1975).

2.4.6 Theorem. *For each* $x \in \{0,1\}^*$ *we have*

(i) $C(x) = K(x) - K^{(2)}(x) + O(K^{(3)}(x))$, *and*
(ii) $K(x) = C(x) + C^{(2)}(x) + O(C^{(3)}(x))$.

Proof. For notation involving error terms see page 75. Also recall from Definition 2.3.2 that x^* is a shortest $\mathbb{U}$-description of x. To save on brackets we write $KC(x)$ instead of $K(C(x))$, etc.

We begin with a brief outline of the proof. We first prove Equation (i) and then obtain Equation (ii) from (i). To prove (i), by Proposition 2.4.5 $C(x) =^+ K(x \mid C(x))$. Note that $K(x \mid C(x)^*) \leq^+ K(x \mid C(x))$ by Fact 2.3.1(ii). In Claim 2.4.7 we show that it is by at most $K^{(2)}C(x)$ smaller. This makes it possible to work with $K(x \mid C(x)^*)$ instead of $K(x \mid C(x))$. Then, applying Levin's Theorem 2.3.6 twice, we obtain that within the small error of $K^{(2)}C(x)$, $C(x)$ can be replaced by

$$K(x \mid C(x)^*) =^+ K(x, C(x)) - KC(x) =^+ K(x) + K(C(x) \mid x^*) - KC(x).$$

Based on this we will be able to reach Equation (i).

2.4.7 Claim. $C(x) \leq^+ K(x \mid C(x)^*) + K^{(2)}C(x)$.
Let $\sigma = C(x)^*$. By Exercise 2.3.5,

$$K(x \mid C(x)) \leq^+ K(x \mid \sigma) + K(\sigma \mid C(x)).$$

We will estimate the second term on the right hand side. For any y we have $K(y^* \mid y) =^+ K(K(y) \mid y)$, since, given y, we can effectively produce a prefix-free description of y^* from one of $K(y)$ (wait till the first $\mathbb{U}$-description of y that has length $K(y)$ comes up), and conversely (take the length). We apply this to $y = C(x)$ and obtain

$$K(\sigma \mid C(x)) = K(\sigma \mid y) =^+ K(K(y) \mid y) \leq^+ K^{(2)}(y) = K^{(2)}C(x).$$

This establishes Claim 2.4.7. Next we provide the main technical result.

2.4.8 Claim. $C(x) = K(x) - KC(x) + O(K^{(2)}C(x))$.
Let $Q(x) = C(x) + KC(x) - K(x)$. To establish the claim, we will show that

$$\forall^\infty x\, [\text{abs}(Q(x)) \leq 5K^{(2)}C(x)]. \tag{2.9}$$

By 2.4.1 $Q(x)$ is bounded from below by a constant. So it suffices to show that $\forall^\infty x\, [Q(x) \leq 5K^{(2)}C(x)]$. We apply Levin's Theorem 2.3.6 twice:

$$K(x \mid C(x)^*) =^+ K(x, C(x)) - KC(x) =^+ K(x) + K(C(x) \mid x^*) - KC(x).$$

Substituting this into Claim 2.4.7 we obtain an estimate for $C(x)$:

$$C(x) \leq^+ K(x) + K(C(x) \mid x^*) - KC(x) + K^{(2)}C(x). \tag{2.10}$$

Our next task is to find a suitable upper bound for the term $K(C(x) \mid x^*)$:

$$\begin{aligned} K(C(x) \mid x^*) &\leq^+ K(C(x) \mid K(x)) \\ &\leq^+ K(\text{abs}(C(x) - K(x))) \\ &\leq^+ K(\text{abs}(C(x) + KC(x) - K(x))) + K^{(2)}C(x). \end{aligned}$$

To obtain the last line, we attached a $\mathbb{U}$-description of $KC(x)$ (and a further bit b) at the beginning of a $\mathbb{U}$-description of $\text{abs}(C(x) - K(x))$ as in the proof of 2.4.4. This causes the extra term $K^{(2)}C(x)$. After rearranging, the estimate (2.10) now turns into

$$Q(x) \leq^+ K(\text{abs}(Q(x))) + 2K^{(2)}C(x).$$

Recall that we already have a constant lower bound for $Q(x)$, so to obtain the upper bound we may assume $Q(x) \geq 0$. Then, since $K(m) \leq^+ 2\log m$ for each $m \in \mathbb{N}$, we have $Q(x) \leq^+ 2\log Q(x) + 2K^{(2)}C(x)$. Since $\frac{1}{2}Q(x) \leq^+ Q(x) - 2\log Q(x)$, this implies $Q(x) \leq^+ 4K^{(2)}C(x)$, which establishes (2.9) and hence Claim 2.4.8.

To proceed we need two facts.

2.4.9 Fact. $K(n+m) = K(n) + O(K(m))$.
For $K(n+m) \leq^+ K(n) + K(m)$ and also $K(n) \leq^+ K(n+m) + K(m)$. It follows that $\text{abs}(K(n+m) - K(n)) \leq^+ K(m)$.

2.4.10 Fact. *Let $f, g\colon \mathbb{N} \to \mathbb{N}$. If $g(x) = f(x) + O(K(f(x)))$, then $K(g(x)) \sim K(f(x))$.*
To see this, we apply K to both sides of the hypothesis. By Fact 2.4.9 we obtain that $K(g(x)) = K(f(x)) + O(\log K(f(x)))$, so $K(g(x)) \sim K(f(x))$.

To establish Equation (i) of Theorem 2.4.6 we want to replace the C's by K's on the right hand side of Claim 2.4.8. We apply Claim 2.4.8 itself for this! Rearranging the claim yields $C(x) + KC(x) = K(x) + O(K^{(2)}C(x))$ Applying Fact 2.4.9 to both sides of this, and using on the right hand side that $K(r) \leq^+ r$ for each $r \in \mathbb{N}$, we obtain

$$KC(x) = K^{(2)}(x) + O(K^{(2)}C(x)). \tag{2.11}$$

This is equivalent to $K^{(2)}(x) = KC(x) + O(K^{(2)}C(x))$, so by Fact 2.4.10, where $f(x) = KC(x)$, we obtain

$$K^{(3)}(x) \sim K^{(2)}C(x). \tag{2.12}$$

Substituting the last two equations into Claim 2.4.8 yields Equation (i).

Next we prove Equation (ii). By Equation (i) with $C(x)$ instead of x,

$$C^{(2)}(x) = KC(x) + O(K^{(2)}C(x)), \tag{2.13}$$

where the error term $O(K^{(3)}C(x))$ has been absorbed into $O(K^{(2)}C(x))$. Using (2.11) and then (2.12), this turns into

$$C^{(2)}(x) = K^{(2)}(x) + O(K^{(3)}(x)).$$

Applying the last equation to $C(x)$ instead of x and using Fact 2.4.10 yields

$$C^{(3)}(x) \sim K^{(2)}C(x). \tag{2.14}$$

Rearranging Claim 2.4.8 yields $K(x) = C(x) + KC(x) + O(K^{(2)}C(x))$. Now we substitute first (2.13) and then (2.14) in order to obtain Equation (ii). This completes the proof of Theorem 2.4.6. □

2.5 Incompressibility and randomness for strings

An object is random if it is disorganized, has no patterns, no regularities. Patterns occuring in a string x yield a description of the string that is shorter than its trivial description (the string itself). For instance, no one would regard x as random if all the bits of x in the even positions are 0. One can use this pattern to show that $C(x) \leq^+ |x|/2$ (see Exercise 2.1.7). On page 99 we will gather further evidence for the thesis that, for strings, the intuitive notion of randomness can be formalized by being hard to describe, that is, incompressibility as defined in 2.1.12. We provide two variants of the concept of incompressibility with respect to C, the first weaker and the second stronger. They yield alternative formal randomness notions for strings.

2.5.1 Definition. The string x is d-incompressible in the sense of K, or *d-incompressible$_K$* for short, if $K(x) > |x| - d$. Else x is called *d-compressible$_K$*.

An incompressibility notion for strings can also be used to study the randomness aspect of an infinite sequence of zeros and ones (i.e., a subset of $\mathbb{N}$) via its finite initial segments. Incompressibility in the sense of K turns out to be the most useful tool; see Section 3.2.

2.5.2 Definition. We say that a string x is *strongly d-incompressible$_K$* if $K(x) > |x| + K(|x|) - d$, namely, $K(x)$ differs by at most d from its upper bound $|x| + K(|x|) + 1$ given by Theorem 2.2.9.

While the intuitive concept of randomness for strings does not involve any number parameter, each of the incompressibility notions for a string is defined in terms of a constant d. It is somewhat arbitrary how large a constant we want to accept when formalizing randomness by incompressibility. Generally, the constant should be small compared to the length of the string. For each fixed d, randomness is an asymptotic concept when interpreted as incompressibility with parameter d. The arbitrariness in the choice of the constant only disappears entirely when we consider infinite sequences of bits, i.e., subsets of $\mathbb{N}$.

It is harder to obtain a formal notion of computational complexity for strings, because the computational complexity of a set Z is given by what an algorithm can do with Z as an oracle, and such an algorithm only makes sense when considered over the whole range of inputs. (A possible measure for the computational complexity of a string y is $C(x) - C(x \mid y)$, taken over a variety of strings x.)

Comparing incompressibility notions

The implications between the incompressibility notions are

$$\text{strongly incompressible}_K \overset{(1)}{\Rightarrow} \text{incompressible}_C \overset{(2)}{\Rightarrow} \text{incompressible}_K.$$

More precisely, for each $p \in \mathbb{N}$ there is a $q \in \mathbb{N}$ such that each strongly p-incompressible$_K$ string is q-incompressible$_C$, and similarly for the second implication.

2.5.3 Remark. For each length, most strings are incompressible in the strongest sense. For it is immediate from Theorem 2.2.26(i) that the number of strongly d-incompressible$_K$ strings of length n is at least $2^n - 2^{\mathbf{c}+n-d}$ where $\mathbf{c}$ is the constant of the theorem. We can make the proportion of strings that are not strongly incompressible$_K$ as small as we wish (independently of n) by choosing d sufficiently large.

The implication (2) holds for a minor change of constants since $C(x) \leq^+ K(x)$. Actually, if x is incompressible$_C$ then all the prefixes of x are incompressible$_K$:

2.5.4 Proposition. *Fix $b \in \mathbb{N}$ such that $C(yz) \leq K(y) + |z| + b$ for each y, z (by 2.4.3). Then for each $d \geq b$, each prefix of a $(d\text{-}b)$-incompressible$_C$ string x is d-incompressible$_K$. That is, $C(x) > |x| - (d - b) \rightarrow \forall y \preceq x\, [K(y) > |y| - d]$.*

Proof. Suppose that $y \preceq x$. Let $x = yz$. If $K(y) \leq |y| - d$ then by the choice of b we have $C(x) \leq K(y) + |x| - |y| + b \leq |x| + b - d$. □

In particular, there are arbitrarily long strings x such that each prefix of x is d-incompressible$_K$. There is no analog of this for C in place of K because for each c, if x has length about 2^c then the C-complexity of some prefix w of x dips below $|w| - c$ (Proposition 2.2.1). This implies that for each c there is a d-incompressible$_K$ string w such that $C(w) \leq |w| - c$. Thus the converse of the implication (2) fails. Next, we address the implication (1).

2.5.5 Proposition. *There is $c \in \mathbb{N}$ such that the following holds. For each d, $K(x) > |x| + K(|x|) - (d - c) \;\rightarrow\; C(x) > |x| - 2d$.*

Proof. We define a bounded request set W to ensure that

$$C(x) \leq |x| - 2d \;\rightarrow\; K(x) \leq |x| + K(|x|) - (d - c), \tag{2.15}$$

where $c - 2$ is the coding constant of W given by Theorem 2.2.17.

> *Construction of W.* At stage s, for each d, each n and each x of length n such that $C_s(x) \leq n - 2d$, enumerate the request $\langle K_s(n) + n - d + 2, x\rangle$ into W unless it is already in W.

Fix $d \in \mathbb{N}$. By Fact 2.1.13 there are no more than 2^{n-2d+1} strings x of length n such that $C(x) \leq n-2d$. So, if $\mathbb{U}(\sigma) = n$, the contribution of σ to the total weight of W (for this d) is at most $2^{-|\sigma|-n+d-2}2^{n-2d+1} = 2^{-|\sigma|-d-1}$. The contribution over all σ is therefore at most $\sum_\sigma 2^{-|\sigma|-d-1}\, [\![\mathbb{U}(\sigma)\downarrow]\!] = \Omega 2^{-d-1}$. The sum over all d is at most 1, hence W is a bounded request set.

Clearly (2.15) is satisfied if we choose the constant c as above. □

The converse of implication (1) fails as well: by Miller (2008), for sufficiently large c, for each Π^0_1 set Q such that $\forall n\, Q \cap \{0,1\}^n \neq \emptyset$ there are infinitely many strings $w \in Q$ such that $K(w) \leq |w| + K(|w|) - c$. For each $d \in \mathbb{N}$, the set of d-incompressible$_C$ strings is a set Q of this kind.

Randomness properties of strings

Incompressibility appears to be the appropriate mathematical definition of randomness for strings because incompressible strings have properties that one commonly associates with randomness. We will provide twofold evidence for this thesis. The statements hold even for the weakest notion, incompressibility in the sense of K.

(a) incompressible strings have only short runs of zeros (i.e. blocks only consisting of zeros), and
(b) zeros and ones occur balancedly in incompressible strings.

We will apply these facts in Proposition 3.2.13 and Theorem 3.5.21 when we consider statistical properties of infinite sequences of bits.

A property G of strings pointing towards organization should be rare in that there are not too many strings of each length with that property. Also, G should be not too hard to recognize: we will mostly consider rare properties G that are also decidable. Having long runs of zeros is such a property, having a large imbalance between zeros and ones is another.

Given a set of strings G, for each n let $G_n = G \cap \{0,1\}^n$. The following lemma shows that for any c.e. set of strings G that is rare in the sense that $\#G_n/2^n$ is bounded by $O(n^{-3})$, the strings in G_n can be compressed to $n - \log n$ in the sense of K. (Recall from page 13 that we use the notation $\log n = \max\{k \in \mathbb{N}\colon\, 2^k \leq n\}$.) This lemma will be applied to prove (a) and (b).

2.5.6 Lemma. *Let G be a c.e. set of strings such that $\#G_n = O(2^{n-3\log n})$ for each n. Then $\forall x \in G_n\, [K(x) \leq^+ n - \log n]$.*

Proof. The idea is that for each string $x \in G$ the length n and its position in the computable enumeration of G_n together provide a short description of x. Instead of giving that description explicitly, we rely on the Machine Existence Theorem 2.2.17.

Since $n - 3\log n > 0$ for $n \geq 10$, there is $k \in \mathbb{N}$ such that for all $n \geq 10$ we have $\#G_n \leq 2^k 2^{n-3\log n}$. Define a bounded request set W as follows: when x is enumerated into G and $|x| = n \geq 10$, put the request

$$\langle n - \log n + k + 1, x\rangle$$

into W. By the hypothesis, the weight contributed by G_n for $n \geq 10$ is at most $2^{-k-1}2^{-n+\log n}2^k 2^{n-3\log n} \leq (n-1)^{-2}/2$ (since $(n-1)^{-1} \geq 2^{-\log n}$). As $\sum_{j>0} 1/j^2 = \pi^2/6 < 2$ this implies that W is a bounded request set. By 2.2.17, $x \in G$ and $|x| = n$ imply $K(x) \leq^+ n - \log n$. □

Our first application of the foregoing lemma is a proof of (a) above: a run of zeros in an incompressible$_K$ string of length n is bounded by $4\log n + O(1)$. We need to show that the relevant property is rare.

2.5.7 Proposition. *Let $n \in \mathbb{N}$ and let x be a string of length n. If there is a run of $4\log n$ zeros in x then $K(x) \leq^+ n - \log n$.*

Proof. Define the computable set G by letting

$$G_n = \{x\colon\, |x| = n \;\&\; x \text{ has a run of } 4\log n \text{ zeros}\}.$$

Such a string x is determined by the least position $i < n$ where such a run of zeros starts and the $n - 4\log n$ bits in the positions outside this run. Thus, $\#G_n \leq n2^{n-4\log n} = O(2^{n-3\log n})$. By Lemma 2.5.6 we may conclude that $K(x) \leq^+ n - \log n$. □

Next we settle (b): strings with a large imbalance between zeros and ones can be compressed in the sense of K. We refer to the brief discussion of probability theory on page 73 and use some of the notation mentioned there. In particular if the sample space is $\{0,1\}^n$, the random variable $S_n(x)$ denotes the number of occurrences of ones in a string x of length n. For $d \in \mathbb{N} - \{0\}$, let

$$A_{n,d} = \{x \in \{0,1\}^n \colon \operatorname{abs}(S_n(x)/n - 1/2) \geq 1/d\}. \tag{2.16}$$

Thus, $A_{n,d}$ is the event that the number of ones differs by at least n/d from the expectation $n/2$. The Chernoff bounds (Shiryayev, 1984, Eqn. (42) on pg. 69) yield

$$P(A_{n,d}) \leq 2e^{-2n/d^2}. \tag{2.17}$$

This considerably improves the estimate $P(A_{n,d}) \leq d^2/(4n)$ after (1.21), page 73, obtained through the Chebycheff inequality. For instance, for $P(A_{1000,10})$ the upper bound is $4.2 \cdot 10^{-9}$ rather than $1/40$.

2.5.8 Theorem. *There is a constant c and a computable function $\lambda d.n_d$ such that $\forall d \in \mathbb{N} - \{0\}\, \forall n \geq n_d\, \forall x \in A_{n,d}\, [K(x) \leq n - \log n + c]$.*

Proof. For each $d \in \mathbb{N} - \{0\}$ let n_d be the least number such that

$$\forall n \geq n_d\, [e^{2n/d^2}/2 \geq n^3].$$

Define the computable set G by letting $G_n = \emptyset$ for $n < 10$, and for $n \geq 10$,

$$G_n = \{x \colon x \in A_{n,d} \text{ where } d \text{ is maximal such that } n_d \leq n\}.$$

Then $\#G_n \leq 2^{n+1}e^{-2n/d^2} \leq 2^{n-3\log n}$. By Lemma 2.5.6 there is c such that $\forall n\, \forall x \in G_n\, [K(x) \leq n - \log n + c]$. Given d, if $n \geq n_d$ and $x \in A_{n,d}$ then $x \in G_n$, so $K(x) \leq n - \log n + c$ as required. □

We may conclude that the occurrences of zeros and ones are asymptotically balanced for b-incompressible$_K$ strings:

2.5.9 Corollary. *Fix b. For each d, for almost every n, each b-incompressible$_K$ string x of length n satisfies* $\operatorname{abs}(S_n(x)/n - 1/2) < 1/d$.

Proof. In the previous notation, if $n > 2^{b+c}$ and $n \geq n_d$ then the required inequality holds, else $x \in A_{n,d}$ and thus $K(x) \leq n - \log n + c \leq n - b$. □

If incompressibility is taken in a stronger sense one also obtains stronger conclusions. For instance, the following exercise shows that for almost all n, if a string x of length n is incompressible in the sense of C then $\operatorname{abs}(S_n(x) - n/2)$ is bounded by $\sqrt{n \ln n}$. Incompressibility in the sense of K merely yields a bound in $o(n)$.

2.5.10 Exercise. For almost all n, if $|x| = n$ and $\operatorname{abs}(S_n(x) - n/2) \geq \sqrt{n \ln n}$ then $C(x) \leq^+ n - \log n$.
Hint. Modify the proof of 2.5.8, using Proposition 2.1.14 rather than the Machine Existence Theorem.

3

MARTIN-LÖF RANDOMNESS AND ITS VARIANTS

Sets computable by a Turing machine are the mathematical equivalent of sets decidable by an algorithm. In this chapter we introduce a mathematical notion corresponding to our intuitive concept of randomness for a set. The intuition is vaguer here than in the case of computability. The central randomness notion will be the one of Martin-Löf (1966). In order to address criticisms to the claim that this notion is the appropriate one, we will consider weaker and stronger randomness notions for sets as alternatives. In this way a (mostly linear) hierarchy of randomness notions emerges. Martin-Löf randomness stands out because we arrive at it from two different directions (Theorem 3.2.9), and also because it interacts best with computability theoretic concepts.

Section 3.1 contains background on randomness and tests. Section 3.2 provides the basics on Martin-Löf randomness, and some interesting further results. In Sections 3.3 and 3.4 we study its interaction with computability theoretic notions via reducibilities and relativization, respectively. In Section 3.5 we introduce weaker variants of Martin-Löf randomness, and in 3.6 stronger variants.

3.1 A mathematical definition of randomness for sets

Our intuitive concept of randomness for a set Z has two related aspects:

(a) Z satisfies no exceptional properties, and
(b) Z is hard to describe.

Towards a mathematical definition of randomness for a set, at first we will consider each of the two aspects separately.

(a) *Z satisfies no exceptional properties.* Think of the set Z as the overall outcome of an idealized physical process that proceeds in time. It produces infinitely many bits. The bits are independent. Zero and one have the same probability. An example is the repeated tossing of a coin. The probability that a string x is an initial segment of Z is $2^{-|x|}$. Exceptional properties are represented by null classes (with respect to the uniform measure λ on Cantor space, Definition 1.9.8). We give examples of exceptional properties $\mathcal{P}$ and $\mathcal{Q}$. The first states that all the bits in even positions are zero:

$$\mathcal{P}(Y) \leftrightarrow \forall i\, Y(2i) = 0. \tag{3.1}$$

The second states there are at least twice as many zeros as ones in the limit:

$$\mathcal{Q}(Y) \leftrightarrow \liminf \#\{i < n\colon\ Y(i) = 0\}/n \geq 2/3. \tag{3.2}$$

The corresponding classes are null. Hence, according to our intuition, they should not contain a random set. In order to obtain a mathematical definition of randomness, we impose extra conditions on the null classes a set must be avoid. Otherwise no set Z would be random at all because the singleton $\{Z\}$ itself is a null class! Some effectivity, or definability, conditions are required for the class (or sometimes a superclass). For instance, one can require that the null class is Π^0_2 (see Definition 1.8.55). The classes given by the properties above are of this type; $\{Y: \mathcal{P}(Y)\}$ is actually Π^0_1.

(b) *Z is hard to describe.* A random object has no patterns, is disorganized. On page 99 we provided evidence that in the setting of finite objects, the intuitive concept of randomness corresponds to being hard to describe. We relied on the fact that there are description systems, called optimal machines, that emulate every other description system of the same type, and in particular describe every possible string. Being hard to describe can be formalized by incompressibility with respect to an optimal machine. Incompressible strings have the properties that one intuitively expects from a random string.

For sets, as for strings our intuition is that some degree of organization makes the object easier to describe. However, we cannot formalize being hard to describe in such a simple way as we did for strings, because each description system only describes countably many sets. To make more precise what we mean by being hard to describe for sets, we recall close descriptions from page 46. The null Π^0_1 classes represent a type of close description; so do the null Π^0_2 classes. A set is hard to describe in a particular sense (say, Π^0_2 classes) if it does not admit a close description in this sense (for instance, it is not in any Π^0_2 null class).

If we want to introduce a mathematical notion of randomness, to incorporate aspect (a) we need a formal condition restricting null classes, and for aspect (b) a formal notion of close description. Both are given by specifying a *test* concept. This determines a mathematical randomness notion: Z is random in that specific sense if it passes all the tests of the given type. Tests are themselves objects that can be described in a particular way; thus only countably many null classes are given by such tests. If $(\mathcal{A}_n)_{n\in\mathbb{N}}$ is a list of all null classes of that kind, then the corresponding class of random sets is $2^{\mathbb{N}} - \bigcup_n \mathcal{A}_n$, which is conull. Remark 2.5.3 is the analog of this for strings. It states that most strings of each length are incompressible.

In the following we give an overview of some test notions. They determine important randomness concepts. A test can be a conceptually simple object such as a Π^0_2 null class. It can also be a more elaborate object such as a sequence of c.e. open sets or a computable betting strategy. The power of tests is directly related to their position in the hierarchy of descriptive complexity for sets and classes (pages 21 and 64). The formal details are provided in later sections of this chapter, and in Chapter 7.

Martin-Löf tests and their variants

Recall from Fact 1.9.9 that a class $\mathcal{A} \subseteq 2^{\mathbb{N}}$ is null iff there is a sequence $(G_m)_{m\in\mathbb{N}}$ of open sets such that $\lim_m \lambda G_m = 0$ and $\mathcal{A} \subseteq \bigcap_m G_m$. Motivated by test concepts from statistics, Martin-Löf (1966) introduced an effective version of this characterization of null classes. He required that the sequence $(G_m)_{m\in\mathbb{N}}$ of open sets be uniformly c.e., and the convergence be effective, namely, for each positive rational δ one can compute m such that $\lambda G_m \leq \delta$. Taking a suitable effective subsequence, one might as well require that $\lambda G_m \leq 2^{-m}$ for each m. A sequence of open sets with these properties will be called a Martin-Löf test, and a set Z is Martin-Löf random if $Z \notin \bigcap_m G_m$ for all Martin-Löf tests $(G_m)_{m\in\mathbb{N}}$. For instance, to see that a Martin-Löf random set does not satisfy the property $\mathcal{P}$ in (3.1), let $G_m = \{Z\colon \forall i < m\, Z(2i) = 0\}$. (A Martin-Löf random set does not satisfy $\mathcal{Q}$ either, but this is postponed to Proposition 3.2.13.) We think of a Martin-Löf test as an effective sequence of c.e. open sets that represent attempts to pin down the unknown set Z by specifying possibilities for Z. These approximations get better and better. Z is Martin-Löf random if it eventually escapes these attempts.

Weaker than Martin-Löf randomness. Schnorr (1971) argued that Martin-Löf's test notion is too strong to be considered algorithmic because one does not know enough about the tests; in particular, while λG_m is a uniformly left-c.e. real number, it may not be computable. He proposed computable test concepts, which will be discussed in Chapter 7. One can go a step further and impose time and space bounds on the computations. Then one obtains randomness notions relevant to the theory of feasible computability, such as being polynomially random.

Stronger than Martin-Löf randomness. From a different point of view, one can maintain that Martin-Löf randomness is too weak as a mathematical notion of randomness. A left-c.e. set can be ML-random. Each set $Y \geq_T \emptyset'$ is Turing equivalent to a ML-random set. According to this viewpoint, these facts are inconsistent with the intuitive idea of a random set. Left-c.e. sets can be approximated fairly well. A random set should not encode arbitrarily complex information.

To get around this, one can proceed to more powerful types of tests. The next stronger notion is weak 2-randomness. Instead of $\forall m\, \lambda G_m \leq 2^{-m}$ one merely asks that $\lim_m \lambda G_m = 0$. These tests are equivalent to null Π^0_2 classes. No weakly 2-random set is Δ^0_2 because each Δ^0_2 set is a Π^0_2 singleton. (In fact, a weakly 2-random set forms a minimal pair with $\emptyset'$.) Even stronger is 2-randomness, Martin-Löf randomness relativized to $\emptyset'$. While these notions are interesting, it turns out that Martin-Löf randomness is the most fruitful one where the interaction of computability and randomness is concerned – precisely because Martin-Löf randomness is not such a strong randomness notion.

For a more radical change away from Martin-Löf randomness, once can move on to higher computability theory. One considers Π^1_1 sets of numbers, an analog of the c.e. sets much higher in the hierarchy of descriptive complexity (Definition 9.1.1). Elements are enumerated at stages that are computable ordinals.

The analog of a Martin-Löf test is a uniformly Π^1_1 sequence $(G_m)_{m\in\mathbb{N}}$ of open sets such that $\lambda G_m \le 2^{-m}$. For an even stronger notion based on higher computability, one can take as tests the null Π^1_1 *classes* $\mathcal{C} \subseteq 2^{\mathbb{N}}$ (9.1.1). Sets Z enter the class at stages that are ordinals computable relative to Z.

The choice of the "superlevel" in the hierarchy of descriptions, namely whether one takes feasible computability, computability, or higher computability, depends on the applications one has in mind. Interestingly, similar patterns emerge at the three superlevels.

Schnorr's Theorem and universal Martin-Löf tests

If the same notion arises from investigations of different areas, it deserves special attention. Among randomness notions, this is the case for Martin-Löf randomness. One obtains the same class when requiring that all the finite initial segments be incompressible in the sense of the prefix-free algorithmic complexity K defined in Section 2.2: a theorem of Schnorr (1973) states that Z is Martin-Löf random if and only if there is $b \in \mathbb{N}$ such that $\forall n\, K(Z\upharpoonright_n) > n - b$. Thus Z is Martin-Löf random if and only if all its initial segments are random as strings! To some extent, the "randomness = incompressibility" paradigm used for finite strings carries over to sets.

Schnorr's Theorem only holds when we formalize randomness of a string x by being b-incompressible in the sense of K (for some b that is small compared to the length of x). We cannot use the plain descriptive complexity C because of its dips at initial segments of a sufficiently long string; see Proposition 2.2.1. However, a similar theorem using a monotonic version Km of string complexity was announced by Levin (1973).

A Martin-Löf test $(U_b)_{b\in\mathbb{N}}$ is called *universal* if $\bigcap_b U_b$ contains $\bigcap_m G_m$ for every Martin-Löf test $(G_m)_{m\in\mathbb{N}}$. Schnorr's Theorem shows that $(\mathcal{R}_b)_{b\in\mathbb{N}}$ is universal, where $\mathcal{R}_b$ is the open set generated by $\{x \in \{0,1\}^*\colon\ K(x) \le |x| - b\}$, the b-compressible$_K$ strings. The existence of a universal test is a further criterion for being natural for a randomness notion. For instance, the argument of Schnorr can be adapted to the Π^1_1 version of Martin-Löf randomness, so a universal test exists as well. Also, there is a largest Π^1_1 null class $\mathcal{Q} \subseteq 2^{\mathbb{N}}$. To be random in this strong sense means to be not in $\mathcal{Q}$ (Theorem 9.3.6).

The initial segment approach

Schnorr's Theorem is our first application of the following:

Characterize $\mathcal{C} \subseteq 2^{\mathbb{N}}$ via the *initial segment complexity* of its members.

This is very useful because one can determine whether $Z \in \mathcal{C}$ by looking at its initial segments, rather than at Z as a whole. We will later provide several other examples of such characterizations. As mentioned above, in Schnorr's Theorem we use K to measure the descriptive complexity of the initial segment $Z\upharpoonright_n$. For other classes one may use different measures, such as the plain complexity C or its conditional variant $\lambda z.\, C(z \mid n)$ where $n = |z|$.

Often membership in $\mathcal{C}$ is determined by a growth condition on the initial segment complexity; 2-randomness (Theorem 3.6.10 below) is somewhat different as one requires that the plain complexity is infinitely often maximal (up to a constant).

3.2 Martin-Löf randomness

We provide the formal details for the discussion in the previous section.

The test concept

In 1.8.22 we introduced the effective listing $([W_e]^{\prec})_{e\in\mathbb{N}}$ of all the c.e. open sets (here W_e is regarded as a subset of $\{0,1\}^*$). A *uniformly c.e. sequence* $(G_m)_{m\in\mathbb{N}}$ *of open sets* is given by a computable function f such that $G_m = [W_{f(m)}]^{\prec}$ for each m. Martin-Löf (1966) effectivized the description of null classes from Fact 1.9.9.

3.2.1 Definition.

(i) A *Martin-Löf test*, or *ML-test* for short, is a uniformly c.e. sequence $(G_m)_{m\in\mathbb{N}}$ of open sets such that $\forall m \in \mathbb{N}\ \lambda G_m \leq 2^{-m}$.

(ii) A set $Z \subseteq \mathbb{N}$ *fails* the test if $Z \in \bigcap_m G_m$, otherwise Z *passes* the test.

(iii) Z is *ML-random* if Z passes each ML-test. Let MLR denote the class of ML-random sets. Let non-MLR denote its complement in $2^{\mathbb{N}}$.

Note that $Z \in \bigcap_m G_m \ \leftrightarrow\ \forall m\, \exists k, s\, [Z\restriction_k] \subseteq G_{m,s}$, so $\bigcap_m G_m$ is a particular kind of Π^0_2 null class. Admitting all the Π^0_2 null classes as tests is equivalent to replacing the condition $\forall m \in \mathbb{N}\ \lambda G_m \leq 2^{-m}$ by the weaker one that $\bigcap_m G_m$ be a null class. In this case, we have more tests and therefore a stronger randomness notion. This notion is called weak 2-randomness, introduced on page 134.

3.2.2 Example. A simple example of a ML-test $(G_m)_{m\in\mathbb{N}}$ is the test showing that a computable set Z is not ML-random: let $G_m = [Z\restriction_m]$. Note that if Z is computable then $\{Z\}$ is a null Π^0_1 class. More generally, for any null Π^0_1 class P there is a ML-test $(G_m)_{m\in\mathbb{N}}$ such that $P = \bigcap_m G_m$: let $(P_s)_{s\in\mathbb{N}}$ be the effective approximation of P by clopen sets from (1.17) on page 55, then $P = \bigcap_s P_s$. Let f be an increasing computable function such that $\lambda P_{f(m)} \leq 2^{-m}$ for each m, and let $G_m = P_{f(m)}$.

We do not require in 3.2.1 that a ML test $(G_m)_{m\in\mathbb{N}}$ satisfy $G_m \supseteq G_{m+1}$ for each m. It would make little difference if we did, since we can replace each G_m by $\bigcap_{i\leq m} G_i$ without changing the described null class $\bigcap_{m\in\mathbb{N}} G_m$. The tests in the examples above have this monotonicity property anyway.

3.2.3 Remark. While each computable set is given by a ML-test, there is no single ML-test that corresponds to the property of being computable, because the class of computable sets is not Π^0_2 by Exercise 1.8.64.

A universal Martin-Löf test

We say that a Martin-Löf test $(U_b)_{b\in\mathbb{N}}$ is *universal* if $\bigcap_b U_b$ contains $\bigcap_m G_m$ for any Martin-Löf test $(G_m)_{m\in\mathbb{N}}$. In other words, $\bigcap_b U_b = 2^{\mathbb{N}} - \mathsf{MLR}$. Martin-Löf (1966) proved that a universal ML-test exists. Fix a listing $(G_k^e)_{k\in\mathbb{N}}$ $(e \in \mathbb{N})$ of all the ML-tests in such a way that G_k^e is c.e. uniformly in e, k.

3.2.4 Fact. *Let* $U_b = \bigcup_{e\in\mathbb{N}} G^e_{b+e+1}$. *Then* $(U_b)_{b\in\mathbb{N}}$ *is a universal ML-test.*

Proof. The sequence $(U_b)_{b\in\mathbb{N}}$ is a ML-test since it is uniformly c.e. and $\lambda U_b \le \sum_e 2^{-(b+e+1)} = 2^{-b}$. For the universality, suppose that Z is not ML-random. Then there is e such that $Z \in \bigcap_k G_k^e$, so $Z \in U_b$ for each b. □

In 1.8.55 we introduced the arithmetical hierarchy for classes of sets. Since $\bigcap_m G_m$ is Π^0_2 for any ML-test $(G_m)_{m\in\mathbb{N}}$, the existence of a universal ML-test implies the following.

3.2.5 Proposition. MLR *is a* Σ^0_2 *class.* □

This shows that Martin-Löf randomness is among the simplest randomness notions as far as the descriptive complexity is concerned; other notions are typically Σ^0_3 or even more complex.

Characterization of MLR *via the initial segment complexity*

Recall from 2.2.4 that $K_M(x) = \min\{|\sigma| \colon M(\sigma) = x\}$ for a prefix-free machine M. The open sets generated by the strings that are b-compressible in the sense of K_M form a monotonic test $(R_b^M)_{b\in\mathbb{N}}$. If M is an optimal prefix-free machine this test turns out to be universal.

3.2.6 Definition. For $b \in \mathbb{N}$ let $R_b^M = [\{x \in \{0,1\}^* \colon K_M(x) \le |x| - b\}]^{\prec}$.

3.2.7 Proposition. $(R_b^M)_{b\in\mathbb{N}}$ *is a ML-test.*

Proof. The condition $K_M(x) \le |x| - b$ is equivalent to

$$\exists\sigma\,\exists s\,[M_s(\sigma) = x \;\&\; |\sigma| \le |x| - b],$$

which is a Σ^0_1-property of x and b. Hence the sequence of open sets $(R_b^M)_{b\in\mathbb{N}}$ is uniformly c.e.

To show $\lambda R_b^M \le 2^{-b}$, let V_b^M be the set of strings in $\{x\colon K_M(x) \le |x| - b\}$ that are minimal under the prefix ordering. Then $\sum_x 2^{-|x|} [\![x \in V_b^M]\!] = \lambda R_b^M$. Note that $2^{-|x^*_M|} \ge 2^b 2^{-|x|}$ for each $x \in V_b^M$, where x^*_M is a shortest M-description of x similar to 2.3.2. Then $1 \ge \sum_x 2^{-|x^*_M|} [\![x \in V_b^M]\!] \ge 2^b \sum_x 2^{-|x|} [\![x \in V_b^M]\!]$, and therefore $\lambda R_b^M \le 2^{-b}$. □

Recall $\mathbb{U}$ is the optimal prefix-free machine defined in 2.2.9, and $K(x) = K_{\mathbb{U}}(x)$.

3.2.8 Definition. For $b \in \mathbb{N}$, let $\mathcal{R}_b = R_b^{\mathbb{U}} = [\{x \in \{0,1\}^* \colon K(x) \le |x| - b\}]^{\prec}$.

Thus, $\mathcal{R}_b$ is the open set generated by the b-compressible$_K$ strings as defined in 2.5.1. The theorem of Schnorr (1973) states that Z is ML-random iff there is b such that each initial segment of Z is b-incompressible in the sense of K. A similar theorem using monotonic string complexity is in Levin (1973).

3.2.9 Theorem. *The following are equivalent for a set Z.*
(i) Z is Martin-Löf random.
(ii) $\exists b \, \forall n \, [K(Z\restriction_n) > n - b]$, that is, $\exists b \, Z \notin \mathcal{R}_b$.

Theorem 3.2.9 simply states that $(\mathcal{R}_b)_{b\in\mathbb{N}}$ is universal. It actually holds for $(R_b^M)_{b\in\mathbb{N}}$, where M is an optimal prefix-free machine, by the same proof.

Proof. The implication (i)$\Rightarrow$(ii) holds because $(\mathcal{R}_b)_{b\in\mathbb{N}}$ is a ML-test by 3.2.7. For (ii)$\Rightarrow$(i), suppose that Z is not ML-random, that is, $Z \in \bigcap_m G_m$ for some ML-test $(G_m)_{m\in\mathbb{N}}$. Since we can replace G_m by G_{2m} we may assume that $\lambda G_m \leq 2^{-2m}$ for each m.

We define a bounded request set L (see 2.2.15). By the representation of c.e. open sets in 1.8.26 we may uniformly in m obtain an effective antichain $(x_i^m)_{i<N_m}$, $N_m \in \mathbb{N}\cup\{\infty\}$, such that $G_m = [\{x_i^m : i < N_m\}]^{\prec}$. Let $L = \{\langle |x_i^m| - m + 1, x_i^m\rangle : m \in \mathbb{N}, i < N_m\}$. The contribution of G_m to the total weight of L is at most 2^{-m-1}, so L is a bounded request set.

Let M_d be the prefix-free machine for L given by the Machine Existence Theorem 2.2.17. Fix $b \in \mathbb{N}$ and let $m = b + d + 1$. Since $Z \in G_m$, we have $x_i^m \prec Z$ for some i. Hence $K(x) \leq |x| - m + 1 + d = |x| - b$ because of the request enumerated for compressing $x = x_i^m$. $\square$

Examples of Martin-Löf random sets

Recall from the introduction to this chapter that for any randomness notion the class of sets that are random in that sense is conull (page 104). However, this gives us no concrete examples of random sets. Because of the "being hard to describe" aspect of randomness, such examples are the more difficult to obtain the stronger the randomness notion becomes. It is still fairly easy to give concrete examples of ML-random sets.

If $(U_m)_{m\in\mathbb{N}}$ is a universal ML-test then for each m, $P_m = 2^{\mathbb{N}} - U_m$ is a Π^0_1 class, $\lambda P_m \geq 1 - 2^{-m}$ and P_m contains only ML-random sets. Our first examples are obtained by applying the basis theorems for Π^0_1 classes to P_1, say.

3.2.10 Examples. *(i) There is a left-c.e. ML-random set.*
(ii) Some ML-random set Z is superlow.

Informally, ML-randomness is not yet a very strong randomness notion. Both (i) and (ii) fail already for our next stronger randomness notion, weak 2-randomness, since a weakly 2-random set forms a minimal pair with $\emptyset'$ (see page 135).

In the following we obtain more explicit examples of left-c.e. Martin-Löf random sets. Recall from (2.4) on page 84 that the halting probability of a prefix-free machine M is the left-c.e. real number $\Omega_M = \lambda[\mathrm{dom}\, M]^{\prec} = \sum_\sigma 2^{-|\sigma|} [\![M(\sigma)\downarrow]\!]$, and that $(\Omega_{M,s})_{s\in\mathbb{N}}$ is a nondecreasing computable approximation of Ω_M, where

$\Omega_{M,s} := \lambda[\text{dom } M_s]^{\prec}$. We identify co-infinite subsets of $\mathbb{N}$ with real numbers in $[0, 1)$ according to Definition 1.8.10. In particular, for a real number r, we let $r \upharpoonright_n$ denote the first n bits of the binary expansion of r.

3.2.11 Theorem. (Chaitin) *The halting probability Ω_R is ML-random for each optimal prefix-free machine R.*

Proof. We will write Ω for Ω_R, and we let $\Omega_s = \Omega_{R,s}$. Let N be the (plain) machine that works as follows on an input x of length n.

(1) Wait for t such that $0.x \le \Omega_t < 0.x + 2^{-n}$.
(2) Output the least string y not in the range of R_t.

If $x = \Omega \upharpoonright_n$ then such a t exists. By stage t all R-descriptions of length $\le n$ have appeared, for otherwise $\Omega \ge \Omega_t + 2^{-n}$. Thus $K(y) > n$ where $y = N(x)$. This implies $\forall n\, [K(\Omega \upharpoonright_n) + c \ge K(N(\Omega \upharpoonright_n)) > n]$ for an appropriate constant c. □

From now on we will write Ω for $\Omega_{\mathbb{U}}$. We let $\Omega_s := \lambda[\text{dom } \mathbb{U}_s]^{\prec}$.

Facts about ML-random sets

Facts derived from Schnorr's Theorem. In the following we think of a set Z as a sequence of experiments with outcomes zero or one, like tossing a coin. The law of large numbers for a set Z says that the number of occurrences of zeros and ones is asymptotically balanced:

3.2.12 Definition. A set Z satisfies the law of large numbers if

$$\lim_n (\#\{i < n : Z(i) = 1\}/n) = 1/2. \tag{3.3}$$

This is one of the simplest criteria for randomness, which is satisfied by any Martin-Löf random set Z. Instead of giving a test directly, we will prove the law of large numbers for Z using Schnorr's Theorem, together with the fact that for large n and any incompressible$_K$ string x of length n, $S_n(x)/n$ gets arbitrarily close to $1/2$ (Corollary 2.5.9).

3.2.13 Proposition. *Each ML-random set Z satisfies the law of large numbers.*

Proof. Fix b such that $Z \notin \mathcal{R}_b$, i.e., each initial segment of Z is b-incompressible$_K$. Apply 2.5.9, noting that $S_n(Z \upharpoonright_n) = \#\{i < n : Z(i) = 1\}$. □

A condition apparently weaker than $\exists b\, Z \notin \mathcal{R}_b$ implies that Z is ML-random:

3.2.14 Proposition. *Suppose there is an infinite computable set R and $b \in \mathbb{N}$ such that $\forall m \in R\, [K(Z \upharpoonright_m) > m - b]$. Then Z is Martin-Löf random.*

Proof. We apply the idea, first used in the proof of 2.4.1, to split off a prefix-free description from a string. The prefix free machine M on input τ first looks for $\sigma \preceq \tau$ such that $\mathbb{U}(\sigma) \downarrow = x$. Next, if $\sigma z = \tau$, it checks whether $|x| + |z|$ is the least number $m \in R$ such that $m \ge |x|$. In this case M outputs xz.

Clearly M is a prefix-free machine. If $K(x) \le |x| - c$ then $K_M(w) \le |w| - c$ for each extension w of x of length the least number $m \ge |x|$ in R. If Z is not

Martin-Löf random, then for each c there is an $x \prec Z$ such that $K(x) \le |x| - c$, so $K(Z\upharpoonright_m) \le^+ m - c$ for $m \in R$ as above, contrary to the hypothesis. □

A *tail* Y of a set Z is what one obtains after taking off an initial segment x. Thus, if $|x| = n$, the tail is the set Y given by $Y(i) = Z(i+n)$. In other words, $Y = f^{-1}(Z)$ where f is the one-one function $\lambda i.\, i + n$.

3.2.15 Proposition. *Suppose Y and Z are sets such that $Y =^* Z$, or Y is a tail of Z. Then Z is ML-random $\Leftrightarrow$ Y is ML-random.*

Proof. Under either hypothesis we have $\forall n\, K(Y\upharpoonright_n) =^+ K(Z\upharpoonright_n)$. Now we apply Schnorr's Theorem 3.2.9. □

The fact that tails of a ML-random set are also ML-random can be strengthened: the pre-image of a ML-random set under a one-one computable function is ML-random. Also, recall from Section 1.3 that, with very few exceptions, classes introduced in computability theory are closed under computable permutations. By the next result this is the case for ML-randomness.

3.2.16 Proposition. *Suppose the one-one function $f\colon \mathbb{N} \to \mathbb{N}$ is computable. If Z is ML-random then so is $Y = f^{-1}(Z)$.*

Proof. For a class $\mathcal{C} \subseteq 2^{\mathbb{N}}$ let $\mathcal{C}^* = \{Z\colon f^{-1}(Z) \in \mathcal{C}\}$. Since f is one-one, we have $\lambda[x]^* = 2^{-|x|}$ for each string x. If $G \subseteq 2^{\mathbb{N}}$ is open, let $(x_i)_{i<N}$ be the minimal strings x such that $[x] \subseteq G$. Then G^* is the disjoint union of the $[x_i]^*$. Thus $\lambda G^* = \lambda G$ for any open set G. Suppose now that $Y \in \bigcap_m G_m$ for some ML-test $(G_m)_{m\in\mathbb{N}}$. Then $(G^*_m)_{m\in\mathbb{N}}$ is a ML-test such that $Z \in \bigcap_m G^*_m$. □

A criterion due to W. Merkle will be useful later on.

3.2.17 Proposition. *The following are equivalent for a set Z.*
(i) Z is not ML-random.
(ii) $Z = z_0 z_1 z_2 \ldots$ for a sequence of strings $(z_i)_{i\in\mathbb{N}}$ such that $\forall i\, K(z_i) \le |z_i| - 1$.
(iii) There is a prefix-free machine M such that $Z = z_0 z_1 z_2 \ldots$, for a sequence of strings $(z_i)_{i\in\mathbb{N}}$ such that $\forall i\, K_M(z_i) \le |z_i| - 1$.

Proof. (i)$\Rightarrow$(ii): We define the sequence $(z_i)_{i\in\mathbb{N}}$ inductively. Since Z is not ML-random, by Theorem 3.2.9 there is a string $z_0 \prec Z$ such that $K(z_0) \le |z_0| - 1$. Suppose $i > 0$ and $z_0, \ldots, z_{i-1}$ have been defined. The tail Y of Z obtained by taking off $z_0 \ldots z_{i-1}$ is not ML-random by Proposition 3.2.15. Thus there is $z_i \prec Y$ such that $K(z_i) \le |z_i| - 1$.
(ii)$\Rightarrow$(iii) is trivial. For (iii)$\Rightarrow$(i), fix n and consider the prefix-free machine N which on input σ first searches for an initial segment $\rho \preceq \sigma$ such that $\mathbb{U}(\rho) = n$ and then looks for $\nu_0, \ldots, \nu_{n-1} \in \mathrm{dom}(M)$ such that $\rho\nu_0 \ldots \nu_{n-1} = \sigma$. If the search is successful it prints $M(\nu_0) \ldots M(\nu_{n-1})$.

Given a string $z_0 \ldots z_{n-1}$, let σ be a concatenation of a shortest $\mathbb{U}$-description of n followed by shortest M-descriptions of $z_0, \ldots, z_{n-1}$. Then $N(\sigma) = z_0 \ldots z_{n-1}$ and hence $K(z_0 \ldots z_{n-1}) \le^+ K(n) + \sum_{i<n} K_M(z_i) \le K(n) + |z_0 \ldots z_{n-1}| - n$. Since $K(n) \le^+ 2\log n$, we obtain $K(z_0 \ldots z_{n-1}) \le^+ |z_0 \ldots z_{n-1}| - (n - 2\log n)$. Then, by Schnorr's Theorem 3.2.9, Z is not ML-random. □

Facts obtained through Solovay tests. Recall that a ML-test is a sequence $(G_m)_{m\in\mathbb{N}}$ of uniformly c.e. open sets such that $\forall m \in \mathbb{N}\ \lambda G_m \le 2^{-m}$, and Z fails the test if $Z \in \bigcap_m G_m$. To define Solovay tests we will broaden the test definition and relax the failure condition. Although Solovay tests are more general than Martin-Löf tests and therefore easier to build, they still determine the same notion of randomness.

3.2.18 Definition. A *Solovay test* is a sequence $(S_i)_{i\in\mathbb{N}}$ of uniformly c.e. open sets such that $\sum_i \lambda S_i < \infty$. Z *fails* the test if $Z \in S_i$ for infinitely many i, otherwise Z *passes* the test.

3.2.19 Proposition. *Z is ML-random $\Leftrightarrow$ Z passes each Solovay test.*

Proof. $\Leftarrow$: Suppose Z fails the ML-test $(G_m)_{m\in\mathbb{N}}$. Each Martin-Löf test is a Solovay test, and since Z fails $(G_m)_{m\in\mathbb{N}}$ as a ML-test (namely, $\forall m\, Z \in G_m$), Z fails it as a Solovay test (namely, $\exists^\infty m\, Z \in G_m$).

$\Rightarrow$: Suppose Z fails the Solovay test $(S_i)_{i\in\mathbb{N}}$, that is, $Z \in S_i$ for infinitely many i. We may omit finitely many of the S_i, and hence assume that $\sum_i \lambda S_i \leq 1$. Let

$$G_m = [\{\sigma\colon [\sigma] \subseteq S_i \text{ for at least } 2^m \text{ many } i\}]^{\prec},$$

then $(G_m)_{m\in\mathbb{N}}$ is a u.c.e. sequence of open sets. Given m, let $(\sigma_k)_{k\in\mathbb{N}}$ be a listing of the minimal strings σ under the prefix ordering such that $[\sigma] \subseteq G_m$. Then

$$1 \geq \sum_i \lambda S_i \geq \sum_i \sum_k \lambda(S_i \cap [\sigma_k]) \geq 2^m \sum_k 2^{-|\sigma_k|} = 2^m \lambda G_m.$$

Thus $\lambda G_m \leq 2^{-m}$, and $(G_m)_{m\in\mathbb{N}}$ is a Martin-Löf test. Since $Z \in \bigcap_m G_m$, Z is not Martin-Löf random. □

We give two applications of Solovay tests. The first is to show that ML-random sets only have short runs of zeros. By Proposition 2.5.7 the length of a run of zeros in an incompressible$_K$ string of length m is bounded by $4\log m + O(1)$. If the run *starts* at the n-th bit position of a Martin-Löf random set, then its length is at most $K(n)$ (for almost all n). This is rather short since $K(n) \leq^+ \log n + 2\log\log n$ by Proposition 2.2.8.

3.2.20 Proposition. *Let Z be ML-random. Then for each r, for almost all n, a run of zeros starting at position n has length at most $K(n) - r$.*

Proof. Recall from (2.6) that $K_t(n)$ is a nonincreasing computable approximation of $K(n)$ at stages t. For $n > 0$, let

$$S_{n,t} = [\{\sigma 0^{\max(K_t(n)-r,0)}\colon |\sigma| = n\}]^{\prec},$$

and let $S_n = \bigcup_t S_{n,t}$. Then $(S_n)_{n\in\mathbb{N}}$ is a uniformly c.e. sequence of open sets. Moreover, $\lambda S_n = 2^{-K(n)+r}$, so that $\sum_n \lambda S_n \leq \Omega 2^r$. The ML-random set Z passes this test, that is, for almost all n we have $Z \notin S_n$. This implies that the runs of zeros in Z satisfy the required bound on the length. □

3.2.21 Proposition. *Z is ML-random $\Leftrightarrow$ $\lim_n K(Z\upharpoonright_n) - n = \infty$.*

Informally, for every set Z the function $\lambda n.\, K(Z\upharpoonright_n)$ avoids being close to n: either for each d there is n such that it dips below $n - d$ (when Z is not ML-random), or for each b eventually it exceeds $n + b$ (when Z is ML-random).

Proposition 3.2.21 will be improved in Theorem 7.2.8: Z is ML-random $\Leftrightarrow$ $\sum_n 2^{-K(Z\upharpoonright_n)+n} < \infty$.

Proof. $\Leftarrow$: This is immediate by Theorem 3.2.9.
$\Rightarrow$: Suppose that there is b such that $\exists^\infty n\, K(Z\upharpoonright_n) - n = b$. Let $S_n = [\{x\colon |x| = n \ \&\ K(x) \le n + b\}]^\prec$, then Z is in infinitely many sets S_n, so we can conclude that Z is not ML-random once we have shown that $(S_n)_{n\in\mathbb{N}}$ is a Solovay test. Clearly $(S_n)_{n\in\mathbb{N}}$ is uniformly c.e. Applying Theorem 2.2.26(i) where $d = K(n)-b$, the number of strings x of length n such that $K(x) \le n+b$ is at most $2^{\mathbf{c}}2^{n-d} = 2^{\mathbf{c}}2^{n-K(n)+b}$. Hence $\lambda S_n \le 2^{\mathbf{c}}2^{-K(n)+b}$, and $\sum_n \lambda S_n < 2^{\mathbf{c}+b}\Omega < \infty$. □

3.2.22 Remark. Let $(S_i)_{i\in\mathbb{N}}$ be a Solovay test. By Fact 1.8.26, each open set S_i is generated by a uniformly computable prefix-free set of strings. We might as well put the basic open cylinders corresponding to the strings in all those prefix-free sets together, because failing the test is equivalent to being in infinitely many intervals. In this way we obtain an equivalent Solovay test where each open set is just a basic open cylinder $[x]$. Thus, we may alternatively represent a Solovay test G by an effective listing of strings $x_0, x_1, \ldots$ (possibly with repetitions) such that $\sum_i 2^{-|x_i|} < \infty$. Usually we will not distinguish between the two types of representation; if we want to stress this difference we will call G an *interval Solovay test.* For instance, in the foregoing proof, the interval Solovay test is an effective listing of the set $\{x\colon K(x) \le |x| + b\}$.

3.2.23 Example. Let B be a c.e. prefix-free set such that $\lambda[B]^\prec < 1$. For each n let S_n be the open set generated by the n-th power B^n of B, namely $S_n = [\{x_1 \ldots x_n\colon \forall i\ x_i \in B\}]^\prec$. Clearly, $\lambda S_n = (\lambda S_1)^n$. Since $\lambda S_1 < 1$, we may conclude that $(S_n)_{n\in\mathbb{N}}$ is a Solovay test. If $\lambda S_1 \le 1/2$ then $(S_n)_{n\in\mathbb{N}}$ is in fact a ML-test.

Tails of sets were defined after Prop. 3.2.13. The following is due to Kučera.

3.2.24 Proposition. *Suppose P is a Π^0_1 class such that $\lambda P > 0$. Then each ML-random set Z has a tail in P.*

Proof. Let $B \subseteq \{0,1\}^*$ be a c.e. prefix-free set such that $[B]^\prec = 2^{\mathbb{N}} - P$. Define the Solovay test $(S_n)_{n\in\mathbb{N}}$ as in 3.2.23. Since Z is ML-random, there is a least $n \in \mathbb{N}$ such that $Z \notin S_n$. If $n > 0$, let $x \prec Z$ be shortest such that $x \in B^{n-1}$, otherwise let $x = \varnothing$. Then the tail Y of Z obtained by taking x off is not in $[B]^\prec$, and hence $Y \in P$. □

In particular, if $P \subseteq \mathsf{MLR}$ (say, $P = 2^{\mathbb{N}} - \mathcal{R}_b$ for some b), then Z is ML-random iff some tail of Z is in P. For, if some tail of Z is in P, then Z is ML-random by Proposition 3.2.15.

Exercises.

3.2.25. Give an alternative proof of Proposition 3.2.20 using 3.2.21.

3.2.26. Use Solovay tests to show that each ML-random set satisfies the law of large numbers.

Left-c.e. ML-random reals and Solovay reducibility

We use the term "real" for real number. Via the identifications in Definition 1.8.10 we can apply definitions about sets of natural numbers to reals.

Solovay reducibility $\leq_S$ is used to compare the randomness content of left-c.e. reals. We obtain two characterizations of the left-c.e. ML-random reals:

(a) They are the reals of the form Ω_R for a optimal prefix-free machine R. Thus, Theorem 3.2.11 accounts for all the left-c.e. ML-random reals in $[0, 1)$.

(b) They are the Solovay complete left-c.e. reals. Thus, they play a role similar to the creative sets for many-one reducibility on the c.e. sets.

We identify the Boolean algebra of clopen sets in Cantor space with the Boolean algebra *Intalg* $[0, 1)_{\mathbb{R}}$ as in 1.8.11. We let α, β, γ denote left-c.e. reals. If γ is left-c.e. then there is a computable sequence of binary rationals $(\gamma_s)_{s\in\mathbb{N}}$ such that $\gamma_s \leq \gamma_{s+1}$ for each s and $\gamma = \sup_s \gamma_s$ (Fact 1.8.15). We say that $(\gamma_s)_{s\in\mathbb{N}}$ is a *non-decreasing computable approximation* of γ.

First we show that adding a left-c.e. real to a left-c.e. ML-random real keeps it ML-random.

3.2.27 Proposition. *Suppose α, β are left-c.e. reals such that $\gamma = \alpha + \beta < 1$. If α or β is ML-random, then γ is ML-random.*

Proof. Suppose that $(G_n)_{n\in\mathbb{N}}$ is a ML-test such that $\gamma \in \bigcap_n G_n$. We show that α is not ML-random. We may assume that $\lambda G_n \leq 2^{-n-1}$, and that β is not a binary rational. Let $(\alpha_s)_{s\in\mathbb{N}}$ and $(\beta_s)_{s\in\mathbb{N}}$ be nondecreasing computable approximations of α, β, and let $\gamma_s = \alpha_s + \beta_s$. We enumerate a ML-test $(H_n)_{n\in\mathbb{N}}$ such that $\alpha \in \bigcap_n H_n$.

> At stage s, if $\gamma_s \in I$ where $I = [x, y)$ is a maximal subinterval of $G_{n,s}$, then put the interval $J = [x - \beta_s - (y - x), y - \beta_s)$ into H_n.

If in fact $\gamma \in I$, then since $\beta_s < \beta < \beta_s + (y - x)$ and $x \leq \gamma$ we have $y - \beta_s > \gamma - \beta_s > \alpha = \gamma - \beta$; also $x - \beta_s - (y - x) \leq x - \beta \leq \alpha$. Thus $\alpha \in J$. The length of an interval J is twice the length of an interval I, so that $\lambda H_n \leq 2^{-n}$. □

The converse of Proposition 3.2.27 was proved by Downey, Hirschfeldt and Nies (2002): if γ is a ML-random left-c.e. real and $\gamma = \alpha + \beta$ for left-c.e. reals α, β, then α or β is ML-random. They also introduced the following algebraic definition of $\leq_S$. It is equivalent to the original definition of Solovay (1975) by Exercise 3.2.33.

3.2.28 Definition. Let $\alpha, \beta \in [0, 1)$ be left-c.e. reals. We write $\beta \leq_S \alpha$ if

$$\exists d \in \mathbb{N}\, \exists \gamma \text{ left-c.e. } \big[2^{-d}\beta + \gamma = \alpha\big].$$

Thus, Solovay completeness in (b) above yields an algebraic characterization of being ML-random within the left-c.e. reals. By Exercise 3.2.33 $\leq_S$ is transitive and implies $\leq_T$.

We proceed to the main result. The implications (iii)⇒(ii)⇒(i) are due to Calude, Hertling, Khoussainov and Wang (2001), and (i)⇒(iii) to Kučera and Slaman (2001).

3.2.29 Theorem. *The following are equivalent for a real $\alpha \in [0,1)$.*

(i) α is left-c.e. and ML-random.

(ii) There is an optimal prefix-free machine R such that $\Omega_R = \alpha$.

(iii) α is Solovay complete, that is, $\beta \le_S \alpha$ for each left-c.e. real β.

Proof. (iii)⇒(ii). Choose $d \in \mathbb{N}$ and a left-c.e. real γ such that $2^{-d}\Omega + \gamma = \alpha$. We define a bounded request set L of total weight α such that the associated prefix-free machine R is optimal. Note that $\Omega_R = \alpha$.

Choose a nondecreasing computable approximation $(\gamma_s)_{s\in\mathbb{N}}$ of γ. We may assume that $\gamma_{s+1} - \gamma_s$ is either 0 or of the form 2^{-n} for some n.

Construction of L. Let $L_0 = \emptyset$.

Stage $s > 0$. If $\mathbb{U}_{s-1}(\sigma)\uparrow$ and $\mathbb{U}_s(\sigma) = y$, put the request $\langle |\sigma| + d, y\rangle$ into L_s unless it is already in L_{s-1}; in this case, to record the increase of Ω in L, put $\langle |\sigma| + d, i\rangle$ into L_s for a number $i > s$ not mentioned so far. If $\gamma_{s+1} - \gamma_s = 2^{-n}$ then put $\langle n, j\rangle$ into L_s for a number $j > s$ not mentioned so far.

It is clear that L is a bounded request set as required.

(ii)⇒(i). Ω_M is left-c.e. for each prefix-free machine M. For an optimal prefix-free machine R, the real Ω_R is ML-random by Theorem 3.2.11.

(i)⇒(iii). Suppose the left-c.e. real $\beta \in [0,1)$ is given. Let $(\beta_s)_{s\in\mathbb{N}}$ and $(\alpha_s)_{s\in\mathbb{N}}$ be nondecreasing computable approximations of β and α, respectively, where $\alpha_s, \beta_s \in \mathbb{Q}_2$. For each parameter $d \in \mathbb{N}$ we build a left-c.e. real γ_d uniformly in d, attempting to ensure that $2^{-d}\beta + \gamma_d = \alpha$. In the end we will define a ML-test $(G_d)_{d\in\mathbb{N}}$. If d is a number such that $\alpha \notin G_d$ then we succeed with γ_d.

The construction with parameter d runs only at *active* stages; 0 is active. If s is active, and t is the greatest active stage less than s, then let $\epsilon_s = 2^{-d}(\beta_s - \beta_t)$. We wish that α increase by an amount of at least ϵ_s in order to record the increase of β. So we put the interval $[\alpha_s, \alpha_s + \epsilon_s)$ into G_d. If $\alpha \notin G_d$, then α will increase eventually, and in that case we have reached the next active stage. If α has increased too much the excess is added to γ. For the formal construction, it is easier to first update γ, and then define the next interval.

Construction for the parameter d.
Stage 0 is declared active. *Let $q_0 = 0$ and $\gamma_{d,0} = 0$.*

Stage $s > 0$. Let $t < s$ be the greatest active stage. If $\alpha_s \ge q_t$ then declare s active, *and do the following.*

(1) *Define $\gamma_{d,s} = \gamma_{d,t} + (\alpha_s - q_t)$.*
(2) *Let $q_s = \alpha_s + 2^{-d}(\beta_s - \beta_t)$. If $q_s > 1$ then stop.*

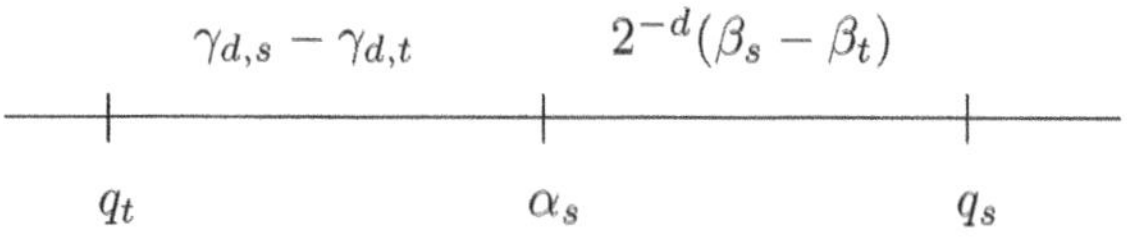

$$\underbrace{q_t \;\longrightarrow\; \alpha_s}_{\gamma_{d,s} - \gamma_{d,t}} \;\underbrace{\longrightarrow\; q_s}_{2^{-d}(\beta_s - \beta_t)}$$

Verification. Let $G_d = \bigcup\{[\alpha_s, q_s): s \text{ active}\}$. Then G_d is (identified with) a c.e. open set in Cantor space uniformly in d, and $\lambda G_d \leq 2^{-d}\beta$. Thus $(G_d)_{d\in\mathbb{N}}$ is a ML-test. Choose d such that $\alpha \notin G_d$, then the construction for parameter d has infinitely many active stages. Inductively, between (1) and (2) of each active stage $s > 0$ we have $2^{-d}\beta_t + \gamma_{d,s} = \alpha_s$, where t is the preceding active stage. Hence $2^{-d}\beta + \gamma_d = \alpha$, where $\gamma_d = \sup_{s \text{ is active}} \gamma_{d,s}$. □

The following fact of Calude, Nies, Staiger and Stephan (2008) can alternatively be proved using Theorem 4.1.11 below.

3.2.30 Proposition. *$\emptyset' \equiv_{wtt} \Omega_R$ for each optimal prefix-free machine R.*

Proof. Clearly $\Omega_R \leq_{wtt} \emptyset'$. To show that $\emptyset' \leq_{wtt} \Omega_R$, define a prefix-free machine M by $M(0^n1) = s$ if $n \in \emptyset'_{\text{at } s}$. Let d be the coding constant for M with respect to R. The reduction procedure is as follows: on input n, using the oracle Ω_R, compute t such that $\Omega_R \restriction_{n+d+1} = \Omega_{R,t} \restriction_{n+d+1}$. Output $\emptyset'_t(n)$.

If n enters $\emptyset'$ at a stage $s > t$ then Ω_R increases by at least $2^{-(n+d+1)}$, contrary to the choice of t. □

3.2.31 Corollary. *Every ML-random left-c.e. set is wtt-complete.* □

We cannot hope to characterize the ML-random ω-c.e. sets as easily: such a set can be superlow by 3.2.10(ii), but it can also be weak truth-table complete.

As an immediate consequence of Theorem 3.2.29, Z is right-c.e. and ML-random iff there is an optimal prefix-free machine R such that $1 - \Omega_R = 0.Z$. How about ML-random reals that are difference left-c.e.? (See (iv) of Definition 1.8.14.) Rettinger (unpublished) has shown that one does not obtain anything new.

3.2.32 Proposition. *Let $r \in [0,1)_{\mathbb{R}}$ be difference left-c.e. and ML-random. Then r is either left-c.e. or right-c.e.*

Proof. Assume r is neither. By Fact 1.8.15, $r = \lim_i q_i$ for an effective sequence $(q_i)_{i\in\mathbb{N}}$ of dyadic rationals such that $\sum_i \text{abs}(q_{i+1} - q_i) < \infty$. For each $m \in \mathbb{N}$ we have $\sup_{i\geq m} q_i > r$, otherwise $r = \lim_{i\geq m} \max\{q_k : m \leq k \leq i\}$, so r is left-c.e. Similarly, $\inf_{i\geq m} q_i < r$ for each m. Thus there are infinitely many i such that $q_i < r < q_{i+1}$. Let $G_i = [q_i, q_{i+1})$ if $q_i < q_{i+1}$ and $G_i = \emptyset$ else. Then $(G_i)_{i\in\mathbb{N}}$ is a Solovay test that succeeds on r. □

3.2.33 Exercise. Check that $\leq_S$ is transitive. Show that $\beta \leq_S \alpha$ iff there is a partial computable $\varphi: \mathbb{Q}_2 \cap [0,\alpha) \to \mathbb{Q}_2 \cap [0,\beta)$ and $c \in \mathbb{N}$ such that $\forall q < \alpha\, [\beta - \varphi(q) < c(\alpha - q)]$. Informally, β is easier to approximate than α. Conclude that $\leq_S$ implies $\leq_T$.

Randomness on reals, and randomness for bases other than 2

Let $\mathcal{X} = [0,1)_{\mathbb{R}} - \mathbb{Q}$ be the space equipped with the subspace topology and Lebesgue measure. One can develop the theory of ML-randomness directly on $\mathcal{X}$, without reference to the representation of reals in base 2. For instance, an open set $U \subseteq \mathcal{X}$ is called computably enumerable if U is an effective union of intervals $(p,q)_{\mathbb{R}} - \mathbb{Q}$ where $p, q \in \mathbb{Q}$, $0 \leq p < q \leq 1$. Based on this one introduces ML-tests on $\mathcal{X}$ by adapting Definition 3.2.1.

3.2.34 Remark. Fix a base $b \in \mathbb{N}$ such that $b \geq 2$, and write $b^{\mathbb{N}}$ for the set of functions $\mathbb{N} \to \{0, \ldots, b-1\}$. Product topology and uniform measure can be defined on the set $b^{\mathbb{N}}$. One can think of the elements of $b^{\mathbb{N}}$ as the overall result of a sequence of experiments with b outcomes, each one occurring with probability $1/b$. For instance if $b = 6$ the experiment could be rolling a dice. Similar to (1.13), define a map

$$F_b\colon \{Z \in b^{\mathbb{N}} : Z \text{ is not eventually periodic}\} \to [0, 1)_{\mathbb{R}} - \mathbb{Q}$$

by $F_b(Z) = \sum_i Z(i)b^{-i-1}$. This map preserves topology and measure in both directions. One can also adapt the definition of ML-randomness in 3.2.1 to base b. It is easy to check that F_b preserves ML-randomness in both directions. Thus, if $Z \in b^{\mathbb{N}}$ is not eventually periodic, then Z is ML-random $\leftrightarrow$ $F_b(Z)$ is ML-random (in the sense of reals) $\leftrightarrow$ the set $F_2^{-1}(F_b(Z))$ is ML-random in the sense of Definition 3.2.1. Thus, ML-randomness is a base-independent concept.

Giving preference to base 2 is not necessary but convenient, because in computability theory one studies subsets of $\mathbb{N}$ rather than functions in $b^{\mathbb{N}}$ for $b > 2$.

A nonempty Π^0_1 *subclass of* MLR *has ML-random measure* ⋆

We consider the uniform measure of nonempty Π^0_1 subclasses of MLR, such as $2^{\mathbb{N}} - \mathcal{R}_1$. This yields examples of right-c.e. ML-random reals other than $1 - \Omega$.

3.2.35 Theorem. *If $P \subseteq \mathsf{MLR}$ is a nonempty Π^0_1 class then λP is ML-random.*

Proof. Note that $\lambda P > 0$, since otherwise $P \cap \mathsf{MLR} = \emptyset$ by 3.2.2. Suppose that λP is not ML-random, and let Z be the co-infinite set identified with the real number λP (namely, $\lambda P = 0.Z$).

Firstly we show that for each $b \in \mathbb{N}$, there is y on P such that $K(y) \leq |y| - b$. We use the fact that an appropriate initial segment x of Z is sufficiently compressible in the sense of K to obtain a b-compressible$_K$ string y that is guaranteed to be on P: if y falls off P then the measure decreases so much that the approximation $0.x$ is wrong.

Let $(P_t)_{t \in \mathbb{N}}$ be the effective approximation of P by clopen sets from (1.17) on page 55. The prefix-free machine M works as follows on an input σ.

(1) Wait for s such that $\mathbb{U}_s(\sigma)\downarrow = x$. Let $n = |x|$. Let $c = \lfloor (n - |\sigma|)/2 \rfloor$.
(2) Wait for $t \geq s$ such that $0.x \leq \lambda P_t < 0.x + 2^{-n}$.
(3) If there is a string y of length $n - c$ such that $\lambda(P_t \cap [y]) \geq 2^{-n}$ then output the leftmost such y.

If $x \prec Z$ and M outputs y, then y is on P, for otherwise $[y] \cap P = \emptyset$ and hence $\lambda(P_t - P) \geq 2^{-n}$ where $n = |x|$. Let d be a coding constant for M. Given $b \in \mathbb{N}$, since we are assuming that λP is not ML-random, there is $x \prec Z$ that is sufficiently compressible in the sense of K, namely, $b + d \leq (n - K(x))/2$, where $n = |x|$; we may also require that $2^{-c} \leq \lambda P$, where $c = \lfloor (n - K(x))/2 \rfloor$. Thus there is some y of length $n - c$ such that $\lambda(P \cap [y]) \geq 2^{-(n-c)}2^{-c} = 2^{-n}$.

If σ is a shortest $\mathbb{U}$-description of x, then M on input σ outputs a string y on P such that $|y| = n - c$. Thus $K_M(y) \leq |y| - c \leq |y| - (b + d)$ and hence $K(y) \leq |y| - b$.

We now iterate the foregoing argument and obtain a sequence of strings $y_0 \prec y_1 \prec \ldots$ such that each y_i is on P and $K(y_i) \leq |y| - i$. Then the set $Y = \bigcup y_i$ is not ML-random and $Y \in P$.

For $S \subseteq 2^{\mathbb{N}}$ let $S^c = 2^{\mathbb{N}} - S$. Let $y_0 = \varnothing$. Suppose $i > 0$ and y_{i-1} has been defined. Then $\lambda(P \cap [y_{i-1}])$ is not ML-random by Proposition 3.2.27: the left-c.e. real number

λP^c is not ML-random, and for each length m, we have $\lambda P^c = \sum_{|y|=m} \lambda(P^c \cap [y])$, so that $\lambda(P^c \cap [y])$ is not ML-random for any y. So we may apply the argument above to the Π^0_1 class $P \cap [y_{i-1}]$ in order to obtain $y_i \succ y_{i-1}$ on P such that $K(y_i) \leq |y_i| - i$. □

3.3 Martin-Löf randomness and reduction procedures

We are mostly interested in the interactions between the degree of randomness and the *absolute* computational complexity of sets; a summary of such interactions will be given in Section 8.6. However, here we address the interaction of randomness with the *relative* computational complexity of sets. Firstly, we consider reducibilities, and then, in the next section, we look at ML-randomness relative to an oracle.

Each set is weak truth-table reducible to a ML-random set

An arbitrarily complex set A can be encoded into an appropriate Martin-Löf random set: there is a ML-random set Z such that $A \leq_{wtt} Z$. This result was obtained independently by Kučera (1985) and by Gács (1986). For instance, there is a ML-random set Z weak truth-table above $\emptyset''$. Thus, ML-random sets can have properties that fail to match our intuition on randomness, a view also supported by the existence of left-c.e. ML-random sets. These particular properties are already incompatible with the somewhat stronger notion of weak 2-randomness introduced in 3.6.1 below. Ultimately, the reason why the coding is possible for ML-random sets, but not for sets satisfying a stronger randomness property, is that MLR contains a Π^0_1 class of positive measure, for instance $2^{\mathbb{N}} - \mathcal{R}_1$. We will provide a mechanism for encoding a set A into members of such a Π^0_1 class. It relies on a simple measure theoretic lemma. Recall from 1.9.3 that for measurable $S \subseteq 2^{\mathbb{N}}$, $\lambda(S|z)$ is the local measure $2^{|z|}\lambda(S \cap [z])$. For each n, λS is the average, over all strings z of length n, of the local measures $\lambda(S|z)$.

3.3.1 Lemma. *Suppose that $S \subseteq 2^{\mathbb{N}}$ is measurable and $\lambda(S|x) \geq 2^{-(r+1)}$ where $r \in \mathbb{N}$. Then there are distinct strings $y_0, y_1 \succeq x$, $|y_i| = |x| + r + 2$, such that $\lambda(S|y_i) > 2^{-(r+2)}$ for $i = 0, 1$.*

Proof. We may assume that $x = \varnothing$. Let y_0 be a string of length $r+2$ such that $\lambda(S|y_0)$ is greatest among those strings. Assume that $\lambda(S|y) \leq 2^{-(r+2)}$ for each $y \neq y_0$ of length $r+2$. Then

$$\lambda S = \sum_{|y|=r+2} \lambda(S \cap [y]) = \lambda(S \cap [y_0]) + \sum_{y \neq y_0 \& |y|=r+2} \lambda(S \cap [y])$$
$$\leq 2^{-(r+2)} + (2^{r+2} - 1)2^{-(r+2)}2^{-(r+2)} < 2^{-(r+1)}. \quad \diamond$$

3.3.2 Theorem. *Let Q be a nonempty Π^0_1 class of ML-random sets. Then for each set A there is $Z \in Q$ such that $A \leq_{wtt} Z \leq_{wtt} A \oplus \emptyset'$. The wtt-reductions are independent A.*

Proof. Since $\lambda Q > 0$, by Theorem 1.9.4 there is a string σ such that $\lambda(Q \mid \sigma) \geq 1/2$, so we may as well assume that $\lambda Q \geq 1/2$. Let f be the function given by $f(0) = 0$ and $f(r+1) = f(r) + r + 2$ (namely, $f(r) = r(r+3)/2$). Let $\widehat{Q}$ be the Π^0_1 class of paths through the Π^0_1 tree

$$T = \{y\colon \forall r\,[f(r) \leq |y| \;\rightarrow\; \lambda(Q|(y\restriction_{f(r)})) \geq 2^{-(r+1)}]\}. \tag{3.4}$$

Note that $\widehat{Q} \subseteq Q$. Since $\lambda Q \geq 1/2$, by Lemma 3.3.1 $\widehat{Q}$ is nonempty.

The set Z will be a member of $\widehat{Q}$. Suppose that so far we have coded $A\restriction_r$ into the initial segment x of Z of length $f(r)$. The idea is to code $A(r)$ into the initial segment of length $k = f(r+1)$, as follows: if $A(r) = 0$, Z takes the leftmost length k extension of x which is on $\widehat{Q}$, otherwise it takes the rightmost one. By the lemma above, these two extensions are distinct. Knowing $Z\restriction_k$, we may enumerate the complement of $\widehat{Q}$ till we see which case applies. In this way we determine $A(r)$.

The details are as follows. We define strings $(x_\tau)_{\tau\in\{0,1\}^*}$ on $\widehat{Q}$ such that $|x_\tau| = f(|\tau|)$. Let $x_\varnothing = \varnothing$. If x_τ has been defined, let $x_{\tau 0}$ be the leftmost y on $\widehat{Q}$ such that $x_\tau \prec y$ and $|y| = f(|\tau|+1)$, and let $x_{\tau 1}$ be the rightmost such y. By Lemma 3.3.1, $x_{\tau 0}$ and $x_{\tau 1}$ exist and are distinct.

For each A, the ML-random set Z coding A simply is the path $\bigcup_{\tau\prec A} x_\tau$ of T determined by A.

Firstly, we describe a reduction procedure for $A \leq_{wtt} Z$, where $f(r+1)$ bounds the use for input r. To determine $A(r)$ let $x = Z\restriction_{f(r)}$ and $y = Z\restriction_{f(r+1)}$. Find s such that

$$\widehat{Q}_s \cap [\{v \succeq x\colon\; |v| = |y| \;\&\; v <_L y\}]^{\prec} = \emptyset, \text{ or}$$
$$\widehat{Q}_s \cap [\{v \succeq x\colon\; |v| = |y| \;\&\; v >_L y\}]^{\prec} = \emptyset.$$

In the first case, output 0. In the second case, output 1.

Next we check that $Z \leq_{wtt} A \oplus \emptyset'$. Suppose that $x = Z\restriction_{f(r)}$ has been determined. If $A(r) = 0$, with $\emptyset'$ as an oracle find the leftmost extension y of x on $\widehat{Q}$ such that $|y| = f(r+1)$. Otherwise, with $\emptyset'$ as an oracle find the rightmost such extension y of x. Then $y = Z\restriction_{f(r+1)}$. Clearly the use on $\emptyset'$ is bounded by a computable function. Also, the wtt-reductions employed do not depend on the particular set A. □

For oracles other than Z, the search in the reduction procedure for $A \leq_{wtt} Z$ may not terminate. Recall from 1.2.20 that a Turing reduction is called a truth-table reduction if it is total for all oracles. In Theorem 4.3.9 below we show that $\emptyset' \not\leq_{tt} Z$ for each ML-random Z. So the wtt-reduction obtained above must be partial for some oracles.

In the exercises we indicate a proof of Theorem 3.3.2 for $Q = 2^{\mathbb{N}} - \mathcal{R}_1$ closer to Kučera's original proof, avoiding the class $\widehat{Q}$. Kučera actually did not use K for his coding. He used an idea taken from the proof of Gödel's incompleteness theorem. A similar coding works for the Π^0_1 class of two-valued d.n.c. functions from Fact 1.8.31. The proof of Gács (1986) used martingales. We will apply his method in Lemma 7.5.6.

Exercises.

3.3.3.$^\diamond$ Show that given an effective listing $(P^e)_{e\in\mathbb{N}}$ of Π^0_1 classes, one may effectively obtain a constant $c \in \mathbb{N}$ such that $\lambda(P^e \cap Q) \leq 2^{-K(e)-c} \;\rightarrow\; P^e \cap Q = \emptyset$.

Hint. Use Theorem 2.2.17 and Remark 2.2.21.

3.3.4. Show that there is $c \in \mathbb{N}$ such that if x is on Q then $\lambda(Q \cap [x]) \geq 2^{-K(x)-c}$.

Now let $h(n) = 2\log(n) + n + c_K$, where by Proposition 2.2.8 the constant c_K is chosen so that $\forall x\,[K(x) < h(|x|)]$. Then $\lambda([x] \cap Q) > 2^{-h(|x|)}$ for every x on Q. Hence there are distinct strings $y_0, y_1 \succ x$ on Q such that $|y_0| = |y_1| = h(|x|)$. Use this to define $(x_\tau)_{\tau \in \{0,1\}^*}$ on Q as before, where $|x_\tau| = h^{(|\tau|)}(0)$.

Autoreducibility and indifferent sets ⋆

A is called *autoreducible* (Trahtenbrot, 1970) if there is a Turing functional Φ such that

$$\forall x\,[A(x) = \Phi(A - \{x\}; x)]. \tag{3.5}$$

Thus one can determine $A(x)$ via queries to A itself, but distinct from x. Intuitively, A is redundant. For example, each set $Y \oplus Y$ is autoreducible via the functional Φ defined by $\Phi(A; 2n + a) = A(2n + 1 - a)$ ($n \in \mathbb{N}, a \in \{0, 1\}$). Thus each many-one degree contains an autoreducible set.

Autoreducibility is a bit like the following hat game. Players sit around a table. Each one has to determine the color of his hat. A player cannot see his own hat, only the hats of the other people. He has to derive his hat color from this information and some known assumptions about the distribution of hat colors.

3.3.5 Proposition. *Some low c.e. set A is not autoreducible.*

Proof sketch. To ensure that A is not autoreducible, we meet the requirements

$$P_e\colon\ \exists x\, \neg A(x) = \Phi_e(A - \{x\}; x).$$

The P_e-strategy is as follows.

(1) Choose a large number x.

(2) When $\Phi_e(A - \{x\}; x) = 0$ enumerate x into A and initialize the strategies for weaker priority requirements P_i, $i > e$. The initialization is an attempt to preserve the computation $\Phi_e(A - \{x\}; x) = 0$.

To make A low we satisfy the usual lowness requirements (1.10) on page 32. This merely needs some extra initialization of the P_e strategies. □

In the following, for a string σ, $i < |\sigma|$, and $h \in \{0, 1\}$, we let $\sigma[i \leftarrow h]$ denote the string where the bit at position i has been changed to h. Below, this notation will be extended in the obvious way to sets Z instead of strings, and to changes of several bits.

Intuitively, being random is opposite to being redundant. Martin-Löf random sets live up to our expectations here by the following fact due to Figueira, Miller and Nies (2009).

3.3.6 Proposition. *No ML-random set is autoreducible.*

Proof. Given a Turing functional Φ, we define a ML-test $(V_m)_{m \in \mathbb{N}}$ in such a way that any set that is autoreducible via Φ fails the test. (In fact a set Z fails the test already if $Z(x) = \Phi(Z - \{x\}; x)$ for infinitely many x.)

Let $V_m = [S_m]^{\prec}$ where $S_0 = \{\varnothing\}$ and, for $m \geq 1$, S_m is the set of minimal strings in

$$\{\sigma\colon \exists x_0 < \ldots < x_m [x_0 = 0 \ \&\ x_m = |\sigma| \ \&$$
$$\forall i < m\, [\Phi_{|\sigma|}(\sigma[x_i \leftarrow 0] \upharpoonright_{x_{i+1}}; x_i) = \sigma(x_i)]]\}.$$

Thus, for each $i < m$, Φ computes $\sigma(x_i)$ using $\sigma \upharpoonright_{x_{i+1}}$ as an oracle, but substituting the answer 0 if the query is $\sigma(x_i)$. Note that $(S_m)_{m\in\mathbb{N}}$ is uniformly c.e., and for each $m \geq 0$ and $\sigma \in S_m$ we have $\lambda(S_{m+1} \mid \sigma) \leq 1/2$. Thus

$$\lambda S_{m+1} = \textstyle\sum_{\sigma\in S_m} 2^{-|\sigma|}\lambda(S_{m+1} \mid \sigma) \leq (\lambda S_m)/2,$$

whence $\lambda S_m \leq 2^{-m}$ for each m. □

Figueira, Miller and Nies (2009) proved that for each ML-random set Z there is an infinite set $I \subseteq \mathbb{N}$ such that Z remains ML-random when some bits with position in I are changed. For a class $\mathcal{C}$, a set $Z \in \mathcal{C}$ and a further set I, we say that I is *indifferent* for Z with respect to $\mathcal{C}$ if each set Y that agrees with Z on $\mathbb{N} - I$ is in $\mathcal{C}$. We want to show that for every ML-random Z there is an infinite set I that is indifferent for Z with respect to being ML-random. To do so, we choose b such that $Z \in P = 2^{\mathbb{N}} - \mathcal{R}_b$ and apply the following result.

3.3.7 Theorem. *Let P be a Π^0_1 class and suppose $Z \in P$ is not autoreducible. Then there is an infinite set $I \leq_T Z'$ that is indifferent for Z with respect to P.*

Proof. Recall from page 48 that for a closed set Q and a string x, we say that x is on Q if $[x] \cap Q \neq \emptyset$. First let us show that there is a number n such that the singleton set $\{n\}$ is indifferent for Z with respect to P. Assume not, then, for each x, one of Z and $Z[x \leftarrow 1 - Z(x)]$ (the set where the bit in position x has been changed) is not in P. This allows us to compute $Z(x)$ from $Z - \{x\}$ as follows: search for $s > x$ such that $Z[x \leftarrow 1] \upharpoonright_s \notin P_s$ or $Z[x \leftarrow 0] \upharpoonright_s \notin P_s$. In the first case output 0, in the second case 1.

An infinite set $I = \{n_0 < n_1 < \ldots\}$ that is indifferent for Z with respect to P can now be determined recursively. Suppose $k \geq 0$ and we already have an indifferent set $\{n_0 < \ldots < n_k\}$. Then Z is a member of the Π^0_1 class

$$Q_k = \{Y : Y \upharpoonright_{n_k+1} = Z \upharpoonright_{n_k+1} \ \&\ \forall a_0, \ldots, a_k \in \{0,1\}\, Y[n_0 \leftarrow a_0, \ldots, n_k \leftarrow a_k] \in P\}.$$

By the argument above let n_{k+1} be an indifferent number for Z with respect to Q_k. Then $n_{k+1} > n_k$ since all the sets $Y \in Q_k$ extend $Z \upharpoonright_{n_k+1}$.

To see that the whole set I is indifferent for Z with respect to P we use that P is closed: suppose Y is obtained from Z by replacing the bit $Z(n_i)$ by a_i. For each k, the set $Y_k = Z[n_0 \leftarrow a_0, \ldots, n_k \leftarrow a_k]$ is in P, and the distance $d(Y_k, Y)$ (see Exercise 1.8.7) is at most $2^{-n_{k+1}}$. Thus $Y \in P$.

Finally, we verify that $I \leq_T Z'$: let $Q_{-1} = P$. To compute $n_0, n_1, \ldots$ recursively from Z', note that if $k \geq -1$, then n_{k+1} is the least n such that

$$\forall s\, (Z[n \leftarrow 0] \upharpoonright_s \in Q_{k,s} \ \&\ Z[n \leftarrow 1] \upharpoonright_s \in Q_{k,s}).$$

Hence n_{k+1} can be computed from an index for the Π^0_1 class Q_k using Z' as an oracle. Next, from n_{k+1} we may find an index for Q_{k+1} using Z as an oracle. □

3.4 Martin-Löf randomness relative to an oracle

Most of the concepts introduced in Chapter 2, and so far in this chapter, are ultimately defined in terms of computations, and hence can be viewed relative to an oracle. In this section we study these concepts in relativized forms. For instance, we interpret the definition of ML-randomness in 3.2.1 relative to an oracle A:

(i) A *ML-test* relative to A is a sequence $(G_m^A)_{m\in\mathbb{N}}$ of uniformly c.e. relative to A open sets such that $\forall m \in \mathbb{N}\ \lambda G_m^A \le 2^{-m}$.
(ii) $Z \subseteq \mathbb{N}$ *fails* the test if $Z \in \bigcap_m G_m^A$.
(iii) Z is *ML-random relative to* A, or *ML-random in* A, if Z passes each ML-test relative to A. MLR^A denotes the class of sets that are ML-random relative to A.

Note that MLR^A is conull and $\mathsf{MLR}^A \subseteq \mathsf{MLR}$ for each A. More generally, for each A, B we have $B \le_T A \ \rightarrow\ \mathsf{MLR}^B \supseteq \mathsf{MLR}^A$: the stronger the oracle is computationally, the stronger are the tests, and therefore the harder for a set to escape them.

In this section we study relative randomness as a binary relation between sets Z and A. An important fact is the symmetry of relative randomness, Theorem 3.4.6. Moreover, in Theorem 5.1.22 of Section 5.1 we consider the situation that some set Z is ML-random in A and also $Z \ge_T A$. We show that this is a strong lowness property of a set A.

In Section 3.6 we fix A; mostly A will be $\emptyset^{(n-1)}$ for some $n > 0$:

3.4.1 Definition. Let $n > 0$. A set Z is called *n-random* if Z is ML-random relative to $\emptyset^{(n-1)}$.

Thus, 1-randomness is the same as ML-randomness. Recall that 1-random sets may fail to match our intuition of randomness because of facts like the Kučera-Gács Theorem 3.3.2, or the existence of a left-c.e. ML-random set. In Section 3.6 we will see that both concerns are no longer valid for 2-randomness.

We can also fix a ML-random set Z and consider the class of oracles A such that Z is ML-random in A. For $Z = \Omega$ this yields the lowness property of being low for Ω, defined in 3.6.17 and studied in Section 8.1.

Relativizing C and K

A Turing functional viewed as a partial map $M : 2^{\mathbb{N}} \times \{0,1\}^* \to \{0,1\}^*$ is called an *oracle machine*. Thus, M is an oracle machine if there is $e \in \mathbb{N}$ such that $M(A, \sigma) \simeq \Phi_e^A(\sigma)$ for each oracle A and string σ. We extend Definition 2.1.2, namely, we let $\mathbb{V}^A(0^{e-1}1\rho) \simeq \Phi_e^A(\rho)$, for each set A, each $e > 0$, and $\rho \in \{0,1\}^*$. We write $C^A(x)$ for the length of a shortest string σ such that $\mathbb{V}^A(\sigma) = x$.

3.4.2 Definition. (i) A *prefix-free oracle machine* is an oracle machine M such that, for each set A, the domain of M^A is prefix-free. We let $\Omega_M^A = \lambda(\operatorname{dom} M^A)$. $K_{M^A}(x)$ denotes the length of a shortest string σ such that $M^A(\sigma) = x$.
(ii) A prefix-free oracle machine R is *optimal* if for each prefix-free oracle machine M there is a constant e_M such that $\forall A \forall x\ [K_{R^A}(x) \le K_{M^A}(x) + e_M]$. Thus, the coding constant is independent of the oracle.

To show that an optimal prefix-free oracle machine exists, we view the proof of Proposition 2.2.7 relative to an oracle A. We may in fact relativize Theorem 2.2.9, using the Recursion Theorem in its version 1.2.10. Thus, there is an optimal prefix-free oracle machine $\mathbb{U}$ such that, where $K^A(x) = K_{\mathbb{U}^A}(x)$, we have $\forall x[K^A(x) \leq K^A(|x|) + |x| + 1]$. We let

$$\Omega^A = \lambda[\operatorname{dom} \mathbb{U}^A]^{\prec}. \tag{3.6}$$

The notation Ω^A is short for $\Omega^A_{\mathbb{U}}$. By the definition of $\mathbb{U}$ as an oracle machine, for each prefix-free oracle machine M there is $d > 1$ such that

$$\forall X \, \forall \rho \, [M^X(\rho) \simeq \mathbb{U}^X(0^{d-1}1\rho)]. \tag{3.7}$$

We say that d is a coding constant for M (with respect to $\mathbb{U}$).

Exercises. For a string α, $C^\alpha(x)$ denotes the length of a shortest string σ such that $\mathbb{V}^\alpha(\sigma) = x$, and $K^\alpha(x)$ denotes the length of a shortest string σ such that $\mathbb{U}^\alpha(\sigma) = x$.

3.4.3. (i) Show that $C(x \mid \alpha) \leq^+ C^\alpha(x)$. (ii) Show that for each n there is α of length n such that $C^\alpha(n) \geq^+ C(n)$ (while $C(n \mid \alpha) = O(1)$).

3.4.4. Show that for each A, B such that $A \leq_T B$ we have $\exists d \forall y \, K^B(y) \leq K^A(y) + d$.

Basics of relative ML-randomness

The relativized form of Theorem 3.2.9 is:

$$Z \text{ is ML-random relative to } A \;\Leftrightarrow\; \exists b \forall n \; K^A(Z \upharpoonright_n) > n - b.$$

In other words, $(\mathcal{R}_b^A)_{b \in \mathbb{N}}$ is a universal ML-test relative to A, where $\mathcal{R}_b^A = [\{x \in \{0,1\}^* : K^A(x) \leq |x| - b\}]^{\prec}$. We obtain examples of sets that are ML-random relative to A by relativizing the results on page 108. In particular, Ω^A defined in (3.6) is ML-random in A.

Fact 3.2.2 that no computable set is ML-random can be relativized:

3.4.5 Fact. *If $Z \leq_T A$ then Z is not ML-random relative to A.*

Proof. Suppose $Z = \Phi^A$ for a Turing reduction Φ. Let $G_m^A = [\Phi^A \upharpoonright_m]$, then $(G_m^A)_{m \in \mathbb{N}}$ is a ML-test relative to A and $Z \in \bigcap_m G_m^A$. □

Symmetry of relative Martin-Löf randomness

Perhaps the most important fact on relative ML-randomness as a relation on sets is the theorem of van Lambalgen (1987) that $A \oplus B$ is ML-random $\Leftrightarrow$ B is ML-random and A is ML-random relative to B. By Proposition 3.2.16 $A \oplus B$ is ML-random $\Leftrightarrow$ $B \oplus A$ is ML-random. So, we also have that $A \oplus B$ is ML-random $\Leftrightarrow$ A is ML-random and B is ML-random relative to A. Thus, relative ML-randomness is a symmetric relationship between sets A and B that are both ML-random: *A is ML-random in B $\Leftrightarrow$ B is ML-random in A*. This is surprising because on the left side we consider the randomness aspect of A and on the right side its computational power.

3.4.6 Theorem. *Let $A, B \subseteq \mathbb{N}$. Then $A \oplus B$ is ML-random $\Leftrightarrow$ B is ML-random and A is ML-random relative to B.*

Proof. $\Rightarrow$: For a string β we let $\mathcal{R}_b^\beta = [\{x \in \{0,1\}^*\colon K^\beta(x) \le |x| - b\}]^\prec$. As in the proof of Proposition 3.2.7 we have $\lambda \mathcal{R}_b^\beta \le 2^{-b}$ for each string β. (We could also use some other universal oracle ML-test here.)

If $A \oplus B$ is ML-random then B is ML-random by Proposition 3.2.16, where the computable one-one function f is given by $f(n) = 2n+1$.

Suppose that A is not ML-random in B. Then $A \in \bigcap_b \mathcal{R}_b^B$. We show that $A \oplus B \in \bigcap_b G_b$ for some ML-test $(G_b)_{b\in\mathbb{N}}$. For each $b, n \in \mathbb{N}$ let

$$G_b(n) = [\{u \oplus \beta\colon |u| = |\beta| = n \ \&\ \exists x \preceq u\, K^\beta(x) \le |x| - b\}]^\prec.$$

Then $G_b(n)$ is a c.e. open set uniformly in b and n, and

$$\lambda G_b(n) \le \textstyle\sum_\beta 2^{-n} \lambda \mathcal{R}_b^\beta \,[\![|\beta| = n]\!] \le 2^{-b}.$$

Clearly $G_b(n) \subseteq G_b(n+1)$ for each n. Let $G_b = \bigcup_n G_b(n)$, then $(G_b)_{b\in\mathbb{N}}$ is a ML-test. If $A \in \bigcap_b \mathcal{R}_b^B$ then for each b there is $x \prec A$ such that $K^B(x) \le |x| - b$, and thus $K^{B\restriction_n}(x) \le |x| - b$ for some $n \ge |x|$. Then $A \oplus B \in G_b(n)$.

$\Leftarrow$: Suppose $A \oplus B$ is not ML-random, then $A \oplus B \in \bigcap_d V_d$ for a ML-test $(V_d)_{d\in\mathbb{N}}$ such that $\lambda V_d \le 2^{-2d}$ for each d. Firstly, we build a Solovay test $(S_d)_{d\in\mathbb{N}}$ in an attempt to show that B is not ML-random. For a string x let $[\emptyset \oplus x]$ denote the clopen set $\{Y_0 \oplus Y_1\colon x \prec Y_1\}$. Let

$$S_d = \bigcup\{[x] : \lambda(V_d \cap [\emptyset \oplus x]) \ge 2^{-d-|x|}\}.$$

By Fact 1.9.16, S_d is a c.e. open set uniformly in d. We claim that $\lambda S_d \le 2^{-d}$. Let (x_i) be a listing of the minimal strings x (under the prefix relation) such that $\lambda(V_d \cap [\emptyset \oplus x]) \ge 2^{-d-|x|}$. Then $S_d = \bigcup_i [x_i]$. Since the sets $V_d \cap [\emptyset \oplus x_i]$ are pairwise disjoint and $\lambda V_d \le 2^{-2d}$, we see that $\sum_i 2^{-d-|x_i|} \le 2^{-2d}$ and hence $\lambda S_d = \sum_i 2^{-|x_i|} \le 2^{-d}$.

If $B \in S_d$ for infinitely many d then B fails the Solovay test, and hence is not ML-random. Now suppose there is d_0 such that $B \notin S_d$ for all $d \ge d_0$. Let

$$H_d(n) = [\{w : |w| = n \ \&\ [w \oplus B\restriction_n] \subseteq V_d\}]^\prec.$$

Then $\lambda H_d(n) \le 2^{-d}$ for each $d \ge d_0$ since $B \notin S_d$. Moreover $H_d(n) \subseteq H_d(n+1)$ for each n. Let $H_d = \bigcup_n H_d(n)$, then $\lambda H_d \le 2^{-d}$, and H_d is a c.e. open set relative to B uniformly in d. Since $A \in H_d$ for each $d \ge d_0$, A is not ML-random relative to B. □

Since MLR^A is conull for each A, the symmetry of relative ML-randomness shows that a ML-random set is ML-random relative to almost every set:

3.4.7 Corollary. *For ML-random A, $\lambda\{B\colon A$ is ML-random in $B\} = 1$.* □

The following is an immediate consequence of the implication "$\Rightarrow$" in Theorem 3.4.6 and Fact 3.4.5. It yields an alternative proof of the Kleene–Post Theorem 1.6.1, and shows that no ML-random set is of minimal Turing degree.

3.4.8 Corollary. *If $A \oplus B$ is ML-random then $A \mid_T B$.* □

The van Lambalgen Theorem holds for k-randomness ($k \ge 1$), as well as for the notions of Π^1_1-ML-randomness and Π^1_1-randomness introduced in Chapter 9. However, it also fails for some important randomness notions, such as Schnorr randomness (3.5.8). See Remark 3.5.22.

3.4.9.$^\diamond$ Exercise. Let $k \geq 1$. Then $A \oplus B$ is k-random $\Leftrightarrow$ B is k-random and A is ML-random relative to $B^{(k-1)}$ (that is, A is k-random relative to B).

Computational complexity, and relative randomness

We prove two results of independent interest that will also be applied later. They are due to Nies, Stephan and Terwijn (2005).

In 3.2.10 we built a (super)low ML-random set by applying the Low Basis Theorem in the version 1.8.38 to the Π^0_1 class $2^{\mathbb{N}} - \mathcal{R}_1$. The proof of 1.8.38 is an explicit construction of the low set. We may also obtain a low ML-random set A by a direct definition: take as A the bits in the odd positions of any ML-random Δ^0_2 set, say Ω. Such a set is low by the following more general fact which also applies to sets A not in Δ^0_2. Recall from Definition 1.5.4 that $\mathrm{GL}_1 = \{A\colon A' \equiv_T A \oplus \emptyset'\}$.

3.4.10 Proposition. *If some Δ^0_2 set Z is ML-random in A then A is in* GL_1.

Proof. Fix a computable approximation $(Z_t)_{t\in\mathbb{N}}$ of Z (see Definition 1.4.1). Let $f\colon \mathbb{N} \to \mathbb{N}$ be the function given by $f(r) = \mu s \forall t \geq s\, [Z_t \upharpoonright r = Z_s \upharpoonright r]$. Recall that $J^A(e) \simeq \Phi^A_e(e)$. Let $\widehat{G}_e$ be the open set, uniformly c.e. in A, given by

$$\widehat{G}_e = \begin{cases} [Z_{s_e} \upharpoonright_{e+1}] & \text{if } s_e \text{ is the stage at which } J^A(e) \text{ converges} \\ \emptyset & \text{if } J^A(e)\uparrow . \end{cases}$$

Let $G_n = \bigcup_{e\geq n} \widehat{G}_e$, then $(G_n)_{n\in\mathbb{N}}$ is a ML-test relative to A. Since $Z \notin \bigcap_n G_n$, only finitely many of the $\widehat{G}_e$ contain Z. Thus $f(e) \geq s_e$ for almost all e such that $J^A(e)\downarrow$. Hence, for almost all e, we have $J^A(e)\downarrow \leftrightarrow J^A_{f(e)}(e)\downarrow$. Since $f \leq_T \emptyset'$, this implies that $A' \leq_T A \oplus \emptyset'$. (Also see Exercise 4.1.14.) □

3.4.11 Corollary. *Suppose the Δ^0_2 set $A = A_0 \oplus A_1$ is ML-random. Then A_0 and A_1 are low.*

Proof. The Δ^0_2 set A_1 is ML-random in A_0 by Theorem 3.4.6. Thus A_0 is in GL_1, and hence A_0 is low. The same argument applies to A_1. □

In 3.4.17 we proved that $\Omega^{\emptyset'}$ is high. So a high set can be in GL_1:

3.4.12 Corollary. *$\Omega^{\emptyset'}$ is in* GL_1.

Proof. $\Omega^{\emptyset'}$ is ML-random relative to $\emptyset' \equiv_T \Omega$, so by Theorem 3.4.6 Ω is ML-random relative to $\Omega^{\emptyset'}$. Thus $\Omega^{\emptyset'}$ is in GL_1 by Proposition 3.4.10. □

Our second result will be applied, for instance, in the proof of Theorem 5.1.19.

3.4.13 Proposition. *Let B be c.e., Z be ML-random, and suppose that $\emptyset' \not\leq_T B \oplus Z$. Then Z is ML-random relative to B.*

Proof idea. Suppose Z is not ML-random relative to B. Thus $Z \in \bigcap_{d \in \mathbb{N}} \mathcal{R}_d^B$. Since $B \oplus Z \not\geq_T \emptyset'$, infinitely many numbers x enter $\emptyset'$ after a stage where Z enters $\mathcal{R}_x^B$ with B correct on the use. This allows us to convert $(\mathcal{R}_d^B)_{d \in \mathbb{N}}$ into an unrelativized ML-test $(S_d)_{d \in \mathbb{N}}$ such that Z fails this test.

Proof details. Let $\mathcal{R}_d^B[s] = [\{x\colon K_s^{B_s}(x) \leq |x| - d\}]^{\prec}$ be the approximation of $\mathcal{R}_d^B$ at stage s. Notice that $\lambda \mathcal{R}_d^B[s] \leq 2^{-d}$ for each s. An enumeration of Z into $\mathcal{R}_d^Y$ at a stage s is due to a computation $\mathbb{U}^Y(\sigma) = z$ where $z \prec Z$ converging at s, and hence has an associated use on the oracle Y. The following function is computable in $B \oplus Z$:

$$f(x) = \mu s.\, Z \in \mathcal{R}_x^B[s] \text{ with use } u \;\&\; B_s \restriction_u = B \restriction_u].$$

Because $B_s \restriction_u = B \restriction_u$ we have $Z \in \mathcal{R}^B[t]$ for all $t \geq s$ (here we need that B is c.e., not merely Δ_2^0). Let $m(x) \simeq \mu s.\, x \in \emptyset'_s$. Then $\exists^\infty x \in \emptyset'\ [m(x) \geq f(x)]$, otherwise one could compute $\emptyset'$ from $B \oplus Z$ because, for almost all x, $x \in \emptyset' \leftrightarrow x \in \emptyset'_{f(x)}$. Let $S_d = \bigcup_{x > d} \mathcal{R}_x^B[m(x)]$. The sequence $(S_d)_{d \in \mathbb{N}}$ is uniformly c.e., and $\mu S_d \leq 2^{-d}$. Also, $Z \in \bigcap_d S_d$ because $m(x) \geq f(x)$ for infinitely many x. This contradicts the assumption that Z is ML-random. □

3.4.14 Exercise. In 3.4.10, if in addition Z is ω-c.e. and A is c.e., then A is superlow.

3.4.15.$^\diamond$ **Problem.** Show that if $Z = X \oplus Y$ is ML-random and $\emptyset' \leq_{wtt} Z$, then X is not superlow. In particular, the sequence of bits of Ω_R in the even positions is low but not superlow for any optimal prefix-free machine R. (The solution to this problem is due to Greenberg and Stephan, 2010.)

The halting probability Ω relative to an oracle ⋆

Recall from (3.6) that $\Omega^A = \lambda[\text{dom}\,\mathbb{U}^A]^{\prec}$. We study the operator $2^{\mathbb{N}} \to \mathbb{R}$ given by $A \mapsto \Omega^A$. The results are from Downey, Hirschfeldt, Miller and Nies (2005). We begin with some observations on the computational complexity of Ω^A.

3.4.16 Fact. *$A' \equiv_T A \oplus \Omega^A$ for each set A.*

Proof. By Proposition 3.2.30 $\emptyset' \equiv_T \Omega$, so $A' \leq_T A \oplus \Omega^A$ by relativization. On the other hand $A' \geq_T A \oplus \Omega^A$ since Ω^A is left-c.e. relative to A. □

3.4.17 Proposition. *If a Δ_2^0 set A is high then Ω^A is high.*

Proof. By the foregoing fact, $\emptyset'' \leq_T A' \equiv_T A \oplus \Omega^A \leq_T (\Omega^A)'$. □

Recall from 3.4.2 that $\Omega_M^A = \lambda(\text{dom}\,M^A)$ for a prefix-free oracle machine M. For a string σ we let $\Omega_M^\sigma = \lambda(\text{dom}\,M^\sigma)$. Some facts below hold for all the operators Ω_M, while at other times it is crucial that the prefix-free machine be optimal.

3.4.18 Proposition. *For each prefix-free oracle machine M, the real numbers $r_0 = \inf\{\Omega_M^X\colon X \in 2^{\mathbb{N}}\}$ and $r_1 = sup\{\Omega_M^X\colon X \in 2^{\mathbb{N}}\}$ are left-c.e.*

Proof. For any rational $q \in [0,1]$ we have $q < r_0 \leftrightarrow$ the Π_1^0 class $\{X\colon \Omega_M^X \leq q\}$ is empty, which is a Σ_1^0 property of q by Fact 1.8.28. Next, $q < r_1 \leftrightarrow \exists \rho\, [q < \Omega_M^\rho]$, which is a Σ_1^0 property of q as well. □

In Theorem 8.1.2 we will show that the infimum r_0 is assumed; in fact, there is a left-Σ^0_2 set A such that $\Omega^A_M = r_0$. Downey, Hirschfeldt, Miller and Nies (2005) proved in their Corollary 9.5 that for $M = \mathbb{U}$ the supremum is assumed by a left-Σ^0_2 set as well. In general, operators Ω_M do not assume their supremum; see Exercise 3.4.21.

An operator $F\colon 2^{\mathbb{N}} \to \mathbb{R}$ is called *lower semicontinuous at* A if for all $\epsilon > 0$ there is $n \in \mathbb{N}$ such that $\forall X \succ A\restriction_n [F(X) > F(A) - \epsilon]$. Dually, F is called *upper semicontinuous at* A if for all $\epsilon > 0$ there is $n \in \mathbb{N}$ such that $\forall X \succ A\restriction_n [F(X) < F(A) + \epsilon]$. Clearly, F is continuous at A iff F is both lower semicontinuous and upper semicontinuous at A.

3.4.19 Fact. *Let M be a prefix-free oracle machine. Then the operator Ω_M is lower semicontinuous at every set $A \in 2^{\mathbb{N}}$.*

Proof. By the use principle, for each set A,

$$\forall \epsilon > 0\, \exists k \in \mathbb{N}\, [\Omega^A_M - \Omega^{A\restriction_k}_M < \epsilon]. \tag{3.8}$$

Hence $\Omega^A_M - \epsilon < \Omega^X_M$ for every $X \succ A\restriction_k$. □

In the following we characterize the class of sets A such that the operator $X \mapsto \Omega^X$ is continuous at A as the 1-generic sets introduced in 1.8.51.

3.4.20 Theorem. *A is 1-generic $\Leftrightarrow$ the operator $X \mapsto \Omega^X$ is continuous at A.*

Proof. $\Rightarrow$: This implication actually holds for Ω_M where M is any prefix-free oracle machine. Since Ω_M is lower semicontinuous at every set, it suffices to show that Ω_M is upper semicontinuous at A. Suppose this fails for the rational ϵ. Let $r \leq \Omega^A_M$ be a rational such that $\Omega^A_M - r < \epsilon$. For each n there is $X \succ A\restriction_n$ such that $\Omega^X_M \geq \Omega^A_M + \epsilon \geq r + \epsilon$. Thus the following c.e. open set is dense along A:

$$S = \{X\colon \exists t\, [\Omega^X_{M,t} \geq r + \epsilon]\},$$

where $\Omega^X_{M,t} = \lambda[\mathrm{dom}\, M^X_t]^{\prec}$. Hence $A \in S$. This implies $\Omega^A_M \geq r + \epsilon > \Omega^A_M$, contradiction.

$\Leftarrow$: We assume that A is not 1-generic and show that there is $\epsilon > 0$ such that $\forall n\, \exists X \succ A\restriction_n [\Omega^X \geq \Omega^A + \epsilon]$. Let the c.e. open S be dense along A but $A \notin S$. Let N^X be the prefix-free oracle machine such that $N^X(\varnothing)\downarrow$ at the first stage t such that $[X\restriction_t] \subseteq S_t$. Let $d > 1$ be the coding constant for N with respect to $\mathbb{U}$ according to (3.7), and let $\epsilon = 2^{-d-1}$. Choose k as in (3.8) for ϵ and $M = \mathbb{U}$. Since N^A is nowhere defined, $\mathbb{U}^{A\restriction k}(0^{d-1}1\rho)\uparrow$ for each string ρ. On the other hand, since S is dense along A, for each $n \geq k$ there is a set $X \succ A\restriction_n$ such that $X \in S$. Then $N^X(\varnothing)\downarrow$, so $\Omega^X \geq \Omega^{A\restriction k} + 2\epsilon \geq \Omega^A + \epsilon$. □

Since the implication from left to right holds for every prefix-free oracle machine M, we have also proved that A is 1-generic $\Leftrightarrow$ for any prefix-free oracle machine M, Ω^X_M is continuous at A. We call a prefix-free oracle machine R uniformly optimal if for each prefix-free oracle machine N there is a fixed α such that $\forall X\, \forall \rho\, [N^X(\rho) \simeq R^X(\alpha\rho)]$. The implication from right to left in 3.4.20 uses that $\mathbb{U}$ is uniformly optimal.

Theorems 5.5.14 and 8.1.2 provide further results about the operators Ω_M.

Exercises.

3.4.21. Give an example of a prefix-free oracle machine M such that $r_1 = \sup\{\Omega_M^X : X \in 2^{\mathbb{N}}\}$ is not assumed.

3.4.22. Show that for a prefix-free oracle machine M, if $\Omega_M^A = \sup\{\Omega_M^X : X \in 2^{\mathbb{N}}\}$, then the operator Ω_M is continuous at A.

3.4.23. Show that the ML-random real $1 - \Omega$ is not of the form Ω^A for any set A.

3.5 Notions weaker than ML-randomness

On page 104 we discussed two criticisms of the notion of ML-randomness.

1. Schnorr (1971) maintained that Martin-Löf tests are too powerful to be considered algorithmic. He suggested to study randomness notions *weaker* than ML-randomness.
2. ML-random sets can be left-c.e., and each set $Y \geq_T \emptyset'$ is Turing equivalent to a ML-random set. This is not consistent with our intuition on randomness. Thus, from an opposite point of view it also makes sense to consider randomness notions *stronger* than ML-randomness.

In this and the next section we vary the notion of a ML-test in order to introduce randomness notions that address these criticisms. Table 3.1 summarizes the three main variants of the concept of a ML-test, and names the corresponding randomness notions. Each time, the tests are u.c.e. sequences of open sets $(G_m)_{m\in\mathbb{N}}$ such that $\bigcap_m G_m$ is a null class, possibly with some extra conditions on the effectivity of a presentation, and how fast λG_m converges to 0.

TABLE 3.1. Variants of the concept of ML-test. Throughout $(G_m)_{m\in\mathbb{N}}$ is a uniformly c.e. sequence of open sets.

Test notion	Definition	Randomness notion
Kurtz test	$(G_m)_{m\in\mathbb{N}}$ is an effective sequence of clopen sets such that $\lambda G_m \leq 2^{-m}$	(weakly random)
Schnorr test	$\lambda G_m \leq 2^{-m}$ is a computable real uniformly in m	Schnorr random
Martin-Löf test	$\forall m\, \lambda G_m \leq 2^{-m}$	Martin-Löf random
Generalized ML-test	$\bigcap_m G_m$ is a null class	weakly 2-random

As before, we say that a set Z *fails* the test if $Z \in \bigcap_m G_m$. Otherwise Z *passes* the test. Schnorr tests are the Martin-Löf tests where λG_m is computable uniformly in m. The corresponding randomness notion is called Schnorr randomness. We begin with the most restricted test concept, Kurtz tests, where the sets G_m are clopen sets given by a strong index for a finite set. The corresponding property will be called weak randomness.

The implications between the notions are

ML-random ⇒ Schnorr random ⇒ weakly random.

The converse implications fail.

Weak randomness

Clopen sets are given by strong indices for finite sets, as explained before 1.8.6.

3.5.1 Definition. A *Kurtz test* is an effective sequence $(G_m)_{m\in\mathbb{N}}$ of clopen sets such that $\forall m\, \lambda G_m \le 2^{-m}$. Z is *weakly random* if it passes each Kurtz test.

Kurtz tests are equivalent to null Π^0_1 classes in a uniform way. Thus, a set is weakly random if and only if it avoids all null Π^0_1 classes.

3.5.2 Fact.

(i) If $(G_m)_{m\in\mathbb{N}}$ is a Kurtz test then $P = \bigcap_m G_m$ is a null Π^0_1 class.
(ii) If P is a null Π^0_1 class then $P = \bigcap_m G_m$ for some Kurtz test $(G_m)_{m\in\mathbb{N}}$.

Proof. (i) The tree $\{x\colon \forall m\ [x] \cap G_m \neq \emptyset\}$ has a c.e. complement in $\{0,1\}^*$. Thus P is a Π^0_1 class (see Definition 1.8.19).
(ii) The test obtained in 3.2.2 is a Kurtz test as required. □

A weakly random set Z is incomputable, otherwise Z would be a member of the null Π^0_1 class given in 3.2.2.

3.5.3 Remark. Weak randomness behaves differently depending on whether the set is of hyperimmune degree or computably dominated.

(i) Each hyperimmune degree contains a weakly 1-generic set by 1.8.50. The law of large numbers (see 3.2.12) may fail for such a set. Weak 1-genericity implies weak randomness. Thus, each hyperimmune degree contains a weakly random set that is far from being random in the intuitive sense.
(ii) If a weakly random set is computably dominated then this set is ML-random, and in fact weakly 2-random.

We discuss (i) here, and postpone (ii) to Proposition 3.6.4.

3.5.4 Fact. *Each weakly 1-generic set is weakly random.*

Proof. Each conull open set D is dense, for otherwise $D \cap [x] = \emptyset$ for some string x, whence $\lambda D \le 1 - 2^{-|x|} < 1$. By the definition, a set G is weakly 1-generic iff G is in every dense c.e. open set. So no weakly 1-generic set is in a null Π^0_1 class. □

3.5.5 Proposition. *The law of large numbers fails for every weakly 1-generic set Z. In fact,* $\liminf_n (\#\{i < n : Z(i) = 1\}/n) = 0$.

Proof. Given $k > 0$, consider the function $f(m) = (k-1)m$. As in the proof of Proposition 1.8.48 let D_f be the dense c.e. open set $[\{\sigma 0^{f(|\sigma|)}\colon \sigma \neq \varnothing\}]^{\prec}$. There is a string σ of length m such that $\sigma 0^{(k-1)m} \prec Z$, so for $n = km$ we have $\#\{i < n : Z(i) = 1\}/n \le 1/k$. □

A similar argument shows that $\limsup_n (\#\{i < n : Z(i) = 1\}/n) = 1$. Thus the number of occurrences of zeros and ones is highly unbalanced. We conclude that weak randomness is indeed too weak to be considered a genuine mathematical randomness notion.

Exercises.

3.5.6. Recall from Theorem 1.8.49 that a weakly 1-generic set can be left-c.e. However, it is far from being c.e.: show that no c.e. set B is weakly random.

3.5.7. We say that Z is *ranked* if Z is a member of a countable Π^0_1 class. (Such a set is far from even being weakly random.) Use Theorem 3.3.7 to show that each ranked set is autoreducible.

Schnorr randomness

Even though Schnorr randomness is a weaker notion, its theory parallels the theory of Martin-Löf randomness. Firstly we study Schnorr tests. Secondly we consider computable measure machines. They are the prefix-free machines with a computable halting probability. We show that each Schnorr test can be emulated by a computable measure machine, which leads to a characterization of Schnorr randomness in terms of the growth of initial segment complexity similar to Theorem 3.2.9. We use this to extend statistical properties of ML-random sets, such as the law of large numbers, to the case of Schnorr random sets.

A main difference between Martin-Löf randomness and Schnorr randomness is that, for the latter, there is no universal test.

Schnorr tests. If $(G_m)_{m\in\mathbb{N}}$ is a universal ML-test, then for each m the Π^0_1 class $P = 2^{\mathbb{N}} - G_m$ is contained in MLR, whence λP is ML-random by Theorem 3.2.35. Thus λG_m is left-c.e. but not computable. Schnorr (1971) introduced a more restricted test concept:

3.5.8 Definition. A *Schnorr test* is a ML-test $(G_m)_{m\in\mathbb{N}}$ such that λG_m is computable uniformly in m. A set $Z \subseteq \mathbb{N}$ *fails* the test if $Z \in \bigcap_m G_m$, otherwise Z *passes* the test. Z is *Schnorr random* if Z passes each Schnorr test.

Each Kurtz test is a Schnorr test, so each Schnorr random set is weakly random. A computable set is not weakly random, and hence not Schnorr random. On the other hand, each Schnorr test $(G_m)_{m\in\mathbb{N}}$ is passed by a computable set: for instance, the Π^0_1 class $2^{\mathbb{N}} - G_1$ contains a computable set by Exercise 1.9.21. We conclude:

3.5.9 Fact. *There is no universal Schnorr test.* □

If λG_m is computable uniformly in m then we have better information about the test than for a ML-test in general, but it does not mean that we "know" the components G_m. There is a c.e. open set R such that λR is computable while $A_R = \{\sigma\colon [\sigma] \subseteq R\}$ is not computable; see Example 1.9.17. The only tests where we know everything are the Kurtz tests, but they do not determine a randomness notion.

By the following we may relax the failure condition in Definition 3.5.8 to the failure condition for a Solovay test that $Z \in G_i$ for infinitely many i.

3.5.10 Fact. *If $\exists^\infty i\,[Z \in G_i]$ for some Schnorr test $(G_i)_{i\in\mathbb{N}}$ then Z is not Schnorr random.*

Proof. Let $\widehat{G}_m = \bigcup_{i>m} G_i$. We show that $(\widehat{G}_m)_{m\in\mathbb{N}}$ is a Schnorr test. Clearly $\lambda\widehat{G}_m \le 2^{-m}$. To see that $\lambda\widehat{G}_m$ is uniformly computable, we apply Fact 1.8.15(iii): given $m, r \in \mathbb{N}$, we compute a rational that is within 2^{-r+1} of $\lambda\widehat{G}_m$. Let $\widehat{G}_{m,s} = \bigcup_{i>m} G_{i,s}$. For each i we may compute s_i such that $\lambda G_i - \lambda G_{i,s_i} \le 2^{-i-r}$. Let $t = \max\{s_i \colon i \le m + r\}$, then

$$\begin{aligned}\lambda\widehat{G}_m - \lambda\widehat{G}_{m,t} &\le \lambda \bigcup_{i>m}(G_i - G_{i,t}) \\ &\le \sum_{i=m+1}^{m+r} 2^{-i-r} + \sum_{i=m+r+1}^{\infty} 2^{-i} \le 2^{-r+1}.\end{aligned}$$

Then Z is not Schnorr random since $Z \in \bigcap_m \widehat{G}_m$. □

On the other hand, by the uniformity of Lemma 1.9.19 we may turn a Schnorr test into one where the measure of the m-th open set is not only uniformly computable, but in fact it *equals* 2^{-m}.

3.5.11 Fact. *For each Schnorr test $(G_m)_{m\in\mathbb{N}}$, one may effectively find a Schnorr test $(\widetilde{G}_m)_{m\in\mathbb{N}}$ such that $\forall m\, G_m \subseteq \widetilde{G}_m$ and $\forall m\, \lambda\widetilde{G}_m = 2^{-m}$.* □

The analog of Proposition 3.2.16 holds by the same proof.

3.5.12 Proposition. *Suppose f is a computable one-one function. If Z is Schnorr random then so is $f^{-1}(Z)$.* □

High degrees and Schnorr randomness. We postpone the proof that there is a Schnorr random, but not ML-random set to Section 7.3, where we actually separate computable randomness from ML-randomness. Computable randomness is a notion defined in terms of computable betting strategies (martingales) which lies properly in between Schnorr and ML-randomness. We will prove in Section 7.3 that each high Turing degree contains a computably random set, and that each c.e. high degree contains a left-c.e. computably random set. If this c.e. degree is Turing incomplete then the set cannot be ML-random by Theorem 4.1.11 below.

On the other hand, if a set is not high, then Schnorr randomness is equivalent to ML-randomness by a result of Nies, Stephan and Terwijn (2005). This is an instance of the heuristic principle that, as we lower the computational complexity of sets, randomness notions tend to coincide. Such an interaction from the computational complexity of sets towards randomness already occurred in Remark 3.5.3(ii), that weak randomness coincides with weak 2-randomness for computably dominated sets.

3.5.13 Proposition. *If Z is Schnorr random and not high, then Z is already Martin-Löf random.*

Proof. Suppose that $(G_m)_{m\in\mathbb{N}}$ is a ML-test such that $Z \in \bigcap_m G_m$. Then the function f given by $f(m) \simeq \mu s.\, Z \in G_{m,s}$ is total. Since $f \le_T Z$ and Z is not high, by Theorem 1.5.19 there is a computable function h not dominated by f, namely, $\exists^\infty m\, h(m) > f(m)$. Let $S_m = G_{m,h(m)}$, then $(S_m)_{m\in\mathbb{N}}$ is a Schnorr (even Kurtz) test such that $Z \in S_m$ for infinitely many m. Then Z is not Schnorr random by Fact 3.5.10. □

Computable measure machines

Recall from (2.4) on page 84 that $\Omega_M = \lambda[\operatorname{dom} M]^{\prec} = \sum_\sigma 2^{-|\sigma|}\,[\![M(\sigma)\downarrow]\!]$ is the halting probability of a prefix-free machine M. Downey and Griffiths (2004) introduced a restricted type of prefix-free machines.

3.5.14 Definition. A prefix-free machine M is called a *computable measure machine* if Ω_M is computable.

(They used the term "computable machines", but we prefer the present term because a machine is the same as a partial computable function.) Similar to Theorem 3.2.9, they characterized Schnorr randomness by a growth condition on the initial segment complexity. In the case of Schnorr randomness one needs an infinite collection of descriptive complexity measures for strings, namely all the functions K_M for a computable measure machine M.

Recall that we approximate $K_M(x)$ by $K_{M,s}(x) := \min\{|\sigma|\colon\ M_s(\sigma) = x\}$. For a computable g such that $\forall n\, g(n) \ge n$ we have the time bounded version of K_M given by $K^g_M(x) = K_{M,g(|x|)}(x)$ (similar to the definition of C^g on page 81).

3.5.15 Proposition. *Suppose that M is a computable measure machine with range $\{0,1\}^*$. Then the function K_M is computable. In fact, there is a computable function g such that $K_M = K^g_M$.*

Proof. Since Ω_M is computable, one can determine $K_M(x)$ as follows: compute the least stage $s = s_x$ such that $\Omega_M - \Omega_{M,s} \le 2^{-n}$, where $n = K_{M,s}(x) < \infty$. If a description of length less than n appears after stage s, then this causes an increase of Ω_M by at least 2^{-n+1}. Thus $K_{M,s}(x) = K_M(x)$.

If we let $g(m) = \max(\{m\} \cup \{s_x\colon\ |x| = m\})$ then $K_M = K^g_M$. □

Our principal tool for building prefix-free machines is the Machine Existence Theorem 2.2.17: from a given bounded request set W one obtains a prefix-free machine M for W such that Ω_M equals $\mathsf{wgt}_W(\{0,1\}^*)$, the total weight of W. If the total weight of W is computable then M is a computable measure machine.

Exercises.

3.5.16. From a computable measure machine M and Ω_M, one may effectively obtain a computable measure machine N such that $\Omega_N = 1$ and $\forall x\, K_N(x) \le K_M(x)$.

3.5.17. Let M be a computable measure machine. Show that from $r \in \mathbb{N}$ one can compute a strong index for the finite set $\{y\colon\ K_M(y) \le r\}$.

Schnorr tests can be emulated by computable measure machines

Downey and Griffiths (2004) proved the analog of Schnorr's Theorem 3.2.9: a set Z is Schnorr random $\Leftrightarrow$ for each computable measure machine M there is b such that each initial segment of Z is b-incompressible in the sense of K_M. First we need a lemma saying that the tests $(R_b^M)_{b\in\mathbb{N}}$ from Definition 3.2.6, where M is a computable measure machine, are Schnorr tests that can emulate any other Schnorr test.

3.5.18 Lemma.

(i) Let M be a computable measure machine. Then $(R_b^M)_{b\in\mathbb{N}}$ is a Schnorr test.
(ii) Let $(G_m)_{m\in\mathbb{N}}$ be a Schnorr test. Then we may effectively obtain a computable measure machine M such that $\bigcap_m G_m \subseteq \bigcap_b R_b^M$.

Proof. (i) We extend the proof of Proposition 3.2.7 that $(R_b^M)_{b\in\mathbb{N}}$ is a ML-test. It suffices to show that λR_b^M is computable uniformly in b. As before, let V_b^M be the set of strings in R_b^M which are minimal under the prefix ordering. Let $V_{b,s}^M$ be the set of minimal strings in $R_{b,s}^M$. Then $\Omega_M - \Omega_{M,s} \geq 2^b(\lambda[V_b^M]^\prec - \lambda[V_{b,s}^M]^\prec)$ for each s. Since Ω_M is computable, by Fact 1.8.15(iv) this shows that $\lambda[V_b^M]^\prec = \lambda R_b^M$ is computable uniformly in b.

(ii) We extend the proof of (ii)$\Rightarrow$(i) in Theorem 3.2.9. Let $(G_m)_{m\in\mathbb{N}}$ be a Schnorr test. We may assume that $\lambda G_m \leq 2^{-2m}$ for each m. Let L be the bounded request set defined as in the proof of Theorem 3.2.9, and let M be the machine obtained from L via the Machine Existence Theorem 2.2.17. By the construction of L, for each $Z \in G_m$ there is $x \prec Z$ such that $K_M(x) \leq |x| - m + 1$. Therefore it suffices to show that the total weight α of L is computable and hence M is a computable measure machine. Recall that the contribution of G_m to the total weight of L is at most 2^{-m-1}. In the notation of the proof of 3.2.9, let

$$L_t = \{\langle |x_i^m| - m + 1, x_i^m\rangle : m \in \mathbb{N} \;\&\; i < \min(N_m, t)\},$$

and let α_t be the total weight of L_t. Given r, similar to the proof of Fact 3.5.10, we compute t such that $\alpha - \alpha_t \leq 2^{-r+1}$: let $G_m(u)$ be the clopen set $[\{x_k^m \colon k < \min(N_m, u)\}]^\prec$. For each i, we may compute s_i such that $\lambda G_i - \lambda G_i(s_i) \leq 2^{-2i-r}$. Let $t = \max\{s_i \colon i \leq r\}$, then $\alpha - \alpha_t \leq \sum_{i\leq r} 2^{-i-r-1} + \sum_{i\geq r+1} 2^{-i-1} \leq 2^{-r+1}$. □

The analog of Theorem 3.2.9 for Schnorr randomness is now immediate.

3.5.19 Theorem. *The following are equivalent.*

(i) Z is Schnorr random.
(ii) For each computable measure machine M, $\exists b \forall n\; K_M(Z\restriction_n) > n - b$, that is, $\exists b\; Z \notin R_b^M$.

Schnorr random sets satisfy the law of large numbers

To obtain statistical properties of incompressible strings we proved Proposition 2.5.7 and Corollary 2.5.9. These results have versions for computable measure machines. They relied on Lemma 2.5.6, so suppose the given set of strings G

in the lemma is computable. In that case, the proof produces a bounded request set W with a computable total weight α. For, we showed that the weight contributed by each G_n, $n \geq 10$, is at most $(n-1)^{-2}/2$. Now $\int_r^\infty x^{-2}dx = 1/r$ and hence $\sum_{n>r+1}(n-1)^{-2} \leq 1/r$. For $r \geq 10$ let the rational α_r be the weight contributed by all the G_n, $10 \leq n \leq r+1$. Since G is computable, α_r can be computed from r, and $\alpha - \alpha_r \leq 1/(2r)$.

The sets G in the proofs of Proposition 2.5.7 and of Corollary 2.5.9 are computable. So we may sharpen these results in the sense that the descriptive complexity can be taken relative to computable measure machines:

3.5.20 Proposition. *There are computable measure machines M and N with the following properties.*

(i) If x is a string of length n with a run of $4\log n$ zeros then $K_M(x) \leq^+ n - \log n$.

(ii) Fix b. For each d, for almost every n, each string x of length n such that $K_N(x) > n - b$ satisfies $\text{abs}(S_n(x)/n - 1/2) < 1/d$. □

This leads to the desired statistical properties of Schnorr random sets.

3.5.21 Theorem. *Let Z be Schnorr random.*

(i) For each n, any run of zeros in $Z\upharpoonright_n$ has length $\leq^+ 4\log n$.

(ii) Z satisfies the law of large numbers in 3.2.12.

Proof. Let M, N be the computable measure machines from the previous proposition. By Theorem 3.5.19, there is $b \in \mathbb{N}$ such that $Z \notin R_b^M$ and $Z \notin R_b^N$. Now (i) and (ii) follow. □

3.5.22 Remark. Stronger results along these lines have been derived. For instance, the law of iterated logarithm holds for Schnorr random sets. This suggests that the randomness notion of Schnorr can be seen as a sufficient formalization of the intuitive concept of randomness as far as statistical properties are concerned. In contrast, Schnorr random sets can have computability theoretic properties that are not consistent with our intuitive notion of randomness. For instance, as Kjos-Hanssen has pointed out, there is a Schnorr random set $A = A_0 \oplus A_1$ such that $A_0 \equiv_T A_1$. To see this, let **a** be a high minimal Turing degree, which exists by the Cooper Jump Inversion Theorem (see Lerman 1983, pg. 207). By Theorem 7.5.9 let $A \in \mathbf{a}$ be Schnorr random. Let $A = A_0 \oplus A_1$, then A_0, A_1 are Schnorr random by 3.5.12, and hence incomputable. Thus $A_0 \equiv_T A_1$. (Since A_0 is not Schnorr random relative to A_1, this example also shows that van Lambalgen's Theorem 3.4.6 fails for Schnorr randomness.)

3.6 Notions stronger than ML-randomness

In this section we mainly study weak 2-randomness and 2-randomness. The implications are

$$\text{2-random} \Rightarrow \text{weakly 2-random} \Rightarrow \text{ML-random}.$$

The converse implications fail. The new notions address the second concern outlined at the beginning of Section 3.5, that for instance left-c.e. sets should not be considered random.

When one is interested in applying concepts related to randomness in computability theory, these notions are less relevant (see the beginning of Chapter 4). Examples like Ω are important for the interaction in this direction. It is at its strongest when one considers the Δ^0_2 sets, and a weakly 2-random set already forms a minimal pair with $\emptyset'$.

In the converse direction, computational complexity is essential for our understanding of weak 2-randomness and 2-randomness. Within the ML-random sets, both notions are characterized by lowness properties: forming a minimal pair with $\emptyset'$ in the former case, and being low for Ω in the latter.

An interesting alternative to deal with the second concern at the beginning of Section 3.5 is Demuth randomness, a further notion between 2-randomness and ML-randomness. It turns out to be incomparable with weak 2-randomness. In Theorem 3.6.25 we show that there is a Demuth random Δ^0_2 set, and in Theorem 3.6.26 that all Demuth random sets are in GL_1.

Weak 2-randomness

The concept of a Martin-Löf test was obtained by effectivizing Fact 1.9.9, that the null classes are the classes contained in $\bigcap_m G_m$ for some sequence of open sets $(G_m)_{m\in\mathbb{N}}$ such that $\lim_m \lambda G_m = 0$. To be a Martin-Löf-test, the sequence $(G_m)_{m\in\mathbb{N}}$ has to be uniformly c.e. and λG_m has to converge to 0 effectively. If one drops the effectivity in the second condition, a strictly stronger randomness notion is obtained.

3.6.1 Definition. A *generalized ML-test* is a sequence $(G_m)_{m\in\mathbb{N}}$ of uniformly c.e. open sets such that $\bigcap_m G_m$ is a null class. Z *fails* the test if $Z \in \bigcap_m G_m$, otherwise Z *passes* the test. Z is *weakly 2-random* if Z passes each generalized ML-test. Let W2R denote the class of weakly 2-random sets.

By Remark 1.8.58, the null Π^0_2 classes coincide with the intersections of generalized ML-tests. To be weakly 2-random means to be in no null Π^0_2 class. This is one of the conceptually simplest randomness notions we will encounter. Note that, by Proposition 1.8.60, these tests are determined by computable sets $S \subseteq \{0,1\}^*$, since each Π^0_2 class is of the form $\{X\colon \exists^\infty n\, S(X\restriction_n)\}$ for such an S.

By the following, the ML-random set Ω is not weakly 2-random.

3.6.2 Proposition. *If Z is Δ^0_2 then $\{Z\}$ is a null Π^0_2 class. In particular, Z is not weakly 2-random.*

Proof. We will turn a computable approximation $(Z_s)_{s\in\mathbb{N}}$ of Z into a generalized ML-test $(V_m)_{m\in\mathbb{N}}$ such that $\{Z\} = \bigcap_m V_m$. We may assume that Z is infinite and $Z_0(m) = 0$ for each m. To enumerate V_m, for each $s > m$, if x is least such that $Z_s(x) \neq Z_{s-1}(x)$, put $[Z_s \restriction x]$ into $V_{m,s}$. To see that $\{Z\} \supseteq \bigcap_m V_m$, note that, if $Z\restriction_k$ is stable from stage t on, then for $m > t$, all the strings enumerated into V_m extend $Z\restriction_k$. Next we show that $Z \in \bigcap_m V_m$. Given m, there is a least x such that $Z_s(x) \neq Z_{s-1}(x)$ for some $s > m$. Let s be the greatest such stage for this x. Then $Z\restriction x = Z_s\restriction x$ is put into V_m at stage s. □

3.6.3 Corollary. *There is no universal generalized ML-test.*

Proof. Let $(G_m)_{m\in\mathbb{N}}$ be a generalized ML-test, and choose m such that the Π^0_1 class $P = 2^{\mathbb{N}} - G_m$ is nonempty. By the Kreisel Basis Theorem 1.8.36, the leftmost path Z of P is left-c.e. Since Z is not weakly 2-random but passes the test, the test is not universal. □

How does one obtain a weakly 2-random set? One way is to take a weakly random computably dominated set, which exists by Theorem 1.8.42. (See Remark 3.5.3 for more details.) Each 2-random set has hyperimmune degree by Corollary 3.6.15 below, so the example we obtain is not 2-random. The result is similar to Proposition 3.5.13.

3.6.4 Proposition. *If a set Z is computably dominated and weakly random then Z already is weakly 2-random.*

Proof. If Z is not weakly 2-random then $Z \in \bigcap_m G_m$ for a generalized ML-test $(G_m)_{m\in\mathbb{N}}$. As in the proof of Proposition 3.5.13, the function f given by $f(m) = \mu s.\, Z \in G_{m,s}$ is total, and $f \le_T Z$. There is a computable function g dominating f. Let $H_m = G_{m,g(m)}$, then $(H_m)_{m\in\mathbb{N}}$ is a Kurtz test such that $Z \in \bigcap_m H_m$. Thus Z is not weakly random. □

Characterizing weak 2-randomness within the ML-random sets. We will later strengthen Proposition 3.6.2. By Exercise 1.8.65, for each Δ^0_2 set A and each Turing functional Φ the class $\{Z\colon \Phi^Z = A\}$ is Π^0_3. In Lemma 5.1.13 below we show that this class is null for incomputable A. Thus, if Z is weakly 2-random and $A \le_T Z, \emptyset'$ then A is computable. In Theorem 5.3.16 we will prove that this property characterizes the weakly 2-random sets within the ML-random sets:

$$Z \text{ is weakly 2-random} \Leftrightarrow Z \text{ is ML-random and } Z, \emptyset' \text{ form a minimal pair.}$$

As a consequence, within the ML-random sets, the weakly 2-random sets are downward closed under Turing reducibility. The proof is postponed because it relies on the cost function method of Section 5.3.

One cannot replace the condition that Z and $\emptyset'$ form a minimal pair by $Z \mid_T \emptyset'$:

3.6.5 Fact. (J. Miller) *Some ML-random set $Z \mid_T \emptyset'$ is not weakly 2-random.*

Proof. Let Ω_0 be the bits of Ω in the even positions, then Ω_0 is low by the comment after Corollary 3.4.11. Let V be a 2-random set such that $\Omega_0 \oplus V \not\ge_T \emptyset'$, which exists since $\{X\colon \Omega_0 \oplus X \not\ge_T \emptyset'\}$ is conull. Let $Z = \Omega_0 \oplus V$, then $Z \mid_T \emptyset'$, Z is ML-random by 3.4.6, and Z does not form a minimal pair with $\emptyset'$. □

On the other hand, we also cannot expect a stronger condition to characterize weak 2-randomness within MLR. For instance, each Σ^0_2 set $B >_T \emptyset'$ bounds a ML-random computably dominated set (and hence a weakly 2-random set) by Exercise 1.8.46.

Exercises. Firstly, compare weak 2-randomness with weak randomness relative to $\emptyset'$. Recall Definition 1.8.59. Each $\Pi^0_1(\emptyset')$ class is a Π^0_2 class by 1.8.67. Thus, each weakly 2-random set is weakly random relative to $\emptyset'$, which yields an alternative proof of Proposition 3.6.2. The converse fails.

3.6.6. A set that is weakly random relative to $\emptyset'$ can fail the law of large numbers. Secondly, show that no Σ^0_2 set is weakly random relative to $\emptyset'$.
3.6.7. If $B \subseteq \mathbb{N}$ is an infinite Σ^0_2 set, then $\{Z\colon B \subseteq Z\}$ is a $\Pi^0_1(\emptyset')$ null class.

3.6.8.$^\diamond$ (Barmpalias, Miller and Nies, 2011) Show that some ML-random set is weakly random in $\emptyset'$ but not weakly 2-random.

3.6.9.$^\diamond$ **Problem.** To what extent does Theorem 3.4.6 hold for weak 2-randomness? (Solved in Barmpalias, Downey and Ng (20xx).)

2-randomness and initial segment complexity

In Section 2.5 we formalized randomness for strings by incompressibility. Theorem 3.2.9 shows that Z is ML-random iff for some b, each $x \prec Z$ is b-incompressible in the sense of K. In this subsection we study 2-randomness, that is, ML-randomness relative to $\emptyset'$ (Definition 3.4.1). The main goal is a characterization based on the incompressibility of initial segments. Relativizing Schnorr's Theorem to $\emptyset'$, a set Z is 2-random iff for some b, each $x \prec Z$ is b-incompressible in the sense of $K^{\emptyset'}$. This is not very satisfying, though: we would rather like a characterization based on the incompressibility of initial segments x with respect to an *unrelativized* complexity measure such as C. One cannot require that for some b, all $x \prec Z$ are b-incompressible in the sense of C. Such sets do not exist because of the complexity dips of C in Proposition 2.2.1. Surprisingly, the weaker condition that

$(\star)$ for some b, *infinitely many* $x \prec Z$ are b-incompressible$_C$

precisely characterizes the 2-random sets. This property was studied by Li and Vitányi (1997). Each set satisfying $(\star)$ is ML-random, because each prefix of an incompressible$_C$ string is incompressible$_K$ (see 2.5.4 for the detailed form of this statement). On the other hand, Ding, Downey and Yu (2004) proved that each 3-random set satisfies $(\star)$. The equivalence of 2-randomness and $(\star)$ is now obtained by extending the arguments in both proofs in such a way that they work with 2-randomness.

Two variants of $(\star)$ will be considered using different notions of incompressibility. The first appears weaker, the second stronger.

1. Recall from page 81 that for a computable function g we defined

$$C^g(x) = \min\{\,|\sigma|\colon\ \mathbb{V}(\sigma) = x \text{ in } g(|x|) \text{ steps}\}. \tag{3.9}$$

The first variant of $(\star)$ is to require that for some b, infinitely many $x \prec Z$ are b-incompressible with respect to C^g, for an appropriate fixed g (which under an additional assumption on $\mathbb{V}$ can be chosen in $O(n^3)$). This is actually equivalent to 2-randomness of Z. Using this fact and that C^g is computable, we give a short proof of the result by Kurtz (1981) that no 2-random set is computably dominated. This shows that 2-randomness is indeed stronger than weak 2-randomness, because a weakly 2-random set can be computably dominated by Proposition 3.6.4. (By Exercise 3.6.22, there is also a weakly 2-random set of hyperimmune Turing degree that is not 2-random.)

2. For the second variant, recall from 2.5.2 that x is strongly d-incompressible$_K$ if $K(x)$ is within d of its maximum possible value $|x| + K(|x|) + 1$, and that being strongly incompressible$_K$ implies being incompressible$_C$. The second variant of $(\star)$ is to require that for some b, infinitely many $x \prec Z$ are strongly b-incompressible$_K$. J. Miller proved that, once again, this seemingly stronger condition is actually equivalent to 2-randomness of Z. See Theorem 8.1.14.

We now prove that Z is 2-random $\Leftrightarrow$ Z it satisfies $(\star)$ above. This is due to Nies, Stephan and Terwijn (2005); the implication "$\Leftarrow$" was also independently and slightly earlier obtained by Miller (2004).

3.6.10 Theorem. *Z is 2-random* $\Leftrightarrow \exists b\, \exists^\infty n\, \big[C(Z\restriction_n) > n - b\big]$.

Proof. $\Rightarrow$: Firstly, we will sketch a proof of the easier result of Ding, Downey and Yu (2004) that any 3-random set Z satisfies $(\star)$. If $(\star)$ fails then $Z \in \bigcap_b V_b$, where $V_b = \bigcup_t P_{b,t}$, and $P_{b,t} = \big\{X\colon \forall n \geq t\, \big[C(X\restriction_n) \leq n - b\big]\big\}$. Note that $\lambda P_{b,t} \leq 2^{-b+1}$ for each t. Hence $\lambda V_b \leq 2^{-b+1}$, as $P_{b,t} \subseteq P_{b,t+1}$ for each t. The class V_b is Σ^0_2. It is not open, but can be enlarged in an effective way to an open class $\widetilde{V}_b$ that is Σ^0_1 relative to C' (the function C is identified with its graph) and has at most twice the measure of V_b. Then Z fails $(\widetilde{V}_{b+2})_{b\in\mathbb{N}}$ which is a ML-test relative to $C' \equiv_T \emptyset''$.

In order to obtain a ML-test relative to $\emptyset'$ we introduce a concept of independent interest. A function $F\colon \{0,1\}^* \to \{0,1\}^*$ is called *compression function* if F is one-one and $\forall x\, [\widehat{F}(x) \leq C(x)]$, where $\widehat{F}(x) = |F(x)|$. For instance, if, for each x, $F(x)$ is a shortest $\mathbb{V}$-description of x, then F is a compression function such that $\widehat{F}(x) = C(x)$. The idea is to replace in the argument sketched above C by $\widehat{F}$ for a compression function F such that $F' \equiv_T \emptyset'$. In this way we obtain a ML-test relative to $\emptyset'$ (and not $\emptyset''$). For the duration of this proof we say that a set Z is *complex for* $\widehat{F}$ if there is $b \in \mathbb{N}$ such that $\widehat{F}(Z\restriction_n) > n - b$ for infinitely many n. If Z is complex for $\widehat{F}$ then $(\star)$ holds. We now extend the argument above to an arbitrary compression function.

3.6.11 Lemma. *Let F be a compression function. Suppose that Z is not complex for $\widehat{F}$, namely,*

$$\forall b\, \exists t\, \forall n \geq t\, \big[\widehat{F}(Z\restriction_n) \leq n - b\big]. \tag{3.10}$$

Then Z is not ML-random relative to F', the jump of F.

Subproof. We identify F with its graph $\{\langle n, m\rangle\colon F(n) = m\}$. Let

$$P_{b,t} = \{X\colon \forall n \geq t\, [\widehat{F}(X\restriction_n) \leq n - b]\},$$

then $P_{b,t}$ is a Π^0_1 class relative to F uniformly in b, t. By (3.10), $Z \in \bigcap_b V_b$ where $V_b = \bigcup_t P_{b,t}$. Note that $\lambda P_{b,t} \leq 2^{-b+1}$ for each t, because as F is one-one, there are fewer than 2^{t-b+1} strings x of length t such that $\widehat{F}(x) \leq t - b$. As $P_{b,t} \subseteq P_{b,t+1}$ for each t this implies $\lambda V_b \leq 2^{-b+1}$. For each b, t, k, using F as an oracle, we can compute the clopen set

$$R_{b,t,k} = \{X\colon \forall n\, [t \leq n \leq k \to \widehat{F}(X\restriction_n) \leq n - b]\}.$$

Since $P_{b,t} = \bigcap_k R_{b,t,k}$, using F' as an oracle, uniformly in b, on input t we can compute $k(t)$ such that

$$\lambda(R_{b,t,k(t)} - P_{b,t}) \leq 2^{-(b+t)}.$$

Let $T_b = \bigcup_t R_{b,t,k(t)}$. Then the T_b are open sets and the corresponding set of strings $\{x : [x] \subseteq T_b\}$ is $\Sigma^0_2(F)$ uniformly in b. Moreover, $V_b = \bigcup_t P_{b,t} \subseteq T_b$ and $\lambda(T_b - V_b) \leq \sum_t 2^{-(b+t)} = 2^{-b+1}$, so $\lambda T_b \leq 4 \cdot 2^{-b}$. Hence Z fails $(T_{b+2})_{b\in\mathbb{N}}$, which is a ML-test relative to F'. ◇

3.6.12 Lemma. *There is a compression function F such that $F' \leq_T \emptyset'$.*

Subproof. (This version is due to L. Bienvenu) A Π^0_1 class $P \subseteq \mathbb{N}^{\mathbb{N}}$ is called bounded if there is a computable function g such that

$$P \subseteq Paths(\{\sigma \in \mathbb{N}^* \colon \forall i < |\sigma| \left[\sigma(i) \leq g(i)\right]\}).$$

In Exercise 1.8.72, the Low Basis Theorem 1.8.37 was extended to nonempty bounded Π^0_1 classes in Baire space. Therefore it suffices to show that the nonempty class $\mathcal{C}$ of compression functions is of this kind: $\mathcal{C}$ is Π^0_1 because F is a compression function iff $\forall n\, [F\restriction_n \in R]$, where R is the computable set

$$\big\{\alpha \in \mathbb{N}^* \colon \forall i,j < |\alpha| \left[i \neq j \to \alpha(i) \neq \alpha(j)\right] \;\&\; \forall i < |\alpha| \left[|\alpha(i)| \leq C_{|\alpha|}(i)\right]\big\}.$$

Here we identify $\mathbb{N}$ with $\{0,1\}^*$ as usual. A compression function F satisfies $|F(x)| \leq C(x) \leq |x| + 1$ for each x, so $\mathcal{C}$ is bounded. ◇

To complete the proof of the implication "⇒", choose F as in Lemma 3.6.12. If Z is 2-random then Z is ML-random relative to F'. By Lemma 3.6.11 Z is complex for $\widehat{F}$, which implies ($\star$).

⇐: We assume that Z is not 2-random and show that ($\star$) fails. We define a machine that attempts to split off a prefix-free description from the input (this idea was first used in the proof of 2.4.1). For instance, there is $d \in \mathbb{N}$ such that

$$C(xy) \leq K(x) + |y| + d \tag{3.11}$$

for all strings x and y: define a machine M which on input τ looks for a decomposition $\tau = \sigma y$ such that $\mathbb{U}(\sigma)\downarrow = x$. If M finds one it outputs xy. (Proposition 2.4.3 gives a sharper bound, but is proved in the same way.) At first, we will work under the stronger hypothesis that Z is not even ML-random: given b, let $x \prec Z$ be such that $K(x) \leq |x| - b - d$. Then, for all y, we have $C(xy) \leq |x| + |y| - b$; in particular, $C(Z\restriction_n) \leq n - b$ for all $n \geq |x|$.

Under the actual hypothesis that Z is not 2-random, we know that, given b, some $x \prec Z$ has a sufficiently short $\mathbb{U}^{\emptyset'}$-description σ. Now we have to look at strings y that are so long that the computation $\mathbb{U}^{\emptyset'}(\sigma)$ is stable by stage $|y|$. The adequate variant of the inequality (3.11) is the following.

3.6.13 Lemma. *There is $d \in \mathbb{N}$ such that $\forall x\, \forall^\infty y\, [C(xy) \leq K^{\emptyset'}(x) + |y| + d]$.*

Subproof. Define a (plain) machine M as follows.

On input τ, let $t = |\tau|$ and search for a decomposition $\tau = \sigma y$ such that $\mathbb{U}^{\emptyset'}(\sigma)[t]\downarrow = x$. If such a decomposition is found, then output xy.

Let d be the coding constant for M with respect to the optimal machine $\mathbb{V}$. Suppose that σ is a shortest $\mathbb{U}^{\emptyset'}$-description of x, and let $t_0 > |\sigma|$ be least such that the computation $\mathbb{U}^{\emptyset'}(\sigma)[t_0]$ is stable. If $|y| \geq t_0$ and $\tau = \sigma y$, then σy is the only decomposition of τ that M can find, because the domain of $\mathbb{U}^{\emptyset'_t}$ is prefix-free for each t. Thus $M(\sigma y) = xy$, whence $C(xy) \leq K^{\emptyset'}(x) + |y| + d$. $\diamond$

We may now argue as before. Given b, let $x \prec Z$ be such that $K^{\emptyset'}(x) \leq |x| - b - d$. By the foregoing Lemma, for almost all y we have $C(xy) \leq |x| + |y| - b$. In particular, $C(Z\upharpoonright_n) \leq n - b$ for almost all n. Thus $(\star)$ fails. $\square$

We will improve Theorem 3.6.10. As already mentioned, using the time bounded version C^g in (3.9) for an appropriate computable g, actually *Z is 2-random iff*

$(\star)_g$ $\qquad \exists b\, \exists^\infty n\, [C^g(Z\upharpoonright_n) > n - b]$.

Clearly $(\star)_g$ is implied by $(\star)$ because $C^g(z) \geq C(z)$ for each z. Under a reasonable implementation of $\mathbb{U}$ by a Turing program we may ensure that $g(n) = O(n^3)$.

3.6.14 Lemma. (Extends Lemma 3.6.13) *There is a computable function g such that, for some $d \in \mathbb{N}$,*

$$\forall x\, \forall^\infty y\, \left[C^g(xy) \leq K^{\emptyset'}(x) + |y| + d\right]. \tag{3.12}$$

Subproof. We determine a bound on the running time of the machine M in the proof of Lemma 3.6.13. On input τ, if $n = |\tau|$, in the worst case, for each $\sigma \preceq \tau$, M has to evaluate $\mathbb{U}^{\emptyset'}(\sigma)[n]$ (which may be undefined). This takes $O(n^2)$ steps. After that, M needs $O(n)$ steps for printing xy. We assume that if no σ is found, M halts with the empty string as an output. So M runs $O(n^2)$ steps on any input of length n.

The copying machine maps each string z to itself. Let $g(n)$ be the maximum of the number of steps it takes $\mathbb{V}$ to simulate both M and the copying machine, for any input of length n. (Namely, $g(n)$ bounds the maximum number of steps it takes $\mathbb{V}(0^{e-1}1\sigma)$ to halt, where e is an index of either machine and $|\sigma| = n$.) If $K^{\emptyset'}(x) \geq |x|$ then (3.12) is satisfied via the copying machine. Otherwise, let σ be a shortest $\mathbb{U}^{\emptyset'}$-description of x. If y is sufficiently long, we have $M(\sigma y) = xy$; since $|\sigma x| \leq |xy|$ we obtain (3.12) via M. $\diamond$

The proof that each Z satisfying $(\star)_g$ is 2-random can be completed as before, using C^g in place of C. $\square$

The following result was originally obtained in a different way by Kurtz (1981).

3.6.15 Corollary. *No 2-random set Z is computably dominated.*

Proof. Choose b so that $(\star)_g$ holds for the computable function g obtained above. Since C^g is computable, the function f defined by

$$G(m) = \mu r.\, \exists G \subseteq \{0, \ldots, r\}\, \left[\#G = m \;\&\; \forall n \in G[C^g(Z\upharpoonright_n) > n - b]\right]$$

is total and computable in Z. Assume for a contradiction that a computable function h dominates f. Then Z is a path of the computable tree
$\{x: \forall m_{|x| \geq h(m)} \exists G \subseteq \{0, \ldots, |x|\} \, [\#G = m \ \& \ \forall n \in G[C^g(x \upharpoonright_n) > n - b]]\}$.
Each path of this tree satisfies $(\star)_g$, and is therefore 2-random. Since its leftmost path is left-c.e. this is a contradiction. □

In fact, Kurtz (1981) proved that each 2-random set Z is c.e. relative to some set $A <_T Z$. Thus $A <_T Z \leq_T A'$, whence Z is of hyperimmune degree by Exercise 1.5.13. For a recent proof of this stronger result see Downey and Hirschfeldt (2010).

3.6.16 Exercise. Show that there is a low compression function for K, namely a one-one function $F\colon \mathbb{N} \to \mathbb{N}$ such that $\forall x \, [|F(x)| \leq K(x)]$ and $\sum_x 2^{-|F(x)|} \leq 1$.

2-randomness and being low for Ω

The following lowness property will be studied in more detail in Section 8.1. Here it allows us to characterize the 2-random sets within the ML-random sets.

3.6.17 Definition. A *is low for* Ω if Ω is ML-random relative to A. The class of sets that are low for Ω is denoted by $\text{Low}(\Omega)$.

In Proposition 8.1.1 we will see that the class $\text{Low}(\Omega)$ does not depend on the particular choice of the optimal prefix-free machine. We make two basic observations. The second observation follows from Proposition 3.4.10.

3.6.18 Fact. *(i)* $\text{Low}(\Omega)$ *is closed downward under Turing reducibility.*
(ii) $\text{Low}(\Omega) \subseteq \text{GL}_1$. □

The proof of Corollary 3.4.12 actually shows that the 2-random set $\Omega^{\emptyset'}$ is low for Ω. The following characterization of 2-randomness is more general.

3.6.19 Proposition. *Z is 2-random $\Leftrightarrow$ Z is ML-random and low for* Ω.

Thus, within the ML-random sets, to be 2-random is equivalent to a lowness property. The same holds for weak 2-randomness, where the lowness property is to form a minimal pair with $\emptyset'$ (see page 135).

Proof. Since $\Omega \equiv_T \emptyset'$ (3.2.30), Z is 2-random $\leftrightarrow$ Z is ML-random relative to $\emptyset'$ $\leftrightarrow$ Z is ML-random relative to Ω. Since Ω is ML-random, by van Lambalgen's Theorem 3.4.6 the latter is equivalent to: Z is ML-random and Ω is ML-random relative to Z. □

By Fact 3.6.18, we have the following.

3.6.20 Corollary. *(i) The 2-random sets are closed downward under Turing reducibility within the ML-random sets. (ii) Each 2-random set is in* GL_1. □

In Theorem 8.1.18 states that each set in $\text{Low}(\Omega)$ is of hyperimmune degree. This yields yet another proof of Corollary 3.6.15.

We discuss to which extent the preceding results hold for weak 2-randomness. By Proposition 3.6.4, a weakly 2-random set can be computably dominated, so 3.6.15 fails. Corollary 3.6.20(i) holds for weakly 2-random sets, but (ii) fails:

Theorem 8.1.19 below shows that no weakly 2-random computably dominated set is in GL_1. On the other hand, by Exercise 3.6.22 there is a weakly 2-random set of hyperimmune degree that is not 2-random.

Exercises. Show the following.

3.6.21. Let A be in Δ^0_2. If Z is 2-random then $Z \triangle A$ is 2-random as well.

3.6.22.$^\diamond$ There is a weakly 2-random set R of hyperimmune degree such that no set in its Turing degree is 2-random.

3.6.23.$^\diamond$ **Problem.** Is there a characterization of weak 2-randomness via the growth of the initial segment complexity? (A partial, affirmative answer is given in (Hölzl, Kräling, Stephan and Wu, 20xx).)

Demuth randomness

The notion was introduced by Demuth (1988) in the language of analysis. His work was made known to a wider audience by Kučera. Like weak 2-randomness, Demuth randomness lies in between 2-randomness and ML-randomness. We show that a Demuth random set can be Δ^0_2, and that all Demuth random sets are in GL_1. This implies that Demuth randomness and weak 2-randomness are incomparable, even up to Turing degree: a Δ^0_2 set is not weakly 2-random; on the other hand, no ML-random set in GL_1 is computably dominated by Theorem 8.1.19, so a weakly 2-random computably dominated set is not Turing equivalent to a Demuth random set. This is an exception to the rule that the randomness notions form a linear hierarchy.

Demuth tests combine features of ML-tests and Solovay tests. We retain the condition of ML-tests that S_m be c.e. open and $\lambda S_m \leq 2^{-m}$, but we relax the uniformity condition: the c.e. index for S_m is now given by an ω-c.e. function (Definition 1.4.6), rather than by a computable function. Informally, we can change the whole open set S_m for a computably bounded number of times. We adopt the failure condition of Solovay tests. In contrast to the previous test notions, one cannot require for all tests that $S_m \supseteq S_{m+1}$ for each m. This will become apparent in the proof of Theorem 3.6.26 below.

3.6.24 Definition. A *Demuth test* is a sequence of c.e. open sets $(S_m)_{m\in\mathbb{N}}$ such that $\forall m\, \lambda S_m \leq 2^{-m}$, and there is an ω-c.e. function f such that $S_m = [W_{f(m)}]^{\prec}$. A set Z *passes* the test if $\forall^\infty m\, Z \notin S_m$. We say that Z is *Demuth random* if Z passes each Demuth test.

Each Demuth test is a Solovay test relative to $\emptyset'$, so, by Proposition 3.2.19 relative to $\emptyset'$, each 2-random set is Demuth random. On the other hand, each ω-c.e. set Z fails to be Demuth random via the test given by $S_n = [Z \upharpoonright_n]$. In particular, this applies to Ω.

The two following results are essentially due to Demuth (1988).

3.6.25 Theorem. *There is a Demuth random Δ^0_2 set Z.*

Proof. We write H_e for $[W_e]^{\prec}$. We use an auxiliary type of tests: a *special test* is a sequence of c.e. open sets $(V_m)_{m\in\mathbb{N}}$ such that $\lambda V_m \leq 2^{-2m-1}$ for each m

and there is a function $g \le_T \emptyset'$ such that $V_m = H_{g(m)}$. Z *passes* the test if $\forall^\infty m\, Z \notin V_m$. Special tests are similar to Demuth tests, but the function g is merely Δ^0_2, while the bound on the measure of the m-th component is tighter than 2^{-m}. It suffices to establish two claims: (1) there is a single special test such that each set passing it is Demuth random, and (2) for each special test there is a Δ^0_2 set that passes it.

Claim 1. *There is a special test $(V_m)_{m\in\mathbb{N}}$ such that each set Z passing the test is Demuth random.*
Z is Demuth random iff for each Demuth test $(U_m)_{m\in\mathbb{N}}$, Z passes the Demuth tests $(U_{2m})_{m\in\mathbb{N}}$ and $(U_{2m+1})_{m\in\mathbb{N}}$. Thus it suffices that $(V_m)_{m\in\mathbb{N}}$ emulate all Demuth tests $(S_m)_{m\in\mathbb{N}}$ such that $\lambda S_m \le 2^{-2m}$ for each m. By Fact 1.4.9, there is a binary function $\widetilde{q} \le_T \emptyset'$ that emulates all ω-c.e. functions. We can stop the enumeration of $H_{\widetilde{q}(e,m)}$ when it attempts to exceed the measure 2^{-2m}; thus there is $q \le_T \emptyset'$ such that $\lambda H_{q(e,m)} \le 2^{-2m}$ for each m, and $H_{q(e,m)} = H_{\widetilde{q}(e,m)}$ if already $\lambda H_{\widetilde{q}(e,m)} \le 2^{-2m}$. Now let

$$V_m = \bigcup_{e<m} H_{q(e,e+m+1)},$$

then $\lambda V_m \le \sum_{e<m} 2^{-2(e+m+1)} \le 2^{-2m-1}$. Clearly, if Z passes this special test $(V_m)_{m\in\mathbb{N}}$ then it passes each Demuth test.

Claim 2. *For each special test $(V_m)_{m\in\mathbb{N}}$ there is a Δ^0_2 set Z such that $Z \notin V_m$ for each m.*

Let $\widehat{V}_m = \bigcup_{i\le m} V_i$. We determine $Z\upharpoonright_m$ by recursion on m using $\emptyset'$ as an oracle. Recall that $\lambda(\mathcal{C} \mid z) = \lambda(\mathcal{C}\cap[z])2^{|z|}$ for each measurable class $\mathcal{C}$ and each string z. For each m we will meet the condition

$$\lambda(\widehat{V}_m \mid Z\upharpoonright_m) \le 1 - 2^{-m-1}. \tag{3.13}$$

Clearly, the condition holds for $m = 0$ as $\lambda\widehat{V}_0 \le 1/2$. Suppose $Z\upharpoonright_m$ has been determined and the condition holds for m. Then using $\emptyset'$ we can determine the least $Z(m) \in \{0,1\}$ such that for $z = Z\upharpoonright_{m+1}$, $\lambda(\widehat{V}_m \mid z) \le 1 - 2^{-m-1}$. Since $\lambda V_{m+1} \le 2^{-2(m+1)-1}$, we have $\lambda(V_{m+1} \mid z) \le 2^{-m-2}$, and hence $\lambda(\widehat{V}_{m+1} \mid z) \le 1 - 2^{-m-1} + 2^{-m-2} = 1 - 2^{-m-2}$, as required.

The two claims establish the theorem. □

3.6.26 Theorem. *Each Demuth random set is in* GL_1.

Proof. We define a Turing functional Θ by $\Theta^Z(m) \simeq \mu s.J^Z_s(m)\downarrow$. We will introduce an ω-c.e. function g and a Demuth test $(S_m)_{m\in\mathbb{N}}$ such that

$$\forall^\infty m[\Theta^Z(m)\downarrow\ \rightarrow\ \Theta^Z(m) \le g(m)]$$

for each Z that passes $(S_m)_{m\in\mathbb{N}}$. That is, g dominates the partial function Θ^Z. Then $J^Z(m)\downarrow\ \leftrightarrow\ J^Z_{g(m)}(m)\downarrow$ for almost all m, whence $Z' \le_T Z \oplus \emptyset'$ (in fact with a computably bounded use on $\emptyset'$).

For each m, let L_m be the open set $\{Z\colon J^Z(m)\downarrow\}$, and let $L_{m,s}$ be the clopen set $\{Z\colon J^Z_s(m)\downarrow\}$. We define an auxiliary clopen set C_m. At stage s we define approximations $g_s(m)$ to $g(m)$ and $C_{m,s}$ to C_m, in such a way that the clopen set $C_{m,s}$ contains the oracles Z such that $g_s(m)$ dominates $\Theta^Z(m)$. Whenever at a stage s the measure of the clopen set $L_{m,s} - C_{m,s-1}$ exceeds 2^{-m}, we put this set into $C_{m,s}$ and increase $g(m)$ to the stage number, so that it also dominates the values $\Theta^Z(m)$ for these newly added oracles Z. These increases of the current approximations to C_m, and hence of $g(m)$, can take place at most 2^m times. Thus g is ω-c.e. and C_m stabilizes, whence $S_m = L_m - C_m$ determines a Demuth test as desired.

Construction of clopen sets $C_m = \bigcup_s C_{m,s}$ and an ω-c.e. function g.
Let $C_{0,0} = \emptyset$ and $g_0(0) = 0$.
Stage $s > 0$. Let $C_{s,s} = \emptyset$ and $g_s(s) = 0$. For each $m < s$,
 if $\lambda(L_{m,s} - C_{m,s-1}) > 2^{-m}$ let $C_{m,s} = L_{m,s}$ and $g_s(m) = s$;
 otherwise, let $C_{m,s} = C_{m,s-1}$ and $g_s(m) = g_{s-1}(m)$.

By this construction, if $\Theta^Z(m)\downarrow$ for $Z \notin S_m$ then $\Theta^Z(m) \le g(m)$. If Z is Demuth random then $Z \notin S_m$ for almost all m, so g dominates Θ^Z. □

In Theorem 8.1.19 we will prove that a computably dominated set in GL_1 is not of d.n.c. degree. Hence no Demuth random set is computably dominated, which strengthens Corollary 3.6.15.

Exercises.

3.6.27. Show that some low ML-random set is not Demuth random.

3.6.28. Suppose Z is Schnorr random relative to $\emptyset'$.
Show that Z is (i) weakly 2-random and (ii) Demuth random.
(iii) Conclude that some weakly 2-random set Y in GL_1 is not 2-random.

3.6.29.$^\diamond$ (Hölzl, Kräling, Stephan and Wu, 20xx) Show that in the Definition 3.6.24 of Demuth randomness, it is sufficient to consider tests $(S_m)_{m\in\mathbb{N}}$ where all the test components are clopen sets given by strong indices for finite sets (namely, $S_m = [D_{f(m)}]^{\prec}$ for some ω-c.e. function f).

3.6.30.$^\diamond$ (Hölzl, Kräling, Stephan and Wu, 20xx) Show that Z is Demuth random $\Leftrightarrow$
$$\exists f\ \omega\text{-c.e. } \exists^\infty m \forall r \ge f(m)\, [C(A\upharpoonright_r) \ge r - m].$$

3.6.31. Show that some 2-random set is Low_2.

4

DIAGONALLY NONCOMPUTABLE FUNCTIONS

We discuss the interactions between computability and randomness. Traditionally the direction is from computability to randomness. In this direction, two types can be distinguished.

Interaction 1a: computability theoretic notions are used to obtain mathematical definitions for the intuitive concept of a random set.

For instance, in Chapter 3 we introduced ML-randomness and its variants using test notions which are based on computable enumerability and other concepts.

Interaction 1b: computational complexity is used to analyze randomness properties of a set.
An example is the result on page 135 that Z is weakly 2-random iff Z is ML-random and $Z, \emptyset'$ form a minimal pair.

The interaction also goes the opposite way.
Interaction 2: concepts related to randomness enrich computability theory.

We have already seen examples of this in Chapter 3: the real number Ω and the operator $X \mapsto \Omega^X$. In Section 5.2 we will study K-triviality, a property of sets that means being far from random. This property turns out to be equivalent to the lowness property of being low for ML-randomness. The class of K-trivial sets has interesting properties from the computability-theoretic point of view. For instance, each K-trivial set is superlow. Many equivalent definitions are known, and all touch in some way upon randomness-related concepts.

In the spirit of Interaction 2, in this chapter we study diagonally noncomputable functions. They arise naturally when one studies ML-random sets. We have briefly considered d.n.c. functions in Remark 1.8.30.

4.0.1 Definition.

(i) A function $f\colon \mathbb{N} \mapsto \mathbb{N}$ is *diagonally noncomputable*, or d.n.c. for short, if $f(e) \neq J(e)$ for any e such that $J(e)\downarrow$.

(ii) A set D has *d.n.c. degree* if there is a d.n.c. function $f \leq_T D$.

A d.n.c. function f is incomputable, for otherwise $f = \Phi_e$ for some e, and hence $f(e) = \Phi_e(e) = J(e)$. We think of f as far from computable since it *effectively* provides the value $f(e)$ as a counterexample to the hypothesis that $f = \Phi_e$.

If Z is ML-random, a finite variant of the function $\lambda n. Z\upharpoonright_n$ is diagonally noncomputable, because $Z\upharpoonright_e = J(e)$ implies $K(Z\upharpoonright_e) \leq^+ 2\log e$. In Section 4.1 we study the sets of d.n.c. degree, or, equivalently, the sets that compute a fixed point free function g (namely $W_{g(x)} \neq W_x$ for each x).

To be of d.n.c. degree is a conull highness property (intuitively speaking, it is weak). Sets of d.n.c. degree are characterized by a property stating that the initial segment complexity grows somewhat fast. Thus the highness property to be of d.n.c. degree can also be interpreted as to be "somewhat random". This characterization is an analog of Schnorr's Theorem 3.2.9. As a consequence, the sets of d.n.c. degree are closed upwards with respect to the preordering $\leq_K$ which compares the degree of randomness of sets (Definition 5.6.1 below.)

In Section 4.2 we discuss Kučera's injury-free solution to Post's problem, a further instance of Interaction 2. It is based on a d.n.c. function $f <_T \emptyset'$, but takes a particularly simple form when f is a finite variant of the function $\lambda n.\, Z\restriction_n$ for a ML-random Δ^0_2 set Z. We use that f is d.n.c. to build a promptly simple set $A \leq_T f$. To make f permit changes of A we threaten that $f(k) = J(k)$ for appropriate numbers k. These methods can be extended to an injury-free construction of a pair of Turing incomparable c.e. sets.

In Section 4.3 we strengthen the concept of a d.n.c. function in various ways. We use the stronger concepts to gain a better understanding of the computational complexity of n-random sets (this is Interaction 1b). Firstly, we show that if a set Z is ML-random and computes a $\{0,1\}$-valued d.n.c. function, then Z already computes the halting problem. Secondly, we introduce a hierarchy of properties of functions strengthening fixed point freeness, and show that an n-random set computes a function at level n of that hierarchy.

4.1 D.n.c. functions and sets of d.n.c. degree

We characterize the sets of diagonally noncomputable degree via a growth condition on the initial segment complexity. Thereafter, we show that the only c.e. sets of diagonally noncomputable degree are the Turing complete ones.

Basics on d.n.c. functions and fixed point freeness

4.1.1 Proposition. *(i) No d.n.c. function f is computable.*
(ii) $\emptyset'$ has d.n.c. degree via some function $f \leq_{tt} \emptyset'$.

Proof. (i) If $f = \Phi_e$ is total, then $f(e) = J(e)$, so f is not d.n.c.
(ii) A $\{0,1\}$-valued d.n.c. function $f \leq_{tt} \emptyset'$ was given in Remark 1.8.30. □

The following provides further examples of d.n.c. functions.

4.1.2 Proposition.

(i) There is $c \in \mathbb{N}$ such that

$$\forall^\infty n\, K(Z\restriction_n) > K(n) + c \;\rightarrow\; Z \text{ has d.n.c. degree}$$

via a function f that agrees with $\lambda n.\, Z\restriction_n$ on almost all n (hence $f \leq_{tt} Z$).

(ii) Each ML-random set Z has d.n.c. degree via a function $f \leq_{tt} Z$.

Proof. (i) Recall from page 12 that we identify strings in $\{0,1\}^*$ with numbers. Let c be a constant such that, for each σ, if $y = J(\mathbb{U}(\sigma))\downarrow$ then $K(y) \leq |\sigma| + c$.

If $Z\restriction_n = J(n)$ and σ is a shortest $\mathbb{U}$-description of n, then $J(\mathbb{U}(\sigma)) = Z\restriction_n$ and hence $K(Z\restriction_n) \le |\sigma| + c = K(n) + c$. Thus $Z\restriction_n = J(n)$ fails for almost all n.
(ii) If Z is ML-random then there is b such that $\forall n\, K(Z\restriction_n) \ge n - b$. Since $K(n) \le^+ 2\log n$, this implies $K(Z\restriction_n) > K(n) + c$ for almost all n. □

Actually, by 8.3.6, every infinite subset of a ML-random set has d.n.c. degree.

By the Recursion Theorem 1.1.5, for each computable function g there is x such that $W_x = W_{g(x)}$. The functions of the following type are far from computable in the sense that this fixed point property fails.

4.1.3 Definition. A function g is *fixed point free* (f.p.f.) if $\forall x\, W_{g(x)} \neq W_x$.

The two notions of being far from computable coincide up to Turing degree by the following result of Kučera (1990).

4.1.4 Proposition. *Let $D \subseteq \mathbb{N}$. Then D has d.n.c. degree $\Leftrightarrow$ D computes a fixed point free function. The equivalence is uniform.*

Proof. It suffices to show that each d.n.c. function computes a fixed point free function and vice versa.
$\Rightarrow$: Suppose that the function f is diagonally noncomputable. We construct a fixed point free function $g \le_T f$. Let

$$\alpha(x) \simeq \text{the first element enumerated into } W_x.$$

Let p be a reduction function for α (see Fact 1.2.15). Thus $\alpha(x) \simeq J(p(x))$ for each x. Let $g \le_T f$ be a function such that $g(x)$ is a c.e. index for $\{f(p(x))\}$. Then $W_x \neq W_{g(x)}$ unless $\#W_x = 1$. If $\#W_x = 1$, then $W_x = \{\alpha(x)\} = \{J(p(x))\}$. Because f is d.n.c., $J(p(x)) \neq f(p(x))$, so $W_x \neq W_{g(x)}$.

$\Leftarrow$: Suppose that the function g is fixed point free. Let h be a computable function such that

$$W_{h(x)} = \begin{cases} W_{J(x)} & \text{if } J(x)\downarrow, \\ \emptyset & \text{otherwise.} \end{cases}$$

If $J(x)\downarrow$ then $W_{g(h(x))} \neq W_{J(x)}$ and hence $g(h(x)) \neq J(x)$. So $f = g \circ h$ is d.n.c. and $f \le_T g$. □

In Definition 4.3.14 we will introduce the hierarchy of n-fixed point free functions ($n \ge 1$). The lowest level $n = 1$ consists of the functions of Definition 4.1.3. The extendability of this definition to higher levels is one of the reasons why we do not define fixed point freeness of a function g by the weaker condition that $\Phi_{g(x)} \neq \Phi_x$.

By Exercise 8.3.6, A has d.n.c. degree iff A computes an infinite subset of a ML-random set.

Exercises. Show the following.

4.1.5. Let f be a d.n.c. function. For each partial computable function ψ, there is a function $\widetilde{f} \le_T f$ such that $\widetilde{f}(e) \neq \psi(e)$ for any e such that $\psi(e)\downarrow$.

4.1.6. No 1-generic set G (see Definition 1.8.51) is of d.n.c. degree. In particular, no 1-generic set computes a ML-random set.

By the following, the properties of functions introduced above are independent of the particular choice of a jump operator, or a universal uniformly c.e. sequence, as far as the Turing degree of the function is concerned.

4.1.7. Suppose that $\widehat{J}$ is a universal partial computable function (Fact 1.2.15). Then each d.n.c. function f is Turing equivalent to a d.n.c. function $\widehat{f}$ with respect to $\widehat{J}$, and vice versa.

4.1.8. Suppose that $(\widehat{W_e})_{e\in\mathbb{N}}$ is a universal uniformly c.e. sequence with the padding property, as in Exercise 1.1.12. Then each f.p.f. function g is Turing equivalent to an f.p.f. function $\widehat{g}$ with respect to $(\widehat{W_e})_{e\in\mathbb{N}}$, and vice versa.

The initial segment complexity of sets of d.n.c. degree

Schnorr's Theorem 3.2.9 states that the ML-random sets are the ones with a nearly maximal growth of the initial segment complexity in the sense of K. The following result of Kjos-Hanssen, Merkle and Stephan (2006) characterizes the sets of d.n.c. degree by a growth condition stating that the initial segment complexity grows somewhat quickly. This yields a further proof besides the one in Exercise 4.1.7 that the property to be of d.n.c. degree does not depend of the somewhat arbitrary definition of the universal partial computable function J.

4.1.9 Theorem. *The following are equivalent for a set Z.*

(i) Z has d.n.c. degree.

(ii) There is a nondecreasing unbounded function $g \le_T Z$ such that $\forall n\,[g(n) \le K(Z\restriction_n)]$.

It is useful to think of the function g in (ii) as slowly growing. By 2.4.2 $K \sim C$, so we could equivalently take the initial segment complexity of Z via C.

Proof. (ii) $\Rightarrow$ (i): The idea is the same as in the proof of Proposition 4.1.2. Suppose (ii) holds via $g \le_T Z$, and let $h(r) = \min\{m : g(m) \ge r\}$. (Think of h as a function that grows quickly.) Note that Z computes the function $f(r) = Z\restriction_{h(r)}$. If $Z\restriction_{h(r)} = J(r)$ and σ is a shortest description of r then $J(\mathbb{U}(\sigma)) = Z\restriction_{h(r)}$ and hence $K(Z\restriction_{h(r)}) \le^+ K(r) = |\sigma| \le^+ 2\log r$. However, by the definition of h we have $K(Z\restriction_{h(r)}) \ge r$. Thus $Z\restriction_{h(r)} = J(r)$ fails for almost all r.

(i) $\Rightarrow$ (ii): Suppose that (ii) fails. Let $\Gamma = \Phi_i$ be a Turing functional such that Γ^Z is total. We show that Γ^Z is not a d.n.c. function, namely, $\Gamma^Z(e) = J(e)$ for some e. For each σ, one can effectively determine an $e(\sigma)$ such that $\Phi_{e(\sigma)}(y) \simeq \Gamma^{\mathbb{U}(\sigma)}(y)$: the number $e(\sigma)$ encodes a Turing program implementing a procedure that on input y first runs $\mathbb{U}$ on the input σ; in case of convergence it takes the output ρ as an oracle string and attempts to compute $\Gamma^\rho(y)$. Let

$$h(r) = \max(\{r\} \cup \{\text{use } \Gamma^Z(e(\sigma)) \colon |\sigma| \le r\}).$$

The function g given by $g(m) = \max\{r\colon h(r) \le m\}$ is computable in Z, nondecreasing and unbounded. If (ii) fails, there is an m such that $\overline{r} = g(m) > K(Z\restriction_m)$. So $\mathbb{U}(\sigma) = Z\restriction_m$ for some σ such that $|\sigma| < \overline{r}$. Let $e = e(\sigma)$. Since

$h(\overline{r}) \leq m$ we have use $\Gamma^Z(e) \leq m$. Thus $J(e) = \Phi_e(e) = \Gamma^{\mathbb{U}(\sigma)}(e) = \Gamma^Z(e)$. □

With a minor modification, the foregoing proof yields a characterization of the sets Z such that there is a d.n.c. function $f \leq_{wtt} Z$. The characterizing condition is that $K(Z\restriction_n)$ be bounded from below by an order function, i.e., a nondecreasing unbounded computable function.

4.1.10 Theorem. *The following are equivalent for a set Z.*

(i) There is a d.n.c. function $f \leq_{wtt} Z$.
(ii) There is an order function g such that $\forall n\,[g(n) \leq K(Z\restriction_n)]$.
(iii) There is a d.n.c. function $f \leq_{tt} Z$.

Proof. (ii)⇒(iii) is proved in the same way as the implication (ii)⇒(i) in Theorem 4.1.9. Notice that if g is computable then $f \leq_{tt} Z$.

For (i)⇒(ii), note that in the proof of (i)⇒(ii) above, if there is a computable bound on the use of Γ, we can choose h and hence g computable. □

A completeness criterion for c.e. sets

The only c.e. sets of d.n.c. degree are the Turing complete ones. This result of Arslanov (1981) builds on work of Martin (1966*b*) and Lachlan (1968).

4.1.11 Theorem. (Completeness Criterion) *Suppose the set Y is c.e.*

(i) There is a d.n.c. function $f \leq_T Y \Leftrightarrow Y$ is Turing complete.
(ii) There is a d.n.c. function $f \leq_{wtt} Y \Leftrightarrow Y$ is wtt-complete.

The result can be viewed as a generalization of the Recursion Theorem 1.1.5 when stated in terms of fixed point free functions: for every function $g <_T \emptyset'$ of c.e. degree there is an x such that $W_{g(x)} = W_x$. For instance, if $A <_T \emptyset'$ is c.e., then the function $p_{\overline{A}}$ has such a fixed point.

Proof idea. If Y is a Turing complete set then Y has d.n.c. degree by Proposition 4.1.1. If Y is *wtt*-complete then there is a d.n.c. function $f \leq_{wtt} Y$ by the same proposition.

Now suppose there is a Turing functional Φ such that $f = \Phi^Y$ is a d.n.c. function. For an appropriate reduction function p (see 1.2.15), when e enters $\emptyset'$ at a stage s such that $\Phi_s^{Y_s}(p(e))\downarrow$, we threaten that this value equal $J(p(e))$. Then Y_s is forced to change below the use of $\Phi_s^{Y_s}(p(e))$. So, once $\Phi_s^{Y_s}(p(e))$ is stable at stage $s(e)$, e cannot enter $\emptyset'$ any more. Since Y is c.e., $s(e)$ is the first stage where $\Phi_s^{Y_s}(p(e))$ converges and $Y_s(x) = Y(x)$ for each x less than the use. So Y can compute such a stage, and $\emptyset' \leq_T Y$. If the use of Φ is bounded by a computable h, then the use of the reduction procedure is bounded by $h(p(e))$ on input e, hence $\emptyset' \leq_{wtt} Y$.

Proof details. We define an auxiliary partial computable function α. By the Recursion Theorem we may assume that we are given a computable reduction function p such that $\alpha(e) \simeq J(p(e))$ for each e (see Remark 4.1.12 below).

Construction of α.
Stage s. If $e \in \emptyset'_{s+1} - \emptyset'_s$ and $y = \Phi_s^{Y_s}(p(e)) \downarrow$, let $\alpha(e) = y$.

We show that $\emptyset' \le_T Y$. Given an input e, using Y as an oracle, we compute the first stage $s = s(e)$ such that $\Phi_s^{Y_s}(p(e)) \downarrow$ and Y is stable below the use of this computation. If e enters $\emptyset'$ at a stage $\ge s(e)$, we define $\alpha(e) = J(p(e)) = f(p(e))$, so f is not a d.n.c. function. Thus $e \in \emptyset' \leftrightarrow e \in \emptyset'_{s(e)}$. □

4.1.12 Remark. We justify the seemingly paradoxical argument where we assume the reduction function p for α is given, even though we are actually *constructing* α. By Fact 1.2.15, from an index e for a partial computable function Φ_e one obtains a reduction function p. Based on p, in our construction we build a partial computable function $\Phi_{g(e)}$. By the Recursion Theorem 1.1.5, there is a fixed point i, namely a partial computable $\alpha = \Phi_i$ such that $\Phi_{g(i)} = \alpha$. So in the interesting case that the given e is such a fixed point, p is a reduction function for $\Phi_{g(e)} = \Phi_e$.

We provide two applications of the Completeness Criterion.
Application 1. By Theorem 3.2.11 the halting probability Ω_R is ML-random for each optimal prefix-free machine R, and by Proposition 3.2.30, Ω_R is *wtt*-complete (we identify Ω_R with its binary representation). We give an alternative proof of the second fact: Since Ω_R is ML-random, a finite variant f of the function $\lambda n. \Omega_R \upharpoonright_n$ is d.n.c., and $f \le_{wtt} \Omega_R$. Also, Ω_R is left-c.e., namely, the set $\{q \in \mathbb{Q}_2 \colon 0 \le q < \Omega_R\}$ is c.e. Thus $\emptyset' \le_{wtt} \{q \in \mathbb{Q}_2 \colon 0 \le q < \Omega_R\} \equiv_{tt} \Omega_R$.

Theorem 4.3.9 below shows that $\emptyset' \not\le_{tt} \Omega_R$. Thus $\{q \in \mathbb{Q}_2 \colon 0 \le q < \Omega_R\}$ is a c.e. set that is weak truth-table complete but not truth-table complete. By Theorem 4.1.10 there is a d.n.c. function $f \le_{tt} \Omega_R$. Thus the Completeness Criterion 4.1.11 has no analog for truth-table reducibility.

The truth-table degree of Ω_R depends on the particular choice of the optimal prefix-free machine: Figueira, Stephan and Wu (2006) have shown that there is a sequence $(R_i)_{i \in \mathbb{N}}$ of such machines such that $\Omega_{R_i} \mid_{tt} \Omega_{R_j}$ for each pair $i \neq j$.

Application 2. Recall from page 32 that a co-infinite c.e. set A is called effectively simple if there is a computable function g such that $\#W_e \ge g(e) \to W_e \cap A \neq \emptyset$. We obtain a short proof of a result due to Martin (1966*b*).

4.1.13 Proposition. *Every effectively simple set A is Turing complete.*

Proof. By Proposition 4.1.4 it suffices to show that A computes a fixed point free function f. Since A is co-infinite, on input e, with A as an oracle one can compute a c.e. index $f(e)$ for the set consisting of the first $g(e)$ elements of $\mathbb{N} - A$. If $W_{f(e)} = W_e$ then $W_e \cap A = \emptyset$ while $\#W_e \ge g(e)$. Hence f is fixed point free. □

Exercises.

4.1.14. Derive Prop. 3.4.10 from the Completeness Criterion 4.1.11 relativized to A.

4.1.15. (Friedberg and Rogers, 1959) Each hypersimple set is *wtt*-incomplete.

4.1.16. If A is effectively simple and not hypersimple then A is *wtt*-complete.

4.2 Injury-free constructions of c.e. sets

One can solve Post's problem and even prove the Friedberg–Muchnik Theorem 1.6.8 avoiding the priority method with injury. These results of Kučera (1986) are interesting because injury makes sets artificial due to the fact that one first fulfills tasks and then gets them undone. This does not happen in Kučera's constructions, even if they are a bit more involved.

The most direct injury-free solution to Post's problem relies on ML-randomness.

Step 1. Begin with Ω_0, the set given by the bits of Ω in the even positions, which is Turing incomplete by Corollary 3.4.8 (and even low by 3.4.11).

Step 2. Build a promptly simple set A (Definition 1.7.9) Turing below Ω_0.

The construction of Kučera (1986) in step 2 works for any Δ^0_2 ML-random set in place of Ω_0; see Remark 4.2.4 below.

There is no injury because in step 1 there are no requirements, and in step 2 we merely meet the prompt simplicity requirements, which cannot be injured. The injury-free solution to Post's problem as it is commonly thought of nowadays is somewhat different. Step 1 uses the Low Basis Theorem. Step 2 is a more general result of independent interest. Its proof needs the Recursion Theorem.

Step 1. Use the Low Basis Theorem 1.8.37 to produce a low set of d.n.c. degree. For instance, take the Π^0_1 class $2^\omega - \mathcal{R}_1$, which by Theorem 3.2.9 consists entirely of ML-random, and hence d.n.c. sets. Or take the Π^0_1 class of $\{0,1\}$-valued d.n.c. functions in Fact 1.8.31. The construction in the proof of Theorem 1.8 is relative to $\emptyset'$. One satisfies the requirements one by one in order. Hence they are not injured.

Step 2. Show that Turing below each Δ^0_2 set Y of d.n.c. degree one can build a promptly simple set.

The original proof in Kučera (1986) uses the Low Basis Theorem in Step 1, but avoids the Recursion Theorem in Step 2, as above. It has been criticized that in the proof of Theorem 1.8.37, injury is merely hidden using $\emptyset'$. Indeed, the effective version of the proof in Theorem 1.8.38 has injury to lowness requirements. This criticism does not apply when we avoid the Low Basis Theorem altogether and rather use Ω_0 as the low set of d.n.c. degree.

Initially Kučera's motivation was to disprove the conjecture in Jockusch and Soare (1972*a*) that each nonempty Π^0_1 class without computable members contains a set $Y <_T \emptyset'$ such that each c.e. set $A \leq_T Y$ is computable. Jockusch then pointed out that Kučera's proof leads to an injury-free solution to Post's problem.

For the injury-free proof of the Friedberg-Muchnik Theorem 1.6.8, Kučera developed his ideas further. Instead of carrying out two independent steps, he let the construction relative to $\emptyset'$ and the effective construction interact via the Double Recursion Theorem 1.2.16. This bears some similarity to the worker's method of Harrington. A so-called level n argument involves interacting priority constructions relative to $\emptyset, \emptyset', \ldots, \emptyset^{(n)}$ (that is, workers at level i for each $i \leq n$), and uses an $(n+1)$-fold Recursion Theorem. See Calhoun (1993).

Each Δ^0_2 set of d.n.c. degree bounds a promptly simple set

We construct a promptly simple set Turing below any given Δ^0_2 set of d.n.c. degree. In the next subsection we describe the orginal argument in Kučera (1986), where the Recursion Theorem is avoided when the given Δ^0_2-set is in fact ML-random. However, the more powerful method of the present construction is used for instance in the injury-free proof of the Friedberg–Muchnik Theorem. The hypothesis that Y be Δ^0_2 is necessary because, say, a weakly 2-random set has d.n.c. degree but forms a minimal pair with $\emptyset'$ (page 135).

4.2.1 Theorem. (Kučera) *Let Y be a Δ^0_2 set of d.n.c. degree. Then there is a promptly simple set A such that $A \leq_T Y$.*

Proof idea: The Completeness Criterion 4.1.11 fails for Δ^0_2 sets because there is a low set of d.n.c. degree. The construction of A can be seen as an attempt to salvage a bit of its proof in the case that Y is Δ^0_2. Where does the proof go wrong? As before, suppose that $f = \Phi^Y$ is a d.n.c. function. A Δ^0_2 set Y cannot compute a stage $s(e)$ such that $\Phi_s^{Y_s}(p(e))$ is stable from $s(e)$ on, only a stage where it has the final value for the first time. The problem is that it may change temporarily after such a stage.

On the other hand, we do not have to code the halting problem into Y. It suffices to code the promptly simple set A we are building. For each e we meet the prompt simplicity requirement from the proof of Theorem 1.7.10

$$PS_e\colon\ \#W_e = \infty \Rightarrow \exists s \exists x\ [x \in W_{e,s} - W_{e,s-1}\ \&\ x \in A_s].$$

At stage s we let x enter A for the sake of a requirement PS_e only if $\Phi^Y(p(e))$ has been stable from stage x to s. If we now threaten that $J(p(e)) = \Phi^Y(p(e))$ then Y has to change to a value not assumed between stage x and s. This allows us to compute A from Y. It also places a strong restriction on PS_e: whenever $\Phi^Y(p(e))$ changes another time at t, then PS_e cannot put any numbers less than t into A at a later stage.

We work with the effective approximation $f_s(x) = \Phi^Y(x)[s]$. (Thus, in contrast to 4.1.11, we actually consider changes of the value $f_s(x)$ rather than changes of the oracle Y.) For an appropriate computable function p, when we want to put a candidate $x \in W_e$ into A for the sake of PS_e, we threaten that $f(p(e))$ equal $J(p(e))$. We only put x into A at stage s if $f_t(p(e))$ has been constant since stage x. (Since $f_t(p(e))$ settles, this holds for large enough x. Thus, PS_e is still able to choose a candidate.) To show $A \leq_T f$, if $f(p(e)) = f_t(p(e))$, then after stage t a number x cannot go into A for the sake of PS_e. See Fig. 4.1.

Proof details. As in the proof of Theorem 4.1.11, we define an auxiliary partial computable function α. By the Recursion Theorem we are given a reduction function p for α, namely, $\forall e\, \alpha(e) \simeq J(p(e))$.

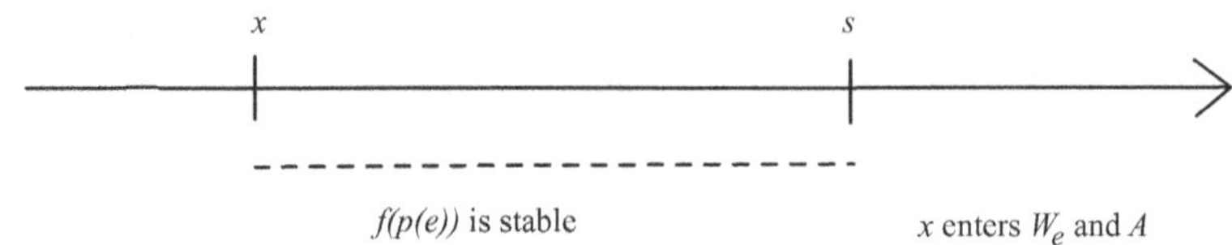

FIG. 4.1. Proof of Kučera's Theorem.

Construction of A and α. Let $A_0 = \emptyset$.

Stage s. For each $e < s$, if PS_e is not satisfied yet, see whether there is an x, $2e \leq x < s$, such that

$$x \in W_{e,s} - W_{e,s-1} \ \& \ \forall t_{x<t<s} \ f_t(p(e)) = f_s(p(e)). \tag{4.1}$$

If so, put x into A_s. Define $\alpha(e) = f_s(p(e))$. Declare PS_e satisfied.

Verification. Clearly A is co-infinite. Choose s_0 such that $\forall s \geq s_0 \ f_s(p(e)) = f(p(e))$. If $x \geq s_0, x \geq 2e$ appears in W_e at stage s then x can be used to satisfy PS_e. Next, we show that $A \leq_T f$. Given an input x, using f compute $t > x$ such that for all e, if $2e \leq x$ then $f_t(p(e)) = f(p(e))$. Then $x \in A \leftrightarrow x \in A_t$, for if we put x into A at a stage $s > t$, then, as we required that $\forall t_{x<t<s} f_t(p(e)) = f_s(p(e))$, the value $\alpha(e) = J(p(e)) = f_s(p(e))$ we define at stage s equals $f(p(e))$. Thus f is not a d.n.c. function, contradiction. □

Variants of Kučera's Theorem

We modify the proof of Kučera's Theorem 4.2.1 in four ways. Firstly, we prove a version for *wtt*-reducibility. Secondly, we work under two stronger hypotheses on the given Δ^0_2 set Y, namely that Y is a $\{0,1\}$-valued d.n.c. function, or that Y is ML-random. In both cases we obtain a uniform version of Kučera's Theorem. Thirdly, we combine the construction with permitting, and finally, in an exercise, we consider the case of two given Δ^0_2 sets Y_0 and Y_1.

4.2.2 Corollary. *Suppose $Y \in \Delta^0_2$ and there is a d.n.c. function $f \leq_{wtt} Y$ (for instance, if Y is ML-random). Then there is a promptly simple set $A \leq_{wtt} Y$.*

Proof. Suppose that h is a computable use bound for the procedure computing f with oracle Y. Then, at the end of the proof of Theorem 4.2.1, to compute $A(x)$ we only need to query Y on numbers less than $h(p(x))$. □

Uniformity considerations for Kučera's Theorem will be important for the injury-free proof of the Friedberg–Muchnik Theorem: given an index k such that $Y = \Phi_k^{\emptyset'}$, can we obtain a c.e. index for A in an effective way? Actually, to build A, we also need to know how to compute the d.n.c. function from Y. This is certainly the case when Y itself is a $\{0,1\}$-valued d.n.c. function:

4.2.3 Corollary. *There is a computable function r such that, for each k, if $Y = \Phi_k^{\emptyset'}$ is total and Y is a $\{0,1\}$-valued d.n.c. function, then $A = W_{r(k)} \leq_{wtt} Y$ and A is promptly simple. An index for the reduction for $A \leq_{wtt} Y$ can be obtained effectively as well.* □

4.2.4 Remark. If Y is Δ_2^0 and ML-random, then a somewhat simpler argument suffices to obtain a promptly simple set $A \leq_{wtt} Y$ (and in fact the use equals the input in the reduction procedure). We do not need the Recursion Theorem or a reduction function. To ensure that $A \leq_{wtt} Y$ we can directly force $Y \upharpoonright_e$ to change by including the string $Y \upharpoonright_e$ in an interval Solovay test G (that is, a certain effective listing of strings defined in 3.2.22). In other words, if Y is ML-random we may let $Y \upharpoonright_e$ play the role of $f(p(e))$ before, so the reduction function p is not needed any longer. Since the procedure to compute the d.n.c. function $\lambda n. Y \upharpoonright_n$ is fixed we obtain A effectively. The reduction procedure to compute A from Y is effectively given by the above, but now it is only correct for almost all inputs because $\sigma \not\preceq Y$ merely holds for almost all σ in G.
Construction of A and G. Let $A_0 = \emptyset$.
Stage $s > 0$. For each $e < s$, if PS_e is not satisfied yet, check whether there is an x, $2e \leq x < s$, such that

$$x \in W_{e,s} - W_{e,s-1} \;\&\; \forall t_{x<t<s}\; Y_t \upharpoonright_e = Y_s \upharpoonright_e . \tag{4.2}$$

If so put x into A. Put the string $Y_s \upharpoonright_e$ into G. Declare PS_e satisfied.

Verification. Given e, choose t_0 such that $\forall s \geq t_0\, Y_s \upharpoonright_e = Y \upharpoonright_e$. If $x \geq t_0$ is enumerated into W_e at a stage s then x can be used to satisfy PS_e, so PS_e is met. G is a Solovay test because the requirement PS_e contributes at most one interval, and this interval has length 2^{-e}.

To see that $A \leq_{wtt} Y$, choose s_0 such that $\sigma \not\preceq Y$ for any σ enumerated into G after stage s_0. Given an input $x \geq s_0$, using Y as an oracle, compute $t > x$ such that $Y_t \upharpoonright_x = Y \upharpoonright_x$. Then $x \in A \;\leftrightarrow\; x \in A_t$, for if we put x into A at a stage $s > t$ for the sake of PS_e then $x > e$, so we list σ in G where $\sigma = Y_s \upharpoonright_e = Y \upharpoonright_e$. This contradicts the fact that $\sigma \not\preceq Y$. □

We have obtained a further uniform version of Theorem 4.2.1 without using the Recursion Theorem.

4.2.5 Corollary. *There is a computable r such that for each e, if $Y = \Phi_e^{\emptyset'}$ is total and ML-random, then $A = W_{r(e)} \leq_{wtt} Y$ and A is promptly simple.* □

As mentioned at the beginning of this section, combining Corollary 3.4.11 (the bits of Ω in an odd position form a Turing incomplete ML-random set Y) with the construction in Remark 4.2.4 (to build a promptly simple set $A \leq_{wtt} Y$), we obtain an injury-free solution to Post's problem that is the simplest known when one also counts the proof that the constructed set is Turing incomplete (or, in fact, low). In comparison, the direct construction of a promptly simple K-trivial set (5.3.11 below) is easier, but it is much harder to verify that the constructed K-trivial set is even Turing incomplete (see from page 201 on). We already compared these construction briefly when we discussed natural solutions to Post's problem on page 34.

We provide two further variants of Kučera's Theorem 4.2.1.

4.2.6 Corollary. *If $Y \in \Delta_2^0$ has d.n.c. degree and C is an incomputable c.e. set, there is a simple set $A \leq_T Y$ such that $A \leq_{wtt} C$.*

Proof. Instead of the prompt simplicity requirements PS_e we now merely meet the requirements $R_e\colon \#W_e = \infty \Rightarrow A \cap W_e \neq \emptyset$. To ensure $A \le_{wtt} C$, we ask that C permit the enumeration of x. Thus, at stage s of the construction, for each $e < s$, if R_e is not satisfied yet, see if there is an x, $2e \le x < s$, such that

$$x \in W_{e,s} \;\&\; C_s \upharpoonright_x \neq C_{s+1} \upharpoonright_x \;\&\; \forall t_{x<t<s}\; f_t(p(e)) = f_s(p(e)).$$

If so then put x into A and define $\alpha(e) = f_s(p(e))$.

If W_e is infinite, then, because C is incomputable, infinitely many numbers x are permitted by C after they enter W_e. So each requirement R_e is met. □

4.2.7 Exercise. No two Δ^0_2 d.n.c. sets form a minimal pair by Kučera (1988). In fact, the proof of 4.2.1 can be adapted to show that there is a promptly simple set below both of them: show that, if Y_0 and Y_1 are Δ^0_2 sets of d.n.c. degree, then there is a promptly simple set $A \le_T Y_0, Y_1$.

On the other hand, a set that is Turing below *all* the Δ^0_2 ML-random sets is computable by Theorem 1.8.39.

An injury-free proof of the Friedberg–Muchnik Theorem ⋆

By Theorem 1.6.8 there are Turing incomparable c.e. sets A, B. An injury-free proof of this result was announced in Kučera (1986) and circulated, but not published.

Proof idea. Let P a nonempty Π^0_1 class, and r be a computable function, such that, if $Y = \Phi_e^{\emptyset'}$ is total and $Y \in P$, then $A = W_{r(e)} \le_{wtt} Y$ and A is not computable. Such a function exists either by Corollary 4.2.3 or 4.2.5. To use 4.2.3 recall that the $\{0,1\}$-valued d.n.c. functions form a Π^0_1 class by Fact 1.8.31; to use 4.2.5 let $P = 2^\omega - \mathcal{R}_1$.

The following attempt looks promising. The Low Basis Theorem, in the version with upper cone avoidance 1.8.39, implies that from any c.e. incomputable set B one may effectively obtain $Y \le_T B \oplus \emptyset' \equiv_T \emptyset'$ such that $Y \in P$ and $B \not\le_T Y$. We start with a *pair* of Δ^0_2 sets $\widetilde{Y}, \widetilde{Z} \in P$, given by indices a, b of reductions from $\emptyset'$, and, applying the function r to these indices we obtain c.e. sets $A \le_{wtt} \widetilde{Y}$ and $B \le_{wtt} \widetilde{Z}$. Now by 1.8.39 we effectively obtain Δ^0_2 sets $Y, Z \in P$ such that $Y \not\ge_T B$ and $Z \not\ge_T A$. By the Double Recursion Theorem 1.2.16 with oracle $\emptyset'$ we may assume that $Y = \widetilde{Y}$ and $Z = \widetilde{Z}$, so that in fact $A \le_T Y$ and $B \le_T Z$. In particular $A \not\ge_T B$ (since not even $Y \ge_T B$), and similarly $B \not\ge_T A$.

In this proof, the letters Y, Z denote either finite strings or infinite sequences of zeros and ones. In the latter case we say that Y (or Z) is total. The problem with the attempt outlined above is that in the proof of Theorem 1.8.39 we needed to know in advance that B is incomputable in order to argue that the parallel search at a stage $2e+2$ (to suitably extend Y or to find a number n such that $P^{2e+1} \cap \{X\colon \Phi_e^X(n)\uparrow\} \neq \emptyset$) terminates. The concern is that B might be computable, in which case the search may fail to terminate. Then Y remains a finite string, and P^{2e+2} remains undefined. The solution here is to keep extending Z while the search proceeds. If it proceeds forever then Z is an infinite sequence over $\{0,1\}$ (that is, a set) which is in P, so B is in fact incomputable. Then the search terminates after all, contradiction.

Proof details. Numbers a, b are given (think of them as Δ^0_2-indices for Y and Z). Let $A = W_{r(a)}$ and $B = W_{r(b)}$. In a construction relative to $\emptyset'$, we build Δ^0_2 sets Y, Z and descending sequences of nonempty Π^0_1 classes $(P^e)_{e\in\mathbb{N}}$ and $(Q^e)_{e\in\mathbb{N}}$ (they correspond to the Π^0_1 classes in the proof of Theorem 1.8.39 defined at even stages). We need to be more specific as to how the parallel search is to be carried out. To do so, we divide stages $i = 0, 1, \ldots$ into substages t.

Construction relative to $\emptyset'$ of Π^0_1 classes P^i, Q^i and strings σ_i on P^i, τ_i on Q^i. Let $\sigma_0 = \tau_0 = \emptyset$ and $P^0 = Q^0 = P$.

Stage $i+1$, $i = 2e$. Let $Q^{i+1} = Q^i$, $t = |\tau_i|$, and $\tau_{i+1,t} = \tau_i$. Go to substage $t+1$.
Substage $t+1$. Check whether

(a) there are σ, k such that $|\sigma|, k < t$, $\sigma_i \prec \sigma$, σ on P^i and $B(k) \neq \Phi^\sigma_e(k)$, *or*

(b) there is $n < t$ such that $P^i \cap \{X\colon \Phi^X_e(n)\uparrow\} \neq \emptyset$.

If neither case applies, or this is the first substage of the current stage, then let $\tau_{i+1,t+1}$ be the leftmost extension of $\tau_{i+1,t}$ of length $t+1$ which is on Q^i. Increment t and proceed to the next substage.
Otherwise, let $\tau_{i+1} = \tau_{i,t}$. If (a) applies let $\sigma_{i+1} = \sigma$ and $P^{i+1} = P^i \cap [\sigma]$. If (b) applies let $\sigma_{i+1} = \sigma_i$ and $P^{i+1} = P^i \cap \{X\colon \Phi^X_e(n)\uparrow\}$. Increment i and proceed to the next stage.
Stage $i+1$, $i = 2e+1$. Proceed in a similar way with the roles of σ_i, τ_i as well as of P^i, Q^i interchanged.

Verification. The construction only needs queries to $\emptyset'$, so it (implicitly) defines computable functions g, h such that

$$Y = \Phi^{\emptyset'}_{g(a,b)} = \bigcup_{i,t} \sigma_{i,t}, \text{ and } Z = \Phi^{\emptyset'}_{h(a,b)} = \bigcup_{i,t} \tau_{i,t}.$$

No matter what a, b are, at least one of Y, Z is total and in P. For Z is extended at the odd stages, and Y at the even stages, both during the first substage. If we get to stage $i+1$ for each i, then both Y and Z are total. Otherwise, say stage $i+1$ is not terminated where $i = 2e$. This makes Z total, hence $Z \in Q_{i+1} \subseteq P$.

By the Double Recursion Theorem 1.2.16 there is a pair of fixed points a, b, and therefore

$$\Phi^{\emptyset'}_{g(a,b)} = \Phi^{\emptyset'}_a \text{ and } \Phi^{\emptyset'}_{h(a,b)} = \Phi^{\emptyset'}_b.$$

We claim that in this case, actually both Y and Z are total. Consider again the case where stage $i+1$, $i = 2e$, is not terminated, whence only $Z = \Phi^{\emptyset'}_b$ is total. Then, since $B = W_{r(b)}$, B is incomputable. By the same argument as in the proof of Theorem 1.8.39, we finish stage $i+1$, contradiction.

Since all stages are terminated, $A \not\leq_T Z$ and $B \not\leq_T Y$. Hence $A \mid_T B$. □

4.3 Strengthening the notion of a d.n.c. function

The computational complexity of a set is related in various ways to its degree of randomness (a summary will be given in Section 8.6). Here we only consider one aspect of the computational complexity: the set computes a function with a fixed point freeness condition.

Sets of PA degree

A highness property of a set D stronger than having d.n.c. degree is obtained when one requires that there be a d.n.c. function $f \leq_T D$ that only takes values in $\{0,1\}$. In a typical argument involving a d.n.c. function f (such as the proof of Theorem 4.1.11) we build a partial computable α and have a reduction function p for α; when we define $\alpha(e) = 0$, say, we know that $f(p(e)) \neq J(p(e)) = \alpha(e)$. If f is $\{0,1\}$-valued, we may in fact conclude that $f(p(e)) = 1$. Thus we may prescribe the value $f(p(e))$, while before we could only avoid a value. Since $f = \Phi^D$ for a given Turing reduction Φ, we can indirectly restrict D.

Up to Turing degree, the $\{0,1\}$-valued d.n.c. functions coincide with the completions of Peano arithmetic PA; see Exercise 4.3.7. This justifies the following terminology.

4.3.1 Definition. We say that a set D has *PA degree* if D computes a $\{0,1\}$-valued d.n.c. function.

The set $\emptyset'$ has PA degree by Remark 1.8.30. Exercise 5.1.15 shows that the class of sets of PA degree is null, so this highness property is indeed much stronger than having d.n.c. degree. Recall from Example 1.8.32 that the $\{0,1\}$-valued d.n.c. functions form a Π^0_1 class. Thus, by the basis theorems of Section 1.8, there is a set of PA degree that is low, and also a set of PA degree that is computably dominated. These examples show that a set can be computationally strong in one sense, but weak in another.

We give two properties that are equivalent to being of PA degree. Both assert that the set is computationally strong in some sense. The result goes back to Kučera (1988).

4.3.2 Theorem. *The following are equivalent for a set D.*

(i) D has PA degree.

(ii) For each partial computable $\{0,1\}$-valued function ψ, there is a total function $g \leq_T D$ that extends ψ, namely, $g(x) = \psi(x)$ whenever $\psi(x)\downarrow$. Moreover, one may choose g to be $\{0,1\}$-valued.

(iii) For each nonempty Π^0_1 class P, there is a set $Z \in P$ such that $Z \leq_T D$. In other words, the sets Turing below D form a basis for the Π^0_1 classes.

Proof. (i)$\Rightarrow$(ii): Suppose $f \leq_T D$ is a $\{0,1\}$-valued d.n.c. function. Let p be a reduction function for ψ, that is, $\forall x\, \psi(x) \simeq J(p(x))$. Then $\forall x\ \neg f(p(x)) = J(p(x))$. So, if $\psi(x)$ converges, then $f(p(x)) = 1 - \psi(x)$. Let $g \leq_T D$ be the $\{0,1\}$-valued function given by $g(x) = 1 - f(p(x))$, then g extends ψ.

(ii)$\Rightarrow$(iii): Let $P = \bigcap_s P_s$ be an approximation of P by a descending effective sequence of clopen sets; see (1.17) on page 52. For a number x (viewed as a binary string), if s is least such that $[xi] \cap P_s = \emptyset$ for a unique $i \in \{0,1\}$, then define $\psi(x) = 1 - i$. (If we see that the extension xi is hopeless, then ψ dictates to go the other way.) On many strings x, the function ψ makes no decision. So let g be a total extension as in (ii), and define Z recursively by $Z(n) = g(Z\restriction_n)$. Then $Z \leq_T D$ and Z is in P.

(iii)⇒(i): This holds because the $\{0,1\}$-valued d.n.c. functions form a Π^0_1 class. □

Exercises.

4.3.3. The equivalence (i)⇔ (ii) above fails for d.n.c. sets without the restriction that the function computed by D is $\{0,1\}$-valued: show that the following are equivalent for a set D. (i) $\emptyset' \leq_T D$. (ii) Each partial computable function ψ can be extended to a (total) function $g \leq_T D$.

4.3.4. (Kučera) Show that for each low set Z there is a promptly simple set A such that $Z \oplus A$ is low.

4.3.5.$^\diamond$ (Jockusch, 1989) In Definition 4.3.1, the condition that D computes a d.n.c. function with range bounded by a constant would be sufficient: suppose that f is a d.n.c. function with bounded range. Show that there is a $\{0,1\}$-valued d.n.c. function $g \leq_T f$.

The next two exercises assume familiarity with Peano arithmetic (see Kaye 1991). The sentences in the language of arithmetic $L(+, \times, 0, 1)$ are effectively encoded by natural numbers, using all the natural numbers. Let α_n be the sentence encoded by n. Let $\dot{n}$ be the effectively given term in the language of arithmetic that describes n.

4.3.6. (Scott) Show that a set D has PA degree ⇔ Peano arithmetic has a complete extension $B \leq_T D$.

4.3.7.$^\diamond$ (Scott) Improve the result of the previous exercise: D has PA degree ⇔ Peano arithmetic has a complete extension $B \equiv_T D$.

Martin-Löf random sets of PA degree

The following theorem of Stephan (2006) shows that there are two types of ML-random sets: the ones that are not of PA degree and the ones that compute the halting problem. However, such a dichotomy only applies for the particular highness property of having PA degree. A ML-random set can satisfy other highness properties, such as being high, without computing $\emptyset'$ (see 6.3.14).

4.3.8 Theorem. *If a ML-random set Z has PA degree then $\emptyset' \leq_T Z$.*

Proof. If the proof seems hard to follow, read the easier proof of Theorem 4.3.9 first, where a similar technique is used.

Let Φ be a Turing functional such that $\Phi^X(n)$ is undefined or in $\{0,1\}$ for each X, n, and Φ^Z is a d.n.c. function. If $\emptyset' \not\leq_T Z$ we build a uniformly c.e. sequence $(C_d)_{d\in\mathbb{N}}$ of open sets such that $\lambda C_d \leq 2^{-d}$ and $Z \in C_d$ for infinitely many d, so Z fails this Solovay test.

We define an auxiliary $\{0,1\}$-valued partial computable function α. By the Recursion Theorem (see Remark 4.1.12) we may assume we are given a reduction function p for α, namely, a computable strictly increasing function p such that $\alpha(e) \simeq J(p(e))$ for each e. Since Φ^Z is $\{0,1\}$-valued, for $r \in \{0,1\}$, if we define $\alpha(x) = 1 - r$, we enforce that $\Phi^Z(p(x)) = r$.

Let $(n_d)_{d>0}$ be defined recursively by $n_1 = 0$ and $n_{d+1} = n_d + d$. The values of α on the interval $I_d = [n_d, n_{d+1})$ are used to ensure that $\lambda C_d \leq 2^{-d}$. When d enters $\emptyset'$ at stage s, consider the set B of strings σ such that $\Phi^\sigma_s \restriction_{p(n_{d+1})}$ converges.

Informally we let C_d be the set of oracles $Y \in [B]^{\prec}$ for which Φ^Y seems to be a $\{0,1\}$-valued d.n.c. function on $p(I_d)$, namely, for all $x \in I_d$, $\Phi^Y(p(x)) \neq J(p(x)) = \alpha(x)$. Now define α in such a way that λC_d is minimal. Then $\lambda C_d \leq 2^{-d}$ because there are 2^d ways to define α on I_d. If $\emptyset' \not\leq_T Z$ then for infinitely many d, $\Phi^Z \upharpoonright_{p(n_{d+1})}$ converges before d enters $\emptyset'$, so $Z \in C_d$.

Construction of a uniformly c.e. sequence of open sets $(C_d)_{d>0}$ and a partial computable function α.

Stage s. Do nothing unless a number $d > 0$ (unique by convention) enters $\emptyset'$ at s. In that case let $B = \{\sigma\colon \Phi_s^\sigma \upharpoonright_{p(n_{d+1})}\downarrow\}$. Let τ_d be a string of length d such that λC_d becomes minimal, where

$$C_d = [\{\sigma \in B\colon \ \forall i < d\ \Phi_s^\sigma(p(n_d + i)) = \tau_d(i)\}]^{\prec}.$$

(Thus $\lambda C_d \leq 2^{-d}$. Note also that C_d is in fact clopen, but we only obtain a c.e. index for it, as we never know whether d will enter $\emptyset'$.) For $i < d$ define $\alpha(n_d + i) = 1 - \tau_d(i)$.

Verification. We show $\exists^\infty d\ Z \in C_d$. Let $g(d) = \mu s.\, \Phi_s^Z \upharpoonright_{p(n_{d+1})}\downarrow$. Since $\emptyset' \not\leq_T Z$ and $g \leq_T Z$, there are infinitely many $d \in \emptyset'$ such that $d \notin \emptyset'_{g(d)}$. If d is such a number then for each $i < d$, $\Phi^Z(p(n_d + i)) \neq J(p(n_d + i)) = \alpha(n_d + i)$ since Φ^Z is d.n.c., so since Φ^Z is $\{0,1\}$-valued, $\Phi^Z(p(n_d + i)) = 1 - \alpha(n_d + i) = \tau_d(i)$. Thus $Z \in C_d$. □

By Proposition 3.2.30 we have $\emptyset' \leq_{wtt} \Omega_R$ for each optimal prefix-free machine R. The following result of Calude, Nies, Staiger and Stephan (2008) shows that this cannot be improved to a truth-table reduction.

4.3.9 Theorem. *Let Φ be a truth-table reduction procedure such that Φ^Z is a $\{0,1\}$-valued d.n.c. function. Then Z is not ML-random. In particular, no ML-random set Z satisfies $\emptyset' \leq_{tt} Z$.*

Proof. The setting is the same as in the proof of Theorem 4.3.8: we define a partial computable function α, and are given a reduction function p such that $\alpha(e) \simeq J(p(e))$. The reduction Φ is total for each oracle Y, and we may also assume Φ^Y is $\{0,1\}$-valued. Therefore we may for each $d > 0$ compute a string τ_d of length d such that λC_d is minimal, where

$$C_d = [\{\sigma\colon \ \forall i < d\ \Phi^\sigma(p(n_d + i)) = \tau_d(i)\}]^{\prec}.$$

For $i < d$, we define $\alpha(n_d + i) = 1 - \tau_d(i)$. Clearly $\lambda C_d \leq 2^{-d}$. Moreover, $Z \in C_d$ for each d because Φ^Z is $\{0,1\}$-valued d.n.c. □

Note that $(C_d)_{d\in\mathbb{N}}$ is a Kurtz test (see Definition 3.5.1). Thus, in fact no weakly random set is truth-table above $\emptyset'$.

Demuth (1988) proved that if Z is ML-random and $\emptyset <_T Y \leq_{tt} Z$, then some $\widetilde{Y} \equiv_T Y$ is ML-random. This also shows that there is no ML-random set $Z \geq_{tt} \emptyset'$.

4.3.10 Exercise. Show that there is a set A such that A and $\emptyset'$ form a minimal pair, but no weakly 2-random set Z computes A.

Turing degrees of Martin-Löf random sets $\star$

We consider ML-randomness in the context of Turing degree structures. Let $\mathcal{D}$ denote the partial order of all Turing degrees, and let $\mathsf{ML} \subseteq \mathcal{D}$ denote the degrees of ML-random sets. We know from Theorem 3.3.2 that the degree of $A \oplus \emptyset'$ contains a ML-random set for each A. However, ML is not closed upwards in $\mathcal{D}$:

4.3.11 Proposition. *For each degree* $\mathbf{c}$, $\{\mathbf{x}\colon \mathbf{x} \geq \mathbf{c}\} \subseteq \mathsf{ML} \Leftrightarrow \mathbf{c} \geq \mathbf{0}'$.

Proof. $\Leftarrow$: This follows from 3.3.2 because each degree $\mathbf{d} \geq \mathbf{0}'$ is in ML.

$\Rightarrow$: Suppose that $C \not\geq_T \emptyset'$. Let P be the $\Pi^0_1(C)$ class of $\{0,1\}$-valued d.n.c. functions f relative to C (i.e., $\forall e \neg f(e) = J^C(e)$). We relativize Theorem 1.8.39 to C. Avoiding the cone above $\emptyset'$, we obtain a set $Y \in P$ such that $Y \oplus C \not\geq_T \emptyset'$. Since $Y \oplus C$ has PA degree, it is not in the same Turing degree as a ML-random set, for otherwise $Y \oplus C \geq_T \emptyset'$ by Theorem 4.3.8. $\square$

This yields a natural first-order definition of $\mathbf{0}'$ in the structure consisting of the partial order $\mathcal{D}$ with an additional unary predicate for ML. Shore and Slaman (1999) proved that the jump operator is first-order definable in the partial order $\mathcal{D}$. Their definition uses metamathematical notions and codings of copies of $\mathbb{N}$ with first-order formulas. Even though later on, Shore (2007) found a proof that does not rely on metamathematics, we still do not know a natural first-order definition of the jump operator (or even of $\mathbf{0}'$) in $\mathcal{D}$.

Next, we study ML within $\mathcal{D}_T(\leq \mathbf{0}')$, the Turing degrees of Δ^0_2 sets. By Exercise 4.2.7, there is an incomputable c.e. set Turing below any pair of ML-random Δ^0_2 sets. So no pair of degrees in $\mathsf{ML} \cap \mathcal{D}_T(\leq \mathbf{0}')$ has infimum $\mathbf{0}$.

4.3.12 Proposition. $\mathsf{ML} \cap \mathcal{D}_T(\leq \mathbf{0}')$ *is not closed upwards in* $\mathcal{D}_T(\leq \mathbf{0}')$. *Also,* $\mathsf{ML} \cap \mathbf{L}$ *is not closed upwards in* $\mathbf{L}$, *the set of low degrees.*

Proof. The $\{0,1\}$-valued d.n.c. functions form a Π^0_1 class, which has a low member D by Theorem 1.8.37. Then the degree of D is not in ML by 4.3.8. On the other hand, by Theorem 4.3.2(iii) there is a ML-random set $Z \leq_T D$. $\square$

4.3.13 Remark. Kučera (1988) proved that there is a minimal pair of sets of PA degree. In fact,

$$\mathbf{0}' = \inf\{\mathbf{a} \vee \mathbf{b}\colon \mathbf{a}, \mathbf{b} \text{ are PA degrees } \& \ \mathbf{a} \wedge \mathbf{b} = \mathbf{0}\}.$$

Thus there also is a natural first-order definition of $\mathbf{0}'$ in the structure consisting of the partial order $\mathcal{D}$ with an additional unary predicate for being a PA degree.

To show that $A \oplus B \geq_T \emptyset'$ for each minimal pair A, B of sets of PA degree, one builds a nonempty Π^0_1 class P without computable members such that $X \oplus Y \geq_T \emptyset'$ whenever $X, Y \in P$ and $X \neq^* Y$. Since there are $X, Y \in P$ such that $X \leq_T A$ and $Y \leq_T B$, this implies $A \oplus B \geq_T \emptyset'$. To obtain P, take a c.e. set $S \equiv_T \emptyset'$ such that $\mathbb{N} - S$ is introreducible. Say, S is the set of deficiency stages for a computable enumeration of $\emptyset'$ (see the comment after Proposition 1.7.6). Let $S = U \cup V$ be a splitting of S into computably inseparable sets (Soare 1987,

Thm. X.2.1) and let $P = \{Z\colon U \subseteq Z \ \&\ V \cap Z = \emptyset\}$. If X, Y are as above, then $X \Delta Y$ is an infinite subset of $\mathbb{N} - S$, so that $\emptyset' \leq_T X \oplus Y$.

Next, one shows that each Σ^0_2 set $C >_T \emptyset'$ bounds a minimal pair A, B of PA sets by a technique similar to the one in the solution to Exercise 1.8.46. Now let $C_0, C_1 >_T \emptyset'$ be a minimal pair of Σ^0_2 sets relative to $\emptyset'$, and let $A_i, B_i \leq_T C_i$ be minimal pairs of sets of PA degree for $i \in \{0,1\}$. Then $\mathbf{0}' = (\mathbf{a}_0 \vee \mathbf{b}_0) \wedge (\mathbf{a}_1 \vee \mathbf{b}_1)$.

Relating n-randomness and higher fixed point freeness

Definition 4.0.1 can be relativized: a function f is d.n.c. relative to C if $f(e) \neq J^C(e)$ for any e such that $J(e)\downarrow$. For $n \geq 1$, we say that f is n-d.n.c. if f is d.n.c. relative to $\emptyset^{(n-1)}$. For example, if $n > 0$ then $\emptyset^{(n)}$ computes a $\{0,1\}$-valued n-d.n.c. function, by relativizing Remark 1.8.30 to $\emptyset^{(n-1)}$.

By Proposition 4.1.2 relativized to $\emptyset^{(n-1)}$, each n-random set computes an n-d.n.c. function. Thus, for this particular aspect of the computational complexity, a higher degree of randomness implies being more complex. The result is not very satisfying because we would prefer highness properties that are not obtained by mere relativization of a highness property to $\emptyset^{(n-1)}$. For this reason, following Kučera (1990), we introduce the hierarchy of n-fixed point free functions. It turns out to coincide up to Turing degree with the hierarchy of n-d.n.c. functions. For sets A, B, let $A \sim_1 B$ if $A = B$ and $A \sim_2 B$ if $A =^* B$. For $n \geq 3$, let $A \sim_n B$ if $A^{(n-3)} \equiv_T B^{(n-3)}$.

4.3.14 Definition. Let $n \geq 1$. We say that a function g is *n-fixed point free* (n-f.p.f. for short) if $W_{g(x)} \not\sim_n W_x$ for each x.

For instance, let $g \leq_T \emptyset''$ be a function such that $W_{g(x)} = \emptyset$ if W_x is infinite and $W_{g(x)} = \mathbb{N}$ otherwise, then g is 2-f.p.f. (By Exercise 4.1.7 in relativized form and a slight modification of Exercise 4.1.8, these properties are independent of the particular choice of jump operator or universal uniformly c.e. sequence, as far as the Turing degree of the function is concerned. In Kučera's notation the hierarchies begin at level 0, but here we prefer notational consistency with the hierarchy of n-randomness for $n \geq 1$.)

We will need the Jump Theorem of Sacks (1963*a*). It states that from a Σ^0_2 set S one may effectively obtain a Σ^0_1 set A such that $A' \equiv_T S \oplus \emptyset'$. (In fact one can also achieve that $C \not\leq_T A$ for a given incomputable Δ^0_2 set C.) For a proof see Soare (1987, VIII.3.1). In fact we need a version of the theorem for the m-th jump.

4.3.15 Theorem. *Let $m \geq 0$. From a Σ^0_{m+1} set S one may effectively obtain a Σ^0_1 set A such that $A^{(m)} \equiv_T S \oplus \emptyset^{(m)}$.*

Proof. The case $m = 0$ is trivial, and the case $m = 1$ is the Jump Theorem itself. For the inductive step, suppose that $m \geq 1$ and S is Σ^0_{m+1}. Then S is a $\Sigma^0_m(\emptyset')$ set. By the result for m relative to $\emptyset'$, using $\emptyset'$ we may obtain a $\Sigma^0_1(\emptyset')$ set B such that $(B \oplus \emptyset')^{(m)} \equiv_T S \oplus \emptyset^{(m+1)}$. Then by the Limit Lemma we may in fact effectively obtain a Σ^0_2 index for B. By the Jump Theorem, we may effectively obtain a c.e. set A such that $A' \equiv_T B \oplus \emptyset'$, and hence $A^{(m+1)} \equiv_T S \oplus \emptyset^{(m+1)}$. □

The main result of this subsection is essentially due to Kučera (1990).

4.3.16 Theorem. *Let $n \geq 1$. Each n-d.n.c. function computes an n-f.p.f. function, and vice versa.*

Proof. The case $n = 1$ is covered by the proof of Proposition 4.1.4, so we may assume that $n \geq 2$. For each set $E \subseteq \mathbb{N}$ and each $i \in \mathbb{N}$, let

$$(E)_i = \{n\colon \langle n, i\rangle \in E\}.$$

1. If an n-d.n.c. function f is given, we define an n-f.p.f. function $g \leq_T f$.

Case $n = 2$. Since the index set $\{e\colon W_e \text{ is finite}\}$ is c.e. relative to $\emptyset'$, there is a Turing functional Φ such that, for each input x, $\Phi^{\emptyset'}(x)$ is the first i in an enumeration relative to $\emptyset'$ such that $(W_x)_i$ is finite if there is such an i, and $\Phi^{\emptyset'}(x)$ is undefined otherwise. Let p be a reduction function (Fact 1.2.15) such that $\Phi^{\emptyset'}(x) \simeq J^{\emptyset'}(p(x))$ for each x, and let $g \leq_T f$ be a function such that $W_{g(x)} = \{\langle y, j\rangle\colon j \neq f(p(x))\}$. If $\Phi^{\emptyset'}(x)$ is undefined then $(W_x)_i$ is infinite for each i while $(W_{g(x)})_{f(p(x))} = \emptyset$, so $W_{g(x)} \neq^* W_x$. If $\Phi^{\emptyset'}(x) = i$ then $(W_{g(x)})_i = \mathbb{N}$ since $f(p(x)) \neq i$, so again $W_{g(x)} \neq^* W_x$.

Case $n = 3$. We modify the foregoing proof. For each set B and each $i \in \mathbb{N}$, let $[B]_{\neq i} = \{\langle n, j\rangle \in B\colon j \neq i\}$. By the finite injury methods of Theorems 1.6.4 and 1.6.8, there is a low c.e. set B such that $(B)_i \not\leq_T [B]_{\neq i}$ for each i. Since B is low, the relation $\{\langle x, i\rangle\colon W_x \leq_T [B]_{\neq i}\}$ is Σ^0_3 (see Exercise 1.5.7, which is uniform), so there is a Turing functional Φ as follows: on input x, $\Phi^{\emptyset''}(x)$ is the first i in an enumeration relative to $\emptyset''$ such that $W_x \leq_T [B]_{\neq i}$ if there is such an i, and $\Phi^{\emptyset''}(x)$ is undefined otherwise. Let p be a reduction function such that $\Phi^{\emptyset''}(x) \simeq J^{\emptyset''}(p(x))$ for each x, and let $g \leq_T f$ be a function such that $W_{g(x)} = [B]_{\neq f(p(x))}$. If $\Phi^{\emptyset''}(x)$ is undefined then $W_x \not\leq_T [B]_{\neq i}$ for each i and hence $W_x \not\leq_T W_{g(x)}$. If $\Phi^{\emptyset''}(x) = i$ then $W_x \leq_T [B]_{\neq i}$. Since $f(p(x)) \neq i$ we have $(B)_i \leq_T W_{g(x)}$, and hence $W_{g(x)} \not\leq_T W_x$ because $(B)_i \not\leq_T [B]_{\neq i}$.

Case $n \geq 4$. Let $m = n - 3$ and $R = \emptyset^{(m)}$. We run a finite injury construction of the kind mentioned in the foregoing proof relative to R, and code R into each $(B)_i$. In this way we obtain a set B that is c.e. in R such that $B' \equiv_T R'$, $R \leq_T (B)_i$ and $(B)_i \not\leq_T [B]_{\neq i}$ for each i.

There is a computable function h such that $W_x^{(m)} = W^R_{h(x)}$, so the relation $\{\langle x, i\rangle\colon W_x^{(m)} \leq_T [B]_{\neq i}\}$ is $\Sigma^0_3(R)$. Since $B' \equiv_T R'$, there is a Turing functional Φ as follows: on input x, $\Phi^{R''}(x)$ is the first i in an enumeration relative to R'' such that $W_x^{(m)} \leq_T [B]_{\neq i}$ if there is such an i, and $\Phi^{R''}(x)$ is undefined otherwise. Let p be a reduction function such that $\Phi^{R''}(x) \simeq J^{R''}(p(x))$ for each x, and let $\widetilde{g} \leq_T f$ be like g before, namely $W^R_{\widetilde{g}(x)} = [B]_{\neq f(p(x))}$. As before one shows that $W^R_{\widetilde{g}(x)} \neq_T W_x^{(m)}$ for each x. Now, by the uniformity of Theorem 4.3.15, there is a function $g \le_T \widetilde{g}$ such that $(W_{g(x)})^{(m)} \equiv_T W^R_{\widetilde{g}(x)} \oplus R$ for each x. Since $R \le_T W^R_{\widetilde{g}(x)}$, this implies that $W^{(m)}_{g(x)} \not\equiv_T W^{(m)}_x$ for each x. (This proof actually works for $n = 3$ as well, in which case we re-obtain the previous one.)

2. If an n-f.p.f. function g is given, we define an n-d.n.c. function $f \le_T g$. We use a result of Jockusch, Lerman, Soare and Solovay (1989) which can also be found in Soare (1987, pg. 273): if ψ is partial computable relative to $\emptyset^{(n-1)}$ then there is a (total) computable function r such that $W_{\psi(x)} \sim_n W_{r(x)}$ whenever $\psi(x)$ is defined. (Say, if $n = 2$, fix a Turing functional Φ such that $\psi = \Phi^{\emptyset'}$, and let $\psi_s(x) \simeq \Phi^{\emptyset'}(x)[s]$. As long as $e = \psi_s(x)$, $W_{r(x)}$ follows the enumeration of W_e.)

Now let $\psi = J^{\emptyset^{(n-1)}}$. If $\psi(x)\downarrow$ then

$$W_{\psi(x)} \sim_n W_{r(x)} \not\sim_n W_{g(r(x))}, \text{ so } \psi(x) \neq g(r(x)).$$

Hence $f = g \circ r$ is n-d.n.c. and $f \le_T g$. □

As a consequence we obtain the following result of Kučera (1990).

4.3.17 Corollary. *If Z is n-random then there is an n-f.p.f. function $g \le_T Z$.*

Proof. By Proposition 4.1.2 relative to $\emptyset^{(n-1)}$, a finite variant f of the function $\lambda n. Z\upharpoonright_n$ is n-d.n.c. There is an n-f.p.f. function $g \le_T f$. □

The converse fails because the n-random degrees are not closed upwards. For $n = 1$ this follows from Proposition 4.3.11, and for $n \ge 2$ it follows because each 2-random set forms a minimal pair with $\emptyset'$. Also note that for each $n > 1$ there is an $(n-1)$-random set $Z \le_T \emptyset^{(n-1)}$. Then Z does not compute an n-f.p.f. function.

Our proof of Corollary 4.3.17 is somewhat indirect as it relies on Schnorr's Theorem 3.2.9 relative to $\emptyset^{(n-1)}$. The proofs in Kučera (1990) are more self-contained (but also longer). The ideas needed in the proof of Theorem 4.3.16 were introduced there. For instance, Kučera's proof of (1.) in the case $n = 2$ is as follows. Given $x \in \mathbb{N}$, using the infinite set Z as an oracle, we may compute $n_x = \mu n. \#Z \cap [0, n) = x$. Let $g \le_T Z$ be a function such that, for each x, $W_{g(x)} = \{\langle y, i\rangle : i < n_x \to i \notin Z\}$. We will show that $W_x \neq^* W_{g(x)}$ for *almost* all x, which is sufficient to establish the theorem.

Since the index set $\{e : W_e \text{ is finite}\}$ is Σ^0_2, there is a computable function h such that $W^{\emptyset'}_{h(x)} = \{i : (W_x)_i \text{ is finite}\}$. We define a ML-test $(G_x)_{x\in\mathbb{N}}$ relative to $\emptyset'$, as follows. On input x, initially let $G_x = \emptyset$. For $x > 0$, once distinct elements $a_0, \ldots, a_{x-1}$ have been enumerated into $W^{\emptyset'}_{h(x)}$, let $G_x = \{Y : \forall i < x\,[Y(a_i) = 1]\}$. Clearly G_x is $\Sigma^0_1(\emptyset')$ uniformly in x, and $\lambda G_x \le 2^{-x}$. Then, since Z is 2-random and $(G_x)_{x\in\mathbb{N}}$ is a Solovay test relative to $\emptyset'$, there is x_0 such that $Z \notin G_x$ for all $x \ge x_0$. Because $\#Z \cap [0, n_x) = x$,

$$\forall x \ge x_0\, W^{\emptyset'}_{h(x)} \neq Z \cap [0, n_x).$$

Assume for a contradiction that $x \ge x_0$ and $W_{g(x)} =^* W_x$. Then $(W_{g(x)})_i =^* (W_x)_i$ for all i. If $i \ge n_x$ this implies that $i \notin W^{\emptyset'}_{h(x)}$. If $i < n_x$ then

$$i \in Z \leftrightarrow (W_{g(x)})_i = \emptyset \leftrightarrow (W_x)_i \text{ is finite } \leftrightarrow i \in W^{\emptyset'}_{h(x)}.$$

Thus $W_{h(x)} = [0, n_x) \cap Z$, a contradiction.

4.3.18 Exercise. Show that there is a weakly 2-random set that does not compute a 2-f.p.f. function. *Hint.* Use Exercise 1.8.46.

5

LOWNESS PROPERTIES AND K-TRIVIALITY

In this chapter and in Chapter 8 we will study lowness properties of a set A. In particular, we are interested in the question of how they interact with concepts related to randomness. The main new notion is the following: A is K-trivial if

$$\forall n \; K(A\upharpoonright_n) \leq K(n) + b$$

for some constant b, namely, up to a constant the descriptive complexity of $A\upharpoonright_n$ is no more than the descriptive complexity of its length. This expresses that A is far from being ML-random, since ML-random sets have a quickly growing initial segment complexity by Schnorr's Theorem 3.2.9. We show that K-triviality coincides with the lowness property of being low for ML-randomness.

Lowness properties via operators. Recall from Section 1.5 that a lowness property of a set A specifies a sense in which A is computationally weak. We always require that a lowness property be closed downward under Turing reducibility. Mostly, computational weakness of A means that A is not very useful as an oracle. Lowness properties of a set A are very diverse, as each one represents only a particular aspect of how information can be extracted from A. They can even exclude each other for incomputable sets.

We have introduced several lowness properties by imposing a restriction on the functions A computes. For instance, A is computably dominated if each function $f \leq_T A$ is dominated by a computable function (Definition 1.5.9). In this chapter we define lowness properties by the condition that the relativization of a class $\mathcal{C}$ to A is the same as $\mathcal{C}$. For example let $\mathcal{C} = \Delta^0_2$. For each A, by the Limit Lemma 1.4.2 relative to A, we have $\mathcal{C}^A = \Delta^0_2(A) = \{X \colon X \leq_T A'\}$. For an operator $\mathcal{C}$ mapping sets to classes, we say that A is *low for* $\mathcal{C}$ if $\mathcal{C}^A = \mathcal{C}$. We write $\text{Low}(\mathcal{C})$ for this class.

The condition $A \in \text{Low}(\mathcal{C})$ means that A is computationally weak in the sense that its extra power as an oracle does not expand $\mathcal{C}$, contrary to what one would usually expect. When $\mathcal{C} = \Delta^0_2$, A is low for $\mathcal{C}$ iff $A' \equiv_T \emptyset'$, that is, A is low in the usual sense of 1.5.2.

In some cases, a lowness property implies that the set is in Δ^0_2. Clearly this is so for the usual lowness. In contrast, being computably dominated is a lowness property such that the only Δ^0_2 sets with that property are the computable ones. One may view the computably dominated sets as a class of the type $\text{Low}(\mathcal{C})$ by letting $\mathcal{C}^X$ be the class of functions dominated by a function $f \leq_T X$.

An operator $\mathcal{C}$ is called *monotonic* if $A \leq_T B \;\rightarrow\; \mathcal{C}^A \subseteq \mathcal{C}^B$. An operator $\mathcal{D}$ is called *antimonotonic* if $A \leq_T B \;\rightarrow\; \mathcal{D}^A \supseteq \mathcal{D}^B$. The operator $X \rightarrow \Delta^0_2(X)$ is monotonic, while MLR, the operator given by Martin-Löf randomness relative to an oracle, is antimonotonic. For both types of operators, the corresponding lowness notion is indeed closed downward under Turing reducibility. For instance, $A \leq_T B$ and $\mathcal{C}^B = \mathcal{C}$ implies $\mathcal{C} \subseteq \mathcal{C}^A \subseteq \mathcal{C}^B = \mathcal{C}$.

Let $\mathcal{D}$ be a randomness notion. Informally, A is in $\text{Low}(\mathcal{D})$ if the computational power of A does not help us to find new regularities in a set that is random in the sense of $\mathcal{D}$. Although $\mathcal{D}^A$ is ultimately defined in terms of computations with oracle A, the operator $\mathcal{D}$ looks at this information extracted from A in a sophisticated way via $\mathcal{D}$-tests relative to A. For this reason, studying the behavior of $\mathcal{D}^A$ often yields interesting results on the computational complexity of A. In this chapter we focus on lowness for ML-randomness. In Chapter 8 we study lowness for randomness notions weaker than ML-randomness.

Existence results and characterization. A computable set satisfies *every* lowness property. Sometimes one is led to define a lowness property and then discovers that *only* the computable sets satisfy it. Consider the operator $\mathcal{C}(X) = \Sigma^0_1(X)$: here $A \in \text{Low}(\mathcal{C})$ implies that A is computable, because both A and $\mathbb{N} - A$ are in $\Sigma^0_1(A) = \Sigma^0_1$.

It can be difficult to determine whether a lowness property only applies to the computable sets. If it does so, this is an interesting fact, especially in the case of a class $\text{Low}(\mathcal{C})$; an example is lowness for computable randomness by Corollary 8.3.11. However, it is also the final result.

On the other hand, if an incomputable set with the property exists, then one seeks to understand the lowness property via some characterization. This is especially useful when the property is given in the indirect form $\text{Low}(\mathcal{C})$. In that case, one seeks to characterize the class by conditions of the following types.

(1) The initial segments of A have a slowly growing complexity.
(2) The functions computed by A are restricted.

A main result of this chapter is a characterization of the first type: A is low for ML-randomness iff A is K-trivial. This is surprising, because having a slowly growing initial segment complexity expresses that A is far from random, rather than computationally weak. This result provides further insight into the class $\text{Low}(\mathsf{MLR})$. For instance, we use it to show that $\text{Low}(\mathsf{MLR})$ induces an ideal in the Turing degrees. This fails for most of the other lowness properties we study.

Theorem 8.3.9 below is a characterization of the second type: a set is low for Schnorr randomness iff it is computably traceable (a strengthening of being computably dominated).

Overview of this chapter. In Section 5.1 we introduce several lowness properties and show their coincidence with $\text{Low}(\mathsf{MLR})$. In Section 5.2 we study K-triviality for its own sake, proving for instance that each K-trivial set is in Δ^0_2. In Section 5.3 we introduce the cost function method. It can be used to build K-trivial sets, or sets satisfying certain lowness properties. Different solutions to Post's problem can be viewed as applications of this method for different cost functions. There is a criterion on the cost function, being nonadaptive, to tell whether this solution is injury-free. Section 5.4 contains the proof that each K-trivial set is low for ML-randomness, introducing the decanter and golden run methods. Section 5.5 applies these methods to derive further properties of the K-trivial sets. In Section 5.6 we introduce the informal concept of weak reducibilities, which are implied by $\leq_T$. We study the weak reducibility $\leq_{LR}$ associated with the class Low(MLR). We also prove the coincidence of two highness properties: we characterize the class $\{C\colon \emptyset' \leq_{LR} C\}$ by a strong domination property.

Some key results of this chapter are non-uniform. For instance, even though every K-trivial set is low, we cannot effectively obtain a lowness index from a c.e. K-trivial set and its constant (Proposition 5.5.5). Corollary 5.1.23 is also non-uniform, as discussed in Remark 5.1.25.

5.1 Equivalent lowness properties

We study three lowness properties that will later turn out to be equivalent: being low for K, low for ML-randomness, and a base for ML-randomness.

The first two properties indicate computational weakness as an oracle. A is low for K if A as an oracle does not help to compress strings any further. A is low for ML-randomness if each ML-random set is already ML-random relative to A. The third property, being a base for ML-randomness, is somewhat different: A is considered computationally weak because the class of oracles computing A looks large to A itself, in the sense that some set $Z \geq_T A$ is ML-random relative to A.

The first two implications are easy to verify: A is low for $K \Rightarrow A$ is low for ML-randomness $\Rightarrow A$ is a base for ML-randomness. The remaining implication is the main result of this section: A is a base for ML-randomness $\Rightarrow A$ is low for K (Theorem 5.1.22).

A fourth equivalent property is lowness for weak 2-randomness. The implication A is low for weak 2-randomness $\Rightarrow A$ is low for K is obtained at the end of this section, the converse implication only in Theorem 5.5.17 of Section 5.5.

Being low for K

5.1.1 Definition. We say that A is *low for K* if there is $b \in \mathbb{N}$ such that

$$\forall y\, [K^A(y) \geq K(y) - b].$$

Let $\mathcal{M}$ denote the class of sets that are low for K.

Each computable set is low for K. Also, $\mathcal{M}$ is closed downward under Turing reducibility, because $B \leq_T A$ implies $\forall y\, [K^B(y) \geq^+ K^A(y)]$ by Exercise 3.4.4.

The notion was introduced by Andrej A. Muchnik in unpublished work dating from around 1999. He built an incomputable c.e. set that is low for K. Later in this section we will reprove his result, but at first in a different, somewhat indirect way: we apply Kučera's construction in 4.2.1 to obtain a c.e. incomputable set A Turing below a low ML-random set. Then, in Corollary 5.1.23 we prove that each set A of this kind is low for K. However, in the proof of Theorem 5.3.35 we also give a direct construction of a set that is low for K, similar to Muchnik's.

5.1.2 Proposition. *Each set $A \in \mathcal{M}$ is generalized low in a uniform way. More precisely, from a constant b one can effectively determine a Turing functional Ψ_b such that, if A satisfies $\forall y\,[K(y) \le K^A(y) + b]$, then $A' = \Psi_b(\emptyset' \oplus A)$.*

Proof idea: Recall that $J^A(e)$ denotes $\Phi_e^A(e)$, and $J_s^A(e)$ denotes $\Phi_{e,s}^A(e)$. For a stage s, using A as an oracle we can check whether $J_s^A(e)\downarrow$. So all we need is a bound on the last stage when this can happen: if such a stage s exists then $K^A(s)$ and hence $K(s)$ is at most $e + O(1)$. Thus $\emptyset'$ can compute such a bound.

Proof details. Let M_c, $c > 1$, be an oracle prefix-free machine such that $M_c^Z(0^e1) \simeq \mu s.\, J_s^Z(e)[s] \downarrow$ for each Z and each e. Thus $M_c^Z(0^e1)$ converges iff $J^Z(e)$ converges. Since $M_c^Z(0^e1) \simeq \mathbb{U}^Z(0^{c-1}10^e1)$, we have $K^Z(M_c^Z(0^e1)) \le e+c+1$ for each Z. Thus, if A is low for K via b then $K(M_c^A(0^e1)) \le e+c+1+b$ for each e. To compute A' from $\emptyset' \oplus A$, given input e use $\emptyset'$ to determine $t = \max\{\mathbb{U}(\sigma) : |\sigma| \le e+c+1+b\}$. Then $J^A(e)\downarrow \leftrightarrow J_t^A(e)\downarrow$, so output 1 if $J_t^A(e)\downarrow$, and 0 otherwise. The reduction procedure was obtained effectively from b. □

For a c.e. set A, modifying the argument improves the result.

5.1.3 Proposition. *Each c.e. set $A \in \mathcal{M}$ is superlow. The reduction procedure for $A' \le_{tt} \emptyset'$ can be obtained in a uniform way.*

Proof. By Proposition 1.4.4 (which is uniform) it suffices to show $A' \le_{wtt} \emptyset'$. Let M_c, $c > 1$, be an oracle prefix-free machine such that for each e,

$$M_c^A(0^e1) \simeq \mu s.\, J_s^A(e)[s]\downarrow \;\;\&\;\; A_s \restriction \mathsf{use}\, J_s^A(e) = A \restriction \mathsf{use}\, J_s^A(e).$$

As before $J^A(e)\downarrow \leftrightarrow M_c^A(0^e1)\downarrow$, in which case $J^A(e)$ has stabilized by stage $M_c^A(0^e1)$. Also $K^Z(M_c^Z(0^e1)) \le e+c+1$ for each Z.

Suppose A is low for K via b. To compute A' from $\emptyset'$, given input e, use $\emptyset'$ to determine $t = \max\{\mathbb{U}(\sigma) : |\sigma| \le e+c+1+b\}$. Then $J^A(e)\downarrow \leftrightarrow J^A(e)\downarrow[t]$. The use of this reduction procedure is bounded by the computable function $\lambda e.\, p(2^{e+c+1+b})$, where p is a reduction function for the partial computable function $\mathbb{U}$. □

Using a more powerful method, in Corollary 5.5.4 we will remove the restriction that A be c.e., by showing that each set $A \in \mathcal{M}$ is Turing below some c.e. set $D \in \mathcal{M}$. This will supersede Proposition 5.1.2 (except for the uniformity statement).

We also postpone the result that $\mathcal{M}$ is a proper subclass of the superlow sets. By the results in Section 5.4, $\mathcal{M}$ induces a proper ideal in the Δ_2^0 Turing degrees, while there are superlow c.e. sets A_0, A_1 such that $\emptyset' \equiv_T A_0 \oplus A_1$, Theorem 6.1.4. Alternatively, there is a superlow c.e. set that is not low for K by Proposition 5.1.20 below; the proof relies on a direct construction.

Exercises. The first two are due to Merkle and Stephan (2007).

5.1.4. Let A be low for K. Suppose $Z \subseteq \mathbb{N}$ and let $Y = Z \triangle A$ be the symmetric difference. Show that $\forall n\, K(Y\restriction_n) =^+ K(Z\restriction_n)$. In particular, if Z is ML-random then Y is ML-random as well by Schnorr's Theorem. (Compare this with 3.6.21.)

5.1.5. Continuing 5.1.4, suppose that A is incomputable and Z is 2-random. Let $Y = Z \triangle A$. Show that $Y \mid_T Z$. (Use that A is Δ_2^0 by 5.2.4(ii) below.)

5.1.6. (Kučera) Show that there is a function $F \leq_{wtt} \emptyset'$ dominating each function that is partial computable in some set that is low for K.

Lowness for ML-randomness

See the introduction to this chapter for background on lowness for operators.

5.1.7 Definition. A is *low for ML-randomness* if $\mathsf{MLR}^A = \mathsf{MLR}$.
Low(MLR) denotes the class of sets that are low for ML-randomness.

Thus, A is low for ML-randomness if MLR^A is as large as possible. This property was introduced by Zambella (1990). He left the question open whether some incomputable set is low for ML-randomness.

The question was answered in the affirmative in work, dating from 1996, of Kučera and Terwijn (1999). They actually built a c.e. incomputable set in Low(MLR). Exercise 5.3.38 below asks for a direct construction of such a set.

By the next fact, the existence of such a set also follows from Muchnik's result that some incomputable c.e. set is low for K.

Our first proof that a c.e. incomputable set exists in Low(MLR) is actually via Kučera's Theorem 4.2.1; see page 170.

5.1.8 Fact. *Each set that is low for K is low for ML-randomness.*

Proof. Recall that Theorem 3.2.9 can be relativized: Z is ML-random relative to $A \Leftrightarrow \forall n\, K^A(Z \upharpoonright_n) \geq^+ n$. Thus the class of ML-random sets MLR can be characterized in terms of K, and MLR^A in terms of K^A. If A is low for K then the function $\text{abs}(K - K^A)$ is bounded by a constant, so $\mathsf{MLR} = \mathsf{MLR}^A$. □

In Corollary 5.1.10 we will characterize Low(MLR) using only effective topology and the uniform measure: $A \in$ Low(MLR) $\Leftrightarrow$ if G is open, c.e. in A, and $\lambda G < 1$, then G is contained in a c.e. open set S (without oracle A) such that $\lambda S < 1$. (In Section 5.6 we will consider several variants of such covering procedures.) The result, due to Kjos-Hanssen (2007), is obtained through the following:

5.1.9 Theorem. *The following are equivalent for a set A.*

(i) A is low for ML-randomness.

(ii) There is a c.e. open set S such that

$$\lambda S < 1 \ \&\ \forall z\, [K^A(z) \leq |z| - 1 \ \rightarrow \ [z] \subseteq S]. \tag{5.1}$$

(iii) For each oracle prefix-free machine M, there is a c.e. open set S such that

$$\lambda S < 1 \ \&\ \forall z\, [K_{M^A}(z) \leq |z| - 1 \ \rightarrow \ [z] \subseteq S]. \tag{5.2}$$

Statement (ii) expresses that A is weak as an oracle in that there are few strings z with a description using A of length $\leq |z| - 1$. Note that the condition (5.1) is equivalent to $\lambda S < 1 \ \&\ \mathcal{R}_1^A \subseteq S$. The characterization of Low(MLR) in (i) $\Leftrightarrow$ (ii) is due to Nies and Stephan (unpublished), who used it to show that the index set $\{e\colon W_e \in \text{Low}(\mathsf{MLR})\}$ is Σ^0_3; see Exercise 5.1.11. This fact also follows from the coincidence of Low(MLR) with $\mathcal{M}$ obtained later on. It is not obvious from the definition of Low(MLR) itself.

Proof. (iii)⇒(ii): trivial.

(ii)⇒(i): If (5.1) holds then non-$\mathsf{MLR}^A \subseteq S$. By Fact 1.8.26, there is a computable antichain B such that $[B]^{\prec} = S$. By Proposition 3.2.15, non-MLR^A is closed under taking off finite initial segments. Thus non-$\mathsf{MLR}^A \subseteq \bigcap_n [B^n]^{\prec}$. By Example 3.2.23, $([B^n]^{\prec})_{n\in\mathbb{N}}$ is a Solovay test. Hence non-$\mathsf{MLR}^A \subseteq$ non-MLR, that is, A is low for ML-randomness.

(i)⇒(iii): Suppose that M is an oracle prefix-free machine for which (5.2) fails. We build a set $Z \in \mathsf{MLR} - \mathsf{MLR}^A$, whence A is not low for ML-randomness. We let $Z = z_0 z_1 z_2 \ldots$ for an inductively defined sequence of strings $z_0, z_1 \ldots$ with the properties (a) and (b) below.

(a) We ensure that $K_{M^A}(z_i) \le |z_i| - 1$. Thus $Z \notin \mathsf{MLR}^A$ by Proposition 3.2.17 relativized to A.

(b) Let H be a c.e. open set such that $\lambda H < 1$ and $2^{\mathbb{N}} - H \subseteq \mathsf{MLR}$. Say, let $H = \mathcal{R}_1 = [\{z\colon K(z) \le |z| - 1\}]^{\prec}$. We will have $[z_0 \ldots z_{n-1}] \not\subseteq H$ for each n, whence $Z \in \mathsf{MLR}$ as $2^{\mathbb{N}} - H$ is closed.

Inductively, suppose we have defined $z_0, \ldots, z_{n-1}$ such that $[w] \not\subseteq H$ where $w = z_0 \ldots z_{n-1}$ (in case $n = 0$, we read this as $w = \emptyset$, so we have $[\emptyset] = 2^{\mathbb{N}} \not\subseteq H$, as required). Let $S = H \mid w = \{Z\colon wZ \in H\}$, then S is c.e. open and $S \neq 2^{\mathbb{N}}$ as H is c.e. open and $[w] \not\subseteq H$. The nonempty Π^0_1 class $2^{\mathbb{N}} - S$ is contained in MLR, and is therefore not null. Hence $\lambda S < 1$. Since (5.2) fails, there is $z = z_n$ such that $K_{M^A}(z_n) \le |z_n| - 1$ and $[z_n] \not\subseteq S$. Thus $[z_0 \ldots z_n] \not\subseteq H$. □

Note that by Fact 1.8.56(ii) relative to an oracle X, a class $G \subseteq 2^{\mathbb{N}}$ is open and c.e. in X iff G is a $\Sigma^0_1(X)$ class.

5.1.10 Corollary. *A is low for ML-randomness ⇔ each $\Sigma^0_1(A)$ class G such that $\lambda G < 1$ is contained in a Σ^0_1 class S such that $\lambda S < 1$.*

Proof. ⇐: For the set $G = \mathcal{R}^A_1$ there is an open c.e. set S, $\lambda S < 1$, such that $G \subseteq S$. Thus (ii) in Theorem 5.1.9 holds.

⇒: Applying the Lebesgue Density Theorem 1.9.4 to $\mathcal{C} = 2^{\mathbb{N}} - G$, we obtain a string σ such that $\lambda(G \mid \sigma) \le 1/2$. Let $H = G \mid \sigma = \{Z\colon \sigma Z \in G\}$. By Fact 1.8.26 relativized to A, let $B \le_T A$ be an antichain in $\{0,1\}^*$ such that $[B]^{\prec} = H$. Then $\{\langle |x| - 1, x\rangle\colon x \in B\}$ is a bounded request set relative to A, so by Theorem 2.2.17 relative to A, there is an oracle prefix-free machine M such that $B = \{x\colon K_{M^A}(x) \le |x| - 1\}$. By (i)⇒(iii) of Theorem 5.1.9, there a Σ^0_1 class $\widehat{S}$ such that $\lambda\widehat{S} < 1$ and $H \subseteq \widehat{S}$. Now $S = \sigma\widehat{S} \cup (2^{\mathbb{N}} - [\sigma])$ is a Σ^0_1 class as required. □

5.1.11 Exercise. Show that the set $\{e\colon W_e \in \mathrm{Low}(\mathsf{MLR})\}$ is Σ^0_3.

When many oracles compute a set

In most cases, computational weakness of a set A means that, in some sense or other, A is not very useful as an oracle. However, another possible interpretation is that A is easy to compute in that the class of oracles computing A is large. Given a set A, let $\mathcal{S}^A = \{Z\colon A \le_T Z\}$. The set A is considered not complex if $\mathcal{S}^A$ is large, and complex if $\mathcal{S}^A$ is small. If smallness merely means being a null class then we can only distinguish between the computable sets A, where $\mathcal{S}^A = 2^{\mathbb{N}}$, and the incomputable sets A, where $\mathcal{S}^A$ is null. In the present form this result appeared in Sacks (1963*b*), but it is in fact an easy consequence of a result in de Leeuw, Moore, Shannon and Shapiro (1956).

5.1.12 Theorem. *A is incomputable $\Leftrightarrow$ the class $\mathcal{S}^A = \{Z\colon A \le_T Z\}$ is null.*

Proof. $\Leftarrow$: If A is computable then $\mathcal{S}^A = 2^{\mathbb{N}}$, so $\lambda\mathcal{S}^A = 1$.
$\Rightarrow$: For each Turing functional Φ let

$$S^A_\Phi = \{Z\colon A = \Phi^Z\}.$$

Thus $\mathcal{S}^A = \bigcup_\Phi S^A_\Phi$. It suffices to show that each S^A_Φ is null. Suppose for a contradiction that $\lambda S^A_\Phi \ge 1/r$ for some $r \in \mathbb{N}$. Then A is a path on the c.e. binary tree

$$T = \{w\colon \lambda[\{\sigma\colon \Phi^\sigma \succeq w\}]^\prec \ge 1/r\}.$$

Each antichain on T has at most r elements, for if $w_0, \ldots, w_r$ is an antichain with $r+1$ elements, we have $r+1$ disjoint sets $[\{\sigma\colon \Phi^\sigma = w_i\}]^\prec$, each of measure at least $1/r$, which is impossible. Then there is n_0 such that for each $n \ge n_0$, $A\upharpoonright_n$ is the *only* string w on T of length n extending $A\upharpoonright_{n_0}$. For otherwise we could pick $r+1$ strings on T branching off A at different levels, and these strings would form an antichain with $r+1$ elements.

Now, to compute $A(m)$ for $m \ge n_0$, wait till some $w \succeq A\upharpoonright_{n_0}$ of length $m+1$ is enumerated into T, and output $w(m)$. □

5.1.13 Remark. For $n > 0$ let

$$S^A_{\Phi,n} = [\{\sigma\colon A\upharpoonright_n \preceq \Phi^\sigma\}]^\prec.$$

Then $S^A_{\Phi,n}$ is $\Sigma^0_1(A)$ uniformly in n. If A is incomputable then $S^A_\Phi = \bigcap_n S^A_{\Phi,n}$ is null, so $\lim_n \lambda S^A_{\Phi,n} = 0$. That is, $(S^A_{\Phi,n})_{n\in\mathbb{N}}$ is a generalized ML-test relative to A (Definition 3.6.1).

If A itself is ML-random, then leaving out the first few components even turns $(S^A_{\Phi,n})_{n\in\mathbb{N}}$ into a ML-test relative to A (Miller and Yu, 2008).

5.1.14 Proposition. *Suppose A is ML-random. Then for each Turing functional Φ there is a constant c such that $(S^A_{\Phi,n+c})_{n\in\mathbb{N}}$ is a ML-test relative to A.*

The easiest proof is obtained by observing that the c.e. supermartingale (see Definition 7.1.5 below) given by $L(x) = 2^{|x|}\lambda[\{\sigma\colon \Phi^\sigma \succeq x\}]^\prec$ is bounded by 2^c for some c along A. The details are provided on page 266.

Exercises.

5.1.15. Show that the sets of PA degree (Definition 4.3.1) form a null class.

5.1.16. Suppose that A and Y are ML-random and $A \le_T Y$. Show that if Y is (i) Demuth random, (ii) weakly 2-random, (iii) ML-random relative to a set C, then A has the same property.

Bases for ML-randomness

We continue to work on the question to what extent high computational complexity of a set A is reflected by smallness of the class $\mathcal{S}^A$ of sets computing A. We have seen that interpreting smallness by being null is too coarse since this makes all the incomputable sets complex. Rather, we will consider the case that $\mathcal{S}^A$ is null in an effective sense relative to A. From Remark 5.1.13 we obtain the following.

5.1.17 Corollary. *Let A be incomputable. If Z is weakly 2-random relative to A then $A \not\leq_T Z$.* □

Thus $\mathcal{S}^A$ looks small to an incomputable A even in a somewhat effective sense: $\mathcal{S}^A$ does not have a member that is weakly 2-random relative to A. On the other hand, by the Kučera-Gács Theorem 3.3.2 $\mathcal{S}^A$ always contains a ML-random set. If A is ML-random then by Proposition 5.1.14 $\mathcal{S}^A$ does not contain a ML-random set relative to A.

Is there any incomputable set A such that $\mathcal{S}^A$ contains a ML-random set relative to A? This property was first studied by Kučera (1993).

5.1.18 Definition. A is a *base for ML-randomness* if $A \leq_T Z$ for some set Z that is ML-random relative to A.

Kučera used the term "basis for 1-RRA". There is no connection to basis theorems. Each set A that is low for ML-randomness is a base for ML-randomness. For, by the Kučera-Gács Theorem 3.3.2 there is a ML-random set Z such that $A \leq_T Z$. Then Z is ML-random relative to A.

We will prove two theorems. They are due to Kučera (1993) and Hirschfeldt, Nies and Stephan (2007), respectively.

Theorem 5.1.19: There is a promptly simple base for ML-randomness.
Theorem 5.1.22: Each base for ML-randomness is low for K.

This completes the cycle: the classes of sets that are low for K, low for ML-randomness, and bases for ML-randomness are all the same! Moreover, this common class reaches beyond the computable sets:

5.1.19 Theorem. *There is a promptly simple base for ML-randomness.*

Proof. Let Z be a low ML-random set, say $Z = \Omega_0$, the bits of Ω in the even positions (see Corollary 3.4.11). By Theorem 4.2.1 there is a promptly simple set $A \leq_T Z$. Then, by 3.4.13, Z is in fact ML-random relative to A. □

For the proof of Theorem 5.1.22 below it might be instructive to begin with a direct construction of a superlow c.e. set A which is not a base for ML-randomness. (Before 5.1.4 we already discussed how to obtain a superlow c.e. set that is not low for K, so this also follows from 5.1.22).

5.1.20 Proposition. *There is a superlow c.e. set A such that for each set Z, if $A \leq_T Z$ then Z is not ML-random relative to A.*

Proof sketch. For each e we build a ML-test $(C^A_{e,d})_{d\in\mathbb{N}}$ relative to A in such a way that for each $j = \langle e, d\rangle$, the following requirement is met:

$$R_j : \forall Z\,[A = \Phi^Z_e \;\Rightarrow\; Z \in C^A_{e,d}].$$

We make A low by meeting the lowness requirements L_e from the proof of Theorem 1.6.4. When $J^A(e)$ newly converges we inititalize the requirements R_j for $j > e$.

The strategy for R_j is as follows. At each stage let k be the number of times R_j has been initialized.

1. Choose a large number m_j.
2. At stage s let $S_{j,s}$ be the clopen set $\{Z\colon\ A\restriction_{m_j} = \Phi^Z_e\restriction_{m_j}[s]\}$. While $\lambda S_{j,s} < 2^{-d-k-1}$, put $S_{j,s}$ into $C^A_{e,d}$ with use m_j on the oracle A. Otherwise put $m_j - 1$ into A, initialize the requirements L_e for $e \geq j$ (we say that R_j acts) and goto 1.

Thus, the strategy for R_j keeps putting $S_{j,s}$ into $C^A_{e,d}$ until this makes the contribution (for this value of k) too large; if it becomes too large, the strategy removes the current contribution by changing A. In the *construction*, at stage s let the requirement of strongest priority that requires attention carry out one step of its strategy.

Verification. Each requirement R_j acts only finitely often: $\lambda S_{j,s}$ can reach 2^{-d-k-1} at most 2^{d+k+1} times, since the different versions of $S_{j,s}$ are disjoint. Thus A is low and each m_j reaches a limit. Next, for each e, d we have $\lambda C^A_{e,d} \leq \sum_k 2^{-d-k-1} \leq 2^{-d}$, so $(C^A_{e,d})_{d\in\mathbb{N}}$ is a ML-test relative to A. If $A = \Phi^Z_e$, for each d let $j = \langle e, d\rangle$ and consider the final value of m_j. Since $A\restriction_{m_j} = \Phi^Z_e\restriction_{m_j}$, Z is in $C^A_{e,d}$. Thus Z is not ML-random relative to A. Finally, there is a computable function f such that $f(e)$ bounds the number of injuries to L_e, so A is superlow. □

5.1.21 Remark. (The accounting method) We describe an important method to show that a c.e. set L of requests is a bounded request set. We associate a request $\langle r, x\rangle$ with an open set C such that $\lambda C \geq 2^{-r}$. The open sets belonging to different requests are disjoint. Then the total weight $\sum_r 2^{-r}\,[\![\langle r, y\rangle \in L]\!]$ is at most the sum of the measures of those sets, and hence at most 1. Informally speaking, we "account" the enumeration of $\langle r, x\rangle$ against the measure of the associated open set. We usually enumerate these open sets actively. When they reach the required measure we are allowed to put the request into L.

We are now ready for the main theorem of this section, which is due to Hirschfeldt, Nies and Stephan (2007).

5.1.22 Theorem. *Each base for ML-randomness is low for K.*

Proof. Suppose that $A \leq_T Z$ for some set Z that is ML-random relative to A. Given a Turing functional Φ, we define an oracle ML-test $(C^X_d)_{d\in\mathbb{N}}$. If $A = \Phi^Z$, we intend to use this test for $X = A$. The goal is as follows: if d is a number such that $Z \notin C^A_d$, then A is low for K via the constant $d + O(1)$. To realize this goal, we build a uniformly c.e. sequence $(L_d)_{d\in\mathbb{N}}$ of bounded request sets. The constructions for different d are independent, but uniform in d, so that for each X, the sequence of open sets $(C^X_d)_{d\in\mathbb{N}}$ is uniformly c.e. in X. (This idea was already used in the proof of Theorem 3.2.29: make an attempt for each d.

Let the construction with parameter d succeed when a ML-random set Z is not in the d-th component of a ML-test. Here, the ML-tests are relative to A, and Z is ML-random in A.) We denote by $C \subseteq \{0,1\}^*$ also the open set $[C]^{\prec}$.

For each computation

$$\mathbb{U}^\eta(\sigma) = y \text{ where } \eta \preceq A$$

(that is, whenever y has a $\mathbb{U}^A$-description σ), we want to ensure that there is a prefix-free description of y not relying on an oracle that is only by a constant longer. Thus we want to put a request $\langle |\sigma| + d + 1, y \rangle$ into L_d.

The problem is that we do not know A, so we do not know which η's to accept; if we accept too many then L_d might fail to be a bounded request set. To avoid this, the description $\mathbb{U}^\eta(\sigma) = y$ first has to prove itself worthy. Once $\mathbb{U}^\eta(\sigma)$ converges, we enumerate open sets $C^\eta_{d,\sigma}$ (if $\mathbb{U}^\eta(\sigma)$ diverges then $C^\eta_{d,\sigma}$ remains empty). We let

$$C^X_d = \bigcup_{\eta \prec X, \sigma \in \{0,1\}^*} C^\eta_{d,\sigma}. \tag{5.3}$$

As long as $\lambda C^\eta_{d,\sigma} < 2^{-|\sigma|-d}$ we think of $C^\eta_{d,\sigma}$ as "hungry", and "feed" it with fresh oracle strings α such that $\eta \preceq \Phi^\alpha_{|\alpha|}$. Since $\lambda C^\eta_{d,\sigma}$ never exceeds $2^{-|\sigma|-d}$, we have $\lambda C^X_d \leq 2^{-d}\Omega^X \leq 2^{-d}$, so $(C^X_d)_{d\in\mathbb{N}}$ is an oracle ML test.

All the open sets $C^\eta_{d,\sigma}$ are disjoint. If $\lambda C^\eta_{d,\sigma}$ exceeds $2^{-|\sigma|-d-1}$ at some stage, then we put the request $\langle |\sigma| + d + 1, y \rangle$ into L_d. As described in Remark 5.1.21, we may account the weight of those requests against the measure of the sets $C^\eta_{d,\sigma}$ because the measure of $C^\eta_{d,\sigma}$ is at least the weight of the request. This shows that each L_d is a bounded request set.

Because Z is ML-random relative to A, there is d such that $Z \notin C^A_d$. This implies that, whenever $\mathbb{U}^\eta(\sigma) = y$ in the relevant case that $\eta \prec A$, then $\lambda C^\eta_{d,\sigma} = 2^{-|\sigma|-d}$, and hence the request $\langle |\sigma| + d + 1, y \rangle$ is enumerated into L_d. For, if $\lambda C^\eta_{d,\sigma} < 2^{-|\sigma|-d}$, then once a sufficiently long initial segment α of Z computes η, we would feed α to $C^\eta_{d,\sigma}$, which would put Z into C^A_d. (Intuitively speaking, lots of sets compute A, so we are able to feed all the sets $C^\eta_{d,\sigma}$ for $\eta \prec A$ and $\sigma \in \mathrm{dom}(\mathbb{U}^\eta)$.)

The open sets C^A_d correspond to the open sets $C^A_{e,d}$ in the proof of Proposition 5.1.20. However, in the present proof, a Turing reduction Φ such that $A = \Phi^Z$ is given in advance, so one does not need the parameter e. In Proposition 5.1.20, we changed A actively to keep $C^A_{e,d}$ small, and if $A = \Phi^Z_e$ then Z was "caught" in $\bigcap_d C^A_{e,d}$. Here, $(C^A_d)_{d\in\mathbb{N}}$ is an ML-test relative to A by definition. The set Z is caught in $\bigcap_d C^A_d$ if A is not low for K as witnessed by a computation $\mathbb{U}^\eta(\sigma) = y$ for $\eta \prec A$.

Construction of c.e. sets $C^\eta_{d,\sigma} \subseteq \{0,1\}^*$, $\sigma, \eta \in \{0,1\}^*$, for the parameter $d \in \mathbb{N}$. Initially, let $C^\eta_{d,\sigma} = \emptyset$.

Stage s. In substages t, $0 \leq t < 2^s$, go through the strings α of length s in lexicographical order. $C^\eta_{d,\sigma}[s,t]$ denotes the approximation at the beginning of substage t of stage s.

If α has been declared used for d, as defined below, then go to the next α. Otherwise, see whether there are σ, η such that

- $\mathbb{U}^{\eta}_{s}(\sigma) \downarrow$ and η is minimal such, namely, $\mathbb{U}^{\eta'}_{s}(\sigma) \uparrow$ for each $\eta' \prec \eta$,
- $\eta \preceq \Phi^{\alpha}_{|\alpha|}$, and
- $\lambda C^{\eta}_{d,\sigma}[s,t] + 2^{-s} \leq 2^{-|\sigma|-d}$.

Choose σ least and put α into $C^{\eta}_{d,\sigma}$. Declare all the strings $\rho \succeq \alpha$ *used for* d.

Verification. For fixed d, the sets $[C^{\eta}_{d,\sigma}]^{\prec}$ $(\sigma, \eta \in \{0,1\}^*)$ are disjoint, because during each stage s we only enumerate unused strings of length s into these sets. Once $\lambda C^{\eta}_{d,\sigma}$ reaches $2^{-|\sigma|-d-1}$, we enumerate $\langle |\sigma| + d + 1, y\rangle$ into L_d. Then L_d is a bounded request set by the accounting method of Remark 5.1.21.

Define C^{X}_{d} by (5.3). Since $(C^{A}_{d})_{d\in\mathbb{N}}$ is a ML-test relative to A, there is d such that $Z \notin C^{A}_{d}$. We verify that L_d works. Suppose that $\mathbb{U}^{A}(\sigma) = y$, and let $\eta \preceq A$ be shortest such that $\mathbb{U}^{\eta}(\sigma) = y$. We claim that $\lambda C^{\eta}_{d,\sigma} = 2^{-|\sigma|-d}$, so that we are allowed to put the required request $\langle |\sigma| + d + 1, y\rangle$ into L_d when the measure of $C^{\eta}_{d,\sigma}$ has reached $2^{-|\sigma|-d-1}$.

Assume for a contradiction that $\lambda C^{\eta}_{d,\sigma} < 2^{-|\sigma|-d}$, and let s be so large that $\mathbb{U}^{\eta}_{s}(\sigma) = y$, $\eta \preceq \Phi^{Z}_{s}$ and $\lambda C^{\eta}_{d,\sigma} + 2^{-s} \leq 2^{-|\sigma|-d}$. Then $\alpha = Z \upharpoonright_{s+1}$ enters $C^{\eta}_{d,\sigma}$ at stage s, unless it enters some $C^{\eta'}_{d,\sigma'}$ instead or is used for d, namely, some $\beta \prec \alpha$ has entered a set $C^{\eta''}_{d,\sigma'}$ at a previous stage. Because $A = \Phi^{Z}$, in any case, $\eta \prec A$, or $\eta' \prec A$, or $\eta'' \prec A$. Thus $Z \in C^{A}_{d}$ contrary to our hypothesis on d. □

The following shows that a certain class defined in terms of plain ML-randomness, rather than relativized ML-randomness, is contained in the c.e. sets that are low for K. In Section 8.5 we will consider subclasses of $\mathcal{M}$ in more detail.

5.1.23 Corollary. *Suppose A is c.e. and there is a ML-random set $Z \geq_T A$ such that $\emptyset' \not\leq_T Z$. Then A is low for K.*

Proof. If A is not low for K, by Theorem 5.1.22, Z is not ML-random relative to A. Then, by Proposition 3.4.13, $\emptyset' \leq_T A \oplus Z \equiv_T Z$. □

We do not know at present whether this containment is strict:

5.1.24 Open question. *If A is c.e. and low for K, is there a ML-random set $Z \geq_T A$ such that $\emptyset' \not\leq_T Z$?*

5.1.25 Remark. It is not hard to see that the proof of Theorem 5.1.22 is uniform, in the sense that if Z is ML-random in A and $\Phi^{Z} = A$, then a constant b such that A is low for K via b can be obtained effectively from an index for Φ and a constant c such that $Z \notin \mathcal{R}^{A}_{c}$. On the other hand, Hirschfeldt, Nies and Stephan (2007) proved that Corollary 5.1.23 is necessarily nonuniform: One cannot effectively obtain a constant b such that A is low for K via b even if one is given c such that $Z \notin \mathcal{R}_c$, Φ, and a lowness index for Z, that is, an index p such that $Z' = \Phi_p(\emptyset')$. (They actually show it for K-triviality instead of being low for K. However, by the proof of Proposition 5.2.3 below, the implication "low for $K \Rightarrow K$-trivial" is uniform in the constants, so the result also applies for the constant via which A is low for K.)

5.1.26 Remark. We describe an unexpected application of Theorem 5.1.22. We say that $\mathcal{S} \subseteq 2^{\mathbb{N}}$ is a *Scott class* (or Scott set) if $\mathcal{S}$ is closed downwards under Turing reducibility, closed under joins, and each infinite binary tree $T \in \mathcal{S}$ has an infinite path in $\mathcal{S}$. Scott classes occur naturally in various contexts, such as the study of models of Peano arithmetic and reverse mathematics. The arithmetical sets form a Scott class. On the other hand, Theorem 1.8.37 in relativized form shows that there is a Scott class consisting only of low sets.

Kučera and Slaman (2007) showed that Scott classes $\mathcal{S}$ are rich: for each incomputable $X \in \mathcal{S}$ there is $Y \in \mathcal{S}$ such that $Y \mid_T X$ (this answered questions of H. Friedman and A. McAllister).

They choose $Y \in \mathsf{MLR}^X$. Then $Y \mid_T X$ unless X is a base for ML-randomness, and hence K-trivial. In that case, they build an infinite computable tree T such that a set $Z \in Paths(T)$ is not K-trivial and satisfies $Z \not\geq_T X$. For the latter they use the Sacks preservation strategy, which relyies on the fact that X is Δ^0_2 (see Soare 1987, pg. 122).

Exercises. Show the following.

5.1.27. If $A \in \text{Low}(\Omega)$ (Definition 3.6.17) and $A \in \Delta^0_2$ then $A \in \text{Low}(\mathsf{MLR})$.

5.1.28. There is an ω-c.e. set A for which Corollary 5.1.23 fails.

5.1.29. Each low set A is a "base for being of PA degree" via a low witness: there is a low set $D \geq_T A$ which is of PA degree relative to A. That is, some $\{0,1\}$-valued function $f \leq_T D$ satisfies $\forall e \, \neg f(e) = J^A(e)$.

5.1.30. If Y and Z are sets such that $Y \oplus Z \in \mathsf{MLR}$, then each set $A \leq_T Y, Z$ is low for K. (For instance, let Y be the bits in an even positions and let Z be the bits in an odd position in the binary representation of Ω. Compare this to Exercise 4.2.7.)

5.1.31.$^\diamond$ Prove Corollary 5.1.23 directly by combining the *proofs* of 5.1.22 and 3.4.13.

Lowness for weak 2-randomness

A recurrent goal of this book is to characterize the class $\text{Low}(\mathcal{C})$ for a randomness notion $\mathcal{C}$. Here and in Section 8.3 we will consider, more generally, lowness for *pairs* of randomness notions such that $\mathcal{C} \subseteq \mathcal{D}$ (relative to each oracle). Relativizing $\mathcal{D}$ to A increases the power of the associated tests, so one would expect that in general $\mathcal{C} \not\subseteq \mathcal{D}^A$. We consider the class of sets A for which, to the contrary, the containment persists when we relativize $\mathcal{D}$.

5.1.32 Definition. (Kjos-Hanssen, Nies and Stephan, 2005) A is in $\text{Low}(\mathcal{C}, \mathcal{D})$ if $\mathcal{C} \subseteq \mathcal{D}^A$. Thus $\text{Low}(\mathcal{C}) = \text{Low}(\mathcal{C}, \mathcal{C})$.

We study classes of the form $\text{Low}(\mathcal{C}, \mathcal{D})$ for various reasons.

(a) The proof techniques suggest so.

(b) We can deal with more than one randomness notion at the same time.

(c) These classes may coincide with interesting computability theoretic classes. For instance, if $\mathcal{C}$ is ML-randomness and $\mathcal{D}$ is Schnorr randomness then $\text{Low}(\mathcal{C}, \mathcal{D})$ coincides with the c.e. traceable sets by Theorem 8.3.3 below.

We frequently use the fact that, if $\mathcal{C} \subseteq \widetilde{\mathcal{C}} \subseteq \widetilde{\mathcal{D}} \subseteq \mathcal{D}$ are randomness notions, then $\mathrm{Low}(\widetilde{\mathcal{C}}, \widetilde{\mathcal{D}}) \subseteq \mathrm{Low}(\mathcal{C}, \mathcal{D})$. That is, the class $\mathrm{Low}(\mathcal{C}, \mathcal{D})$ is enlarged by either decreasing $\mathcal{C}$ or increasing $\mathcal{D}$. In particular, both $\mathrm{Low}(\mathcal{C})$ and $\mathrm{Low}(\mathcal{D})$ are contained in $\mathrm{Low}(\mathcal{C}, \mathcal{D})$.

Weak 2-randomness was introduced in Definition 3.6.1, and $\mathsf{W2R} \subseteq \mathsf{MLR}$ are the classes of weakly 2-random and ML-random sets, respectively. The following result is due to Downey, Nies, Weber and Yu (2006).

5.1.33 Theorem. $\mathrm{Low}(\mathsf{W2R}, \mathsf{MLR}) = \mathrm{Low}(\mathsf{MLR})$.

Thus $\mathrm{Low}(\mathsf{W2R}) \subseteq \mathrm{Low}(\mathsf{MLR})$. Theorem 5.5.17 below shows the converse containment, so that actually $\mathrm{Low}(\mathsf{W2R}) = \mathrm{Low}(\mathsf{MLR})$.

Proof. Suppose that $A \notin \mathrm{Low}(\mathsf{MLR})$. By the characterization of $\mathrm{Low}(\mathsf{MLR})$ in 5.1.9, there is no c.e. open set R such that $\lambda R < 1$ and $\forall z\, [K^A(z) \leq |z| - 1 \rightarrow [z] \subseteq R]$). Claim 5.1.34 below, which is a consequence of this failure of (5.1), is used to show that $\mathsf{W2R} \not\subseteq \mathsf{MLR}^A$. The argument extends the one for the implication (i)⇒(iii) in the proof of Theorem 5.1.9. Recall from Definition 1.9.3 that, for a measurable class $V \subseteq 2^{\mathbb{N}}$ and a string w, the local measure $\lambda(V \mid w)$ is $2^{|w|}\lambda(V \cap [w])$.

5.1.34 Claim. *Suppose (5.1) on page 167 fails for A. Let β, γ be rationals such that $\beta < \gamma < 1$. For each c.e. open set V and each string w, if $\lambda(V \mid w) \leq \beta$, there is z such that $K^A(z) \leq |z| - 1$ and $\lambda(V \mid wz) \leq \gamma$.*

Subproof. Assume that no such z exists, and consider the c.e. set of strings

$$G = \{z\colon\ \lambda(V \mid wz) > \gamma\}.$$

Whenever $K^A(z) \leq |z| - 1$ then $z \in G$. Let $S = [G]^{\prec}$. Let $(y_i)_{i<N}$, $N \leq \infty$ be a listing of the minimal strings in G under $\preceq$, so that $S = \bigcup_{i<N}[y_i]$. Now

$$\beta \geq \lambda(V \mid w) \geq \textstyle\sum_{i<N} 2^{-|y_i|}\lambda(V \mid wy_i) \geq \lambda S \cdot \gamma.$$

Thus $1 > \beta/\gamma \geq \lambda S$, whence (5.1) holds via S, contradiction. $\diamond$

We build a set $Z \in \mathsf{W2R}$ that is not ML-random relative to A. Let $(G^e_n)_{n\in\mathbb{N}}$ be a listing of all generalized ML-tests (Definition 3.6.1) with no assumption on the uniformity in e. We define Z by finite extensions (somewhat similar to Theorem 1.6.1), defeating the tests $(G^e_n)_{n\in\mathbb{N}}$ one by one. Claim 5.1.34 ensures that we can choose the extensions in such a way that $Z \notin \mathsf{MLR}^A$.

As in Theorem 5.1.9, we define a sequence of strings $z_0, z_1, \ldots$ such that $K^A(z_i) \leq |z_i| - 1$. Then $Z = z_0z_1z_2\ldots$ is not ML-random relative to A by Proposition 3.2.17 relativized to A. In step e we define z_e, and, to ensure that $Z \in \mathsf{W2R}$, we also define a number n_e such that $Z \notin G^e_{n_e}$. At the beginning of step e, we have defined $z_0, \ldots, z_{e-1}$ and $n_0, \ldots, n_{e-1}$. We let $H_e = \bigcup_{i<e} G^i_{n_i}$ and $w_e = z_0 \ldots z_{e-1}$. We ensure inductively that for each e

$$\lambda(H_e \mid w_e) \leq \gamma_e := 1 - 2^{-e}. \tag{5.4}$$

Note that w_0 is the empty string and $H_0 = \emptyset$, so that (5.4) holds for $e = 0$. In step $e \geq 0$, we choose n_e so large that

$$\lambda(G^e_{n_e}) \leq 2^{-|w_e|-e-2}.$$

Then $\lambda(G^e_{n_e}|w_e) \leq 2^{-(e+2)}$. Since $H_{e+1} = H_e \cup G^e_{n_e}$,

$$\lambda(H_{e+1} \mid w_e) \leq \gamma_e + 2^{-(e+2)} < \gamma_{e+1}.$$

By Claim 5.1.34 for $V = H_{e+1}$, $w = w_e$, $\beta = \gamma_e + 2^{-(e+2)}$, and $\gamma = \gamma_{e+1} > \beta$, we can choose $z = z_e$ such that $K^A(z) \leq |z| - 1$ and $\lambda(H_{e+1} \mid w_e z) \leq \gamma_{e+1}$. Thus (5.4) holds for $e+1$ where $w_{e+1} = w_e z$.

If $Z \in G^e_{n_e}$, there is $m > n_e$ such that $[w_m] \subseteq G^e_{n_e} \subseteq H_m$ as $G^e_{n_e}$ is open. However, since $\lambda(H_m|w_m) < 1$ by (5.4), we have $[w_m] \not\subseteq H_m$ for each m. Thus $Z \notin G^e_{n_e}$. □

5.2 K-trivial sets

We say that a set A is K-trivial if up to an additive constant the function $\lambda n. K(A\restriction_n)$ grows no faster than the function $\lambda n.K(n)$. Thus, $A\restriction_n$ has no more information than its length has. It is easily verified that each set that is low for K is K-trivial, so by Section 5.1, page 170, there is a promptly simple K-trivial set. In Proposition 5.3.11 we will give a direct construction of such a set, after introducing the cost function method.

We already stated in the introduction to this chapter that the sets that are low for ML-randomness (or low for K) actually coincide with the K-trivial sets. This we will prove in Section 5.4. Here we study K-triviality for its own sake, mostly by combinatorial means. Given the equivalence with the lowness properties of Section 5.1, this leads to results involving those properties which would be hard to obtain if their definitions were used directly. For instance, it is not too difficult to show that each K-trivial set is Δ^0_2. A direct proof of this result is possible, but more difficult, for the sets that are low for ML-randomness (Nies, 2005a). A further example of this is the closure of the K-trivial sets under the operation $\oplus$ on sets, where no direct proof is known using the definition of Low(MLR).

Some results of this section will be improved in the Sections 5.4 and 5.5 using "dynamic" methods, such as the golden run. For instance, once we know that being K-trivial is the same as being low for K, we may conclude from Proposition 5.1.2 that the K-trivial sets are not only Δ^0_2, but in fact low. Alternatively, in Corollary 5.5.4 we use the golden run method to show directly that each K-trivial set is superlow, and hence ω-c.e.

Basics on K-trivial sets

Each prefix-free description of a string y also serves as a description for $|y|$. Thus $K(|y|) \leq^+ K(y)$ for each y. Clearly $\forall n\, K(n) =^+ K(0^n)$, so the following property of a set A expresses that the K-complexity of the initial segments of A grows as slowly as the one of a computable set.

5.2.1 Definition. A is K-trivial via $b \in \mathbb{N}$ if $\forall n\ K(A\restriction_n) \leq K(n) + b$.
Let $\mathcal{K}$ denote the class of K-trivial sets.

Although we have defined the class of K-trivial sets in terms of the standard prefix-free universal machine, it actually does not depend on this particular choice: a change of the universal machine would merely lead to a different constant b.

The intuitive meaning of K-triviality is to be far from ML-random. By Theorem 3.2.9 A is ML-random iff $\forall n\ K(A\upharpoonright_n) \geq n - c$ for some c. Thus, A is ML-random if for each n, the number $K(A\upharpoonright_n)$ is within $K(n)+c+1 \leq^+ 2\log n$ of its upper bound $n+K(n)+1$ given by Theorem 2.2.9. In contrast, A is K-trivial if $K(A\upharpoonright_n)$ is within a constant of its lower bound $K(n)$.

We say that A is *C-trivial* if $\forall n\, C(A\upharpoonright_n) \leq C(n) + b$ for some $b \in \mathbb{N}$ (Chaitin, 1976). Computable sets are both K-trivial and C-trivial. Chaitin proved that there are no further C-trivial sets (Theorem 5.2.20 below). He still managed to show that all K-trivial sets are in Δ^0_2. Solovay (1975) constructed an incomputable K-trivial set. These results will be covered in the present and in the next subsection.

Intuitively, an incomputable K-trivial set exists because both sides of the defining inequality $\forall n\, K(A\upharpoonright_n) \leq K(n) + b$ are noncomputable, and also because we do not ask for uniformity in the inequality. In particular, we do not require that a short description of $A\upharpoonright_n$ can be obtained from a short description of n. Chaitin's argument gets around this for C-trivial sets because the computable upper bound $C(n) \leq 1+\log(n+1)$ is attained in each interval $[2^i-1, 2^{i+1}-1)$. So, roughly speaking, one can replace the right hand side $C(n)+b$ in the definition of C-triviality by such a computable bound.

5.2.2 Fact. *Each computable set A is K-trivial.*

Proof. On input σ, the prefix-free machine M attempts to compute $n = \mathbb{U}(\sigma)$, and outputs $A\upharpoonright_n$. Then $\forall n\, K_M(A\upharpoonright_n) \leq K(n)+b$ where b is the coding constant for M. Alternatively, by Example 2.2.16, $W = \{\langle n+1, A\upharpoonright_n\rangle : n \in \mathbb{N}\}$ is a bounded request set, so $\forall n\, K(A\upharpoonright_n) \leq^+ K(n)$. □

By Theorem 5.1.19, there is a promptly simple base for ML-randomness, and by Theorem 5.1.22 each base for ML-randomness is low for K. Then the following implies that there is a promptly simple K-trivial set.

5.2.3 Proposition. (Extends 5.2.2) *Each set that is low for K is K-trivial.*

Proof. We actually show that there is a fixed d such that, if A is low for K via a constant c, then A is K-trivial via the constant $c+d$. Let M be the oracle prefix-free machine such that $M^X(\sigma) \simeq X\upharpoonright_{\mathbb{U}(\sigma)}$ for each X, σ, and let d be the coding constant for M. Then $K^X(X\upharpoonright_n) \leq K(n)+d$ for each X and n. Hence $K(A\upharpoonright_n) \leq K^A(A\upharpoonright_n) + c \leq K(n)+c+d$ for each n. □

K-trivial sets are Δ^0_2

The following theorem of Chaitin (1976) shows that the K-trivial sets are rare: for each constant b there are at most $O(2^b)$ many. As a consequence, each K-trivial set is Δ^0_2. In Corollary 5.5.4 we improve this: each K-trivial set is superlow.

However, the work in this section is not wasted, since the dynamic methods used there rely on a computable approximation of the set.

5.2.4 Theorem.

(i) There is a constant $\mathbf{c} \in \mathbb{N}$ *such that for each* b*, at most* $2^{\mathbf{c}+b}$ *sets are* K*-trivial with constant* b*.*

(ii) Each K*-trivial set is* Δ^0_2*.*

Proof. (i). The paths of the Δ^0_2 tree

$$T_b = \{z : \forall u \le |z| \; [K(z \upharpoonright_u) \le K(u) + b]\}$$

coincide with the sets that are K-trivial via the constant b. If $\mathbf{c}$ is the constant of Theorem 2.2.26(ii), then for each n the size of the level $\{z \in T_b\colon |z| = n\}$ is at most $2^{\mathbf{c}+b}$.

(ii). By (i) each path of T_b is isolated, and hence Δ^0_2 by Fact 1.8.34 relativized to $\emptyset'$. □

We give an affirmative answer to a question of Kučera and Terwijn (1999). After building a c.e. incomputable set in Low(MLR), they asked whether each set in Low(MLR) is in Δ^0_2. The result was first obtained in a direct manner; see Nies (2005*a*). Here we use the foregoing result and Theorem 5.1.22.

5.2.5 Corollary. Low(MLR) $\subseteq \Delta^0_2$.

Proof. If $A \in$ Low(MLR) then A is a base for ML-randomness, and hence low for K by Theorem 5.1.22. Each set that is low for K is K-trivial. Then, by the foregoing theorem, A is in Δ^0_2. □

By Theorem 5.2.4 the class $\mathcal{K}$ of K-trivial sets can be represented as an ascending union of finite classes $(Paths(T_b))_{b\in\mathbb{N}}$. Note that we do *not* obtain a uniform listing of Δ^0_2-indices (i.e., of total Turing reductions to $\emptyset'$) for $\mathcal{K}$. An obvious attempt would be to represent a path A of T_b by a string $\sigma \in T_b$ such that A is the only path extending σ. However, the property of a string σ to be on T_b and have a unique path above it is not known to be Δ^0_2. Nevertheless, using other methods and the fact that each K-trivial set is ω-c.e., we will see in Theorem 5.3.28 that there is such a listing, which even includes the constants for K-triviality.

As in 1.4.5 let V_e be the e-th ω-c.e. set.

5.2.6 Fact. $\{e\colon V_e \in \mathcal{K}\}$ *is* Σ^0_3.

Proof. $V_e \in \mathcal{K}$ is equivalent to $\exists b \forall n \forall s \exists t > s \, [K_t(V_{e,t} \upharpoonright_n) \le K_t(n) + b]$. □

As a consequence, the index set $\{e\colon W_e$ is K-trivial$\}$ is Σ^0_3 as well. Since each finite set is K-trivial, there is a uniformly c.e. listing of all the c.e. K-trivial sets, using Exercise 1.4.22. Then the index set is Σ^0_3-complete by Exercise 1.4.23. Note that $\mathcal{K}$ is also Σ^0_3 as a class, by a proof similar to the proof of Fact 5.2.6.

Exercises.

5.2.7. Show that A is low for K $\Leftrightarrow$ $\forall n \, K(A \upharpoonright_n) \le^+ K^A(n)$.

5.2.8. Let A be a c.e. set that is *wtt*-incomplete. Show that $\exists^\infty n \, K(A \upharpoonright_n) \le^+ K(n)$ and $\exists^\infty n \, C(A \upharpoonright_n) \le^+ C(n)$.

5.2.9. Give a "far-from-random" analog of Proposition 3.2.14. Let $R = \{r_0 < r_1 < \ldots\}$ be an infinite computable set, and let $A \subseteq \mathbb{N}$ be any set.
(i) For all n, $K(r_n) =^+ K(n)$ and $K(A\restriction_n) \leq^+ K(A\restriction_{r_n})$
(ii) If $b \in \mathbb{N}$ and $\forall n\, K(A\restriction_{r_n}) \leq K(r_n) + b$, then A is K-trivial via $b + O(1)$.

5.2.10.$^\diamond$ Using Theorem 5.2.4(i), devise a strategy to build a c.e. set that is not K-trivial. Use it to show that some superlow c.e. set A is not K-trivial.

The number of sets that are K-trivial for a constant b $\star$

Let $G(b)$ be the number of sets that are K-trivial via b. By Theorem 5.2.4 $G(b) = O(2^b)$. The actual values $G(b)$ depend on the choice of an optimal prefix-free machine. However, in Proposition 5.2.11 below we derive machine-independent lower bounds for G which are rather close to the upper bound $O(2^b)$, for instance $\lfloor \epsilon 2^b/b^2 \rfloor$ for some $\epsilon > 0$. The finite sets alone are sufficient to obtain these lower bounds. The next result, Theorem 5.2.12 due to J. Miller, states that $\sum_b G(b)/2^b < \infty$. This shows that the lower bound $\lfloor \epsilon 2^b/b^2 \rfloor$ is rather tight; for instance, it cannot be improved to $\lfloor \epsilon 2^b/b \rfloor$ for any $\epsilon > 0$. On the other hand, the upper bound $O(2^b)$ is not tight since $\lim_b G(b)/2^b = 0$.

Recall that $T_b = \{x : \forall y \preceq x\ K(y) \leq K(|y|) + b\}$. Thus $G(b) = \#\mathit{Paths}(T_b)$.

5.2.11 Proposition. *Let $D : \mathbb{N} \to \mathbb{N}$ be a nondecreasing function which is computably approximable from above and satisfies $\sum_b 2^{-D(b)} < \infty$. Then there is $\epsilon > 0$ such that $\forall b\ G(b) \geq \lfloor \epsilon 2^{b-D(b)} \rfloor$.*

For instance, if $D(b) = 2\log b$, we obtain the lower bound $\lfloor \epsilon 2^b/b^2 \rfloor$ for $G(b)$.

Proof. Note that $\forall b\, K(b) \leq^+ D(b)$ by Proposition 2.2.18, and hence $\forall x\, K(x) \leq^+ |x| + D(|x|)$. We may increase D by a constant without changing the validity of the conclusion, so let us assume that in fact $\forall x\, K(x) \leq |x| + D(|x|)$.

There is a constant $r \in \mathbb{N}$ such that, for each string x and each $m \in \mathbb{N}$, $K(x0^m) \leq K(x) + K(|x|+m) + r$, and for each $x' \prec x$, $K(x') \leq K(x) + K(|x'|) + r$, (since x' can be computed from x and $|x'|$). Thus, if $b \geq r$, then for each x such that $K(x) \leq b - r$, the set $x0^\infty$ is K-trivial via b. If $|x| \leq (b-r) - D(b-r)$ then $K(x) \leq b - r$. So the number of such x is at least $\lfloor 2^{b-r-D(b-r)} \rfloor$. Thus $G(b) \geq \lfloor 2^{-r} 2^{b-D(b)} \rfloor$ for $b \geq r$. Since this inequality holds vacuously for $b < r$, the proof is complete. □

5.2.12 Theorem. $\sum_b G(b)/2^b < \infty$.

Proof. Let $F(b,n) = \#\{x\colon |x| = n\ \&\ K(x) \leq K(n) + b\}$.

Claim 1. $\sum_b G(b)/2^b \leq \liminf_n \sum_b F(b,n)/2^b$.

Since $G(b)$ is finite for each b, we may choose n_b so large that each $A \in \mathit{Paths}(T_b)$ is the only path of T_b extending $A\restriction_{n_b}$. Given k, let $m_k = \max_{b \leq k} n_b$. Then for all $n \geq m_k$ we have that $\sum_{b=0}^k G(b)/2^b \leq \sum_{b=0}^k F(b,n)/2^b$. Thus $\sum_b G(b)/2^b \leq \liminf_n \sum_b F(b,n)/2^b$.

Claim 2. *There is $c \in \mathbb{N}$ such that $\sum_b F(b,n)/2^b \leq c$ for each n.*

Let $S_d = \{x\colon K(x) = K(|x|) + d\}$, and let $\widetilde{F}(d,n) = \#\{x\colon |x| = n \ \&\ x \in S_d\}$. Since $\{0,1\}^*$ is partitioned into the sets S_d, for each n we have

$$\textstyle\sum_d \widetilde{F}(d,n) 2^{-K(n)-d} = \sum_{|x|=n} 2^{-K(x)}.$$

If $K(x) \le K(|x|) + b$ then $x \in S_d$ for a unique $d \le b$. Thus for each n

$$\begin{aligned}\sum_b F(b,n)2^{-K(n)-b} &= \sum_b \sum_{K(x)\le K(n)+b} 2^{-K(n)-b}\\ &= \sum_d \sum_{K(x)=K(n)+d} \sum_{b\ge d} 2^{-K(n)-b}\\ &= 2\sum_d \widetilde{F}(d,n)2^{-K(n)-d}\\ &= 2\sum_{|x|=n} 2^{-K(x)}.\end{aligned}$$

It now suffices to show that $\sum_{|x|=n} 2^{-K(x)} = O(2^{-K(n)})$, for then Claim 2 follows after multiplying by $2^{K(n)}$. Recall from the Coding Theorem 2.2.25 that for a prefix-free machine M we defined $P_M(y) = \lambda[\{\sigma\colon M(\sigma) = y\}]^{\prec}$, and that $P_{\mathbb{U}}(y) \sim 2^{-K(y)}$. Let M be the machine from the proof of Theorem 2.2.26 given by $M(\sigma) = |\mathbb{U}(\sigma)|$. Then $P_M(n) = \sum_{|x|=n} P_{\mathbb{U}}(x) \sim \sum_{|x|=n} 2^{-K(x)}$. By 2.2.25, $P_M(n) = O(2^{-K(n)})$. Thus $\sum_{|x|=n} 2^{-K(x)} = O(2^{-K(n)})$ as required. □

The sequence $(G(b)/2^b)_{b\in\mathbb{N}}$ converges to 0 rather slowly:

5.2.13 Proposition. *There is no computable function $h\colon \mathbb{N} \to \mathbb{Q}_2$ such that $\lim_b h(b) = 0$ and $\forall b\,[G(b)/2^b \le h(b)]$. In particular, the function G is not computable.*

Proof. Assume that such a function h exists. We enumerate a bounded request set L. We assume that an index d for a machine M_d is given, thinking of M_d as a prefix machine for L (see Remark 2.2.21).
Construction of L. Compute the least $b \ge d$ such that $h(b) < 2^{-d}$. Now enumerate L in such a way that there are 2^{b-d} K-trivial sets for the constant b; this is a contradiction since in that case $G(b)/2^b \ge 2^{-d}$. For each string x of length $b-d$ let $A_x = x0^\infty$. Whenever $s > 0$ is a stage such that $K_s(n) < K_{s-1}(n)$ (possibly $K_{s-1}(n) = \infty$), then for each such x put the requests $\langle K_s(n) + b - d, A_x\restriction_n\rangle$ into L.
Verification. The weight we put into L for each A_x is at most $\Omega 2^{b-d}$, so the total weight of L is at most Ω, no matter what d is. If M_d, $d > 1$, is in fact a machine for L, then $K_{M_d}(A_x\restriction_n) \le K(n) + b - d$ for each x, n, whence A_x is K-trivial via b. □

Since each A_x is finite, the Proposition remains valid when one replaces G by the function G_{fin}, where $G_{\mathsf{fin}}(b)$ is the number of finite sets that are K-trivial via b.

Exercises.

5.2.14. There is $r_0 \in \mathbb{N}$ such that, if A is K-trivial via the constant c, then xA is K-trivial via $2K(x) + c + r_0$, for each string x.

5.2.15.$^\diamond$ Improve 5.2.13: a function h with the properties there is not even Δ^0_2. In particular $G \not\le_T \emptyset'$. The same is true for G_{fin}.

5.2.16$^\diamond$ **Problem.** It is not hard to verify that $G \le_T \emptyset^{(3)}$. Show that $G \le_T \emptyset^{(2)}$ (for any choice of an optimal machine). See Barmpalias and Sterkenburg (20xx).

Closure properties of $\mathcal{K}$

We show that $\mathcal{K}$ induces an ideal in the Δ^0_2 weak truth-table degrees. The closure under $\oplus$ was proved by Downey, Hirschfeldt, Nies and Stephan (2003).

5.2.17 Theorem. *If $A, B \in \mathcal{K}$ then $A \oplus B \in \mathcal{K}$. More specifically, if both A and B are K-trivial via b, then $A \oplus B$ is K-trivial via $3b + O(1)$.*

Proof idea: By Exercise 5.2.9, it is sufficient to show

$$\forall n\, K(A \oplus B \restriction_{2n}) \le K(n) + 3b + O(1).$$

The set $S_{n,r} = \{x \colon |x| = n \;\&\; K(x) \le r\}$ is c.e. uniformly in n and r. If $r = K(n) + b$, then by Theorem 2.2.26(ii) we have $\#S_{n,r} \le 2^{\mathbf{c}+b}$ for the constant $\mathbf{c}$. So we may define a prefix-free machine describing a *pair* of strings in $S_{n,r}$ with $K(n)+O(1)$ bits. It has to describe n only once, using a shortest $\mathbb{U}$-description σ. Thereafter, it specifies via two numbers $i, j < 2^{\mathbf{c}+b}$ in which position the strings $A\restriction_n$ and $B\restriction_n$ appear in the enumeration of $S_{n,r}$ where $r = |\sigma| + b$.

Proof details. Let M be the prefix-free machine which works as follows. On input $0^b1\rho$ search for $\sigma \preceq \rho$ such that $\mathbb{U}(\sigma)\downarrow = n$. If $\rho = \sigma\alpha\beta$ where α, β are strings of length $\mathbf{c} + b$, then let $i, j < 2^{\mathbf{c}+b}$ be the numbers with binary representations 1α and 1β. Search for strings x and y, the i-th and the j-th element in the computable enumeration of $S_{n,r}$, respectively. If x and y are found output $x \oplus y$.

For an appropriate string $\rho = \sigma\alpha\beta$ of length $K(n)+2(\mathbf{c}+b)$ we have $M(0^b1\rho) = A \oplus B\restriction_{2n}$. Hence $K(A \oplus B\restriction_{2n}) \le K(n) + 3b + O(1)$. □

Is the class of K-trivial sets closed downward under Turing reducibility? In Section 5.4, with considerable effort, we will answer this question in the affirmative by showing that $\mathcal{K} = \mathcal{M}$. In contrast, an easier fact follows straight from the definitions, downward closure under weak truth-table reducibility. This already implies that $\mathcal{K}$ is closed under computable permutations.

5.2.18 Proposition.

(i) Let A be K-trivial via b. If $B \le_{wtt} A$ then B is K-trivial via a constant d determined effectively from b and the wtt reduction.

(ii) If A is K-trivial then $\emptyset' \not\le_{wtt} A$.

Proof. (i) Suppose $B = \Gamma^A$, where Γ is a *wtt* reduction procedure with a computable bound f on the use. Then, for each n,

$$K(B\restriction_n) \le^+ K(A\restriction_{f(n)}) \le^+ K(f(n)) \le^+ K(n).$$

(ii) follows from (i) because Ω is not K-trivial, and $\Omega \le_{wtt} \emptyset'$ by 1.4.4. □

5.2.19 Exercise. If A and B are K-trivial and $0.C = 0.A + 0.B$, then C is K-trivial.

C-trivial sets

In the next two subsections we aim at understanding K-triviality by varying it. We say that A is *C-trivial* if $\exists b \forall n\, C(A\upharpoonright_n) \le C(n) + b$ (see page 177). A modification of the proof of Fact 5.2.2 shows that each computable set is C-trivial. By the following result of Chaitin (1976) there are no others.

5.2.20 Theorem.

(i) For each b at most $O(b^2 2^b)$ sets are C-trivial with constant b.
(ii) Each C-trivial set is computable.

Proof. We first establish the fact that for each string x, there are no more than $O(b^2 2^b)$ many M-descriptions of x that are at most $C(x) + b$ long. Thus, surprisingly, the number of such descriptions depends only on b, not on x.

5.2.21 Lemma. *Given a machine M, we can effectively find $d \in \mathbb{N}$ such that, for all b, x,*

$$\#\{\sigma\colon\ M(\sigma) = x \ \&\ |\sigma| \le C(x) + b\} < 2^{b+2\log b+d+5} = O(b^2 2^b). \tag{5.5}$$

Subproof. The argument is typical for the theory of descriptive string complexity: if there are too many M-descriptions of x, we can find a $\mathbb{V}$-description of x that is shorter than $C(x)$, contradiction.

Recall from page 13 that $\mathsf{string}(b)$ is the string identified with b, which has length $\log(b+1)$. Let $\widehat{b} = 0^{|\mathsf{string}(b)|}1\mathsf{string}(b)$ so that $|\widehat{b}| \le 2\log b + 3$. We define a machine R. By the Recursion Theorem (with a parameter for M) we may assume that we are effectively given a coding constant $d > 0$ for R, that is, $\Phi_d = R$.

R: For each b, each $m \ge 2\log b + d + 4$ and each x, if there are $2^{b+2\log b+d+5}$ strings σ of length at most $m + b$ such that $M(\sigma) = x$, let $R(\widehat{b}\rho) = x$, for the leftmost ρ of length $m - 2\log b - d - 4$ such that $\widehat{b}\rho$ is not yet in the domain of R. Note that $|\widehat{b}\rho| < m - d$.

This definition of R is consistent: for each b, m, there are $2^{m+b+1} - 1$ strings of length at most $m+b$, so at most $2^{m+b+1}/2^{b+2\log b+d+5} = 2^{m-2\log b-d-4}$ strings x can have a sufficient number of M-descriptions to get an R-description $\widehat{b}\rho$.

Suppose (5.5) fails for b, x, and let $m = C(x)$. Then $2^{m+b+1} \ge 2^{b+2\log b+d+5}$, so $m \ge 2\log b+d+4$, hence we ensure that $C_R(x) < m-d$ and therefore $C(x) < m$, contradiction. ◇

(i) We apply the lemma to count the strings z of length n such that $C(z) \le C(n)+b$. Similar to the proof of Theorem 2.2.26(i), consider the machine M given by $M(\sigma) \simeq |\mathbb{V}(\sigma)|$. Each shortest $\mathbb{V}$-description of such a z is an M-description of n that has length at most $C(n)+b$, so by the lemma there are at most r many, where $r = O(b^2 2^b)$ is independent of n.

Similar to the proof of Theorem 5.2.4(i), consider the tree

$$T_b^C = \{z\colon\ \forall u \le |z|\, [C(z\upharpoonright_u) \le C(u) + b]\},$$

which contains at most r strings of each length n. Each set that is C-trivial for b is a path of T_b^C, so there are at most r such sets.

(ii) Let A be C-trivial via b. The tree T_b^C is merely Δ_2^0, so the fact that A is an isolated path is not sufficient to show that A is computable. Instead, let $\widetilde{S}_b \supseteq T_b^C$ be the c.e. tree $\{z\colon \forall u \le |z| \, [C(z\restriction_u) \le 1 + \log(u+1) + b]\}$. For almost every i, there is a string y of length i such that $C(y) = |y| + 1$ by Exercise 2.1.19 (ii), and therefore (with the usual identifications from page 13) there is a number u, $2^i - 1 \le u < 2^{i+1} - 1$, such that $C(u) = 1 + \log(u+1)$. Let

$$S_b = \{z\colon \exists w \succeq z \, [|w| = 2|z| \; \& \; w \in \widetilde{S}_b]\},$$

then there are at most r strings on S_b at almost every level $k = 2^i - 1$ of S_b. The set A is a path of S_b, so there is a string $z \prec A$ such that for each $k > |z|$ of the form $2^i - 1$, $A\restriction_k$ is the only string in S_b extending z. Since the tree S_b is c.e., we can enumerate it till this string appears. Hence A is computable. □

Exercises.

5.2.22. We say that A is low for C if $\exists b \forall y \, [C^A(y) \ge C(y) - b]$. Show that each set that is low for C is computable.

5.2.23. Explain why the proof of Theorem 5.2.20(ii) cannot be adapted to show that each K-trivial set is computable.

5.2.24. (Loveland, 1969) Show that the following are equivalent for a set A.

(i) A is computable.

(ii) There is $d \in \mathbb{N}$ such that $\forall n \, K(A\restriction_n | \, n) \le d$.

(iii) There is $b \in \mathbb{N}$ such that $\forall n \, C(A\restriction_n | \, n) \le b$.

Replacing the constant by a slowly growing function ⋆

We replace the right hand side $K(n) + O(1)$ in the definition of K-triviality by $K(n) + p(K(n)) + O(1)$, where p is a function we think of as a slowly growing. For each function p, let $\mathcal{K}_p$ denote the class of sets A such that

$$\forall n \; K(A\restriction_n) \le^+ K(n) + p(K(n)).$$

We prove a result of Stephan which shows that, if $p\colon \mathbb{N} \to \mathbb{N}$ is unbounded and computably approximable from above (see Definition 2.1.15), the class $\mathcal{K}_p$ is much larger than $\mathcal{K}$. An example of such a function was given in Proposition 2.1.22: $p(n) = \overline{C}(n) = \min\{C(m)\colon m \ge n\}$. The approximability condition is needed, since Csima and Montalbán (2005) defined a nondecreasing unbounded function f such that A is K-trivial $\leftrightarrow$ $\forall n \; K(A\restriction_n) \le^+ K(n) + f(n)$, which implies that A is K-trivial $\leftrightarrow$ $\forall n \; K(A\restriction_n) \le^+ K(n) + f(K(n))$.

5.2.25 Theorem. *Suppose p is a function such that $\lim_n p(n) = \infty$ and p is computably approximable from above. Then there is a Turing complete c.e. set E such that each superset of E is in $\mathcal{K}_p$.*

Proof. For a set $X \subseteq \mathbb{N}$ and $n \in \mathbb{N}$, we write $\overline{X}$ for $\mathbb{N} - X$ and $X \cap n$ for $X \cap [0, n)$. The function $g(n) = \lfloor p(K(n))/2 \rfloor$ is computably approximable from above via $g_s(n) = \lfloor p_s(K_s(n))/2 \rfloor$. The idea is to make the complement of the c.e. set E very thin, so that for every $A \supseteq E$ and each n, a description of n and very little extra information yields a description of $A \upharpoonright_n$. We enumerate E as follows. At stage s,

(1) for each $r < s$ in increasing order, if $|\overline{E}_{s-1} \cap r| > g_s(r)$, put the greatest element of $\overline{E}_{s-1} \cap r$ into E_s.
(2) If $i \in \emptyset'_s - \emptyset'_{s-1}$ then put the i-th element of $\overline{E}_{s-1}$ into E.

Firstly, we verify that E is Turing complete. By the coding in (2), it is sufficient to show that E is co-infinite. We prove by induction on $q \in \mathbb{N}$ that $\#(\overline{E} \cap n) = q$ for some n. This is trivial for $q = 0$. If $q > 0$, by the inductive hypothesis there is m such that $\#(\overline{E} \cap m) = q - 1$; let t be the least stage such that $E_t \cap m = E \cap m$. Choose $k \geq m, t$ such that $g_s(r) > q$ for each $k \geq r$ and each s. If $q \in \emptyset'_s - \emptyset'_{s-1}$ for some s, let v be the number enumerated in (2) at stage s, else let $v = 0$. Let $n \geq \max\{v+1, k+1\}$ be least such that $n - 1 \notin E_{k+1}$. If $\#(\overline{E} \cap n - 1) > q - 1$ we are done. Suppose otherwise, that is, $(m, n-1) \subseteq E$. Then $n - 1 \notin E$, since $n - 1$ is not enumerated via (2), nor can any number $r \geq n$ in (1) demand that $\#(\overline{E} \cap r) < q$. Thus $\#(\overline{E} \cap n) = q$.
Secondly, we show that each set $A \supseteq E$ is in $\mathcal{K}_p$. We introduce a prefix-free machine M such that $K_M(A \upharpoonright_n) \leq K(n) + p(K(n)) + 1$ for each n. On input $0^{|\tau|}1\tau\sigma$, M first attempts to compute $n = \mathbb{U}(\sigma)$. In case of convergence, it waits for the least stage s such that $\#(\overline{E}_s \cap n) \leq |\tau|$. Interpreting τ as the bits of A in the positions where membership has not yet been determined, it outputs a string y of length n such that $y(i) = 1$ if $i \in E_s$, and $y(i) = \tau(j)$ if i is the j-th number not in E_s and $j < |\tau|$. Suppose σ is a shortest string such that $\mathbb{U}(\sigma) = n$. If we choose an appropriate τ of length $g(n)$ then $M(0^{|\tau|}1\tau\sigma) = A \upharpoonright_n$, and the length of this M-description is at most $2g(n) + 1 + |\sigma| \leq p(|\sigma|) + 1 + |\sigma|$. □

5.3 The cost function method

We introduce an important method to build a set A satisfying a lowness property. It was first used by Kučera and Terwijn (1999) to build an incomputable c.e. set that is low for ML-randomness, and later, in more explicit form, by Downey, Hirschfeldt, Nies and Stephan (2003) to build an incomputable c.e. K-trivial set.

We define a *cost function*, a computable function c that maps a pair x, s of natural numbers where $x < s$ to a nonnegative binary rational. At stage s, we interpret $c(x, s)$ as the cost of a potential enumeration of x into A. The set A has to *obey* this cost function in the sense that the sum of the costs of all enumerations is finite (if several numbers are enumerated at the same stage we only count the least one). This restrains the enumeration into A, so via a cost function construction one can build a c.e. set A satisfying a specific lowness property. Under some extra condition on the cost function one can make A incomputable, and even promptly simple (see 1.7.9).

Using the cost function method, we will

(I) directly build a promptly simple K-trivial set A;

(II) rephrase the construction of a promptly simple set A that is weak truth-table reducible to a given ML-random Δ^0_2 set Y; this is Kučera's Theorem in the restricted version of Remark 4.2.4.

In both cases, since the total cost of the enumerations is finite, we can define an auxiliary c.e. object that in some sense has a finite weight. In (I) this object is a bounded request set showing that A is K-trivial, while in (II) it is a Solovay test needed to show that $A \leq_T Y$. Table 5.1 on page 200 lists some important examples of cost functions.

Recall that $\{Y\}$ is a Π^0_2 class for each Δ^0_2 set Y. A cost function construction allows us to extend (II): for each Σ^0_3 null class $\mathcal{C}$, there is a promptly simple set A such that $A \leq_T Y$ for every ML-random set $Y \in \mathcal{C}$. This is applied in Theorem 8.5.15 to obtain interesting classes contained in the c.e. K-trivial sets.

Most cost functions $c(x, s)$ will be non-increasing in x and nondecreasing in s. That is, at any stage larger numbers are no cheaper, and a number may become more expensive at later stages.

Note that we have already proved the existence of a promptly simple set that is low for K and hence K-trivial (see the comment before Proposition 5.2.3). However, the cost function construction in (I) gives a deeper insight into K-triviality. Indeed, we will prove that each K-trivial set can be viewed as being built via such a construction. For this, it will be necessary to extend the cost function method to Δ^0_2 sets: one now considers the sum of the costs $c(x, s)$ of changes $A_s(x) \neq A_{s-1}(x)$. This characterization via a cost function shows that each K-trivial set A is Turing below a c.e. K-trivial set C, where C is the change set of A defined in the proof of the Limit Lemma 1.4.2. The only known proof of this result is the one relying on cost functions.

To build a promptly simple set A, we meet the prompt simplicity requirements PS_e in the proof of Theorem 1.7.10. Such a requirement acts at most once, and is typically allowed to incur a cost of up to 2^{-e}. In that case, the sum of the costs is finite, that is, A obeys the cost function. Instead of 2^{-e} we could use any other nonnegative quantity $f(e) \in \mathbb{Q}_2$, as long as the function f is computable and $\sum_e f(e) < \infty$. We are able to meet each requirement PS_e, provided that the cost function satisfies the *limit condition*, namely, for each $e \in \mathbb{N}$, almost all x cost at most 2^{-e} at all stages $s > x$.

We sketch the construction for (I) above. The *standard* cost function

$$c_{\mathcal{K}}(x, s) = \sum_{w=x+1}^{s} 2^{-K_s(w)}$$

satisfies the limit condition. Whenever a computable enumeration of a set A obeys this cost function, we can build a bounded request set L showing that A is K-trivial. The set L yields descriptions of the initial segments of A and keeps up with the changes of A. Let $p \in \mathbb{N}$ be a constant such that the total cost S of all enumerations is at most 2^p. If x is the least number entering A at stage s, then all the initial segments $A_s \upharpoonright_w$, $x < w \leq s$, need new descriptions via the prefix-free machine obtained from L. Thus, for each w, $x < w \leq s$, we put a request $\langle K_s(w) + p + 1, A_s \upharpoonright_w \rangle$ into L. The weight contributed to L is $2^{-p-1}c(x, s)$. In total, the contributed weight is at most $2^{-p-1}S \leq 1/2$. (The other half is needed for new descriptions of $A_s \upharpoonright_w$ when $K_s(w) < K_{s-1}(w)$. Details are supplied when we prove Theorem 5.3.10 below.)

In the applications (I) and (II) of the method, the cost function is given in advance. We will also consider the case that $c(x, s)$ depends on A_{s-1}. Such a cost function, called adaptive, is necessary for a direct construction of a promptly simple set that is low for K (5.3.34). Also adaptive is the cost function used by Kučera and Terwijn (1999) to build an incomputable c.e. set that is low for ML-randomness (5.3.38). The latter two cost function constructions merely prove the existence of a promptly simple set in the class $\mathcal{K}$, given that being K-trivial is equivalent to being low for K. However, it is still instructive to study these direct constructions because they expose different aspects of $\mathcal{K}$.

Adaptive cost functions can be used to hide injury to requirements. For instance, in Remark 5.3.37 we reformulate the construction of a low simple set (which has injury to the lowness requirements) in the language of an adaptive cost function. However, we also argue that a cost function given in advance cannot be used in that way, so the constructions based on non-adaptive cost functions, such as (I) and (II) above, can be considered injury-free.

The basics of cost functions

5.3.1 Definition. A *cost function* is a computable function

$$c : \mathbb{N} \times \mathbb{N} \rightarrow \{x \in \mathbb{Q}_2 \colon x \geq 0\}.$$

We say that c satisfies the *limit condition* if $\lim_x \sup_{s>x} c(x, s) = 0$, that is, $\forall e \, \forall^\infty x \, \forall s > x \, [c(x, s) \leq 2^{-e}]$.

We say that c is *monotonic* if $c(x + 1, s) \leq c(x, s) \leq c(x, s + 1)$ for each $x < s$, namely, $c(x, s)$ does not decrease when we enlarge the interval $[x, s)$.

In the following we will usually only define the values of a cost function $c(x, s)$ for $x < s$, and let $c(x, s) = 0$ for $x \geq s$.

We already discussed an important example of a cost function, the one for building a K-trivial set. We use the convention that $2^{-\infty} = 0$.

5.3.2 Definition. The *standard cost function* $c_{\mathcal{K}}$ is given by

$$\boxed{c_{\mathcal{K}}(x, s) = \textstyle\sum_{w=x+1}^{s} 2^{-K_s(w)}.}$$

5.3.3 Lemma. *(i) $c_{\mathcal{K}}$ is monotonic. (ii) $c_{\mathcal{K}}$ satisfies the limit condition.*

Proof. (i) Immediate.
(ii) Given $e \in \mathbb{N}$, since $\sum_w 2^{-K(w)} \leq 1$, there is an x_0 such that $\sum_{w \geq x_0} 2^{-K(w)} \leq 2^{-e}$. Hence $c_{\mathcal{K}}(x, s) \leq 2^{-e}$ for all $x \geq x_0$ and all $s > x$. □

Recall from Definition 1.4.1 that a computable approximation $(A_s)_{s \in \mathbb{N}}$ of a Δ^0_2 set A is a computable sequence of (strong indices for) finite sets such that $A(x) = \lim_s A_s(x)$.

5.3.4 Definition. We say that a computable approximation $(A_s)_{s \in \mathbb{N}}$ *obeys* a cost function c if

$$S = \sum_{x,s} c(x, s) \, [\![x < s \; \& \; x \text{ is least s.t. } A_{s-1}(x) \neq A_s(x)]\!] < \infty. \tag{5.6}$$

If the computable approximation of a Δ^0_2 set A is clear from the context, we will also say that the set A obeys the cost function.

We think of a cost function as a description of a class of Δ^0_2 sets: those sets with an approximation obeying the cost function. For instance, the standard cost function describes the K-trivial sets. This is somewhat similar to a sentence in some formal language describing a class of structures.

We proceed to the general existence theorem. A cost function with the limit condition has a promptly simple "model".

5.3.5 Theorem. *Let c be a cost function with the limit condition. Then there is a promptly simple set A with a computable enumeration $(A_s)_{s\in\mathbb{N}}$ obeying c. Moreover, $S \le 1/2$ in (5.6), and we obtain A uniformly in c.*

Proof. We meet the prompt simplicity requirements from the proof of 1.7.10

$$PS_e\colon\ \#W_e = \infty \ \Rightarrow\ \exists s\, \exists x\, [x \in W_{e,\mathsf{at}\ s} \ \&\ x \in A_s]$$

(where $W_{e,\mathsf{at}\ s} = W_{e,s} - W_{e,s-1}$). We define a computable enumeration $(A_s)_{s\in\mathbb{N}}$ as follows.

> Let $A_0 = \emptyset$. At stage $s > 0$, for each $e < s$, if PS_e has not been met so far and there is $x \ge 2e$ such that $x \in W_{e,\mathsf{at}\ s}$ and $c(x,s) \le 2^{-e}$, put x into A_s. Declare PS_e met.

The computable enumeration $(A_s)_{s\in\mathbb{N}}$ obeys the cost function, since at most one number is put into A for the sake of each requirement. Thus, the sum S in (5.6) is bounded by $\sum_e 2^{-e} = 2$.

If W_e is infinite, there is an $x \ge 2e$ in W_e such that $c(x,s) \le 2^{-e}$ for all $s > x$, because c satisfies the limit condition. We enumerate such an x into A at the stage $s > x$ where x appears in W_e, if PS_e has not been met yet by stage s. Thus A is promptly simple.

If we modify the construction so that each requirement PS_e is only allowed to spend 2^{-e-2}, we have ensured that $S \le 1/2$. Clearly the construction of A is uniform in an index for the computable function c. □

The c.e. change set $C \ge_T A$ for a computable approximation $(A_s)_{s\in\mathbb{N}}$ of a Δ^0_2 set A was introduced in the proof of the Limit Lemma 1.4.2: if $s > 0$ and $A_{s-1}(x) \ne A_s(x)$ we put $\langle x, i\rangle$ into C_s, where i is least such that $\langle x,i\rangle \notin C_{s-1}$. If A is ω-c.e. via this approximation then $C \ge_{tt} A$. The following will be used in Corollary 5.5.3, that each K-trivial set is Turing below a c.e. K-trivial set.

5.3.6 Proposition. *Suppose c is a cost function such that $c(x,s) \ge c(x+1,s)$ for each x, s. If a computable approximation $(A_s)_{s\in\mathbb{N}}$ of a set A obeys c, then the corresponding computable enumeration of the change set $C \ge_T A$ obeys c as well.*

Proof. Since $x < \langle x,i\rangle$ for each x, i, we have $C_{s-1}(x) \ne C_s(x) \to A_{s-1}\upharpoonright_x \ne A_s\upharpoonright_x$ for each x, s. Then, since $c(x,s)$ is nonincreasing in x, the sum in (5.6) for C does not exceed the sum for A. □

Exercises.

5.3.7. There is a computable enumeration $(A_s)_{s\in\mathbb{N}}$ of $\mathbb{N}$ in the order $0, 1, 2, \ldots$ (i.e., each A_s is an initial segment of $\mathbb{N}$) such that $(A_s)_{s\in\mathbb{N}}$ does not obey $c_{\mathcal{K}}$.

5.3.8. Show the converse of Theorem 5.3.5 for a monotonic cost function c: if a computable approximation $(A_s)_{s\in\mathbb{N}}$ of an incomputable set A obeys c, then c satisfies the limit condition.

5.3.9. We view each Φ_i as a (possibly partial) computable approximation (A_s) by letting $A_s \simeq D_{\Phi_i(s)}$. V_e is the e-th ω-c.e. set (1.4.5). Let c be a cost function with $\forall x\, c(x, s) \geq 2^{-x}$. Show that $\{e\colon$ some total computable approximation of V_e obeys $c\}$ is Σ^0_3.

A cost function criterion for K-triviality

We give a general framework for the cost function construction of a promptly simple K-trivial set. This construction was already explained on page 185 in the introduction to this section.

5.3.10 Theorem. *Suppose a computable approximation $(A_s)_{s\in\mathbb{N}}$ of a set A obeys the standard cost function $c_{\mathcal{K}}(x, s) = \sum_{x<w\leq s} 2^{-K_s(w)}$. Then A is K-trivial.*

Proof. By the hypothesis the total cost S of all changes defined in (5.6) is finite. First suppose that $S \leq 1$. We enumerate a bounded request set W at stages s: put the request $\langle K_s(w) + 1, A_s\restriction_w\rangle$ into W whenever $w \leq s$ and

(a) $K_s(w) < K_{s-1}(w)$, or

(b) $K_s(w) < \infty$ & $A_{s-1}\restriction_w \neq A_s\restriction_w$.

Requests enumerated because of (a) contribute at most $\Omega/2$ to W, since for each w and each value $K_s(w)$ there is at most one such request. Suppose now that a request $\langle K_s(w) + 1, A_s\restriction_w\rangle$ is enumerated at stage s because of (b). Then $w > x$ where x is least such that $A_{s-1}(x) \neq A_s(x)$. Thus the term $2^{-K_s(w)}$ occurs in the sum $c_{\mathcal{K}}(x, s)$, and hence in the sum S. If we assume $S \leq 1$, the contribution of such requests is at most $1/2$. Thus W is a bounded request set.

Let M_d be the prefix-free machine for W obtained by the Machine Existence Theorem 2.2.17. We claim that $K(A\restriction_w) \leq K(w)+d+1$ for each w. Given w, let s be greatest such that $s = 0$ or $A_{s-1}\restriction_w \neq A_s\restriction_w$. If $s > 0$ then the requests in (b) at stage w cause $K_u(A\restriction_w) \leq K_s(w)+d+1$ for some $u > s$. If $K_s(w) = K(w)$, we are done. Otherwise, the inequality is caused by a request in (a) at the greatest stage $t > s$ such that $K_t(w) < K_{t-1}(w)$ (this includes the case $s = 0$ as $K_0(w) = \infty$).

More generally, suppose $S \leq 2^p$ where $p \in \mathbb{N}$. We now put requests of the form $\langle K_s(w) + p + 1, A_s\restriction_w\rangle$ into W, and argue as before. □

Since $c_{\mathcal{K}}$ satisfies the limit condition by Lemma 5.3.3, the proof of Theorem 5.3.5 provides a direct construction for the following.

5.3.11 Proposition. *There is a promptly simple K-trivial set A.* □

As one would expect, obeying the standard cost function restricts the number of changes in a computable approximation. Let r be a constant such that $K(y) \leq 2\log y+r$ for each y, and let $h(y) = \min\{s\colon\ K_s(y) \leq 2\log y + r\}$.

5.3.12 Proposition. *If a computable approximation $(A_s)_{s\in\mathbb{N}}$ of a set A obeys the cost function $c_{\mathcal{K}}$, then for each y, $A_s(y)$ for $s \geq h(y)$ changes at most $O(y^2)$ times. In particular, A is ω-c.e.*

Proof. Given $y < s$, when $A_{s-1}(y) \neq A_s(y)$, the sum S in (5.6) increases by at least $2^{-K_s(y)}$. Since $K_s(y) \leq 2\log(y) + r$, we have $2^{-K_s(y)} \geq \epsilon y^{-2}$ for some fixed $\epsilon > 0$. As $S < \infty$, the required bound on the number of changes follows. □

Cost functions and injury-free solutions to Post's problem

We now have two injury-free solutions of Post's problem: Kučera's solution in Section 4.2, and the construction of a promptly simple K-trivial set in Proposition 5.3.11 (we will show in 5.5.4 that each K-trivial set is low). The two solutions are closely related. Firstly, Kučera's Theorem in the restricted version of Remark 4.2.4 yields a base for ML-randomness and hence a K-trivial set. Secondly, the proof in Remark 4.2.4 can be rephrased as a cost function construction, as already discussed in (II) at the beginning of this section. Given a ML-random Δ^0_2 set Y, we want to build a promptly simple set $A \leq_{wtt} Y$ using the construction in Theorem 5.3.5. The cost function c_Y depends on a computable approximation of Y. Let $c_Y(x,s) = 2^{-x}$ for each $x \geq s$. If $x < s$, and $e < x$ is least such that $Y_{s-1}(e) \neq Y_s(e)$, let

$$c_Y(x,s) = \max(c_Y(x,s-1), 2^{-e}). \qquad (5.7)$$

This makes all the numbers $x < s$ inaccessible to PS_j for $j > e$. Clearly c_Y satisfies the limit condition, because if $e < x < s$, then $c_Y(x,s) \leq 2^{-e}$ is equivalent to $\forall t_{x \leq t < s}\, Y_t \upharpoonright_e = Y_s \upharpoonright_e$. Therefore the construction in the proof of Theorem 5.3.5 for $c = c_Y$ reproduces the construction of the promptly simple set A in 4.2.4.

5.3.13 Fact. (Greenberg and Nies, 20xx) *Suppose Y is a ML-random Δ^0_2 set and $(A_s)_{s\in\mathbb{N}}$ is a computable approximation of a set A obeying c_Y. Then $A \leq_{wtt} Y$ with use function bounded by the identity.*

Proof. We modify the argument in Remark 4.2.4. We build an interval Solovay test G (see 3.2.22) as follows: when $A_{s-1}(x) \neq A_s(x)$ and $c_Y(x,s) = 2^{-e}$, we list the string $Y_s \upharpoonright_e$ in G. Then G is indeed an interval Solovay test since the computable approximation of A obeys c_Y.

Choose s_0 such that $\sigma \not\prec Y$ for each σ listed in G after stage s_0. To show $A \leq_{wtt} Y$, given an input $x \geq s_0$, using Y as an oracle, compute $t > x$ such that $Y_t \upharpoonright_x = Y \upharpoonright_x$. We claim that $A(x) = A_t(x)$. Otherwise $A_s(x) \neq A_{s-1}(x)$ for some $s > t$. Let $e \leq x$ be the largest number such that $Y_r \upharpoonright_e = Y_t \upharpoonright_e$ for all r, $t < r \leq s$. If $e = x$ then $c_Y(x,s) \geq c_Y(x,0) = 2^{-x}$. If $e < x$ then $Y(e)$ changes in the interval $(t,s]$ of stages, so $c_Y(x,s) \geq 2^{-e}$. Hence, by the choice of t, we list an initial segment of $Y_t \upharpoonright_e = Y \upharpoonright_e$ in G at stage $s \geq s_0$, contradiction. □

5.3.14 Remark. It is instructive to compare the standard cost function $c_{\mathcal{K}}$ with the cost function c_Y. Let us see how each of them restricts a prompt simplicity requirement PS_e when we build in 5.3.5 a promptly simple set A obeying the cost function. (We meet (I) or (II) as outlined at the beginning of this section.) Let c be one of the two cost functions. A number x can enter A for the sake of PS_e only if $c(x, s) \leq 2^{-e}$. Firstly, consider $c_{\mathcal{K}}(x, s) = \sum_{x<y\leq s} 2^{-K_s(y)}$. Note that $c_{\mathcal{K}}(x, s) = 0$ for $x \geq s$, and if at a stage $t > s$ we have $c(x, t) > 2^{-e}$ we may as well assume that the entire interval $[x, t)$ has become unusable for PS_e (the numbers with short descriptions at stage t might be close to t). So PS_e will have to look for future candidates x among the numbers $\geq t$.

For the cost function c_Y, whenever $Y \upharpoonright_e$ changes at stage t, all the numbers $< t$ become unusable for PS_e. For $c_{\mathcal{K}}$ this process of intervals becoming unusable can be repeated at most 2^e times, as each time Ω increases by more than 2^{-e}. In contrast, for c_Y the process of intervals becoming unusable can be repeated as often as $Y \upharpoonright_e$ changes.

In Definition 8.5.3 we will introduce benign cost functions to capture this behavior of $c_{\mathcal{K}}$. Often, results involving $c_{\mathcal{K}}$ hold more generally for benign cost functions. If Y is ω-c.e. then c_Y is benign.

Construction of a promptly simple Turing lower bound

We will prove a useful variant of Fact 5.3.13. By Proposition 3.6.2, $\{Y\}$ is a Π^0_2 class for any Δ^0_2 set Y. Hirschfeldt and Miller (unpublished) showed in 2006 that instead of the class $\{Y\}$ one can take any null Σ^0_3 class $\mathcal{H}$ and still obtain a promptly simple *Turing* lower bound for all its ML-random members. In Theorem 8.5.15 we will see interesting examples of such classes $\mathcal{H}$, for instance the class of ω-c.e. sets. On the other hand, frequently $\mathcal{H}$ is a highness property of sets, such as being uniformly a.e. dominating (Definition 5.6.26).

5.3.15 Theorem. *From a null Σ^0_3 class $\mathcal{H}$ one can effectively obtain a promptly simple set A such that $A \leq_T Y$ for each ML-random set $Y \in \mathcal{H}$.*

Proof. Firstly, given a description of $\mathcal{H}$ as a Σ^0_3 class, we define a cost function c with the limit condition such that every Δ^0_2 set A obeying c is a Turing lower bound for the ML-random sets in $\mathcal{H}$. Secondly, we use that, by Theorem 5.3.5 we can effectively obtain a computable enumeration $(A_s)_{s\in\mathbb{N}}$ of a promptly simple set A obeying c.

Let us first work under the stronger assumption that $\mathcal{H}$ is a Π^0_2 class. By Remark 1.8.58 we have $\mathcal{H} = \bigcap_x V_x$ for an effective sequence $(V_x)_{x\in\mathbb{N}}$ of Σ^0_1 classes such that $V_{x+1} \subseteq V_x$ for each x. By (1.16) on page 54 let $(V_{x,s})_{x,s\in\mathbb{N}}$ be an effective double sequence of clopen sets such that $V_{x,s} = \emptyset$ for $x \geq s$, $V_{x,s} \subseteq V_{x,s+1}$ for each x, s and $V_x = \bigcup_s V_{x,s}$. Then the cost function $c(x, s) = \lambda V_{x,s}$ satisfies the limit condition in Definition 5.3.1 because $\lim_x \lambda V_x = 0$.

Suppose $(A_s)_{s\in\mathbb{N}}$ is a computable approximation of a set A obeying c. To show $A \leq_T Y$ for each ML-random set $Y \in \mathcal{H}$, as in the proof of Fact 5.3.13 we enumerate an interval Solovay test G. When $A_s(x) \neq A_{s-1}(x)$ for $s > x$, list in G all the strings σ of length s such that $[\sigma] \subseteq V_{x,s}$. As before, G is an interval Solovay test by the hypothesis that the approximation of A obeys c.

Showing $A \leq_T Y$ is similar to the proof of 5.3.13. Choose s_0 such that $\sigma \not\preceq Y$ for any σ enumerated into G after stage s_0. Given an input $x \geq s_0$, using Y as an oracle compute $t > x$ such that $[Y \upharpoonright_t] \subseteq V_{x,t}$. We claim that $A(x) = A_t(x)$. Otherwise $A_s(x) \neq A_{s-1}(x)$ for some $s > t$, which would cause the strings σ of length s such that $[\sigma] \subseteq V_{x,s}$ to be listed in G, contrary to $Y \in V_{x,t}$. (In general there is no computable bound for the use t on Y; see Exercise 5.3.19.)

Intuitively, we enumerate a Turing functional Γ such that $A = \Gamma^Y$ (see Section 6.1). At stage t we define $\Gamma^Y(x) = A_t(x)$ for all Y in $V_{x,t}$. When $A_s(x) \neq A_{s-1}(x)$ for $s > t$ we have to remove all these oracles by declaring them non-random.

If $\mathcal{H}$ is a Σ^0_3 class we slightly extend the argument: by Remark 1.8.58 we have $\mathcal{H} = \bigcup_i \bigcap_x V^i_x$ for an effective double sequence $(V^i_x)_{i,x\in\mathbb{N}}$ of Σ^0_1 classes such that $V^i_x \supseteq V^i_{x+1}$ for each i, x and $V^i_x = 2^{\mathbb{N}}$ for $i > x$. Then the cost function $c(x,s) = \sum_i 2^{-i}\lambda V^i_{x,s}$ is $\mathbb{Q}_2$-valued and satisfies the limit condition: given k, there is x_0 such that

$$\forall i \leq k+1 \ \forall x \geq x_0 \ \lambda V^i_x \leq 2^{-k-1}.$$

Then $c(x,s) \leq 2^{-k}$ for all $x \geq x_0$ and all s, since the total contribution of terms $2^{-i}\lambda V^i_{x,s}$ for $i \geq k+2$ to $c(x,s)$ is bounded by 2^{-k-1}.

Suppose we are given a computable approximation of a set A obeying c. For each i, when $A_s(x) \neq A_{s-1}(x)$ we list the strings σ of length s such that $[\sigma] \subseteq V^i_{x,s}$ in a set G_i. Then G_i is an interval Solovay test by the definition of c. If $Y \in \mathcal{H}$ we may choose i such that $Y \in \bigcap_x V^i_x$. If Y is also ML-random then we use G_i as before to argue that $A \leq_T Y$. □

As an application we characterize weak 2-randomness within ML-randomness. This unpublished 2006 result of Hirschfeldt and Miller was promised on page 135.

5.3.16 Theorem. *Let Z be ML-random. Then the following are equivalent:*

(i) Z is weakly 2-random.

(ii) Z and $\emptyset'$ form a minimal pair.

(iii) There is no promptly simple set $A \leq_T Z$.

Proof. (i) ⇒ (ii): Suppose the Δ^0_2 set A is incomputable and $A = \Phi^Z$ for some Turing functional Φ. Since $\{A\}$ is Π^0_2, the class $\{Y \colon \Phi^Y = A\}$ is Π^0_2 by Exercise 1.8.65. Also, this class is null by Lemma 5.1.13. Thus Z is not weakly 2-random.

(ii) ⇒ (iii): Trivial.

(iii) ⇒ (i): Suppose Z is not weakly 2-random, then Z is in a null Π^0_2 class $\mathcal{H}$. By Theorem 5.3.15 there is a promptly simple set $A \leq_T Z$. □

Exercises.

5.3.17. Show that Theorem 5.3.15 fails for null Π^0_3 classes.

5.3.18. Suppose $\mathcal{H} = \{Y\}$ for a Δ^0_2 set Y. Recall the representation of the Π^0_2 class $\mathcal{H}$ given by Proposition 3.6.2, namely $\mathcal{H} = \bigcap_x V_x$ for a sequence of uniformly Σ^0_1 classes $(V_x)_{x\in\mathbb{N}}$ where, whenever $s > x$ and r is least such that $Y_s(r) \neq Y_{s-1}(r)$, we put $Y_s \upharpoonright_{r+1}$ into $V_{x,s}$. We may suppose that $\forall x\, Y_x(x) \neq Y_{x+1}(x)$. Let c be the cost function from the proof of Theorem 5.3.15. Show that $c_Y = c$.

5.3.19. Explain why we merely obtain a Turing reduction in Theorem 5.3.15, and not a weak truth-table reduction as in Remark 4.2.4.

5.3.20. (i) Show that if Y and Z are sets such that $Y \oplus Z$ is ML-random and Y is weakly 2-random (for instance, if $Y \oplus Z$ is weakly 2-random), then Y, Z form a minimal pair. (ii) Use this to show there is a minimal pair of computably dominated *low*$_2$ sets.

K-trivial sets and Σ_1-induction ⋆

In this subsection we assume familiarity with the theory of fragments of Peano arithmetic; see Kaye (1991). We ask how strong induction axioms are needed to prove that there is a promptly simple K-trivial set. All theories under discussion will contain a finite set of axioms PA^- comprising sufficiently much of arithmetic to formulate number-theoretic concepts, carry out the identifications of binary strings with numbers on page 13, and formulate and verify some basics on c.e. sets, machines and K. $I\Sigma_1$ is the axiom scheme for induction over Σ_1 formulas. Simpson showed that the proof of the Friedberg-Muchnik Theorem 1.6.8 can be carried out in $I\Sigma_1$. See Mytilinaios (1989).

The scheme $I\Delta_0$ is induction over Δ_0 formulas. $B\Sigma_1$ is $I\Delta_0$ together with collection for Σ_1 formulas. $B\Sigma_1$ states for instance that $f([0, x])$ is bounded for each Σ_1 function f and each x. All formulas may contain parameters. Note that $I\Sigma_1 \vdash B\Sigma_1$ but not conversely (see Kaye 1991).

Hirschfeldt and Nies proved the following.

5.3.21 Theorem. *$I\Sigma_1 \vdash$ "there is a promptly simple K-trivial set".*

Proof sketch. It is not hard, if tedious, to verify that the proofs of Theorems 5.3.5 and 5.3.10 can be carried out within $I\Sigma_1$. Thus it suffices to prove from $I\Sigma_1$ that $c_{\mathcal{K}}$ satisfies the limit condition in Definition 5.3.1. Let $\mathcal{M} \models I\Sigma_1$. Consider the Σ_1 formula $\varphi(m, e)$ given by

$$\exists u \left[|u| = m + 1 \;\&\; \forall i \, (0 \le i < m \rightarrow c_{\mathcal{K}}(u_i, u_{i+1}) > 2^{-e}) \right].$$

Suppose the limit condition fails for $c_{\mathcal{K}}$ via $e \in \mathcal{M}$. Then, using $I\Sigma_1$, we have $\mathcal{M} \models \forall m \, \varphi(m, e)$. Now let m be $2^e + 1$ (in $\mathcal{M}$), and let $u \in \mathcal{M}$ be a witness for $\mathcal{M} \models \varphi(m, e)$. For each $i < m$, we have in $\mathcal{M}$ a clopen set C_i such that $\lambda C_i = c_{\mathcal{K}}(u_i, u_{i+1})$ and $C_i \cap C_j = \emptyset$ for $i \neq j$. Thus, in $\mathcal{M}$, we have

$$1 \ge \textstyle\sum_{0 \le i \le 2^e} c_{\mathcal{K}}(u_i, u_{i+1}) \ge (2^e + 1)2^{-e} > 1,$$

contradiction. □

Note that we have actually shown that $I\Sigma_1$ suffices for the "benignity" property of $c_{\mathcal{K}}$ described in Remark 5.3.14.

The weaker axiom scheme $B\Sigma_1$ is not sufficient to prove that there is a promptly simple K-trivial set. If $\mathcal{M} \models I\Delta_0$, we say that $A \subseteq \mathcal{M}$ is *regular* if for each $n \in \mathcal{M}$, $A \upharpoonright_n$ is a string of $\mathcal{M}$ (i.e., $A \upharpoonright_n$ corresponds to an element of $\mathcal{M}$ via the usual identifications defined for $\mathbb{N}$ on page 12). Each K-trivial set $A \subseteq \mathcal{M}$ is regular because for each n there is in $\mathcal{M}$ a prefix-free description of $A \upharpoonright_n$. But there is a model $\mathcal{M} \models B\Sigma_1$ in which each regular c.e. set A is computable; see Chong and Yang (2000).

Hájek and Kučera (1989) formulated and proved in $I\Sigma_1$ a version of the solution to Post's problem from Kučera (1986) which uses the Low Basis Theorem (see page 150).

In this context it would be interesting to know to what extent the proof of 5.3.13 can be carried out in $I\Sigma_1$ provided that $Y(e)$ changes sufficiently little, say $O(2^e)$ times, and whether the alternative solution, avoiding the Low Basis Theorem and using the even bits of Ω, can be carried out in $I\Sigma_1$. It is not known whether $I\Sigma_1$ proves that each K-trivial set A is Turing incomplete (see page 202).

Avoiding to be Turing reducible to a given low c.e. set

For many classes of c.e. sets studied in computability theory, given a Turing incomplete c.e. set B, there is a set A in the class such that $A \not\leq_T B$. This is the case, for instance for the class of low c.e. sets (and even of superlow c.e. sets), because there are (super)low c.e. sets A_0, A_1 such that $A_0 \oplus A_1 \equiv_T \emptyset'$. For lowness, this follows from the Sacks Splitting Theorem (1.6.10). To extend it to superlowness, see Theorem 6.1.4.

The class of c.e. K-trivial sets is different: some *low*$_2$ c.e. set B is Turing above all the K-trivial sets by a result of Nies (see Downey and Hirschfeldt 2010, or Barmpalias and Nies 2011). However, as shown in Nies (2002), one can still build a K-trivial set that is not Turing reducible to a given *low* c.e. set B. More generally, this holds for any class of c.e. sets obeying a fixed cost function with the limit condition. To show this, we extend the construction in the proof of Theorem 5.3.5, by combining it with a method to certify computations that rely on a given low c.e. set B as an oracle. This is known as the guessing method of Robinson (1971). He introduced it to show that for each pair of c.e. Turing degrees $\mathbf{b} < \mathbf{a}$ such that $\mathbf{b}$ is low, there exist incomparable low c.e. Turing degrees $\mathbf{a}_0, \mathbf{a}_1$ such that $\mathbf{a}_0 \vee \mathbf{a}_1 = \mathbf{a}$ and $\mathbf{b} < \mathbf{a}_0, \mathbf{a}_1$.

5.3.22 Theorem. *Let c be a cost function satisfying the limit condition. Then for each low c.e. set B, there is a c.e. set A obeying c such that $A \not\leq_T B$.*

Proof. Recall that $\mathbb{N}^{[e]}$ denotes the set of numbers of the form $\langle y, e\rangle$. We meet the requirements

$$P_e : A \neq \Phi_e^B,$$

by enumerating a number $x \in \mathbb{N}^{[e]}$ into A when $\Phi_e^B(x) = 0$. The problem is that B may change below the use after we do this, allowing the output of $\Phi_e^B(x)$ to switch to 1. To solve this problem, we use the lowness of B to guess at whether a computation $\Phi_e^B(x)[s] = 0$ is correct. Finitely many errors can be tolerated. We ask questions about the enumeration of A (involving B), in such a way that the answer "yes" is $\Sigma_1^0(B)$. Since $\Sigma_1^0(B) \subseteq \Delta_2^0$, we have a computable approximation to the answers. Which computable enumeration of A should we use? We may assume that one is *given*, by the Recursion Theorem! Formally, we view a computable enumeration (Definition 1.1.15) as an index for a partial computable function $\mathcal{A}$ defined on an initial segment of $\mathbb{N}$ such that, where $\mathcal{A}(t)$ is interpreted as a strong index (Definition 1.1.14) for the part of A enumerated by stage t, we have $\mathcal{A}(s) \subseteq \mathcal{A}(s+1)$ for each s. Thus we allow partial enumerations. We write $\mathcal{A}_t$ for $\mathcal{A}(t)$. Given any (possibly partial) computable enumeration $\widetilde{\mathcal{A}}$, we effectively produce an enumeration $\mathcal{A}$, asking $\Sigma_1^0(B)$-questions about the given enumeration $\widetilde{\mathcal{A}}$. We must show that $\mathcal{A}$ is total in the interesting case that $\mathcal{A} = \widetilde{\mathcal{A}}$ (by the Recursion Theorem), because in this case these questions are actually about $\mathcal{A}$.

The $\Sigma^0_1(B)$-question for requirement P_e is as follows:

Is there a stage s and $x \in \mathbb{N}^{[e]}$ such that $\widetilde{\mathcal{A}}$ is defined up to $s-1$, and

(i) $\Phi^B_e(x) = 0[s]$ & $B_s \upharpoonright \mathsf{use}\, \Phi^B_e(x)[s] = B \upharpoonright \mathsf{use}\, \Phi^B_e(x)[s]$ *(B is stable up to the use of the computation), and*

(ii) $c(x, s) \leq 2^{-(e+n)}$*?*

Here $n = \#(\mathbb{N}^{[e]} \cap \widetilde{\mathcal{A}}(s-1))$ is the number of enumerations for the sake of P_e prior to s.

Since B is low, there is a total computable function $g(e, s)$ such that $\lim g(e, s) = 1$ if the answers is "yes", and $\lim g(e, s) = 0$ otherwise. (The function $g(e, s)$ actually depends on a further argument which we supress, an index for $\widetilde{\mathcal{A}}$.)

Construction. Let $\mathcal{A}_0 = \emptyset$. At stage $s > 0$, we attempt to define $\mathcal{A}_s$, assuming that $\mathcal{A}_{s-1}$ has been defined already.

At the beginning of the stage we let $D = \mathcal{A}_{s-1}$. Then, for $e = 0, \ldots, s-1$, we do the following: we check whether there is an $x < s$, $x \in \mathbb{N}^{[e]}$ satisfying

$$\Phi^B_e(x) = 0[s] \;\&\; c(x, s) \leq 2^{-(e+n)}, \tag{5.8}$$

where $n = \#(\mathbb{N}^{[e]} \cap \mathcal{A}_{s-1})$. Now the answer to the $\Sigma^0_1(B)$ question above for e seems to be "yes"; so we choose x least and search for the least $t \geq s$ such that $g(e, t) = 1$, or $B_t \upharpoonright_u \neq B_s \upharpoonright_u$, where $u = \mathsf{use}\, \Phi^B_e(x)[s]$. In the first case, we put x into D. If we get done with all $e < s$, we let $\mathcal{A}_s = D$. Otherwise the search does not end for some $e < s$, and $\mathcal{A}_s$ is undefined.

This search is essential for the Robinson guessing method. If there is an apparent contradiction between what we see at stage s (a computation with oracle B of a certain kind) and the prediction (that there is no such computation), we look ahead till the apparent contradiction is reconciled, either by a change of B destroying the computation, or by a change of the prediction to "yes". We will argue that this only works for c.e. sets B.

Verification. We may assume that $\mathcal{A} = \widetilde{\mathcal{A}}$ by the Recursion Theorem.

5.3.23 Claim. *The function $\mathcal{A}$ is total.*

Inductively assume that $\mathcal{A}_{s-1}$ is defined if $s > 0$. Since $\mathcal{A} = \widetilde{\mathcal{A}}$ and by the correctness of $\lim_t g(e, t)$, the search at stage s ends for each e. So we define $\mathcal{A}_s$.

5.3.24 Claim. *$(\mathcal{A}_s)_{s \in \mathbb{N}}$ obeys the cost function c.*

At stage s, suppose x is least s.t. $\mathcal{A}_{s-1}(x) \neq \mathcal{A}_s(x)$. We enumerate x for the sake of some requirement P_e, which so far has enumerated n numbers. Then $c(x, s) \leq 2^{-(e+n)}$, hence $S \leq \sum_{e,n \in \mathbb{N}} 2^{-(e+n)} = 4$ where S is defined in (5.6).

Let $A = \bigcup_s \mathcal{A}_s$.

5.3.25 Claim. *Each requirement P_e is met.*

Assume for a contradiction that $A = \Phi_e^B$. First suppose that $\lim_s g(e,s) = 1$. Choose witnesses x, s for the affirmative answer to the $\Sigma_1^0(B)$ question for P_e and let $u = \text{use } \Phi_e^B(x)[s]$. Since $B \upharpoonright_u$ does not change after s, we search for t till we see $g(e,t) = 1$. Then P_e enumerates x at stage s.

Now consider the case $g(e,s) = 0$ for all $s \geq s_0$. Then P_e does not enumerate any numbers into A after stage s_0. Suppose it has put n numbers into A up to stage s_0. Since $A = \Phi_e^B$, there is $x \in \mathbb{N}^{[e]}$ and $s \geq s_0$ such that $\Phi_e^B(x) = 0[s]$ and $c(x,s) \leq 2^{-(e+n)}$. So the answer to the $\Sigma_1^0(B)$ question for P_e is "yes", contradiction. □

If B is merely a Δ_2^0 set, the argument in the proof of Lemma 5.3.25 breaks down in the case $\lim_s g(e,s) = 1$. Otto can now present the correct computation $\Phi_e^B(x) = 0$ at a stage s where the $g(e,s)$ has not yet stabilized. To fool us when we try to reconcile the apparent contradiction, he temporarily changes B below the use at stage $t > s$ while keeping $g(e,t) = 0$, and we do not put x into A at s. At a later stage the correct computation $\Phi_e^B(x) = 0$ will return, but now he has increased the cost of x above $2^{-(e+n)}$, so we have lost our opportunity.

Indeed, Kučera and Slaman (20xx) have shown that some low set B is Turing above all the c.e. sets that are low for K (and hence above all the c.e. K-trivial sets by Section 5.4). In fact they prove this for any class inducing a Σ_3^0 ideal in the c.e. degrees such that there is a function $F \leq_T \emptyset'$ that dominates each function partial computable in a member of the class. The class of sets that are low for K is of this kind by Exercise 5.1.6.

5.3.26 Exercise. In addition to the hypotheses of Theorem 5.3.22, let E be a c.e. set such that $E \not\leq_T B$. Then there is a c.e. set A obeying c such that $A \not\leq_T B$ and $A \leq_T E$.

Necessity of the cost function method for c.e. K-trivial sets

In the following two subsections, we study aspects of the cost function method peculiar to K-triviality. Later on we will see that Theorem 5.3.10, the cost function criterion for K-triviality, is actually a characterization: a Δ_2^0 set A is K-trivial iff some computable approximation of A obeys the standard cost function (as defined in 5.3.4). Thus each K-trivial set can be thought of as being constructed via the cost function method with $c_{\mathcal{K}}$. However, we cannot expect that *every* computable approximation of a K-trivial set obeys the standard cost function, for instance because for an appropriate computable enumeration $0, 1, 2, \ldots$ of $\mathbb{N}$, the total cost of changes is infinite (see Exercise 5.3.7). The total cost is only finite when one views a given approximation in "chunks". One introduces an appropriate computable set E of stages. In the new approximation changes are only reviewed at stages in E. In the calculation of the total cost in (5.6), only the change at the least number counts at such a stage.

In this subsection we only consider the c.e. K-trivial sets. It is harder to show the necessity of the cost function method for all the K-trivial sets: this requires the golden run method of Section 5.4, and is postponed to Theorem 5.5.2.

5.3.27 Theorem. *The following are equivalent for a c.e. set A.*

(i) A is K-trivial.

(ii) Some computable enumeration $(\widetilde{A}_i)_{i\in\mathbb{N}}$ of A obeys the standard cost function $c_{\mathcal{K}}$.

Proof. (ii) $\Rightarrow$ (i): This follows from Theorem 5.3.10.
(i) $\Rightarrow$ (ii): Suppose that the c.e. set A is K-trivial via b. Then there is an increasing computable function f such that $K(A\restriction_n) \le K(n) + b$ $[f(s)]$ for each s and each $n < s$, i.e., the inequality holds at stage $f(s)$. The set of stages $E = \{s_0 < s_1 < \ldots\}$ is obtained by iterating f: let

$$s(0) = 0 \text{ and } s(i+1) = f(s(i)). \tag{5.9}$$

Let $\widehat{c}_{\mathcal{K}}(x, s(i)) = \sum_y 2^{-K_{s(i+1)}(y)} [\![x < y \le s(i)]\!]$, i.e., $K(y)$ is computed only at stage $s(i+1)$. Let x_i be the least number $x < s(i)$ such that $A_{s(i)}(x) \neq A_{s(i+1)}(x)$. In the following, we only consider numbers i, j such that x_i (or x_j) is defined. Using a variant of the accounting method in Remark 5.1.21, we will show that

$$\widehat{S} = \sum_i \widehat{c}_{\mathcal{K}}(x_i, s(i)) [\![x_i \text{ is defined}]\!] \le 2^b. \tag{5.10}$$

For each y such that $x_i < y \le s(i)$, at stage $s = s(i+1)$ there is a $\mathbb{U}$-description of $A_s\restriction_y$ of length $\le K_s(y) + b$. Let E_i be the set of descriptions for such y. The key fact is that, since $(A_s)_{s\in\mathbb{N}}$ is a computable *enumeration*, the strings $A_s\restriction_y$ described at different stages $s(i), s(j)$ are distinct, so that $E_i \cap E_j = \emptyset$. We will account the cost of an A-change between stages $s(i)$ and $s(i+1)$ against $\lambda[E_i]^{\prec}$. For each y, $x_i < y \le s(i)$, if $s = s(i+1)$, since $K_s(A_s\restriction_y) \le K_s(y) + b$ we have $2^{-K_s(y)} \le 2^b 2^{-K_s(A_s\restriction_y)}$, and hence, by taking the sum over all y such that $x_i < y \le s(i)$, we have $\widehat{c}_{\mathcal{K}}(x_i, s(i)) \le 2^b\lambda[E_i]^{\prec}$. Since $\sum_i \lambda[E_i]^{\prec} \le 1$, this implies $\sum_i \widehat{c}_{\mathcal{K}}(x_i, s(i)) \le 2^b$, that is, (5.10). Now let $\widetilde{A}_i = A_{s(i+1)} \cap [0, i)$. For each $x < i$,

$$c_{\mathcal{K}}(x, i) = \textstyle\sum_{x<y\le i} 2^{-K_i(y)} \le \widehat{c}(x, s(i)).$$

Note that if $i > 0$ and $i-1 \in A_{s(i+1)}$, then $i-1 \in \widetilde{A}_i - \widetilde{A}_{i-1}$, which may contribute a cost of $c_{\mathcal{K}}(i-1, i) = 2^{-K_i(i)}$. Thus we have

$$\textstyle\sum_{x,i} c_{\mathcal{K}}(x, i) [\![i > 0 \ \& \ x \text{ least s.t. } \widetilde{A}_{i-1}(x) \neq \widetilde{A}_i(x)]\!] \le \widehat{S} + \Omega \le 2^b + 1. \qquad \square$$

Listing the (ω-c.e.) K-trivial sets with constants

We provide a presentation of the class of ω-c.e. K-trivial sets. In the remark after Theorem 5.5.2, we will learn that each K-trivial set is ω-c.e., so we actually have a presentation of the entire class $\mathcal{K}$. As for every class of ω-c.e. sets that contains the finite sets and has a Σ^0_3 index set, there is a uniformly ω-c.e. listing $(A_e)_{e\in\mathbb{N}}$ of the ω-c.e. sets in $\mathcal{K}$ (Exercise 1.4.22). Here we show that there is a listing which includes the witnesses for the Σ^0_3 statement, namely, for each e, a constant via which A_e is K-trivial. The result is due to Downey, Hirschfeldt, Nies and Stephan (2003). We give a simpler proof using ideas developed in the proofs of Theorems 5.3.10 and 5.3.27. Recall from 1.4.5 that V_e is the e-th ω-c.e. set.

5.3.28 Theorem. *There is an effective sequence $(B_e, d_e)_{e \in \mathbb{N}}$ of ω-c.e. sets and of constants such that each B_e is K-trivial via d_e, and each K-trivial set occurs in the sequence. Furthermore, there is an effective sequence of the c.e. K-trivial sets with constants.*

Proof. We effectively transform a pair A, b of an ω-c.e. set and a constant into an ω-c.e. set $\widetilde{A}$ and a constant d such that $\widetilde{A}$ is K-trivial via d and $A = \widetilde{A}$ in case that A is in fact K-trivial via b. (But even then, d may be larger than b.) To obtain the required listing $(B_e, d_e)_{e \in \mathbb{N}}$, for each $e = \langle i, b \rangle$, let $A = V_i$, apply this transformation to the pair A, b, and let $B_e = \widetilde{A}$ and $d_e = d$.

We define a sequence of stages $s(i)$ the same way we did before Theorem 5.3.27, except that now the sequence breaks off if A is not K-trivial via b. The computable approximation of $\widetilde{A}$ follows the approximation of A, but is updated only at such stages. In more detail, for each s, let

$$f(s) \simeq \mu t > s. \forall n < s \; K_t(A_t \upharpoonright_n) \leq K_t(n) + b, \text{ and let}$$

$$s(0) = 0 \text{ and } s(i+1) \simeq f(s(i)). \tag{5.11}$$

Let E be the (possibly finite) set of stages of the form $s(i)$. Because "$t = f(s)$" is computable uniformly in A, b, we can compute E uniformly. Therefore the following is a computable approximation:

$$\widetilde{A}_u(x) = A_{\max(E \cap \{0,\dots,u\})}.$$

(Thus, if E is finite the final value is $A_{\max(E)}$.) We define a prefix-free machine M such that

$$\forall s \forall w < s \, K_M(\widetilde{A}_s \upharpoonright_w) \leq K_s(w) + b + 2. \tag{5.12}$$

Let r be the coding constant for M according to the Machine Existence Theorem, and let $d = b + 2 + r$. Then $\widetilde{A}$ is K-trivial via d, as required.

To meet (5.12), we have to provide a new M-description of $A_s \upharpoonright_w$ in two cases, which are the same as in the proof of Theorem 5.3.10:

(a) $K_s(w) < K_{s-1}(w)$, or

(b) $K_s(w) < \infty \;\&\; \widetilde{A}_{s-1} \upharpoonright_w \neq \widetilde{A}_s \upharpoonright_w$.

Here (a) includes the case that $K_{s-1}(w) = \infty$. To ensure (5.12) in case (a), we use strings beginning in 00 as M-descriptions. Thus, if $\mathbb{U}_s(\sigma) = w$ where $|\sigma| = K_s(w) < K_{s-1}(w)$, we declare

$$M(00\sigma) = A_s \upharpoonright_w.$$

In case (b), since $s(0) = 0$ and $\widetilde{A}$ only changes at stages in E, we have $s = s(i+1)$ for some i. There are two subcases.

(b1) $s(i) \leq w < s(i+1) = s$. For each w there is at most one such i (this is a key point). So in that case, to meet (5.12) we declare

$$M_s(10\sigma) = \widetilde{A}_s\upharpoonright_w,$$

for any string σ of length $K_s(w)$ such that $\mathbb{U}_s(\sigma) = w$.

(b2) Otherwise, i.e., $w < s(i)$. By the definitions, $\widetilde{A}_s\upharpoonright_w = A_s\upharpoonright_w$ and there is a $\mathbb{U}$-description σ of $A_s\upharpoonright_w$ at stage s such that $|\sigma| \leq K_s(w) + b$. So to meet (5.12) it suffices to copy $\mathbb{U}$, that is, to let

$$M_t(11\sigma) \simeq \mathbb{U}_t(\sigma) \text{ for all } t \text{ and all } \sigma.$$

To obtain a listing of the c.e. K-trivial sets with constants, we carry out the same proof based on the indexing $(W_i)_{i\in\mathbb{N}}$ of the c.e. sets. □

Let C be an index set for a class of c.e. sets, namely $e \in C$ & $W_e = W_i \to i \in C$ (Definition 1.4.18). We say that C is *uniformly* Σ^0_3 if there is a Π^0_2 relation P such that $e \in C \leftrightarrow \exists b\, P(e,b)$ and there is an effective sequence $(e_n, b_n)_{n\in\mathbb{N}}$ such that $P(e_n, b_n)$ and $\forall e \in C\, \exists n\, W_e = W_{e_n}$. For instance, let $P(e,b)$ be $\forall n\, \forall s\, \exists t > s\, [K_t(W_{e,t}\upharpoonright_n) \leq K_s(n)+b]$. Then Theorem 5.3.28 shows the following (also see Exercise 5.3.32).

5.3.29 Corollary. *The class of c.e. K-trivial sets has a uniformly Σ^0_3 index set.* □

Exercises.

5.3.30. Let $(B_e)_{e\in\mathbb{N}}$ be as in Theorem 5.3.28. Show that there is a computable binary function f such that $B_{f(i,j)} = B_i \oplus B_j$ for each $i, j \in \mathbb{N}$.

5.3.31. Let $Q(e,b)$ be the Π^0_2 relation $W_e \cup W_b = \mathbb{N}$ & $W_e \cap W_b = \emptyset$, so that W_e is computable iff $\exists b\, Q(e,b)$. Show that Q does not serve as a Π^0_2 relation via which the index set of the class of computable sets is uniformly Σ^0_3.

5.3.32. (Stephan) Show that an index set C is uniformly Σ^0_3 $\Leftrightarrow$ its class is uniformly c.e., namely, there is a computable sequence of indices $(e_n)_{n\in\mathbb{N}}$ such that $i \in C \leftrightarrow \exists n\, W_i = W_{e_n}$. For instance, by Exercise 1.4.22, if C is a Σ^0_3 index set of a class that contains all finite sets, then C is uniformly Σ^0_3.

5.3.33. Use 5.3.32 to show that $C = \{e\colon W_e \text{ is infinite}\}$ is not uniformly Σ^0_3.

Adaptive cost functions

A modification of the proof of Theorem 5.3.11 yields a direct argument that some promptly simple set is low for K. Understanding this transition from K-triviality to being low for K may be helpful for the proof in Section 5.4 that the two classes are equal (in particular, for Lemma 5.4.10).

Given a prefix-free oracle machine M and a Δ^0_2 set A with a computable approximation $(A_s)_{s\in\mathbb{N}}$, consider the cost function

$$c_{M,A}(x,s) = \sum_\sigma 2^{-|\sigma|}\, [\![M^A(\sigma)[s-1]\downarrow \ \&\ x < \mathsf{use}\ M^A(\sigma)[s-1]]\!]. \tag{5.13}$$

Note that $c_{M,A}(x,s)$ is the measure of the M^A-descriptions at stage $s-1$ that are threatened by a potential change $A_s(x) \neq A_{s-1}(x)$. In other words, $c_{M,A}(x,s)$ is the maximum decrease of Ω^A_M that can be caused by $A_s(x) \neq A_{s-1}(x)$. In contrast to the previous examples, this cost function is *adaptive*, namely,

$c_{M,A}(x,s)$ depends on A_{s-1}. (It would thus be more accurate to use the notation $c_M(x,s;A_{s-1})$, but this is too cumbersome.) The computable approximation of A determines whether such a function is non-decreasing in s, and whether the limit condition holds.

If $M = \mathbb{U}$, a set that obeys the cost function it determines is low for K.

5.3.34 Proposition. *Suppose that $A(x) = lim_s\, A_s(x)$ for a computable approximation $(A_s)_{s\in\mathbb{N}}$ that obeys $c_{\mathbb{U},A}$, namely, there is $u \in \mathbb{N}$ such that*

$$S = \sum_{x,s} c_{\mathbb{U},A}(x,s)\, [\![s > 0 \ \& \ x \text{ is least s.t. } A_{s-1}(x) \neq A_s(x)]\!] \leq 2^u. \qquad (5.14)$$

Then A is low for K.

Proof. The proof is somewhat similar to the proof of Theorem 5.3.10. As before we enumerate a bounded request set W:

> at stage $s > 0$, put the request $\langle |\sigma|+u+1, y\rangle$ into W if $\mathbb{U}^A(\sigma)[s] = y$ newly converges, that is, $\mathbb{U}^A(\sigma)[s] = y$ but $\mathbb{U}^A(\sigma)[s-1]\uparrow$.

To show that A is low for K, it suffices to verify that W is indeed a bounded request set. Suppose a request $\langle |\sigma| + u + 1, y\rangle$ is put into W at a stage s via a newly convergent computation $\mathbb{U}^A(\sigma)[s] = y$. Let $w = \mathsf{use}\ \mathbb{U}^A(\sigma)[s]$.

Stable case. $\forall t > s\ A_s\restriction_w = A_t\restriction_w$. The contribution to W of such requests is at most $\Omega^A/2^{u+1}$, since at most one request is enumerated for each description $\mathbb{U}^A(\sigma) = y$ once $A\restriction_w$ is stable.

Change case. $\exists t > s\ A_s\restriction_w \neq A_t\restriction_w$. Choose t least, and $x < w$ least such that $A_{t-1}(x) \neq A_t(x)$. Then $2^{-|\sigma|}$ is part of the sum $c_{\mathbb{U},A}(x,t)$, which is part of S. Since A obeys its cost function $c_{\mathbb{U},A}$ and by the choice of u, the total contribution to W in this case is at most $1/2$. □

We apply the criterion to give a direct proof of Muchnik's result, presumably close to his original proof.

5.3.35 Theorem. *There is a promptly simple set A that is low for K.*

Proof. The construction is the one from the proof of Theorem 5.3.5 where a requirement PS_j can spend at most 2^{-j}. The only difference is that we now use the cost function $c_{\mathbb{U},A}$. (This is allowed since for $s > 0$, $c_{\mathbb{U},A}(x,s)$ is defined in terms of A_{s-1}.) The set A is low for K by Proposition 5.3.34. By the same argument as in the previous proof, we know that A is promptly simple, once we have shown the following.

5.3.36 Claim. *$c_{\mathbb{U},A}$ satisfies the limit condition.*

Given $e \in \mathbb{N}$, we will find m such that $\sup_{s>m} c_{\mathbb{U},A}(m,s) \leq 2^{-e}$. Note that if $\sigma \in \operatorname{dom} \mathbb{U}^A$ and t is a stage such that $\mathbb{U}^A(\sigma)[t]\downarrow$ and each requirement PS_j for $j \leq |\sigma|$ has ceased to act, then the computation $\mathbb{U}^A(\sigma)[t]$ is stable. Let $\sigma_0, \ldots, \sigma_{k-1} \in \operatorname{dom} \mathbb{U}^A$ be strings such that, where $\alpha = \sum_i 2^{-|\sigma_i|}$, we have

TABLE 5.1. Overview of cost functions to build a c.e. incomputable set.

Cost function	Definition	Purpose	Ref.
$c_{\mathcal{K}}(x,s)$	$\sum_{x<w\le s} 2^{-K_s(w)}$	build a K-trivial	5.3.2
c_Y	$c_Y(x,s)=\max(c_Y(x,s-1),2^{-e})$ where $Y_{s-1}(e)\neq Y_s(e)$	build a set below a Δ^0_2 set $Y\in \mathsf{MLR}$	(5.7), pg. 189
$c(x,s)$	$\lambda V_{x,s}$, where V_x is uniformly Σ^0_1 and $\mathcal{H}=\bigcap_x V_x$ is null	build a set below $\mathcal{H}\cap\mathsf{MLR}$	5.3.15
$c_{\mathbb{U},A}(x,s)$	$\sum_\sigma 2^{-\lvert\sigma\rvert}$ $[\![\mathbb{U}^A(\sigma)[s-1]\downarrow$ & $x<\mathsf{use}\ \mathbb{U}^A(\sigma)[s-1]]\!]$	build a set that is low for K	(5.13), pg. 198

$\Omega^A-\alpha\le 2^{-e-1}$. We may choose a stage $m\ge e+1$ such that all the computations $\mathbb{U}^A(\sigma_i)[m]$ are stable and no PS_j, $j\le e+1$ acts from stage m on. At each stage $s>m$ we have $\Omega^A[s]-\Omega^A\le 2^{-e-1}$, since the most $\Omega^A[t]$ can decrease (because computations $\mathbb{U}^A(\sigma)$ are destroyed by enumerations into A) is $\sum_{j>e+1}2^{-j}=2^{-e-1}$. Thus, $\Omega^A[s]-\alpha\le 2^{-e}$ for each $s>m$. Now, for $s>m$, the sum in (5.13) defining $c_{\mathbb{U},A}(m,s)$ refers to computations other than $\mathbb{U}^A(\sigma_i)[m]$ as their use is at most m. So $c_{\mathbb{U},A}(m,s)\le\Omega^A[s]-\alpha\le 2^{-e}$. This proves the claim. □

5.3.37 Remark. Recall that a cost function is called *adaptive* if the cost at stage s depends on A_{s-1}. If the underlying cost function is adaptive then a cost function construction must be regarded as having injury. For instance, during the construction of a low simple set in Theorem 1.6.4, the lowness requirements

$$L_e\colon\ \exists^\infty s\, J^A(e)[s-1]\downarrow\ \Rightarrow\ J^A(e)\downarrow$$

are injured. The following adaptive cost function encodes the restraint imposed by L_e: if $J^A(e)$ newly converges at stage $s-1$, define

$$c(x,s)=\max\{c(x,s-1),2^{-e}\}$$

for each $x<\mathsf{use}\ J^A(e)[s-1]$. If A is enumerated in such a way that the total cost of changes is finite, then L_e is injured only finitely often. Thus A is low.

In contrast, a cost function c given in advance cannot be used to hide injury, because to encode a restraint that is in force at the beginning of stage s we have to know A_{s-1}. The cost functions in the first three rows of Table 5.1 are non-adaptive. In particular, so is the cost function used to build a promptly simple Turing lower bound in Theorem 5.3.15. Thus, the corresponding constructions are injury-free.

5.3.38 Exercise. (Kučera and Terwijn, 1999). Give a direct construction of a promptly simple set A that is low for ML-randomness.

5.4 Each K-trivial set is low for K

In Section 5.1 we studied three equivalent lowness properties: being low for K, being low for ML-randomness, and being a base for ML-randomness. It is easily shown that a set that is low for K is K-trivial (5.2.3). We provide the remaining implication. Thus, these three equivalent lowness properties also coincide with K-triviality, a property expressing that the set is far from being Martin-Löf random.

This implication has a complicated history. Downey, Hirschfeldt, Nies and Stephan (2003) showed that each K-trivial set is Turing incomplete. These ideas were later on explained through the decanter model, which appeared in Downey, Hirschfeldt, Nies and Terwijn (2006). Nies in 2002 combined the decanter model with a new technique called the golden run method in order to show that the K-trivial sets are closed downward under $\leq_T$. Hirschfeldt and Nies together used the golden run method to show the stronger result that K-triviality implies being low for ML-randomess. This result appeared in Nies (2005*b*). For even more on the story see Nies (2011).

5.4.1 Theorem. *Each K-trivial set is low for K.*

First we prove the easier result of Downey, Hirschfeldt, Nies and Stephan (2003) that each K-trivial set is Turing incomplete, thereby introducing the *decanter method.*

Combining this with a further new technique, the *golden run method*, yields the full result. By Proposition 5.1.2 and Theorem 5.2.4(ii) each set that is low for K is low, so the full result indeed implies lowness and hence Turing incompleteness for a K-trivial set. In Section 5.5, the core of the combined decanter and golden run methods will be isolated in a powerful (if technical) result, the Main Lemma 5.5.1. It can be used, for instance, to show that Theorem 5.3.10 in fact provides a characterization of the K-trivial sets: A is K-trivial iff some computable approximation of A obeys the standard cost function $c_{\mathcal{K}}$.

Introduction to the proof

We outline the proof of the implication "K-trivial $\Rightarrow$ low for K". We also introduce some terminology and auxiliary objects that will be used in the detailed proof. We go through stronger and stronger intermediate results, showing that a K-trivial set is *wtt*-incomplete, then Turing incomplete, and then low. Each step introduces new techniques. Important comments are made in Remarks 5.4.3 and 5.4.4.

Throughout, we fix a constant b such that the given set A is K-trivial via b, that is, $\forall n\, K(A\upharpoonright_n) \leq K(n) + b$. By Theorem 5.2.4(ii) we may also fix a computable approximation $(A_s)_{s\in\mathbb{N}}$ of A.

1. No K-trivial set A is weak truth-table complete.

This was already proved in Proposition 5.2.18(ii). Here we assume $\emptyset' \leq_{wtt} A$ and obtain a contradiction. The very basic idea how to use the K-triviality of A is the following: we choose a number n and give it a short description. The opponent Otto claims that A is K-trivial, so he has to respond by giving a short description of $A\upharpoonright_n$. If he later needs to change $A\upharpoonright_n$, he has to provide a description of the new $A\upharpoonright_n$. Via the short description of n we have made such changes expensive for him. In this way we limit the changes of A to an extent contradicting its weak truth-table completeness.

As an extreme case, let us assume that for each n, his description of $A\upharpoonright_n$ is no longer than our description of n. We choose n large enough so that we can force Otto to change $A\upharpoonright_n$, using that A is *wtt*-complete (see below for details). We issue a description of n that has length 0, and wait for the stage where he provides a description of $A\upharpoonright_n$ that also has length 0. He has now wasted *all* his capital: if we force him to change $A\upharpoonright_n$, then he has to give up on providing descriptions.

The bounded request set L, and the constant d. Actually, Otto can choose his descriptions by a constant longer. To counter this, we force him to change A more often. His constant is known to us in advance. We issue descriptions of numbers n by enumerating requests of the form $\langle r, n\rangle$ into a bounded request set L. By the Recursion Theorem (see Remark 2.2.21), we may assume an index d is given such that M_d is a machine for L. To respond to our enumeration of a request $\langle r, n\rangle$ into L, he has to provide a description of $A\upharpoonright_n$ that has length at most $r+b+d$. For at least 2^{b+d} times, we want $A\upharpoonright_n$ to change after he has given such a description. In fact we want $A\upharpoonright_n$ not to change back, but rather to a configuration not seen before. In that case, if our description of n is of length $r=0$ (say), his total capital spent is at least $(2^{b+d}+1)2^{-(b+d)} > 1$, contradiction.

The hypothesis $\emptyset' \leq_{wtt} A$ can be used to force Otto into making sufficiently many changes. We build a c.e. set B. By the Recursion Theorem, we are given a c.e. index for B, and hence a many-one reduction showing that $B \leq_m \emptyset'$ (1.2.2). Combining this with the fixed *wtt*-reduction of $\emptyset'$ to A, we are given a Turing reduction Γ and a computable function g such that $B = \Gamma^A$ and $\forall x\, \mathsf{use}\, \Gamma^A(x) \leq g(x)$. (In fact we have used the Double Recursion Theorem 1.2.16, because we also needed the constant d in advance. This slight technical complication will disappear when we proceed to showing that all K-trivial sets are low.)

Construction of L and the c.e. set B. Let $c = 2^{b+d}$, and $n = g(c)$. We put the single request $\langle 0, n\rangle$ into L. From now on, at each stage t such that $B\upharpoonright_c = \Gamma^A\upharpoonright_c [t]$ and $K_t(A_t\upharpoonright_n) \leq b+d$, we force $A\upharpoonright_n$ to change to a configuration not seen before by putting into B the largest number less than c which is not yet in B.

In the fixed point case we have $B = \Gamma^A$, so we can force c such changes. Since all the $A\upharpoonright_n$-configurations are different, the measure of their descriptions is at least $(c+1)2^{-(b+d)} > 1$, which is impossible.

2. No K-trivial set A is Turing complete.

We assume for a contradiction that A is Turing complete. As before, we build the c.e. set B and, by the Recursion Theorem, we are given a Turing functional Γ such that $B = \Gamma^A$. Let $\gamma^A(m) = \mathsf{use}\, \Gamma^A(m)$ for each m. We have no computable bound on $\gamma^A(m)$ any longer. If we try to run the previous strategy with n greater than the current use $\gamma^A(m)$ for a number m we can put into B, Otto can beat us as follows: *first* he changes $A\upharpoonright_n$; when a new computation $\Gamma^A(m)$ appears, then $\gamma^A(m) > n$. Only then he gives a description of the new version of $A\upharpoonright_n$. Such a premature A-change deprives us of the ability to later cause changes of $A\upharpoonright_n$ by putting m into B.

To solve the problem we start a new attempt whenever $\Gamma^A(m)$ changes. But we have to be careful to avoid spending everything on a failed attempt, so we will put into L requests of the type $\langle r, n\rangle$ for large r and lots of numbers n. Each time we put such a request we wait for the $A\upharpoonright_n$ description before putting in the next one. If $\Gamma^A(m)$ changes first then r is increased. Since Γ^A is total, $\Gamma^A(m)$ must settle eventually, so r settles as well. On premature $A\upharpoonright_n$-change we have only wasted 2^{-r}, but then we increase r. (This is our first example of a *controlled risk strategy*. Our risk is only 2^{-r} for each attempt. We choose the successive values of r so large that their combined risks are tolerable.) On the other hand, if $\Gamma^A(m)$ remains stable for long enough, we can build up in small portions as much as we would have put in one big chunk. So eventually we can make an A-change sufficiently expensive for Otto.

It is instructive to consider the hypothetical case $b = d = 0$ once again. Recall from Definition 2.2.15 that the weight of $E \subseteq \mathbb{N}$ with regard to L is defined by

$$\mathsf{wgt}_L(E) = \textstyle\sum_{\rho,x} 2^{-(\rho)_0} [\![\rho \in L \;\&\; x \in E \;\&\; (\rho)_1 = x]\!].$$

The following procedure enumerates a set F_2 of weight $p = 3/4$ such that for each $n \in F_2$, there are $\mathbb{U}$-descriptions of two versions $A\upharpoonright_n$. It maintains auxiliary sets F_0 and F_1.

Procedure $P_2(p)$.

(1) Choose a large number m.
(2) WAIT for $\Gamma^A(m)$ to converge.
(3) Let $r \in \mathbb{N}$ be such that the procedure has gone through (2) for $r-2$ times. Pick a large number n. Put $\langle r_n, n\rangle$ into L where $r_n = r$, and put n into F_0. WAIT for a stage t such that $K(A\upharpoonright_n)[t] \le r+b+d = r$, and transfer n from F_0 to F_1. (If M_d is a machine for L, then t exists.) IF $\mathsf{wgt}_L(F_2 \cup F_1) \le p$ GOTO (3).
    If the expression $\Gamma^A(m)$ changes during this loop then all the numbers currently in F_1 have obtained their A-change, so put F_1 into F_2 and declare $F_1 = \emptyset$. (We call this an *early promotion* of the numbers in F_1. This important topic is pursued further in Remark 5.4.3.) Also, the number n in F_0 is *garbage* as we cannot any longer hope to get descriptions for two $A\upharpoonright_n$ configurations, so declare $F_0 = \emptyset$. GOTO (2).
(4) Put m into B. This provides the A-change for the elements now in F_1, so put F_1 into F_2 and declare $F_1 = \emptyset$. (We call this a *regular promotion*.)

To see that L is a bounded request set, consider what happens to a number n after a request $\langle r_n, n\rangle$ has entered L.

- For each r there is at most one n such that $r_n = r$ and n is not promoted to F_1. The total weight of such numbers n is thus at most $\sum_{r\ge 3} 2^{-r} = 1/4$.

- The weight of the numbers n that get into F_1 is at most $p = 3/4$, since at each stage we have $\mathsf{wgt}_L(F_2 \cup F_1) \leq p$, and we empty a set F_1 into F_2 before we start a new one (and F_2 never gets emptied).

Since Γ^A is total, the procedure reaches (4), which causes $\mathsf{wgt}(F_2) = 3/4$. This is a contradiction, since for each $n \in F_2$, two configurations $A\restriction_n$ have descriptions of length r_n. ◇

We now remove the hypothesis that $b = d = 0$. We allow a weight of $1/4$ of numbers n that are garbage. This extra "space" in L is obtained by forcing $2^{b+d+1} - 1$ many $A\restriction_n$-changes when the request is not garbage (that is, almost twice as many as in the *wtt* case). Throughout, we will let

$$\boxed{\boldsymbol{k} = 2^{b+d+1}.}$$

To force these changes we use $\boldsymbol{k} - 1$ levels of procedures P_i ($2 \leq i \leq \boldsymbol{k}$). A procedure at level $i > 2$ calls one at level $i - 1$. The bottom procedure P_2 acts as above. An argument similar to the one just used shows that the total weight of numbers that are not garbage does not exceed $1/2$. Before giving more details in Fact 5.4.2, we introduce two technical concepts.

Stages when A looks K-trivial. The construction is restricted to *stages*, defined exactly as before Theorem 5.3.27 (except that the given set A was c.e. then). Thus, for each s, let $f(s) = \mu t > s\, \forall n < s\; K(A\restriction_n) \leq K(n) + b\; [t]$. Let $s(0) = 0$ and $s(l + 1) = f(s(l))$. We write *stage* (in italics) when we mean a stage of this type.

Definition of i-sets. The following device keeps track of the number of times the opponent has had to give new descriptions of strings $A\restriction_n$: the number n is in an i-set if this has happened i times. Thus, for $1 \leq i \leq \boldsymbol{k}$, we say that a finite set $E \subseteq \mathbb{N}$ is an *i-set* at *stage* t if, for all $n \in E$, at some *stage* $u < t$ we enumerated a request $\langle r_n, n\rangle$ into L, and now there are i distinct strings z of the form $A_v\restriction_n$ for some *stage* v, $u \leq v \leq t$, such that $K_v(z) \leq r_n + b + d$. A c.e. set with an enumeration $E = \bigcup E_t$ is an i-set if E_t is an i-set at each *stage* t.

Note that $\mathsf{wgt}_L(E) = \sum\{2^{-r_n} : n \in E\}$ for each $E \subseteq \mathbb{N}$. Since Otto has to match our description of n by descriptions of strings of length n that are by at most $b + d$ longer, we have the following.

5.4.2 Fact. *If the c.e. set E is a $\boldsymbol{k}$-set, where $\boldsymbol{k} = 2^{b+d+1}$, then* $\mathsf{wgt}_L(E) \leq 1/2$.

Proof. For all $n \in E$, there is a request $\langle r_n, n\rangle$ in L and there are $\boldsymbol{k}$ distinct strings z of length n such that $K(z) \leq r_n + b + d$. Hence

$$1 \geq \Omega = \lambda[\mathrm{dom}\,\mathbb{U}]^{\prec} \geq \boldsymbol{k} \textstyle\sum_{n\in E} 2^{-(r_n+b+d)} = \boldsymbol{k} 2^{-(b+d)} \mathsf{wgt}_L(E).$$

Because $\boldsymbol{k} = 2^{b+d+1}$, this implies $\mathsf{wgt}_L(E) \leq 1/2$. □

A procedure P_i ($2 \leq i \leq \boldsymbol{k}$) enumerates an i-set F_i, beginning with $F_i = \emptyset$. If P_i is initialized then F_i is emptied. The construction begins by calling $P_{\boldsymbol{k}}$, which calls $P_{\boldsymbol{k}-1}$ lots of times, and so on down to P_2, which also enumerates L

and F_0, F_1, as above. Each procedure P_i is called with parameters p, α, where $\alpha = 2^{-m}$ for some m, $p \in \mathbb{Q}_2$ and $p \geq \alpha$.

- The *goal* p is the weight P_i needs its set to reach to be able to return.
- The *garbage quota* α is how much garbage P_i is allowed to produce.

The garbage quota of a procedure is closely related to the goals of the sub-procedures it calls, because the quantity of garbage produced when they are cancelled does not exceed their goal.

In the following construction, the garbage of a run $P_i(p, \alpha)$ is the weight of the numbers that are eventually at level $i-2$ (and never promoted to level $i-1$).

Procedure $P_i(p, \alpha)$ ($2 \leq i \leq \boldsymbol{k}$, $p \in \mathbb{Q}_2$, $\alpha \leq p$ of the form 2^{-l}).

(1) Choose a large number m.
(2) WAIT for $\Gamma^A(m) \downarrow$.
(3) Let v be the number of times P_i has gone through (2).

Case $i = 2$. Pick a large number n. Put $\langle r_n, n \rangle$ into L and n into F_0, where $2^{-r_n} = 2^{-v}\alpha$. WAIT for a *stage* t such that $n < t' < t$ for some *stage* t' and $K_t(n) \leq r_n + d$, and put n into F_1. (If M_d is a machine for L, then t exists.)

Case $i > 2$. CALL $P_{i-1}(2^{-v}\alpha, \alpha')$, where $\alpha' = \min(\alpha, 2^{-i-2}2^{-w_{i-1}})$, and w_{i-1} is the number of P_{i-1} procedures started so far in the construction.

IF $\mathsf{wgt}_L(F_i \cup F_{i-1}) < p$ GOTO 3.

If the expression $\Gamma^A(m)$ changes during this loop, cancel the run of all sub-procedures, and GOTO (2). (*Comment:* despite the cancellation, what is in F_{i-1} now is an i-set because of this very change.) Put F_{i-1} into F_i, and declare $F_{i-1} = \emptyset$. Also declare $F_{i-2}, \ldots, F_0 = \emptyset$ (without putting their content into any set, as it is garbage).

(4) Put m into B. (*Comment:* this forces A to change below $\gamma(m) < \min(F_{i-1})$, and hence makes F_{i-1} an i-set, as we assume inductively that F_{i-1} is an $i-1$-set.) As above, put F_{i-1} into F_i and then declare $F_{i-1} = \emptyset$.

5.4.3 Remark. The comment on cancellation in step (3) contains a key idea for subsequent constructions. When we have to cancel the sub-procedures of a run $P_i(q, \beta)$, what they are working on becomes garbage, but *not* F_{i-1}. For F_{i-1} already is an i-1-set, so all we need is another A-change, which is provided here by the cancellation itself (early promotion), as opposed to being caused actively in step (4) once the run reaches its goal (regular promotion). The only difference between the two types of promotion is that in the first case F_{i-1} has not reached the target weight $2^{-v}\beta$. (There is no garbage at level $\boldsymbol{k}$ -1 because the run of $P_{\boldsymbol{k}}$ is never cancelled.)

Construction of L and B. Call $P_{\boldsymbol{k}}(3/4, 1/4)$.

Verification. Let C_i be the c.e. set consisting of the numbers which are in F_i at some stage. Thus $C_{\boldsymbol{k}} = F_{\boldsymbol{k}}$, and for $i < \boldsymbol{k}$, $C_{i-1} - C_i$ is the set of numbers that

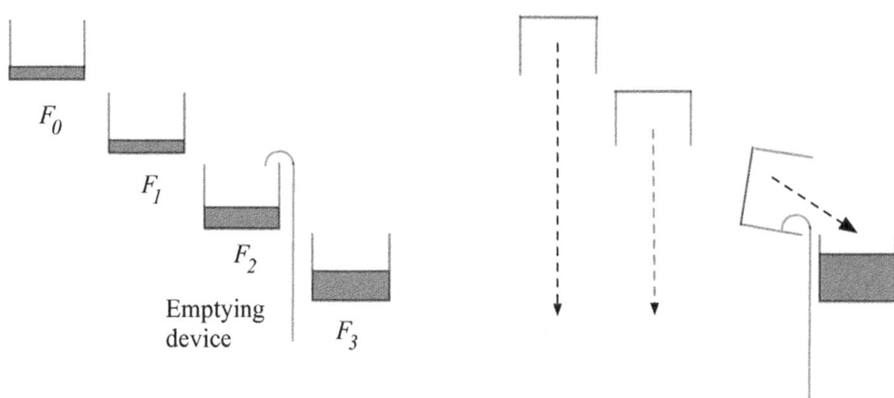

FIG. 5.1. Decanter model. Right side: premature A-change. F_2 is emptied into F_3, while F_0 and F_1 are spilled on the floor.

get stuck at level $i-1$. To show that L is a bounded request set, note that by the choice of the garbage quotas, at each stage, $C_{i-1} - C_i$ has weight at most 2^{-i-2} for $1 \le i < k$, and the weight of $C_k \cup C_{k-1} = F_k \cup F_{k-1}$ is at most $3/4$. (We have to be careful here: we need to show that L is a bounded request set for *each* d, so we have to accommodate the case that the construction breaks off, which can happen if d is not a fixed point. This is why we insist that the argument work at each stage.)

Since Γ^A is total, each procedure reaches (4) unless it is cancelled. In particular, $P_{\boldsymbol{k}}(3/4, 1/4)$ returns a $\boldsymbol{k}$-set of weight $3/4$, contrary to Fact 5.4.2. This contradiction shows that $\emptyset' \not\le_T A$. ◇

The decanter model. We visualize this construction by an arrangement of $\boldsymbol{k}+1$ levels of decanters $F_{\boldsymbol{k}}, F_{\boldsymbol{k}-1}, \ldots, F_0$, where $F_{\boldsymbol{k}}$ is at the bottom. See Figure 5.1 for the (hypothetical) case $\boldsymbol{k} = 3$. The decanters hold amounts of a precious liquid – how about 1955 Biondi-Santi Brunello wine?

We put F_{i-1} above F_i so that F_{i-1} can be emptied into F_i. At any stage F_i is an i-set. Procedure $P_i(p, \alpha)$ wants F_i to reach weight p, by filling F_{i-1} to the quantity of at most p needed, and then emptying it into F_i. Emptying corresponds to adding one more $A\restriction_n$-change for each n in F_{i-1}.

The use of $\Gamma^A(m)$ acts as the emptying device, which, besides being activated on purpose, may go off finitely often by itself. This is how we visualize an A-change. It is premature if the device goes off by itself. When F_{i-1} is emptied into F_i then $F_{i-2}, \ldots, F_0$ are spilled on the floor.

We suppose that a bottle of wine corresponds to a weight of 1. We first pour wine into the highest decanter F_0. We want to ensure that a weight of at most $1/4$ of wine we put into F_0 does not reach $F_{\boldsymbol{k}}$. Recall that the parameter α is the amount of garbage $P_i(p, \alpha)$ allows. If v is the number of times the emptying device has gone off by itself, then for $i > 2$, P_i lets P_{i-1} fill F_{i-1} in portions of

$2^{-v}\alpha$. For then, when F_{i-1} is emptied into F_i, a quantity of at most $2^{-v}\alpha$ can be lost because it is in higher decanters $F_{i-2}, \ldots, F_0$.

The procedure P_2 is a special case, but limits its garbage in the same way: it puts requests $\langle r_n, n\rangle$ into L where $2^{-r_n} = 2^{-v}\alpha$. Once it sees the corresponding description of $A\upharpoonright_n$, it empties F_0 into F_1. However, if the hook $\gamma^A(m)$ belonging to P_2 moves before this, F_0 is spilled on the floor while F_1 is emptied into F_2.

5.4.4 Remark. If A is merely Δ^0_2 then $A\upharpoonright_n$ can return to a previous value. However, in the definition of $\boldsymbol{k}$-sets we need $\boldsymbol{k}$ distinct values. Why can we still conclude that $P_{\boldsymbol{k}}(3/4, 1/4)$ returns a $\boldsymbol{k}$-set of weight $3/4$?

A run of P_2 is the result of recursive calls of runs $P_{\boldsymbol{k}}, P_{\boldsymbol{k}-1}, \ldots, P_3$; there are numbers $m_{\boldsymbol{k}} < m_{\boldsymbol{k}-1} < \ldots < m_2$ where m_i belongs to the run of P_i. Consider a number n such that the run of P_2 puts a request $\langle r_n, n\rangle$ into L at a *stage* s. Then $n \geq \gamma^A(m_i)$ for each i. The weight of numbers n that do not reach $F_{\boldsymbol{k}}$ is at most $1/4$. Suppose now that n reaches $F_{\boldsymbol{k}}$. Then for each $i \geq 2$, $A\upharpoonright_{\gamma(m_i)}$ is stable from s to the *stage* t when n reaches F_{i-1}, otherwise n would become garbage. So when n is promoted from F_{i-1} to F_i, the current value $A\upharpoonright_n$ has not been seen from s to t.

3: Each K-trivial set is superlow.

The construction showing that each K-trivial set A is Turing incomplete is sequential at each level: at most one procedure runs at any particular level. To show that A is superlow, and even low for K, we use a construction where at each level procedures can run in parallel.

Recall that $A' = \operatorname{dom} J^A$. A procedure $P_i(p, \alpha)$ at level $i \geq 2$ (with goal p and garbage quota α, of the form 2^{-l}, enumerating a set F_i) attempts to provide a computable approximation for A', making a guess at each stage. Initially, the guess is "divergent". When $J^A(e)$ newly converges, P_i calls a subprocedure $Q_{i-1,e}(q, \beta)$, where $q = 2^{-e-1}\alpha$, in order to test the stability of this computation. When the run of $Q_{i-1,e}(q, \beta)$ returns an $(i-1)$-set $G_{i-1,e}$ of weight q, then P_i makes a guess that $J^A(e)$ converges.

Stable case. If A does not change later below the use of $J^A(e)$, the guess is correct. However, $G_{i-1,e}$ is now garbage, since its elements will not reach the i-set F_i. This is a one-time event for each e, so the total garbage produced in this way is at most $\sum_e 2^{-e-1}\alpha = \alpha$, as required. (Note that this garbage due to the failure of A to change is of a new type. It was not present in the previous constructions where we could force A to change.)

Change case. If A changes later below the use of $J^A(e)$, then all the numbers in $G_{i-1,e}$ are put into F_i, so $\mathsf{wgt}_L(F_i)$ increases by the fixed quantity $2^{-e-1}\alpha$. In this case P_i changes its guess back to "divergent". If this happens r times, where $r = p2^{e+1}/\alpha$, then the run of P_i reaches its goal p.

We consider the hypothetical case that $b = d = 0$ in some more detail.

Procedure $P_2(p, \alpha)$ ($p \in \mathbb{Q}_2$, $\alpha \leq p$ of the form 2^{-l}).
It enumerates a set F_2, beginning with $F_2 = \emptyset$.

At *stage* s, declare $e = s$ available (availability is a local notion for each run of a procedure). For each $e \leq s$, do the following.

P1$_e$ IF e is available and $J^A(e)\downarrow [s]$, CALL the procedure $Q_{1,\mathbf{e}}(2^{-e-1}\alpha, \beta)$, where $\beta = \min(2^{-e-1}\alpha, 2^{-v-1})$ and v is the number of runs of Q_1 type procedures started so far in the construction. Declare e unavailable.

P2$_e$ IF e is unavailable due to a run $Q_{1,e}(q, \beta)$ (which may have returned already) and $A_s\restriction_w \neq A_{s-1}\restriction_w$ where w is the use of the associated computation $J^A(e)$, declare e available.

(a) Put $G_{1,e}$ into F_2.

(b) IF $\mathsf{wgt}_L(F_2) < p$ and the run $Q_{1,e}$ has not returned yet, CANCEL this run;
ELSE RETURN the set F_2, CANCEL all the runs of its subprocedures and END this run of P_2.

Note that we have combined steps (3) and (4) of the corresponding procedure $P_2(p)$ on page 203, because we cannot any longer force A to change.

Procedure $Q_{1,e}(q, \beta)$ ($\beta \leq q$, both of the form 2^{-l} for some l).
It enumerates a set $G_{1,e}$, beginning with $G_{1,e} = \emptyset$. It behaves similarly to step (3) in the procedure $P_2(p)$ on page 203.

Q1 Pick a number n larger than any number used so far. Put $\langle r_n, n\rangle$ into L, where $2^{-r_n} = \beta$. WAIT for a *stage* $t > n$ such that $K_t(n) \leq r_n + d$. (If M_d is a machine for L then t exists.)

Q2 Put n into $G_{1,e}$. IF $\mathsf{wgt}_L(G_{1,e}) < q$ GOTO (Q1). ELSE RETURN the set $G_{1,e}$.

A tree of runs. Next, we remove the extra hypothesis that $b = d = 0$. We have $2\boldsymbol{k} - 2$ levels of runs of procedures $Q_1, P_2, Q_2, \ldots, P_{\boldsymbol{k}-1}, Q_{\boldsymbol{k}-1}, P_{\boldsymbol{k}}$. The runs are now arranged as a tree, where the successor relation is given by recursive calls. The root of the tree is the single run of procedure $P_{\boldsymbol{k}}$, which calls procedures $Q_{\boldsymbol{k}-1,e}$ for $e \in \mathbb{N}$. Each one of them calls a procedure $P_{\boldsymbol{k}-1}$, and so on, till we reach the leaves, consisting of runs of procedures $Q_{1,e}$ which behave as outlined above. The procedures $Q_{j,e}$, $j > 1$ behave similar to $Q_{1,e}$, but in (Q1) the enumeration of a request $\langle r_n, n\rangle$ into L is replaced by a recursive call of a procedure P_j. See Fig. 5.2.

Recall that a run of P_i enumerates an i-set F_i, and a run of $Q_{i-1,e}$ enumerates an $(i-1)$-set $G_{i-1,e}$. To continue the decanter model, if $2 \leq i < \boldsymbol{k}$ and a run $Q_{i,e'}$ calls P_i, then F_i can be emptied into $G_{i,e'}$. For $2 \leq i \leq \boldsymbol{k}$, when a run P_i calls $Q_{i-1,e}$, then $G_{i-1,e}$ can be emptied into F_i. If $G_{i-1,e}$ reaches the weight required by $Q_{i-1,e}$ and the A-change does not occur, then this decanter is stuck – its content is garbage and will never be used again.

To verify that L is a bounded request set, for $2 \leq i \leq \boldsymbol{k}$, as before we let C_i be the c.e. set of numbers which are in a set F_i at some stage, and we let D_{i-1} be the c.e. set of numbers which are in a set $G_{i-1,e}$ at some stage. We let C_1 be the right domain of L. Numbers start out in C_1, and may successively be promoted

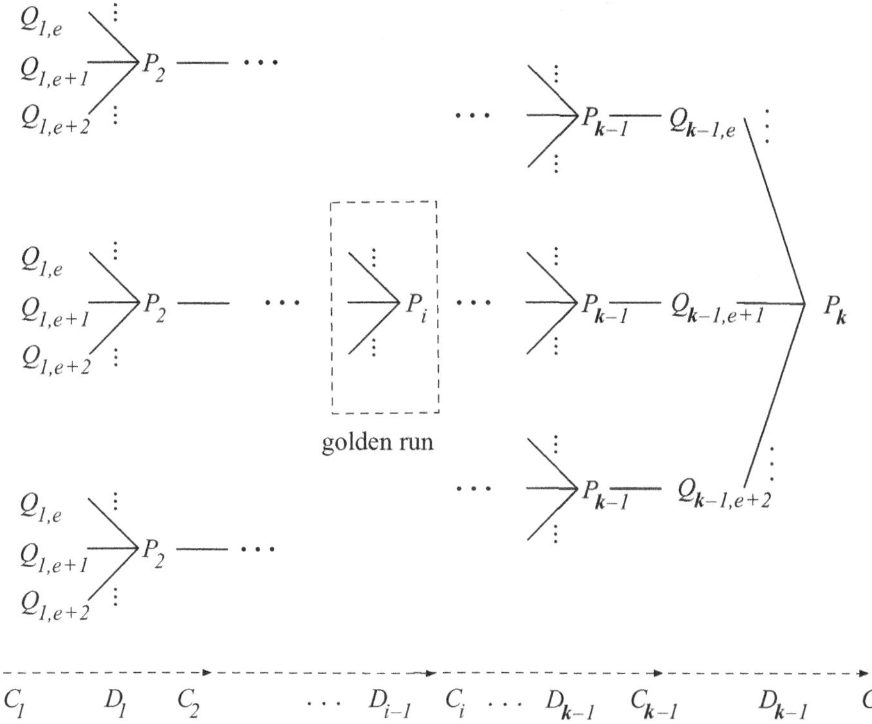

FIG. 5.2. Tree of runs. At the bottom we indicate the promotion of numbers.

to $D_1, C_2, \ldots, D_{\boldsymbol{k}-1}, C_{\boldsymbol{k}}$. For $1 \le j < \boldsymbol{k}$, $C_j - D_j$ contains the garbage produced when a run $Q_{j,e}$ is cancelled, and for $2 \le i \le \boldsymbol{k}$, $D_{i-1} - C_i$ is the new type of garbage which is produced when A fails to change after a run $Q_{i-1,e}$ returns.

The golden run. As the result of a premature A-change, a run may be cancelled by runs of procedures which precede this run on the tree. The initial procedure $P_{\boldsymbol{k}}$ is not cancelled and never returns, since we start it with goal $3/4$ while a $\boldsymbol{k}$-set has weight at most $1/2$ by Fact 5.4.2. So there must be a *golden run* of a procedure $P_i(p, \alpha)$: it is not cancelled, does not reach its goal, but all the subprocedures $Q_{i-1,e}$ it calls either reach their goals or are cancelled. This run P_i builds a computable approximation for A': as explained above, given e, the change case occurs fewer than $r = p2^{e+1}/\alpha$ times, because otherwise P_i would reach its goal. Thus P_i's guess at whether $J^A(e)$ converges can change at most $2r$ times. Hence A is superlow.

The computable approximation of A' was not obtained effectively, since we needed to know which run is golden. This nonuniformity cannot be avoided by Corollary 5.5.5 below.

4. Each K-trivial set is low for K.

Now a run $P_i(p, \alpha)$ that does not reach its goal tries to build a bounded request set W showing that A is low for K. It emulates the argument of Proposition 5.3.34 that each Δ^0_2 set A obeying $c_{\mathbb{U},A}$ is low for K. If the run is golden then it succeeds. The hypothesis in 5.3.34 that A obeys $c_{\mathbb{U},A}$ is replaced by the hypothesis that the run $P_i(p, \alpha)$ does not reach its goal.

When a computation $\mathbb{U}^A(\sigma) = y$ newly converges, $P_i(p, \alpha)$ calls a procedure $Q_{i-1,\sigma}(q, \beta)$ with the goal $q = 2^{-|\sigma|-1}\alpha$. As in the foregoing construction, it is necessary to call in parallel procedures based on different inputs σ. The procedure $Q_{i-1,\sigma}$ works in the same way as $Q_{i-1,e}$ before: unless it is cancelled, it returns an $(i-1)$-set $G_{i-1,\sigma}$ of weight q. The numbers in $G_{i-1,\sigma}$ are no less than the use of the computation $\mathbb{U}^A(\sigma)$. Let $v \in \mathbb{N}$ be least such that $2^v \geq p/\alpha$. When the run $Q_{i-1,\sigma}(q, \beta)$ returns, $P_i(p, \alpha)$ hopes that the computation $\mathbb{U}^A(\sigma) = y$ is stable, so it puts a request $\langle |\sigma| + v + 2, y \rangle$ into W.

Stable case: A does not change later below the use of $\mathbb{U}^A(\sigma)$. The enumeration into W causes $K(y) \leq^+ |\sigma|$, so if σ is a shortest $\mathbb{U}^A$-description of y, we obtain $K(y) \leq^+ K^A(y)$. The set $G_{i-1,\sigma}$ is now garbage. The total garbage produced in this way is at most $\Omega^A\alpha \leq \alpha$, as required. The contribution to W is at most $1/2$.

Change case: A changes later below the use of $\mathbb{U}^A(\sigma)$. Then all numbers in $G_{i-1,\sigma}$ are put into F_i, so $\mathsf{wgt}_L(F_i)$ increases by at least $q = 2^{-|\sigma|-1}\alpha$. Since $2^v \geq p/\alpha$, the total weight contributed to W in this way stays below $1/2$, or else the run of P_i reaches its goal p.

If the run $P_i(p, \alpha)$ is golden, it succeeds in showing that A is low for K: it is not cancelled, and does not reach its goal (so W is a bounded request set). Each run $Q_{i-1,\sigma}$ returns unless cancelled, necessarily by an A-change below the use of the computation $\mathbb{U}^A(\sigma)$ it is based on. Thus $P_i(p, \alpha)$ puts the corresponding request into W at a *stage* when $\mathbb{U}^A(\sigma)$ is stable.

It is instructive to compare (a) the proof of Theorem 5.1.22 (on bases for ML-randomness) with (b) the proof of Theorem 5.4.1. In both cases, we show that a given set A is low for K (in (b) also a computable approximation of A is given). The construction for a fixed parameter d in Theorem 5.1.22 is somewhat similar to a golden run $P_i(p, \alpha)$. In (a) one feeds hungry sets $C^\eta_{d,\sigma}$ on ce $\mathbb{U}^\eta(\sigma) = y$ is defined. In (b) subprocedures $Q_{i-1,\sigma}$ are called once $\mathbb{U}^A(\sigma)[s] = y$ is defined. The measure $2^{-|\sigma|-d-1}$ that $C^\eta_{d,\sigma}$ has to reach so that one can put the request $\langle |\sigma|+d+1, y \rangle$ into L_d corresponds to the goal $q = 2^{-|\sigma|-1}\alpha$ the subprocedure has to reach, so that it can return and the request $\langle |\sigma| + v + 2, y \rangle$ can go into W. However, the verifications that L_d and W are bounded request sets are completely different.

Exercises.

5.4.5. Get hold of a bottle of 1955 Biondi-Santi Brunello wine. Why is a single bottle sufficient for any of the decanter constructions?

5.4.6. Sketch a tree of runs as in Figure 5.2 for the case that $k = 4$. The types of procedures are Q_1, P_2, Q_2, P_3, Q_3, and P_4.

The formal proof of Theorem 5.4.1

We retain the definitions of the previous subsection, in particular of *stages* $s(l)$ and i-sets (page 204). We modify the computable approximation of A so that $A_v(x) = A_{s(l)}(x)$ for all x, v, l such that $s(l) \leq v < s(l+1)$.

Firstly, we outline the procedures in more detail. We indicate why no run exceeds its garbage quota. We allow a run of a procedure to "overfulfill" its goal up to at most the double amount. The reason is that immediately before the run returns, it may receive too much from the run of a subprocedure it has called, and which causes it to return. (We thank R. Solovay for pointing out the necessity of this.)

Recall that a run $P_i(p, \alpha)$ enumerates a set F_i, and a run $Q_{j,\sigma}(q, \beta)$ enumerates a set $G_{j,\sigma}$. We will write $Q_{j,\sigma yw}(q, \beta)$ instead of $Q_{j,\sigma}(q, \beta)$ when the associated computation is $\mathbb{U}^A(\sigma) = y$ and its use is w. For $2 \le i \le \boldsymbol{k}$ we let C_i be the c.e. set of numbers which are in a set F_i at some stage, and we let D_{i-1} be the c.e. set of numbers which are in a set $G_{i-1,\sigma}$ at some stage. Note that sets of type F_i and $G_{j,\sigma}$ are local in that they belong to individual runs of the corresponding procedures. Just as for runs, there may be several such sets at any stage. The indices are merely used to indicate the procedures they belong to. In contrast, the c.e. sets C_i and D_j (used only in the verification) are global.

For $j = 1$, a run $Q_{j,\sigma yw}(q, \beta)$ chooses a large number n, puts a request $\langle r, n\rangle$ into L, where $2^{-r} = \beta$, and waits for $K_t(n) \le r + d$ at a later *stage* t, so that it can put n into $G_{1,\sigma}$. This is repeated until it has reached its goal.

For $j > 1$, while it has not reached its goal q, the run $Q_{j,\sigma yw}(q, \beta)$ keeps calling a single procedure $P_j(\beta, \alpha)$ (for decreasing garbage quotas α) and waits until this procedure returns a set F_j, at which time it puts F_j into $G_{j,\sigma}$. Thus the amount of garbage left in $C_j - D_j$ is produced during a single run of a procedure P_j which does not reach its goal β, and hence is bounded by β.

A run $P_i(p, \alpha)$ calls procedures $Q_{j,\sigma yw}(2^{-|\sigma|-1}\alpha, \beta)$ for appropriate values of β. The weight left in $D_{i-1} - C_i$ by all the returned runs of Q_{i-1}-procedures which never receive an A-change adds up to at most $\Omega^A\alpha$ since this is a one-time event for each σ. (Keep in mind that $Q_{j,\sigma yw}$ may actually return a weight of up to $2^{-|\sigma|}\alpha$.) The runs of procedures Q_{i-1} which are cancelled and have so far enumerated a set G do not contribute to the garbage of $P_i(p, \alpha)$, since G goes into F_i upon cancellation.

To assign the garbage quotas we need some global parameters. At any substage of *stage* s, let

$$\alpha_i^* = 2^{-(2i+5+n_{P,i})}, \tag{5.15}$$

where $n_{P,i}$ is the number of runs of P_i-procedures started in the construction prior to this substage of *stage* s. Furthermore, let

$$\beta_j^* = 2^{-(2j+4+n_{Q,j})}, \tag{5.16}$$

where $n_{Q,j}$ is the number of runs of Q_j-procedures started so far. Then the sum of all the values of α_i^* and β_j^* is at most $1/4$. When P_i is called at a substage of *stage* s, its parameter α is at most α_i^*. Similarly, the parameter β of Q_j is at most β_j^*. This ensures $\mathsf{wgt}_L(C_1 - C_{\boldsymbol{k}}) \le 1/4$. We start the construction by calling $P_{\boldsymbol{k}}(3/4, 1/4)$, so $\mathsf{wgt}_L(C_{\boldsymbol{k}}) \le 3/4$. In this way we will show that L is a bounded request set.

We now give the formal description of the procedures and the construction. We check by induction on levels that no procedure reaches more than twice its goal.

Procedure $P_i(p, \alpha)$ $(1 < i \le \boldsymbol{k}, p \in \mathbb{Q}_2, p \ge \alpha = 2^{-l}$ for some $l)$.
It enumerates a set F_i, beginning with $F_i = \emptyset$.
At *stage* s, declare each σ of length s *available* (recall that availability is a local notion for each run of a procedure). Go through the σ such that $|\sigma| \le s$ in lexicographical order doing the following.

P1$_\sigma$ IF σ is available, and $\mathbb{U}^A(\sigma)[s] = y$ for some $y < s$, let w be the use of this computation, and CALL the procedure $Q_{i-1,\sigma y w}(2^{-|\sigma|-1}\alpha, \beta)$, where $\beta = \min(2^{-|\sigma|-2}\alpha, \beta^*_{i-1})$, and β^*_{i-1} is defined in (5.16).
Declare σ *unavailable.*

P2$_\sigma$ IF σ is unavailable due to a (possibly returned) run $Q_{i-1,\sigma y w}(q, \beta)$ and $A_s\restriction_w \neq A_{s-1}\restriction_w$, declare σ *available.*

(a) Say the run is *released.* Put $G_{i-1,\sigma}$ into F.
IF $\mathsf{wgt}_L(F) < p$ GOTO (b).
ELSE RETURN the set F, CANCEL all the runs of subprocedures and END this run of P_i. Assuming that $G_{i-1,\sigma}$ already was an $(i-1)$-set at the last *stage*, F is an i-set, as we will verify below. Inductively we have $\mathsf{wgt}_L(G_{i-1,\sigma}) \le 2^{-|\sigma|}\alpha \le p$, so that $\mathsf{wgt}_L(F) \le 2p$.

(b) IF the run $Q_{i-1,\sigma y w}$ has not returned yet, CANCEL this run and all the runs of subprocedures it has called.

The procedure $Q_{j,\sigma y w}(q, \beta)$ $(0 < j < \boldsymbol{k}, q = 2^{-l}, \beta = 2^{-r}$ for some $r > l)$.
It enumerates a set $G = G_{j,\sigma}$. Initially $G = \emptyset$.

Q1 *Case* $j = 1$. Pick a number n larger than any number used so far. Put $\langle r_n, n \rangle$ into L, where $2^{-r_n} = \beta$, and GOTO (Q2).
Case $j > 1$. CALL $P_j(\beta, \alpha)$, $\alpha = \min(\beta, \alpha^*_j)$, where α^*_j is defined in (5.15) and GOTO (Q2).

Q2 *Case* $j = 1$. WAIT for a *stage* $t > n$ such that $n < t' < t$ for some *stage* t' and $K_t(n) \le r_n + d$. (If M_d is a machine for L then t exists.) Put n into G. (By the definition of *stages*, $K(A\restriction_n) \le K(n) + b\,[t]$, so G remains a 1-set.)
Case $j > 1$. WAIT till $P_j(\beta, \alpha)$ returns a set F'. Put F' into G (G remains a j-set, assuming inductively that the sets F' are j–sets).
In any case, IF $\mathsf{wgt}_L(G) < q$ then GOTO (Q1). ELSE RETURN the set G. In this case, if $j = 1$ then $\mathsf{wgt}_L(G) = q$. If $j > 1$ then inductively we have $\mathsf{wgt}_L(F') \le 2\beta \le q$, and hence $\mathsf{wgt}_L(G) \le 2q$.

Construction. Begin by calling $P_{\boldsymbol{k}}(3/4, \alpha^*_{\boldsymbol{k}})$ at *stage* 0. At each *stage*, descend through the levels of procedures of type $P_{\boldsymbol{k}}, Q_{\boldsymbol{k}-1} \ldots P_2, Q_1$. At each level start or continue finitely many runs of procedures. Do so in some effective order, say from left to right on that level of the tree of runs of procedures, so that the values α^*_i and β^*_j are defined at each substage. Since one descends through the levels, a possible termination of a procedure in (P2$_\sigma$.b) occurs before the procedure can act.

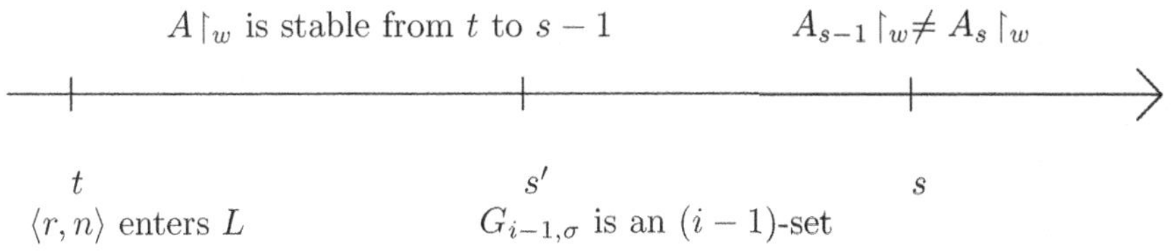

FIG. 5.3. Illustration of Lemma 5.4.7.

Verification. Before Lemma 5.4.9, we do not assume that M_d is a machine for L. C_1, the right domain of L, is enumerated in (Q1). For $1 \le j < \boldsymbol{k}$, let $D_{j,t}$ be the union of sets $G_{j,\sigma}$ enumerated by runs $Q_{j,\sigma yw}$ up to the end of stage t. Let $C_{i,t}$ be the union of sets F_i enumerated by runs P_i $(1 < i \le \boldsymbol{k})$ by the end of stage t.

5.4.7 Lemma. *The c.e. sets C_i, $2 \le i \le \boldsymbol{k}$, and D_i, $1 \le i < \boldsymbol{k}$, are i-sets.*

Subproof. By the comments in ($\mathrm{P2}_\sigma$) and (Q2) above, D_1 is a 1-set. Suppose $2 \le i \le k$, and assume inductively that D_{i-1} is an $i-1$-set. To see that C_i (and hence D_i in case $i < k$) is an i-set, assume that during *stage* s a number n enters C_i. Thus $A_{s-1}\upharpoonright_n \neq A_s\upharpoonright_n$. Among other things we will have to verify that this A-change is not a change back to a previous configuration (see Remark 5.4.4).

For some run of P_i, at $P2_\sigma(a)$ for some σ, the number n is in a set $G_{i-1,\sigma}[s]$ and enters the set F_i of that run. Let s' be the last *stage* before s. No Q_{i-1}-type procedure has been active yet at s, so the run of $Q_{i-1,\sigma yw}$ enumerating $G_{i-1,\sigma}[s]$ was already present at s', and hence $n \in G = G_{i-1,\sigma}[s']$. Also $\min G > w$, and inductively G was an $(i-1)$-set already at s'. Thus at a *stage* $t \le s'$, $\langle r, n\rangle$ was enumerated into L by a subprocedure of type Q_1 of this run $Q_{i-1,\sigma yw}$, and there are $i-1$ distinct strings z of the form $A_v\upharpoonright_n$ for some *stage* v, $t \le v < s$ such that $K_v(z) \le r + c$. Moreover, $n < s'$ and hence $K_s(A_s\upharpoonright_n) \le r + c$ by the definition of *stages*, the wait in (Q2) and because $n \in D_1$. Also, $A\upharpoonright_w$ did not change from t to $s-1$, else the run of $Q_{i-1,\sigma yw}$ would have been canceled before s. Since $A_{s-1}\upharpoonright_w \neq A_s\upharpoonright_w$, we have a new string $z = A_s\upharpoonright_n$ as required to show that C_i is an i-set. See Fig. 5.3. ◇

Next we verify that L is a bounded request set. First we show that no run of a procedure exceeds its garbage quota. Since the runs are arranged as a tree, each number n enumerated into the right domain of L at a leaf during a *stage* t corresponds to a unique run of a procedure at *each* level at the same *stage*. We say n *belongs* to this run.

5.4.8 Lemma. *(a) Let $1 \le j < \boldsymbol{k}$. At each stage, the weight of the numbers in $C_j - D_j$ which belong to a run $Q_{j,\sigma yw}(q,\beta)$ is at most β.*

(b) Let $1 < i \le \boldsymbol{k}$. At each stage, the weight of the numbers in $D_{i-1} - C_i$ which belong to a run $P_i(p,\alpha)$ is at most α.

Subproof. (a) For $j = 1$ the upper bound β on the weight holds since the run has at most one number n in $C_1 - D_1$ at any given stage. So, if the run gets stuck waiting at (Q2), it has left a weight of β in $C_1 - D_1$.

If $j > 1$, all the numbers as in (a) of the lemma belong to a single run of a procedure $P_j(\beta, \alpha)$ called by $Q_{j,\sigma yw}(q, \beta)$, because, once such a run returns a set F', this set is put into $G_{j,\sigma}$ and hence into D_j. As long as the run of P_j has not returned, it has not reached its goal β. Thus the weight of such numbers is at most β at any stage of the run of $Q_{j,\sigma yw}$.

(b) Suppose that n belongs to a run $P_i(p, \alpha)$ and $n \in D_{i-1,t}$ for a *stage* t. Then n is in the set $G_{i-1,\sigma}[t]$ of a run $Q_{i-1,\sigma yw}(2^{-|\sigma|-1}\alpha, \beta)$ called by P_i. We claim that, *if n does not reach C_i then no further procedure $Q_{i-1,\sigma y' w'}(q', \beta')$ is called during the run of P_i after t.*

Firstly assume that $A_s \upharpoonright_w \neq A_{s-1} \upharpoonright_w$ for some *stage* $s > t$. The only possible reason that n does not reach C_i is that the run of P_i did not need n to reach its goal in (P2$_\sigma$), in which case the run of P_i has returned by the end of *stage* s.

Secondly, assume there is no such s. Then the run of P_i, as far as it is concerned with σ, keeps waiting at $(P2_\sigma)$, and σ does not become available again. This proves the claim.

As a consequence, for each σ there is at most one run $Q_{i-1,\sigma yw}(2^{-|\sigma|-1}\alpha, \beta)$ called by $P_i(p, \alpha)$ which leaves numbers in $D_{i-1} - C_i$. The weight of the set of these numbers is at most $2^{-|\sigma|}\alpha$. If the run of P_i returns at a *stage* s then the sum of the weights of such numbers is bounded by the value of $\Omega^A\alpha$ at the last *stage* before s. Otherwise it is bounded by $\Omega^A\alpha$. ◇

By the foregoing lemma and the definitions of the values α_i^*, β_j^* at substages, at each stage we have

$$\mathsf{wgt}_L(C_1 - C_k) \leq \sum_{j=1}^{k-1} \mathsf{wgt}_L(C_j - D_j) + \sum_{i=2}^{k} \mathsf{wgt}_L(D_{i-1} - C_i) \leq 1/4.$$

Also $\mathsf{wgt}_L(C_k) \leq 3/4$ since we call P_k with a goal of 3/4. Thus $\mathsf{wgt}_L(C_1) \leq 1$. Since C_1 is the right domain of L, we may conclude that L is a bounded request set.

From now on we may assume that M_d is a machine for L, using the Recursion Theorem as explained in Remark 2.2.21.

5.4.9 Lemma. *There is a run of a procedure P_i, called a* golden run, *such that*

(i) the run is not cancelled,

(ii) each run of a procedure $Q_{i-1,\sigma yw}$ called by P_i returns unless cancelled, and

(iii) the run of P_i does not return.

Subproof. Assume that no such run exists. We claim that *each run of a procedure returns unless cancelled.* This yields a contradiction, since we call P_k with a goal of 3/4, this run is never cancelled, but if it returns, it has enumerated a weight of 3/4 into C_k, contrary to Fact 5.4.2.

To prove the claim, we carry out an induction on levels of procedures of type $Q_1, P_2, Q_2, \ldots, Q_{k-1}, P_k$. Suppose the run of a procedure is not cancelled.

$Q_{j,\sigma yw}(q, \beta)$: If $j = 1$, the wait at (Q2) always succeeds because M_d is a machine for L. If $j > 1$, inductively each run of a procedure P_j called by $Q_{j,\sigma yw}$ returns, as it is not cancelled. In either case, each time the run is at (Q2), the weight of D increases by β. Therefore $Q_{j,\sigma yw}$ reaches its goal and returns.

$P_i(p, \alpha)$: The run satisfies (i) by hypothesis, and (ii) by inductive hypothesis. Thus, (iii) fails, that is, the run returns. $\diamond$

5.4.10 Lemma. *A is low for K.*

Subproof. Choose a golden run of a procedure $P_i(p, \alpha)$ as in Lemma 5.4.9. We enumerate a bounded request set W showing that A is low for K. Let $v \in \mathbb{N}$ be least such that $p/\alpha \leq 2^v$. At *stage* s, when a run $Q_{i-1,\sigma y w}(2^{-|\sigma|-1}\alpha, \beta)$ returns, put $\langle |\sigma| + v + 2, y\rangle$ into W. We prove that W is a bounded request set, namely, $S_W = \sum_r 2^{-r} [\![\langle r, z\rangle \in W]\!] \leq 1$. Suppose $\langle r, z\rangle$ enters W at *stage* s due to a run $Q_{i-1,\sigma y w}(2^{-|\sigma|-1}\alpha, \beta)$ which returns.

Stable case. The contribution to S_W of those requests $\langle r, z\rangle$ where $A\restriction_w$ is stable from s on is bounded by $2^{-(v+2)}\Omega^A$, since for each σ such that $\mathbb{U}^A(\sigma)$ is defined, this can only happen once.

Change case. Now suppose that $A\restriction_w$ changes after *stage* s. Then the set G returned by $Q_{i-1,\sigma y w}$, with a weight of at least $2^{-|\sigma|-1}\alpha$, went into the set F_i of the run $P_i(p, \alpha)$. Since this run does not reach its goal $p \leq 2^v\alpha$,

$$\sum_s \sum_{\sigma,y,w} 2^{-|\sigma|-1} [\![Q_{i-1,\sigma y w} \text{ returns at } s \ \&\ \exists t > s\ A_t\restriction_w \neq A_{t-1}\restriction_w]\!] < 2^v. \quad (5.17)$$

Thus the contribution of the corresponding requests to S_W is less than $1/2$.

Let M_e be the machine for W according to the Machine Existence Theorem. *We claim that $K(y) \leq K^A(y) + v + e + 2$ for each y.* Suppose that s is the least *stage* such that a stable computation $\mathbb{U}^A_s(\sigma) = y$ appears, where σ is a shortest $\mathbb{U}^A$-description of y. Let w be the use of this computation. Then σ is available at s: otherwise some run $Q_{i-1,\sigma y' w'}$ is waiting to be released at (P2$_\sigma$). In that case, $A\restriction_{w'}$ has not changed since that run was started at a *stage* $< s$. Then $w = w'$ and $y = y'$, contrary to the minimality of s. So we call $Q_{i-1,\sigma y w}$. Since $A\restriction_w$ is stable and the run of P_i is not cancelled, this run is not cancelled, so it returns by (ii) of Lemma 5.4.9. At this *stage* we put the request $\langle |\sigma| + v + 2, y\rangle$ into W, which causes $K(y) \leq K^A(y) + v + e + 2$ as required. $\diamond$

This concludes the proof of Theorem 5.4.1. $\square$

As already mentioned, the proof of Lemma 5.4.10 extends the proof of Proposition 5.3.34, where we showed that any Δ^0_2 set A obeying a cost function $c_{\mathbb{U},A}$ is low for K. Instead of the hypothesis that A obey $c_{\mathbb{U},A}$, we now have (5.17). The new complications arise because we can only issue a request when the run of the corresponding procedure of type Q has returned.

5.5 Properties of the class of K-trivial sets

Theorem 5.3.10 states that, if some computable approximation of a set A obeys the standard cost function c_K, then A is K-trivial. We will prove the converse: each K-trivial set A has a computable approximation that obeys c_K. This was not too hard to show for c.e. sets A in Theorem 5.3.27. Without the restriction to being c.e., it seems to require the full power of the golden run method. Luckily, it suffices to extend the proof of Theorem 5.4.1 slightly. In the Main Lemma 5.5.1 we rephrase a central aspect of the proof in a more general language.

While the cost function method is a general method to build K-trivial (and other) sets, the Main Lemma is our principal method to prove restricting results on K-trivial sets. We will show that each K-trivial set A is Turing below a c.e. K-trivial set, and that each K-trivial set is superlow. No proof of these results is known using the equivalent property to be low for ML-randomness. (The closure under $\oplus$ in Theorem 5.2.17 is another example where the definition via K-triviality appears to be essential.) We obtain the first result as a corollary to the characterization of K-triviality by the standard cost function; the second follows from the first because each c.e. set that is low for K is superlow.

As further applications of the Main Lemma, we prove that if a Δ^0_2 set A is K-trivial then Ω^A is left-c.e. (the converse follows from Theorem 5.1.22), and that each K-trivial set is low for weak 2-randomness. We also show in Corollary 5.5.6 that any proof of Theorem 5.4.1 is necessarily nonuniform.

A Main Lemma derived from the golden run method

The golden run method works for any prefix-free oracle machine M: we have only used that the machine is optimal near the end of the proof in Lemma 5.4.10 when we verified that A is low for K. In (5.13) on page 198 we introduced the adaptive cost function $c_{M,A}(x,r)$, the measure of the M^A-descriptions at stage $r-1$ that are threatened by a potential change $A_{r-1}(x) \neq A_r(x)$. The Main Lemma extracts a computable sequence of stages $0 = q(0) < q(1) < \ldots$ from a golden run, and uses a variant $\widehat{c}_{M,A}$ of this cost function: the measure of the M^A-descriptions existing at stage $q(r+1)$ but with use $\leq q(r)$ that are threatened by a potential change of $A(x)$. The quantity $\widehat{S}$ bounds the sum of measures of descriptions where the computations are actually destroyed by stage $q(r+2)$. The Main Lemma states that $\widehat{S}$ is finite. In typical applications we will define an appropriate M and use the finiteness of $\widehat{S}$ to prove the desired restriction on A.

5.5.1 Main Lemma. *Let A be K-trivial. Fix a computable approximation $(A_s)_{s\in\mathbb{N}}$ of A. Let M be a prefix-free oracle machine. Then there is a constant $u \in \mathbb{N}$ and a computable sequence of* stages $q(0) < q(1) < \ldots$ *such that*

$$\widehat{S} = \sum_{x,r} \widehat{c}_{M,A}(x,r) \, [\![x \text{ is least s.t. } A_{q(r+1)}(x) \neq A_{q(r+2)}(x)]\!] < 2^u, \tag{5.18}$$

$$\textit{where } \widehat{c}_{M,A}(x,r) = \sum_{\sigma} 2^{-|\sigma|} \left[\!\!\left[\begin{array}{l} M^A(\sigma)[q(r+1)] \downarrow \ \& \\ x < \mathsf{use}\ M^A(\sigma)[q(r+1)] \leq q(r) \end{array} \right]\!\!\right]. \tag{5.19}$$

The intuition is that a golden run $P_i(p,\alpha)$ calls runs $Q_{i-1,\sigma y w}(2^{-|\sigma|-1}\alpha, \beta)$ at *stages* $v < q(r+1)$, where the use w of the underlying computation $M^A(\sigma)[t]$ is at most $q(r)$. By stage $q(r+1)$ they have returned (unless cancelled), and they

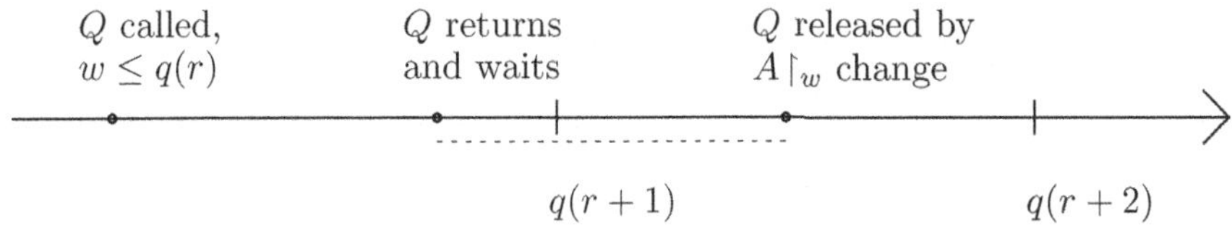

FIG. 5.4. Illustration of the Main Lemma.

are waiting up to stage $q(r+2)$ for an $A \upharpoonright_w$ change in order to be released (see Figure 5.4). The overall weight that the run $P_i(p,\alpha)$ receives in this way is at least $\alpha\widehat{S}/2$. Since the run is golden, we have $\alpha\widehat{S}/2 < p$, so if u is chosen so that $p/\alpha \leq 2^{u-1}$, then $\widehat{S} < 2^u$.

Proof. We extend the proof of Theorem 5.4.1 with the machine M in place of $\mathbb{U}$. Choose a golden run $P_i(p,\alpha)$ by Lemma 5.4.9.

Claim. *For each* stage *s, there is a* stage *$t > s$ such that, for all $\sigma < s$, if $M^A(\sigma)[t] \downarrow$ with use $w \leq s$ then a run $Q_{i-1,\sigma y w}$ has returned by the end of stage t and is not released yet, that is, the run waits at (P2$_\sigma$).*

Let $r \geq s$ be the least *stage* by which $A_r \upharpoonright_s$ has settled. A run $Q_{i-1,\sigma y w}$ such that $w \leq s$ is never cancelled after *stage* r. Therefore it returns by property (ii) of a golden run. This proves the claim.

Let $q(0) = 0$. If $s = q(r)$ has been defined, let $q(r+1)$ be the least *stage* $t > s$ such that the condition of the claim holds (this can be determined effectively).

Let $u \in \mathbb{N}$ be least such that $p/\alpha \leq 2^{u-1}$. We show that $\widehat{S} < 2^u$. Suppose x is least such that $A_{q(r+1)}(x) \neq A_{q(r+2)}(x)$. Then $A_{v-1}(x) \neq A_v(x)$ for some stage v such that $q(r+1) < v \leq q(r+2)$. At stage $q(r+1)$, $\widehat{c}_{M,A}(x,r)$ equals the measure of the descriptions σ such that a run of a procedure $Q_{i-1,\sigma y w}$, $x < w \leq q(r)$ is waiting at (P2$_\sigma$) to be released. Such a run is released at a *stage* v', $q(r+1) < v' \leq v$, in which case a weight that is at least its goal $\alpha \cdot 2^{-|\sigma|-1}$ is added to the set F enumerated by $P_i(p,\alpha)$. So an overall weight of at least $\alpha \cdot \widehat{c}_{M,A}(x,r)/2$ is added to F. Thus $\alpha\widehat{S}/2 < p$ and hence $\widehat{S} < 2^u$, otherwise the run $P_i(p,\alpha)$ would reach its goal and return. □

The standard cost function characterizes the K-trivial sets

We remove the hypothesis that A be c.e. from Theorem 5.3.27.

5.5.2 Theorem. (Extends 5.3.27) *The following are equivalent for a set A.*

(i) A is K-trivial.

(ii) There is a computable approximation $(\widetilde{A}_r)_{r\in\mathbb{N}}$ of A that obeys the standard cost function $c_{\mathcal{K}}$ defined in (5.6).

Note that by (ii) and Fact 5.3.12, every K-trivial set A is ω–c.e. We will later improve this in two ways. Firstly, in Corollary 5.5.4 we show that A is superlow. Secondly, in 5.5.11 we find a computable approximation of A that changes as little as desired.

Proof. (ii) $\Rightarrow$ (i): this is Theorem 5.3.10.

(i) ⇒ (ii): Fix some computable approximation $(A_s)_{s\in\mathbb{N}}$ of A. We apply the Main Lemma to the prefix-free oracle machine M such that $M^X(\sigma)$ attempts to compute $\mathbb{U}(\sigma)$ (without using the oracle). In case the computation converges, at this same stage it reads $y = \mathbb{U}(\sigma)$ bits of the oracle set, thereby making the use y. Then it stops with output y.

Let $0 = q(0) < q(1) < \ldots$ be the computable sequence of stages obtained from the Main Lemma. Since $\mathsf{use}\, M^A(\sigma)[s] \simeq \mathbb{U}(\sigma)[s]$ for each σ, s, we have

$$\begin{aligned}\widehat{c}_{M,A}(z,r) &= \sum_\sigma 2^{-|\sigma|}\, [\![M^A(\sigma)[q(r+1)]\downarrow \;\& \\ &\qquad\qquad z < \mathsf{use}\, M^A(\sigma)[q(r+1)] \le q(r)]\!] \\ &= \sum_\sigma 2^{-|\sigma|}\, [\![z < \mathbb{U}(\sigma)[q(r+1)] \le q(r)]\!] \\ &\ge \sum_y 2^{-K_{q(r+1)}(y)}\, [\![z < y \le q(r)]\!].\end{aligned}$$

The remainder of the proof is similar to the proof of Theorem 5.3.27. As in that proof, we write $\widehat{c}_{\mathcal{K}}(z, q(r))$ for the sum expression in the last line displayed above. Let $\widetilde{A}_r = A_{q(r+2)} \cap [0, r)$. For each $x < r$ we have

$$c_{\mathcal{K}}(x,r) = \textstyle\sum_{x<y\le r} 2^{-K_r(y)} \le \widehat{c}_{\mathcal{K}}(x, q(r)).$$

Since $\widehat{c}_{\mathcal{K}}(z, q(r)) \le \widehat{c}_{M,A}(z,r)$, and using (5.18),

$\sum_{r,x} c_{\mathcal{K}}(x,r)\, [\![r > 0 \;\&\; x \text{ is least s.t. } \widetilde{A}_{r-1}(x) \ne \widetilde{A}_r(x)]\!]$

$\le \sum_{r,x} \widehat{c}_{\mathcal{K}}(x, q(r))\, [\![r > 0 \;\&\; x \text{ is least s.t. } A_{q(r+1)}(x) \ne A_{q(r+2)}(x)]\!] \;\le\; \widehat{S} + 1.$

Here $\widehat{S}$ is defined in the Main Lemma, and the extra term $+1$ is necessary for the same reason as in Theorem 5.3.27 (if $r-1 \in A_{q(r+2)}$ then $r-1 \in \widetilde{A}_r - \widetilde{A}_{r-1}$). Since $\widehat{S} < \infty$, the computable approximation $(\widetilde{A}_r)_{r\in\mathbb{N}}$ of A obeys $c_{\mathcal{K}}$. □

The characterization of K-triviality by the standard cost function has interesting consequences. Firstly, being K-trivial is intrinsically related to being c.e.:

5.5.3 Corollary. *For each K-trivial set A, there is a c.e. K-trivial set $C \ge_{tt} A$.*

Proof. Some computable approximation $(A_s)_{s\in\mathbb{N}}$ of A obeys $c_{\mathcal{K}}$. By Proposition 5.3.12, there is a computable function h such that for each y, $A_s(y)$ for $s \ge h(y)$ changes at most $O(y^2)$ times. In particular, A is ω-c.e. By Proposition 5.3.6 the c.e. change set $C \ge_{tt} A$ obeys $c_{\mathcal{K}}$. Then C is K-trivial. □

5.5.4 Corollary. *Each K-trivial set A is superlow, that is, $A' \le_{tt} \emptyset'$.*

Proof. Since superlowness is downward closed under $\le_T$, it suffices to show that the K-trivial c.e. set $C \ge_{tt} A$ obtained in Corollary 5.5.3 is superlow. But C is low for K by Theorem 5.4.1, and hence superlow by Proposition 5.1.3.

One can also directly spell out the golden run construction sketched on page 207, or use the Main Lemma to obtain a stronger result, Proposition 5.5.12. □

On the other hand, Exercise 5.2.10 shows that some c.e. superlow set is not K-trivial. Alternatively, in Theorem 6.1.4 we will prove that there are superlow c.e. sets A_0,A_1 such that $A_0 \oplus A_1$ is Turing complete. Then at least one among the sets A_0, A_1 is not K–trivial.

A *lowness index* for a set X is a number i such that $X' = \Phi_i(\emptyset')$. Results obtained via the golden run method tend to be nonuniform since one would have to know which run is golden in order to determine the required object (such as a lowness index for A, or a bounded request set that demonstrates A is low for K). The following results show that this non-uniformity is necessary. Corollary 5.5.4 is nonuniform even when the conclusion is merely lowness.

5.5.5 Proposition. *One cannot effectively obtain a lowness index for A from a pair (A, b), where A is a c.e. set that is K-trivial via b.*

Proof. Otherwise, by Theorem 5.3.28 there would be an effective sequence $(B_r, i_r)_{r \in \mathbb{N}}$ of all the c.e. K-trivial sets and lowness indices for them. But such a sequence does not exist, since a straightforward extension of Theorem 5.3.22 yields a set $C \in \mathcal{K} = \mathcal{M}$ not Turing below any set B_r. To prove this, one meets requirements $P_{\langle e,r \rangle}: A \neq \Phi_e^{B_r}$, and asks $\Sigma_1^0(B_r)$ questions for requirement $P_{\langle e,r \rangle}$. Since the lowness index for B_r is given effectively, there is a total computable function $g(e, r, s)$ such that $\lim g(e, r, s) = 1$ if the answers is "yes", and $\lim g(e, r, s) = 0$ otherwise. □

As a consequence, Theorem 5.4.1 is nonuniform as well.

5.5.6 Corollary. *One cannot effectively obtain a constant d such that A is low for K via d from a pair (A, b), where A is a c.e. set that is K-trivial via b.*

Proof. If a c.e. set A is low for K, then, by the uniformity of Proposition 5.1.3, a lowness index for A can be computed from a c.e. index for A and a constant via which A is low for K. □

The nonuniformity in the particular proof of Theorem 5.4.1 given above is easily detected: the constant via which A is low for K given by that proof is $v + e + 2$, and both numbers e and v depend on what the golden run is. This actually proves that we cannot determine a golden run effectively from the given objects in the construction.

In the following we summarize the properties of the class of K-trivial sets within the Turing degrees. Ideals in uppersemilattices were defined in 1.2.27, and the effective listing $(V_e)_{e \in \mathbb{N}}$ of the ω-c.e. sets in 1.4.5.

5.5.7 Theorem.

(i) The K-trivial sets induce an ideal $\boldsymbol{K}$ in the Turing degrees.

(ii) $\boldsymbol{K}$ is the downward closure of its c.e. members.

(iii) Each K-trivial set is superlow. $\boldsymbol{K}$ is a Σ_3^0 ideal in the ω-c.e. T-degrees.

(iv) $\boldsymbol{K}$ is nonprincipal.

Proof. Theorems 5.4.1, 5.2.17 and Corollary 5.5.3 imply (i) and (ii). Fact 5.2.6 and Corollary 5.5.4 imply (iii). For (iv), assume that there is a degree $\mathbf{b}$ such that $\boldsymbol{K} = [\mathbf{0}, \mathbf{b}]$. Then $\mathbf{b}$ is c.e. by Corollary 5.5.3, and $\mathbf{b}$ is low by Theorem 5.5.4. By Theorem 5.3.22 there is $\mathbf{a} \in \boldsymbol{K}$ such that $\mathbf{a} \not\leq \mathbf{b}$, contradiction. □

Nies proved that each proper Σ^0_3 ideal in the c.e. Turing degrees has a low_2 upper bound (see Downey and Hirschfeldt 2010). Then, by Corollary 5.5.3, there is a low_2 c.e. set E such that $A \leq_T E$ for each K-trivial set A.

Kučera and Slaman (20xx) built a low set B Turing above all the (c.e.) K-trivial sets. By Theorem 5.3.22, such a set is not computably enumerable. Also see before 5.3.26.

5.5.8.$^\diamond$ **Problem.** Are there c.e. Turing degrees $\mathbf{a}, \mathbf{b}$ such that $\boldsymbol{K} = [\mathbf{0}, \mathbf{a}] \cap [\mathbf{0}, \mathbf{b}]$? Such a pair of degrees is called an exact pair for the ideal. Each Σ^0_3-ideal with a bound below $\mathbf{0}'$ has an exact pair in the Δ^0_2 Turing degrees by Shore (1981).

5.5.9 Exercise. There is no cost function that is non-increasing in the first argument and characterizes the superlow sets in the sense of Theorem 5.5.2.

5.5.10 Exercise. (i) There is a 1-generic K-trivial set.
(ii) No Schnorr random set is K-trivial.

The number of changes ⋆

Each K-trivial set A is superlow, that is, A' is ω-c.e. We ask to what extent the number of changes in a computable approximation for A can be minimized. Thereafter we do the same for A'.

5.5.11 Proposition. *Let $A \in \mathcal{K}$. For each order function h there is a computable approximation $(\widetilde{A}_r)_{r\in\mathbb{N}}$ of A such that $\widetilde{A}_r(y)$ changes at most $h(y)$ times.*

Proof. Proposition 5.3.12 and Theorem 5.5.2 imply this result for some order function h in $O(y^2)$. For the full result we modify the proof of (i)⇒(ii) in Theorem 5.5.2. Let $\widetilde{h}(x) = \lfloor h(x)^{1/4} \rfloor$. Let g be an increasing computable function such that $x \leq g(\widetilde{h}(x))$ for each x, and there is c such that $K_{g(y)}(y) \leq 2\log y + c$ for each y. (For instance, if $h(x) = \log x$ then $g(y) \geq 2^{y^4}$.)

We apply the Main Lemma to the prefix-free oracle machine M such that $M^X(\sigma)$ attempts to compute $y = \mathbb{U}(\sigma)$; in the case of convergence, it reads $g(y)$ bits of the oracle, thereby making the use $g(y)$, and then halts with output y. Let $\widetilde{A}_r(x) = A_{q(r+2)}(x)$ if $g(\widetilde{h}(x)) \leq q(r)$, and $\widetilde{A}_r(x) = 0$ otherwise. Clearly $(\widetilde{A}_r)_{r\in\mathbb{N}}$ is a computable approximation of A. Suppose that $r > 0$ and $\widetilde{A}_{r-1}(x) \neq \widetilde{A}_r(x)$. Then $A_{q(r+1)}(x) \neq A_{q(r+2)}(x)$ and, by the definition of g, $x \leq g(\widetilde{h}(x)) \leq q(r)$ and $K_{q(r)}(\widetilde{h}(x)) \leq 2\log \widetilde{h}(x) + c$. Thus $M^A(\sigma)[q(r+1)]\downarrow$ with use at most $q(r)$ for some σ such that $|\sigma| \leq^+ 2\log h(x)$. So this change contributes at least $\epsilon \widetilde{h}(x)^{-2}$ to $\widehat{S}$ for some $\epsilon > 0$ independent of x. Then, since $\widehat{S} < \infty$, the number of changes of $\widetilde{A}_r(x)$ is $O(\widetilde{h}^2(x))$. Since $\widetilde{h}^2(x) \leq \sqrt{h(x)}$, this means that for almost all x the number of changes is at most $h(x)$. We are done after mending the approximation $(\widetilde{A}_r)_{r\in\mathbb{N}}$ for finitely many x. □

As a consequence, we can choose a polynomial time truth-table reduction in Corollary 5.5.3, namely, the truth table for input x can be computed on a Turing machine in time polynomial in $\log x$. Moreover, for a given order function h such that $h(x) \leq \log x$ and $h(x)$ itself can be computed in time polynomial in $\log x$, we can ensure there are at most $h(x)$ queries. To see this, choose a computable approximation $(\widetilde{A}_r)_{r\in\mathbb{N}}$ of A for h as above, let C be its change set. Note that the reduction procedure of A to C in more detail is as follows: *on input x, let i^* be the greatest $i < h(x)$ such that $\langle x, i\rangle \in C$. If i^* is even, then $A(x) = 1 - A_0(x)$. If i^* is odd or there is no such i^*, then $A(x) = A_0(x)$.* Since $h(x) \leq \log x$, all the queries $\langle x, i\rangle$ for $i < h(x)$ and hence the whole truth table can be determined in time polynomial in $\log x$.

As a further application of the Main Lemma, we strengthen Corollary 5.5.4: for each K-trivial set A, the number of changes in an appropriate computable approximation to $A'(m)$ is $O(m \log^2 m)$. We could base the definition of the Turing jump on some other universal Turing program, because the result holds in fact for each set Z that is c.e. relative to A.

5.5.12 Proposition. (Extends 5.5.4) *Let A be K-trivial. Then, for each e, the set $Z = W_e^A$ is ω-c.e., and in fact there is a computable approximation $(Z_r)_{r\in\mathbb{N}}$ of Z such that for each m, $Z_r(m)$ changes at most $O(m\log^2 m)$ times.*

Proof. Recall that $W_e^A = \operatorname{dom}\Phi_e^A$. We apply the Main Lemma 5.5.1 to the oracle prefix-free machine given by $M^X(\sigma) \simeq \Phi_e^X(\mathbb{U}(\sigma))$. By Proposition 2.2.8, there is a computable function p and a constant c such that for each m, $\mathbb{U}_{p(m)}(\sigma_m) = m$ for some $\mathbb{U}$-description σ_m such that $|\sigma_m| \leq \log m + 2\log\log m + c$. To define the computable approximation $(Z_r)_{r\in\mathbb{N}}$, if $r \geq p(m)$ and

$$\Phi_e^A(m)[q(r+1)]\downarrow \text{ with use } \leq q(r)$$

(and hence $M^A(\sigma_m)[q(r+1)]\downarrow$ with use $\leq q(r)$), then declare $Z_r(m) = 1$, otherwise $Z_r(m) = 0$.

If $Z_r(m) = 1$ and A changes below the use of $\Phi_e^A(m)[q(r+1)]$ between the stages $q(r+1)$ and $q(r+2)$, then $2^{-|\sigma_m|} \sim 1/(m\log^2 m)$ is added to $\widehat{S}$. Thus $\lim_r Z_r(m) = Z(m)$ and the number of changes of $Z_r(m)$ is in $O(m\log^2 m)$. □

For each order function h, the class of c.e. sets such that A' can be approximated with at most $h(x)$ changes contains a promptly simple set by Theorem 8.4.29 and Corollary 8.4.35. However, if h grows sufficiently slowly, this class is a proper subclass of the c.e. K-trivial sets by Theorems 8.5.1, 8.4.34, and Corollary 8.5.5. Thus, for a c.e. K-trivial set A in general, the number of changes in an approximation to A' cannot be decreased arbitrarily.

5.5.13 Exercise. (Barmpalias, Downey and Greenberg, 2009) Let h be an order function such that $\sum_m 1/h(m) < \infty$. Extend 5.5.12 to the case of the function h instead of $\lambda m.m\log^2 m$.

Ω^A *for K-trivial A*

In our next application of the Main Lemma, we characterize the K-trivial sets in terms of the operator Ω: if A is Δ_2^0, then A is K-trivial $\Leftrightarrow$ Ω^A is left-c.e. (the implication "⇐" follows from Theorem 5.1.22). The results in this subsection are due to Downey, Hirschfeldt, Miller and Nies (2005).

By Definition 3.4.2, $\Omega_M^A = \lambda[\operatorname{dom} M^A)]^{\prec}$ for a prefix-free oracle machine M. We view $\mathbb{U}$ as a universal prefix-free oracle machine (as explained after Definition 3.4.2) and write Ω^A for $\Omega_{\mathbb{U}}^A$.

5.5.14 Theorem. *Let A be a set in Δ_2^0. Then the following are equivalent.*

(i) A is K-trivial.
(ii) $A \le_T \Omega^A$.
(iii) Ω^A is left-c.e.
(iv) Ω_U^A is left-c.e. for every optimal prefix-free oracle machine U.
(v) Ω_M^A is difference left-c.e. for every prefix-free oracle machine M.

Proof. (i)⇒(v): We will define left-c.e. real numbers α, β such that $\Omega_M^A = \alpha - \beta$. Applying the Main Lemma 5.5.1 to the machine M we obtain a computable sequence of stages $0 = q(0) < q(1) < \ldots$; for the duration of this proof, we will write $M^A(\sigma)\downarrow [q(r+1)]$ if $M^A(\sigma)$ converges at stage $q(r+1)$ with use $\le q(r)$, and $M^A(\sigma)\uparrow [q(r+1)]$ otherwise. The idea is roughly that α is the sum of the Ω_M^A-increases, while β is the sum of the Ω_M^A-decreases, measured only at stages of the form $q(r)$. More precisely, for each σ, r, let

$$\alpha_{\sigma,r} = \begin{cases} 2^{-|\sigma|} & \text{if } M^A(\sigma)\uparrow [q(r)] \ \& \ M^A(\sigma)\downarrow [q(r+1)] \\ 0 & \text{otherwise} \end{cases}$$

$$\beta_{\sigma,r} = \begin{cases} 2^{-|\sigma|} & \text{if } M^A(\sigma)\downarrow [q(r+1)] \ \& \ M^A(\sigma)\uparrow [q(r+2)] \\ 0 & \text{otherwise.} \end{cases}$$

Since the sum $\widehat{S}$ defined in (5.18) is finite, $\forall^\infty r\, [\alpha_{\sigma,r} = \beta_{\sigma,r} = 0]$ for each σ. Let

$$\alpha = \textstyle\sum_\sigma \sum_r \alpha_{\sigma,r} \text{ and } \beta = \sum_\sigma \sum_r \beta_{\sigma,r},$$

then $\beta \le \widehat{S} < \infty$. Thus $\alpha - \beta = \sum_\sigma \sum_r (\alpha_{\sigma,r} - \beta_{\sigma,r})$. (We use the fact that the difference operation: $\mathbb{R} \times \mathbb{R} \to \mathbb{R}$ is continuous. So far we have not excluded the possibility that $\alpha = \infty$.) Only the last convergent computation with input σ counts in Ω_M^A, hence

$$\begin{aligned} \Omega_M^A &= \sum_\sigma 2^{-|\sigma|} [\![\exists r\, (\alpha_{\sigma,r} \neq 0 \ \& \ \forall t \ge r\ \beta_{\sigma,t} = 0)]\!] \\ &= \sum_\sigma \sum_r (\alpha_{\sigma,r} - \beta_{\sigma,r}) \\ &= \alpha - \beta. \end{aligned}$$

In particular $\alpha < \infty$. Clearly α and β are left-c.e.

(v)⇒(iv): If U is optimal then Ω_U^A is ML-random (even relative to A) and difference left-c.e., so by Proposition 3.2.32, Ω_U^A is left-c.e. or right-c.e. Now Ω_U^A is certainly left-c.e. relative to A. So if Ω_U^A is right-c.e., we have $\Omega_U^A \leq_T A$, contrary to Fact 3.4.5. Thus Ω_U^A is left-c.e.

(iv)⇒(iii): immediate.

(iii)⇒(ii): If Ω^A is left-c.e. then $\Omega^A \equiv_T \emptyset'$ by Corollary 3.2.31. Thus $A \leq_T \Omega^A$ since we are assuming that A is Δ_2^0.

(ii) ⇒ (i): If $A \leq_T \Omega^A$ then A is a base for ML-randomness, and hence K-trivial by Theorem 5.1.22. □

The hypothesis that $A \in \Delta_2^0$ is needed because there is a ML-random set Z such that Ω^Z is left-c.e. by Theorem 8.1.2.

The difference left-c.e. real numbers form a subfield D of $\mathbb{R}$ by 1.8.16. Since $\mathcal{K}$ induces an ideal in the Turing degrees, the K-trivial real numbers form a subfield of $\mathbb{R}$ as well (see before 1.8.16). By the following, this field is in fact a subfield of D.

5.5.15 Corollary. *For each K-trivial set A, the real number $0.A$ is difference left-c.e.*

Proof. Let M be the prefix-free oracle machine such that $M^Y(\sigma)\downarrow \leftrightarrow \exists n\,[\sigma = 0^n 1 \;\&\; Y(n) = 1]$. Then $\Omega_M^Y = 0.Y$ for each Y. Thus $0.A$ is difference left-c.e. by Theorem 5.5.14. □

By a result of Raichev (2005), all these fields are real closed, that is, each polynomial of odd degree with coefficients in the field assumes the value 0.

5.5.16 Exercise. Let U be a universal prefix-free machine and let A be a set (not necessarily Δ_2^0). Then A is K-trivial $\Leftrightarrow$ $A \leq_T \Omega_U^A$.

Each K-trivial set is low for weak 2-randomness

In Theorem 5.1.33 we proved that $\text{Low}(\mathsf{W2R}, \mathsf{MLR}) = \text{Low}(\mathsf{MLR})$, which implies that $\text{Low}(\mathsf{W2R}) \subseteq \text{Low}(\mathsf{MLR})$. Here we show that actually $\text{Low}(\mathsf{W2R}) = \text{Low}(\mathsf{MLR})$. This is surprising because the two randomness notions are quite different. Two proofs of the containment $\text{Low}(\mathsf{MLR}) \subseteq \text{Low}(\mathsf{W2R})$ were found independently. The first, due to Nies, uses the Main Lemma 5.5.1. The second, due to Kjos-Hanssen, Miller and Solomon (2011), relies on Theorem 5.6.9 below and will be discussed after having proved that theorem.

In Remark 5.6.19 we will briefly address lowness for 2-randomness.

5.5.17 Theorem. *The following are equivalent for a set A:*

(i) A is low for generalized ML-tests in a uniform way. That is, given a null $\Pi_2^0(A)$ class U, one can effectively obtain a null Π_2^0 class $V \supseteq U$.

(ii) A is low for weak 2-randomness.

(iii) A is K-trivial.

Note that the condition (i) appears to be stronger than (ii). It states that each test for weak 2-randomness relative to A can be covered by a test for unrelativized weak 2-randomness. The difference between being lowness for a test notion and lowness for the corresponding randomness notion will be discussed after Definition 8.3.1.

Proof. (i) $\Rightarrow$ (ii): Immediate.
(ii) $\Rightarrow$ (iii): This follows from Theorem 5.1.33 and the fact that each set that is low for ML-randomness is K-trivial.

(iii) $\Rightarrow$ (i): In fact, if A is K-trivial, we obtain a conclusion that appears even stronger than (i) as it applies to $\Pi^0_2(A)$ classes in general:

(i′) *Given a $\Pi^0_2(A)$ class U, one can effectively obtain a Π^0_2 class $V \supseteq U$ such that $\lambda V = \lambda U$.*

In (i′) it actually suffices to cover $\Sigma^0_1(A)$ classes.

5.5.18 Lemma. (J. Miller) *Suppose that for each $\Sigma^0_1(A)$ class G there is a Π^0_2 class S such that $S \supseteq G$ and $\lambda S = \lambda G$. Then (i′) holds for A.*

Subproof. Let $N^X_e = [W^X_e]^{\prec}$. Then $(N^X_e)_{e \in \mathbb{N}}$ is an effective listing of the $\Sigma^0_1(X)$ classes. Let

$$N^X = \bigcup_e 0^e 1 N^X_e$$

(where for $\mathcal{C} \subseteq 2^{\mathbb{N}}$ and $x \in \{0,1\}^*$ we denote by $x\mathcal{C}$ the class $\{xZ : Z \in \mathcal{C}\}$). The class N^X is an effective disjoint union of all the N^X_e.

By the hypothesis for $G = N^A$ there is a Π^0_2 class $S \supseteq N^A$ of the same measure as N^A. For each $i \in \mathbb{N}$ let $S_i = \{Z\colon 0^i 1 Z \in S\}$, then S_i is a Π^0_2 class uniformly in i. Given a $\Pi^0_2(A)$ class U, one can effectively obtain a computable function p such that $U = \bigcap_e N^A_{p(e)}$ by Remark 1.8.58. Further, $S_{p(e)} \supseteq N^A_{p(e)}$ and $2^{-p(e)-1}\lambda(S_{p(e)} - N^A_{p(e)}) \leq \lambda(S - N^A) = 0$, hence $\lambda S_{p(e)} = \lambda N^A_{p(e)}$. Let $V = \bigcap_e S_{p(e)}$, then $V \supseteq U$ and the Π^0_2 class V was obtained effectively from U. Also $\lambda V = \lambda U$ since $V - U \subseteq \bigcup_e (S_{p(e)} - N^A_{p(e)})$. ◇

To conclude the proof of (iii) $\Rightarrow$ (i′), we show that for each K-trivial set A the hypothesis of Lemma 5.5.18 holds. By Fact 1.8.26 relative to A there is a c.e. index d such that $G = [W^A_d]^{\prec}$, and W^X_d is a prefix-free set for each X. Consider a prefix-free oracle machine M^X such that, for each σ,

$$M^X(\sigma)\downarrow \;\leftrightarrow\; \sigma \in W^X_d \tag{5.20}$$

with the same use on both sides. By Main Lemma 5.5.1 there is a constant $u \in \mathbb{N}$ and a computable sequence of stages $q(0) < q(1) < \ldots$ such that (5.18) holds. We define a sequence $(U_n)_{n \in \mathbb{N}}$ such that U_n is a Σ^0_1 class uniformly in n, and $S = \bigcap_n U_n$ is the required Π^0_2 class. The idea is to put into U_n all $[\sigma]$ such that the computation $M^A(\sigma)$ converges with a use of at most $q(r)$ at a stage $q(r+1)$ where $r + 1 \geq n$. If the computation is stable then $\sigma \in W^A_d$. The measure of the computations that are not stable tends to 0 with n because of (5.18).
Here are the details on how to enumerate the Σ^0_1 classes U_n: *for each r,*

($\star$) *if $M^A(\sigma)[q(r+1)]\downarrow$ and* use *$M^A(\sigma)[q(r+1)] \leq q(r)$,*
put $[\sigma]$ into $U_{n,r+1}$ for each $n \leq r+1$.

Clearly $G \subseteq U_n$ for each n, since $\sigma \in W_d^A$ implies that for some r there is a stable computation $M^A(\sigma)[q(r)]$. Thus $G \subseteq S$. To show that $\lambda(S - G) = 0$, note that for each $\epsilon > 0$ there is n such that $\widehat{S}_n \leq \epsilon$, where

$$\widehat{S}_n = \sum_{x,k} \widehat{c}_M(x,k) \, [\![n+1 \leq k \text{ \& } x \text{ is least s.t. } A_{q(k+1)}(x) \neq A_{q(k+2)}(x)]\!],$$

and $\widehat{c}_{M,A}(x,k)$ is defined in (5.19) on page 216. Suppose there is a number r such that the computation $M^A(\sigma)[q(r+1)]$ with use at most $q(r)$ has put $[\sigma]$ into $U_{n,r+1}$. If $[\sigma] \not\subseteq G$ then there is $k \geq n+1, r$ such that this computation is destroyed by an A-change at a stage $t \in [q(k+1), q(k+2))$. Then $2^{-|\sigma|}$ is included in the sum $\widehat{S}_n$. Since

$$U_n - G \subseteq \bigcup \{[\sigma] \colon \sigma \text{ is as in } (\star) \text{ \& } [\sigma] \not\subseteq G\}$$

this implies that $\lambda(U_n - G) \leq \epsilon$. Hence $\lambda S = \lambda G$. □

5.5.19$^\diamondsuit$ **Problem.** Characterize lowness for Demuth randomness. (See Downey and Ng 2009, together with Bienvenu, Downey *et al.* 20xx.)

5.6 The weak reducibility associated with Low(MLR)

A reducibility is a preordering on $2^{\mathbb{N}}$ that specifies a way to compare sets with regard to their computational complexity (Section 1.2). Informally, we will say that a reducibility $\leq_W$ is *weak* if $A \leq_T B$ implies $A \leq_W B$ (as opposed to strong reducibilities like $\leq_{tt}$ that imply $\leq_T$). We also ask that $\leq_W$ is Σ^0_n for some n as a relation on sets (often $n = 3$), and $X' \not\leq_W X$ for each set X. Thus, we want $\leq_W$ to be somewhat close to $\leq_T$; for instance, arithmetical reducibility, defined by $X \leq_{ar} Y \leftrightarrow \exists n \, X \leq_T Y^{(n)}$, does not qualify. In general, there are no reduction procedures for a weak reducibility.

Two of the classes introduced in Section 5.1, being low for K and being low for ML-randomness, form the least degrees of weak reducibilities denoted $\leq_{LK}$ and $\leq_{LR}$, respectively. For further examples of weak reducibilities, see $\leq_{\text{cdom}}$ in Exercise 5.6.8 (each A-computable function is dominated by a B-computable function) and $\leq_{JT}$ in Definition 8.4.13. An overview of weak reducibilities is given in Table 8.3 on page 363.

The intuition for a weak reducibility relation $A \leq_W B$ is that B can only understand a small aspect of the abilities of A. For instance, $A \leq_{\text{cdom}} B$ means that B can emulate the growth of functions computed by A, and $A \leq_{JT} B$ that B can approximate the values of J^A. In contrast, if $A \leq_T B$ then B can emulate everything A does.

The K-trivial sets form the least degree of a preordering $\leq_K$.

5.6.1 Definition. For sets A, B, let

(i) $A \leq_{LK} B \quad \leftrightarrow \quad \exists d \, \forall y \, [K^B(y) \leq K^A(y) + d]$
(ii) $A \leq_{LR} B \quad \leftrightarrow \quad \mathsf{MLR}^B \subseteq \mathsf{MLR}^A$
(iii) $A \leq_K B \quad \leftrightarrow \quad \exists b \forall n \, K(A\restriction_n) \leq K(B\restriction_n) + b.$

Informally, $A \leq_{LK} B$ means that up to a constant, one can compress a string with B as an oracle at least as well as with A, and $A \leq_{LR} B$ means that, whenever A can find "regularities" in a set Z, then so can B. By these definitions, A is low for K iff $A \leq_{LK} \emptyset$, and A is low for ML-randomness iff $A \leq_{LR} \emptyset$.

For each X we have $X' \not\leq_{LR} X$ because Ω^X is not ML-random relative to X' but Ω^X is ML-random relative to X. If A is low for ML-randomness relative to X, namely, $\mathsf{MLR}(A \oplus X) = \mathsf{MLR}(X)$, then $A \leq_{LR} X$. The converse fails in general; see Remark 5.6.24(iii) for a counterexample. In particular, $\oplus$ does not determine a join in the LR-degrees.

By Exercise 3.4.4, $\leq_T$ implies $\leq_{LK}$. The converse fails since some incomputable set is low for K. By Schnorr's Theorem 3.2.9 in relativized form, $\leq_{LK}$ implies $\leq_{LR}$. Note that, while the definition of $\leq_{LK}$ is in Σ^0_3 form, the definition of $\leq_{LR}$ is not even in arithmetical form as it involves universal quantification over sets. Nonetheless, in Theorem 5.6.5 we will show that $\leq_{LR}$ is in fact equivalent to $\leq_{LK}$. This yields an alternative proof of the result in Section 5.1 that the sets that are low for K coincide with the sets that are low for ML-randomness.

Further results in this section show that $\leq_{LR}$ is rather different from $\leq_T$. For instance, the set $\{Z\colon Z \leq_{LR} \emptyset'\}$ is uncountable. (This is possible only because there are no reduction procedures.) The equivalence relation $\equiv_{LR}$ is somewhat close to $\equiv_T$: for example, $A \equiv_{LR} B$ implies that A is K-trivial in B and B is K-trivial in A (5.6.20), and therefore $A' \equiv_{tt} B'$. Thus each LR-degree is countable, while the initial interval below the LR-degree of $\emptyset'$ is uncountable.

Each weak reducibility $\leq_W$ determines a lowness property $\{A\colon A \leq_W \emptyset\}$ and, dually, a highness property $\{C\colon \emptyset' \leq_W C\}$. We will characterize the highness property associated with $\leq_{LR}$ by a domination property, being uniformly a.e. dominating.

5.6.2 Remark. Note that A is K-trivial iff $A \leq_K \emptyset$. The preordering $\leq_K$ compares the degree of randomness of sets: if $A \leq_K B$ then B is "at least as random" as A. As an alternative, Miller and Yu (2008) compared the degree of randomness of sets by introducing van Lambalgen reducibility:

$$A \leq_{vL} B \;\leftrightarrow\; \forall Y \in \mathsf{MLR}\,[A \in \mathsf{MLR}^Y \rightarrow B \in \mathsf{MLR}^Y].$$

Intuitively, B is ML-random in at least as many sets $Y \in \mathsf{MLR}$ as A is (it is not known whether the condition that $Y \in \mathsf{MLR}$ actually makes a difference). Interestingly, they showed that $\leq_K$ implies $\leq_{vL}$ on MLR (see Exercise 8.1.17).

The least $\leq_{vL}$ degree consists of the sets that are not ML-random, because any ML-random set A is ML-random in Ω^A by Theorem 3.4.6. If A and B are ML-random then $A \leq_{vL} B \leftrightarrow A \geq_{LR} B$ by 3.4.6., so $\leq_{vL}$ is actually just another way to look at $\geq_{LR}$ on the ML-random sets.

Preorderings coinciding with LR-reducibility

We characterized the class Low(MLR) by effective topology in Corollary 5.1.10. This is a special case of characterizing the underlying weak reducibility: $A \leq_{LR} B$ means that we can cover each $\Sigma^0_1(A)$ class of (uniform) measure < 1 by a $\Sigma^0_1(B)$

class of measure < 1. We will encounter several variants of such *covering procedures.* For instance, in Lemma 5.6.4 we cover "small" A-c.e. sets by "small" B-c.e. sets, and in Theorem 5.6.9 we cover each $\Pi^0_2(A)$ class by a $\Pi^0_2(B)$ class of the same measure. Theorem 5.5.17 already involved the special case of this covering procedure where $B = \emptyset$. A summary of the covering procedures in this chapter is given in Table 5.2 on page 230. Lowness for $\mathcal{C}$-null classes (8.3.1) is another example. See 1.8.59 for the definition of $\Sigma^0_n(C)$ classes and $\Pi^0_n(C)$ classes.

5.6.3 Lemma. *$A \le_{LR} B \Leftrightarrow$ each $\Sigma^0_1(A)$ class G such that $\lambda G < 1$ is contained in a $\Sigma^0_1(B)$ class S such that $\lambda S < 1$.*

Proof. While a mere relativization of Theorem 5.1.9 and Corollary 5.1.10 is not enough, it suffices to extend the notation in their statements and proofs. The statement "A is low for ML-randomness" becomes "$A \le_{LR} B$", and "S is open and c.e." becomes "S is a $\Sigma^0_1(B)$ class". □

Kjos-Hanssen, Miller and Solomon (2011) proved that $\le_{LR}$ is equivalent to $\le_{LK}$. We mostly follow the exposition of this result due to Simpson (2007).

For a function $f\colon \mathbb{N} \to \mathbb{N}$, a set $I \subseteq \mathbb{N}$ is called *f-small* if $\sum_{n \in I} 2^{-f(n)} < \infty$. (Equivalently, $\mu_f(I)$ is finite where μ_f is the measure on $\mathcal{P}(\mathbb{N})$ given by $\mu_f(\{n\}) = 2^{-f(n)}$.) The main work is in the following lemma, where the idea is once again that $A \le_{LR} B$ means we can cover A-c.e. objects by B-c.e. objects of the same type. We only need one implication; for the converse see 5.6.7.

5.6.4 Lemma. *$A \le_{LR} B \Rightarrow$ for each computable function f, each f-small A-c.e. set I is contained in an f-small B-c.e. set R.*

Proof. We use a fact from elementary analysis: for a sequence of real numbers $(a_n)_{n \in \mathbb{N}}$ such that $0 \le a_n < 1$ for each n, we have

$$\sum_{n=0}^{\infty} a_n < \infty \;\leftrightarrow\; \prod_{n=0}^{\infty} (1 - a_n) > 0. \tag{5.21}$$

To see this let $g(x) = -\ln(1-x)$ for $x \in [0,1)_{\mathbb{R}}$. Since $e^y \ge 1 + y$ for each $y \in \mathbb{R}$, we have $x \le g(x)$ for each x. On the other hand $g'(x) = 1/(1-x)$, so $g'(0) = 1 = \lim_{x \to 0}(g(x) - g(0))/x$. Hence there is $\epsilon > 0$ such that $g(x) \le 2x$ for each $x \in [0, \epsilon)$.

If the sequence $(a_n)_{n \in \mathbb{N}}$ does not converge to 0 then both sides in (5.21) are false. Otherwise, we may as well assume that $a_n < \epsilon$ for each n. Then

$$\begin{aligned}
\textstyle\prod_{n=0}^{\infty}(1-a_n) > 0 \;&\leftrightarrow\; \exists c \in \mathbb{R}\, \forall k \;\; \ln \textstyle\prod_{n=0}^{k}(1-a_n) > c \\
&\leftrightarrow\; \exists d \in \mathbb{R}\, \forall k \;\; \textstyle\sum_{n=0}^{k} -\ln(1-a_n) < d \\
&\leftrightarrow\; \exists d \in \mathbb{R}\, \forall k \; \textstyle\sum_{n=0}^{k} a_n < d \\
&\leftrightarrow\; \textstyle\sum_{n=0}^{\infty} a_n < \infty.
\end{aligned}$$

We use (5.21) to infer Lemma 5.6.4 from the implication "$\Rightarrow$" of Lemma 5.6.3. We may assume $f(n) > 0$ for each n, for we can replace f by $\lambda n. f(n) + 1$ without changing the notion of f-smallness. For each n, let $g(n) = \sum_{i<n} f(i)$, and let E_n be the clopen set $\{Z\colon \exists i \in [g(n), g(n+1))\, [Z(i) \neq 0]\}$. Note that

$\lambda E_n = 1 - 2^{-f(n)}$. Also $\lambda(\bigcap_{n \in X} E_n) = \prod_{n \in X} \lambda E_n$ for each $X \subseteq \mathbb{N}$. (This is easy to check for finite X; in the general case, $\lambda(\bigcap_{n \in X} E_n) = \lim_m \lambda(\bigcap_{n \in X, n \le m} E_n) = \lim_m \prod_{n \in X, n \le m} \lambda E_n = \prod_{n \in X} \lambda E_n$. We use that the events E_n are independent in the language of probability theory.)

Since I is A-c.e., the class $P = \bigcap_{n \in I} E_n$ is $\Pi^0_1(A)$. For we may assume that $I \neq \emptyset$, so there is a function $h \le_T A$ with range I; then $P = \{Z \colon \forall k\, Z \in E_{h(k)}\}$. Since I is f-small, by (5.21) we have $\lambda P = \prod_{n \in I}(1 - 2^{-f(n)}) > 0$. By Lemma 5.6.3 choose a $\Pi^0_1(B)$ class $Q \subseteq P$ such that $\lambda Q > 0$. Let $R = \{n \colon Q \subseteq E_n\}$. Then R is B-c.e., $I \subseteq R$, and R is f-small, by (5.21) and because

$$\prod_{n \in R}(1 - 2^{-f(n)}) = \prod_{n \in R} \lambda(E_n) = \lambda \bigcap_{n \in R} E_n \ge \lambda Q > 0,$$

by the independence of the events E_n. □

5.6.5 Theorem. *For each pair of sets A, B we have $A \le_{LK} B \Leftrightarrow A \le_{LR} B$.*

Proof. $\Rightarrow$: This follows from the relativized form of Schnorr's Theorem 3.2.9: $Z \in \mathsf{MLR}^B \leftrightarrow \forall n\, K^B(Z \restriction_n) \ge^+ n \rightarrow \forall n\, K^A(Z \restriction_n) \ge^+ n \leftrightarrow Z \in \mathsf{MLR}^A$.

$\Leftarrow$: Let f be the computable function given by $f(\langle r, y \rangle) = 2^r$. The set $I = \{\langle |\sigma|, y \rangle \colon \mathbb{U}^A(\sigma) = y\}$ is a bounded request set relative to A and hence f-small. So by Lemma 5.6.4, I is contained in an f-small B-c.e. set $\widetilde{R}$. Let $R \subseteq \widetilde{R}$ be a bounded request set relative to B such that $\widetilde{R} - R$ is finite. Then, applying to R the Machine Existence Theorem 2.2.17 relative to B, we may conclude that $\forall y\, K^B(y) \le^+ K^A(y)$. □

Exercises.

5.6.6. Show that $A \le_{LK} B \Leftrightarrow \forall n\, K^B(A \restriction_n) \le^+ K^A(n)$.

5.6.7. Prove the converse of Lemma 5.6.4.

5.6.8. Consider the weak reducibility defined by $A \le_{\text{cdom}} B$ if each A-computable function is dominated by a B-computable function. Show that $\emptyset' \le_{\text{cdom}} C \Leftrightarrow \emptyset' \le_T C$.

A stronger result under the extra hypothesis that $A \le_T B'$

If one can approximate A computably in B, then $A \le_{LR} B$ implies that each $\Sigma^0_1(A)$ class G can be approximated from above by a sequence of uniformly $\Sigma^0_1(B)$ classes $(U_n)_{n \in \mathbb{N}}$, in the sense that $G \subseteq U_n$ and $\lambda(U_n - G)$ tends to 0. In other words, $G \subseteq S$ for a $\Pi^0_2(B)$ class S of the same measure as G. This result of Kjos-Hanssen, Miller and Solomon (2011) will be applied in Theorem 5.6.30 to prove the coincidence of two highness properties.

Note that $A \le_{LR} B$ does not in general imply $A \le_T B'$. (For instance, let $B = \emptyset'$, and note that $\#\{A \colon A \le_{LR} \emptyset'\} = 2^{\aleph_0}$ by Theorem 5.6.13 proved shortly.) However, $A \le_{LR} B$ does imply $A \le_T B'$ if B is low for Ω (Corollary 8.1.10), and thus for almost all sets B in the sense of the uniform measure.

5.6.9 Theorem. *The following are equivalent for sets A and B.*

(i) $A \le_{LR} B$ and $A \le_T B'$.

(ii) For each $\Sigma^0_1(A)$ class G there is a $\Pi^0_2(B)$ class $S \supseteq G$ such that $\lambda S = \lambda G$.

(iii) Given a $\Pi^0_2(A)$ class U one can effectively obtain a $\Pi^0_2(B)$ class $V \supseteq U$ such that $\lambda V = \lambda U$.

By Exercise 5.6.10 $A \leq_{LR} B$ and $A \leq_T B'$ implies $A' \leq_T B'$. Thus the theorem characterizes the intersection of the weak reducibility relations $A \leq_{LR} B$ and $A' \leq_T B'$ for sets A and B.

The implication (iii)⇒(i) in Theorem 5.5.17 states that each K-trivial set is low for generalized ML-tests. Since each set in Low(MLR) is Δ^0_2, the implication (i)⇒(iii) in the foregoing theorem for $B = \emptyset$ shows that each set in Low(MLR) is low for generalized ML-tests. Thus we have obtained an alternative proof of this implication (iii)⇒(i) in Theorem 5.5.17: instead of using the Main Lemma 5.5.1 it relies on Lemma 5.6.4.

Proof. (ii)⇒(iii): this is similar to the proof of Lemma 5.5.18, replacing Π^0_2 classes by $\Pi^0_2(B)$ classes.

(iii)⇒(i): $A \leq_{LR} B$ by Lemma 5.6.3. To show $A \leq_T B'$, recall from page 12 that $\leq_L$ is the lexicographical order on sets, and note that $U = \{Z\colon Z \leq_L A\}$ is a $\Pi^0_1(A)$ class such that $\lambda U = 0.A$. Choose a $\Pi^0_2(B)$ class $V \supseteq U$ such that $\lambda V = \lambda U$. By Exercise 1.9.22 λV is left-$\Pi^0_2(B)$, and hence $0.A = \lambda V$ is right-c.e. relative to B'. Applying the same argument to $\mathbb{N} - A$ instead of A shows that $1 - 0.A$ is right-c.e. relative to B'. Therefore $A \leq_T B'$.

(i)⇒(ii): Although the two proofs were found independently, the present proof follows the same outline as the proof of (iii)⇒(i) in Theorem 5.5.17. (In fact, that proof shows (ii) of the present theorem under the hypothesis that A is K-trivial in B, which is stronger than (i) of the present theorem by 6.3.16.)

Recall that $W^X_e = \operatorname{dom} \Phi^X_e$. Let d be a c.e. index such that $G = [W^A_d]^{\prec}$ and W^X_d is an antichain for each oracle X. The plan is as follows:

(a) translate the setting of $\Sigma^0_1(A)$ classes into the setting of A-c.e. sets,
(b) apply Lemma 5.6.4 to obtain a B-c.e. set,
(c) translate the setting of B-c.e. sets back into the setting of $\Sigma^0_1(B)$ classes to define a sequence $(U_n)_{n\in\mathbb{N}}$ of uniformly $\Sigma^0_1(B)$ sets such that for the $\Pi^0_2(B)$ class $S = \bigcap_n U_n$ we have $S \supseteq G$ and $\lambda S = \lambda G$.

We carry out the steps in detail.
(a) The set $I = \{\langle A\restriction_u, x\rangle\colon \Phi^A_d(x)\downarrow \text{ with use } u\}$ is A-c.e.
(b) Note that $\sum_{\sigma,x} 2^{-|x|} [\![\langle\sigma, x\rangle \in I]\!] = \lambda G \leq 1$, so by Lemma 5.6.4 for the function $f(\langle\sigma, x\rangle) = 2^{-|x|}$, there is a B-c.e. set $R \supseteq I$ such that

$$\sum_{\sigma,x} 2^{-|x|} [\![\langle\sigma, x\rangle \in R]\!] < \infty.$$

Since $A \leq_T B'$ there is a B-computable approximation $(A_r)_{r\in\mathbb{N}}$ of A. For each n let

$$F_n = \{\langle\sigma, x\rangle \in R\colon \exists r \geq n\, [\sigma \prec A_r \ \&\ \Phi^\sigma_{d,r}(x)\downarrow \text{ with use } |\sigma|]\}.$$

Then the sequence $(F_n)_{n\in\mathbb{N}}$ is uniformly c.e. in B and $I \subseteq F_n \subseteq R$ for each n. To see that $I = \bigcap_n F_n$, given $\langle\sigma, x\rangle$, let n be a stage such that $A_r\upharpoonright_{|\sigma|} = A_n\upharpoonright_{|\sigma|}$ for each $r \geq n$. If $\langle\sigma, x\rangle \in F_n$ then $\sigma \prec A$ and hence $\langle\sigma, x\rangle \in I$.

(c) Let $U_n = [\{x\colon \exists\sigma\, \langle\sigma, x\rangle \in F_n\}]^{\prec}$ be the open set generated by the projection of F_n on the second component (these open sets correspond to the sets U_n in the proof of (iii)⇒(i) of Theorem 5.5.17). Since $I \subseteq F_n$ for each n, we have $G \subseteq S := \bigcap_n U_n$. Note that S is $\Pi^0_2(B)$. To show $\lambda S = \lambda G$, fix $\epsilon > 0$. Let k be so large that

$$\textstyle\sum_{\sigma,x} 2^{-|x|}[\![\langle\sigma, x\rangle \in R \;\&\; \langle\sigma, x\rangle \geq k]\!] < \epsilon,$$

and let n be so large that, for all $\langle\sigma, x\rangle < k$, if $\langle\sigma, x\rangle \notin I$ then $\langle\sigma, x\rangle \notin F_n$. If $Y \in U_n - G$ then there is $\langle\sigma, x\rangle \in F_n - I$ such that $x \prec Y$, so $\lambda(U_n - G) \leq \sum_{\sigma,x} 2^{-|x|}[\![\langle\sigma, x\rangle \in F_n - I]\!] \leq \sum_{\sigma,x} 2^{-|x|}[\![\langle\sigma, x\rangle \in R \;\&\; \langle\sigma, x\rangle \geq k]\!] < \epsilon$. □

5.6.10 Exercise. Show that (i) in Theorem 5.6.9, namely $A \leq_{LR} B$ and $A \leq_T B'$, implies $A' \leq_T B'$.

The size of lower and upper cones for $\leq_{LR}$ ⋆

If "size" is interpreted by "measure" then $\leq_{LR}$ behaves like $\leq_T$ in that each lower cone and each nontrivial upper cone is null (the second result is an analog of Theorem 5.1.12). In contrast, if size means cardinality, some $\leq_{LR}$ lower cone has the maximum possible size $2^{\aleph_0}$, while of course each Turing lower cone is countable.

5.6.11 Fact. *The class $\{X\colon X \leq_{LR} A\}$ is null for each A.*

Proof. If $X \in \mathsf{MLR}^A$ then $X \not\leq_{LR} A$ since X is not ML-random relative to itself. This suffices since MLR^A is conull. □

5.6.12 Fact. *If $A \notin$ Low(MLR) then $\{Z\colon A \leq_{LR} Z\}$ is null.*

TABLE 5.2. Overview of the covering procedures. Sets A and B are given.

Condition described	Given	Covered by	Reference
$A \leq_{LR} \emptyset$	$\Pi^0_2(A)$ class	Π^0_2 class of the same measure	5.5.17
$A \leq_{LR} B$	non-conull $\Sigma^0_1(A)$ class	non-conull $\Sigma^0_1(B)$ class	5.6.3
$A \leq_{LR} B$	(f-small A-c.e. set)	(f-small B-c.e. set)	5.6.4
$A \leq_{LR} B$ & $A \leq_T B'$	$\Pi^0_2(A)$ class	$\Pi^0_2(B)$ class of the same measure	5.6.9
$\emptyset' \leq_{LR} B$	Σ^0_2 class	$\Pi^0_2(B)$ class of the same measure	5.6.30

Proof. (Stephan) Let $X \in \mathsf{MLR} - \mathsf{MLR}^A$. If $Z \in \mathsf{MLR}^X$ then $X \in \mathsf{MLR}^Z$ by Theorem 3.4.6, so $A \not\leq_{LR} Z$. Again this suffices since MLR^X is conull. □

Note that $Z \leq_{LR} \emptyset'$ iff each 2-random set is in MLR^Z. This class is much larger than the class of Δ^0_2 sets by a result of Barmpalias, Lewis and Soskova (2008).

5.6.13 Theorem. $\#\{Z\colon Z \leq_{LR} \emptyset'\} = 2^{\aleph_0}$.

Proof. We prove the result for the equivalent reducibility $\leq_{LK}$. We may assume that $\Omega^X \leq 3/4$ for each X by arranging that the oracle prefix-free machine M_2 never halts (see Theorem 2.2.9). For the duration of this proof, contrary to the definition on page 14, we will write $\mathbb{U}^\alpha(\sigma) = y$ if $\mathbb{U}^\alpha_{|\alpha|}(\sigma) = y$, namely the computation halts within $|\alpha|$ steps. Let $\Omega^\alpha = \sum_\sigma 2^{-|\sigma|}[\![\mathbb{U}^\alpha(\sigma)\downarrow \ \& \ |\sigma| \leq |\alpha|]\!]$.

5.6.14 Fact. *Let* $k \in \mathbb{N}$ *and* $\gamma \in \{0,1\}^*$. *Then* $\exists \alpha \succeq \gamma \forall \beta \succeq \alpha\, [\Omega^\beta - \Omega^\alpha \leq 2^{-k}]$.

Subproof. If the fact fails for k, γ, we can build a sequence $\gamma \preceq \alpha_0 \prec \alpha_1 \prec \ldots$ such that $\Omega^{\alpha_{i+1}} - \Omega^{\alpha_i} > 2^{-k}$ for each k, which contradicts $\Omega^{\bigcup_i \alpha_i} \leq 1$. ◇

We define a Δ^0_2 tree $B \subseteq 2^{\mathbb{N}}$ such that $\#\mathit{Paths}(B) = 2^{\aleph_0}$ and $Z \leq_{LK} \emptyset'$ for each $Z \in \mathit{Paths}(B)$. We represent B by a monotonic one-one Δ^0_2 function $F\colon \{0,1\}^* \to \{0,1\}^*$ in the sense that $B = \{\alpha\colon \exists \eta\ \alpha \preceq F(\eta)\}$. Thus, the class $\mathit{Paths}(B)$ is perfect, and the range of F is the set of branching nodes of B. We build a single bounded request set L relative to $\emptyset'$ that can simulate the $\mathbb{U}^Z$-descriptions for any path Z of B. For this we have to choose the branching nodes sufficiently far apart. Define F inductively:

Let $F(\varnothing)$ be the least string α such that $\forall \beta \succeq \alpha\, [\Omega^\beta - \Omega^\alpha < 2^{-4}]$. If $n > 0$ and $F(\eta)$ has been defined for all strings η of length $n-1$, then, for $a \in \{0,1\}$, let $F(\eta a)$ be the least string $\alpha \succeq F(\eta)a$ such that $\forall \beta \succeq \alpha\, [\Omega^\beta - \Omega^\alpha < 2^{-2n-4}]$.

Note that $F \leq_T \emptyset'$ since the function $\alpha \to \Omega^\alpha$ is computable. Now let $L = \bigcup_n L_n$, where L_n is the set of pairs $\langle |\sigma|, y\rangle$ such that $\mathbb{U}^{F(\eta)}(\sigma) = y$, for some η of length n such that $\mathbb{U}^{F(\eta')}(\sigma)\uparrow$ for each $\eta' \prec \eta$. Clearly L is c.e. in $\emptyset'$. For any set $S \subseteq \mathbb{N} \times \{0,1\}^*$, let $\mu(S) = \sum_r 2^{-r}\, [\![\langle r, y\rangle \in S]\!]$. We have to show that $\mu(L) \leq 1$. Now $\mu(L_0) \leq 3/4$, and for $n > 0$, $\mu(L_n) \leq 2^n 2^{-2n-2} = 2^{-n-2}$, since for each string η of length $n-1$, if $\gamma = F(\eta)$, then for each extension $\delta_a = F(\eta a)$, $a \in \{0,1\}$, we have $\Omega^{\delta_a} - \Omega^\gamma \leq 2^{-2n-2}$. Thus $\mu(L) \leq 3/4 + \sum_{n>0} 2^{-n-2} \leq 1$.

If $Z \in \mathit{Paths}(B)$ and $\mathbb{U}^Z(\sigma) = y$ then $\langle |\sigma|, y\rangle \in L$. Thus $\forall y\, [K^Z(y) \leq K^{\emptyset'}(y) + d]$ where d is the coding constant for L, and hence $Z \leq_{LK} \emptyset'$. □

Stronger results have been obtained: $\#\{Z\colon Z \leq_{LR} B\} = 2^{\aleph_0}$ whenever $B \notin \mathrm{GL}_2$ (Barmpalias, Lewis and Soskova, 2008) and whenever $B \in \Delta^0_2 - \mathrm{Low}(\mathsf{MLR})$ (Barmpalias, 2010). On the other hand, if $B \in \mathrm{Low}(\Omega)$ then $Z \leq_{LR} B$ implies $Z \leq_T B'$ by Corollary 8.1.10, so $\#\{Z\colon Z \leq_{LR} B\} \leq \aleph_0$.

The following lowness property may be interesting: A is an *LR-base for ML-randomness* if $A \leq_{LR} Z$ for some $Z \in \mathsf{MLR}^A$. If A is ML-random then A is not such a base because $Z \in \mathsf{MLR}^A$ implies $A \in \mathsf{MLR}^Z$, so $A \not\leq_{LR} Z$. Barmpalias, Lewis and Stephan (2008) have built a perfect Π^0_1 class P such that $Y \leq_{LR} \emptyset'$ for each $Y \in P$, and also

$P \cap \text{Low}(\mathsf{MLR}) = \emptyset$. Barmpalias has pointed out that P contains a set A that is low for Ω by 8.1.3, so A is an LR-base for ML-randomness while $A \notin \text{Low}(\mathsf{MLR})$. This contrasts with Theorem 5.1.22.

5.6.15.$^{\diamond}$ **Exercise.** (Barmpalias, Lewis and Stephan 2008, Thm. 16)
Show that for each incomputable set W there is set $A \equiv_{LR} W$ such that $A \mid_T W$.

5.6.16.$^{\diamond}$ **Exercise.** (Barmpalias, Miller and Nies 2011, Thm. 5.2)
Show that some weakly 2-random set Z satisfies $Z \leq_{LR} \emptyset'$.

Operators obtained by relativizing classes

We have introduced preorderings associated with the classes $\text{Low}(\mathsf{MLR})$, $\mathcal{M}$ (being low for K) and $\mathcal{K}$ (the K-trivial sets). Now we will view these classes relative to an oracle X, thereby turning them into operators $\mathcal{C}\colon \mathcal{P}(\mathbb{N}) \to \mathcal{P}(\mathcal{P}(\mathbb{N}))$. The classes $\mathcal{M}$, $\text{Low}(\mathsf{MLR})$, and $\mathcal{K}$, respectively, yield the operators

$$\begin{aligned}
\mathcal{M}(X) &= \{A\colon \exists d \forall y\, K^X(y) \leq K^{A\oplus X}(y) + d\} = \{A\colon A \oplus X \equiv_{LK} X\},\\
\text{Low}(\mathsf{MLR}^X) &= \{A\colon \mathsf{MLR}^{A\oplus X} = \mathsf{MLR}^X\} = \{A\colon A \oplus X \equiv_{LR} X\}, \text{ and}\\
\mathcal{K}(X) &= \{A\colon \exists b \forall n\, [K^X(A\restriction_n) \leq K^X(n) + b]\}.
\end{aligned}$$

(Note that $K^X(A \oplus X \restriction_{2n}) =^+ K^X(A\restriction_n)$ and $K^X(2n) =^+ K^X(n)$, so in the definition of $\mathcal{K}(X)$ we could as well replace A by $A \oplus X$ for a proper relativization.) The classes $\mathcal{M}(X)$ and $\text{Low}(\mathsf{MLR}^X)$ are closed downward under $\leq_T$ for each X. All the three operators $\mathcal{C}$ are degree invariant, namely, $X \equiv_T Y$ implies $\mathcal{C}(X) = \mathcal{C}(Y)$. This is clear for $\mathcal{M}$ and $\text{Low}(\mathsf{MLR})$, since $\leq_T$ implies $\leq_{LR}$.

In fact, the three operators coincide, because the results that the corresponding classes coincide can be relativized. This yields further information on $\leq_{LR}$. For instance, each LR-degree is countable. The only known proof of this result is via relativized K-triviality. Note that $X \equiv_{LR} Y$ implies that $\mathcal{K}(X) = \mathcal{K}(Y)$. Therefore $\mathcal{C}(X)$ only depends on the LR-degree of X.

We summarize some of the previous results in their relativized forms.

5.6.17 Theorem.

(i) *$\mathcal{K}(X)$ is closed under $\oplus$ for each X.*
(ii) *There is a c.e. index e such that, for each X, $W_e^X \in \mathcal{K}(X)$ and $W_e^X \not\leq_T X$, and therefore $X <_T X \oplus W_e^X \in \mathcal{K}(X)$.*
(iii) *We have $\mathcal{M}(X) = \text{Low}(\mathsf{MLR}^X) = \mathcal{K}(X)$; in particular, $\mathcal{K}(X)$ is closed downward under $\leq_T$.*

Proof. It is sufficient to note that the proofs of 5.2.17, 5.3.11, 5.1.22 and 5.4.1 can be carried out relative to the oracle X. □

In contrast, the following is more than a relativization of the statements of 5.5.3 and 5.5.4. Such a relativization would merely yield relativized truth-table reductions, where the oracle X is used first to compute the truth table and then to answer queries in the truth-table. To prove versions of these corollaries for $\mathcal{K}(X)$ with plain truth-table reductions, we have to look at their proofs.

5.6.18 Theorem.

(i) $A \in \mathcal{K}(X) \;\Rightarrow\; A \le_{tt} C$ *for some* $C \in \mathcal{K}(X)$ *which is c.e. in* X.
(ii) $A \in \mathcal{K}(X) \;\Rightarrow\; A' \le_{tt} X'$.

Proof. (i) By Proposition 5.3.12 relative to X, there is an X-computable approximation $(A_s)_{s\in\mathbb{N}}$ such that $A_s(y)$ changes at most $O(y^2)$ times. Then the change set C in the proof of Corollary 5.5.3 is in $\mathcal{K}(X)$, and $C \ge_{tt} A$.
(ii) The proof of Corollary 5.5.4 relative to X shows that the change set C for A satisfies $(C \oplus X)' \le_{tt} X'$, because in the proof of Proposition 5.1.3 relative to X, the use of the weak truth-table reduction for $A' \le_{wtt} \emptyset'$ is bounded by a computable function independent of X. Since $A \le_{tt} C$ we have $A' \le_m C' \le_{tt} X'$.
Alternatively, one can view the proof of Proposition 5.5.12 relative to X. □

5.6.19 Remark. Note that Z is 2-random relative to A iff Z is ML-random in A' (see 3.4.9). Thus, A is low for 2-randomness $\Leftrightarrow A' \equiv_{LR} \emptyset'$. By Theorem 5.6.17 this class also coincides with $\{A\colon\ A' \in \mathcal{K}(\emptyset')\}$. Each low set is low for 2-randomness. See Exercise 5.6.23 for more.

Studying $\le_{LR}$ ***by applying the operator*** $\mathcal{K}$

Let $\mathcal{S}_{LR}$ be the operator given by

$$\mathcal{S}_{LR}(X) = \{A\colon\ A \le_{LR} X\}.$$

If $A \oplus X \le_{LR} X$ then $A \le_{LR} X$, so $\mathcal{K}(X) \subseteq \mathcal{S}_{LR}(X)$ for each X because $\mathcal{K}(X) = \text{Low}(\text{MLR}^X)$. Note that $\mathcal{K}(X)$ can be proper subclass of $\mathcal{S}_{LR}(X)$: for instance $\mathcal{S}_{LR}(\emptyset')$ is uncountable by Theorem 5.6.13 while $\mathcal{K}(X)$ is countable for each X (also see 5.6.24(iii) below). In contrast, we have:

5.6.20 Lemma. $A \equiv_{LR} B \;\Leftrightarrow\; A \in \mathcal{K}(B)\ \&\ B \in \mathcal{K}(A)$.

Proof. $\Leftarrow$: This is immediate since $\mathcal{K}(X) \subseteq \mathcal{S}_{LR}(X)$ for each X.
$\Rightarrow$: Note that $K^A(A\restriction_n) \le^+ K^A(n)$ for each n. By Theorem 5.6.5 $A \equiv_{LK} B$, so we may replace the oracle A by B in this inequality, which shows that $A \in \mathcal{K}(B)$. Similarly $B \in \mathcal{K}(A)$. □

Together with Theorem 5.6.18(ii) we obtain:

5.6.21 Corollary. *For each pair of sets* A, B, *if* $A \equiv_{LR} B$ *then* $A' \equiv_{tt} B'$. *In particular, each LR-degree is countable.* □

This contrasts with Theorem 5.6.13 that the LR-lower cone below $\emptyset'$ is of cardinality $2^{\aleph_0}$. Since the union of fewer than $2^{\aleph_0}$ countable sets has cardinality less than $2^{\aleph_0}$, the cone below the degree of $\emptyset'$ in the LR-degrees is of cardinality $2^{\aleph_0}$. In the Turing degrees each lower cone $\{\mathbf{b}\colon\ \mathbf{b} \le \mathbf{a}\}$ is countable (Fact 1.2.26).

Exercises. Show the following.
5.6.22. (Barmpalias and Nies) The LR-degrees of the ML-random and the c.e. sets have a singleton as an intersection, namely, the LR-degree of Ω: if Z is ML-random, A is c.e., and $Z \equiv_{LR} A$, then $\emptyset' \in \mathcal{K}(Z)$. In particular, $Z \equiv_{LR} \emptyset'$.
5.6.23. (i) Some nonlow c.e. set A is low for 2-randomness (See 5.6.19.).
(ii) If A is low for 2-randomness then $A'' \equiv_{tt} \emptyset''$ (A is superlow$_2$)

Comparing the operators $\mathcal{S}_{LR}$ and $\mathcal{K}$ $\star$

Recall from the chapter introduction that an operator $\mathcal{S}\colon \mathcal{P}(\mathbb{N}) \to \mathcal{P}(\mathcal{P}(\mathbb{N}))$ is *monotonic* if $X \le_T Y \;\to\; \mathcal{S}(X) \subseteq \mathcal{S}(Y)$ for each X, Y. This property is stronger than degree invariance. We say that an operator $\mathcal{S}$ is closed under $\oplus$ if $\mathcal{S}(X)$ is closed under $\oplus$ for each X, and $\mathcal{S}$ is Σ^0_n if the relation "$Z \in \mathcal{S}(X)$" is Σ^0_n.

5.6.24 Remark. (i) The operator $\mathcal{K}$ is Σ^0_3 and closed under $\oplus$. Shore provided an example of non-monotonicity for $\mathcal{K}$. By Theorem 5.3.11, let A be a promptly simple set in $\mathcal{K}(\emptyset) = \mathcal{K}$. Then A is low cuppable, i.e. there is a low c.e. set G such that $\emptyset' \equiv_T A \oplus G$, by Theorem 6.2.2 below. Hence $A \in \mathcal{K}(\emptyset) - \mathcal{K}(G)$, for otherwise $A \oplus G \in \mathcal{K}(G)$ and therefore $\emptyset'' \equiv_T (A \oplus G)' \le_T G' \equiv_T \emptyset'$ by Theorem 5.6.18(ii), which is a contradiction.

(ii) By Theorem 5.6.5, the lower cone operator $\mathcal{S}_{LR}$ is Σ^0_3 as well, but unlike $\mathcal{K}$, $\mathcal{S}_{LR}$ is monotone. The same example can be used to show that $\mathcal{S}_{LR}$ is not closed under $\oplus$: since $A \in \mathcal{K} = \text{Low}(\mathsf{MLR})$ we have $A \le_{LR} G$. Trivially $G \le_{LR} G$. $A \oplus G \le_{LR} G$ would imply $\emptyset' \equiv_{LR} G$ (since $A \oplus G \equiv_T \emptyset' \ge_T G$), which contradicts Corollary 5.6.21 because G is low. Alternatively, pick $B \in \mathcal{S}_{LR}(\emptyset') - \mathcal{K}(\emptyset')$, then $B \oplus \emptyset' \not\le_{LR} \emptyset'$. By these examples, $\le_{LR}$ behaves differently from $\le_T$ in a further aspect: the operation $\oplus$ does not determine a supremum in the $\le_{LR}$-degrees.

(iii) We have also obtained a low c.e. set G such $\mathcal{K}(G)$ is a proper subclass of $\mathcal{S}_{LR}(G)$, because $A \in \mathcal{S}_{LR}(G) - \mathcal{K}(G)$.

Slaman (2005) studied monotonic operators $\mathcal{S}$ that are Borel (in the sense that the relation "$Z \in \mathcal{S}(X)$" is Borel) and for each X, $\mathcal{S}(X) \neq \mathcal{P}(\mathbb{N})$ contains X, is closed downward under $\le_T$, and closed under $\oplus$. He proved that every such operator is given by (possibly transfinite) iterates of the jump on an upper cone in the Turing degrees. Some possibilities for such an operator $\mathcal{S}(X)$ are $\{Y\colon Y \le_T X\}$, $\{Y\colon Y \le_T X'\}$, or $\{Y\colon \exists n \in \mathbb{N}\; Y \le_T X^{(n)}\}$. Slaman's result can also be used to derive the properties in Remark 5.6.24. The operator $\mathcal{K}$ is Σ^0_3 and hence Borel. By Theorem 5.6.18, $\mathcal{K}$ is not given by iterates of the jump, so $\mathcal{K}$ fails to be monotonic. $\mathcal{S}_{LR}$ is not given by iterates of the jump because $X' \notin \mathcal{S}_{LR}(X)$ for each X; see the beginning of this section. Since $\mathcal{S}_{LR}(X)$ is downward closed under $\le_T$ for each X, this yields a further, indirect, proof that $\mathcal{S}_{LR}(X)$ is not for all X closed under $\oplus$.

5.6.25 Exercise. Show that if $A \in \mathcal{K}$ and X is ML-random, then $A \in \mathcal{K}(X)$.

Uniformly almost everywhere dominating sets

A highness property of a set C expresses that C is close to being Turing above $\emptyset'$. Our first example of such a property was that C is *high*$_1$, namely, $\emptyset'' \le_T C'$; equivalently, some function $f \le_T C$ dominates all the computable functions (Theorem 1.5.19). Dobrinen and Simpson (2004) introduced a stronger highness property when they investigated the strength of axiom systems in the language of second-order arithmetic. In the following, we say that a property $\mathcal{Q}$ of a set holds for *almost every* Z if the class $\{Z\colon \mathcal{Q}(Z)\}$ is conull.

5.6.26 Definition. We say that a set C is *uniformly almost everywhere dominating*, or *u.a.e.d.* for short, if there is a function $f \le_T C$ such that for all e,

$$\text{for almost every } Z \left[\Phi_e^Z \text{ total } \rightarrow \Phi_e^Z \text{ is dominated by } f\right].$$

Clearly each u.a.e.d. set is high. The word "uniform" reminds us of the fact that the same f works for almost all Z (and all e). However, in Proposition 5.6.31 we will show that the seemingly weaker notion of being almost everywhere dominating, where f can depend on Z and e, is in fact equivalent.

It is not necessary to require domination of the total Φ_e^Z for all Turing functionals Φ_e as a particular one is sufficient: for each Z, e, n, let

$$\Theta^{0^e 1Z}(n) \simeq \Phi_e^Z(n).$$

5.6.27 Lemma. *Suppose a function $f \le_T C$ dominates Θ^Y for almost every Y. Then C is uniformly almost everywhere dominating.*

Proof. Fix e. Then $\lambda\{Z\colon \Phi_e^Z \text{ total } \ \& \ \Phi_e^Z \text{ not dominated by } f\}$
$\le 2^{e+1}\lambda\{Y\colon \Theta^Y \text{ total} \ \& \ \Theta^Y \text{ not dominated by } f\} = 0.$ □

We will keep Θ fixed for the rest of this section.

Our main goal is to show that $C \ge_{LR} \emptyset'$ if and only if C is uniformly almost everywhere dominating. This is an analog for highness properties of the coincidences of lowness properties obtained in Section 5.1: $C \ge_{LR} \emptyset'$ is the dual of $C \le_{LR} \emptyset$, namely, $C \in$ Low(MLR). Unlike the case of Low(MLR), the property characterizing $C \ge_{LR} \emptyset'$ is not directly related to randomness. Rather, it is a domination property defined in terms of the uniform measure. It is instructive to begin with the special case $C = \emptyset'$, due to Kurtz (1981), which will be used in the next subsection to prove the full result.

5.6.28 Proposition. *$\emptyset'$ is uniformly almost everywhere dominating.*

Proof. The class $\{Y\colon \Theta^Y\!\upharpoonright_{n+1}\downarrow\}$ is Σ^0_1 uniformly in n. Hence $\emptyset'$ can determine the greatest q_n of the form $i2^{-n}$, $0 \le i < n$, such that $q_n < \lambda\{Y\colon \Theta^Y\!\upharpoonright_{n+1}\downarrow\}$. Also, $\emptyset'$ can determine $t_n \in \mathbb{N}$ such that $q_n \le \lambda[\{\sigma\colon |\sigma| = t_n \ \& \ \Theta^\sigma_{t_n}\!\upharpoonright_{n+1}\downarrow\}]^\prec$. Define a function $f \le_T \emptyset'$ as follows. On input n, using $\emptyset'$, determine t_n and output the maximum of values $\Theta^\sigma_{t_n}(n)$ over all such strings σ of length t_n.

For each n let $C_n = \{Y\colon \Theta^Y\!\upharpoonright_{n+1}\downarrow \ \& \ \Theta^Y(n) > f(n)\}$. Note that $\lambda C_n < 2^{-n}$ by the definition of q_n. Hence $2^{-k} \ge \lambda \bigcup_{n>k} C_n$ for each k. The intersection of the classes $\bigcup_{n>k} C_n$ over all k is $\{Y\colon \Theta^Y \text{ is total and not dominated by } f\}$, which is a null class. □

Dobrinen and Simpson (2004) pointed out that $\emptyset' \le_T C \Rightarrow C$ is u.a.e.d. $\Rightarrow$ C is *high*$_1$, and asked whether the class of u.a.e.d. sets coincides with one of these two extremes. The answer is negative. The first examples of u.a.e.d. sets $C \not\ge_T \emptyset'$ appeared in Cholak, Greenberg and Miller (2006). One of their constructions uses a method inspired by the forcing method from set theory, and builds a u.a.e.d. set C such that $C \not\ge_T E$ for a given incomputable set E. The other construction yields a c.e. set C. Here we first prove that C is u.a.e.d. iff $C \ge_{LR} \emptyset'$. Then, in

Section 6.3, we give examples of sets $C \geq_{LR} \emptyset'$ such that $C <_T \emptyset'$. Such a set C can be chosen to be either c.e., or ML-random.

Binns, Kjos-Hanssen, Lerman and Solomon (2006) showed that some high set is not almost everywhere dominating. In Theorem 8.4.15 we will prove a stronger result: $\emptyset' \leq_{LR} C$ implies that C is superhigh, namely, $\emptyset'' \leq_{tt} C'$. This property only depends on the Turing degree of C. The superhigh degrees form a proper subclass of the high degrees.

$\emptyset' \leq_{LR} C$ *if and only if* C *is uniformly a.e. dominating*

The first step towards proving this equivalence was made by Dobrinen and Simpson (2004). They expressed the property of being u.a.e.d. in terms of a covering procedure (see before Lemma 5.6.3, or Table 5.2 on page 230): each Σ^0_2 class, and in fact, each Π^0_3 class (by Exercise 5.6.32) can be covered by a $\Pi^0_2(C)$ class of the same measure. This states indeed that C is computationally strong since the $\Pi^0_2(C)$ class emulates the given Π^0_3 class. In comparison, (ii) of Theorem 5.5.17 says that A is weak, because each $\Pi^0_2(A)$ class can be emulated by a Π^0_2 class.

Since the domain of a Turing functional is a Π^0_2 class, the notation becomes simpler when one rephrases the covering procedure using complements.

5.6.29 Lemma. *C is uniformly a.e. dominating $\Leftrightarrow$ for each Π^0_2 class H there is a $\Sigma^0_2(C)$ class $L \subseteq H$ such that $\lambda L = \lambda H$.*

Proof. $\Rightarrow$: Suppose C is u.a.e.d. via the strictly increasing function $f \leq_T C$. Let $H = \bigcap_n S_n$ where the S_n are Σ^0_1 classes uniformly in n. Let Ψ be the Turing functional such that $\Psi^Z(n) \simeq \mu t.\, Z \in S_{n,t}$ for each Z, n. The required $\Sigma^0_2(C)$ class is

$$L = \{Z \colon \exists m\, [\forall n < m\, \Psi^Z(n)\downarrow \;\&\; \forall n \geq m\, \Psi^Z(n)[f(n)]\downarrow\,]\}.$$

Note that $Z \in H \;\leftrightarrow\; \Psi^Z$ is total, so $L \subseteq H$. Whenever Ψ^Z is total and f dominates Ψ^Z then $Z \in L$, so $\lambda L = \lambda H$. Note that (after fixing f) the class L was obtained uniformly from H.

$\Leftarrow$: Let Θ be the Turing functional of Lemma 5.6.27 and let H be the Π^0_2 class $\{Z \colon \Theta^Z \text{ total}\}$. Choose a $\Sigma^0_2(C)$ class $L \subseteq H$ such that $\lambda L = \lambda H$. Thus $L = \bigcup_n P_n$ where P_n is $\Pi^0_1(C)$ uniformly in n and $P_n \subseteq P_{n+1}$ for each n. Uniformly in n we have a C-computable sequence $(P_{n,t})_{t\in\mathbb{N}}$ of clopen sets such that $P_n = \bigcap_t P_{n,t}$, by (1.17) on page 55.

We define a function $f \leq_T C$ that dominates the total function Θ^Z for almost every Z. On input n, using C as an oracle find $t \in \mathbb{N}$ such that

$$\forall \sigma\, [(|\sigma| = t \;\&\; [\sigma] \subseteq P_{n,t}) \;\rightarrow\; \Theta^\sigma_t(n)\downarrow\,].$$

Note that t exists since $P_n \subseteq H$. Let $f(n)$ be the maximum of all the values $\Theta^\sigma_t(n)$ where $|\sigma| = t$ and $[\sigma] \subseteq P_{n,t}$. Almost every set $Z \in H$ is in P_k for some k. In that case $\Theta^Z(n) \leq f(n)$ for all $n \geq k$. □

The coincidence result is due to Kjos-Hanssen, Miller and Solomon (2011).

5.6.30 Theorem. $\emptyset' \leq_{LR} C \;\Leftrightarrow\;$ *C is uniformly a.e. dominating.*

Proof. $\Rightarrow$: Firstly, $\emptyset'$ is u.a.e.d. by 5.6.28, so by Lemma 5.6.29 every Π^0_2 class contains a $\Sigma^0_2(\emptyset')$ class of the same measure (also see Exercise 5.6.33).

Secondly, we apply the implication (i)$\Rightarrow$(iii) in Theorem 5.6.9 for $A = \emptyset'$ and $B = C$. Note that (i) holds by our hypothesis $\emptyset' \leq_{LR} C$, so taking complements in (iii) we obtain that each $\Sigma^0_2(\emptyset')$ class contains a $\Sigma^0_2(C)$ class of the same measure.

We may conclude that each Π^0_2 class contains a $\Sigma^0_2(C)$ class of the same measure. Hence C is uniformly a.e. dominating by a further application of Lemma 5.6.29.

$\Leftarrow$: Each $\Pi^0_1(\emptyset')$ class is a Π^0_2 class by Exercise 1.8.67, so by Lemma 5.6.29 each $\Pi^0_1(\emptyset')$ class contains a $\Sigma^0_2(C)$ class of the same measure. The implication (ii)$\Rightarrow$(i) of Theorem 5.6.9 now shows that $\emptyset' \leq_{LR} C$. □

In the following we consider a further property of a set C introduced by Dobrinen and Simpson (2004) that appears to be weaker than being uniformly a.e. dominating. We say that C is *almost everywhere dominating* if for almost every set Z, every function $g \leq_T Z$ is dominated by a function $f \leq_T C$. The difference is that now the dominating function $f \leq_T C$ can depend on the oracle Z and on g. However, by the following result of Kjos-Hanssen (2007), together with the foregoing Theorem, this property is equivalent to being u.a.e.d.

5.6.31 Proposition. *$\emptyset' \leq_{LR} C \Leftrightarrow C$ is almost everywhere dominating.*

Proof. $\Rightarrow$: Immediate by Theorem 5.6.30.

$\Leftarrow$: By Lemma 5.6.3 it suffices to show that each $\Pi^0_1(\emptyset')$ class H of positive measure contains a $\Sigma^0_2(C)$ class L of positive measure (and hence a $\Pi^0_1(C)$ class of positive measure). We proceed similar to the implication "$\Rightarrow$" of Lemma 5.6.29. Let $H = \bigcap_n S_n$, where S_n is uniformly Σ^0_1, and let the functional Ψ be defined as before. Since C is a.e. dominating, there is e such that $0 < \lambda\{Z \colon \Phi^C_e \text{ dominates} \Psi^Z\}$. Define a $\Sigma^0_2(C)$ class L as before, but based on $f = \Phi^C_e$. Then $L \subseteq H$ and $\lambda L > 0$. □

Exercises.

5.6.32. Show that if C is u.a.e.d., then in fact each Π^0_3 class G can be covered by a $\Pi^0_2(C)$ class of the same measure.

5.6.33. Give a direct proof (not via Proposition 5.6.28) for the fact used in the proof of Theorem 5.6.30 that each Π^0_2 class H contains a $\Sigma^0_2(\emptyset')$ class L of the same measure. Thus Z is weakly 3-random iff Z is weakly 2-random relative to $\emptyset'$.

6

SOME ADVANCED COMPUTABILITY THEORY

We outline the plan for the next three chapters. In the present chapter we mainly develop computability theory, having its interaction with randomness in mind. Chapter 7 is mostly about randomness. We discuss betting strategies, and introduce the powerful notion of martingales as their mathematical counterpart. Like Chapter 5, Chapter 8 is on the interactions of computability and randomness. It combines the technical tools of the foregoing two chapters in order to study lowness properties and highness properties of sets. For instance, in Theorem 8.3.9, lowness for Schnorr randomness is characterized by a computability theoretic property called computable traceability. The proof relies on computable martingales.

A main technical advance of this chapter is a method for building Turing functionals: they can be viewed as ternary c.e. relations of a particular kind. An appropriate language is introduced. As a first application of this method, we build superlow c.e. sets A_0 and A_1 such that $A_0 \oplus A_1$ is Turing complete.

Prompt simplicity is a highness property within the c.e. sets we introduced in 1.7.9. (However, this property is compatible with being low for ML-randomness by Theorem 5.1.19.) In Theorem 6.2.2 we use the language for building Turing functionals to characterize the degrees of promptly simple sets as the low cuppable degrees. The promptly simple degrees also coincide with the non-cappable degrees, as mentioned at the beginning of Section 1.7. Historically, these results of Ambos-Spies, Jockusch, Shore and Soare (1984) constituted the first example of a coincidence result for degree classes that had been studied separately before. In Chapter 5 we obtained such coincidence results for the K-trivial degrees and for the degrees of uniformly a.e. dominating sets.

C.e. operators are functions of the form $Y \to W_e^Y$, first mentioned in (1.4) on page 10. Here we use the new language in order to build such operators. We also present inversion results for c.e. operators V that are increasing in the sense of $\leq_T$. For instance, there is a c.e. set C, as well as a ML-random Δ^0_2 set C, such that $V^C \equiv_T \emptyset'$. If V is the c.e. operator corresponding to the construction of a c.e. K-trivial set, then $C \equiv_{LR} \emptyset'$, whence C is uniformly a.e. dominating by Theorem 5.6.30. Inversion results can be used to separate highness properties. For instance, we will build a superhigh set that is not uniformly a.e. dominating.

The language for building Turing functionals introduced here will also be used in the proof of Theorem 8.1.19 that a computably dominated set in GL_1 does not have d.n.c. degree, and when we show near the end of Section 8.4 that the strongly jump traceable c.e. sets form a proper subclass of the c.e. K-trivial sets.

6.1 Enumerating Turing functionals

By Remark 1.6.9, constructions using the priority method can be viewed as a game between us and an evil opponent called Otto. It is important that Otto and we have the same types of objects. For instance, he enumerates the c.e. sets W_e, and we also build c.e. sets. How about the Turing functionals Φ_e? In the proofs of results such as the Friedberg–Muchnik Theorem 1.6.8, the functionals belonged to him while we had to diagonalize against them. In subsequent constructions we often build a set Turing below another, so we have to provide a Turing functional. Usually we give an informal procedure that relies on an oracle Y. Then, by the oracle version of the Church–Turing thesis a Turing program P_e formalizing the procedure exists; this determines the required Turing functional Φ_e. We have already introduced functionals in such a way, for instance in the proof of the implication from right to left of Proposition 1.2.21, and in the proof of Theorem 1.7.2. In more complex constructions it is convenient to view a functional belonging to us as a particular kind of a c.e. relation, since its enumeration can be made part of the construction.

Basics and a first example

A computation $\Phi^Y(x) = w$ relies on the finite initial segment η of the oracle such that $|\eta| = \mathsf{use}\, \Phi^Y(x)$ (recall from 1.2.18 that the use is $1 + q$ where q is the maximum oracle question asked during the computation). We will view the convergence of such a computation as the enumeration of a triple $\langle \eta, x, w\rangle$ into a c.e. set, where η is the initial segment of the oracle, x is the input and w the output. We choose η shortest possible, namely, no substring of η yields a halting computation on input x.

6.1.1 Fact. *The Turing functionals Φ_e correspond in a canonical way to the c.e. sets $\Gamma \subseteq \{0,1\}^* \times \mathbb{N} \times \mathbb{N}$ such that*

(F1) $\langle \eta, x, v\rangle, \langle \eta, x, w\rangle \in \Gamma \ \rightarrow\ v = w$ (output uniqueness), *and*
(F2) $\langle \eta, x, v\rangle, \langle \rho, x, w\rangle \in \Gamma \ \rightarrow\ (\eta = \rho \ \vee\ \eta \mid \rho)$ (oracle incomparability).

Proof. Given Φ_e, enumerate Γ as follows. Run the computation of the Turing program P_e with η on the oracle tape on input x. If an oracle question q is asked such that $q \geq |\eta|$ then stop. Otherwise, if the computation halts and the largest oracle question has been $|\eta| - 1$, or no questions have been asked and $\eta = \varnothing$, then put $\langle \eta, x, w\rangle$ into Γ. Clearly Γ has the properties (F1) and (F2).

Conversely, given Γ we use the oracle version of the Church–Turing thesis to obtain a Turing program P_e and hence Φ_e. In the informal procedure, suppose the oracle is Y, and the input is x. Whenever $\langle \eta, x, w\rangle$ is enumerated into Γ, test whether $\eta \prec Y$. If so, output w and stop. □

Terminology used previously only for the functionals Φ_e can now be applied to any Turing functional Γ as in Fact 6.1.1. For instance, $\Gamma^Y(x) = w$ means that $\Phi_e^Y(x) = w$ for the corresponding Φ_e, which is equivalent to $\langle \eta, x, w\rangle \in \Gamma$ for some $\eta \prec Y$. In this case, we let $\mathsf{use}\, \Gamma^Y(x) \simeq \mathsf{use}\, \Phi_e^Y(x) = |\eta|$. We write $\gamma^Y(x)$

= use $\Gamma^Y(x)$. More generally, if an upper case Greek letter denotes a functional, the corresponding lower case Greek letter denotes its use function.

If a Δ^0_2 set A is given by a computable approximation $(A_s)_{s\in\mathbb{N}}$, we let $\Gamma^A(x)[s] = w$ if $\langle \eta, x, w\rangle \in \Gamma_s$ for some $\eta \prec A_s$.

Theorem 1.7.2 states that for each c.e. incomputable set C, there is a simple set $A \le_{wtt} C$. In the proof we implicitly built a Turing functional Γ to show that $A \le_{wtt} C$. Let us reformulate the construction in this new language to define Γ explicitly.

Construction of A and Γ. Let $A_0 = \emptyset$ and $\Gamma_0 = \emptyset$.

Stage $s > 0$.
1. For each $i < s$, if $A_{s-1} \cap W_{i,s-1} = \emptyset$, and there is $x \in W_{i,s}$ such that $x \ge 2i$ and $\Gamma^C(x)$ is currently undefined (i.e., there is no $\langle \eta, x, w\rangle \in \Gamma$ such that $\eta \prec C_s$), then enumerate the least such x into A_s.

2. For each $x < s$, if $\Gamma^C(x)$ is currently undefined, define it with use $\gamma^C(x) = x$ and output $A_s(x)$ (i.e., enumerate $\langle C_s\upharpoonright_x, x, A_s(x)\rangle$ into Γ).

Verification. Clearly A is co-infinite and (F1) holds. For (F2), if $\eta \neq \rho$ and we define $\Gamma^\eta(x)$ and $\Gamma^\rho(x)$, then $\eta \mid \rho$. (Here we have used that C is c.e.: if $C\upharpoonright_m$ changes, it can never turn back to the previous configuration.) Also $\Gamma^C = A$ because at each stage, we define $\Gamma^C(x)$ with use x if it is undefined, and once $C\upharpoonright_x$ has settled such a definition is permanent. To show that each requirement S_i is met, suppose that W_i is infinite. Then there are x and s such that $x \in W_{i,s}$ and $\Gamma^C(x)$ is undefined in phase 1 of stage s because $C_s\upharpoonright_x \neq C_{s-1}\upharpoonright_x$ by the same argument as before. Thus we put x into A, thereby meeting S_i. □

6.1.2 Exercise. Recall from Proposition 3.2.30 that Ω is *wtt*-complete. Given a c.e. set A, specify a Turing functional Γ such that $A = \Gamma(\Omega)$.

C.e. oracles, markers, and a further example

Sometimes we build a Turing functional when we are only interested in c.e. oracles C (equipped with a computable enumeration $(C_s)_{s\in\mathbb{N}}$). Then it suffices to specify the use $\gamma_s(x)$ and a value v at each stage $s > x$, because the oracle string is $C_s\upharpoonright_{\gamma_s(x)}$. It is understood that, if we specify $\gamma_s(x)$ and v, we put $\langle C_s\upharpoonright_{\gamma_s(x)}, x, v\rangle$ into Γ. For output uniqueness (F1), the value v has to be kept the same unless $C\upharpoonright_{\gamma_s(x)}$ changes. The use $\gamma_s(x)$ is pictured as a *movable marker* (Soare, 1987). Its value is always nondecreasing in both s and x. In order to achieve oracle incomparability (F2) we ensure that

$$\gamma_s(x) > \gamma_{s-1}(x) \rightarrow C_s\upharpoonright_{\gamma_{s-1}(x)} \neq C_{s-1}\upharpoonright_{\gamma_{s-1}(x)}, \tag{6.1}$$

because then $\eta = C_s\upharpoonright_{\gamma_s(x)}$ satisfies $\eta \mid \rho$ for any ρ such that $\langle \rho, x, w\rangle \in \Gamma_{s-1}$ (a c.e. set never turns back). To make Γ^C total, we also require that

$$\forall x\ \lim_s \gamma_s(x) < \infty, \tag{6.2}$$

so that for each x there is v such that $\langle \eta, x, v\rangle \in \Gamma$, where $\gamma(x) = \lim_s \gamma_s(x)$ and $\eta = C\upharpoonright_{\gamma(x)}$. Thus $\Gamma^C(x)\downarrow$.

It would be safest to keep $\gamma_s(x)$ from changing all-together, as in the rephrased proof of Theorem 1.7.2 above. We usually "correct" the output of $\Gamma^C(x)$ after putting $\gamma(x) - 1$ into C. In more complex constructions, this conflicts with constraints on C, so we have to choose $\gamma(x)$ larger than these constraints. Certain strategies declare $\gamma(x)$ to be undefined. It is redefined at some later stage to be a *large* number y, that is, $y - 1$ is larger than any number mentioned so far. (In particular, y is not yet in C.) Informally, this process is called *lifting the use*. To achieve (6.2) we must ensure that $\gamma(x)$ is lifted only finitely often.

6.1.3 Remark. In the next two applications of the method we build reductions to a set C of the form $E_0 \oplus E_1$. A proper formal treatment would require to view functionals as c.e. sets of quadruples $\langle \eta_0, \eta_1, x, v \rangle$, where η_i denotes an oracle string that is an initial segment of E_i. Similarly, in the treatment via markers we would have to keep the left use $\gamma_0(x)$ and the right use $\gamma_1(x)$ separate. We can usually avoid this by requiring that the use be the same on both sides (an exception is the proof of Theorem 6.3.1 below). This common use will also be denoted by $\gamma(x)$. A change of either of the two sets below $\gamma(x)$ allows us to redefine the computation $\Gamma^{E_0 \oplus E_1}(x)$.

The Sacks Splitting Theorem 1.6.10 implies that there are low c.e. sets A_0 and A_1 such that $\emptyset' \leq_T A_0 \oplus A_1$. Bickford and Mills (1982) strengthened this:

6.1.4 Theorem. *There are superlow c.e. sets A_0, A_1 such that $\emptyset' \leq_T A_0 \oplus A_1$.*

Proof. We enumerate the sets A_0 and A_1, and a Turing functional Γ such that $\emptyset' = \Gamma(A_0 \oplus A_1)$. The use of $\Gamma(A_0 \oplus A_1; p)$ is denoted by $\gamma(A_0 \oplus A_1; p)$ and pictured as a movable marker. For the duration of this proof, k and l denote numbers in $\{0, 1\}$, p and q denote numbers in $\mathbb{N}$, and $[p, k]$ stands for $2p + k$. Also, $j(X, n)$ stands for use $J^X(n)$. To avoid that $J^{A_k}(p)$ change too often, we ensure that at each stage s, for each p and k such that $[p, k] \leq s$, we have

$$J^{A_k}(p)[s]\downarrow \;\rightarrow\; \gamma(A_0 \oplus A_1)([p, k]) > j(A_k, p)[s] \tag{6.3}$$

To do so, when $J^{A_k}(p)[s]$ newly converges for the least $[p, k]$, we change A_{1-k}. (This idea to change the other side stems from the proof of the Sacks Splitting Theorem 1.6.10.)

Construction of A_0, A_1 and Γ. Let $A_0 = A_1 = \emptyset$. Define $\Gamma(A_0 \oplus A_1; 0) = 0$ with use 2.

Stage $s > 0$. Define $\Gamma(A_0 \oplus A_1; s) = 0$ with large use. Do the following.

(a) If there is $[p, k]$ such that $J^{A_k}(p)[s-1]\uparrow$ and $J^{A_k}(p)\downarrow$ at the beginning of stage s, then choose $[p, k]$ least. Put $\gamma(A_0 \oplus A_1; [p, k]) - 1$ into A_{1-k} and redefine $\Gamma(A_0 \oplus A_1; q)$, $s \geq q \geq [p, k]$, with value $\emptyset'_s(q)$ and large use.

(b) If $n \in \emptyset'_s - \emptyset'_{s-1}$ then put $\gamma(A_0 \oplus A_1; n) - 1$ into A_0.

A typical set-up is shown in Figure 6.1, where $J^{A_1}(p)$ converged after $J^{A_0}(p)$.

Claim 1. *The condition (6.3) holds for each s.*

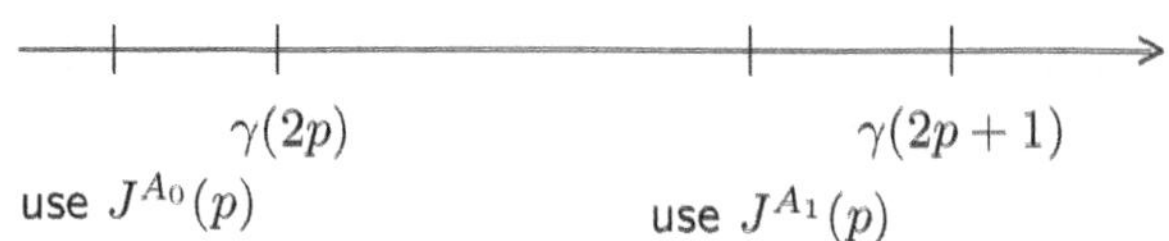

FIG. 6.1. Typical set-up in the proof of 6.1.4.

We use induction on s. The condition holds for $s = 0$. If $s > 0$, we may suppose there is a least $[p, k]$ such that $J^{A_k}(p)[s-1] \uparrow$ and $J^{A_k}(p)[s] \downarrow$ (else there is nothing to prove), in which case we put $y = \gamma(A_0 \oplus A_1; [p, k]) - 1$ into A_{1-k}.

- If $[q, l] < [p, k]$ then $y \geq \gamma(A_0 \oplus A_1; [q, l]) > j(A_l, q)[s]$ by the inductive hypothesis, so (6.3) remains true for $J^{A_l}(q)[s]$.
- Since we enumerate y into A_{1-k}, the computation $J^{A_k}(p)[s]$ remains convergent, so we ensure that $\gamma(A_0 \oplus A_1; [p, k]) > j(A_k, p)[s]$.
- For $[q, l] > [p, k]$, the use of computations $J^{A_l}(q)[s]$ is below the new value of $\gamma([q, l])$. ◇

Claim 2. *A_0 and A_1 are superlow.*
Similar to the construction of a superlow simple set in 1.6.5, if $J^{A_k}(p)[s] \downarrow$ let $f_k(p, s) = 1$, and otherwise let $f_k(e, s) = 0$. We define a computable function h such that $h([p, k])$ bounds the number of times $J^{A_k}(p)$ becomes defined. Then $b_k(p) = 2h([p, k]) + 1$ bounds the number of times $f_k(p, s)$ changes.

By (6.3), $J^{A_0}(p)$ becomes undefined at most $2p$ times due to a change of $\emptyset' \upharpoonright_{2p}$. Otherwise, $J^{A_k}(p)$ becomes undefined only when some computation $J^{A_{1-k}}(q)$ becomes defined where $[q, 1-k] < [p, k]$. Thus the function h given by $h(0) = 1$ and $h([p, k]) = 1 + 2p + \sum_q h([q, 1-k]) \, [\![[q, 1-k] < [p, k]]\!]$ is as desired. ◇

By the proof of Claim 2 each marker $\gamma(A_0 \oplus A_1; m)$ reaches a limit. Therefore $\emptyset' = \Gamma(A_0 \oplus A_1)$. □

6.1.5 Remark. In contrast to the case of the Sacks Splitting theorem, we cannot achieve that the use function of Γ is computably bounded for the oracle $A_0 \oplus A_1$: Bickford and Mills (1982) showed that each superlow c.e. set A is non-cuppable in the c.e. *wtt*-degrees, namely, if $\emptyset' \leq_{wtt} A \oplus W$ for a c.e. set W then already $\emptyset' \leq_{wtt} W$. Thus some low c.e. set is not superlow. One can also obtain such a set by a direct construction (see Exercise 1.6.7).

6.2 Promptly simple degrees and low cuppability

We say that a c.e. set A is *cuppable* if $\emptyset' \leq_T A \oplus Z$ for some c.e. set $Z \not\geq_T \emptyset'$. Informally speaking, A is helpful for Z to compute $\emptyset'$. This class contains incomputable sets, and even high sets, by a result in Miller (1981) with Harrington.

Variants of this concept are obtained by asking that the set $Z \not\geq_T \emptyset'$ be in some other class instead of the c.e. sets. For instance, A is *low cuppable* if $\emptyset' \leq_T A \oplus Z$

for some low c.e. set Z. Being non-cuppable in a particular sense is a lowness property within the c.e. sets. In Remark 8.5.16 we show that any c.e. set that is not cuppable by a ML-random set $Z \not\geq_T \emptyset'$ is K-trivial.

The main result of this section, due to Ambos-Spies *et al.* (1984), is that A has promptly simple Turing degree iff A is low cuppable. Thus the corresponding lowness property, to be not low cuppable, is equivalent to being cappable (see page 37). On the other hand, it is known that the non-cuppable c.e. sets form a proper subclass of the cappable sets.

C.e. sets of promptly simple degree

By Definition 1.7.9, a co-infinite c.e. set E is promptly simple if it has a computable enumeration $(E_s)_{s\in\mathbb{N}}$ such that for each e the requirement

$$PS_e\colon\ \#W_e = \infty \Rightarrow \exists s > 0\, \exists x\, [x \in W_{e,\mathrm{at}\ s} \cap E_s]$$

is met. We say that A has *promptly simple degree* if $A \equiv_T E$ for some promptly simple set E, and A has *promptly simple wtt-degree* if $A \equiv_{wtt} E$ for some promptly simple set E. We prove an auxiliary result of Ambos-Spies *et al.* (1984) characterizing the c.e. sets of promptly simple degree in terms of so-called delay functions.

6.2.1 Theorem. *Let A be c.e. Then the following are equivalent.*

(i) A has promptly simple degree.

(ii) A has promptly simple wtt-degree.

(iii) There is a computable delay function d *with $\forall s\, d(s) \geq s$ such that for each e*

$$PSD_e\colon\ \#W_e = \infty \ \Rightarrow\ \exists s \exists x\, \left[x \in W_{e,\mathrm{at}\ s}\ \&\ A_{d(s)}\restriction_x \neq A_s\restriction_x\right]. \qquad (6.4)$$

If $A_{d(s)}\restriction_x \neq A_s\restriction_x$ we say that A *promptly permits below* x.

Proof. (i)$\Rightarrow$(iii): we prove (iii) under the weaker hypothesis that there is a promptly simple set $E \leq_T A$. Suppose E is promptly simple via a computable enumeration $(E_s)_{s\in\mathbb{N}}$, and $E = \Phi^A$ for some Turing functional Φ. As in the proof of Theorem 1.7.14, we enumerate auxiliary sets G_e uniformly in e. By the Recursion Theorem we are given a computable function g such that $G_e = W_{g(e)}$ for each e (this is similar to the proof of 1.7.14). We show that PSD_e is met for each e, where the delay function is

$$d(s) = \mu t > s.\, \forall e < s\, \left[W_{g(e),t}\restriction_s = G_{e,t}\restriction_s\ \&\ \forall y < s\, E_t(y) = \Phi^A(y)[t]\right].$$

The set G_e is used to achieve the prompt A-permitting necessary to meet PSD_e.

Construction of sets G_e, $e \in \mathbb{N}$. Let $G_{e,0} = \emptyset$ for each e.

Stage $s > 0$. If there is $e < s$ such that PSD_e is not met, $W_{g(e),s-1} = G_{e,s-1}$, and there is $x \in W_{e,\mathrm{at}\ s}$ and $y \notin E_s \cup G_{e,s-1}$, $y < s$ such that use $\Phi^A(y)[s] \leq x$, then let e be least, let $\langle x, y\rangle$ be least for e, and put y into $G_{e,s}$. (We attempt to meet PSD_e.)

Verification. Consider PSD_e. If W_e is infinite then $G_e = W_{g(e)}$ is infinite as well. Since E is promptly simple via the given enumeration, there is a stage s where we attempt to meet PSD_e via x, y, and $y \in W_{g(e),t}$ for some $t \geq s$ such that $y \in E_t$. Since $E_s(y) = 0 = \Phi^A(y)[s]$ and use $\Phi^A(y)[s] \leq x$, by the definition of the function d this means that $A_{d(s)} \upharpoonright_x \neq A_s \upharpoonright_x$.

(iii)⇒(ii): We use the methods of Theorem 1.7.3 to build a promptly simple set $E \equiv_{wtt} A$. We meet the requirements PS_e above.

Construction of E. Let $E_0 = \emptyset$.

Stage $s > 0$.

- For each $e < s$, if PS_e is not yet met and there is $x \geq 3e$ such that $x \in W_{e,\text{at } s}$, then check whether $A_{d(s)} \upharpoonright_x \neq A_s \upharpoonright_x$. If so, put x into E_s.
- If $y \in A_{\text{at } s}$, put $p_{\overline{E}_{s-1}}(3y)$ into E.

Clearly E is promptly simple and $E \leq_{wtt} A$. Note that $\#E \cap [0, 3i) \leq 2i$ for each i, since at most i elements less than $3i$ enter due to the requirements PS_e and at most i for the coding of A into E. Hence $p_{\overline{E}}(y) \leq 3y$ for each y. Then $E_s \upharpoonright_{3y+1} = E \upharpoonright_{3y+1}$ implies $A_s(y) = A(y)$, so that $A \leq_{wtt} E$. □

A c.e. degree is promptly simple iff it is low cuppable

We now prove the main result of this section, which is due to Ambos-Spies, Jockusch, Shore and Soare (1984). It was applied already in Remark 5.6.24.

6.2.2 Theorem. *Let A be c.e. Then the following are equivalent.*

(i) A has promptly simple degree.
(ii) A is low cuppable, namely, there is a low c.e. set Z such that $\emptyset' \leq_T A \oplus Z$.

Proof. (i) ⇒ (ii):
Idea. We enumerate Z and build a Turing functional Γ such that $\emptyset' = \Gamma^{A \oplus Z}$. To ensure Z is low, we meet the usual requirements from the proof of 1.6.4

$$L_e : \exists^\infty s\, J^Z(e)[s] \downarrow \;\Rightarrow\; J^Z(e) \downarrow.$$

To aid in meeting L_e we enumerate a c.e. set G_e. As in Theorem 1.6.4, when $J^Z(e)$ converges we would like to preserve Z up to the use of this computation. However, now we also have to maintain the correctness of Γ. When a number enters $\emptyset'$ this may force us to change Z. We would like to achieve the situation where $\gamma^{A \oplus Z}(e) >$ use $J^Z(e)$ after $\emptyset' \upharpoonright_e$ has settled, for in that case we do not have to change Z below the use of $J^Z(e)$. For this we use the hypothesis that A has promptly simple degree via a delay function d as in Theorem 6.2.1. When we see a new computation $J^Z(e)$ at stage s, it will generally be the case that $x = \gamma(e) \leq$ use $J^Z(e)[s]$. In this case we put x into G_e hoping that A will promptly permit below x.

If A does so, we may compute the stage r by when it has permitted, and we only resume the construction then. The A-change allows us to lift the use $\gamma_r(e)$. Nothing happened from stage r to $s-1$. In particular, Z was not enumerated, so

$J^Z(e)[s-1]$ remains unchanged at stage r. Any computation $\Gamma^{A\oplus Z}(e)[t]$ defined at a stage $t \geq r$ has use $\gamma_t(e) > r \geq \mathsf{use}\ J^Z(e)[r]$. We say that we have *cleared* the computation $J^Z(e)$ of the use $\gamma(e)$.

If A does not permit promptly then instead we put x into Z at stage s, and lift $\gamma(e)$ via this Z-change. Then the next computation $\Gamma^{A\oplus Z}(e)$ will have a larger use. However, this case only applies finitely often: otherwise, G_e is infinite, so eventually we would get the prompt permission, contradiction. (Putting x into Z is often called the "capricious destruction of the computation $J^Z(e)[s]$". The actual purpose is that we threaten to make G_e infinite each time we give up on this computation.)

Details. As in the proof of Theorem 1.7.14, we are given a computable function g and think of $W_{g(e)}$ as G_e by the Recursion Theorem.

Construction of Z, Γ, and an auxiliary u.c.e. sequence $(G_e)_{e\in\mathbb{N}}$.
Let $Z_0 = \emptyset$ and $G_{e,0} = \emptyset$ for each e. Let $\Gamma = \emptyset$.

Stage $s > 0$.

(1) *Coding $\emptyset'$.* If $v < s$ is the last stage carried out and $i \in \emptyset'_s - \emptyset'_v$, put $\gamma_{s-1}(i)-1$ into Z_s, and declare $\gamma(k)$ undefined for $k \geq i$.

(2) *Defining Γ.* Let $i < s$ be least such that $\gamma(i)$ is now undefined. Define the $\gamma(j)$ for $i \leq j < s$ in a monotonic fashion, with large values (say, define $\gamma(j) = N+1+j$ where N is the largest number mentioned so far in the construction), and define $\Gamma^{A\oplus Z}(j)$ with use $\gamma(j)$ and the correct output $\emptyset'_s(j)$.

(3) *Attempt at clearing $J^Z(e)$.* Let $e < s$ be least such that

$G_{e,s} = W_{g(e),s}$,
$J^Z(e)[s]\downarrow$,
$\Gamma^{A\oplus Z}(e)[s-1]\downarrow$, and
$x = \gamma_{s-1}(e) \leq \mathsf{use}\ J^Z(e)[s]$.

Enumerate x into $G_{e,s}$, and search for the least $t \geq s$ such that $x \in W_{g(e),t}$. (Such a t exists if actually $G_e = W_{g(e)}$. Otherwise the construction might get stuck here.) Let $r = d(t)$ (where d is the delay function via which A has promptly simple degree).

Case 1. $A_r \upharpoonright_x \neq A_t \upharpoonright_x$. End stage s, declare $\gamma(e)$ undefined and jump to stage r.
Case 2. Otherwise. Put $x-1$ into Z_s, declare $\gamma(e)$ undefined and go to the next stage, $s+1$.

Claim. *Let $e \in \mathbb{N}$. Then (a) L_e is met, and (b) $\lim_s \gamma_s(e)\downarrow$.*
Assume inductively that the claim holds for all $i < e$. To show (a) for e, choose a stage s_0 such that $\emptyset' \upharpoonright_e$ is stable at s_0, and $\gamma_s(i)$ has reached its limit at s_0 for all $i < e$. If the hypothesis of L_e holds, then G_e is infinite, so at some stage $s \geq s_0$ of the construction we choose e in (3) and are in Case 1. Let r be as in the construction. Then $J^Z(e)[s]$ is preserved till r since the construction only resumes at stage r. Also, the new value $\gamma_r(e)$ is above the use of $J^Z(e)[r-1]$, so since $\emptyset' \upharpoonright_e$ is stable, $J^Z(e)[r-1]$ is stable. Thus L_e is met.

For (b), choose $s_1 \geq s_0$ such that either $J^Z(e)$ is stable from s_1 on, or $J^Z(e)[s]\uparrow$ for all $s \geq s_1$, and also $\emptyset'(e)$ is stable at s_1. Then $\gamma(e)$ reaches its limit by stage s_1. $\diamond$

By the remarks above, (b) implies that $\Gamma^{A \oplus Z}$ is total. Also $\emptyset' = \Gamma^{A \oplus Z}$ since we keep each $\Gamma^{A \oplus Z}(i)$ correct at each stage $s > i$.

(ii) $\Rightarrow$ (i):
Idea. We use the lowness of Z via the Robinson guessing method already applied in the proof of Theorem 5.3.22. Fix a Turing reduction for $\emptyset' \leq_T A \oplus Z$. We enumerate a c.e. set C and a Turing functional Δ. By the Double Recursion Theorem 1.2.16, we may assume that we are given in advance a Turing functional Γ such that $C = \Gamma(A \oplus Z)$, and a reduction function p for Δ in the sense of Fact 1.2.15. Keep in mind that Γ now belongs to Otto, while Δ is ours.

In more detail, given indices for a c.e. set $\widetilde{C}$ and a Turing functional $\widetilde{\Delta}$, we effectively obtain a many-one reduction showing $\widetilde{C} \leq_m \emptyset'$ and a reduction function p for $\widetilde{\Delta}$ because Proposition 1.2.2 and Fact 1.2.15 are uniform. The reduction for $\emptyset' \leq_T A \oplus Z$ is fixed in advance, so combining it with the many-one reduction we obtain a Turing functional Γ such that $\widetilde{C} = \Gamma(A \oplus Z)$. Based on Γ and p we run the construction to build C and Δ. By the Double Recursion Theorem we may assume that $C = \widetilde{C}$ and $\Delta = \widetilde{\Delta}$. Then $C = \Gamma(A \oplus Z)$ and p is a reduction function for Δ.

The so-called length-of-agreement function associated with Γ is

$$\ell_\Gamma(t) = \max\{y \colon C\upharpoonright_y = \Gamma^{A \oplus Z}\upharpoonright_y [t]\}.$$

We enumerate a number $w \in \mathbb{N}^{[e]}$ into C to enforce the prompt change of A needed in order to meet the requirement PSD_e. Suppose w is not in C yet and $\Gamma^{A \oplus Z}(w) = 0$. If we put w into C, then one of A or Z must change below the use $\gamma^{A \oplus Z}(w)$ in order to correct the Γ computation at input w. When $x \geq \gamma^{A \oplus Z}(w)[s]$ enters W_e at stage s, a reasonable strategy would be to enumerate w into C at the same stage, hoping that the correction of $\Gamma^{A \oplus Z}(w)$ will be via A. In this case we would have a prompt permitting by A via the delay function

$$d(s) = \min\{t > s \colon \ell_\Gamma(t) \geq s\}.$$

Namely, $d(s)$ is the first stage greater than s where the length of agreement is at least s again.

If the correction is always via a change of Z, we waste all the numbers w in $\mathbb{N}^{[e]}$ without ever getting the desired prompt permission. To avoid this, before enumerating w we ask whether Z is stable up to the use $u = \gamma^{A \oplus Z}(w)[s]$. This is possible since Z is low. We define a new value of the auxiliary functional $\Delta^Z(e)$ with the same use u, which allows us to control the jump J^Z at the input $p(e)$. Only if the answer is "yes" do we put w into C at s. If $Z\upharpoonright_u$ changes after all then the $\Delta^Z(e)$ computation goes away. We have to try again when the next (sufficiently large) number enters W_e, using the next $w \in \mathbb{N}^{[e]}$. We can afford finitely many incorrect answers. Eventually the answer will be correct, so we get the desired prompt permission by A.

Details. Since Z is low, by the Limit Lemma 1.4.2 there is a computable binary function g such that $Z'(m) = \lim_s g(m,s)$ for each m.

Construction of C and Δ, given Γ and the reduction function p. Let $C_0 = \emptyset$.

Stage $s > 0$. For each e, if the requirement PSD_e has not yet been met, let $w = \min(\mathbb{N}^{[e]} - C_s)$. If $\ell_\Gamma(s) > w$ and $\Delta^Z(e)\,[s]$ is undefined, then see if there is an $x \in W_{e,s} - W_{e,s-1}$ such that $x \geq u = \gamma^{A\oplus Z}(w)[s]$. If so, define $\Delta^Z(e) = 0$ with use u. Search for the least $t \geq s$ such that

$$Z_t \upharpoonright_u \neq Z_s \upharpoonright_u \text{ or } g(p(e),t) = 1.$$

(In the fixed point case, one of the two alternatives applies, since p is a reduction function for Δ.)

If $g(p(e),t) = 1$ do the following: enumerate w into C (at this stage s). If $A_{d(s)} \upharpoonright_u \neq A_s \upharpoonright_u$, then we got the prompt permission, so PSD_e is met. Otherwise $Z_{d(s)} \upharpoonright_u \neq Z_s \upharpoonright_u$, so we may declare $\Delta^Z(e)\,[d(s)]$ undefined. (Later we may try again with the next w in $\mathbb{N}^{[e]}$ and a new x.)

Verification. Assume for a contradiction that W_e is infinite but PSD_e is not met.

Case 1: $\Delta^Z(e)$ is defined permanently from some stage s on. This is so because at s we put the number w into C when $Z_s \upharpoonright_u$ was stable already, where $u = \gamma^{A\oplus Z}(w)[s]$. Then $A_{d(s)} \upharpoonright_x \neq A_s \upharpoonright_x$, so PSD_e is met, contradiction.

Case 2: $\Delta^Z(e)$ is undefined at infinitely many stages. Since $\Delta^Z(e) \simeq J^Z(p(e))$, this implies $p(e) \notin Z'$, so $\lim_s g(p(e),s) = 0$. Thus $\mathbb{N}^{[e]} \cap C$ is finite; let $s_0 = d(t)$ where t is the greatest stage such that $t = 0$ or an element of $\mathbb{N}^{[e]}$ enters C. Let $w = \min(\mathbb{N}^{[e]} - C)$ and suppose that $\Gamma^{A\oplus Z}(w)$ has settled by stage $s_1 \geq s_0$. Let $s \geq s_1$ be least such that some $x \geq u = \gamma^{A\oplus Z}(w)[s_1]$ enters W_e at stage s. Then $\Delta^Z(e)[s]$ is undefined (else PSD_e would have been met by stage $d(t)$). So we define $\Delta^Z(e)[s] = 0$ with use u. Since $Z \upharpoonright_u$ is stable from s_1 on, this contradicts the case hypothesis. □

In the proof of (i)⇒(ii) the number of injuries to L_e depends on the element of G_e that is promptly permitted by A. Thus the set Z we build is not superlow for any obvious reason. We say that A is *superlow cuppable* if there is a superlow c.e. set Z such that $\emptyset' \leq_T A \oplus Z$. Indeed, Diamondstone (2009) has proved that some promptly simple set fails to be superlow cuppable. Also see Exercise 8.5.24.

6.3 C.e. operators and highness properties

For each c.e. operator W we will build a c.e. set C, and also a ML-random set C, such that $W^C \oplus C \equiv_T \emptyset'$. If W is the c.e. operator given by the construction of a low simple set, then $C <_T \emptyset'$ and C is high. For a stronger result, we take the operator given by the cost function construction of a c.e. K-trivial set, (5.3.11) and obtain a set $C <_T \emptyset'$ such that $C \equiv_{LR} \emptyset'$.

The basics of c.e. operators

Definitions of c.e. sets can usually be viewed relative to an oracle C. For instance, we have done this for the halting problem $\emptyset'$: in Definition 1.2.9 we defined

$C' = \text{dom}(J^C) = \{e\colon e \in W_e^C\}$. Given a particular definition of a c.e. set A and an oracle C, we will write A^C for the set defined in the same way relative to C. In this way, the definition of A yields a c.e. operator. For a further example, let $A = \{q \in \mathbb{Q}_2\colon q < \Omega\}$. Then the corresponding c.e. operator is given by $A^C = \{q \in \mathbb{Q}_2\colon q < \Omega^C\}$.

Consider the case that the definition of A is a construction by stages. Such a definition can be viewed relative to an oracle C, but some extra care is needed, because the enumeration into A^C at a stage s now requires us to specify a use. By convention, at stage s, queries to the oracle C are less than s. So this use will always be defined to be s.

We provide some more detail on the construction in the proof of Theorem 1.6.4 relative to an oracle C. It now describes an enumeration of a co-infinite set A^C relative to C that is both simple relative to C and low relative to C, namely $(A^C \oplus C)' \equiv_T C'$. To ensure A^C is simple relative to C, we meet the requirements $S_i^C\colon \#W_i^C = \infty \Rightarrow W_i^C \cap A^C \neq \emptyset$. To make A^C low relative to C, we meet $L_e : \exists^\infty s\, J(A^C \oplus C; e)[s-1] \downarrow \Rightarrow J(A^C \oplus C; e) \downarrow$. If x enters W_e^C at stage s then the use of the newly convergent computation $\Phi_e^C(x)$ is at most s. We may now put x into A^C with use s in order to meet S_e^C. Thus, only for the relevant oracles C do we put x into A^C.

How do we build a c.e. operator W? Recall from (1.4) on page 10 that a c.e. operator is given by a Turing functional Φ_e, mapping each oracle C to the set $W^C = W_e^C = \text{dom}(\Phi_e^C)$. We describe such a Turing functional using the language introduced after Fact 6.1.1. Thus, when building a c.e. operator A, we implicitly enumerate a Turing functional Γ as in 6.1.1 and let $A^C = \text{dom}(\Gamma^C)$. Since the output is not relevant, we may always enumerate triples of the form $\langle \eta, x, 0\rangle$ into Γ. At stage s we have $|\eta| = s$. We think of η as a possible value of $C \upharpoonright_s$.

For instance, the functional Γ describing the jump operator is given by enumerating at stage s into Γ all triples $\langle \eta, e, 0\rangle$ such that $|\eta| = s$ and $J^\eta(e)[s] \downarrow$ while $J^\eta(e)[s-1] \uparrow$. The construction of a low simple set A relative to C can be seen as the implicit enumeration of a Turing functional as well. At stage $s > 0$ we consider all the strings η such that $|\eta| = s$ and carry out the construction above with $C \upharpoonright_s = \eta$. To put x into A^η means to put $\langle \eta, x, 0\rangle$ into Γ.

By Fact 6.1.1, to each c.e. set Γ with the properties (F1) and (F2) corresponds a Turing functional Φ_e. Hence $W_e^C = \text{dom}(\Phi_e^C)$ equals A^C, so any c.e. operator we build is included in the list $(W_e)_{e\in\mathbb{N}}$.

Given a c.e. operator W, usually we avoid naming the underlying functional $\Phi = \Phi_e$ explicitly, but if $i \in W^C$ we may need the use of the computation $\Phi^C(i)$. We let

$$\mathsf{use}\ (i \in W^C) \simeq \mathsf{use}\ \Phi^C(i).$$

When we say that *i enters W^C at stage s* we mean that $\Phi^C(i)$ newly converges at s. For a Δ_2^0 set C given by a computable approximation, we write

$$\mathsf{use}\ (i \in W^C[s]) \simeq \mathsf{use}\ \Phi^C(i)[s].$$

The intuition is as follows: a number i enters W^C at stage s with a certain use. From then on it stays in W^C as long as the approximation to C does not change below that use. The set W^C is Σ^0_2 via the approximation $(W^C[s])_{s\in\mathbb{N}}$ in the sense of Proposition 1.4.17, since $W^C(i) = \liminf_s W^C(i)[s]$.

Pseudojump inversion

The jump operator is increasing in the sense that $C' \geq_T C$ for each C. This fails for a c.e. operator W in general. For instance, let $W^C = \{q \in \mathbb{Q}_2 \colon q < \Omega^C\}$, then $W^C \not\geq_T C$ unless C is a K-trivial (see Corollary 5.5.16). However, the operator $V^C = W^C \oplus C$ is increasing. We call V a *pseudojump operator.* A pseudojump operator can differ a lot from the usual jump operator, even if it is strictly increasing. For instance, the construction of a low simple set determines a pseudojump operator V such that $V^C >_T C$ and $(V^C)' \equiv_T C'$ for each C.

Recall that the jump operator is degree invariant and even monotonic (see 1.2.14 and the discussion on page 34). Degree invariance fails for a pseudojump operator in general. For a trivial counterexample, we view the construction in the proof of Theorem 1.6.8 relative to an oracle. We now build c.e. operators A and B such that $A^C \oplus C \mid_T B^C \oplus C$ for each C. Let

$$W^C = \begin{cases} A^{C-\{0\}} & \text{if } 0 \in C, \\ B^{C-\{0\}} & \text{else.} \end{cases}$$

The operator given by $V^C = W^C \oplus C$ is not degree invariant since $V^{\{0\}} \mid_T V^{\{1\}}$. By Downey, Hirschfeldt, Miller and Nies (2005) the operator $C \to \Omega^C$ is not degree invariant.

The pseudojump inversion theorem, due to Jockusch and Shore (1984), states that no matter what W is, there exists a c.e. set C such that $W^C \oplus C \equiv_T \emptyset'$. This result can be used, for instance, to build a high c.e. set $C <_T \emptyset'$. For, a c.e. set C is high iff $\emptyset'$ is low relative to C. Let W be the c.e. operator given by the construction of a low simple set, then $C <_T \emptyset'$ and $\emptyset'$ is low relative to C.

6.3.1 Theorem. *For each c.e. operator W there is a c.e. set C such that $W^C \oplus C \equiv_T \emptyset'$.*

Proof idea. We enumerate a Turing functional Γ such that $\emptyset' = \Gamma(W^C \oplus C)$. We also ensure that W^C is Δ^0_2 (and not merely Σ^0_2, which is automatic), so that $W^C \oplus C \leq_T \emptyset'$ by the Limit Lemma 1.4.2. The requirements for $W^C \in \Delta^0_2$ are

$$Q_e : \exists^\infty s > 0 \left[e \in W^C(e)[s-1]\right] \Rightarrow e \in W^C.$$

If we meet all the Q_e then W^C is in Δ^0_2 via the computable approximation $\lambda e, s.\, W^C(e)[s]$. Note that the Q_e are similar to the lowness requirements L_e from the proof of Theorem 1.6.4, but with the c.e. operator W instead of the jump operator.

The construction parallels the one in the proof of (i)$\Rightarrow$(ii) of Theorem 6.2.2 that each c.e. set A of promptly simple degree can be cupped to $\emptyset'$ by a low c.e. set Z. In both cases we build a reduction Γ computing $\emptyset'$ with the appropriate oracle. Here we want to achieve $W^C \oplus C \equiv_T \emptyset'$, there we ensured $A \oplus Z \equiv_T \emptyset'$. The set C here corresponds to Z there. Both C and Z are built by us. The set W^C plays the role of the given set A there.

Recall the discussion in Remark 6.1.3. Here we let $\gamma_s(e) > e$ denote the use on the right side and view it as a movable marker. It matters that the actual use on the left side is $e+1$. However, to simplify notation we artificially increase this use to $\gamma(e)$.

To maintain $\emptyset' = \Gamma(W^C \oplus C)$, when i enters $\emptyset'$ we put $\gamma(i) - 1$ into C. The strategy for Q_e is *not* to restrain W^C (which is impossible), but rather to stabilize $W^C(e)$, by ensuring for all stages s and for each $e < s$ that

$$e \in W^C[s] \to \gamma_s(e) > \mathsf{use}\ (e \in W^C[s]). \tag{6.5}$$

For then, if $i \geq e$ enters $\emptyset'$, the Γ-correction via a C-change does not remove e from W^C. Thus e can leave W^C at most e times, till $\emptyset' \upharpoonright_e$ is stable. Hence Q_e is met. The potential problem is that oracle incomparability (F2) in Fact 6.1.1 might fail for Γ because $W^C(e)$ can change back and forth. But e can only leave W^C at a stage s when $C \upharpoonright_r$ changes, where $r = \mathsf{use}\ (e \in W^C[s])$. So by (6.5), without violating (F2) we can put a new computation into Γ when e reappears in W^C.

It may help our understanding to introduce further requirements

$$T_i : \emptyset'(i) = \Gamma(W^C \oplus C; i).$$

When i enters $\emptyset'$ then T_i puts $\gamma(i) - 1$ into C, but otherwise it wants to keep $\gamma(i)$ from changing. Q_e is allowed to move $\gamma(e)$, which injures T_i for $i \geq e$. If T_i enumerates into C and this removes e from W^C, then Q_e is injured for $e > i$. This is reflected by the priority ordering $Q_0 < T_0 < Q_1 < T_1 < \ldots$

Construction of C and Γ. Let $C_0 = \emptyset$ and $\Gamma_0 = \emptyset$.
Stage $s > 0$.

(1) *Coding $\emptyset'$.* If $i \in \emptyset'_s - \emptyset'_{s-1}$, put $\gamma_{s-1}(i) - 1$ into C_s and declare $\gamma(k)$ undefined for $k \geq i$.

(2) *Undefining Γ.* In the remaining instructions of stage s, we do not change C, so we already know $W^C \oplus C[s]$. If $e \in W^C[s] - W^C[s-1]$, then declare $\gamma(k)$ undefined for $k \geq e$.

(3) *Defining Γ.* Let $i < s$ be least such that $\gamma(i)$ is now undefined. Define $\gamma(j)$, $i \leq j < s$ in a monotonic fashion and with large values (in particular, greater than j), and define $\Gamma(W^C \oplus C; j)$ with use $2\gamma(j)$, and output $\emptyset'_s(j)$. (That is, put $\langle W^C \oplus C \upharpoonright_{2\gamma(j)}, j, \emptyset'_s(j)\rangle$ into Γ.)

Verification. The condition (6.5) holds by construction. Next, we prove that W^C is a Δ^0_2 set, and each $\gamma(e)$ reaches its limit.

6.3.2 Claim. *For each e, (a) Q_e is met, and (b) $\lim_s \gamma_s(e)\downarrow$.*

Assume inductively that the claim holds for each $i < e$. Choose a stage s_0 such that $e \in \emptyset'$ implies $e \in \emptyset'_{s_0}$, and for all $i < e$, $\gamma_s(i)$ has reached its limit at s_0.

(a) If e enters W^C at a stage $s_1 \geq s_0$ then $\gamma_s(e) > \mathsf{use}\,(e \in W^C[s])$ for $s \geq s_1$, so a later enumeration into C will not remove e from W^C.
(b) By (a) let $t \geq s_0$ be a stage by which $W^C(e)$ has reached its limit. Then $\gamma_s(e) = \gamma_t(e)$ for all $t \geq s$. ◇

Next we show that Γ is a Turing functional. The output uniqueness (F1) in Fact 6.1.1 follows from the coding in (1): suppose $\langle \eta, i, v\rangle \in \Gamma_t$, so that $|\eta| = 2\gamma_t(i)$. We only change the output when i enters $\emptyset'$ at stage s. In that case $\gamma_{s-1}(i) - 1$ entered C, so we do not put a further triple $\langle \eta, i, w\rangle$ into Γ.

Fix e. For each stage u, let $\eta_u = W^C \oplus C \restriction_{2\gamma(e)} [u]$. In order to establish oracle incomparability (F2) it suffices to show the following.

6.3.3 Claim. *Let $s < t$ be stages. Then $\eta_s = \eta_t$ or $\eta_s \mid \eta_t$.*

We may assume that $|\eta_s| \neq |\eta_t|$. Then $\gamma_s(e) < \gamma_t(e)$. If $C_s \restriction_{\gamma_s(e)} \neq C_t \restriction_{\gamma_s(e)}$ we are done, so assume otherwise. The following figure illustrates the argument.

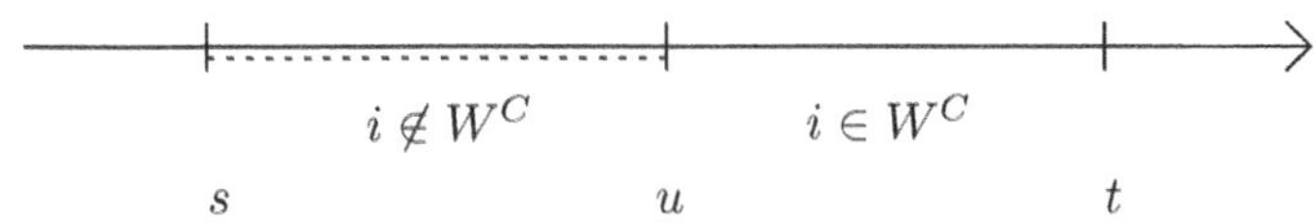

We declare $\gamma(e)$ undefined at some stage u, $s < u \leq t$. Thus some $i \leq e$ enters W^C at stage u. We may assume i is chosen minimal. If $i \in W^C[s]$ then $\gamma_s(i) > \mathsf{use}\,(i \in W^C[s])$, so i cannot leave W^C before stage u by the hypothesis that $C_s \restriction_{\gamma_s(e)} = C_t \restriction_{\gamma_s(e)}$. So $i \notin W^C[s]$. Next, $i \in W^C[t]$ because $\gamma(j)$, $j < i$, remains unchanged from s to t by the minimality of i, and $\gamma_u(i) > \mathsf{use}\,(i \in W^C[u])$, so an enumeration of $\gamma(k)$, $k \geq i$, at a stage $\geq u$ does not remove i from W^C. Thus $i \notin W^C[s]$ and $i \in W^C[t]$, whence $\eta_s \mid \eta_t$ since $\gamma(i) > i$ at each stage. ◇

At each stage u, we ensure that $\langle \eta_u, e, \emptyset'_u(e)\rangle \in \Gamma$ where $\eta_u = W^C \oplus C \restriction_{2\gamma(e)} [u]$. By Claim 6.3.2, $\gamma(e)$ and $W^C \restriction_{\gamma(e)}$ settle. Thus $\emptyset' = \Gamma(W^C \oplus C)$. □

Applications of pseudojump inversion

A set C is called high if $\emptyset'' \leq_T C'$ (1.5.18). For instance, $\Omega^{\emptyset'}$ is high by 3.4.17. So far we have not given any example of a Turing incomplete high Δ^0_2 set.

6.3.4 Corollary. *There is a high c.e. set $C <_T \emptyset'$.*

Proof. We obtain a c.e. set C by applying Theorem 6.3.1 to the c.e. operator W given by the construction of a low simple set in the proof of Theorem 1.6.4. Then $\emptyset'$ is low relative to C, that is, $(\emptyset' \oplus C)' \equiv_T C'$. Since C is c.e., this implies $\emptyset'' \equiv_T C'$. Also, W^C is simple relative to C. Hence $\mathbb{N} - W^C$ is not c.e. relative to C, and therefore $W^C \not\leq_T C$ by Proposition 1.2.8. Since $W^C \in \Delta^0_2$, this shows that $C <_T \emptyset'$. □

A direct proof of 6.3.4 uses the priority method with infinite injury (see Soare 1987). A requirement for $\emptyset' \not\leq_T C$ may be injured infinitely often by a stronger priority requirement for making C high. While the construction becomes more complex, this method is also more flexible than the one in the foregoing proof: for instance we can achieve $C \not\geq_T B$ for a given incomputable c.e. set B. A particular implementation of the method is by using trees of strategies. This even yields a minimal pair of high c.e. sets (see Exercise 7.5.12).

By Definitions 1.5.2 and 1.5.18, a set Z is called low_n if $Z^{(n)} \equiv_T \emptyset^{(n)}$, and Z is $high_n$ if $Z^{(n)} \geq_T \emptyset^{(n+1)}$. The downward closed classes low_n and non-$high_n$ form a hierarchy displayed in (1.8) on page 28. We will see that Theorem 6.3.1 can be used to separate all the classes of c.e. degrees given by this hierarchy: for each $n \geq 0$ there is a c.e. set in $low_{n+1} - low_n$, and a c.e. set in $high_{n+1} - high_n$.

For each e we let $V_e^X = W_e^X \oplus X$. Notice that Theorem 6.3.1 is uniform, namely, there is a computable function $\widetilde{f}$ such that, given a c.e. operator $W = W_e$, the c.e. set $C = W_{\widetilde{f}(e)}$ satisfies $W^C \oplus C \equiv_T \emptyset'$. The construction can be viewed relative to an oracle Z, so we may regard C as a c.e. operator. Here we adapt the notation slightly in the statement and proof by joining the oracle Z to the set C^Z we construct. In this way we obtain a computable function f such that, for each oracle Z,

$$V_e(V_{f(e)}(Z)) \equiv_T Z'. \tag{6.6}$$

The classes low_n and $high_n$ can be relativized to an oracle Z:

$$\begin{aligned} low_n^Z &= \{A\colon\ (A \oplus Z)^{(n)} \equiv_T Z^{(n)}\}, \text{ and} \\ high_n^Z &= \{A\colon\ (A \oplus Z)^{(n)} \geq_T Z^{(n+1)}\}. \end{aligned}$$

The proof of Corollary 6.3.4 shows that, if $V_e^Z \in low_1^Z - low_0^Z$ for each Z, then the set C obtained via Theorem 6.3.1 is high but not Turing complete. This is (i) of the following Lemma for $n = 0$.

6.3.5 Lemma. *Let $n \geq 0$.*
(i) $\forall X\, [V_e(X) \in low_{n+1}^X - low_n^X] \ \Rightarrow\ \forall Y\, [V_{f(e)}(Y) \in high_{n+1}^Y - high_n^Y]$.
(ii) $\forall Y\, [V_e(Y) \in high_{n+1}^Y - high_n^Y] \ \Rightarrow\ \forall X\, [V_{f(e)}(X) \in low_{n+2}^X - low_{n+1}^X]$.

Proof. (i) Given Y, let $X = V_{f(e)}(Y)$. By hypothesis $V_e(X)^{(n+1)} \equiv_T X^{(n+1)}$ while $V_e(X)^{(n)} >_T X^{(n)}$. Also $V_e(X) = V_e(V_{f(e)}(Y)) \equiv_T Y'$ by the definition of f. Thus $Y^{(n+2)} \equiv_T V_{f(e)}(Y)^{(n+1)}$ while $Y^{(n+1)} >_T V_{f(e)}(Y)^{(n)}$, as required.
(ii) is similar: given X, let $Y = V_{f(e)}(X)$. By hypothesis $V_e(Y)^{(n+1)} \equiv_T Y^{(n+2)}$ while $V_e(Y)^{(n)} <_T Y^{(n+1)}$. Also $V_e(Y) \equiv_T X'$. Thus $X^{(n+2)} \equiv_T V_e(Y)^{(n+1)} \equiv_T V_{f(e)}(X)^{(n+2)}$ while $X^{(n+1)} \equiv_T V_e(Y)^{(n)} <_T V_{f(e)}(X)^{(n+1)}$, as required. □

6.3.6 Theorem. *All the classes of c.e. sets in the hierarchy (1.8)*

$$\textit{computable} \subseteq low_1 \subseteq low_2 \subseteq \ldots \subseteq \text{non-}high_2 \subseteq \text{non-}high_1 \subseteq \{Z\colon\ Z \not\geq_T \emptyset'\}$$

are distinct. In fact, there is a uniformly c.e. sequence $(A_k)_{k \in \mathbb{N}^+}$ such that for each $i \in \mathbb{N}$ we have $A_{2i+1} \in low_{i+1} - low_i$ and $A_{2i+2} \in high_{i+1} - high_i$.

Proof. Let e be such that W_e is the c.e. operator given by the construction of a low simple set (see page 248). For $k \geq 1$ let $A_k = V_{f^{(k-1)}(e)}(\emptyset)$. (For instance,

$A_1 = V_{f^{(0)}(e)}(\emptyset) = W_e \oplus \emptyset$ is low but not computable, and $A_2 = V_{f(e)}(\emptyset)$ is high but not Turing complete.) By Lemma 6.3.5 the sequence $(A_k)_{k\in\mathbb{N}^+}$ is as required. □

The operators W of interest in Theorem 6.3.1 satisfy $W^X <_T X'$ for each X, so C is automatically incomputable. However, one can also meet additional requirements in the construction to ensure this actively. One can even build C of promptly simple degree. This extension will be needed for the next theorem.

6.3.7 Exercise. Show that one can choose the set C in Theorem 6.3.1 incomputable, and even of promptly simple degree.

6.3.8 Theorem. *There is a c.e. set that is not low$_n$ or high$_n$ for any n.*

Proof. Let f be as in (6.6) for the c.e. operator W_e given by the extended construction of Exercise 6.3.7. By the Recursion Theorem 1.1.5 relative to an oracle, there is e such that $V_e(X) = V_{f(e)}(X)$ for each X. Let $V = V_e$. Then

$$V(V(X)) \equiv_T X'$$

for each X. Moreover $X <_T V(X) <_T V(V(X))$ since we used the extended construction. (Such a c.e. operator V is called a half-jump.)

If $n \geq 1$ then $\emptyset^{(n)} \equiv_T V^{(2n)}(\emptyset) <_T V^{(2n+1)}(\emptyset) <_T V^{(2n+2)}(\emptyset) \equiv_T \emptyset^{(n+1)}$. Since $V^{(2n+1)}(\emptyset) \equiv_T (V(\emptyset))^{(n)}$, this shows that $\emptyset^{(n)} <_T (V(\emptyset))^{(n)} <_T \emptyset^{(n+1)}$, so $V(\emptyset)$ is a c.e. set as desired. □

No half-jump V is degree invariant (see page 34). The foregoing proof shows in fact that $X^{(n)} <_T (V(X))^{(n)} <_T X^{(n+1)}$ for each X. On the other hand, Downey and Shore (1997) have proved that if an increasing c.e. operator V is degree-invariant then V is relatively *low*$_2$ or relatively *high*$_2$ on an upper cone. That is, there is a set D such that either $V(X)'' \equiv_T X''$ for all $X \geq_T D$, or $V(X)'' \equiv_T X'''$ for all $X \geq_T D$.

Inversion of a c.e. operator via a ML-random set

The following new result is an analog of Theorem 6.3.1.

6.3.9 Theorem. *For each c.e. operator W and each Π^0_1 class P of positive measure there is a set $C \in P$ such that $W^C \oplus C \equiv_{wtt} \emptyset'$.*

6.3.10 Corollary. *For each c.e. operator W there is a ML-random set C such that $W^C \oplus C \equiv_{wtt} \emptyset'$.*

Proof. We apply the theorem to the Π^0_1 class $P = \{Z : \forall n\, K(Z\restriction_n) \geq n\}$. Note that $\lambda P \geq 1/2$ by 3.2.7, and each member of P is ML-random by 3.2.9. □

This inversion result has similar applications as Theorem 6.3.1: there is a high ML-random set $C <_T \emptyset'$ as in Corolllary 6.3.4, and in fact the hierarchy in Theorem 6.3.6 can be separated by ML-random Δ^0_2 sets. This was shown by Kučera (1989) in a direct way. Downey and Miller (2006) proved a stronger result already announced by Kučera: for each Σ^0_2 set $S \geq \emptyset'$ there is a ML-random Δ^0_2 set C such that $C' \equiv_T S$. This is analogous to the Jump Theorem of Sacks (1963*a*) discussed on page 160.

Proof of Theorem 6.3.9. By Theorem 1.9.4 we may suppose that $\lambda P \geq 1/2$. We build a set $C \in P$ in such a way that $W^C \oplus C \equiv_{wtt} \emptyset'$. For $W^C \leq_{wtt} \emptyset'$, we will provide an effective approximation to W^C. In the construction we have a strategy Q_e (without an explicit requirement) making a guess at whether $e \in W^C$. It defines a Π^0_1 class P^e such that either $e \in W^Y$ for no $Y \in P^e$, or for all. This resembles the strategy to guess the value of $J^Y(e)$ in the proof of the Low Basis Theorem 1.8.37. As in that proof, Q_e may change from the guess that $e \notin W^C$ to the guess that $e \in W^C$. Thereafter it only returns to the guess that $e \notin W^C$ when it is initialized, which may happen finitely often. At stage s, Q_e determines a clopen set $P^{e,s}$. The actual Π^0_1 class on which Q_e succeeds is $P^e = \bigcap_{s \geq s_1} P^{e,s}$, where s_1 is the first stage from which on Q_e is no more initialized and does not change its guess. At each stage s, $\tau_{e,s}$ is an approximation of $W^C \upharpoonright_{e+1}$ given by the guesses of the strategies Q_i for $i \leq e$.

As in the proof of Theorem 6.3.1, we build a functional Γ, and a strategy T_e is responsible for $\emptyset'(e) = \Gamma(W^C \oplus C; e)$. At stage s this strategy determines an approximation $\sigma = \sigma_{e,s}$ to $C \upharpoonright_{e+1}$, and defines $\Gamma(\tau_{e,s} \oplus \sigma_{e,s}; e) = \emptyset'_s(e)$. In the end we verify that $\sigma_e = \lim_s \sigma_{e,s}$ exists and show that $C = \bigcup_e \sigma_e$ is as required.

The framework for the construction is as follows: for each stage s, strategies Q_e, T_e for $e \leq s$ act at substages $e = 0, \ldots, s$. The strategy Q_e defines $P^{e,s}$ and $\tau_{e,s}$. The strategy T_e defines $\sigma_{e,s}$. Both $\tau_{e,s}$ and $\sigma_{e,s}$ have length $e+1$.

The strategies choose their objects in such a way that $P^{0,s} \supseteq \ldots \supseteq P^{s,s}$ and $\sigma_{0,s} \prec \ldots \prec \sigma_{s,s}$. More precisely, Q_e chooses $P^{e,s}$ inside $P^{e-1,s} \cap [\sigma_{e-1,s}]$, and T_e defines $\sigma_{e,s}$ as the leftmost string σ of length $e+1$ such that $[\sigma] \cap P^{e,s} \neq \emptyset$. The priority ordering is $Q_0 < T_0 < Q_1 < T_1 < \ldots$, since for $e > 0$, Q_e works in the environment $[\sigma_{e-1,s}]$ given by T_{e-1}, and for each e, T_e has to work in the environment $P^{e,s}$ prescribed by Q_e.

We also build a c.e. open set U containing the garbage. If e enters $\emptyset'$, then T_e puts $[\sigma_{e,s-1}]$ into U because there is an incorrect Γ-definition involving this string. Before we begin stage s, the garbage produced at previous stages has been removed, that is, we only consider strings σ such that $[\sigma] \cap P_s - U_{s-1} \neq \emptyset$. Then $\lambda U \leq 1/4$ since $|\sigma_{e,s}| = e+1$, and T_e contributes no string for $e \in \{0, 1\}$ and at most one string for $e \geq 2$. Thus $P_s - U_{s-1} \neq \emptyset$ for each s because we assume that $\lambda P \geq 1/2$.

An important idea in the construction is automatic Γ-recovery. While Q_e is in a phase guessing that $e \notin W^C$, as more and more strings σ of length $e+1$ leave P^e, the approximation $\sigma_{e,s}$ moves to the right. Suppose $\sigma = \sigma_{e,s}$ leaves P^e at stage $s' > s$. When e enters $\emptyset'$ after stage s', such a used string σ will not be removed via enumeration into U. But the now incorrect $\Gamma(\tau \oplus \sigma; e)$ definition is not a problem anyway: the strategy T_e can only return to σ at a stage $t > s$ if some Q_i, $i \leq e$, has changed its guess. In that case $\tau_{e,s} \neq \tau_{e,t}$, so that the definition of $\Gamma(e)$ made at stage s does not apply any longer. This is similar to the argument in Claim 6.3.3 above.

Construction of clopen sets $P^{e,s}$, strings $\tau_{e,s}$ and $\sigma_{e,s}$, a Turing functional Γ and a c.e. open set U. Let $U_{-1} = \emptyset$.

Stage $s \ge 0$. At substage -1, let $P^{-1,s} = P_s - U_{s-1}$ and let $\sigma_{-1,s} = \tau_{-1,s} = \varnothing$. Carry out substage e for $e = 0, 1, \ldots, s$.

Substage e.

Q_e: If $G \neq \emptyset$, where

$$G = P^{e-1,s} \cap [\sigma_{e-1,s}] \cap [\{\rho \colon |\rho| = s \ \& \ e \notin W_s^{\rho}\}]^{\prec},$$

then let $P^{e,s} = G$ and $\tau_{e,s} = \tau_{e-1,s}0$. Otherwise, let $P^{e,s} = P^{e-1,s} \cap [\sigma_{e-1,s}]$ and $\tau_{e,s} = \tau_{e-1,s}1$.

T_e: If $e \ge 2$ and $e \in \emptyset'_s - \emptyset'_{s-1}$ then put $[\sigma_{e,s-1}]$ into U.
Let $\sigma_{e,s}$ be the leftmost string $\sigma \in \{\sigma_{e-1,s}0, \sigma_{e-1,s}1\}$ such that $P^{e,s} \cap [\sigma] \neq \emptyset$. Enumerate $\langle \tau_{e,s} \oplus \sigma_{e,s}, e, \emptyset'_s(e) \rangle$ into Γ.

Verification. We have $\lambda U \le 1/4$ as explained above, and thus $\forall s\, P_s - U_{s-1} \neq \emptyset$.

6.3.11 Claim. *Let* $e \ge -1$. *(i)* $\tau_e = \lim_s \tau_{e,s}$ *exists. (ii)* $\sigma_e = \lim_s \sigma_{e,s}$ *exists.*

The claim holds trivially for $e = -1$. Now suppose that $e \ge 0$. Inductively, let s_0 be the stage by which $\tau_{e-1,s}$ and $\sigma_{e-1,s}$ have reached their limits. Let P^{e-1} be the Π^0_1 class $\bigcap_{s \ge s_0} P^{e-1,s}$. Then $P^{e-1} \cap [\sigma_{e-1}] \neq \emptyset$ by the definition of σ_{e-1} at stage t.
(i) If $P^{e-1} \cap [\sigma_{e-1}] \cap \{Y \colon e \notin W^Y\} \neq \emptyset$, then from $s_1 := s_0$ on we define $\tau_{e,s} = \tau_{e-1}0$. Otherwise, from some $s_1 \ge s_0$ on we define $\tau_{e,s} = \tau_{e-1}1$.
(ii) Let $P^e = \bigcap_{s \ge s_1} P^{e,s}$ where s_1 is as in (i) above. Then $P^e \cap [\sigma_{e-1}] \neq \emptyset$. Thus there is a least $a \in \{0,1\}$ such that $P^e \cap [\sigma_{e-1}a] \neq \emptyset$, and we eventually define $\sigma_{e,s}$ to be $\sigma_{e-1}a$. ◇

Let $C = \bigcup_e \sigma_e$, then $C \in P^e$ for each e. Thus $W^C = \bigcup_e \tau_e$.

6.3.12 Claim. Γ *is a Turing functional such that* $\emptyset' - \{0,1\} = \Gamma(W^C \oplus C)$.

We verify the conditions (F1) and (F2) in Fact 6.1.1.We only enumerate triples $\langle \eta, e, v \rangle$ into Γ such that $|\eta| = 2e+2$. Thus oracle incomparability (F2) is trivially satisfied. For output uniqueness (F1), we give a formal argument for the automatic Γ-recovery mentioned above. Assume for a contradiction that at stages $s < t$ we enumerate $\langle \eta, e, v \rangle$ and $\langle \eta, e, w \rangle$ into Γ, respectively, where $v \neq w$. Let $\eta = \tau \oplus \sigma$, so $\tau_{e,s} = \tau_{e,t} = \tau$ and $\sigma_{e,s} = \sigma_{e,t} = \sigma$. We have $e \in \emptyset'_r - \emptyset'_{r-1}$ for some r such that $s < r \le t$, causing the output w to be 1 at stage t, while at stage s the output is $v = 0$. This implies $\sigma_{e,r-1} \neq \sigma$, otherwise T_e would have enumerated $[\sigma]$ into U at stage r, contrary to $\sigma_{e,t} = \sigma$. So there is a least $j \le e$ such that for some u, $s < u \le t$, we have

(a) $\tau_{j,u-1} \neq \tau_{j,u}$, or
(b) $\tau_{j,u-1} = \tau_{j,u}$ and $\sigma_{j,u-1} \neq \sigma_{j,u}$.

If (a) holds then by the minimality of j we have $\tau_{j,s}(j) = 0$ and $\tau_{j,t}(j) = 1$. If (b) holds then $P^{j,p}$ is stable for $s < p \le t$, so by the choice of $\sigma_{j,p}$ as the leftmost extension of $\sigma_{j-1,p}$ on $P^{j,p}$ we have $\sigma_{j,s}(j) = 0$ and $\sigma_{j,t}(j) = 1$. Both possibilities contradict $\tau_{e,s} = \tau_{e,t} =$ and $\sigma_{e,s} = \sigma_{e,t}$. So Γ is a Turing functional. Now $\emptyset' - \{0,1\} = \Gamma(W^C \oplus C)$ is immediate by the enumeration into Γ of the strategies T_e. ◇

The use of $\Gamma(W^C \oplus C; e)$ is bounded by $2e + 2$. Thus $\emptyset' \leq_{wtt} W^C \oplus C$. The proof of Claim 6.3.11 allows us to compute recursively a bound on the number of times $\tau_{e,s}$ and $\sigma_{e,s}$ can change. Thus $W^C \oplus C \leq_{wtt} \emptyset'$. □

An alternative proof of Theorem 6.3.9 due to Kučera is sketched in Simpson (2007). It uses ideas from the proof of Theorem 3.3.2 in a construction relative to $\emptyset'$. The construction in the present proof is not relative to $\emptyset'$, and thereby provides directly a computable approximation to $W^C \oplus C$. Further, it yields weak truth-table equivalence $W^C \oplus C \equiv_{wtt} \emptyset'$. We do not know whether the conclusion in Theorem 6.3.1 can be improved to weak truth-table equivalence. The proof of 6.3.1 suggests a negative answer.

Separation of highness properties

The inversion theorems can be used to separate highness properties implied by *high*$_1$, both via c.e. sets and via ML-random Δ^0_2 sets. We will consider two properties between *high*$_1$ and $\{C\colon \emptyset' \leq_T C\}$. The property of being uniformly a.e. dominating was already introduced in Definition 5.6.26. In Theorem 5.6.30 we showed that C is uniformly a.e. dominating iff $C \geq_{LR} \emptyset'$. Mohrherr (1986) introduced a further property and separated it from highness within the c.e. sets.

6.3.13 Definition. We say that a set C is *superhigh* if $\emptyset'' \leq_{tt} C'$.

It is equivalent to require $\emptyset'' \leq_{wtt} C'$. For by Exercise 1.4.8 due to Mohrherr (1984), if E is a set such that $\emptyset' \leq_{tt} E$, then $X \leq_{wtt} E$ implies $X \leq_{tt} E$.

A further equivalent formulation is $\forall x\, [\emptyset''(x) = \lim_s f(x, s)]$ for some function $f \leq_T C$ such that the number of changes of $f(x, s)$ is bounded by a computable function $g(x)$.

In the framework of weak reducibilities, superhighness is the highness property dual to superlowness (see Table 8.3 on page 363). The basic idea is to apply pseudojump inversion to the c.e. operator W given by the construction of a c.e. incomputable set with a lowness property to obtain a set $C <_T \emptyset'$ that satisfies the dual highness property. In Theorem 6.3.14 we extend this method in order to separate highness properties.

The implications between these highness properties are the following: for each set C,

$$\emptyset' \leq_T C \;\Rightarrow\; \emptyset' \leq_{LR} C \;\Rightarrow\; C \text{ is superhigh} \;\Rightarrow\; C \text{ is high.} \tag{6.7}$$

Only the second implication is nontrivial; it will be proved in Theorem 8.4.17. In Proposition 8.5.11 below we show that *high*$_1$ is a null class.

6.3.14 Theorem. *The converse implications in (6.7) fail. Moreover, a counterexample C can be chosen to be either c.e., or ML-random and Δ^0_2.*

Proof. We apply the inversion results Theorem 6.3.1 and Corollary 6.3.10. Throughout, C denotes either a c.e. set, or a ML-random Δ^0_2 set.

1. To obtain a set $C \geq_{LR} \emptyset'$ such that $C <_T \emptyset'$, let W be the c.e. operator given by the construction of an incomputable K-trivial set in Proposition 5.3.11. By the inversion results, there is a set C such that $W^C \oplus C \equiv_T \emptyset'$ and $W^C \not\leq_T C$. Hence $C <_T \emptyset'$. Moreover, $\emptyset' \in \mathcal{K}(C)$, whence $\emptyset' \leq_{LR} C$.

2. To build a superhigh set C such that $\emptyset' \not\leq_{LR} C$, we apply one of the inversion results to the c.e. operator given by the construction in Proposition 5.1.20 of a c.e. superlow set that is not a base for ML-randomness. (The solution to Exercise 5.2.10 could be used as well.) If we carry out this construction relative to C we obtain a binary function $q \leq_T C$ such that $e \in (W^C \oplus C)' \leftrightarrow \lim_s q(e,s) = 1$ for each e.

By the remark at the end of the proof of 5.1.20, the number of changes in this approximation is bounded by a computable function (and not merely by a function computable in C). Therefore $(W^C \oplus C)' \leq_{tt} R \leq_m C'$ where R is the change set for q as in the proof of Proposition 1.4.4. If C is obtained via one of the inversion results then $\emptyset' \equiv_T W^C \oplus C$, so that $\emptyset'' \leq_{tt} C'$.

The set $\emptyset'$ is not low for ML-randomness relative to C (see page 232). Since C is Δ^0_2 this implies $\emptyset' \not\leq_{LR} C$.

3. To obtain a set C that is high but not superhigh, let W be the c.e. operator given by the construction of a low but not superlow set in the solution on page 385 to Exercise 1.6.7. If C is obtained via one of the inversion results, then on the one hand $\emptyset'$ is low relative to C, namely $(\emptyset' \oplus C)' \equiv_T C'$. On the other hand, $\emptyset'$ is not superlow relative to C, namely, $(\emptyset' \oplus C)' \not\leq_{tt(C)} C'$; here $\leq_{tt(C)}$ denotes the reducibility where C can be used as an oracle to compute the truth table. Since C is Δ^0_2 we have $\emptyset'' \equiv_m (\emptyset' \oplus C)'$, so this implies $\emptyset'' \not\leq_{tt} C'$. □

In Remark 5.6.24 we compared the operators $\mathcal{K}(X)$ and $\mathcal{S}_{LR}(X) = \{A\colon A \leq_{LR} X\}$. Let us now compare the classes $\{C\colon \emptyset' \in \mathcal{K}(C)\}$ and $\{C\colon \emptyset' \leq_{LR} C\}$. The former is not closed upward under $\leq_T$. The two classes coincide on the Δ^0_2 sets but can be separated by a Σ^0_2 set. From (iii) below we obtain a further proof that the operator $\mathcal{K}$ is not monotonic.

6.3.15 Proposition.

(i) If $\emptyset' \in \mathcal{K}(C)$ then $\emptyset' \leq_{LR} C$.

(ii) If C is Δ^0_2, then, conversely, $\emptyset' \leq_{LR} C$ implies $\emptyset' \in \mathcal{K}(C)$.

(iii) There is a c.e. set D and a Σ^0_2 set S, $S' \equiv_T \emptyset''$, $D \leq_T S$, such that $\emptyset' \in \mathcal{K}(D)$ but $\emptyset' \notin \mathcal{K}(S)$.

(iv) If S is as in (iii) then $\emptyset' \leq_{LR} S$ while $\emptyset' \notin \mathcal{K}(S)$.

Proof. (i) $\emptyset' \in \mathcal{K}(C) \leftrightarrow \emptyset' \oplus C \in \mathcal{K}(C) \leftrightarrow \emptyset' \oplus C \equiv_{LR} C \rightarrow \emptyset' \leq_{LR} C$.

(ii) If C is Δ^0_2 then $\emptyset' \leq_{LR} C$ implies that $\emptyset' \oplus C \equiv_{LR} C$.

(iii) We combine the argument in Remark 5.6.24 with pseudojump inversion. Let W be the c.e. operator given by the construction of a promptly simple K-trivial set in Proposition 5.3.11. Let D be a c.e. set such that $W^D \oplus D \equiv_T \emptyset'$, then $\emptyset' \in \mathcal{K}(D)$. Because $\emptyset'$ is of promptly simple degree relative to D, by

Theorem 6.2.2 relativized to D there is a set $S \geq_T D$ such that $\emptyset' \oplus S \equiv_T D' \equiv_T S'$. Since $D \geq_{LR} \emptyset'$ we have $S \geq_{LR} \emptyset'$. But also $\emptyset' \notin \mathcal{K}(S)$, for otherwise $\emptyset' \oplus S \in \mathcal{K}(S)$ and hence $(\emptyset' \oplus S)' \leq_T S'$, contrary to the fact that $\emptyset' \oplus S \equiv_T S'$.

Since S is c.e. in D, we have that S is Σ^0_2 and $S' \equiv_T \emptyset''$.

(iv) See the proof of (iii). □

6.3.16 Exercise. Find sets A, B such that $A \leq_{LR} B$, $A \leq_T B'$, and $A \notin \mathcal{K}(B)$.

6.3.17.$^\diamond$ **Problem.** Decide whether $\emptyset' \leq_{LR} C \Leftrightarrow \exists B \leq_T C\, [\emptyset' \in \mathcal{K}(B)]$.

Minimal pairs and highness properties ⋆

Recall minimal pairs for Turing reducibility from Definition 1.7.13. Given a highness property $\mathcal{H}$, it is interesting to determine whether there is a minimal pair of sets A, B satisfying $\mathcal{H}$: as we remarked already at the beginning of Section 1.5, this would show that $\mathcal{H}$ is not too close to being Turing above $\emptyset'$. In fact, for all the highness properties we study there is such a minimal pair.

Next, one can ask whether there is in $\mathcal{H}$ a minimal pair of sets of a particular kind, for instance of c.e. sets, or of ML-random sets. Lachlan (1966) proved that there is a minimal pair of c.e. high sets (see the comment after 6.3.4). Shore (unpublished) and Ng (2008*b*) independently improved this by showing that there is a minimal pair of c.e. superhigh sets.

6.3.18.$^\diamond$ **Problem.** Is there a minimal pair of c.e. sets that are u.a.e.d.?

A minimal pair of high ML-random sets A, B can be obtained as follows: the 2-random set $A = \Omega^\Omega$ is high by Proposition 3.4.17. Let $B = \Omega$, then $A \oplus B$ is ML-random by Theorem 3.4.6, and A, B form a minimal pair by Exercise 5.3.20. On the other hand, the superhigh sets are contained in a null Σ^0_3 class by 8.5.12 below. So by Theorem 5.3.15 there is a promptly simple set Turing below all the ML-random superhigh sets (see page 357 for details).

By Remark 4.3.13 there is a minimal pair of sets of PA degree. However, we cannot find such a pair among the PA sets of c.e. or of ML-random degree, because they compute $\emptyset'$ by Theorems 4.1.11 and 4.3.8 respectively.

A condition on a highness property $\mathcal{H}$ stronger than containing a minimal pair is that for *each* set F there is $B \in \mathcal{H}$ such that F and B form a minimal pair. Simpson observed that the class of u.a.e.d. sets satisfies this condition. For Cholak, Greenberg and Miller (2006) proved that for each incomputable set E, there is a u.a.e.d. set $B \not\geq_T E$. Their forcing construction can be extended slightly so that $B \not\geq_T E_i$ for each i, where $(E_i)_{i \in \mathbb{N}}$ is any sequence of incomputable sets. We may assume that F is incomputable. Let $(E_i)_{i \in \mathbb{N}}$ be a list of the incomputable sets computed by F. Then B and F form a minimal pair.

7

RANDOMNESS AND BETTING STRATEGIES

Schnorr (1971) criticized that Martin-Löf randomness is too strong a notion to be considered algorithmic. He suggested a weaker notion, nowadays called Schnorr randomness (Section 3.5). In this chapter we study randomness notions in between ML-randomness and Schnorr randomness.

The main notion is computable randomness. Recall that we identify subsets Z of $\mathbb{N}$ with infinite sequences of bits. For $n \geq 0$, we refer to $Z(n)$ as the bit of Z in position n. Our test concept is (the mathematical counterpart of) a computable betting strategy B that tries to gain capital along Z by predicting $Z(n)$ after having seen $Z(0), \ldots, Z(n-1)$. The set Z is computably random if each such betting strategy B fails in the sense that the capital $B(Z \upharpoonright_n)$ is bounded from above. Our test concept can be viewed as a special case of the martingale notion from probability theory (see Shiryayev 1984).

If it is allowed to leave $B(x)$ undefined for strings x that are not prefixes of Z, we obtain a notion called partial computable randomness.

If B may also bet on the bits in an order it chooses, we obtain the even stronger notion of Kolmogorov–Loveland (KL) randomness. Each ML-random set is KL-random. It is unknown at present whether the two notions coincide. In this case, ML-randomness could in fact be characterized by a computable test concept. (Most researchers think at present that KL-randomness is weaker than ML-randomness.)

The approach to randomness via unpredictability is useful because we have an intuitive understanding of betting strategies. On the other hand, this approach is still within the framework given in the introduction to Chapter 3. A betting strategy is a particular type of test, and the null class it describes is the class of sets on which it succeeds.

Schnorr did not even view computable randomness as the right answer to his critic, because a computable betting strategy B may succeed on a set very slowly, so slowly that $B(Z \upharpoonright_n)$ cannot be bounded from below by an order function for infinitely many n. If our tests are computable betting strategies together with such an order function and we require this lower bound on the capital for infinitely many n, we re-obtain Schnorr randomness defined in 3.5.8, as we will see in Section 7.3. This gives further evidence that Schnorr randomness is the right answer.

A questions we ask frequently in this book is how the strength of a randomness notion is reflected by the growth rate of the initial segment complexity. For instance, for a computably random set it can be as low as $O(\log n)$, but not for a partial computably random set. KL-random sets behave more like ML-random sets. Their growth rate is closer to the growth rate for ML-random sets, where a lower bound of $\lambda n.n - b$ for some b is given by Schnorr's Theorem.

We also show that the computational complexity is more varied for weaker randomness notions. For instance, there is a computably random set in each high degree.

We have seen in the previous chapters that Martin-Löf-randomness interacts well with computability. Not only is Martin-Löf randomness defined as an algorithmic notion, it also leads to objects such as Ω, classes such as Low(MLR), and methods such as the elegant solution to Post's problem outlined in Remark 4.2.4. There are fewer interactions of this kind for Schnorr randomness and the other randomness notions studied in this chapter. However, in Section 8.3 we will characterize the corresponding lowness notions and see that they determine classes of considerable interest in computability theory. For instance, a set is low for Schnorr randomness if and only if it is computably traceable, a property stronger than being computably dominated.

7.1 Martingales

We introduce martingales as a mathematical counterpart of the intuitive concept of a betting strategy. At first we do not make any assumptions on the effectivity. It takes some effort to develop the basics, but we will be rewarded with simple proofs of facts that have been used already, such as Proposition 5.1.14. The advantage of martingales over Martin-Löf tests and their variants is that martingales have useful built-in algebraic properties. For instance, we can define a new martingale as the sum of given ones.

Formalizing the concept of a betting strategy

Imagine a gambler in a casino is presented with bits of a set Z in ascending order. So far she has seen $x \prec Z$, and her current capital is $B(x) \geq 0$. She bets an amount α, $0 \leq \alpha \leq B(x)$, on her prediction that the next bit will be 0, say. Then the bit is revealed. If she was right, she wins α, else she loses α. Thus,

$$B(x0) = B(x) + \alpha \text{ and } B(x1) = B(x) - \alpha,$$

and hence $B(x0) + B(x1) = B(x) + \alpha + B(x) - \alpha = 2B(x)$. The same considerations apply if she bets that the next value will be 1: the betting is fair in that the expected capital after the next bet is equal to the current capital.

7.1.1 Definition. A *martingale* is a function $B\colon \{0,1\}^* \to \mathbb{R}^+ \cup \{0\}$ that satisfies for every $x \in \{0,1\}^*$ the equality

$$B(x0) + B(x1) = 2B(x). \tag{7.1}$$

For a martingale B and a set Z, let

$$B(Z) = \sup_n B(Z\upharpoonright_n).$$

We say that the martingale B *succeeds* on Z if the capital it reaches along Z is unbounded, that is, $B(Z) = \infty$. Let $\mathsf{Succ}(B) = \{Z\colon\ B \text{ succeeds on } Z\}$.

Martingales are equivalent to measures on $2^{\mathbb{N}}$ via the measure representations of Definition 1.9.2. We prefer martingales here because they formalize betting strategies, for which we have an intuitive understanding.

7.1.2 Fact. *Let $B : \{0,1\}^* \to \mathbb{R}^+ \cup \{0\}$ and let $r = \lambda x.2^{-|x|}B(x)$. Then*

$$B \text{ is a martingale } \Leftrightarrow r \text{ is a measure representation.}$$

Proof. Given $x \in \{0,1\}^*$, let $n = |x|$. Then

$$\begin{aligned}(7.1) \ &\Leftrightarrow\ 2^{-(n+1)}B(x0) + 2^{-(n+1)}B(x1) = 2^{-n}B(x)\\ &\Leftrightarrow\ r(x0) + r(x1) = r(x).\end{aligned}$$ □

7.1.3 Example. Recall that $r(x) = 2^{-|x|}$ is the representation of the uniform measure λ on $2^{\mathbb{N}}$. It corresponds to the martingale B with constant value 1. Note that B formalizes the strategy where the player bets the amount $\alpha = 0$ on any prediction.

More generally, for a measurable class $\mathcal{C} \subseteq 2^{\mathbb{N}}$, one defines the measure $\lambda_{\mathcal{C}}$ by $\lambda_{\mathcal{C}}(\mathcal{A}) = \lambda(\mathcal{C} \cap \mathcal{A})$ for measurable A. Its representation is $r(x) = \lambda(\mathcal{C} \cap [x])$. Then

$$B_{\mathcal{C}}(x) = 2^{|x|}r(x) = \lambda(\mathcal{C} \mid x) \tag{7.2}$$

is a martingale by Fact 7.1.2. $B_{\mathcal{C}}(x)$ is a conditional probability, namely the chance to get into $\mathcal{C}$ when starting from x. In particular, its initial capital is $\lambda\mathcal{C}$. If $[x] \subseteq \mathcal{C}$ then $B_{\mathcal{C}}(x) = 1$. If $\mathcal{C}$ is closed the converse implication holds as well, because the only conull closed class is $2^{\mathbb{N}}$.

7.1.4 Example. Given a string x, the elementary martingale E_x starts with capital 1. It bets its whole capital on the next bit of x till x is exhausted. Thus, E_x is the martingale given by

$$E_x(y) = \begin{cases} 2^{|y|} & \text{if } y \preceq x \\ 2^{|x|} & \text{if } x \prec y \\ 0 & \text{if } x \mid y. \end{cases}$$

E_x coincides with $2^{|x|}B_{\mathcal{C}}$ for $\mathcal{C} = [x]$.

Note that $B_{\mathcal{C}}$ is bounded by 1 and hence does not succeed on any set. It is used to build up more complex martingales via infinite sums, which may then succeed on some set Z. If $\lambda\mathcal{C} > 0$ then $B_{\mathcal{C}}(x)$ gets arbitrarily close to 1 for appropriate x by Theorem 1.9.4.

Supermartingales

We often work with a notion somewhat broader than martingales.

7.1.5 Definition. A *supermartingale* is a function $S\colon \{0,1\}^* \to \mathbb{R}^+ \cup \{0\}$ that satisfies for every $x \in \{0,1\}^*$ the inequality

$$S(x0) + S(x1) \leq 2S(x). \tag{7.3}$$

Thus, the average of $S(x0)$ and $S(x1)$ is at most $S(x)$. This implies that

$$S(x0) \leq S(x) \text{ or } S(x1) \leq S(x). \tag{7.4}$$

Extending Definition 7.1.1, for a set Z we let $S(Z) = \sup_n S(Z \upharpoonright_n)$. We say S *succeeds* on Z if $S(Z) = \infty$, and $\mathsf{Succ}(S)$ denotes the class of sets Z on which S succeeds.

A supermartingale can still be viewed as the mathematical counterpart of a betting strategy. If the supermartingale inequality (7.3) is proper, i.e.,

$$d(x) = S(x) - 1/2(S(x0) + S(x1)) > 0, \tag{7.5}$$

the gambler first donates the amount $d(x)$ to charity. Thereafter she bets with the rest $S(x) - d(x)$. Of course, she could as well keep $d(x)$ and never bet with it. Thus, each supermartingale is below a martingale with the same start capital:

7.1.6 Proposition. *For each supermartingale S there is a martingale B such that $B(\varnothing) = S(\varnothing)$ and $B(x) \geq S(x)$ for each x.*

Proof. Let $B(x) = S(x) + \sum_{z \prec x} d(z)$, where the function d is defined by (7.5). Clearly $\forall x\, B(x) \geq S(x)$ and $B(\varnothing) = S(\varnothing)$. We verify the martingale equality for B by induction on $|x|$:

$$\begin{aligned}
1/2(B(x0) + B(x1)) &= 1/2(S(x0) + S(x1)) + \sum_{z \preceq x} d(z) \\
&= 1/2(S(x0) + S(x1)) + d(x) + \sum_{z \prec x} d(z) \\
&= S(x) + \sum_{z \prec x} d(z) \\
&= B(x).
\end{aligned}$$

Some basics on supermartingales

In the following fact, parts (i) and (ii) are used to assemble complex (super)martingales from simple ones, and (iii) is a localization principle which allows us to transfer results on supermartingales we have proved in $\{0,1\}^*$ to a set of strings of the form $\{x\colon x \succeq v\}$.

7.1.7 Fact.

(i) If $\alpha \in \mathbb{R}^+$ and B and C are (super)martingales then so are αB and $B + C$.

(ii) If N_i is a (super)martingale for each $i \in \mathbb{N}$ and $\sum_i N_i(\varnothing) < \infty$ then $N = \sum_i N_i$ is a (super)martingale.

(iii) If B is a (super)martingale and $v \in \{0,1\}^*$ then $\lambda x.B(vx)$ is a (super)martingale.

Proof. (i) and (iii) are easily verified. For (ii), it suffices to show that $N(x) < \infty$ for each x. We use induction on $|x|$. For $x = \varnothing$ this is a hypothesis. Now $N_i(xa) \leq 2N_i(x)$ for each i and each $a \in \{0,1\}$. So $N(x) < \infty$ implies $N(xa) < \infty$. □

Measure representations can be used via Fact 7.1.2 to generalize the supermartingale inequality (7.3).

7.1.8 Lemma. *(i) Let S be a supermartingale and let $x_0, x_1 \ldots$ be a finite or infinite antichain of strings. Then $\sum_i 2^{-|x_i|} S(x_i) \leq S(\varnothing)$.*
(ii) If S is a martingale and $\bigcup_i [x_i] = 2^{\mathbb{N}}$ then equality holds.

Proof. (i) By 7.1.6 we may assume that S is a martingale. Let r be the corresponding measure representation: $r(x) = 2^{-|x|} S(x)$. Then $\sum_i r(x_i) \leq r(\varnothing)$ since the clopen sets $[x_i]$ are pairwise disjoint. This implies the required inequality.
(ii) is clear by the definition of measure representations. □

The lemma can be generalized using the localization principle 7.1.7(iii): for each string v, if $x_0, x_1 \ldots$ is a finite or infinite antichain of strings extending v then $\sum_i 2^{-|x_i|-|v|} S(x_i) \leq S(v)$.

A frequently used fact is that, if $b > S(\varnothing)$, then $S(\varnothing)/b$ bounds the measure of the class of sets along which S reaches at least b. For instance, if $S(\varnothing) = 1$ then S reaches the capital 4 along at most 1/4 of the sets.

7.1.9 Proposition. *Let S be a supermartingale, $b \in \mathbb{R}$, and $S(\varnothing) < b$. Then*

$$\boxed{\lambda\{Z \colon \exists n\, S(Z\upharpoonright_n) \geq b\} \leq S(\varnothing)/b.}$$

Proof. Let $(x_i)_{i<N}$ be the antichain of all minimal strings x under the prefix ordering such that $S(x) \geq b$ (here $N \in \mathbb{N} \cup \{\infty\}$). By 7.1.8(i), $\sum_{i<N} 2^{-|x_i|} S(x_i) \leq S(\varnothing)$. Then
$\lambda\{Z \colon \exists n\, S(Z\upharpoonright_n) \geq b\} = \sum_{i<N} 2^{-|x_i|} \leq \sum_{i<N} 2^{-|x_i|} S(x_i)/b \leq S(\varnothing)/b$. □

Exercises.

7.1.10. Show the following. (i) Generalized martingales, where the range can be all of $\mathbb{R}$, form a vector space over $\mathbb{R}$. (ii) If $x \neq y$ then E_x and E_y are linearly independent. (iii) Let B be a martingale, $B(\varnothing) = 1$, and let $n \in \mathbb{N}$. Suppose $B(z) = B(z\upharpoonright_n)$ for each z such that $|z| \geq n$. Show that B is a convex linear combination of the E_x for $|x| = n$.

7.1.11. Define a martingale S such that the $\leq$ relation in 7.1.9 becomes an equality for infinitely many $b \in \mathbb{N}$.

Sets on which a supermartingale fails

Our examples of computably random sets will often be sets on which a single sufficiently powerful supermartingale S fails. One way to obtain a set on which S fails is provided by the following immediate consequence of (7.4).

7.1.12 Fact. *For each supermartingale S there is a set Z, called the leftmost non-ascending path of S, such that $\forall n\, S(Z\upharpoonright_n) \geq S(Z\upharpoonright_{n+1})$.* □

7.1.13 Remark. We discuss a further way to obtain a set on which S fails. Proposition 7.1.9 yields a closed class of positive measure where S does not succeed: let $b > S(\varnothing)$ and

$$T_b^S = \{x \colon \forall y \preceq x\, [S(y) \leq b]\}.$$

Then $\lambda Paths(T_b^S) > 1 - S(\varnothing)/b > 0$ by 7.1.9. For instance, S does not succeed on the leftmost path of T_b^S. Note that $Paths(T_b^S)$ is nonempty even for $b = S(\varnothing)$, since it contains the leftmost non-ascending path of S.

Recall that $\mathbb{Q}_2$ is the set of rationals of the form $z2^{-n}$ where $z \in \mathbb{Z}$ and $n \in \mathbb{N}$. The *undergraph* of a supermartingale S is the set of pairs

$$\{\langle x, q\rangle : q \in \mathbb{Q}_2 \ \& \ q < S(x)\}.$$

When we apply computability theoretic notions to a supermartingale we actually mean its undergraph. For instance, we write $S \leq_T X$ to express that X computes the undergraph of S.

7.1.14 Exercise. (Savings Lemma) Show that for each supermartingale S there is a supermartingale $R \leq_T S$ such that $\lim_n R(Z\upharpoonright_n) = \infty$ for each $Z \in \mathsf{Succ}(S)$; in fact $R(y) \geq R(x) - 2$ for every pair of strings $y \succeq x$. If S is a martingale then so is R.

Characterizing null classes by martingales

By Proposition 1.9.9 a class $\mathcal{A} \subseteq 2^{\mathbb{N}}$ is null if and only if there is a sequence of open sets $(G_i)_{i\in\mathbb{N}}$ such that $\lambda G_i \leq 2^{-i}$ and $\mathcal{A} \subseteq \bigcap_i G_i$. Martingales provide a characterization of null classes as well, by a result known as Ville's Theorem. In fact, sequences of open sets can be converted into martingales, and conversely. Later, we will consider the case that $(G_i)_{i\in\mathbb{N}}$ is a ML-test, or even a Schnorr test, and study the effectiveness condition that ensues for the corresponding martingale.

Recall from Example 7.1.3 that any measurable class $\mathcal{C} \subseteq 2^{\mathbb{N}}$ determines a martingale $B_{\mathcal{C}}$ where $B_{\mathcal{C}}(x) = \lambda(\mathcal{C} \mid x)$.

7.1.15 Proposition. *Let $\mathcal{A} \subseteq 2^{\mathbb{N}}$. Then*

$$\begin{aligned} \mathcal{A} \text{ is null } &\Leftrightarrow \mathcal{A} \subseteq \mathsf{Succ}(B) \text{ for some martingale } B \\ &\Leftrightarrow \mathcal{A} \subseteq \mathsf{Succ}(S) \text{ for some supermartingale } S. \end{aligned}$$

In fact,

(i) If $\mathcal{A} \subseteq \bigcap_i G_i$, where $(G_i)_{i\in\mathbb{N}}$ is a sequence of open sets such that $\lambda G_i \leq 2^{-i}$, then $\mathcal{A} \subseteq \mathsf{Succ}(B)$ for $B = \sum_i B_{G_i}$.

(ii) If S is a supermartingale, $S(\varnothing) \leq 1$, and $\mathcal{A} \subseteq \mathsf{Succ}(S)$ then $\mathcal{A} \subseteq \bigcap_i G_i$, where $G_i = \{Z\colon \exists x \prec Z\ [S(x) > 2^i]\}$. Further, G_i is open and $\lambda G_i \leq 2^{-i}$ for each i.

Proof. (i) Note that $\sum_i B_{G_i}(\varnothing) \leq 2$, so B is a martingale by Fact 7.1.7. We have $B_{G_i}(\sigma) = 1$ whenever $[\sigma] \subseteq G_i$. Thus if $Z \in \bigcap_i G_i$ then $B(Z\upharpoonright_n)$ is unbounded.
(ii) Clearly G_i is open; $\lambda G_i \leq 2^{-i}$ follows from Proposition 7.1.9. □

7.2 C.e. supermartingales and ML-randomness

In the following we impose effectivity conditions on supermartingales. In this section we consider computably enumerable supermartingales. They yield a further characterization of ML-randomness besides Theorem 3.2.9, which helps us to understand the growth rate of the function $\lambda n.K(Z\upharpoonright_n) - n$ for a ML-random set Z. In the next section we will proceed to computable supermartingales.

Computably enumerable supermartingales

7.2.1 Definition. A supermartingale L is called computably enumerable if $L(x)$ is a left-c.e. real number uniformly in x. Equivalently, the undergraph $U = \{\langle x, q\rangle : q \in \mathbb{Q}_2 \;\&\; q < L(x)\}$ is computably enumerable. A c.e. index for U serves as an index for L. We write $L_s(x) > q$ if $\langle x, q\rangle \in U_s$.

We provide a version of Fact 7.1.7 for c.e. supermartingales.

7.2.2 Fact. *(i) If $\alpha > 0$ is left-c.e. and B and C are c.e. (super)martingales, then so are αB and $B + C$.*

(ii) If $(N_i)_{i\in\mathbb{N}}$ is a uniformly c.e. sequence of (super)martingales and $\sum_i N_i(\varnothing) < \infty$, then $N = \sum_i N_i$ is a c.e. (super)martingale.

Proof. (i) is easily verified. For (ii), N is a (super)martingale by 7.1.7. To see that N is c.e., note that for each $q \in \mathbb{Q}_2$ we have $N(x) > q \leftrightarrow$
$\exists n \exists s \, \exists q_0 \ldots q_{n-1} \in \mathbb{Q}_2 \left[\sum_{i<n} q_i > q \;\&\; \forall i < n \; N_{i,s}(x) > q_i\right]$. □

7.2.3 Definition. A *supermartingale approximation* is a uniformly computable sequence $(L_s)_{s\in\mathbb{N}}$ of $\mathbb{Q}_2$-valued supermartingales such that $L_{s+1}(x) \geq L_s(x)$ for each x, s. We say that $(L_s)_{s\in\mathbb{N}}$ is an *approximation of* L if $L(x) = \sup_s L_s(x)$ for each x. In this case, L is a c.e. supermartingale.

For instance, if R is c.e. open and $R = \bigcup_s [R_s]^{\prec}$ as in (1.16) then the martingale $L = B_R$ from Example 7.1.3 has a supermartingale approximation given by $L_s = B_{[R_s]^{\prec}}$. Thus B_R is computably enumerable.

7.2.4 Fact. *Each c.e. supermartingale L has a supermartingale approximation $(L_s)_{s\in\mathbb{N}}$. Moreover, it can be obtained uniformly.*

Proof. Let $U \subseteq \{0,1\}^* \times \mathbb{Q}_2$ be the undergraph of L. At stage s let the variables q, q' range over numbers of the form $i2^{-s}$. Let $\widetilde{L}_s(y)$ be the largest q such that $\langle y, q\rangle \in U_s$. Define $L_s(x)$ by induction on $|x|$. If $L_s(x)$ has been defined for all x such that $|x| < n$, let $L_s(x0)$ be the largest $q \leq \widetilde{L}_s(x0)$ such that $q \leq 2L_s(x)$. Then, let $L_s(x1)$ be the largest $q' \leq \widetilde{L}_s(x1)$ such that $q + q' \leq 2L_s(x)$. □

7.2.5 Exercise. Show that there is a uniformly c.e. listing $(S_e)_{e\in\mathbb{N}}$ of all the c.e. supermartingales S such that $S(\varnothing) \leq 1$.

Characterizing ML-randomness via c.e. supermartingales

We provide an effective version of Proposition 7.1.15.

7.2.6 Proposition. *The following are equivalent for a set Z.*

(i) Z is ML-random.

(ii) No c.e. martingale succeeds on Z.

(iii) No c.e. supermartingale succeeds on Z.

Proof. (ii) $\Rightarrow$ (i): We show that if $(G_i)_{i\in\mathbb{N}}$ is a ML-test then the martingale $B = \sum_i B_{G_i}$ from the proof of Proposition 7.1.15(i) is computably enumerable.

Note that $(B_{G_i})_{i\in\mathbb{N}}$ is a uniformly c.e. sequence of martingales. For each i we have $B_{G_i}(\emptyset) = \lambda G_i \le 2^{-i}$. Hence B is c.e. by Fact 7.2.2. If $Z \in \bigcap_i G_i$ then B succeeds on Z.
(i) $\Rightarrow$ (iii): Given is a c.e. supermartingale S, we may assume that $S(\emptyset) \le 1$. Then $(G_i)_{i\in\mathbb{N}}$ is a ML-test where $G_i = \{Z\colon \exists x \prec Z\, [S(x) > 2^i]\}$ as in 7.1.15(ii). If S succeeds on Z then Z fails this test. □

We are now in the position to prove Proposition 5.1.14 that if A is ML-random, then for each Turing functional Φ there is $c \in \mathbb{N}$ such that $(S^{\Phi}_{A,n+c})_{n\in\mathbb{N}}$ is a Martin-Löf test relative to A. Let

$$L(x) = 2^{|x|}\lambda[\{\sigma\colon \Phi^{\sigma} \succeq x\}]^{\prec}.$$

Then $L(x0) + L(x1) \le 2L(x)$ for each x. Further, $\{\langle q, x\rangle\colon q \in \mathbb{Q}_2 \ \&\ q < L(x)\}$ is c.e., that is, L is a c.e. supermartingale. Since A is ML-random, by Proposition 7.2.6 there is c such that $L(x) \le 2^c$ for each $x \prec A$. Letting $x = A\restriction_{n+c}$ we obtain $\lambda S^{\Phi}_{A,n+c} \le 2^{c-|x|} = 2^{-n}$. □

Universal c.e. supermartingales

In analogy to the concept of a universal ML-test, we say that a c.e. supermartingale S is *universal* if $\mathsf{non\text{-}MLR} = \mathsf{Succ}(S)$, that is, for each Z,

$$Z \text{ is ML-random} \Leftrightarrow S(Z) < \infty.$$

The implication "$\Rightarrow$" holds for any c.e. supermartingale S by Proposition 7.2.6. We provide some examples of universal supermartingales.

1. Let $(U_m)_{m\in\mathbb{N}}$ be a universal ML-test. Then $B = \sum_m B_{U_m}$ is a universal c.e. martingale: B is c.e. as in Proposition 7.2.6, and $\mathsf{non\text{-}MLR} = \bigcap_m U_m \subseteq \mathsf{Succ}(S)$ as in (i) of Proposition 7.1.15.
2. Consider the martingale

$$\mathsf{FS} = \sum_y 2^{-K(y)} E_y. \tag{7.6}$$

Note that $2^{-K(y)}E_y$ has the supermartingale approximation $2^{-K_s(y)}E_y$ uniformly in y. Then, by Fact 7.2.2(ii), FS is a c.e. martingale. We will verify its universality in Theorem 7.2.8.

7.2.7 Exercise. A c.e. supermartingale S is called *multiplicatively optimal* if for each c.e. supermartingale B there is $c \in \mathbb{N}$ such that $\forall x\, B(x) \le cS(x)$. Show:
(i) $\sum_e 2^{-e}S_e$ is a multiplicatively optimal supermartingale, where $(S_e)_{e\in\mathbb{N}}$ is the uniformly c.e. listing of all c.e. supermartingales S with $S(\emptyset) \le 1$ from Exercise 7.2.5.
(ii) Each multiplicatively optimal supermartingale is universal.
(iii) The martingale FS is not multiplicatively optimal. (In fact, by Levin 1973 no c.e. martingale is multiplicatively optimal. Also see Downey and Hirschfeldt 2010.)

The degree of nonrandomness in ML-random sets ⋆

A ML-random set Z can retain nonrandom features (see page 117). If S is a universal c.e. supermartingale, the real number $S(Z)$ indicates how much nonrandomness is still present in Z. We will discuss more general ways to assign a value $r(Z) \in [0, \infty]$ to a set Z in such a way that $r(Z)$ indicates the degree of nonrandomness of Z. In each case Z ML-random if and only if $r(Z) < \infty$, and the larger $r(Z)$, the less random is Z. Any universal c.e. supermartingale is a function of that kind. Actually, we have studied such functions earlier. Schnorr's Theorem 3.2.9 states that Z is ML-random iff $Z \notin \mathcal{R}_b$ for some b, and our intuition is that the larger b must be chosen, the less random is Z. To obtain a function r as above, let $\mathsf{CPS}(Z) = \sup_n 2^{n-K(Z\upharpoonright_n)}$. Then

$$\mathsf{CPS}(Z) < 2^b \leftrightarrow \forall n\, K(Z\upharpoonright_n) > n - b \leftrightarrow Z \notin \mathcal{R}_b.$$

In particular, Z is ML-random $\leftrightarrow$ $\mathsf{CPS}(Z) < \infty$.

The function CPS is somewhat crude for gauging nonrandomness. The following function due to Miller and Yu (2008) is finer. One takes the *sum* of all the $2^{n-K(Z\upharpoonright_n)}$ rather than their supremum. Thus, let

$$\mathsf{JM}(Z) = \sum_n 2^{n-K(Z\upharpoonright_n)}.$$

Let us compare JM with the universal c.e. martingale FS defined in (7.6). Fix Z and let $n \in \mathbb{N}$. If $y = Z\upharpoonright_n$ then $2^{n-K(Z\upharpoonright_n)} = 2^{-K(y)} E_y(y)$. Since $E_y(z) = 2^{|y|}$ for any $z \succeq y$, it follows that for each k,

$$\sum_{n \le k} 2^{n-K(Z\upharpoonright_n)} = \sum_{y \preceq Z\upharpoonright_k} 2^{-K(y)} E_y(Z\upharpoonright_k) \le \mathsf{FS}(Z\upharpoonright_k).$$

Hence $\mathsf{JM}(Z) \le \mathsf{FS}(Z)$ for each Z.

We summarize the preceding discussion. The implication (iv)⇒(i) shows that the c.e. martingale FS is universal.

7.2.8 Theorem. *The following are equivalent for a set Z.*

(i) Z is ML-random.
(ii) $\mathsf{CPS}(Z) < \infty$.
(iii) $\mathsf{JM}(Z) < \infty$.
(iv) $\mathsf{FS}(Z) < \infty$.

Proof. The implications (iv)⇒ (iii) ⇒ (ii) hold since $\mathsf{CPS}(Z) \le \mathsf{JM}(Z) \le \mathsf{FS}(Z)$. The proof of (ii)⇒(i) coincides with the proof of (ii)⇒(i) in Theorem 3.2.9. The implication (i)⇒(iv) follows from 7.2.6 since FS is a c.e. martingale. □

Of particular interest is the implication (i)⇒(iii), due to Miller and Yu (2008), that $\mathsf{JM}(Z) < \infty$ for each ML-random set Z. They call this result the "Ample Excess Lemma" because it says that $K(Z\upharpoonright_n)$ exceeds n considerably for large n.

In the following let $f, g\colon \mathbb{N} \to \mathbb{Z}$. To provide more detail, a function f such that $\sum_n 2^{-f(n)} < \infty$ cannot grow too slowly. (For monotonic functions, the borderline is somewhere between $\lambda n. \log n$ and $\lambda n. 2\log n$, as the sum diverges for the former, and converges for the latter.) For ML-random Z, the function $f_Z(n) = K(Z\upharpoonright_n) - n$ has a positive value for almost all n, and does not grow too slowly. Theorem 3.2.9 merely states that there is b such that $f_Z(n) \ge -b$ for all n. Proposition 3.2.21, that $\lim_n f_Z(n) = \infty$, is already a stronger condition on the growth rate of f_Z. The condition $\mathsf{JM}(Z) < \infty$ is even stronger and clearly implies Proposition 3.2.21. (If there is d such that $f_Z(n) = d$ for infinitely many n, then $\mathsf{JM}(Z) = \infty$ since each such d contributes 2^{-d} to the sum $\mathsf{JM}(Z)$.)

In this context, Miller and Yu (20xx) have proved that if $\sum_n 2^{-g(n)} < \infty$ then there is a ML-random set Z such that $\forall n\, f_Z(n) \leq^+ g(n)$. Thus, for a ML-random set Z in general, we cannot prove a stronger growth condition on f_Z than $\mathsf{JM}(Z) < \infty$.

7.2.9 Exercise. (Miller and Yu, 20xx) (i) Show that if $\sum_n 2^{-f(n)} < \infty$ then there is a function $g \in \Delta^0_2(f)$ such that $\lim_n(f(n) - g(n)) = \infty$ and $\sum_n 2^{-g(n)} < \infty$.
(ii) Use the result of Miller and Yu (20xx) mentioned above and (i) to show that for each ML-random set Y there is a ML-random set $Z <_K Y$ (see 5.6.1 for $\leq_K$).

7.3 Computable supermartingales

We will study a randomness notion where the tests are the computable betting strategies. This notion lies properly in between Martin-Löf randomness and Schnorr randomness defined in 3.5.8.

7.3.1 Definition. (i) A supermartingale L is called *computable* if $L(x)$ is a computable real number uniformly in x. Equivalently, the undergraph $\{\langle x, q\rangle : q \in \mathbb{Q}_2 \ \&\ q < L(x)\}$ is computable.
(ii) We say that Z is *computably random* if no computable martingale succeeds on Z. The class of computably random sets is denoted by CR.

For each computable supermartingale S there is a computable martingale M such that $\forall x\, M(x) \geq S(x)$ by the proof of Proposition 7.1.6. Thus Z is computably random iff no computable supermartingale succeeds on Z.

Each computable supermartingale is a c.e. supermartingale, so by 7.2.6 each ML-random set is computably random. In the next sections we will give various examples that the converse implication fails. This was first proved by Schnorr (1971, Satz 7.2). In the present section we characterize Schnorr randomness in terms of martingales and develop some basics on computable supermartingales.

Schnorr randomness and martingales

It is instructive to see a simple proof that any Schnorr test can be emulated by a computable martingale. Thereafter, we proceed to a stronger result: Z is not Schnorr random iff some computable martingale succeeds quickly on Z. This will be used in Theorem 7.5.10 to show that some Schnorr random set is not computably random.

7.3.2 Proposition. *Every computably random set is Schnorr random.*

Proof. It suffices to show that if $(G_i)_{i\in\mathbb{N}}$ is a Schnorr test, then the martingale $B = \sum_i B_{G_i}$ from the proof of Proposition 7.1.15 is computable, for B succeeds on any set in $\bigcap_i G_i$. By Fact 1.9.18 (which is uniform), $B_{G_i}(x) = 2^{|x|}\lambda(G_i \cap [x])$ is a computable real number uniformly in x and i. Moreover, $B_{G_i}(x) \leq 2^{-i+|x|}$, so $2^{-k} \geq \sum_{i>|x|+k} B_{G_i}(x)$ for each k. Given a string x and $k \in \mathbb{N}$, for each $i \leq |x| + k$ we may compute $q_{i,x} \in \mathbb{Q}_2$ such that $0 \leq B_{G_i}(x) - q_{i,x} < 2^{-k-i-1}$. Then $0 \leq B(x) - \sum_{i\leq|x|+k} q_{i,x} < 2^{-k+1}$. □

Below we will apply the fact that, if α_i is a computable real number uniformly in $i \in \mathbb{N}$ and $0 \le \alpha_i \le 2^{-i}$ for each i, then $\sum_i \alpha_i$ is computable (Exercise 1.8.18).

7.3.3 Theorem. *The following are equivalent for a set Z.*

(i) $Z \in \bigcap_i G_i$ for a Schnorr test $(G_i)_{i\in\mathbb{N}}$.

(ii) There is a computable martingale B and an order function h such that $\exists^\infty n\, B(Z\restriction_n) \ge h(n)$.

(iii) There is a computable martingale D and a strictly increasing computable function g such that $\exists^\infty k\, D(Z\restriction_{g(k)}) \ge k$.

Proof. (iii)⇒(ii): If D and g are as in (iii), let $h(n) = \max\{l\colon\ g(l) \le n\}$. Then $h(g(k)) = k$ for each k, so (ii) holds via D and h.

(ii)⇒(i): We may assume that $B(\varnothing) \le 1$. For each $i \in \mathbb{N}$ let S_i be the computable set of minimal strings x (under the prefix relation) such that $B(x) \ge h(|x|) \ge 2^i$, and let $G_i = [S_i]^\prec$. Then $Z \in \bigcap_i G_i$, and $\lambda G_i \le 2^{-i}$ by Proposition 7.1.9. It remains to show that λG_i is computable uniformly in i. Given r, we will compute a rational $q_r \le \lambda G_i$ such that $\lambda G_i - q_r \le 2^{-r}$. Compute the least p such that $h(p) \ge r$. Then $B(x) \ge r$ for each $x \in S_i$ of length at least p, so $\lambda[\{x \in S_i\colon\ |x| \ge p\}]^\prec \le 2^{-r}$ by Proposition 7.1.9. Thus $q_r = \lambda[\{x \in S_i\colon\ |x| < p\}]^\prec$ is a rational as required.

(i)⇒(iii): By Fact 1.8.26, uniformly in i there is a computable antichain $S_i \subseteq \{0,1\}^*$ such that $G_i = [S_i]^\prec$. We will apply the following lemma to $R = \bigcup_i S_i$. It introduces a version of interval Solovay tests (3.2.22) for Schnorr randomness.

7.3.4 Lemma. *Let R be a computable set of strings such that $\sum_{x\in R} 2^{-|x|}$ is finite and a computable real. If $\exists^\infty m\, Z\restriction_m \in R$ then (iii) holds.*

Subproof. There is an order function f such that, for each n,

$$\textstyle\sum_x 2^{-|x|}\, [\![x \in R\ \&\ |x| \ge n]\!] \le 2^{-2f(n)}.$$

Let $U_k = \{x \in R\colon\ f(|x|) \ge k\}$. We think of f as a function that grows slowly, so U_k is small. Then we can uniformly in k construct a computable martingale D_k that has a start capital of at most 2^{-k+1} and reaches at least $2^{f(|x|)}$ on every string $x \in U_k$. Now $D = \sum_k D_k$ is a computable martingale as required.

We provide the details. For $r \in \mathbb{N}$ let $\alpha_r = \sum_x 2^{f(|x|)-|x|}\, [\![x \in R\ \&\ f(|x|) = r]\!]$. Then $\alpha_r \in \mathbb{Q}_2$ and $\alpha_r = 2^r \sum_x 2^{-|x|}\, [\![x \in R\ \&\ f(|x|) = r]\!] \le 2^r 2^{-2r} = 2^{-r}$.

For each string x, let

$$p_x = 2^{f(|x|)-|x|}.$$

Note that $\sum_x p_x\, [\![x \in U_k]\!] = \sum_{r\ge k} \alpha_r$ is a computable real uniformly in k. More generally, the following is a computable real, uniformly in k and a string w:

$$F_k(w) = 2^{|w|} \textstyle\sum_x p_x\, [\![w \preceq x\ \&\ x \in U_k]\!].$$

Let $E_k(w) = \sum_v 2^{f(|v|)}\, [\![v \prec w\ \&\ v \in U_k]\!]$, and let

$$D_k(w) = E_k(w) + F_k(w).$$

Then $D_k(w)$ is a computable real uniformly in k. Informally, $E_k(w)$ is the amount D_k has achieved already, while $F_k(w)$ prepares for future achievements.

We check that the martingale property holds for D_k. First suppose that $w \in U_k$. Since $E_k(wa) = E_k(w) + 2^{|w|}p_w$ ($a \in \{0,1\}$) and $F_k(w0) + F_k(w1) = 2(F_k(w) - 2^{|w|}p_w)$, we have

$$(E_k(w0) + E_k(w1))/2 = E_k(w) + 2^{|w|}p_w,$$
$$(F_k(w0) + F_k(w1))/2 = F_k(w) - 2^{|w|}p_w,$$

so the martingale property holds. The case that $w \notin U_k$ is simpler since we do not need the term $2^{|w|}p_w$ in either equation.

Since $\sum_{r \geq k} \alpha_r \leq 2^{-k+1}$, we have $F_k(w) \leq 2^{-k+1}2^{|w|}$. Also $E_k(w) = 0$ for $k \geq |w|$, so $D = \sum_k D_k$ is a computable martingale.

To show (iii), let $g(k) = \min\{n\colon f(n) > k\}$. Given $k_0 \in \mathbb{N}$, let m be least such that $k = f(m) \geq k_0$ and $Z\upharpoonright_m \in R$. Then $Z\upharpoonright_m \in U_k$. If $x \in U_k$ then $D_k(z) \geq 2^k$ for each $z \succeq x$. As $g(k) > m$, we have in fact $D_k(Z\upharpoonright_{g(k)}) \geq D_k(Z\upharpoonright_m) \geq 2^k$. ◇

Note that the set $R = \bigcup_i S_i$ is computable since $x \in S_i$ implies $|x| \geq i$. Further, the real number $\sum_{x \in R} 2^{-|x|}$ is computable because $\lambda G_i \leq 2^{-i}$ and λG_i is computable uniformly in i. If $Z \in \bigcap_i G_i$ then for each i there is m such that $Z\upharpoonright_m \in S_i$, and $m \geq i$ since $\lambda G_i \leq 2^{-i}$. Thus $\exists^\infty m\, Z\upharpoonright_m \in R$. □

Exercises.

7.3.5. For a prefix-free machine M let FS_M be the martingale $\sum_y 2^{-K_M(y)}E_y$ (see page 266). Show that if M is a computable measure machine then FS_M is computable. Use this to give yet another proof of Proposition 7.3.2.

7.3.6. Use Lemma 7.3.4 to give another proof of Theorem 3.5.21 that each Schnorr random set satisfies the law of large numbers.

7.3.7.◊ Show that Z is Schnorr random $\Leftrightarrow$ $\liminf_n K_M(Z\upharpoonright_n) - n = \infty$ for each computable measure machine M. *Hint.* Adapt the proof of Proposition 3.2.21.

Preliminaries on computable martingales

If V is a computable martingale then by Fact 1.8.15(iv) one can approximate each value $V(x) \in \mathbb{R}$ by a rational up to an arbitrary precision. Nonetheless, it would be desirable to have complete information about these values. For this reason we will mostly work with computable (super)martingales B taking values in $\mathbb{Q}_2$. By the following fact of Schnorr (1971), this restriction does not affect the notion of computable randomness: if V is a computable martingale and $V(Z) = \infty$ then $B(Z) = \infty$ for a computable $\mathbb{Q}_2$-valued martingale B.

7.3.8 Proposition. *For each computable martingale V there is a $\mathbb{Q}_2$-valued computable martingale B such that $V(x) \leq B(x) \leq V(x) + 2$ for each x.*

The constant 2 is chosen for notational simplicity, and could be replaced by any rational $\epsilon > 0$.

Proof. Suppose that at a string y the martingale V bets $r(y) = V(y0) - V(y)$ on the prediction 0. We allow a negative value $r(y)$ here, which means that V

actually bets $-r(y)$ on 1. Since $V(x)$ is a computable real number uniformly in x, there is a computable function $f\colon \{0,1\}^* \to \mathbb{Q}_2$ such that $\mathrm{abs}(r(y) - f(y)) \le 2^{-|y|-2}$ for each y.

We assign B a start capital of $B(\varnothing) \in [V(\varnothing) + 1/2, V(\varnothing) + 1]$. This implies $0 \le V(\varnothing) + 1 - B(\varnothing) \le 1/2$. In order to define B on strings other than $\varnothing$, we replace r by f. Recursively we define

$$B(ya) = B(y) + (-1)^a f(y)$$

for each string y and each $a \in \{0,1\}$. Then the martingale equality holds. To show that $V(x) \le B(x) \le V(x) + 2$ for each x, note that, where y ranges over $\{0,1\}^*$, and a over $\{0,1\}$, we have $V(x) = V(\varnothing) + \sum_{y,a} (-1)^a r(y)\ [\![ya \preceq x]\!]$ and similarly $B(x) = B(\varnothing) + \sum_{y,a}(-1)^a f(y)\ [\![ya \preceq x]\!]$, so that

$$\begin{aligned}\mathrm{abs}(V(x) + 1 - B(x)) &= \mathrm{abs}(V(\varnothing) + 1 - B(\varnothing) \\ &\qquad + \sum_{y,a}(-1)^a (r(y) - f(y))\ [\![ya \preceq x]\!]) \\ &\le 1/2 + \sum_{j<|x|} 2^{-j-2} \le 1.\end{aligned}$$

Thus $V(x) \le B(x) \le V(x) + 2$ for each x. □

7.3.9 Proposition. *Let S be a $\mathbb{Q}_2$-valued supermartingale. Then for each string z and each $u > |z|$, one may compute relative to S the leftmost w of length u such that S does not increase along w, namely,*

$$\forall n \ \left[|z| \le n < u \to S(w\!\upharpoonright_n) \ge S(w\!\upharpoonright_{n+1})\right].$$

We say that w is the leftmost non-ascending string of length u above z given by S.

Proof. The following procedure computes w relative to S.

1. Let $w := z$.
2. FOR $n = |z|$ to $u - 1$ DO: IF $S(w0) \le S(w)$ let $w := w0$ ELSE $w := w1$.
3. Output w. □

7.3.10 Corollary. *For a $\mathbb{Q}_2$-valued supermartingale S the leftmost non-ascending path in the sense of 7.1.12 is computable in S.* □

7.3.11 Exercise. A computable supermartingale S does not succeed on some computable set. In particular, there is no universal computable supermartingale.

7.4 How to build a computably random set

We will follow two approaches for building a computably random set Z that is not ML-random. Both actually yield a stronger result by combining Schnorr's original proof with new ideas. Thus, beyond separating the two notions, we show that the computably random sets are more diverse than the ML-random sets.

(I) We build a computably random Z that has a slowly growing initial segment complexity, say $K(Z\upharpoonright_n) \leq^+ 3\log n$ for each n.

(II) Turing below each high set C we build a computably random set Z.

The set Z in approach (I) is not ML-random because each ML-random set Z satisfies $\forall n\, K(Z\upharpoonright_n) \geq^+ n$ by Theorem 3.2.9. In approach (II), if the high set C is c.e. and $C <_T \emptyset'$, then Z is not ML-random because it is not even of d.n.c. degree: otherwise C is of d.n.c. degree and hence Turing complete by the completeness criterion 4.1.11.

By Theorem 3.5.13, a computably random set that is not high is already ML-random. Theorem 7.5.9 is a result stronger than (II) that complements 3.5.13: each high degree contains a computably random set that is not ML-random.

Besides computable randomness we will consider partial computable randomness, which lies strictly in between ML-randomness and computable randomness. The tests are now $\mathbb{Q}_2$-valued partial computable martingales B. The fact that $B(x)$ may be undefined for some strings x gives the betting strategy B a considerable advantage. It can wait as long as it wishes before it decides on its bet at a position. As before, B succeeds on Z if $\sup_n B(Z\upharpoonright_n) = \infty$; in particular, B must be defined for all the initial segments of Z. In Section 7.5, as a preparation to Theorem 7.5.9, we give a direct construction of a computably random but not partial computably random set.

One can also separate all three randomness notions mentioned above by considering the initial segment complexity. This extends approach (I). It turns out that a partial computably random set Z cannot satisfy $\forall n\ K(Z\upharpoonright_n) \leq^+ c\log n$ for a constant c (Theorem 7.6.7), while a computably random set can do so. A slightly larger bound, such as $\lambda n.\log^2 n$, is consistent with being partial computably random, but still not with being ML-random.

We summarize the implications between the randomness notions:

$$\begin{aligned}\text{ML-random} &\Rightarrow \text{partial computably random}\\ &\Rightarrow \text{computably random}\\ &\Rightarrow \text{Schnorr random.}\end{aligned}$$

The converse implications fail. A set that is Schnorr random but not computably random is obtained in Theorem 7.5.10, a variant of Theorem 7.5.9.

Three preliminary theorems: outline

The results in this and the next section will be achieved in small steps. In each of three preliminary theorems, we prove that there is a computably random set Z with certain additional properties. Each theorem introduces a new main idea. Eventually all the ideas are combined to prove Theorem 7.5.9, due to Nies, Stephan and Terwijn (2005), that each high degree contains a computably random set that is not ML-random.

The proofs share a template. In the simplest case, a supermartingale L is introduced such that for each total B_k, for an appropriate $c > 0$ we have $\forall x\, [c \cdot$

$L(x) \geq B_k(x)]$. Then a set Z is built on which L does not succeed. Thus Z is computably random.

We describe the three preliminary theorems.

- In Theorem 7.4.8 we carry out approach (I) on page 271. We introduce the template and the idea of copying partial computable martingales.
- In Theorem 7.5.2 we carry out approach (II). The idea is to use the highness of C to build a supermartingale $L \leq_T C$ that dominates each B_k up to a multiplicative constant.
- In Theorem 7.5.7 we build a computably random set that is not partial computably random. We introduce the idea of encoding the definition of L into Z.

The template will be used to separate all the randomness notions that can be characterized via martingales, that is, all the notions between Schnorr randomness and ML-randomness.

Partial computable martingales

First we provide the formal definition of a concept already mentioned above.

7.4.1 Definition. A *partial computable martingale* is a partial computable function $B: \{0,1\}^* \to \{x \in \mathbb{Q}_2 : x \geq 0\}$ such that $\mathrm{dom}(B)$ is closed under prefixes, and for each $x \in \mathrm{dom}(B)$, $B(x0)$ is defined iff $B(x1)$ is defined, in which case the martingale equality $B(x0) + B(x1) = 2B(x)$ holds.

By this definition, partial computable martingales are always $\mathbb{Q}_2$-valued, while we allow computable supermartingales with values in $\mathbb{R}$. This fine point rarely matters because of Proposition 7.3.8.

For a function f, we write $f(x)\downarrow$ to denote that $f(x)$ is defined, and $f(x)\uparrow$ otherwise. Definition 7.1.1 can be adapted to partial computable martingales. Thus, we let $B(Z) = \sup\{B(Z\upharpoonright_n) : B(Z\upharpoonright_n)\downarrow\}$, and B succeeds on Z if $B(Z) = \infty$. This is only possible if B is defined along Z, that is, $B(Z\upharpoonright_n)\downarrow$ for each n.

7.4.2 Definition. Z is *partial computably random* if no partial computable martingale succeeds on Z. The class of partial computably random sets is denoted by PCR.

Each partial computable martingale B yields a c.e. supermartingale $\widehat{B}$ in the sense of Definition 7.2.1: let $\widehat{B}(x) = B(x)$ if the latter is defined, and otherwise $\widehat{B}(x) = 0$. Note that B and $\widehat{B}$ succeed on the same sets. Hence each ML-random set Z is partial computably random by Proposition 7.2.6.

For the template and its applications we will need some more notation. Via the effective identifications of $\{0,1\}^*$ and $\{q \in \mathbb{Q}_2 : q \geq 0\}$ with $\mathbb{N}$ we may view each Φ_k as a partial function $\{0,1\}^* \to \{q \in \mathbb{Q}_2 : q \geq 0\}$. Let B_k be the partial computable function that copies Φ_k as long as it looks like a martingale:

$B_k(\varnothing) \simeq \Phi_k(\varnothing)$, and

$B_k(x) = \Phi_k(x)$ if $x = ya$ for $a \in \{0,1\}$, $B_k(y)$ has been defined already, and $\Phi_k(y0) + \Phi_k(y1) = 2\Phi_k(y)$ (all defined).

Then $(B_k)_{k\in\mathbb{N}}$ is an effective listing of all the partial computable martingales. Let

$$\mathsf{TMG} = \{k \colon B_k \text{ is total}\}.$$

We frequently need a partial computable martingale $B_{k,n}$, $n \in \mathbb{N}$, which copies B_k, but only for strings x such that $|x| \geq n$, and with some scaling. For shorter strings it returns the value 1.

7.4.3 Fact. *There is an effective listing $(B_{k,n})_{k,n\in\mathbb{N}}$ of partial computable martingales such that*

(i) $B_{k,n}(x) = 1$ *for* $|x| < n$.
(ii) For each k, n there is a constant $c > 0$ such that, for each x, if $B_k(x)\downarrow$ then $B_{k,n}(x)\downarrow$ and $B_k(x) \leq c \cdot B_{k,n}(x)$.

Proof. For $|x| \geq n$, if $B_k(x\upharpoonright_n) = 0$ let $B_{k,n}(x) = 0$. Otherwise let $B_{k,n}(x) \simeq B_k(x)/B_k(x\upharpoonright_n)$. Then (ii) holds via $c = \max\{B_k(y) \colon |y| = n\}$. □

7.4.4 Definition. For each n let G_n be the supermartingale such that $G_n(x) = 1$ for $|x| < n$ and $G_n(x) = 0$ else.

Exercises. Show the following.

7.4.5. PCR is closed under finite variants.

7.4.6. For each partial computable martingale B there is a computable set E such that either $\exists n\, B(E\upharpoonright_n)\uparrow$ or $\forall n\, B(E\upharpoonright_{n+1}) \leq B(E\upharpoonright_n)$.

A template for building a computably random set

We apply the template already mentioned above in all the subsequent constructions of a set Z that is computably random, or satisfies some variant of this property. First a definition.

7.4.7 Definition. Let L and B be supermartingales. We say that L *multiplicatively dominates* B if there is $c \in \mathbb{N}$ such that $B(x) \leq c \cdot L(x)$ for each x.

In the template, for each k one introduces a supermartingale V_k that copies B_k with certain restrictions. Then one lets $L = \sum_k 2^{-k} V_k$, and Z is some set on which L does not succeed. For instance, Z could be the leftmost non-ascending path of L. Enough copying of B_k to V_k is carried out so that L multiplicatively dominates each total B_k. Hence Z is computably random.

To ensure that L is $\mathbb{Q}_2$-valued and for other purposes, one defines a sequence $0 = n_0 < n_1 < \ldots$ and lets V_k copy not B_k but rather $B^*_k := B_{k,n_{k+1}}$ defined in Fact 7.4.3 (B_k is only copied for strings of length $\geq n_{k+1}$). Then, if $|x| \in I_k = [n_k, n_{k+1})$, all the V_j for $j \geq k$ together contribute 2^{-k+1} to $L(x)$. Thus $L(x)$ is in $\mathbb{Q}_2$, and only depends on the values $V_j(x)$ for $j < k$.

To summarize, the template consists of the following three steps.

1. Define the sequence $0 = n_0 < n_1 < \ldots$, which determines the supermartingales $B_k^* = B_{k,n_{k+1}}$.
2. Introduce the supermartingales V_k that copy the B_k^* with certain constraints. In this way obtain the supermartingale $L = \sum_k 2^{-k} V_k$.
3. Define Z in such a way that $L(Z) < \infty$.

Computably random sets and initial segment complexity

We carry out approach (I) on page 272, building a computably random set Z of slowly growing initial segment complexity. The construction works best for the conditional initial segment complexity $K(Z\upharpoonright_n \mid n)$. Note that by Section 2.3

$$K(Z\upharpoonright_n) \le^+ K(Z\upharpoonright_n \mid n) + K(n) \le^+ K(Z\upharpoonright_n \mid n) + 2\log n,$$

so $K(Z\upharpoonright_n)$ also grows quite slowly. However, we need to distinguish the two types of initial segment complexity since we will also consider sublogarithmic upper bounds for $K(Z\upharpoonright_n \mid n)$.

Our first application of the template is to build a computably random Z such that $K(Z\upharpoonright_n \mid n) \le^+ h(n)$ for each order function h. We begin with the simpler case where h is fixed. V_k copies B_k^* in the case that B_k is total. If not, $V_k(x) = 1$ for $|x| < n_{k+1}$, and $V_k(x) = 0$ otherwise. The initial segment complexity grows slowly if we choose the n_k sufficiently far apart. If $|x| < n_{k+1}$ then $L(x)$ only depends on $B_i(x)$ for $i < k$; the set Z is the leftmost non-ascending path of L, so to determine $Z\upharpoonright_{n_{k+1}}$ we only need the information which ones among these B_i are total. This amount of information is small compared to $|x|$.

7.4.8 Theorem. *For each order function h there is a computably random set Z such that $\forall^\infty n\, K(Z\upharpoonright_n \mid n) \le h(n)$.*

Proof. *Step 1.* Define the computable sequence $0 = n_0 < n_1 < \ldots$ by

$$n_{k+1} = \mu n > n_k\, [h(n) > 4(k+1)]. \tag{7.7}$$

(The slower h grows, the sparser is the sequence. For instance, if $h(n) = \log n$, then $n_k = 2 \times 16^k$ for $k > 0$.) Let

$$I_k = [n_k, n_{k+1}).$$

Sometimes we write $\mathsf{Int}(z)$ for the number k such that $|z| \in I_k$. Let

$$B_k^* = B_{k,n_{k+1}}. \tag{7.8}$$

Step 2. Let $V_k = B_k^*$ if B_k^* is total, and $V_k = G_{n_{k+1}}$ otherwise (see 7.4.4 for G_n). Then each V_k is a supermartingale. Define a supermartingale L by

$$L = \sum_k 2^{-k} V_k. \tag{7.9}$$

Note that $L(\varnothing) = 2$. Once again, we draw attention to the following.

7.4.9 Fact. *If $|x| \in I_k$ we have $V_j(x) = 1$ for all $j \geq k$, so the total contribution of all such V_j to $L(x)$ is 2^{-k+1}. In particular, L is $\mathbb{Q}_2$-valued.* ◇

Step 3. Let Z be the leftmost non-ascending path of L, i.e., Z is leftmost such that $\forall n\, L(Z\restriction_{n+1}) \leq L(Z\restriction_n)$. Then $L(x) \leq 2$ for all $x \prec Z$, that is, $L(Z) \leq 2$. By Fact 7.4.3(ii), L multiplicatively dominates each total B_k. So Z is computably random.

It remains to show that $K(Z\restriction_n | n) \leq h(n)$ for almost all n. To do so, when $n \in I_k$ we compute $Z\restriction_n$ from n and $\mathsf{TMG}\restriction_k$ (recall that $\mathsf{TMG} = \{e\colon B_e \text{ total}\}$).

7.4.10 Compression Lemma. *There is a partial computable function F such that $\forall k \forall n \in I_k\, \left[Z\restriction_n = F(n, \mathsf{TMG}\restriction_k)\right]$.*

Subproof. We describe a procedure for F on inputs n and τ which in the relevant case that $\tau = \mathsf{TMG}\restriction_k$ simply provides more formal details in the definition of Z. For other strings τ it may attempt to compute undefined values of partial martingales, and thus may fail to terminate.

(1) Compute k such that $n \in I_k$.

(2) Let $z = \varnothing$. FOR $p = 0$ to $n - 1$ DO:

 (a) Let r be such that $p \in I_r$. Write
 $$S_r(x) \simeq 2^{-r+1} + \textstyle\sum_i 2^{-i} B_i^*(x)[\![i < r \;\&\; \tau(i) = 1]\!].$$

 (b) Let $z := z0$ if $S_r(z0) \leq S_r(z)$ and $z := z1$ else.
 (If $S_r(z0)$ is undefined then the procedure does not return.)

(3) Output z.

If $n \in I_k$ and $\tau = \mathsf{TMG}\restriction_k$ then S_r is defined for all z such that $|z| < n_{k+1}$, so the procedure returns $Z\restriction_n$. ◇

The length of a prefix-free description of $\mathsf{TMG}\restriction_k$ is bounded by $2k + O(1)$. Via F such a description can be turned into a description of $Z\restriction_n$ given n. Since $h(n) \geq 4k$ for $n \in I_k$, this shows that $K(Z\restriction_n | n) \leq h(n)$ for almost all n. □

We will refine the argument to obtain a result of Merkle (2003), anticipated in earlier works, which states that there is a computably random set Z of an even more slowly growing initial segment complexity. The quantifiers in Theorem 7.4.8 can be interchanged, resulting in a set Z with initial segment complexity dominated by all order functions. Such sets, called facile sets, will be studied in Section 8.2, page 319. The sequence (n_k) is no longer computable because to determine n_{k+1} one has to take into account all order functions Φ_i for $i < k$. To describe $Z\restriction_n$ for $n \in I_k$ one has to know which among the Φ_i for $i < k$ are order functions.

7.4.11 Theorem. (Extends 7.4.8) *There is a computably random facile set Z, namely, $\forall^\infty n\, K(Z\restriction_n | n) \leq h(n)$ for each order function h.*

Exercise 8.6.6 below shows that a computably random set can in fact be jump traceable. This lowness property, introduced in 8.4.1, implies being facile.

Proof. Let $\mathsf{Ord} = \{i\colon \Phi_i \text{ is an order function}\}$. We modify the sequence $n_0 < n_1 < \ldots$ as follows. We let $n_0 = 0$ and

$$n_{k+1} = \mu n > n_k \left[\min\{\Phi_i(n)\colon i < k \;\&\; i \in \mathsf{Ord}\} > 4(k+1)\right]. \tag{7.10}$$

Now we define L by (7.9), but based on this new sequence. As before, the leftmost non-ascending path Z of L is computably random. It remains to verify that the initial segment complexity of Z grows slowly. We modify Lemma 7.4.10.

7.4.12 Compression Lemma. *There is a partial computable ternary function F such that $\forall k \forall n \in I_k \left[Z\restriction_n = F(n, \mathit{TMG}\restriction_k, \mathsf{Ord}\restriction_k)\right]$.*

Subproof. The procedure for F has the inputs n, τ and ρ. We modify step (1) in the proof of 7.4.10. It may now fail to terminate if $\rho \not\prec \mathsf{Ord}$.

(1′) Attempt to compute numbers $0 = n_0 < n_1 < \ldots$ by

$$n_{k+1} \simeq \mu n > n_k \left[\min\{\Phi_i(n)\colon i < k \;\&\; \rho(i) = 1\} > 4(k+1)\right]$$

till k is found such that $n \in I_k = [n_k, n_{k+1})$.

The remaining steps are as before. ◇

There is a prefix-free description of the pair $\langle \mathsf{TMG}\restriction_k, \mathsf{Ord}\restriction_k\rangle$ of length at most $3k + O(1)$. If $n \in I_k$ then F can turn such a description into a description of $Z\restriction_n$ given n. If $h = \Phi_i$ is an order function then $h(n) \geq 4k$ for all $k > i$ and $n \in I_k$. Thus $K(Z\restriction_n \mid n) \leq h(n)$ for almost all n. □

7.4.13 Remark. Since TMG and Ord are Π^0_2 sets we have $Z \leq_T \emptyset''$ by 7.4.12. We modify the construction in the proof of Theorem 7.4.8 to obtain a set $Z \leq_T \emptyset'$. We define the computable sequence $n_0 < n_1 < \ldots$ to be a bit sparser: $n_0 = 0$ and

$$n_{k+1} = \mu n > n_k \left[h(n) > 4(k+1)^2\right].$$

Recall that $\mathsf{Int}(z)$ denotes the number k such that $|z| \in I_k$. Let

$$V_k(x) = \begin{cases} B^*_k(x) & \text{if } \mathsf{Int}(x) < k \text{ or } \forall y\, [\mathsf{Int}(y) = \mathsf{Int}(x) \rightarrow B^*_k(y)\downarrow] \\ 0 & \text{otherwise,} \end{cases}$$

and define L and Z as before but using these new versions of the V_k. Then $Z \leq_T \emptyset'$ since the function $\lambda k, x.V_k(x)$ is computable in $\emptyset'$. To describe $Z\restriction_n$ given $n \in I_k$ in a prefix-free way, we need to know, for each $i < k$, which (if any) of the intervals I_r, $r \leq k$, is the last such that $B^*_i(x)\downarrow$ for all x with $\mathsf{Int}(x) = r$. This requires at most $4k \log k$ bits.

7.4.14 Exercise. There is a computably random facile set $Z \leq_T \emptyset'$.
Hint. Apply an argument similar to the one in 7.4.13 as well to the order functions.

The case of a partial computably random set

A modification of the foregoing proof yields a set Z that is partial computably random (see 7.4.2), at the cost of an initial segment complexity that grows faster by a factor of $\log n$. By Theorem 7.6.7 below this cannot be improved, because a growth rate of $O(\log n)$ implies that the set is not partial computably random.

7.4.15 Theorem. *There is a partial computably random set Z such that $\forall^\infty n\, K(Z\upharpoonright_n \mid n) \le h(n)\log n$ for each order function h.*

Note that $K(x \mid n) \le^+ K(x) \le^+ K(x \mid n) + K(n) \le^+ K(x \mid n) + 2\log n$ by Section 2.3, so we could as well state that $\forall^\infty n\, K(Z\upharpoonright_n) \le h(n)\log n$ for each order function h. We merely keep conditional complexity to be notationally consistent with Theorem 7.4.11.

Proof. We define the sequence $(n_k)_{k\in\mathbb{N}}$ by (7.10). Let

$$V_k(x) = \begin{cases} B_k^*(x) & \text{if } B_k^*(x)\downarrow \\ 0 & \text{otherwise.} \end{cases}$$

As before, let $L = \sum_k 2^{-k} V_k$, and let Z be the leftmost non-ascending path of L. Then Z is partial computably random: suppose $B_k^*(Z\upharpoonright_n)\downarrow$ for each n, then $V_k(Z\upharpoonright_n) = B_k^*(Z\upharpoonright_n)$ is bounded along Z.

In the proof of the foregoing theorem, to determine $Z(n)$ from $z = Z\upharpoonright_n$ for $n \in I_k$, we merely required a yes/no information for each $i < k$: whether B_i is total, and whether Φ_i is an order function. Now we need to know $V_i(z0)$, which depends on whether $B_i^*(z0)$ (or equivalently, $B_i^*(z1)$) is defined. This is a considerably larger amount of information: we need the value of p_i, where p_i is the least $p \le n$ such that $B_i^*(Z\upharpoonright_p)\uparrow$ if there is such a p, and $p_i = n+1$ else. The length of a prefix-free description of p_i is bounded by $2\log n + O(1)$. Since $h(n) \ge 4k$ for each order function h, almost all k and all $n \in I_k$, we need at most $h(n)\log n$ bits.

7.4.16 Compression Lemma. *There is a partial computable ternary function F such that*

$$\forall k \forall n \in I_k\ [Z\upharpoonright_n = F(n, m, \mathsf{Ord}\upharpoonright_k)],$$

where $m = \langle p_0, \ldots, p_{k-1}\rangle$, and the $p_i \le n+1$ are defined as above.

Proof. The modified procedure to compute F is as follows.

(1) Attempt to compute numbers $0 = n_0 < n_1 < \ldots$ by
$$n_{k+1} \simeq \mu n > n_k\ [4(k+1) < \min\{\Phi_i(n) \colon i < k \ \&\ \rho(i) = 1\}]$$
till k is found such that $n \in I_k = [n_k, n_{k+1})$.

(2) Let $z = \varnothing$. FOR $l = 0$ to $n-1$ DO:

 (a) Let r be such that $l \in I_r$. For each $i \le r$ and $a \in \{0,1\}$, attempt to compute $V_i(za)$: if $p_i \le |z|+1$ let $V_i(za) = 0$, otherwise let $V_i(za) \simeq B_i^*(za)$. Now calculate $S_r(za) \simeq \sum_{i<r} 2^{-i} V_i(za)$.

 (b) Let $z := z0$ if $S_r(z0) \le S_r(z)$, and let $z := z1$ otherwise. (If $S(z0)$ is undefined then the procedure does not return.)

(3) Output z.

As explained above, if $n \in I_k$, m is as above, and $\rho = \mathsf{Ord}\upharpoonright_k$, the procedure returns $Z\upharpoonright_n$. Since $K(\rho) \le^+ 2k$, with the remarks above this shows that $K(Z\upharpoonright_n \mid n) \le h(n)\log n$ for almost all n. □

7.4.17 Remark. A slight modification of the proof yields a left-c.e. set Z in Theorem 7.4.15. The V_k are uniformly c.e. supermartingales, hence L is computably enumerable. Recall from Remark 7.1.13 that $T_2^L = \{x\colon \forall y \preceq x\, [L(y) \le 2]\}$. Since L is c.e. and $L(\varnothing) \le 2$, $Paths(T_2^L)$ is a nonempty Π_1^0 class. The leftmost path Z of T_2^L is left-c.e. by Fact 1.8.36, and Z is partial computably random. To show the bound on the initial segment complexity, in the procedure for F we replace step (2.b) by

(2.b′) Let $z := z0$ if $S_r(z0) \le 2$, and $z := z1$ otherwise.

7.5 Each high degree contains a computably random set

Having studied the initial segment complexity of computably random sets, we now turn to their computational complexity. We carry out the approach (II) on page 271. Given a high set C, our first goal is to build a computably random set $Z \le_T C$. Later on, we will achieve that $Z \equiv_T C$, and we will also ensure that Z is not partial computably random. Each Schnorr random set of non-high degree is ML-random by 3.5.13, so this characterizes the Turing degrees of sets that are computably random but not partial computably random (or ML-random).

Martingales that dominate

Theorem 1.5.19 states that C is high iff there is a function $f \le_T C$ dominating every computable function. From such an f we will obtain a $\mathbb{Q}_2$-valued supermartingale $L \le_T C$, $L(\varnothing) \le 2$, that multiplicatively dominates each computable martingale. If Z is the leftmost path on $T_2^L = \{x\colon \forall y \preceq x\, [L(y) \le 2]\}$ then Z is computably random and $Z \le_T C$.

7.5.1 Domination Lemma. *If C is high then there is a $\mathbb{Q}_2$-valued supermartingale $L \le_T C$ that multiplicatively dominates each computable martingale.*

Proof. Only steps 2 and 3 of the template on page 274 are needed here (think of n_e as being e). The partial computable martingales $B_{k,n}$ were defined in 7.4.3.

Suppose $f \le_T C$ dominates each computable function. For each e, let

$$g_e(n) \simeq \mu t\, [\forall x\, (|x| \le n \to\ B_{e,e+1}(x)[t]\downarrow)], \text{ and let}$$

$$V_e(x) = \begin{cases} B_{e,e+1}(x) & \text{if } |x| \le e \text{ or } \forall m\ \big[e \le m \le |x|) \to g_e(m) \le f(m) + e\big] \\ 0 & \text{else.} \end{cases}$$

As usual let $L(x) = \sum_e 2^{-e} V_e(x)$. Clearly L is a supermartingale. Further, L is $\mathbb{Q}_2$-valued by the standard argument: for $e > |x|$ we have $V_e(x) = 1$, so the tail of the sum due to the $e > |x|$ equals $2^{-|x|}$. Note that $L \le_T C$ since $V_e \le_T C$ uniformly in e.

If B_e is total then there is m such that $g_e(n) \le f(n)$ for all $n \ge m$. By the Padding Lemma 1.1.3 we may choose e' such that the Turing program $P_{e'}$ does exactly the same as P_e and $e' > \max\{g_e(n)\colon n < m\}$. Thus $g_e(n) = g_{e'}(n) \le f(n) + e'$ for *all* n. (This saves on notation here: else in the definition of L one would have to consider pairs consisting of a martingale and a constant.) It follows that $B_e(x) \le 2^{e'} L(x)$ for each x. □

The following preliminary result is now immediate.

7.5.2 Theorem. *For each high set C there is a computably random set $Z \le_T C$.*

Proof. Let L be as in Lemma 7.5.1. Note that $L(\varnothing) \le 2$. As in Remark 7.1.13, let Z be the leftmost path on the tree $T_2^L = \{x\colon \forall y \preceq x\, [L(y) \le 2]\}$. Then $Z \le_T L \le_T C$ and Z is computably random. □

Each high c.e. degree contains a computably random left-c.e. set

Eventually we want to improve the conclusion in Theorem 7.5.2 to $Z \equiv_T C$. This is easier when C is c.e., and in this case we even obtain a left-c.e. set Z. The proof deviates from the template.

7.5.3 Theorem. *For each high c.e. set C, there is a computably random left-c.e. set $Z \equiv_T C$.*

Proof. We need a version of the Domination Lemma 7.5.1 for the c.e. setting.

7.5.4 Lemma. *Suppose the set C is high and also c.e. Then there is a supermartingale L as in 7.5.1 such that, in addition, L is computably enumerable.*

Subproof. We refine the proof Lemma 7.5.1 in order to obtain a supermartingale approximation $(L_s)_{s\in\mathbb{N}}$ for L (see Definition 7.2.3). Suppose Δ is a Turing functional such that $\widetilde{f} = \Delta^C$ dominates every computable function. We will replace $\widetilde{f}$ by an even larger function computable in C which can be approximated in a non-decreasing way: let

$$f_s(x) = \max\nolimits_{t\le s}\Delta^C(x)[t], \text{ and } f(x) = \lim\nolimits_s f_s(x).$$

Note that $f(x) < \infty$ since Δ^C is total. To compute $f(x)$ relative to C, let s be least such that $\Delta^C(x)[s]\downarrow$ with C_s correct up to u the use of this computation and output $f_s(x)$. Then $f_t(x) = f(x)$ for each stage $t > s$ (here we have used that C is computably enumerable). Now let $L \le_T C$ be as in the proof of Lemma 7.5.1. To obtain the supermartingale approximation $(L_s)_{s\in\mathbb{N}}$, let

$$V_{e,s}(x) = \begin{cases} B_e(x) & \text{if } |x| < e \text{ or } \forall m\, [e \le m \le |x| \to g_e(m) \le f_s(m+e)] \\ 0 & \text{else.} \end{cases}$$

Let $L_s(x) = \sum_e 2^{-e} V_{e,s}(x)$. ◇

As in Remark 7.4.17, *Paths*(T_2^L) is a nonempty Π_1^0 class. So the leftmost path of T_2^L is left-c.e. and computable in C. To prove Theorem 7.5.3, we have to ensure that, conversely, C can be computed from the leftmost path on a variant of T_2^L. This is achieved by adding a c.e. supermartingale N to L. The following isolates the argument.

7.5.5 Lemma. *Let L be a $\mathbb{Q}_2$-valued c.e. supermartingale with an approximation $(L_s)_{s\in\mathbb{N}}$, and suppose that $L(\lambda) \le 2$. Then for each c.e. set C there is a $\mathbb{Q}_2$-valued c.e. supermartingale $N \le_T C$ such that $N(\lambda) \le 2$, and $C \le_{wtt} Z$ for the leftmost path Z of $T = T_4^{L+N} = \{x\colon \forall y \preceq x\ (L+N)(y) \le 4\}$.*

Note that Z exists by (7.4). In our case, since L is $\mathbb{Q}_2$-valued and $L \leq_T C$, we have $L + N \leq_T C$. Thus $Z \leq_T C$ as well, which establishes Theorem 7.5.3.

To prove the Lemma, we build N via a supermartingale approximation $(N_s)_{s\in\mathbb{N}}$. We may assume that $\#(C_{s+1} - C_s) = 1$ for each stage s. At stage $s+1$, let Z_s be the leftmost path of length $s+3$ of $T_s = \{x\colon \forall y \preceq x\ (L_s + N_s)(y) \leq 4[s]\}$. Since $L_s(y)$ and $N_s(y)$ are nondecreasing in s, the approximation $(Z_s)_{s\in\mathbb{N}}$ only moves to the right, so Z is left-c.e.

We devote 2^{-n} of the initial capital of N to code into Z whether $n \in C$. Recall that G_n is the supermartingale given by $G_n(x) = 1$ for $|x| < n$ and $G_n(x) = 0$ otherwise. Let $N_0 = \sum_n 2^{-n} G_n$, so that $N_0(x) = 2^{-|x|}$. When n enters C at stage s, then N_{s+1} withdraws at the string $z = Z_s \restriction_n$ the capital 2^{-n}. Thereafter it doubles its capital $n+3$ times along Z_s, reaching the value 8 at $Z_s \restriction_{2n+3}$. More formally, let

$$F_{n,s}(x) = \begin{cases} 2^{|x|-n} & \text{if } x \preceq Z_s \restriction_{2n+3} \ \&\ n \leq |x| \\ 0 & \text{else,} \end{cases}$$

$$N_{s+1} = N_s + 2^{-n} F_n.$$

Then $G_n + F_n$ is a supermartingale for each n. Hence, by Fact 7.1.7(ii), $(N_s)_{s\in\mathbb{N}}$ is a supermartingale approximation, and $N(x) = \sup_s N_s(x)$ defines a c.e. supermartingale. To see that N is $\mathbb{Q}_2$-valued and $N \leq_T C$, given x, compute relative to C a stage s such that $C \restriction_{|x|+1}$ is stable. Then $N(x) = N_s(x)$ by the definition of the $F_{n,s}$.

To show that $C \leq_{wtt} Z$, given n, compute t such that $Z_t \restriction_{2n+3} = Z \restriction_{2n+3}$. Then $C(n) = C_t(n)$, for if n is enters C at a stage $s > t$, this causes $(L+N)(Z_s \restriction_{2n+3}) \geq 4$, so $Z_t \restriction_{2n+3}$ is not on the leftmost path of T. □

A computably random set that is not partial computably random

Coding a set C into the leftmost path of a Π^0_1 class works when the set is c.e., but for coding an arbitrary set a different idea is needed. Suppose that as before L is a supermartingale dominating all the partial computable supermartingales. We introduce a new method for coding information into a set Z such that $L(Z) < \infty$. Given a string x which already encodes k bits of information, one finds distinct strings $y_0, y_1 \succ x$ such that along each y_a, L increases by at most a factor of $1 + 2^{-k}$. If Z extends x, one may code one further bit a into Z by letting Z extend y_a. Such strings $y_0, y_1 \succ x$ exist by the following Lemma. It is an analog of Lemma 3.3.1 used for the proof of the Kučera–Gács Theorem 3.3.2.

7.5.6 Lemma. *Let L be a supermartingale. Then for each string x and each number k, there are distinct strings $y_0, y_1 \succ x$ of length $|x| + k + 1$ such that L does not grow beyond $(1 + 2^{-k})L(x)$ along each y_a, $a = 0, 1$. That is,*

$$\forall z\, [x \preceq z \preceq y_a \rightarrow L(z) \leq (1 + 2^{-k})L(x)]. \tag{7.11}$$

Proof. Because of the localization principle 7.1.7(iii) we may suppose $x = \varnothing$. Let y_0 be a string of length $k+1$ along which L does not increase. Let $z_0, \ldots, z_{n-1}$ be the strings z of length $k+1$ that are minimal under the prefix relation with $L(z) > (1 + 2^{-k})L(\varnothing)$. Assume for a contradiction that for each $y \neq y_0$ of length $k+1$ there is $i < n$ such that $z_i \preceq y$. Then $\sum_{i<n} 2^{-|z_i|} = 1 - 2^{-(k+1)}$. By Lemma 7.1.8(i)

$$L(\varnothing) \geq \sum_{i<n} 2^{-|z_i|} L(z_i) > (1+2^{-k})L(\varnothing)(1-2^{-(k+1)}) > L(\varnothing),$$

contradiction. □

We let

$$y^L_{x,k+1,0} \text{ and } y^L_{x,k+1,1} \tag{7.12}$$

be the leftmost and the rightmost extension of x with length $|x|+k+1$ such that (7.11) holds; these strings are distinct by Lemma 7.5.6. If L is a $\mathbb{Q}_2$-valued supermartingale then $y^L_{x,k,0}$ and $y^L_{x,k,1}$ can be computed from x, k and the values $L(z)$ for $|z| \leq |x|+k+1$.

7.5.7 Theorem. *There is a computably random set Z that is not partial computably random.*

Proof. Let $n_0 = 0$ and $n_{k+1} = n_k + k + 2$. As usual, let $I_k = [n_k, n_{k+1})$. Let

$$u_k = n_k + k + 1,$$

and $u_{-1} = 0$. We define supermartingales V_k ($k \in \mathbb{N}$) almost as in 7.4.8:

$$V_k = B^*_k \text{ if } B_k \text{ is total, and } V_k = G_{n_{k+1}} \text{ otherwise,}$$

where $B^*_k = B_{k,n_{k+1}}$. As always let $L = \sum_k 2^{-k} V_k$. Then L multiplicatively dominates each total martingale B_k.

The set Z is defined by recursion. The bit positions in $[n_k, u_k)$ are used to code $\mathsf{TMG}(k)$ into Z. One lets $Z(u_k) = 0$ unless this makes the value of L increase, in which case one chooses the value $Z(u_k) = 1$. This decision can be predicted by a suitable partial computable martingale D, which therefore may double its capital. The detailed definition of Z follows:

- If $x = Z\upharpoonright_{n_k}$ has been defined let $Z\upharpoonright_{u_k} = y^L_{x,k+1,\mathsf{TMG}(k)}$.
- If $z = Z\upharpoonright_{u_k}$ has been defined let $Z(u_k) = 0$ if $L(z0) \leq L(z)$, and $Z(u_k) = 1$ otherwise.

A typical portion of Z looks like this:

...010	1	10111000	1	0100...
	u_6	n_7	u_7	n_8

(the bit in position u_7 is 1, for instance, because $L(z0) > L(z)$ for $z = Z\upharpoonright_{u_7}$).

To show that Z is computably random, it suffices to verify that L does not succeed on Z. We need the following fact from analysis.

7.5.8 Fact. *If $(a_k)_{k\in\mathbb{N}}$ is a sequence of nonnegative real numbers and $s = \sum_k a_k < \infty$, then $\prod_k (1+a_k) < \infty$.*

Proof. $e^s = \sup_k \prod_{i=0}^k e^{a_i}$, and $1 + a_i \leq e^{a_i}$. Thus $\prod_k(1+a_k) \leq e^s < \infty$. ◇

We apply this fact to $a_k = 2^{-k}$. As $L(\varnothing) \leq 2$ we have $L(Z\upharpoonright_n) \leq 2\prod_{k<n}(1+2^{-k})$ for each n. So $L(Z) = \sup_n L(Z\upharpoonright_n) < \infty$.

We define a partial computable martingale D that succeeds on Z. It can predict each value $Z(u_k)$, and hence bets all of its current capital on that value. The prediction is based on $\mathsf{TMG}\upharpoonright_k$, since $\mathsf{TMG}\upharpoonright_k$ determines the values $L(x)$ for $|x| \in I_k$. D merely "reads" Z at the bit positions $j < u_{k-1}$ not of the form u_l without betting. $\mathsf{TMG}\upharpoonright_k$ can be recovered from $Z\upharpoonright_{u_{k-1}}$ by the coding.

For the formal definition of D, we begin with a Turing functional Δ with computably bounded use. We will verify by induction on k that

$$\Delta^{Z\upharpoonright_{u_k}} = \mathsf{TMG}\upharpoonright_{k+1}. \tag{7.13}$$

(For instance, the string $Z\upharpoonright_{u_0}$ suffices to determine whether B_0 is total.) Below, the definition of Δ is in italics while the rest is explanatory. Note that $\Delta^{\varnothing} = \varnothing$ for any Turing functional Δ.

1. If $|z| = u_k$ and $\tau = \Delta^{z\upharpoonright_{u_{k-1}}}$ has been defined, where $|\tau| = k$, then write

$$S_k = 2^{-k+1} + \sum_{i<k} 2^{-i} B_i^* \,[\![\tau(i) = 1]\!]. \tag{7.14}$$

Note that $S_0(w) = 2$ for each w. Also (7.13) holds for $k = 0$, since $\Delta^{Z\upharpoonright_1} = \mathsf{TMG}\upharpoonright_1$ by the definition of Z. Suppose $k > 0$ and (7.13) holds for $k-1$. Thus $\Delta^{z\upharpoonright_{u_{k-1}}} = \tau = \mathsf{TMG}\upharpoonright_k$ for $z = Z\upharpoonright_{u_k}$. Hence all the computations in the sum $S_k(w)$ are defined for $|w| \in I_k$, and $S_k(w) = L(w)$.

2. Let $x = Z\upharpoonright_{n_k}$ and attempt to compute $y_a = y^{S_k}_{x,k+1,a}$ for $a = 0, 1$. If y_a is defined and equal to z, define $\Delta^z(k) = a$. Then (7.13) holds for k by the definition of Z. This completes the definition of Δ. Clearly the properties in Fact 6.1.1 are satisfied. For oracle incomparability (F2) note that we only make a definition $\Delta^z(k) = a$ if $|z| = u_k$.

Now we define a partial computable supermartingale D in such a way that $D(Z\upharpoonright_{u_k+1}) = 2^k$ for each k. Let $D(\varnothing) = 1$. If $D(z)\downarrow$ for a string z such that $|z| = u_k$, wait for $\Delta^z = \tau$, define S_k by (7.14), and attempt to compute $S_k(z0)$. If this computation converges and $S_k(z0) \le S_k(z)$, let $D(z0) = 2D(z)$ and $D(z1) = 0$, otherwise let $D(z1) = 2D(z)$ and $D(z0) = 0$. Extend D to all the strings $w \succeq za$, $|w| \le u_{k+1}$, by defining $D(w) = D(za)$. □

An alternative proof of the result will be obtained through Theorem 7.6.7 below.

A strictly computably random set in each high degree

We combine the ideas from the proofs of the previous three preliminary theorems to obtain the result of Nies, Stephan and Terwijn (2005) that there is a computably random, but not partial computably random set in each high degree. While the Domination Lemma 7.5.1 cannot be applied as it is, we still rely on a function $f \le_T C$ dominating each computable function. The set TMG we coded into Z in the proof of Theorem 7.5.7 will be replaced by an auxiliary set $Q \equiv_T C$ that tells us whether a function like g_e (the running time of B_e) in the proof of the Domination Lemma is dominated by f at certain values. In this way the question whether B_e is total is divided into infinitely many subquestions whether $B_e(x)$ is defined for all x such that $|x| \in I_r$. This is similar to the transition from Theorem 7.4.8 to Remark 7.4.13.

7.5.9 Theorem. *For each high set C there is a computably random, but not partial computably random set $Z \equiv_T C$. Moreover, given an order function h, one can achieve that $\forall^\infty n\, K(Z\restriction_n | n) \le h(n)$.*

Proof. Let $n_0 = 0$. For $k \ge 0$, let

$$n_{k+1} = \mu n > n_k + k + 1.\, h(n) > 4(k+1).$$

Let $I_k = [n_k, n_{k+1})$, $u_k = n_k + k + 1$, and $u_{-1} = 0$. As in the proof of the Domination Lemma 7.5.1, choose a function $f \le_T C$ which dominates each computable function, and for each e let $g_e(n) \simeq \mu t. \forall x\, [|x| \le n \to B_{e,t}(x)\downarrow]$. Let $Q = \{\langle 0, r\rangle \colon r \in C\} \cup$

$$\{\langle e, r\rangle \colon e > 0\ \&\ \forall p \le r\, [g_e(n_{\langle e+p+1, e+p+1\rangle}) \le f(e+p)]\}.$$

Clearly $Q \equiv_T C$. We say B_e is *active* in I_k if $k \le e+1$, or $\langle e, r\rangle \in Q$ for the maximal r such that $r = 0$ or $\langle e, r\rangle < k$. In the second case $\langle e, r+1\rangle \ge k$, so that $n_{\langle e+r+1, e+r+1\rangle} > n_k$. Hence $B_e(w)\downarrow$ for each w such that $|w| < n_{k+1}$.

Let $V_0 = 1$. For $e > 0$, as always let $B_e^* = B_{e,n_{e+1}}$. Let

$$V_e(x) = \begin{cases} B_e^*(x) & \text{if } |x| \in I_k\ \&\ B_e \text{ is active in } I_k \\ 0 & \text{else.} \end{cases}$$

As usual $L(x) = \sum_e 2^{-e} V_e(x)$ is a $\mathbb{Q}_2$-valued supermartingale. Given x, k, the strings $y^L_{x,k+1,0}$ and $y^L_{x,k+1,1}$ are defined in (7.12). The set Z is given by the following recursion.

- If $x = Z\restriction_{n_k}$ has been defined let $Z\restriction_{u_k} = y^L_{x,k+1,Q(k)}$.
- If $z = Z\restriction_{u_k}$ has been defined let $Z\restriction_{n_{k+1}}$ be the leftmost non-ascending string above z given by L (see 7.3.9).

We show that Z is computably random. Firstly, we verify that $L(Z) < \infty$. Note that $L(Z\restriction_n) \le 2\prod_{k\le n}(1 + 2^{-k})$ by the choice of the strings in (7.12) and since L is non-ascending along Z on bit positions in the intervals $[u_k, n_{k+1})$. By Fact 7.5.8 $\prod_k(1 + 2^{-k}) < \infty$. So $L(Z) = \sup_n L(Z\restriction_n) < \infty$.

Secondly, we show that L multiplicatively dominates each total B_e. The function v given by $v(i) = g_e(n_{\langle i+1, i+1\rangle})$ is computable, so that $\forall i \ge i_0\, [v(i) \le f(i)]$ for some i_0. By the Padding Lemma 1.1.3 we may choose e' such that the Turing program $P_{e'}$ behaves like P_e and $e' > i_0$. Hence, for all p, we have $v(e'+p) = g_{e'}(n_{\langle e'+p+1, e'+p+1\rangle}) \le f(e'+p)$, so $\langle e', r\rangle \in Q$ for e' and all r. Then $B_{e'}$ is active in all intervals I_r, whence $V_{e'} = B^*_{e'}$. Since $B_e = B_{e'}$, this shows that L multiplicatively dominates B_e. (The Padding Lemma is more than a notational convenience here: it saves us from keeping track of the point from which on f exceeds g_e.)

Since $C \equiv_T Q$, to show $C \equiv_T Z$ it suffices to verify that $Q \equiv_T Z$. For $Q \leq_T Z$ we introduce a Turing functional Δ and verify inductively that

$$\Delta^{Z\restriction_{u_k}} = Q\restriction_{k+1} \tag{7.15}$$

for each k. As always $\Delta^{\varnothing} = \varnothing$.

1. If $|z| = u_k$ and $\tau = \Delta^{z\restriction_{u_{k-1}}}$ has been defined where $|\tau| = k$, let

$$E(\tau) = \{e\colon 0 < e < k \ \&\ \forall p\,[\langle e, p\rangle < k \to \tau(\langle e, p\rangle) = 1]\}$$

and

$$S_k = 2^{-k+1} + \sum_e 2^{-e} B_e^* \,[\![e \in E(\tau)]\!]. \tag{7.16}$$

Note that (7.15) holds for $k = 0$ by the definition of Z (in which case $S_k(x) = 2$ for each x). If $k > 0$ and (7.15) holds for $k-1$, then for $z = Z\restriction_{u_k}$ all the computations in the sum $S_k(w)$, $|w| \in I_k$, are defined, and $S_k(w) = L(w)$. This holds because inductively $\tau = Q\restriction_k$, so $e \in E(\tau)$ implies that B_e is active in I_k.

2. Let $x = Z\restriction_{n_k}$ and calculate $y_a = y^{S_k}_{x,k+1,a}$ for $a = 0, 1$. If y_a is defined and equal to z then define $\Delta^z(k) = a$. Then (7.15) holds for k by the definition of Z. This completes the definition of Δ and shows that $\Delta^Z = Q$.

For $Z \leq_T Q$ it suffices to recall that, if $\tau = Q\restriction_k$ and S_k is determined as above, then $S_k(w) = L(w)$ whenever $|w| \in I_k$, so that the recursive definition of Z can be computed with Q as an oracle.

To show that Z is not partial computably random, we use a partial computable martingale D as in the proof of Theorem 7.5.7, but based on the new background definitions. The proof that $\forall^\infty n\, K(Z\restriction_n|\, n) \leq h(n)$ is as in Theorem 7.4.8, except that we now compute $Z\restriction_n$ ($n \in I_k$) from n and $\tau = Q\restriction_k$. So we modify the procedure for F in Lemma 7.4.10: in step (2a) we define S_r by (7.16), namely, we let $S_r = 2^{-r+1} + \sum_e 2^{-e} B_e^* \,[\![e \in E(\tau\restriction_r)]\!]$. □

A strictly Schnorr random set in each high degree

Some Schnorr random set is not computably random. In fact, a set separating the two randomness notions exists in each high Turing degree. The result is again due to Nies, Stephan and Terwijn (2005). We modify the proof of Theorem 7.5.9. The main change is in the definition of Z.

7.5.10 Theorem. *For each high set C there is a Schnorr random set $Z \equiv_T C$ such that Z is not computably random.*

Proof. We are not attempting to achieve a slowly growing initial segment complexity, so we may drop the order function and let $n_0 = 0$, $u_k = n_k + k + 1$, and $n_{k+1} = u_k + 1$. We define the computable supermartingale L as before. While we mostly determine $Z(u_k)$ in such a way that $L(Z\restriction_{n_{k+1}}) \leq L(Z\restriction_{u_k})$, for a sparse set of numbers k we guarantee that $Z(u_k) = 0$. This set is so sparse that no computable martingale D succeeds on Z in the strong sense of Theorem 7.3.3(iii).

On the other hand we are able to build a computable martingale that uses these guaranteed zeros to succeed on Z.

Let $\theta_e(m) \simeq \mu s.\, \Phi_e(m)[s]\downarrow$, and let

$$\psi(e,m) \simeq u_{\langle\langle e,\theta_e(m)\rangle,m\rangle+1}.$$

Note that ψ is one-one, $\psi(e,m) \geq u_{m+1} > m$ for all pairs e,m in its domain, and its range is computable. We define a computable function p by the recursion

$$p(r) = \begin{cases} p(i)+1 & \text{if } i < r \;\&\; \exists e < \log p(i) - 1\, [\psi(e,i) = r], \\ r+4 & \text{otherwise.} \end{cases}$$

Informally, if $\psi(e,i) = r$ for small e, then instead of taking the value $r+4$ the function goes back to $p(i)+1$ (usually i is small compared to r). The second condition ensures that $\log p(r) > 1$ for each r.

By induction on n, we show that p attains every value $\leq n$ only finitely often: if $r \geq n$ and $p(r) = n$ then $p(r) \neq r+4$, so $p(r) = p(i)+1$ for some i. Hence $p(i) < n$ and the inductive hypothesis applies.

As before, we fix a function $f \leq_T C$ that dominates each computable function. We may assume without loss of generality that Φ_0 is total and $f(m) \geq \psi(0,m)$ for all m. We define a quickly growing function $h \leq_T C$ by

$$h(m) = \max\{\psi(e,m)\colon\ \psi(e,m)\downarrow \leq f(m) \;\&\; e < \log(p(m)) - 1\}.$$

The definition of Z in the proof of Theorem 7.5.9 is modified as follows.

- If $v = Z\restriction_{n_k}$ is defined, let $Z\restriction_{u_k} = y^L_{v,k+1,Q(k)}$.
- If $z = Z\restriction_{u_k}$ is defined, let $Z(u_k) = 0$ if $u_k \in \mathrm{ran}(h)$ or $L(z0) \leq L(z)$, and let $Z(u_k) = 1$ otherwise.

We define the set Q as before. To show that $Z \equiv_T Q$, one modifies the corresponding proof above, taking into account that $h \leq_T C$. To see that Z is not computably random, we define a computable martingale N that succeeds on Z. Given i, a finite set that is uniformly computable in i is determined by

$$G_i = \{\psi(e,i)\colon\ \psi(e,i)\downarrow \wedge\, e < \log p(i) - 1\}.$$

Since ψ is one-one, we have $G_i \cap G_j = \emptyset$ for $i \neq j$. Further, $\psi(0,i) \in G_i$, and $\psi(0,i)$ is a number u_k such that $Z(u_k) = 0$.

The computable betting strategy $\mathcal{S}$ starts with capital 1. It has a parameter i; initially $i = u_0$. Let r be the bit position it currently bets on. $\mathcal{S}$ only risks capital if $r \in G_i$. If $\mathcal{S}$ has lost m times while betting on positions in G_i, it bets an amount of $2^m/p(i)$ on its prediction that $Z(y) = 0$. If it loses this bet, it stays with i. If it wins the bet, it has gained on the points of G_i in total the amount $1/p(i)$; it updates i to the current value of r.

The following formal description of $\mathcal{S}$ determines a computable martingale.

The betting strategy $\mathcal{S}$. The strategy is presented with a sequence of bits. We denote its capital by N; initially $N = 1$. The strategy $\mathcal{S}$ declares its bets, and in response the casino updates N. Let $i = u_0 = 1$ and $r = -1$.

(1) Let $c = 1/(2p(i))$.
(2) Let $r := r + 1$. IF $r \notin G_i$ GOTO 2.
(3) Let $c := 2c$. Bet the amount $\min(N, c)$ that the next bit is 0.
(4) Request the next bit b. IF $b = 0$ (and hence N is updated to $N + c$), let $i := r$ and GOTO 1.
(5) (Now N has been updated to $N - c$.) IF $N > 0$ GOTO 3, ELSE END.

Suppose now that the strategy is fed the bits of Z. Given i, let m be the number of times $\mathcal{S}$ bets at (3) for this value of i. Since $Z(r) = 0$ for some $r \in G_i$, we have $m \leq \#G_i \leq \log p(i) - 1$. The maximum value of c is $2^m/p(i) \leq 1/2$, so there is always some capital left for $\mathcal{S}$ to bet with. $\mathcal{S}$ loses $1/p(i), 2/p(i), 4/p(i)$ but then wins $2^m/p(i)$. So it wins $1/p(i)$ on Z with parameter i.

If $\mathcal{S}$ updates its parameter i to r in (4) then $r = \psi(e, i)$ for some $e < \log p(i) - 1$, so $p(r) = p(i) + 1$. After going back to (1) for t times, $\mathcal{S}$ has a capital of at least $\sum_{j=0}^{t-1} 1/(u_0 + j + 4)$, which is unbounded in t. Thus $\mathcal{S}$ succeeds on Z.

Assume for a contradiction that Z is not Schnorr random. Then, by Theorem 7.3.3, there is a total martingale B_j and a computable strictly increasing function g such that $B_j(Z \upharpoonright_{g(m)}) > m$ for infinitely many m. Since p takes each value only finitely often, for almost all m we have $h(\log\log m) > g(m)$.

We claim that $L(Z \upharpoonright_{g(m)}) = O(\log m)$ for each $m > 1$: in an interval of the form $[n_k, u_k)$, L can increase its capital by a factor of at most $1 + 2^{-k}$. Since $\prod_k (1 + 2^{-k}) < \infty$, overall the capital increases only by a constant factor. At a position u_k it can only increase its capital if $u_k \in \operatorname{ran}(h)$. For almost all m, there are at most $\log\log(m)$ such u_k below $g(m)$. So even if L doubles the capital each time, it will only achieve an increase by a factor of $O(\log m)$. This proves the claim.

In the proof of Theorem 7.5.9 we showed that L multiplicatively dominates each total B_i. Thus $B_i(Z \upharpoonright_{g(m)}) = O(\log m)$ for each $m > 1$, contradiction. □

Nies, Stephan and Terwijn (2005) proved Theorem 7.5.3 by extending the methods of Theorem 7.5.9. In a similar vein, they proved that in 7.5.10, if the high set C is c.e., the Schnorr random set Z can be chosen left-c.e.

Exercises.

7.5.11. Show that if C is hyperimmune, there is a set $Z \equiv_T C$ such that Z is weakly random but not Schnorr random. (This complements 3.6.4.)

7.5.12.$^\diamond$ Remark 7.4.17 contains an example of a left-c.e. partial computably random set with a slowly growing initial segment complexity. As a further way to separate this randomness notion from ML-randomness, show that there is a Turing minimal pair Z_0, Z_1 of left-c.e. partial computably random sets.
Hint. Modify the usual construction of a minimal pair of high c.e. sets via a tree of strategies already mentioned after 6.3.4; see for instance Soare (1987, XIV.3.1).

7.5.13.$^\diamond$ **Problem.** Decide whether for each high c.e. set C there is a left-c.e. set $Z \equiv_T C$ such that Z is partial computably random. (Also see Ex. 8.6.6.)

7.6 Varying the concept of a betting strategy

We consider two variants of betting strategies: selection rules and nonmonotonic betting strategies. The former are weaker, the latter stronger than betting strategies with the same type of effectivity condition.

Instead of betting, a selection rule selects from a set Z (viewed as a sequence of bits) a subsection of bits that "count". It succeeds on the set if the law of large numbers fails for this selected subsequence. This notion can be traced back to von Mises (1919).

A nonmonotonic betting strategy bets on the bits in an order it determines.

Basics of selection rules

A selection rule ρ processes a set Z bit by bit. For each bit position n it makes a decision whether to select the bit in position n based on what it has seen so far. Thus, ρ maps a string x of length n (thought of as a possible initial segment $Z\restriction_n$) to a bit.

Recall from 1.7.1 that $p_E(n)$ denotes the n-th element of a set $E \subseteq \mathbb{N}$.

7.6.1 Definition.

(i) A *selection rule* is a partial function $\rho\colon \{0,1\}^* \to \{0,1\}$ with a domain closed under prefixes.

(ii) Given a set Z, let $E = \{n\colon \rho(Z\restriction_n) = 1\}$ (the set of positions selected from Z by ρ). The *set selected* from Z by ρ is $S = \{k\colon Z(p_E(k)) = 1\}$.

For instance, suppose ρ selects a bit position if the last pair of bits it has read is 01. If $Z = 110110110\ldots$ then $E = \{3m+1\colon m > 0\}$ and $S = \mathbb{N}$.

Stochasticity

Recall from page 109 that a set Z satisfies the law of large numbers if

$$\lim_n \#\{i < n\colon Z(i) = 1\}/n = 1/2. \tag{7.17}$$

This is one of the simplest criteria a set Z must fulfill according to our intuitive notion of randomness. The criterion is not sufficient by far, for instance because the computable set $\{2i\colon i \in \mathbb{N}\}$ satisfies the law of large numbers. In the following we discuss a stronger test concept that nonetheless retains the idea to check whether the occurrences of zeros and ones are balanced. We say that a selection rule ρ *succeeds* on Z if $\rho(Z\restriction_n)$ is defined for each n, and the set S selected from Z fails the law of large numbers (7.17). To obtain a notion of stochasticity, we will specify a set of allowed selection rules, and require that none of them succeeds on Z. For instance, if we require the rules to be computable, we obtain a notion introduced by Church (1940).

7.6.2 Definition. Z is *computably stochastic* (or Church stochastic) if no computable selection rule ρ suceeds on Z.

7.6.3 Fact. *If Z is computable then Z is not computably stochastic.*

Proof. Let $b = 1$ if Z is infinite and $b = 0$ else. Define a computable selection rule ρ by $\rho(x) = 1$ if $Z(|x|) = b$. Then the set selected from Z by ρ is $\mathbb{N}$. □

If a computable selection rule ρ succeeds on a set Z then $\rho(Z\upharpoonright_n)$ is defined for each n. By the main result of this section, Theorem 7.6.7 below, extra power is gained if we allow ρ to be undefined on strings that are not initial segments of Z.

7.6.4 Definition. Z is *partial computably stochastic* (or Church–Mises–Wald stochastic) if no partial computable selection rule ρ succeeds on Z.

The two notions just introduced correspond to computable randomness and partial computable randomness. Martingales as tests are strictly more powerful than the selection rules with the same effectivity condition. For instance, the computably random sets form a proper subclass of the computably stochastic sets. In fact, Merkle *et al.* (2006, Thm. 30) proved that some partial computably stochastic set is not even weakly random.

Anbos-Spies *et al.* (1996) called a (possibly partial) martingale B *simple* if for some $q \in (0,1)_{\mathbb{Q}}$, the set of betting factors $\{B(x0)/B(x)\colon x \in \{0,1\}^*\}$ is contained in $\{1, q, 1-q\}$. They showed that Z is computably stochastic iff no simple computable martingale succeeds on Z, and Z is partial computably stochastic iff no simple partial computable martingale succeeds on Z.

Stochasticity and initial segment complexity

We think of a K-trivial set A as far from random because $K(A\upharpoonright_n)$ grows as slowly as possible. If we relax the growth condition somewhat, we can still expect the set to have strong nonrandom features. Such a condition is being $O(\log)$ bounded. Recall that $\log n$ denotes the largest $k \in \mathbb{N}$ such that $2^k \le n$.

7.6.5 Definition. A string x is $O(\log)$ *bounded* via $b \in \mathbb{N}$ if $C(x) < b\log(|x|)$. A set Z is $O(\log)$ *bounded* via $b \in \mathbb{N}$ if almost every initial segment of Z is $O(\log)$ bounded via b, that is, $\forall^\infty n\, C(Z\upharpoonright_n) < b\log n$.

By Corollary 2.4.2 $C(x) \le^+ K(x) \le^+ C(x) + 2\log|x|$, so the class of $O(\log)$ bounded sets would remain the same if we defined it using K instead of C. For an example of an $O(\log)$ bounded set, let R be ML-random and consider the set $Z = \{2^i\colon R(i) = 1\}$. Then Z is $O(\log)$ bounded via $b = 2$. (Note that by Schnorr's Theorem 3.2.9 we also have $\log n \le^+ K(Z\upharpoonright_n)$ for each n, so Z is not K-trivial.)

7.6.6 Fact. *Let $b \in \mathbb{N}$. For each n, fewer than n^b strings of length n are $O(\log)$ bounded via b.*

Proof. It suffices to note that $\#\{\sigma\colon |\sigma| < b\log n\} < 2^{b\log n} \le n^b$. □

Merkle (2003) proved the following.

7.6.7 Theorem. *An $O(\log)$ bounded set Z is not partial computably stochastic.*

On the other hand, Z may be computably random by Theorem 7.4.11 and Exercise 8.2.31 that each facile set is $O(\log)$ bounded. Thus we have another proof of Theorem 7.5.7.

We need some preparations to prove 7.6.7. The following notation will be used throughout. For an interval $I = [n, m)$ let $Z \upharpoonright I = Z(n) \ldots Z(m-1)$. Moreover,

(a) $m_0 < m_1 < \ldots$ is a certain computable sequence of numbers,
(b) $u_p\colon = \sum_{i<p} m_i$, and
(c) $I^p\colon = [u_p, u_{p+1})$.

Thus $\#I^p = m_p$. The following lemma has to be refined to be of use in the proof of Theorem 7.6.7. However, in its present simpler form it serves us better as an introduction to the techniques.

7.6.8 Lemma. *Under the following hypotheses a set Z is not partial computably stochastic:*

- *m_p, u_p and I^p are as above, and, in addition, $10u_p \le m_p$ for each p (for instance, this is satisfied when $m_p = 11^p$);*
- *there is a uniformly c.e. sequence $(T^p)_{k\in\mathbb{N}}$ of strings of length m_p such that the following two properties hold:*
  (i) $\forall^\infty p\,[Z \upharpoonright I^p \in T^p]$, and
  (ii) $\exists^\infty p\,[\#T^p < 0.2m_p]$.

The lemma is motivated by the case that there is a rational $\delta < 1$ such that $\exists^\infty p\ C(Z \upharpoonright I^p) < \delta \log m_p$. (Roughly, this condition strengthens being $O(\log)$ bounded.) The hypotheses of the lemma are met for $T^p = \{z\colon |z| = m_p \;\&\; C(z) < \delta \log m_p\}$. In fact, for almost all p we have $\#T^p \le 2^{\delta \log m_p} \le (m_p)^\delta < 0.2m_p$.

Proof idea. We build selection rules ρ_a for $a \in \{0,1\}$, where ρ_a selects a bit position when it is guessing that the value of the given set at that position is a. Each bit position is selected by either ρ_0 or ρ_1. The bit positions in I^p will be selected in such a way that for infinitely many p, for one of the selection rules ρ_a the quotient of the number of selected bits up to u_{p+1} with value a by all selected bits is at least 4/7. Then this selection rule succeeds on Z, so Z is not partial computably stochastic.

The strings in T^p act as "advisors" for ρ_a. By (i) there is a number p_0 such that $Z \upharpoonright I^p \in T^p$ for each $p \ge p_0$. Given such a p, to make guesses on bit positions in I_p, the selection rule ρ_a hires the first advisor that appears in the given computable enumeration of T^p. It starts at position u_p, and believes the predictions of its current advisor as long as they turn out to be correct when the bit is revealed. When the prediction is a, it also selects that position.

The first time an advisor is wrong, he is fired, and ρ_a hires the next one in the computable enumeration of T^p. When a new advisor has been hired, ρ_a continues at the position following the one where the old advisor was fired, as long as I_p is not exhausted. The number of incorrect predictions ρ_a makes in this way is no more than the number of possible advisors. Thus, if by (ii) $p \ge p_0$ is one of the infinitely many numbers such that $\#T^p < 0.2m_p$, then one of the ρ_a will succeed in selecting a high fraction of bits with value a in I^p. We have to ensure that the sizes of the intervals I^p grow sufficiently fast so that the imbalance obtained on I^p outweighs the selections at all the previous intervals.

Proof details. Suppose (i) holds for all $p \geq p_0$. Fix $a \in \{0,1\}$. On input a string y, the following procedure attempts to compute $\rho_a(y) \in \{0,1\}$, where $\rho_a(y) = 1$ means that the next bit is selected.

(1) Let p be so that $|y| \in I_p$. IF $p < p_0$, output 0 and END.
(2) Let $i = u_p$ and WAIT for the first element v to be enumerated into T^p.
(3) (a) IF $i = |y|$ THEN: IF $v(i) = a$, output 1, ELSE output 0; END.
    (b) ELSE: IF $v(i) \neq y(i)$, let v be the next element enumerated into T^p (if necessary wait here for v to appear).
(4) Increment i and goto (3).

If $y \prec Z$ then we never get stuck when waiting for a new advisor v at (3b), because $Z \upharpoonright I^p \in T^p$ for $p \geq p_0$. So the procedure ends. Moreover, each position of Z is selected by either ρ_0 or ρ_1. Let us say the position i is *selected correctly* if for some $a \in \{0,1\}$, i is selected by ρ_a (i.e., $\rho_a(Z\upharpoonright_i) = 1$) and indeed $Z(i) = a$. By (ii) there are infinitely many $p \geq p_0$ such that $\#T^p \leq 0.2m_p$. Fix such a p. Firstly, we will bound from above the number of positions $i < u_{p+1}$ selected *incorrectly*.

(a) For $i \in I^p$, each time a position is selected incorrectly, the advisor v is changed. Hence fewer than $0.2m_p$ such positions are selected incorrectly.
(b) For $i < u_p$, at most $0.1m_p$ positions are selected at all since $10u_p \leq m_p$.

Alltogether, at most $0.3m_p$ positions are selected incorrectly.

Secondly, by (a) again, at least $0.8m_p$ positions $i \in I^p$ are selected correctly as $\#I^p = m_p$, so for some $a \in \{0,1\}$ at least $0.4m_p$ positions $i \in I^p$ are selected by ρ_a when $Z(i) = a$, and the remaining among those selected correctly are not selected by ρ_a. We now have an imbalance for this a: among the bits up to u_{p+1} selected by ρ_a, the fraction of the ones with value a is at least 4/7. Moreover, for one of ρ_0, ρ_1, this holds for infinitely many p such that $\#T^p \leq 0.2m_p$. So Z is not partial computably stochastic. □

In Lemma 7.6.8 we had in mind as the sets T^p the strings y of length m_p such that $C(y) < \delta \log m_p$ where $\delta < 1$ is fixed. Let us see how close we can get to the hypothesis of Lemma 7.6.8 under our actual assumption that Z is $O(\log)$ bounded via b, assuming also that the sequence $m_0 < m_1 < \ldots$ grows fast enough. Let $A_p = \{w\colon\ |w| = m_p \ \&\ C(w) < (b+1)\log m_p\}$. By Fact 7.6.6,

$$\#A_p < (m_p)^{b+1}. \tag{7.18}$$

7.6.9 Lemma. *Suppose $u_p = \sum_{i<p} m_i \leq m_p$ for each p (for instance, this is the case if $m_p = 2^p$). If Z is $O(\log)$ bounded via b then $\forall^\infty p\ Z \upharpoonright I^p \in A_p$.*

Proof. Since the sequence $(u_p)_{p\in\mathbb{N}}$ is computable, there is a machine M such that $M(\sigma)$ is the string obtained by removing the first u_p bits from $\mathbb{V}(\sigma)$ in the case that $\mathbb{V}(\sigma)\downarrow = y$ and $|y| = u_{p+1}$. If $\mathbb{V}(\sigma) = Z \upharpoonright u_{p+1}$ then $M(\sigma) = Z \upharpoonright I^p$. For each p we have $u_{p+1} \leq 2m_p$. Thus, for almost all p,

$$C_M(Z \restriction I^p) < C(Z \restriction u_{p+1}) \leq b \log u_{p+1} \leq b(1 + \log m_p).$$

This implies $C(Z \restriction I^p) < (b+1) \log m_p$ for almost all p. (Using $b+1$ instead of b we have beaten the coding constant for M in the first inequality.) □

Proof of Theorem 7.6.7. It is tempting to use selection rules as in the proof of Lemma 7.6.8, taking the strings in A_p as advisors. The problem is that $\#A_p$ is merely bounded by $(m_p)^{b+1}$, not by $0.2m_p$ as in Lemma 7.6.8. Let

$$k = b + 2.$$

To get around this problem we use k levels of selection rules $\rho_{a,s}$ for each a. The selection rule $\rho_{a,1}$ is pictured at the bottom. The variables s and t range over $\{1, \ldots, k\}$. We define the sequence $(m_p)_{p \in \mathbb{N}}$ by $m_0 = 1$ and, for $p > 0$,

$$m_p := 10ku_p,$$

where as always $u_p = \sum_{i<p} m_i$. We also use the notation $l_p := m_p/k = 10u_p$. We split $I^p = [u_p, u_{p+1})$ into consecutive subintervals J_s^p of length l_p, for $s \in \{1, \ldots, k\}$, namely, $J_s^p = [u_p + (s-1)l_p, u_p + sl_p)$. See Fig. 7.1.

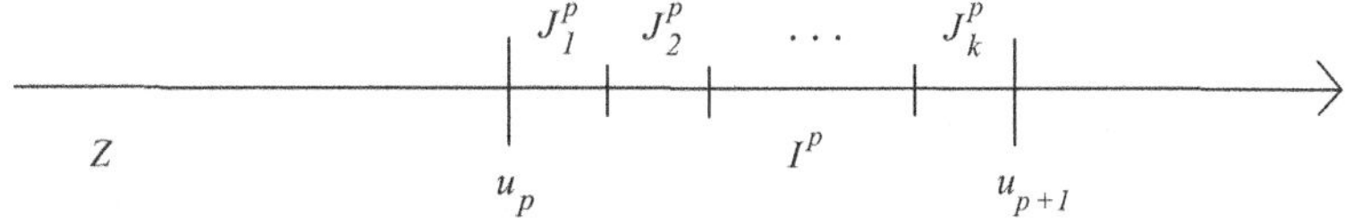

FIG. 7.1. Splitting I^p.

For each $s \in \{1, \ldots, k\}$, we let $\widehat{s} = k + 1 - s$, and similarly for $t \in \{1, \ldots, k\}$. The selection rule $\rho_{a,s}$, $a \in \{0, 1\}$, $s \in \{1, \ldots, k\}$, selects positions in intervals of the form $J_{\widehat{s}}^p$. Thus $\rho_{a,1}$ selects from the last subinterval, and $\rho_{a,k}$ selects from the first. Which set of advisors should $\rho_{a,s}$ use? For $s \in \{1, \ldots, k\}$, consider a string z of length $(\widehat{s} - 1)l_p$. We let (see Figure 7.2)

$$T_s^p(z) = \{v : |v| = l_p \text{ and there are at least } (0.2l_p)^{s-1} \text{ strings } y \text{ such that } zvy \in A^p\}.$$

The string z is thought of as the bits of Z in the positions from u_p to $u_p + (\widehat{s}-1)l_p$. The selection rule has read z (without selecting any bits), and uses the strings in $T_s^p(z)$ as its advisors. Note that $T_s^p(z)$ is c.e. uniformly in p, s and z. Consider the extreme cases: if $s = 1$ then $|z| = (\widehat{s} - 1)l_p = (k - 1)l_p$, so $y = \varnothing$, and thus $T_1^p(z) = \{v\colon |v| = l_p \;\&\; zv \in A_p\}$. On the other hand, if $s = k$ we have $z = \varnothing$, and therefore

$$T_k^p(z) = \{v : |v| = l_p \text{ and there are at least } (0.2l_p)^{k-1} \text{ strings } y \text{ such that } vy \in A^p\}. \tag{7.19}$$

For each $p \in \mathbb{N}$, $s \in \{1, \ldots, k\}$, let

$$z_{p,s} = Z \restriction [u_p, u_p + (\widehat{s} - 1)l_p).$$

(In particular, $z_{p,1} = Z \upharpoonright [u_p, u_p + (k-1)l_p)$ and $z_{p,k} = \varnothing$.) Let us show that for some $t \in \{1, \ldots, k\}$, for infinitely many p the set of advisors when $\rho_{a,t}$ is processing Z has a size of less than $0.2l_p$. Thereafter, we can argue as in the proof of Lemma 7.6.8. The idea is that otherwise, because there are k levels, for almost every p the size of A_p exceeds the bound $(m_p)^{k-1}$ in (7.18).

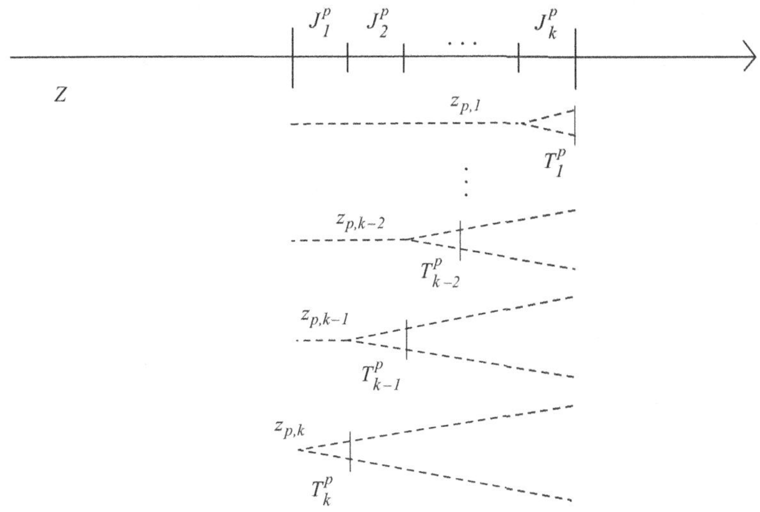

FIG. 7.2. The sets $T^p_s(z_{p,s})$.

7.6.10 Lemma. *There is* $t \in \{1, \ldots, k\}$ *such that*

(i)$_t$ $\forall^\infty p \;\; Z \upharpoonright J^p_t \in T^p_t(z_{p,t})$.

(ii)$_t$ $\exists^\infty p \;\; \#T^p_t(z_{p,t}) < 0.2l_p$.

Proof. (i)$_1$ states that $Z \upharpoonright I^p \in A_p$ for almost all p, that is, $C(Z \upharpoonright I^p) < (b+1)\log m_p$ (since $k = b+2$), which holds by Lemma 7.6.9. If (ii)$_1$ is satisfied as well, we are done. Otherwise, for almost all p we have $\#T^p_1(z_{p,1}) \geq 0.2l_p$, which means that (i)$_2$ holds. More generally, if (i)$_t$ holds for some $t < k$, then either (ii)$_t$ is satisfied as well and we are done, or (i)$_{t+1}$ holds. Suppose that, iterating the argument, we proceed all the way to (i)$_k$. If (ii)$_k$ fails then, where $\epsilon = (0.2/k)^k$, for almost all p by (7.19) we have

$$\#A^p \geq (0.2l_p)^{k-1} \#T^p_k(\varnothing) \geq (0.2l_p)^k = (0.2m_p/k)^k = \epsilon(m_p)^k \geq (m_p)^{k-1},$$

contrary to (7.18). So (ii)$_k$ holds. □

Fix $a \in \{0,1\}$. It remains to describe the selection rules $\rho_{a,s}$ in detail. We adapt the procedure in the proof of Lemma 7.6.8. Now $\rho_{a,s}$ does within the intervals J^p_s what ρ_a did in the intervals I^p, and $\rho_{a,s}$ uses the strings in $T^p_s(z)$ as sets of advisors for an appropriate z derived from the input.

On input y the procedure attempts to compute $\rho_{a,s}(y)$ as follows.

(1) Let p, t be so that $|y| \in J^p_t$. IF $p < p_0$ or $s \neq t$, output 0 and END. ELSE let $u = u_p + (\widehat{s} - 1)l_p$ and $z = y \restriction [u_p, u)$.
(2) Let $i = u$ and WAIT for the first element v to be enumerated into $T^p_s(z)$.
(3) (a) IF $i = |y|$ THEN: IF $v(i) = a$, output 1, ELSE output 0; END.
 (b) ELSE: IF $v(i) \neq y(i)$, let v be the next element enumerated in $T^p_s(z)$ (if necessary, wait here for v to appear).
(4) Increment i and GOTO (3).

Let t be as in Lemma 7.6.10. As in the proof of 7.6.8, we may argue that for some $a \in \{0, 1\}$ the selection rule $\rho_{a,t}$ succeeds on Z. We use that now $m_p = 10ku_p$, so we obtain the imbalance even though $\rho_{a,t}$ only selects bits in subintervals of the form $J^p_{\widehat{t}}$. □

The method used is related to the decanter/golden run methods in Section 5.4. In both cases there is a constant number of levels. Lemma 7.6.10 is analogous to Lemma 5.4.9 that the golden run exists.

7.6.11$^\diamond$ **Problem.** Decide whether the weaker hypothesis suffices in Theorem 7.6.7 that $\forall r \in S\, [C(Z \restriction_r) \le b \log r]$ for some infinite computable set S and $b \in \mathbb{N}$.

Nonmonotonic betting strategies

So far, our strategies have processed the bits of a set in the usual ascending order. A strategy becomes more powerful if we allow it to bet in an order it chooses. This leads to the notion of Kolmogorov–Loveland randomness (7.6.19) which is not known to differ from ML-randomness. First we study the simplest case of a nonmonotonic betting strategy. A permutation betting strategy bets on the bits in the order $\pi(0), \pi(1), \pi(2) \ldots$ where π is a computable permutation of $\mathbb{N}$. Thus, all the bits are considered in an order which does not depend on the particular given set. For most types of betting strategies this yields no advantage over the monotonic case, because, as mentioned in Section 1.3, classes we study are usually closed under computably permutations. We proved this closure property for ML-randomness (3.2.16) and for Schnorr randomness (3.5.12) by directly inspecting the definitions via (variants of) ML-tests. No such definition has been given for computable randomness, so we will have to work harder to establish its closure under computable permutations (Theorem 7.6.24). In contrast, the class PCR of partial computably random sets is one of the few interesting classes where this closure property fails (Corollary 7.6.15). Thus, the following randomness notion introduced by Miller and Nies (2006) is strictly stronger than partial computable randomness.

7.6.12 Definition. Z is *permutation random* if $Z \circ \pi$ is partial computably random for each computable permutation π. A *permutation betting strategy* is a pair $P = (\pi, B)$ consisting of a computable permutation and a partial computable martingale. We say that P succeeds on Z if $Z \circ \pi \in \mathsf{Succ}(B)$.

The permutation random sets form the largest subclass of PCR that is closed under computable permutations. Kastermans and Lempp (2010) constructed a permutation random set that is not ML-random. It is not known whether such a set can be left-c.e.; it is also open whether a permutation random set must be of d.n.c. degree.

Muchnik's splitting technique

We give two applications of a technique due to Andrej Muchnik (see Muchnik, Semenov and Uspensky 1998, Thm. 9.1) to show that a set Z satisfies a property indicating non-randomness.

Let S be an infinite co-infinite computable set. Split Z into sets Z_0 and Z_1, the bits with positions in S and in $\mathbb{N} - S$, respectively. In both applications, the hypothesis states that for each $r \in \{0, 1\}$, for each $m \in \mathbb{N}$, the set Z_r satisfies a "local" non-randomness condition $\mathcal{P}_{m,r}$. These conditions are uniformly $\Sigma_1(Z_r)$; let $t_{m,r}$ be the stage by which we discover that $\mathcal{P}_{m,r}$ holds for Z_r. We may choose p such that $t_{m,p} \geq t_{m,1-p}$ for infinitely many m.

Theorem 7.6.13, due to Merkle *et al.* (2006), is our first application of this technique. We use Z_p as an oracle set to build a test for Z_{1-p}.

Theorem 7.6.14, for which the technique was introduced by Muchnik, is our second application. It uses the oracle set Z_p in a more specific way: the required bits of Z_p are read by a permutation betting strategy in order to succeed on Z.

7.6.13 Theorem. *Suppose $Z = Z_0 \oplus Z_1$ and neither Z_0 nor Z_1 is ML-random. Then Z_{1-p} is not Schnorr random relative to Z_p for some $p \in \{0, 1\}$.*

Proof. Suppose that for each $r \in \{0, 1\}$, $(G^r_m)_{m \in \mathbb{N}}$ is a ML-test such that $Z_r \in \bigcap_m G^r_m$. (The local conditon $\mathcal{P}_{m.r}$ states that $Z_r \in G^r_m$.)

1. Let $G^r_{m,s}$ be the approximation of G^r_m at stage s by strings of length at most s as in (1.16) on page 54.
2. Let $t_{m,r} = \mu t.\, Z_r \in [G^r_{m,t}]^{\prec}$. Note that $\lambda m.\, t_{m,r} \leq_T Z_r$.
3. Let $p \in \{0, 1\}$ be least such that $\exists^\infty m\, [t_{m,p} \geq t_{m,1-p}]$.
4. Let C^p_m be the clopen set $[G^{1-p}_{m,t_{m,p}}]^{\prec}$.
5. Let $V_n = \bigcup_{m>n} C^p_m$. Then $Z_{1-p} \in \bigcap_n V_n$ by the choice of p.

We show that $(V_n)_{n \in \mathbb{N}}$ is a Schnorr test relative to Z_p. Clearly, V_n is uniformly c.e. relative to Z_p and $\lambda V_n \leq 2^{-n}$. Also, λV_n is uniformly computable relative to Z_p, because for $k > n$ the rational $\alpha_k = \lambda \bigcup_{n<m\leq k} C^p_m$ is computable in Z_p, and $\lambda V_n - \alpha_k \leq \lambda \bigcup_{m>k} C^p_m \leq 2^{-k}$; now apply 1.8.15(iv) relative to Z_p. □

The second application shows that for a permutation random set the initial segment complexity (in the sense of C, and hence also in the sense of K) has to be large at infinitely many positions in a given infinite computable set. From this we will conclude that PCR is not closed under computable permutations.

7.6.14 Theorem. (Andrej Muchnik) *Suppose Z is a set such that, for some strictly increasing computable function g with $g(1) > 1$, we have $\forall k\, [C(Z \upharpoonright_{g(k)}) \le^+ g(k) - k]$. Then Z is not permutation random.*

Proof. First we show that there is a strictly increasing computable h such that, where $(I_n)_{n\in\mathbb{N}}$ are the consecutive intervals of $\mathbb{N}$ of length $h(n)$, we have

$$\forall n\, C(Z \upharpoonright I_n) \le^+ h(n) - n.$$

Let $h(0) = 1$ and $h(n) = g(n + \widehat{h}(n)) - \widehat{h}(n)$ where $\widehat{h}(n) = \sum_{i<n} h(i)$. Similar to the proof of Lemma 7.6.9, let M be a machine such that $M(\sigma)$ is the string obtained by removing the first $\widehat{h}(n)$ bits from y in the case that $\mathbb{V}(\sigma) = y$ and $|y| = g(n + \widehat{h}(n)) = \widehat{h}(n+1)$. Then

$$C_M(Z \upharpoonright I_n) \le C(Z \upharpoonright_{\widehat{h}(n+1)}) \le^+ \widehat{h}(n+1) - \widehat{h}(n) - n,$$

and hence $C(Z \upharpoonright I_n) \le^+ h(n) - n$.

Let d be such that $\forall n\, C(Z \upharpoonright I_n) \le h(n) - n + d$. Let $b_{m,r} = 2m + r - d$ for $m \in \mathbb{N}$ and $r \in \{0,1\}$, then $C(Z \upharpoonright I_{2m+r}) \le h(2m + r) - b_{m,r}$. (This is the local conditon $\mathcal{P}_{m.r}$, where the splitting is given by $S = \bigcup_m I_{2m}$.) We describe permutation betting strategies Q_r ($r \in \{0,1\}$) one of which succeeds on Z. While Q_0 is monotonic, Q_1 processes the intervals in the order $I_1, I_0, I_3, I_2, \ldots$. Within each interval, Q_1 bets monotonically.

We describe the strategies Q_r informally. The start capital of either one is 2^{d+2}. Suppose $m \in \mathbb{N}$ and on a set Y, Q_r has already processed all the intervals I_k for $k < 2m$.

(1) Read $x = Y \upharpoonright I_{2m+r}$. WAIT for t such that $C_t(x) \le |x| - b_{m,r}$.

(2) Let $E = E_{m,r,t}$ be the clopen set

$$[\{y\colon\ |y| = h(2m + 1 - r)\ \&\ C_t(y) \le |y| - b_{m,1-r}\}]^{\prec}.$$

Note that $\lambda E \le 2^{-b_{m,1-r}+1}$. Now bet on the bits in I_{2m+1-r} in monotonic order as follows: for $|z| < h(2m+1-r)$ and $a \in \{0,1\}$ bet $\lambda(za \mid E)$ on za. That is, bet locally like B_E in (7.2).

Since $\sum_{m,r} 2^{-b_{m,1-r}+1} \le 2^{d+2}$, sufficient capital is available along Y.

We consider the behavior of the strategies on the given set Z. Let

$$t_{m,r} = \mu t.\, C_t(Z \upharpoonright I_n) \le h(n) - n + d \text{ where } n = 2m + r.$$

Let $p \in \{0,1\}$ be such that $t_{m,p} \ge t_{m,1-p}$ for infinitely many m. Then Q_p succeeds on Z because for each such m we have $Z \upharpoonright I_{2m+1-p} \in E_{m,p,t}$ for $t = t_{m,p}$. So Q_p increases its capital by 1 for this m. □

7.6.15 Corollary. *Some partial computably random set Z is not permutation random.*

Proof. By Theorem 7.4.15 there is a partial computably random set Z such that $\forall m\, C(Z \upharpoonright_m) \le^+ m/2$. Then the hypothesis of the foregoing theorem is satisfied via $g(k) = 2k$, so Z is not permutation random. □

Kolmogorov–Loveland randomness

Most of the material in this subsection is due to Muchnik, Semenov and Uspensky (1998) or Merkle, Miller, Nies, Reimann and Stephan (2006).

A nonmonotonic computable betting strategy determines the next position of a set Y where a bet is placed from the previous positions and the values of Y at these positions. For instance, initially the strategy bets at position 3; if $Y(3) = 0$ it next bets at position 1, otherwise at 2, and so on.

The history up to the n-th bet of betting positions and values is given by a finite assignment $\alpha = (\langle d_0, r_0\rangle, \ldots, \langle d_{n-1}, r_{n-1}\rangle)$ as defined in Section 1.3. Thus all $d_i \in \mathbb{N}$ are distinct and $r_i \in \{0,1\}$. We let $\mathrm{dom}(\alpha)$ denote the set $\{d_0, \ldots, d_{n-1}\}$, called the set of *positions* of α. We write $\alpha \lhd Y$ if $Y(d_i) = r_i$ for each $i < n$. Let FinA denote the set of finite assigments, and let the variable α range over FinA. We fix some effective encoding of finite assignments by natural numbers.

7.6.16 Definition. A *partial computable scan rule* is a partial computable function $S\colon \mathsf{FinA} \to \mathbb{N}$ such that $S(\alpha) \notin \mathrm{dom}(\alpha)$ whenever $S(\alpha)$ is defined.

In the example above we have $S(\emptyset) = 3$, $S(\langle 3, 0\rangle) = 1$, and $S(\langle 3, 1\rangle) = 2$. Each one-one computable function f determines a partial computable scan rule S_f that picks the positions $f(0), f(1), \ldots$; formally, $S_f(\alpha) = f(n)$ if $\mathrm{dom}(\alpha) = \{f(0), \ldots, f(n-1)\}$.

For the remainder of this section, the variables Y, Z denote either a string or an infinite sequence of zeros and ones. To indicate that the latter case applies, we say that Y is a *set.* The finite assigment after n bets is given by a Turing functional Θ_S. Let $\Theta_S^Y(0) = \varnothing$. If $\Theta_S^Y(n) = \alpha$ has been defined, let $\Theta_S^Y(n+1) \simeq \alpha\widehat{\ }\langle S(\alpha), Y(S(\alpha))\rangle$.

Let $S(Y)$ be the sequence Z of bits such that $Z(n)$ is the second component of the last entry of $\Theta_S^Y(n+1)$ if defined. A partial computable scan rule S is called *total* if $S(Y)$ is an infinite sequence (that is, a set) for each set Y.

7.6.17 Remark. If S is total we may assume that $S(\alpha)$ is defined for each $\alpha \in \mathsf{FinA}$: test for $i = 0, 1, \ldots$ whether $S(\alpha \upharpoonright_i)\downarrow$. Either (1) we reach $i = |\alpha|$ or (2) we find $i < |\alpha|$ such that $S(\alpha \upharpoonright_i) \downarrow \neq (\alpha_i)_0$; otherwise, $S(Y)$ fails to be a set for each set $Y \rhd \alpha \upharpoonright_{i+1}$. In case (2) we may vacuously extend S by defining $S(\alpha) = \min(\mathbb{N} - \mathrm{dom}(\alpha))$ without changing $S(Y)$ for any set Y.

7.6.18 Remark. For each n, by definition $S(Y)\upharpoonright_n$ is given by $\Theta_S^Y(n)$. Conversely, $\Theta_S^Y(n)$ is determined by $S(Y)\upharpoonright_n$. This is clear for $n = 0$. If we have already found $\alpha = \Theta_S^Y(n)$ and $d = S(\alpha)$ is defined, then $\Theta_S^Y(n+1) = \alpha\langle d, S(Y)_n\rangle$.

The main notion of this subsection was introduced independently by Kolmogorov (1963) and Loveland (1966).

7.6.19 Definition. A *Kolmogorov–Loveland (KL) betting strategy* is a pair $\rho = (S, B)$ where S is a partial computable scan rule, and B is a partial computable martingale; ρ *succeeds* on a set Z if $S(Z)$ is a set on which B succeeds.

We say that Z is *Kolmogorov-Loveland (KL-)random* if no KL-betting strategy succeeds on it.

Thus, Z is KL-random if $S(Z)$ is partial computably random whenever it is a set. By Exercise 7.6.25 we may assume that S and B are total.

7.6.20 Proposition. *Each Martin-Löf random set Z is KL-random.*

Proof. We use ideas from the proof of Proposition 3.2.16. Given a partial computable scan rule S, for each string x let $C_x = \{Y \in 2^{\mathbb{N}} \colon x \preceq S(Y)\}$. Note that C_x is a c.e. open set uniformly in x. Moreover, either $C_{x0} = C_{x1} = \emptyset$ (this can only happen if S is partial) or these two sets split C_x into parts of the same measure. Then by induction on $|x|$ we have $\lambda C_x \leq 2^{-|x|}$ for each x. If G is c.e. open then so is $G^* = \bigcup_{[x]\subseteq G} C_x$, and $\lambda G^* \leq \lambda G$.

Now suppose the KL-betting strategy (S, B) succeeds on Z. We may assume that $B(\varnothing) \leq 1$. By Proposition 7.1.9 the c.e. open set $G_d = \{X \colon \exists r\, B(X\restriction_r) > 2^{-d}\}$ satisfies $\lambda G_d \leq 2^{-d}$. Thus Z fails the Martin-Löf test $(G^*_d)_{d\in\mathbb{N}}$. □

Each KL-betting strategy has an "Achilles heel": it does not succeed on some c.e. set by Exercise 7.6.26. However, by Merkle *et al.* (2006) there are two KL-betting strategies such that on each c.e. set one of them succeeds. In fact, it is an open question whether there are two KL-betting strategies such that on each set that is not ML-random one of them succeeds.

Although we do not know whether each KL-random set is already Martin-Löf random, various results of Merkle *et al.* (2006) show that KL-randomness is at least closer to ML-randomness than the previously encountered notions (except for, possibly, permutation randomness).

7.6.21 Theorem. *If $Z = Z_0 \oplus Z_1$ is KL random, then Z_p is Martin-Löf random for some $p \in \{0, 1\}$.*

Sketch of proof. Suppose otherwise, then by Theorem 7.6.13 Z_{1-p} is not Schnorr random relative to Z_p for some $p \in \{0,1\}$, say $p = 0$. By Proposition 7.3.2 relative to Z_0 we may choose a Turing functional Ψ such that Ψ^{Z_0} is a $\mathbb{Q}_2$-valued martingale that succeeds on Z_1. We use Ψ to build a KL-betting strategy that succeeds on Z. We may assume $\Psi^{\sigma}_{|\sigma|}(y)\downarrow$ implies $\Psi^{\sigma}_{|\sigma|}(x)\downarrow$ for each $x \prec y$. The strategy reads more and more bits on the Z_0-side in ascending order without betting until a new computation $q = \Psi^{\sigma}_{|\sigma|}(x0)$ appears, where $\sigma \prec Z_0$ has been scanned on the Z_0-side, and $x \prec Z_1$ on the Z_1-side. Then it bets $q - \Psi^{\sigma}_{|\sigma|}(x)$ on the prediction that the n-th bit on the Z_1-side is 0. Each bit on the Z_1-side is scanned eventually, so the strategy succeeds on Z.

In more detail, the sequence of scanned bits has the form $\sigma_0 b_0 \sigma_1 \ldots b_{n-1}\sigma_n$, where the σ_i are strings of consecutive oracle bits on the Z_0-side and the b_j are consecutive bits on the Z_1-side. To formalize the above by a KL-betting strategy (S, B), for instance let $n = 1$. The finite assigment corresponding to $\sigma_0 b_0 \sigma_1$, where $k_i = |\sigma_i|$, is

$$\alpha = (\langle 0, \sigma_{0,0}\rangle, \ldots, \langle 2(k_0 - 1), \sigma_{0,k_0-1}\rangle, \langle 1, b_0\rangle, \langle 2k_0, \sigma_{1,0}\rangle, \ldots, \langle 2(k_0 + k_1 - 1), \sigma_{1,k_1-1}\rangle,$$

and we define $S(\alpha) = 3$ since $q = \Psi^{\sigma_0\sigma_1}_{k_0+k_1}(b_0 0)$ newly converges. Also $B(w) = 1$ for $w \preceq \sigma_0$, $B(w) = \Psi^{\sigma_0}(b_0)$ for $\sigma_0 b_0 \preceq w \preceq \sigma_0 b_0 \sigma_1$, and $B(\sigma_0 b_0 \sigma_1 0) = q$. □

The following is an effective version of the Hausdorff dimension of the class $\{Z\}$. See for instance Downey et al. (2006).

7.6.22 Definition. Let $\dim(Z) = \liminf_n K(Z\restriction_n)/n$.

Note that by Corollary 2.4.2 we could as well take C instead of K. By Schnorr's Theorem 3.2.9 and Proposition 4.1.2 we have:

7.6.23 Corollary. *If Z is KL-random then $\dim(Z) \geq 1/2$, and Z has d.n.c. degree via a finite variant of the function $\lambda n.\, Z\restriction_n$.* □

In particular, each left-c.e. KL-random set is *wtt*-complete by the completeness criterion 4.1.11. It is not known whether such a set is already ML-random. An extension of the argument in Theorem 7.6.21 shows that $\dim(Z) = 1$ for each KL-random set Z. Further, if Z is Δ^0_2 then both Z_0 and Z_1 are ML-random. See Merkle *et al.* (2006).

We say that a set Z is *KL-stochastic* if, for each partial computable scan rule S, if $S(Z)$ is a set then $S(Z)$ is partial computably stochastic. Merkle *et al.* (2006) showed that actually $\dim(Z) = 1$ for each KL-stochastic set. On the other hand, there is a nonempty Π^0_1 class of KL-stochastic sets that are not even weakly random.

The following implies that CR is closed under computable permutations.

7.6.24 Theorem. *Suppose that S is a partial computable scan rule that scans all positions for each set Y, namely, $S(Y)$ is a set and $\bigcup_n \operatorname{dom} \Theta^Y_S(n) = \mathbb{N}$. Then, if Z is computably random, so is $S(Z)$.*

Sketch of proof. S is total, so by 7.6.17 we may assume that $S(\alpha)$ is defined for each $\alpha \in \mathsf{FinA}$. Suppose the computable martingale B succeeds on $S(Z)$. We may assume that $B(x) > 0$ for each x and, by the Savings Lemma 7.1.14, that $\lim_n B(S(Z)\restriction_n) = \infty$. We define a computable sequence $0 = n_0 < n_1 < \ldots$ by

$$n_{i+1} = \mu n > n_i.\, \forall y\, [|y| = n \;\rightarrow\; \exists r\, [0, n_i) \subseteq \operatorname{dom} \Theta^y_S(r)]. \tag{7.20}$$

That is, n_{i+1} is the least $n > n_i$ such that for each y of length n all the positions in $[0, n_i)$ have been scanned. For some $p \in \{0, 1\}$ we may assume that the KL-betting strategy $\rho = (S, B)$ only bets on positions in $[n_{2k+p}, n_{2k+p+1})$, because the product over all l of the betting factors $B(S(Z)\restriction_{l+1})/B(S(Z)\restriction_l)$ is infinite. Say $p = 0$.

We inductively define a computable martingale D that succeeds on Z. Let $D(\varnothing) = B(\varnothing)$. Suppose $|x| = n_{2k}$ and $D(x)$ has been defined. Let $m = n_{2k+2} - n_{2k}$. We will define $D(xv)$ for each string v of length at most m.

For each string u of length m, let r_u be least such that $[0, n_{2k+1}) \subseteq \operatorname{dom} \Theta^{xu}_S(r_u)$. Note that $n_{2k+2} \geq r_u \geq n_{2k+1}$. Let

$$w_u = S(xu) \restriction r_u. \tag{7.21}$$

To proceed with the definition of D we need to verify the following.

Claim. $D(x) = \sum_{|u|=m} 2^{-m} B(w_u)$.

Then, recalling the martingales E_u from 7.1.4, define for $|v| \leq m$

$$D(xv) = \sum_{|u|=m} 2^{-m} B(w_u) E_u(v). \tag{7.22}$$

For $v = \emptyset$ the right hand side is $D(x)$ by the claim. For $|v| = m$ we have the value $B(w_v)$. Since $\lim_n B(S(Z)\restriction_n) = \infty$ this shows $D(Z) = \infty$.

To prove the claim, let G be the prefix-free set $\{w_u : |u| = m\}$. Since $S(\alpha)$ is defined for each $\alpha \in \mathsf{FinA}$, from a string w we can determine the finite assignment $\alpha = \alpha_w$, $|\alpha| = |w|$ that induces w in the sense that $w(i) = S(\alpha \restriction_i)$ for $i < |w|$ (as in Remark 7.6.18). Let q_w be the number of positions of α_w in $[n_{2k}, n_{2k+2})$. Then

$$D(x) = \textstyle\sum_{w \in G} 2^{-q_w} B(w).$$

To see this, apply Lemma 7.1.8(ii) to the "local" martingale that starts with the capital $D(x)$ and bets like B along each α_w, except that for $k > 0$ it ignores the positions less than n_{2k}. On these positions B does not bet anyway, since all the positions less than n_{2k-1} have been visited already after processing x, namely, $[0, n_{2k-1}) \subseteq \mathrm{dom}\, \Theta_S^x$. This leaves q_w positions.

For each $w \in G$ there are 2^{m-q_w} strings u of length m such that $\alpha_w \triangleleft xu$, or equivalently $w_u = w$. Hence $D(x) = \sum_{w \in G} 2^{-q_w} B(w) = \sum_{|u|=m} 2^{-m} B(w_u)$. This proves the claim. □

By the solution to Exercise 7.6.25, each KL-betting strategy can be replaced by a KL-betting strategy (S, B) with total S and B that on every set scans at least as many positions as the given one. Thus, by the foregoing theorem, to be more powerful than a monotonic betting strategy, a KL-betting strategy needs to avoid scanning positions on some sets (actually, on a class of sets of non-zero measure by Exercise 7.6.28).

Exercises.

7.6.25. (Merkle) Show that, if Z is not KL-random, then some KL-betting strategy (S, B) with total scan rule S and total B suceeds on Z.

7.6.26. (i) Show that for each computable set E and each partial computable scan rule S, there is a c.e. set A such that either $S(A)$ is not a set or $S(A) = E$.
(ii) Use 7.4.6 to infer that each KL-betting strategy (S, B) does not succeed on some c.e. set A.

7.6.27. Extend Schnorr's Theorem to finite assignments:
Z is ML-random $\Leftrightarrow \exists b \forall \alpha \triangleleft Z [K(\alpha) > |\alpha| - b]$.

7.6.28. Strengthen 7.6.24: if a KL-betting strategy (S, B) succeeds on a computably random set Z, then the probability that S scans all places of a set Y is less than 1.

8

CLASSES OF COMPUTATIONAL COMPLEXITY

A lowness property of a set specifies a sense in which the set is computationally weak. In this chapter we complete our work on lowness properties and how they relate to randomness relative to an oracle. Recall that Low(MLR) denotes the sets that are low for ML-randomness. In Chapter 5 we showed that this property coincides with a number of other lowness properties, such as being low for K. It also coincides with K-triviality, a property that expresses being far from ML-random. We will carry out similar investigations for two classes that are variants of Low(MLR). We give examples of incomputable sets in each class and prove the coincidence with classes that arise from a different context.

(1) By Definition 3.6.17, Low(Ω) is the class of sets A such that Ω is ML-random relative to A. Clearly, Low(MLR) is contained in Low(Ω). Note that the class Low(Ω) is conull because each 2-random set is low for Ω. We think of being low for Ω as a weak lowness property. However, it is very restrictive within the Δ^0_2 sets: each Δ^0_2 set that is low for Ω is a base for ML-randomness, and hence is in Low(MLR) by 5.1.22.

We prove that each nonempty Π^0_1 class contains a set that is low for Ω. Thereafter we show that Low(Ω) coincides with the class of sets A that are weakly low for K, namely, $\exists b \, \exists^\infty y \, [K^A(y) \geq K(y) - b]$. This yields a further characterization of 2-randomness via the initial segment complexity.

(2) The second variant of Low(MLR) is the class Low(SR), the sets that are low for Schnorr randomness. In Theorem 3.5.19 we characterized Schnorr randomness in terms of the initial segment complexity given by computable measure machines. In this chapter we show that Low(SR) coincides with the class of sets that are low for computable measure machines. This corresponds to the coincidence of Low(MLR) and being low for K.

More importantly, the sets in Low(SR) can be characterized by being computably traceable, a restriction on the functions they compute. This characterization is a further example of the interactions between randomness and computability in both directions. A computability theoretic property characterizes lowness for Schnorr randomness, but also, Schnorr randomness helps us to understand what this property means.

The general idea of traceability of a set A is that the possible values of each function computed by A are in finite sets which, together with a bound on their size, can be determined in some effective way from the argument. The set A is computably traceable (Terwijn and Zambella, 2001) if for each function $f \leq_T A$, each value $f(n)$ is in a finite set $D_{g(n)}$, where g is a computable function depending on f; furthermore, there is a computable bound on the size $\#D_{g(n)}$ independent of f (see Definition 8.2.15 below for the details). If we merely require that the trace be uniformly c.e., we obtain the weaker notion of c.e. traceability. A set is computably traceable iff it is c.e. traceable and computably dominated.

Broadly speaking, lowness for Schnorr randomness and computable traceability coincide because they are linked by the idea of a covering procedure (see page 226). A is low for Schnorr randomness iff A is low for Schnorr null classes, that is, for each Schnorr test $(G_m)_{m\in\mathbb{N}}$ relative to A there is a Schnorr test $(H_k)_{k\in\mathbb{N}}$ such that $\bigcap_m G_m$ is covered by $\bigcap_k H_k$. A is computably traceable if for each $f \leq_T A$ we can trace, or "cover", the possible values $f(n)$ by a finite set obtained without the help of A. The proof of this coincidence shows how a covering procedure of one type can be transformed into a covering procedure of the other type.

How about being low for computable randomness? Surprisingly, the only sets of this kind are the computable ones. Recall from Definition 5.1.32 that for randomness notions $\mathcal{C} \subseteq \mathcal{D}$, we denote by $\mathrm{Low}(\mathcal{C},\mathcal{D})$ the class of sets A such that $\mathcal{C} \subseteq \mathcal{D}^A$. We determine the lowness notions for SR and CR by first characterizing $\mathrm{Low}(\mathcal{C},\mathcal{D})$ for each pair $\mathcal{C} \subset \mathcal{D}$ among the three randomness notions under discussion. We show that $\mathrm{Low}(\mathsf{MLR},\mathsf{SR})$ coincides with being c.e. traceable, $\mathrm{Low}(\mathsf{CR},\mathsf{SR})$ with being computably traceable, and $\mathrm{Low}(\mathsf{MLR},\mathsf{CR})$ with being low for K. If A is low for computable randomness then A is computably dominated and Δ^0_2, so A is computable. For the same reason, although $\mathsf{MLR} \subseteq \mathsf{SR}$, the corresponding lowness properties exclude each other for incomputable sets.

For the classes $\mathrm{Low}(\mathsf{MLR})$ and $\mathrm{Low}(\Omega)$, no characterization in purely computability theoretic terms is known. However, both classes are contained in interesting computability theoretic classes: each set in $\mathrm{Low}(\mathsf{MLR})$ is superlow by Corollary 5.5.4, and each incomputable set that is low for Ω is of hyperimmune degree by Theorem 8.1.18 below. In particular, the classes $\mathrm{Low}(\mathsf{SR})$ and $\mathrm{Low}(\Omega)$ exclude each other for incomputable sets, strengthening the aforementioned similar fact about $\mathrm{Low}(\mathsf{SR})$ and $\mathrm{Low}(\mathsf{MLR})$.

A further example of a traceability property of a set A is jump traceability with an order function g as a bound (Nies, 2002). As for c.e. traceability, the trace for J^A is a uniformly c.e. sequence $(T_n)_{n\in\mathbb{N}}$ such that $\#T_n \leq g(n)$ for each n. We ask that $J^A(n) \in T_n$ whenever $J^A(n)$ is defined. If g grows suffciently fast then there is a perfect class of jump traceable sets with bound g. On the c.e. sets this property coincides with superlowness. Unlike the case of computable traceability, or c.e. traceability, it now matters which order function we choose as a bound. We will show that if an order function h grows much slower than g, the class for the bound h is properly included in the class for the bound g.

We say that A is strongly jump traceable if for *each* order function g there is a c.e. trace for J^A with bound g. Figueira, Nies and Stephan (2008) built a promptly simple strongly jump traceable set. Their hope was that, at least for the c.e. sets, this property might be a computability theoretic characterization of $\mathrm{Low}(\mathsf{MLR})$. However, Cholak, Downey and Greenberg (2008) proved that

TABLE 8.1. Properties in the same row coincide (except for the facile sets).

Low(MLR)	Low for K	K-trivial	
Low for Ω	Weakly low for K		
Low(SR)	Low for computable measure machines	(facile)	Computably traceable
	Lowly for C		strongly jump tr.

the strongly jump traceable c.e. sets form a proper subclass of the c.e. sets in Low(MLR).

Table 8.1 summarizes the classes studied in this chapter. The first row illustrates some of the coincidences of Chapter 5, the second row relates to (1) above, the third to (2) above, and the fourth to strong jump traceability. Classes in the same row coincide, disregarding the class of facile sets. Properties in the same column are analogous.

The analog of the coincidence of lowness for ML-randomness and being a base for ML-randomness fails in the domain of Schnorr randomness: Hirschfeldt, Nies and Stephan (2007) proved that the class of bases for Schnorr randomness is larger than Low(SR). Similarly, the analog of the coincidence of lowness for Schnorr randomness and computable traceability fails in the domain of ML-randomness, because the strongly jump traceable c.e. sets form a proper subclass of the c.e. sets in Low(MLR).

8.1 The class Low(Ω)

In 3.6.17 we introduced the weak lowness property Low(Ω). In Fact 3.6.18 we observed that $\text{Low}(\Omega) \subseteq \text{GL}_1$, and that Low($\Omega$) is closed downward under Turing reducibility (actually, it is easily seen to be closed downward under $\leq_{LR}$). We now study the class Low(Ω) in more detail. We will frequently use the fact that in the definition of Low(Ω) one can replace Ω by any left-c.e. ML-random set.

8.1.1 Proposition. *The following are equivalent for a set A.*

(i) Some left-c.e. set is ML-random relative to A.
(ii) A is low for Ω.
(iii) Every left-c.e. ML-random set is ML-random relative to A.

Proof. The implications (iii)⇒(ii) and (ii)⇒(i) are trivial. For the implication (i)⇒(iii), suppose that Z is left-c.e. and ML-random relative to A. If Y is left-c.e. and ML-random then by Theorem 3.2.29 we have $0.Z \leq_S 0.Y$, which implies that $0.Z \leq_{S^A} 0.Y$, where $\leq_{S^A}$ denotes Solovay reducibility relativized to A. Thus $\beta \leq_{S^A} 0.Z \leq_{S^A} 0.Y$ for any real number β that is left-c.e. in A. Hence Y is ML-random relative to A by Theorem 3.2.29 relativized to A. □

The Low(Ω) *basis theorem*

We give some examples of sets in Low(Ω). By 5.1.27 the only Δ^0_2 sets in Low(Ω) are the ones in Low(MLR). By 3.6.19 each 2-random set is low for Ω.

The Low(Ω) basis theorem states that each nonempty Π^0_1 class P contains a set in Low(Ω). For instance, there is a set $A \in$ Low(Ω) in the Π^0_1 class of two-valued d.n.c. functions (see Fact 1.8.31). This yields a further type of sets in Low(Ω): the Turing degree of A does not contain a ML-random set, for otherwise $A \geq_T \emptyset'$ by Theorem 4.3.8, contrary to the fact that Low(Ω) $\subseteq$ GL$_1$. Also, there is a ML-random set $Z \leq_T A$ by Theorem 4.3.2, so A is not in Low(MLR).

The Low(Ω) basis theorem can be derived easily from the following result of Downey, Hirschfeldt, Miller and Nies (2005). Recall from Definition 3.4.2 that $\Omega^A_M = \lambda(\text{dom}\, M^A)$ for a prefix-free oracle machine M.

8.1.2 Theorem. *For each prefix-free oracle machine M and each nonempty Π^0_1 class P there is a left-Σ^0_2 set $Z \in P$ such that $\Omega^Z_M = \inf\{\Omega^X_M \colon X \in P\}$, which is a left-c.e. real number.*

Proof. Recall from 1.8.3 that $T_P = \{x \in \{0,1\}^* \colon P \cap [x] \neq \emptyset\}$ is the tree corresponding to P. We label each $x \in T_P$ with the real number

$$r_x = \inf\{\Omega^X_M \colon X \in P \cap [x]\}.$$

If $x \notin T_P$ we let $r_x = \infty$. Clearly, $r_x = \min(r_{x0}, r_{x1})$. For $q \in \mathbb{Q}_2$, the class

$$S_{x,q} = P \cap \{X \succ x \colon \Omega^X_M \leq q\}$$

is Π^0_1 uniformly in x, q. Then $L(r_x) = \{q \in \mathbb{Q}_2 \colon q < r_x\}$ consists of those q such that $S_{x,q} = \emptyset$. Now r_x is the infimum of all Ω^X_M such that $X \in P \cap [x]$, so it is the supremum of those q such that $S_{x,q} = \emptyset$. In particular, each r_x is left-c.e. by Fact 1.8.28.

Note that $r_{x0} > r_{x1} \leftrightarrow L(r_{x0}) \supset L(r_{x1}) \leftrightarrow \exists q \in \mathbb{Q}_2[S_{x0,q} = \emptyset \;\&\; S_{x1,q} \neq \emptyset]$, even in the case that $P \cap [x0] = \emptyset$ (namely $r_{x0} = \infty$). This is a Σ^0_2 property of x, and hence Σ^0_1 in $\emptyset'$. Therefore the class $P \cap \{X \colon \forall n\, [r_{X\upharpoonright(n+1)} \leq r_{X\upharpoonright n}]\}$ is $\Pi^0_1(\emptyset')$. Let Z be its leftmost path, then Z is left-c.e. in $\emptyset'$ by the Kreisel Basis Theorem 1.8.36, that is, Z is left-Σ^0_2.

We verify that $\Omega^Z_M = r_\varnothing$. By definition of Z we have $r_\varnothing = r_x$ for each $x \prec Z$. Let Ω^x_M denote $\lambda(\text{dom}\, M^x)$, then $\Omega^x_M \leq r_x = r_\varnothing$ for each $x \prec Z$. But clearly $\lim_n \Omega^{Z\upharpoonright n}_M = \Omega^Z_M$, as $\Omega^Z_M - \Omega^{Z\upharpoonright n}_M$ is the measure of the M^Z-descriptions with use greater than n. Thus $\Omega^Z_M = r_\varnothing$ as required. □

If M is the optimal oracle prefix-free machine $\mathbb{U}$, we obtain a set Z that is low for Ω, because the left-c.e. real number Ω^Z is ML-random relative to Z. Thus:

8.1.3 Corollary. *Each nonempty Π^0_1 class contains a left-Σ^0_2 set $Z \in$ Low(Ω).*

We mention a further consequence of Theorem 8.1.2: there is a left-c.e. real number r with a preimage under the operator Ω of positive measure. (Thus the operator Ω is rather different from the jump operator in that for many sets X, Ω^X has no information at all about X.) First we observe that such an r is necessarily left-c.e. For $r \in [0,1]_{\mathbb{R}}$ let $G_r = \{X : \Omega^X = r\}$.

8.1.4 Proposition. *If $\lambda G_r > 0$ then r is left-c.e.*

Proof. By Theorem 1.9.4 there is $\sigma \in 2^{<\omega}$ such that the operator Ω maps more than half of the sets that extend σ to r, namely, $\lambda\{X \in [\sigma]\colon \Omega^X = r\} > 2^{-|\sigma|-1}$. For each $s \in \mathbb{N}$, since $\Omega_s^X \in \mathbb{Q}_2$ for each X, we can compute the largest $q_s \in \mathbb{Q}_2$ such that $\lambda\{X \in [\sigma]\colon \Omega_s^X \geq q_s\} > 2^{-|\sigma|-1}$. Then $r = \lim_s q_s$. Since the sequence $(q_s)_{s\in\mathbb{N}}$ is non-descending, this shows that r is left-c.e. □

8.1.5 Lemma. *Let $r \in [0,1]_\mathbb{R}$ be left-c.e. Then*

$$\lambda G_r > 0 \Leftrightarrow G_r \text{ contains a ML-random set.}$$

Proof. The implication "⇒" is trivial because the class MLR is conull.

For the implication "⇐", suppose that X is ML-random and $\Omega^X = r$. Let R be the set such that $r = 0.R$. Now R is ML-random in X, so by the van Lambalgen Theorem 3.4.6, X is ML-random in R. But $R \equiv_T \emptyset'$ because R is ML-random and left-c.e., so X is 2-random. By Exercise 8.1.8 below, G_r is a Π^0_2 class containing X, so we may conclude that $\lambda G_r > 0$. □

8.1.6 Corollary. *There is a real number r such that $\lambda G_r > 0$: if P is a Π^0_1 class such that $\emptyset \neq P \subseteq$ MLR, then $r = \min\{\Omega^X\colon X \in P\}$ is such a number.*

Proof. Immediate from Theorem 8.1.2 and Lemma 8.1.5. □

Exercises.

8.1.7. Show that not every nonempty Π^0_1 class P contains a Σ^0_2 set in Low(Ω).

8.1.8. (Bienvenu) Show that for each Δ^0_2 real number $r \in [0,1)$, the class $G_r = \{X\colon \Omega^X = r\}$ is Π^0_2.

Being weakly low for K

By Theorem 5.1.22, A is low for $K \Leftrightarrow A \in$ Low(MLR). Here we prove an analogous result of Miller (2009). Both sides of the equivalence are weakened. (Bienvenu, and Downey, 2009) gave an alternative proof of the implication ⇐ using Solovay functions.

We say that A is *weakly low for K* if $\exists b\, \exists^\infty y\, [K^A(y) \geq K(y) - b]$.

8.1.9 Theorem. *A is weakly low for $K \Leftrightarrow A$ is low for Ω.*

Proof. ⇒: Suppose that A is not low for Ω, that is, for each b there is m such that $K^A(\Omega\restriction_m) \leq m - b$. We want to show that for each b, for almost all y, there is a $\mathbb{U}^A$-description that is by $b + O(1)$ shorter than a $\mathbb{U}$-description σ of y.

If for some sufficiently large m, A "knew" a stage t such that $\Omega - \Omega_t \leq 2^{-m}$, it could ensure this by starting a bounded request set L at t, and enumerating a request $\langle |\sigma| - m, y\rangle$ whenever a computation $\mathbb{U}(\sigma) = y$ converges after t.

The set A actually does not know such a stage t, but for each b there is an m such that $\Omega\restriction_m$ has a $\mathbb{U}^A$-description τ of length at most $m - b$. This information about Ω is sufficient: at stage s, if there is a new $\mathbb{U}^A$-description τ of $\Omega_s\restriction_m$, we start a bounded request set L_τ, which is enumerated similar to the set L described above, but only at stages $t \geq s$, and only as long as $\Omega_t\restriction_m = \Omega_s\restriction_m$. The c.e. index for L_τ can be computed from A, hence we obtain prefix-free descriptions relative to A by concatenating τ and the descriptions for the prefix-free machine for L_τ. If $\mathbb{U}^A(\tau) = \Omega \upharpoonright_m$ and $|\tau| \leq m - b$, then the new prefix-free description of y has a length of at most $(m - b) + (|\sigma| - m) = |\sigma| - b$.

Construction relative to A of bounded request sets L_τ.
Initially, let $L_\tau = \emptyset$ for all $\tau \in \{0,1\}^*$.

If $\mathbb{U}_s^A(\tau)$ newly converges at stage s with output $\Omega_s \upharpoonright_m$, start $P(\tau, s)$.
Procedure $P(\tau, s)$: at stage $t \geq s$, if $\Omega_t \upharpoonright_m = \Omega_s \upharpoonright_m$ and a computation $\mathbb{U}_t(\sigma) = y$ newly converges (necessarily $|\sigma| \geq m$), then put the request $\langle |\sigma| - m, y\rangle$ into L_τ.
The set L_τ is a bounded request set because we only enumerate into it as long as Ω has not increased by more than 2^{-m} since stage s. By the uniformity of the Machine Existence Theorem 2.2.17, we have a prefix-free machine M_τ for each L_τ, and an index for M_τ as a partial computable function can be computed by A. Thus, letting $M(\tau\sigma) \simeq M_\tau(\sigma)$ we define a prefix-free machine M relative to A.

To show that A is not weakly low for K, given b choose m, and s least for m, such that $\Omega_s \upharpoonright_m = \Omega \upharpoonright_m$ and $\mathbb{U}_s^A(\tau) = \Omega \upharpoonright_m$ for some τ of length at most $m - b$. Then the procedure $P(\tau, s)$ is started (at stage s). If a computation $\mathbb{U}(\sigma) = y$ converges after s, the enumeration of L_τ ensures that $M(\tau\rho) = y$ for some ρ such that $|\rho| \leq |\sigma| - m$. Thus $K^A(y) \leq^+ (m - b) + (|\sigma| - m) = |\sigma| - b$.

$\Leftarrow$: Suppose that A is not weakly low for K. Let $R \neq 2^{\mathbb{N}}$ be a c.e. open set such that $2^{\mathbb{N}} - R \subseteq \mathsf{MLR}$ (for instance, let R be the component $\mathcal{R}_1$ of the universal ML-test in Definition 3.2.8). Let Z be the leftmost member of the Π_1^0 class $2^{\mathbb{N}} - R$. Then Z is left-c.e. and ML-random. We show that Z is not ML-random relative to A, whence A is not low for Ω by 8.1.1.

The set Z will fail a ML-test $(\widehat{G}_e)_{e\in\mathbb{N}}$ relative to A. We first build a ML-test $(G_e)_{e\in\mathbb{N}}$ relative to A such that $G_e \subseteq R$ and $\lambda G_e \leq 2^{-e-1}$ for each e. While Z passes this test, for each e we copy sufficiently much of R into G_e to ensure that

$$(\circ) \qquad [p, 0.Z) \subseteq G_e \text{ for some } p \in \mathbb{Q}_2 \text{ such that } p < 0.Z$$

(we identify $2^{\mathbb{N}} - \{Z\colon Z \text{ co-infinite}\}$ and $[0,1)_{\mathbb{R}}$ according to 1.8.10 and 1.8.11). It now suffices to enlarge G_e to at most twice its size: let

$$\widehat{G}_e = \bigcup\{[p, p+2\epsilon)\colon\ p, \epsilon \in \mathbb{Q}_2 \ \&\ p, \epsilon > 0 \ \&\ p + 2\epsilon \leq 1 \ \&\ [p, p+\epsilon) \subseteq G_e\}.$$

Then $\lambda\widehat{G}_e \leq 2\lambda G_e$ and $\widehat{G}_e$ is c.e. relative to A uniformly in e. Thus $(\widehat{G}_e)_{e\in\mathbb{N}}$ is a ML-test relative to A and $Z \in \bigcap_e G_e$ by $(\circ)$.

By Fact 1.8.26 there is a computable prefix-free set B such that $[B]^{\prec} = R$. We will define sets $B_e \subseteq B$ that are uniformly c.e. relative to A. Thereafter we let $G_e = [B_e]^{\prec}$. For each m, any string of length m in B is enumerated into B_e as long as the number of such strings does not exceed $2^{m-K^A(m)-e-1}$. If $K^A(m)$ decreases we are allowed to enumerate more strings into B_e. More formally, we define an A-computable enumeration $(B_{e,s})_{s\in\mathbb{N}}$ of B_e. The finite set $B_{e,s}$ is given by requiring that, for each m,

$$B_{e,s} \cap \{0,1\}^m = B_t \cap \{0,1\}^m,$$

where $t \le s$ is the greatest number such that $\#(B_t \cap \{0,1\}^m) \le 2^{m-K_s^A(m)-e-1}$.

Note that $\lambda G_e = \sum_m 2^{-m} \#(B_e \cap \{0,1\}^m) \le 2^{-e-1} \sum_m 2^{-K^A(m)} \le 2^{-e-1}$.

Claim. *For each e, for sufficiently large m we have $B \cap \{0,1\}^m = B_e \cap \{0,1\}^m$.*

To see this let $b_m = \#(B \cap \{0,1\}^m)$. By Fact 2.2.20 there is $c \in \mathbb{N}$ such that

$$K(m) - c \le m - \log b_m$$

for each m. Since A is not weakly low for K, there is $m_0 \in \mathbb{N}$ such that for each $m > m_0$,

$$K^A(m) \le K(m) - c - e - 2.$$

Thus $m - K^A(m) - e - 2 \ge \log b_m$ for all $m > m_0$, which implies that $2^{m-K_s^A(m)-e-1} \ge b_m$ (since $2^{\log b_m} \ge b_m/2$ by our definition of the logarithm), and hence $B \cap \{0,1\}^m = B_e \cap \{0,1\}^m$.

We use the claim to verify the property (∘) for $p = 0.Z \upharpoonright_{m_0}$. The basic open cylinder $[Z \upharpoonright_{m_0}]$ is identified with the interval $[p, p + 2^{-m_0})$, which contains $[p, 0.Z)$. Since $Z \notin R$ we have $[Z \upharpoonright_{m_0}] \cap [x] = \emptyset$ for each $x \in B$ such that $|x| \le m_0$. Now $[p, 0.Z) \subseteq R$ by the definition of Z. Since strings in B of length no more than m_0 do not help to cover $[p, 0.Z)$ and all longer strings in B are enumerated into B_e, this implies that $[p, 0.Z) \subseteq G_e$. □

The foregoing theorem has interesting consequences. The first is presented here, the second in Theorem 8.1.14. For background on the first, see the discussion after Theorem 5.6.13.

8.1.10 Corollary. (J. Miller) *If $A \le_{LR} B$ and B is low for Ω then $A \le_T B'$. In particular, if B is low for Ω then $\{X \colon X \le_{LR} B\}$ is countable.*

Proof. Note that $A \le_{LK} B$ by Theorem 5.6.5. Hence, for each n we have

$$K^B(A \upharpoonright_n) \le^+ K^A(A \upharpoonright_n) =^+ K^A(n) \le^+ K(n).$$

The set B is weakly low for K by Theorem 8.1.9, so $S = \{n \colon K(n) \le K^B(n) + d\}$ is infinite for some $d \in \mathbb{N}$. Then, for some $b \in \mathbb{N}$, the set A is a path of the tree

$$V = \{z \colon \forall n \le |z| \; [n \in S \rightarrow K^B(z \upharpoonright_n) \le K^B(n) + b]\}.$$

Note that $S \le_T B'$, and therefore $V \le_T B'$. By Theorem 2.2.26(ii) relativized to B, for each $n \in S$ there are only $O(2^b)$ strings of length n on V. Hence A is an isolated path of V, which implies that $A \le_T V \le_T B'$. □

Note that the set A in 8.1.10 is low for Ω, and hence in GL_1 by Fact 3.6.18(ii). Therefore, from $A \le_T B'$ we can conclude that $A' \le_T A \oplus \emptyset' \le_T B'$. Thus, for sets A and B in the conull class Low(Ω), the weak reducibility $\le_{LR}$ is well-behaved in that $A \le_T B \rightarrow A \le_{LR} B \rightarrow A' \le_T B'$.

In Theorem 5.6.9, we obtained statements equivalent to the condition that $A \le_{LR} B$ and $A \le_T B'$. Hence all these hold under the hypotheses of Corollary 8.1.10. Thus, we can also, more generally, invoke 5.6.9 via Exercise 5.6.10 to show that $A \le_{LR} B$ and $A \le_T B'$ implies $A' \le_T B'$ for each A and B.

Exercises.

8.1.11. If $A \in \Delta^0_2$ is weakly low for K then A is already low for K.

8.1.12. (Miller, Yu) If $X \oplus Y$ is 2-random then X and Y form a minimal pair with respect to $\leq_{LR}$. Namely, $C \leq_{LR} X, Y$ implies $C \in \text{Low}(\mathsf{MLR})$ for each set C.

8.1.13.$^\diamond$ **Problem.** Show that, if $\{X\colon X \leq_{LR} B\}$ is countable, then B low for Ω.

2-randomness and strong incompressibility$_K$

In Theorem 3.6.10 we proved that Z is 2-random iff for some b, infinitely many $x \prec Z$ are b-incompressible$_C$. In our second application of Theorem 8.1.9, we show that one can equivalently require that infinitely many initial segments are strongly incompressible in the sense of K (Definition 2.5.2). This also yields a new proof of the implication "$\Rightarrow$" in Theorem 3.6.10. The result is due to Miller (2009).

8.1.14 Theorem. *Z is 2-random* $\Leftrightarrow \exists b\, \exists^\infty n\, \left[K(Z\restriction_n) \geq n + K(n) - b\right]$.

Proof. $\Leftarrow$: Each strongly incompressible$_K$ string is incompressible$_C$ in a sense made precise in Proposition 2.5.5. So the right hand side implies that for some b', infinitely many strings $x \prec Z$ are b'-incompressible$_C$. Therefore Z is 2-random by Theorem 3.6.10.

$\Rightarrow$: Unlike the proof of 3.6.10, this proof is based on lowness properties. A 2-random set is low for Ω and hence weakly low for K. We will show that, if Z is ML-random and weakly low for K, then Z satisfies the right hand side.

8.1.15 Remark. A c.e. operator L is called a *request operator* if L^X is a bounded request set relative to X for each set X. The proof of the Machine Existence Theorem 2.2.17 can be viewed relative to an oracle: from a request operator L one can effectively obtain a prefix-free oracle machine $M = M_d$ ($d > 1$) such that $\forall X\, \forall r, y\left[\langle r, y\rangle \in L^X \leftrightarrow \exists w\, [|w| = r \;\&\; M^X(w) = y]\right]$. In particular, if $\langle r, y\rangle \in L^X$ then $K^X(y) \leq r + d$. As before, we call d a coding constant for L. Remark 2.2.21 remains valid for request operators: suppose we build a c.e. operator L based on a given parameter $d \in \mathbb{N}$ in such a way that for each choice of d, L is a request operator. Then, by the oracle version of the Recursion Theorem we may assume that the coding constant d for L is given in advance.

The following lemma says that, for ML-random sets, the slower the initial segment complexity grows, the stronger is the set computationally.

8.1.16 Lemma. *If B is ML-random then $K^B(n) \leq^+ K(B\restriction_n) - n$ for each n.*

Subproof. By Theorem 7.2.8 there is $c \in \mathbb{N}$ such that $\sum_n 2^{n-K(B\restriction_n)} \leq c$. Define a request operator L as follows: for each X, n, at each stage $s > 0$ such that $K_s(X\restriction_n) < K_{s-1}(X\restriction_n)$ put the request $\langle K_s(X\restriction_n) - n + c + 1, n\rangle$ into L^X_s, as long as this does not make the total weight of L^X exceed 1. By the choice of c, in enumerating L^B the total weight never threatens to exceed 1, which implies the required inequality. (Note here that $2^{-K(x)+1} \geq \sum\{2^{-K_s(x)}\colon s \in \mathbb{N}\}$.) $\diamond$

To conclude the proof of 8.1.14, Z is low for Ω by Proposition 3.6.19, and hence weakly low for K by Theorem 8.1.9. Then, by the lemma, for infinitely many n we have $K(n) \leq^+ K^Z(n) \leq^+ K(Z\restriction_n) - n$. $\square$

8.1.17$^{\diamond}$ **Exercise.** (Miller and Yu, 2008) (i) Show that

$$K^B(m) =^+ \min\{K(B\upharpoonright_{\langle m,i\rangle}) - \langle m,i\rangle : i \in \mathbb{N}\}$$

for each ML-random set B. (ii) Deduce that $\leq_K$ implies $\geq_{LK}$ on MLR. (iii) Further, for each $n > 1$, within MLR the class of n-random sets is closed downwards under $\leq_{LR}$, and closed upwards under $\leq_K$.

Hint. (i) For the inequality "$\leq^+$" generalize 8.1.16. The converse inequality "$\geq^+$" holds for every set X: define an appropriate prefix-free machine that simulates a computation $\mathbb{U}^X(\rho) = m$ by processing both input bits and oracle bits. By modifying $\mathbb{U}$ so that it reads further oracle bits if necessary, you may assume that the use of such a computation is of the form $\langle m,i\rangle$ for some i.

Each computably dominated set in Low(Ω) *is computable*

By Proposition 1.5.12, each low (and in fact each Δ^0_2) computably dominated set is computable. We say that the two lowness properties are orthogonal. We will prove that Low(Ω) and being computably dominated are orthogonal. By Exercise 5.1.27 the only Δ^0_2 sets in Low(Ω) are the ones in Low(MLR), so this result is rather different from Proposition 1.5.12. Since each 2-random set is in Low(Ω), we obtain an alternative proof of Corollary 3.6.15 that no 2-random set is computably dominated.

8.1.18 Theorem. (Miller and Nies, unpublished) *Suppose that A is low for Ω and computably dominated. Then A is computable.*

Proof. We will construct a computable binary tree T such that for some fixed k, for each n the level $T_n = \{\sigma \in T : |\sigma| = n\}$ has at most k elements. The set A will be a path of T. Any path of T is isolated, and hence computable.

Fix $b \in \mathbb{N}$ such that $\Omega \notin \mathcal{R}^A_b$. Then the function g defined by

$$g(r) = \mu t \geq r. \forall m < r\, [K^A_t(\Omega_t\upharpoonright_m) > m - b]$$

is total and $g \leq_T A$. Since A is computably dominated, there is a computable function f such that $f(r) \geq g(r)$ for each r. We define T level by level. Let $n > 0$, and suppose T_{n-1} has been defined. In the beginning we let T_n be the set of extensions of length n of strings in T_{n-1}. We obtain the final T_n by a thinning-out procedure consisting of at most 2^n rounds. Given σ of length n, we use f in an attempt to rule out that $\sigma \prec A$. If we succeed, we remove σ from T_n. We challenge σ at stage r by causing $K^\sigma(\Omega_r\upharpoonright_m) \leq m - b$ for some m which exceeds n by a constant to be determined later. Then, by the definition of f, we have $\sigma \not\prec A$ unless $\Omega\upharpoonright_m$ changes before stage $f(r)$. In that case we plan to account our investment to ensure that $K^\sigma(\Omega_r\upharpoonright_m)$ is small against Otto's investment to increase Ω. We build a request operator L (see 8.1.15) to make $K^\sigma(\Omega_r\upharpoonright_m)$ small. As usual, our investment into a single L^σ is c times his, for a constant $c \in \mathbb{N}$ known to us in advance. In each round, if c strings of length n are challenged (in a sequential fashion) and survive, then Otto has to increase Ω each time, so his investment is as large as ours.

We actually ask that $k = 2c$ strings survive on T_n for the procedure to continue, for we want at our disposal an extra weight of $1/2$ in each set L^X. We need this extra weight, for if σ is challenged and $\Omega\restriction_m$ does not change then σ is removed from T_n, but the request due to σ already is in L^X. Given an oracle X, this can happen at most once for each length n, for $\sigma = X\restriction_n$: either it happens when σ is removed from T_n, or when we are done with level n. The rounds are necessary because we are only allowed to contribute 2^{-n-1} to L^X in that way. (This an example of a controlled risk strategy. Such strategies were already used in Section 5.4.) The thinning-out procedure can have at most 2^n rounds for level n because each time Ω increases by 2^{-n}.

Construction of the levels T_n and the request operator L, given $d \in \mathbb{N}$.
Let $k = 2^{b+d+1}$. Let $T_0 = \{\varnothing\}$.

Suppose $n > 0$ and we have completed the definition of T_{n-1} by stage s. Let $m = n+b+d+1$. Begin with $T_{n,s} = \big\{\rho a\colon\ \rho \in T_{n-1,s}\ \&\ a \in \{0,1\}\big\}$. Carry out the thinning-out procedure, starting at stage s.

(1) IF $\#T_{n,s} \le k$, let $T_n = T_{n,s}$ and return.
ELSE start a new round by declaring all strings $\sigma \in T_{n,s}$ *available.*

(2) IF no string $\sigma \in T_{n,s}$ is available GOTO (1).

(3) Let σ be the leftmost available string. Put the request $\langle n{+}1, \Omega_s\restriction_m\rangle$ into L^σ and declare σ *unavailable.*

(4) WAIT for the first stage $r > s$ such that $K^\sigma(\Omega_s\restriction_m) \le n+d+1 = m-b$.
(If d is indeed a coding constant for L then r exists.)
IF $\Omega_{f(r)}\restriction_m = \Omega_s\restriction_m$, remove σ at stage r. That is, let $T_{n,r} = T_{n,s} - \{\sigma\}$.
(We will show that this cannot happen if $\sigma \prec A$.)

(5) Let $s := r$ and GOTO (2).

Verification. At first we do not assume that $d \in \mathbb{N}$ is a coding constant for L.

Claim 1. *Suppose $s < t$ are stages such that the procedure for n is at* (1) *at stage s and first reaches* (1) *again at stage t. If $\#T_{n,t} > k$ then $\Omega_t - \Omega_s \ge 2^{-n}$.*
If σ is declared unavailable at (3) during stage $u \in [s,t]$, and σ is not removed when the procedure reaches (4) at stage $v > u$, then $\Omega_v\restriction_m \neq \Omega_u\restriction_m$. At least $k+1$ strings are declared unavailable at some stage v, $s \le v \le t$. Therefore $\Omega_t - \Omega_s \ge k2^{-m} = 2^{-n}$. (We need here that $\Omega\restriction_m$ change more than k times, for when the procedure for n begins at stage r, Ω_r could already be very close to $2^{-m} + 0.\Omega_r\restriction_m$. In the binary representation of Ω_r there could be a long sequence of ones from the position m on.)

Claim 2. *L is a request operator.*
Fix $X \subseteq \mathbb{N}$. A request $\langle n+1, \Omega_s\restriction_m\rangle$ is put into L^X at (3) during a stage s because of some $\sigma = X\restriction_n$ for $n > 0$. If σ is next declared available at stage $t > s$ then Ω has increased by 2^{-n}, twice the weight of our request. The contribution of such requests to the total weight of L^X is therefore bounded by $1/2$. If σ is not declared available again (possibly because we have finished level n), we have a one-time contribution of 2^{-n-1} for this length n. Thus the total weight of L^X is at most $1/2 + \sum_{n>0} 2^{-n-1} \le 1$, and Claim 2 is proved.

From now one we may assume d is a coding constant for L. Then a stage r as in (4) always exists. As a consequence, by Claim 1, the thinning-out procedure for T_n returns (after passing (1) for at most 2^n times).

Claim 3. *A is computable.*
It suffices to show that A is a path of T since each path of T is isolated and hence computable. We must verify that for each $n > 0$ the string $\sigma = A\restriction_n$ is not removed from T_n. Assume σ is chosen at stage s and removed at $r > s$. Then $\Omega_{f(r)}\restriction_m = \Omega_s\restriction_m$. By the definition of g and since $g(r) \le f(r)$, there is a stage t, $r \le t \le f(r)$, such that $K_t^A(\Omega_s\restriction_m) > m - b$. This contradicts the fact that already $K_r^\sigma(\Omega_s\restriction_m) \le n + d + 1 = m - b$. □

A related result on computably dominated sets in GL_1

Each 2-random set is generalized low (i.e., in GL_1) by Corollary 3.6.20. Lewis, Montalban and Nies (2007) were interested in the question whether this is still true for weakly 2-random sets. As a counterexample, they found a ML-random computably dominated set that is not in GL_1. Later it turned out that actually *no* ML-random computably dominated set is in GL_1! (This was already mentioned after Proposition 3.6.4.)

To see this, let us replace in Theorem 8.1.18 being in Low(Ω) by the weaker property of being in GL_1. Not all computably dominated sets in GL_1 are computable. For, by Exercise 8.2.20 below, there is a perfect Π^0_1 class of incomputable sets that are in GL_1, and such a class has a perfect subclass of computably dominated sets by Theorem 1.8.44.

8.1.19 Theorem. (Miller and Nies, unpublished) *Suppose A is in GL_1 and computably dominated. Then A is not of d.n.c. degree.*

Proof idea. Let $h \le_T A$. We want to show that $h(x) = J(x)$ for some x. Let Ψ be a Turing functional such that $A' = \Psi(A \oplus \emptyset')$. The function g defined by

$$g(r) = \mu t \ge r.\, \forall k < r\, [A'_t(k) = \Psi(A \oplus \emptyset'; k)[t]]$$

is total and computable in A. Since A is computably dominated there is a computable function f such that $f(r) \ge g(r)$ for each r.

We build a Turing functional Γ, and by the Recursion Theorem we are given in advance a reduction function p such that $\Gamma^X(e) \simeq J^X(p(e))$ for each X, e. We look for a string α and a stage t such that $\Psi(\alpha \oplus \emptyset', p(0))\downarrow [t] = 0$. In case $\alpha \prec A$ the opponent Otto has already declared his use on the A-side to be $|\alpha|$, while we may still define $\Gamma^\sigma(0)$ for many pairwise incomparable strings $\sigma \succeq \alpha$ in an attempt to rule out $\sigma \prec A$. The only possible reason we cannot do this is that $\emptyset'$ changes below the use of $\Psi(\alpha \oplus \emptyset', p(0))$ by a stage computed using f. Then we may take a one-time action based on $\langle \alpha, t\rangle$ to ensure $h(x) = J(x)$ for some x.

Proof details. Fix $h \le_T A$, and let Θ be a Turing functional such that $h = \Theta^A$. To show that there is x such that $J(x) = h(x)$, besides Γ we build a Turing functional Δ which is only relevant without an oracle (i.e., Δ is treated as a partial computable function). Then, by Fact 1.2.15 and the Double Recursion Theorem, we are given in advance computable functions p and q such that

$$\forall\sigma\,\forall e\,[\Gamma^\sigma(e) \simeq J^\sigma(p(e))] \text{ and } \forall j\,[\Delta(j) \simeq J(q(j))].$$

For Γ we only need the input 0: we define $\Gamma^\sigma(0)$ for lots of strings σ. We write $\overline{0}$ for $p(0)$. Since $A'(\overline{0}) = \Psi(A \oplus \emptyset'; \overline{0})$ we can wait for a computation $\Psi(\alpha \oplus \emptyset'; \overline{0})$ at a stage $t > \overline{0}$. Let $j = \langle \alpha, t\rangle$. We attempt to rule out any string $\sigma \succeq \alpha$ such that $y = \Theta^\sigma(q(j))$ converges by defining $\Gamma^\sigma(0)$ at a stage $s > t$. We wait for the response $J^\sigma(\overline{0})\downarrow$ at stage $r > s$, and then look ahead till stage $f(r)$: if $\emptyset'$ has changed below the use of $\Psi(\alpha \oplus \emptyset'; \overline{0})$, then because this is a one-time event for $j = \langle \alpha, t\rangle$, we are allowed to use up j as an input for Δ. We define $\Delta(j) = y$. Then $x = q(j)$ is as required. If $\emptyset'$ has not changed then $\sigma \not\prec A$.

Construction of a Turing functional Γ and a partial computable function Δ, given computable functions p and q.
At stage $t > \overline{0} = p(0)$, if there is a new convergence $\Psi(\alpha \oplus \emptyset', \overline{0})[t] = 0$, where α is shortest for this computation, let m be its use on the $\emptyset'$-side, let $j = \langle \alpha, t\rangle$, and start the procedure $P(\alpha, t)$ which runs at stages $s > t$. Several procedures may run in parallel.

Procedure $P(\alpha, t)$:

(1) IF at stage $s > t$ there is a string $\sigma \succeq \alpha$ such that $\Gamma^\sigma(0)\uparrow$ and $y = \Theta^\sigma_s(q(j))\downarrow$, let σ be the least such and define $\Gamma^\sigma(0) = 1$ (the output is irrelevant).
(2) WAIT for the first stage $r > s$ such that $J^\sigma(\overline{0})\downarrow$. IF $\emptyset'_{f(r)}\restriction_m \neq \emptyset'_t\restriction_m$, define $\Delta(j) = y$ and END the procedure $P(\alpha, t)$, ELSE GOTO (1).

Verification. We claim that there is x such that $J(x) = h(x)$. First we show that $\Gamma^A(0)\downarrow$. Otherwise, let t be the least stage $> \overline{0}$ such that the computation $\Psi(A \oplus \emptyset', \overline{0})[t] = 0$ is stable. Let α be the initial segment of the oracle A used by this computation, and let m be its use on the $\emptyset'$ side. We start procedure $P(\alpha, t)$ at stage t. Let $j = \langle \alpha, t\rangle$, and let $\sigma \prec A$ be least such that $\Theta^\sigma(q(j))\downarrow$. Whenever we enter (2) for some σ', we get back to (1) because $\emptyset'_t\restriction_m$ is already stable. So, eventually it is the turn of σ, and we define $\Gamma^\sigma(0) = 1$, contradiction.

We may conclude that we define $\Gamma^\sigma(0) = 1$ for some least $\sigma \prec A$ in (1) of some procedure $P(\alpha, t)$. We use the notation of the construction for this procedure: let $j = \langle \alpha, t\rangle$, $x = q(j)$ and suppose $r > s$ is least such that $J^\sigma_r(\overline{0})\downarrow$. By the definition of g and since $\overline{0} < r$ and $g(r) \le f(r)$, there is a stage v, $r \le v \le f(r)$, such that $A'(\overline{0}) = \Psi(A \oplus \emptyset'; \overline{0})[v]$. But we only start a procedure $P(\alpha, t)$ if $\Psi_t(\alpha \oplus \emptyset', \overline{0}) = 0$. Thus $\emptyset'_{f(r)}\restriction_m \neq \emptyset'_t\restriction_m$ and we cause $J(x) = \Delta(j) = y = h(x)$. □

8.1.20 Exercise. (Greenberg) Let $\mathcal{C}$ be the class of computably dominated sets, and let $\mathcal{D}$ be the sets of d.n.c. degree. Show that the intersection of any two of the classes GL_1, $\mathcal{C}$, and $\mathcal{D}$ contains a perfect class (although the intersection of all three is empty by 8.1.19). *Hint.* For $\mathcal{C} \cap \mathrm{GL}_1$ one needs function trees; see Exercise 8.2.20.

8.2 Traceability

A domination property of a set A, such as being computably dominated, restricts the growth of the functions f computed by A. In contrast, a traceability property states that each value $f(n)$ is in a finite set T_n that can be determined from n in an effective way (see the chapter introduction for more background). We will study four traceability properties. The first two are covered in this section: c.e. traceability and computable traceability, where only total functions computed by A are traced. Each c.e. traceable set is array computable, a weak lowness property introduced in 8.2.7 below. Each computably traceable set is computably dominated.

In Section 8.4 we study jump traceability and strong jump traceability, where also functions that are partial computable in A are traced.

In this section and Section 8.4 we consider traceability for its own sake. In Sections 8.3 and 8.5 we will see that traceability is closely related to lowness for randomness notions.

Recall that an order function is a nondecreasing unbounded computable function. *From now on we will assume that order functions only have positive values.*

8.2.1 Definition. Let h be an order function. A *trace with bound* h is a sequence $(T_n)_{n\in\mathbb{N}}$ of nonempty sets such that $\#T_n \le h(n)$ for each n. We say that $(T_n)_{n\in\mathbb{N}}$ is a trace for the function $f\colon \mathbb{N} \to \mathbb{N}$ if $f(n) \in T_n$ for each n. We call $(T_n)_{n\in\mathbb{N}}$ a *c.e. trace* if $(T_n)_{n\in\mathbb{N}}$ is uniformly computably enumerable.

The following notion was first studied by Ishmukametov (1999).

8.2.2 Definition. We say that a set A is *c.e. traceable* if there is an order function h such that each function $f \le_T A$ has a c.e. trace with bound h.

The point is that all the functions $f \le_T A$ can be traced via the same bound. It suffices to require that there is n_0 such that $f(n) \in T_n$ for all $n \ge n_0$, for we can modify the trace by letting $T_n = \{f(n)\}$ for $n < n_0$. This fact will be used without mentioning, not only for c.e. traceability, but also for its variants encountered later on.

We give an overview of this section. After the basics on c.e. traceability, we introduce the stronger notion of computable traceability, where one requires that the trace be given by a computable sequence of strong indices for finite sets. We build a perfect class of computably traceable sets; in particular, there are $2^{\aleph_0}$ of them. Then we show that being computably traceable is equivalent to lowness for computable measure machines, a property analogous to being low for K. Under the assumption that the set already is computably dominated, the property is also equivalent to being facile, a weakening of being K-trivial. (See Table 8.1 on page 303.)

Our interest in these traceability notions stems in part from the fact that they can be used to characterize lowness for randomness notions. In Section 8.3 we will show for instance that Low(SR) coincides with the class of computably traceable sets. The only known proof that there is an incomputable set in Low(SR) is by actually building a computably traceable set and then using this coincidence.

C.e. traceable sets and array computable sets

For c.e. traceability, an arbitrary order function q can be taken as a bound for the c.e. traces, no matter how slowly it grows. This result of Terwijn and Zambella (2001) relies on the fact that one traces only total functions.

8.2.3 Theorem. *A c.e. traceable set is c.e. traceable via every order function.*

Proof. Recall that on page 67 we defined a computable injection $D\colon \mathbb{N}^* \to \mathbb{N}$ by $D(m_0, \ldots, m_{k-1}) = \prod_{i=0}^{k-1} p_i^{m_i+1} = \alpha$ where p_i is the i-th prime number. We will write $\alpha(i)$ for m_i. If $f \in \mathbb{N}^{\mathbb{N}}$ and $k \in \mathbb{N}$, we will write $f\upharpoonright_k$ for $D(m_0, \ldots, m_{k-1})$, where $m_j = f(j)$ for $j < k$.

Suppose that the set A is c.e. traceable via some order function h. Suppose an order function q is given. Let r be the computable function given by

$$r(n) = 1 + \max\{i\colon\ q(i) < h(n+1)\}.$$

(For instance, if $h(n) = n + 1$ and $q(i) = 1 + \log i$, then $r(n) = 2^n$.)

Now suppose $f \le_T A$. We will define a c.e. trace $(T_i)_{i \in \mathbb{N}}$ with bound q for f. Since f is total we may define a function $\widetilde{f} \le_T A$ by $\widetilde{f}(n) = f\upharpoonright_{r(n)}$. There is a c.e. trace $(\widetilde{T}_n)_{n \in \mathbb{N}}$ for $\widetilde{f}$ with bound h. We may assume that, for each $\alpha \in \widetilde{T}_n$,

$$\alpha \in \mathrm{ran}(D)\ \&\ |D^{-1}(\alpha)| = r(n).$$

For each i, let $T_i = \{\alpha(i)\colon\ \alpha \in \widetilde{T}_{n_i}\}$, where n_i is the least n such that $q(i) < h(n+1)$. Since $r(n_i) > i$, we have $|D^{-1}(\alpha)| > i$, so that $f(i) \in T_i$. Moreover, $\#T_i \le \#\widetilde{T}_{n_i} \le h(n_i) \le q(i)$ for each i. □

The preceding argument can be adapted to a range of contexts where only total functions are traced. One can vary the type of total functions being traced (8.2.29), the effectivity condition on the traces (8.2.15), or on the bounds.

8.2.4 Corollary. *For each order function h there is a c.e. trace $(V_n)_{n \in \mathbb{N}}$ with bound h that is universal in the sense that, if A is c.e. traceable and $f \le_T A$, then $f(n) \in V_n$ for almost all n.*

Proof. Let g be the order function such that $g(n) = \max\{i\colon\ i^2 \le h(n)\}$. There is a uniformly c.e. listing $(S_n^i)_{i,n \in \mathbb{N}}$ of all the c.e. traces with bound g. Let $V_n = \bigcup_{i \le g(n)} S_n^i$, then $\#V_n \le h(n)$ for each n. By 8.2.3 $(V_n)_{n \in \mathbb{N}}$ is universal. □

It is instructive to observe that a constant bound for the c.e. traces is not sufficient unless the set is computable.

8.2.5 Proposition. *Suppose there is $c \in \mathbb{N}$ and a c.e. trace $(T_n)_{n \in \mathbb{N}}$ for the function $\lambda n.\, A\upharpoonright_n$ such that $\forall n\, \#T_n \le c$. Then A is computable.*

Proof. The string $A\upharpoonright_n$ can be determined from n and the information that it is the i-th element of T_n (in the order of the given computable enumeration). Preceding a shortest $\mathbb{V}$-description of n by 0^i1 shows that $\forall n\, C(A\upharpoonright_n) \le^+ C(n)$. Thus A is C-trivial, and hence computable by Theorem 5.2.20. □

Being c.e. traceable is incompatible with the conull highness property of having d.n.c. degree.

8.2.6 Proposition. *If A has d.n.c. degree then A is not c.e. traceable.*

Proof. Assume for a contradiction that A is c.e. traceable. By Theorem 4.1.9 there is a nondecreasing unbounded function $g \le_T A$ such that $\forall n\, [g(n) \le K(A \upharpoonright_n)]$. Let $f \le_T A$ be the function given by $f(r) = A \upharpoonright_{h(r)}$ where $h(r) = \min\{n\colon g(n) \ge r\}$. If $n = h(r)$ then $f(r) = A \upharpoonright_n$, so $K(f(r)) \ge g(n) \ge r$ for each r. By Theorem 8.2.3 there is a trace $(T_r)_{r \in \mathbb{N}}$ for f such that $\#T_r \le r+1$ for each r. But $f(r) \in T_r$ implies $K(f(r)) \le^+ 4 \log r$, contradiction. □

Kjos-Hanssen, Merkle and Stephan (2011) proved a stronger result: A is c.e. traceable $\Leftrightarrow$ there is an order function q showing that each $f \le_T A$ strongly fails to be a d.n.c. function, in the sense that for each k we have $k \le \#\{e < q(k)\colon f(e) = J(e)\}$. They also proved that A is not of d.n.c. degree $\Leftrightarrow$ there is an order function h such that for each $f \le_T A$, for infinitely many x we can trace $f(x)$ by a c.e. trace with bound h.

We will compare c.e. traceability with the following weak lowness property introduced by Downey, Jockusch and Stob (1990).

8.2.7 Definition. A is called *array computable* if there is a single function $g \le_{wtt} \emptyset'$ dominating every function computable in A.

Every array computable set is in GL_2 by Exercise 1.5.21. Most of the lowness properties we study imply being array computable. For instance, each set in $\mathrm{Low}(\Omega)$ is array computable by Exercise 8.3.5. Thus the class of array computable sets is conull.

8.2.8 Corollary. *Each c.e. traceable set A is array computable.*

Proof. Recall from Exercise 1.4.7 that a function g is ω-c.e. iff $g \le_{wtt} \emptyset'$. By Corollary 8.2.4 let $(V_n)_{n \in \mathbb{N}}$ be a universal c.e. trace, and let $g(n) = \max V_n$. Then $g \le_{wtt} \emptyset'$ and g dominates any A–computable function. (We can choose $(V_n)_{n \in \mathbb{N}}$ with bound an arbitrary order function h. Then the number of changes in an effective approximation to g is bounded by h.) □

The converse of 8.2.8 fails because the class of c.e. traceable sets is null. However, Ishmukhametov (1999) proved that each c.e. array computable set is c.e. traceable (see Exercise 8.4.26). The proof is similar to (and easier than) the implication "$\Leftarrow$" in proof of Theorem 8.4.23 below, which states that each c.e. superlow set is jump traceable (Definition 8.4.1), and hence c.e. traceable. To summarize, the implications between these lowness properties on the c.e. sets are (also see Figure 8.2 on page 362)

$$\text{superlow} \Rightarrow \text{c.e. traceable} \Leftrightarrow \text{array computable} \Rightarrow \mathrm{Low}_2.$$

Some low c.e. set is not array computable by Downey, Jockusch and Stob (1990). See Exercise 8.2.12.

Exercises. Show the following.

8.2.9. Each computably dominated set A is array computable.

8.2.10.$^\diamond$ (Ng, Nies) Each superlow set C is array computable.
Hint. Apply Exercise 1.5.5, and use a construction similar to the one in 1.5.16.

8.2.11. Each K-trivial set A is c.e. traceable.

8.2.12. Some low c.e. set A is not c.e. traceable. Explain why your construction does not make A superlow.

8.2.13. (Nies, 2002) By 1.5.8, a set A is in Low_2 iff $\mathrm{Tot}^A = \{e\colon \Phi_e^A \text{ total}\}$ is Σ_3^0. Show that a c.e. traceable Δ_2^0 set A is uniformly in Low_2 in that, from a computable approximation $(A_s)_{s\in\mathbb{N}}$ of A, one can effectively obtain a Σ_3^0-index for Tot^A.

8.2.14.$^\diamond$ **Problem.** Show that some weakly 2-random set is not array computable.

Computably traceable sets

Terwijn and Zambella (2001) introduced a further notion of traceability. One has complete information about the components of the traces. Recall from 1.1.14 that $(D_n)_{n\in\mathbb{N}}$ is an effective listing of the finite subsets of $\mathbb{N}$.

8.2.15 Definition. A trace $(T_n)_{n\in\mathbb{N}}$ is called *computable* if there is a computable function g such that $T_n = D_{g(n)}$ for each n. We say that A is *computably traceable* if there is an order function h such that each function $f \leq_T A$ has a computable trace with bound h.

The following fact of Kjos-Hanssen, Nies and Stephan (2005) characterizes the computably traceable sets within the class of c.e. traceable sets.

8.2.16 Proposition. *A is computably traceable $\Leftrightarrow$ A is c.e. traceable and computably dominated.*

Proof. $\Rightarrow$: Each computable trace $(D_{g(n)})_{n\in\mathbb{N}}$ is a c.e. trace. If it is a trace for a function f then f is dominated by the computable function $\lambda n. \max D_{g(n)}$.

$\Leftarrow$: Suppose A is c.e. traceable via the bound h. Let $f \leq_T A$, and let $(T_n)_{n\in\mathbb{N}}$ be a c.e. trace for f with bound h. Let $p(n) = \mu s. f(n) \in T_{n,s}$, then $p \leq_T A$. Since A is computably dominated, we may choose a computable function r such that $r(n) \geq p(n)$ for each n. Then $(T_{n,r(n)})_{n\in\mathbb{N}}$ is a computable trace for f with bound h. □

Recall from page 56 that a nonempty closed class $\mathcal{C} \subseteq 2^{\mathbb{N}}$ is called perfect if the corresponding tree $T_{\mathcal{C}} = \{x\colon [x] \cap \mathcal{C} \neq \emptyset\}$ has no isolated paths. Terwijn and Zambella (2001) built a perfect class $\mathcal{C}$ of computably traceable sets. Since a perfect class has cardinality $2^{\aleph_0}$, this implies that some computably traceable set is not computable. In fact we will obtain such a set in Δ_3^0.

8.2.17 Theorem. *There is a perfect class $\mathcal{C}$ of computably traceable sets. Moreover, the corresponding tree $\{x\colon [x] \cap \mathcal{C} \neq \emptyset\}$ is Δ_3^0.*

The proof is similar to the proof of Theorem 1.8.44, yet we cannot expect to make $\mathcal{C}$ a subclass of an arbitrary given Π_1^0 class without computable members. For instance, the class $2^{\mathbb{N}} - \mathcal{R}_1$ of ML-random sets does not even contain a c.e. traceable set by Proposition 8.2.6. However, we can ensure that $\mathcal{C}$ is a subclass of a given perfect class P such that T_P is computable.

Proof. We need an alternative notation for the tree representation of perfect classes given by 1.8.3. A *function tree* is a map $F\colon \{0,1\}^* \to \{0,1\}^*$ such that for each σ, both $F(\sigma 0)$ and $F(\sigma 1)$ properly extend $F(\sigma)$, and $F(\sigma 0) <_L F(\sigma 1)$.

Function trees correspond to binary trees without dead branches or isolated paths; if F is a function tree, the corresponding binary tree is $B_F = \{x\colon \exists \sigma\, [x \preceq F(\sigma)]\}$. Note that $B_F \equiv_T F$. Let $Paths(F) = \{Z\colon \forall n \exists \sigma\, [Z \upharpoonright_n \preceq F(\sigma)]\}$. The perfect classes are precisely the ones of the form $Paths(F)$ for a function tree F.

We define a $\emptyset''$-computable sequence of computable function trees $(F^e)_{e \in \mathbb{N}}$ such that $Paths(F^{e+1}) \subseteq Paths(F^e)$ for each e. For F^0 we may choose an arbitrary computable function tree, for instance the identity function on $\{0,1\}^*$. The class $\mathcal{C}$ will be a perfect subclass of $\bigcap_e Paths(F^e)$.

For each e, we let F^{e+1} copy F^e on strings of length up to e. For each σ of length $e+1$, all paths Y of F^{e+1} extending $F^{e+1}(\sigma)$ behave in the same way with regard to Φ_e: either Φ_e^Y is partial for each such Y, or there is a single computable trace for each Φ_e^Y with bound $h(n) = 2^n$.

Inductive definition of the computable function trees F^e for $e \in \mathbb{N}$.

Step $e+1$. Suppose that F^e has been determined. For each ν such that $|\nu| \le e$, let $F^{e+1}(\nu) = F^e(\nu)$.
Let S_e be the set of strings σ of length $e+1$ such that

$$\forall \eta \succeq \sigma\, \forall n\, \exists \rho \succeq \eta\, [\Phi_{e,|\rho|}^{F^e(\rho)}(n)\downarrow]. \tag{8.1}$$

For each σ of length $e+1$ we define F^{e+1} on all strings extending σ:

If $\sigma \notin S_e$, let $\eta \succeq \sigma$ and $n \in \mathbb{N}$ be a pair of witnesses for the failure of (8.1), and define $F^{e+1}(\sigma\alpha) = F^e(\eta\alpha)$ for each string α. This ensures that $\Phi^Y(n)\uparrow$ for each path Y of F^{e+1} extending $F^{e+1}(\sigma)$.

If $\sigma \in S_e$, define $F^{e+1}(\sigma\alpha)$ by induction on $n = |\alpha|$: let $F^{e+1}(\sigma) = F^e(\sigma)$. If $F^{e+1}(\sigma\alpha) = F^e(\eta)$ has been defined, then for each $r \in \{0,1\}$, let $F^{e+1}(\sigma\alpha r) = F^e(\rho)$, where $\rho \succeq \eta r$ is least such that $\Phi_{e,|\rho|}^{F^e(\rho)}(n)\downarrow$. Note that ρ exists since $\sigma \in S_e$. *This completes the inductive definition.*

For each e, F^{e+1} is computable. Moreover, using $\emptyset''$ as an oracle, we can compute S_e, and hence find a Turing program computing F^{e+1}. We define a function $G \le_T \emptyset''$ by $G(\sigma) = F^{|\sigma|}(\sigma)$. Then G is a function tree since $F^{e+1}(\nu) = F^e(\nu)$ for $|\nu| = e$, and $F^{e+1}(\nu 0) <_L F^{e+1}(\nu 1)$. Thus $\mathcal{C} = Paths(G)$ is perfect.

We show that each $Y \in \mathcal{C}$ is computably traceable with bound $\lambda n.2^n$. Given e, let σ be the string of length $e+1$ such that $F^{e+1}(\sigma) \prec Y$. If $\sigma \notin S_e$ then Φ_e^Y is partial. Otherwise let g be the computable function such that

$$D_{g(n)} = \{\Phi_{e,|F^{e+1}(\sigma\alpha)|}^{F^{e+1}(\sigma\alpha)}(n)\colon |\alpha| = n\}.$$

Then $\#D_{g(n)} \le 2^n$ for each n, and $(D_{g(n)})_{n \in \mathbb{N}}$ is a computable trace for Φ_e^Y.

Note that $G \equiv_T \operatorname{ran} G \equiv_T \{x\colon [x] \cap \mathcal{C} \neq \emptyset\}$ by Exercise 8.2.19, so the set $\{x\colon [x] \cap \mathcal{C} \neq \emptyset\}$ is Δ_3^0. □

8.2.18 Corollary. *There is a computably traceable set A in Δ_3^0 that is not computable.*

Proof. Let the function tree G be as above. We define inductively a Δ_3^0 sequence of strings $\sigma_0 \prec \sigma_1 \prec \ldots$ such that $|\sigma_e| = e$, and let $A = \bigcup_e G(\sigma_e)$. If σ_e has been determined, using $\emptyset''$ as an oracle, check whether $G(\sigma_e 0)(k) = 1 \leftrightarrow \Phi_e(k) = 1$ for each $k < |G(\sigma_e 0)|$. If so, let $\sigma_{e+1} = \sigma_e 1$. Otherwise, let $\sigma_{e+1} = \sigma_e 0$. Then $A \neq \{k \colon \Phi_e(k) = 1\}$ for each e, so A is incomputable. □

Stephan and Yu (2006) built a $\{0,1\}$-valued partial computable function ψ with co-infinite domain such that the perfect Π^0_1 class P of total $\{0,1\}$-valued extensions of ψ consists of sets that are neither c.e. traceable nor of d.n.c. degree. By Theorem 1.8.44, P has a perfect subclass of computably dominated sets. In particular, there is a perfect class of computably dominated sets that are neither of d.n.c. degree nor computably traceable.

They also showed that being computably dominated and not of d.n.c. degree is equivalent to being low for weak 1-genericity (see 1.8.47). Note that by Theorem 8.1.19 this class contains the sets in GL_1 (Definition 1.5.4) that are computably dominated.

Some, but not all, computably traceable sets are in GL_1. See the discussion after Corollary 8.4.7.

Exercises.

8.2.19. Show that $F \equiv_T \operatorname{ran} F$ for each function tree F.

8.2.20. Show that there is a perfect Π^0_1 class $P \subseteq \mathrm{GL}_1$. Conclude that a computably dominated set in GL_1 can be incomputable.

Refine this further by also ensuring that P has no computable member; hence GL_1 has a perfect subclass of computably dominated sets.

8.2.21. Let h be an order function. Let $(I_n)_{n\in\mathbb{N}}$ be the sequence of consecutive intervals such that $\#I_n = h(n)$. Show that A is c.e. traceable [computably traceable] $\Leftrightarrow$ for each $f \leq_T A$ there is a partial computable [a computable] function ψ such that $\forall n\, \exists x \in I_n\, \psi(x) = f(x)$.

Lowness for computable measure machines

Recall from 3.5.14 that a prefix-free machine M is called a computable measure machine (c.m.m.) if its halting probability Ω_M is computable. We will say that an oracle prefix-free machine is a *computable measure machine relative to A* if Ω^A_M is computable in A. Downey, Greenberg, Mikhailovich and Nies (2008) have shown that for each such M there is an oracle prefix-free machine $\widehat{M}$ such that $\widehat{M}^A = M^A$ and $\Omega^X_{\widehat{M}}$ is computable in X for each X.

We introduce an analog of being low for K in the setting of such machines. Since there is no universal computable measure machine we need an extra quantification.

8.2.22 Definition. A is *low for computable measure machines* if for each computable measure machine M relative to A, there is a computable measure machine N and a number $b \in \mathbb{N}$ such that $\forall y\, [K^A_M(y) \geq K_N(y) - b]$.

The following result is due to Downey *et al.* (2008).

8.2.23 Theorem. *A is computably traceable $\Leftrightarrow$ A is low for c.m.m.*

Proof. $\Rightarrow$: For the duration of this proof, a machine will be called *finite* if its domain is finite. Such a machine is given by a finite set of pairs of numbers, so it can be effectively encoded by a number.

Suppose M is an oracle prefix-free machine such that Ω_M^A is A-computable. We actually build a c.m.m. N such that $\forall y\,[K_M^A(y) \ge K_N(y) - 1]$. For $t < u$, let $M_{[t,u)}^A$ denote the finite machine obtained by restricting M^A to computations that converge at a step $s \in [t, u)$. We partition M^A into finite machines of this type: let

$$t_n = \mu t.\, \Omega_M^A - \Omega_{M,t}^A \le 2^{-2n},$$

and let $f\colon \mathbb{N} \to \mathbb{N}$ be the function such that $f(n)$ is the code for $M_{[t_n, t_{n+1})}^A$. Since Ω_M^A is A-computable, $\lambda n.t_n \le_T A$ and hence $f \le_T A$. Let $(D_{g(n)})_{n\in\mathbb{N}}$ be a computable trace for f with bound $\lambda n.2^n$. We may assume that $D_{g(n)}$ only contains (codes for) finite machines S such that $\Omega_S \le 2^{-2n}$. We combine all the finite machines in this computable trace to build a bounded request set: let $L = \bigcup_n L_n$ where $L_n = \{\langle |\sigma| + 1, y\rangle \colon \exists S \in \bigcup_{i\le n} D_{g(i)}\,[S(\sigma) = y]\}$. Let $w_n \in \mathbb{Q}_2$ be the total weight of L_n, and let w be the total weight of L. Then $(w_n)_{n\in\mathbb{N}}$ is a nondecreasing computable sequence and

$$w - w_n \le (1/2)\sum_{m>n} 2^m 2^{-2m} = 2^{-n-1}.$$

Thus w is a computable real number by Fact 1.8.15. Also, $w_0 \le 1/2$ since $\#D_{g(0)} = 1$, so $w \le 1$. If N is the prefix-free machine obtained from L through the Machine Existence Theorem 2.2.17, then $\Omega_N = w$, so N is a computable measure machine. Clearly N is as required.

$\Leftarrow$: Suppose that $f \le_T A$. Let M be the prefix-free machine defined by

$$M(0^{|x|}1x) = f(x).$$

Then $K_M(f(x)) \le^+ 2\log x$ for each x. Since the strings of length n contribute $2^n 2^{(-2n+1)} = 2^{-n-1}$ to the measure of $\operatorname{dom} M$, this measure is computable. In particular, M is a c.m.m. relative to A. Hence there is a c.m.m. N such that $\forall y\, K_N(y) \le^+ K_M(y)$. By Exercise 3.5.17 a strong index for the finite set $\{y\colon K_N(y) \le r\}$ can be computed from r. Let g be the computable function such that $D_{g(x)} = \{y\colon K_N(y) \le 3\log x\}$. Note that $\#D_{g(x)} \le 2x^3 + 1$. Moreover, $f(x) \in D_{g(x)}$ for almost all x. Thus, a finite variant of $(D_{g(x)})_{x\in\mathbb{N}}$ is a computable trace for f with bound $\lambda x.2x^3 + 1$. □

Exercises.

8.2.24. Show that a variation of the preceding result yields a characterization of the c.e. traceable sets in terms of oracle prefix-free machines: A is c.e. traceable $\Leftrightarrow$ for each c.m.m. M relative to A, there is b such that $\forall y\,[K_M^A(y) \ge K(y) - b]$.

8.2.25. We say that A is Schnorr trivial (Downey, Griffiths and LaForte, 2004) if for each c.m.m. M there is a c.m.m. N such that $\forall n\, K_N(A\restriction_n) \le^+ K_M(n)$. Show that if A is low for c.m.m. then A is Schnorr trivial.

Facile sets as an analog of the K-trivial sets ⋆

We will characterize the computably traceable sets within the class of computably dominated sets by a property defined in terms of a growth condition on the initial segment complexity. The material in this subsection is due to unpublished work of Kjos-Hanssen and Nies dating from 2005.

8.2.26 Definition. Z is *facile* if for each order function h,

$$\forall n\, [K(Z\restriction_n \mid n) \leq^+ h(n)].$$

By Theorem 7.4.11 a facile set can be computably random. By the next two facts, similar to the K-trivial sets, the facile sets determine an ideal in the weak truth-table degrees (see Definition 1.2.27 for ideals).

8.2.27 Proposition. *If A is facile and $B \leq_{wtt} A$ then B is facile as well.*

Proof. Suppose that $B = \Phi^A$ and $\forall n\, [\mathsf{use}\, \Phi^A(n) \leq f(n)]$ for a computable f. We may assume that f is strictly increasing. A prefix-free description of $B\restriction_n$ given n may be obtained from a prefix-free description σ of $A\restriction_{f(n)}$ given n via the binary machine M defined by $M(\sigma, n) \simeq \Phi^{\mathbb{U}^2(\sigma,n)}\restriction_n$. Thus,

$$K(B\restriction_n \mid n) \leq^+ K(A\restriction_{f(n)} \mid n) \leq^+ K(A\restriction_{f(n)} \mid f(n))$$

(for the second inequality, we have used that f is computable and strictly increasing). If h is an order function, then $\widehat{h}(r) = h(\min\{n\colon\ f(n) \geq r\})$ is an order function as well (without the extra condition that $\widehat{h}(0) > 0$). Also, $\widehat{h}(f(n)) = h(n)$ for each n. Since $K(A\restriction_r \mid r) \leq^+ \widehat{h}(r)$ for each r, we obtain that $K(B\restriction_n \mid n) \leq^+ h(n)$ for each n. ☐

8.2.28 Proposition. *If A and B are facile, then $A \oplus B$ is facile.*

Proof. It suffices to observe that, for each n, we have
$K(A \oplus B\restriction_{2n} \mid 2n) =^+ K(A \oplus B\restriction_{2n} \mid n) \leq^+ K(A\restriction_n \mid n) + K(B\restriction_n \mid n)$. ☐

We will show that each c.e. traceable set A is facile. In fact, a weaker property suffices. We say that set A is *weakly c.e. traceable* if Definition 8.2.2 holds restricted to the functions $f \leq_T A$ that are bounded from above by a computable function. As before, the choice of the bound for the trace does not matter as long as it is an order function, because in the proof of Theorem 8.2.3, if f bounded by a computable function then so is $\widetilde{f}$.

8.2.29 Theorem. *A is weakly c.e. traceable $\Leftrightarrow$ each set $Z \leq_T A$ is facile.*

Proof. $\Rightarrow$: Suppose $Z \leq_T A$, and an order function h is given. The function $\lambda n. Z\restriction_n$ is computable in A, and dominated by the function $\lambda n. 2^n$. Therefore it has a c.e. trace $(T_n)_{n\in\mathbb{N}}$ with bound h. Given n, we may describe $z = Z\restriction_n$ by the number $i \leq h(n)$ such that z is the i-th string enumerated into T_n. Since $K(i) \leq^+ 2\log i$ we have $K(z \mid n) \leq^+ h(n)$.

$\Leftarrow$: Suppose that $f \leq_T A$ and $\forall n\, f(n) \leq b(n)$ where b is an order function. For $r \in \mathbb{N}$, let $n_r = \sum_{i<r} b(i)$. Define a set $Z \leq_T A$ in such a way that the number of ones in the interval $[n_r, n_{r+1})$ equals $f(r)$: for $j < b(r)$, let

$$Z(n_r + j) = \begin{cases} 1 & \text{if } j < f(r) \\ 0 & \text{else.} \end{cases}$$

Let h be the order function given by $h(m) = 1 + \lfloor r/2 \rfloor$ for $m \in [n_r, n_{r+1})$. Since Z is facile, $K(Z\restriction_{n_r} \mid n_r) \leq^+ h(n_r) = 1 + \lfloor r/2 \rfloor$, and because $\lambda r. n_r$ is computable, we obtain $\forall^\infty r\, K(Z\restriction_{n_r} \mid r) \le r$, and hence $\forall^\infty r\,[K(Z\restriction_{n_r}) \le r + 2\log r]$. This leads to the desired trace $(T_r)_{r\in\mathbb{N}}$ for f: given r, for each σ such that $|\sigma| \le r + 1 + 2\log(r+1)$, if $\mathbb{U}(\sigma) = z$ where $|z| = n_{r+1}$, then enumerate $\#\{i \in [n_r, n_{r+1})\colon z(i) = 1\}$ into T_r. Clearly $f(r) \in T_r$ for almost all r. While $(T_r)_{r\in\mathbb{N}}$ depends on the sequence (n_r) and hence on the computable bound b for f, the *size* $\#T_r$ is bounded by 2^{cr} for a fixed c, independent of such a computable bound for f. □

We are now able to characterize computable traceability within the class of computably dominated sets.

8.2.30 Corollary. *Let A be computably dominated. Then*
A is computably traceable $\Leftrightarrow$ A is facile.

Proof. By 1.5.11, $Z \le_T A \Leftrightarrow Z \le_{wtt} A$ for each Z. By 8.2.27, being facile is closed downward under $\le_{wtt}$, so by Theorem 8.2.29 A is weakly c.e. traceable iff A is facile. Finally, by 8.2.16, for computably dominated sets A we have:
A is computably traceable $\Leftrightarrow$ A c.e. traceable $\Leftrightarrow$ A weakly c.e. traceable. □

Franklin and Stephan (2010) have given an alternative characterization of this kind: let A be computably dominated. Then A is computably traceable $\Leftrightarrow$ A is Schnorr trivial (as defined in Exercise 8.2.25). Note that "$\Rightarrow$" holds for each set by 8.2.23 and 8.2.25. On the other hand, a Schnorr trivial set can be Turing complete by Downey, Griffiths and LaForte (2004).

Each K-trivial set is c.e. traceable by Exercise 8.2.11, and hence facile. The converse fails. For instance, no K-trivial set is Schnorr random, otherwise, being nonhigh, it would already be ML-random (5.5.10). So the facile computably random set of Theorem 7.4.11 is not K-trivial. Alternatively, there are $2^{\aleph_0}$ many computably traceable and hence facile sets, while each K-trivial set is Δ^0_2.

Exercises. Show the following.

8.2.31. Each facile set Z is $O(\log)$ bounded (as defined in 7.6.5, page 289).

8.2.32. If Z is facile then $\exists^\infty n\, K(Z\restriction_n) < h(n)$ for each order function h. (By 4.1.10, there is a d.n.c. function $f \le_{wtt} Z \Leftrightarrow$ there is an order function h such that $K(Z\restriction_n) \ge h(n)$ for each n. Thus there is no such f for a facile set.)

To summarize, the hierarchy of growth conditions on the initial segment complexity is K-trivial $\Rightarrow$ facile $\Rightarrow$ $O(\log)$ bounded $\Rightarrow$ not ML-random. The converse implications fail. A growth condition strictly in between the last two was introduced in 7.6.22: $\dim(Z) = \liminf_n K(Z\restriction_n)/n < 1$. For more details, see for instance Section 15 of Downey, Hirschfeldt, Nies and Terwijn (2006).

8.3 Lowness for randomness notions

The main results of this section are the following.

- Lowness for Schnorr randomness coincides with computable traceability.
- Only the computable sets are low for computable randomness.

First we study lowness for pairs of randomness notions (Definition 5.1.32). Recall that if $\mathcal{C} \subseteq \mathcal{D}$ then A is in $\mathrm{Low}(\mathcal{C},\mathcal{D})$ if $\mathcal{C} \subseteq \mathcal{D}^A$. If $\mathcal{C} \subseteq \widetilde{\mathcal{C}} \subseteq \widetilde{\mathcal{D}} \subseteq \mathcal{D}$ then $\mathrm{Low}(\widetilde{\mathcal{C}},\widetilde{\mathcal{D}}) \subseteq \mathrm{Low}(\mathcal{C},\mathcal{D})$. It will require some effort to prove the following:

(1)	Low(MLR, SR)	$\Leftrightarrow$	c.e. traceable	(Thm. 8.3.3)
(2)	Low(CR, SR)	$\Leftrightarrow$	comp. traceable	(Thm. 8.3.7)
(3)	Low(MLR, CR)	$\Leftrightarrow$	low for K	(Thm. 8.3.10).

The implications "$\Leftarrow$" are the easier ones. For (3), we have already seen in Fact 5.1.8 that each set that is low for K is low for ML-randomness. For (1) and (2), we use that the relevant traceability notion can be expressed in terms of computable measure machines (Exercise 8.2.24 and Theorem 8.2.23, respectively). Then we apply Theorem 3.5.19 that Schnorr randomness can be described by such machines. In the case of (2), this argument actually establishes that each computably traceable set is in Low(SR); see Proposition 8.3.2.

The implications "$\Rightarrow$" are harder. We derive (2) from (1) and Lemma 8.3.8 that each set in Low(CR, SR) is computably dominated. To prove (1) and (3), we build a test in the sense of $\mathcal{D}^A$, and use the fact that no set Z on which the test succeeds is in $\mathcal{C}$. We first derive from this fact a condition on finite objects (we call such a condition combinatorial). This enables us to obtain the desired lowness property; see Lemmas 8.3.4 and 8.3.13, respectively. For instance, in (3), to show that each set in $A \in \text{Low}(\mathsf{MLR}, \mathsf{CR})$ is low for K, we build a $\mathbb{Q}_2$-valued martingale $L^A \leq_T A$. By the hypothesis, $\mathsf{Succ}(L^A) \subseteq \text{non-}\mathsf{MLR}$. From this we derive in Lemma 8.3.13 a combinatorial condition stating, roughly, that $[x] \subseteq \mathcal{R}_1$ for any string x such that $N(x)$ exceeds a certain value 2^d.

The technique to derive a combinatorial condition from the hypothesis that A is low for some randomness notion(s) was first used in (i)$\Rightarrow$(iii) of Theorem 5.1.9. The proof is always by contraposition: if the condition fails, one can build a set Z that shows A does not satisfy the lowness property. The condition is combinatorial because at some point the attempted definition of Z by finite extensions must stop.

The characterizations of Low(SR) and Low(CR) are now easily obtained. By (2), $A \in \text{Low}(\mathsf{SR})$ implies that A is computably traceable, and the converse also holds. By (2) and (3), $A \in \text{Low}(\mathsf{CR})$ iff A is computable, because computably traceable sets are computably dominated while sets that are low for K are Δ^0_2.

Lowness for $\mathcal{C}$-null classes

A randomness notion $\mathcal{C}$ is usually introduced by specifying a notion of a $\mathcal{C}$-test. We say that $\mathcal{S} \subseteq 2^{\mathbb{N}}$ is a *$\mathcal{C}$-null class* if there is a $\mathcal{C}$-test such that all the sets in $\mathcal{S}$ fail the test. For instance, $\mathcal{S}$ is a Schnorr null class if there is a Schnorr test $(G_m)_{m\in\mathbb{N}}$ such that $\mathcal{S} \subseteq \bigcap_m G_m$.

A $\mathcal{C}$-test is called *universal* if the sets that fail the test form the largest $\mathcal{C}$-null class. If there is a universal $\mathcal{C}$-test, then $\mathcal{S}$ is a $\mathcal{C}$-null class iff $\mathcal{S} \cap \mathcal{C} = \emptyset$.

The following is a lowness property defined in terms of a covering procedure (see page 227).

8.3.1 Definition. We say that A is *low for $\mathcal{C}$-null classes* if each $\mathcal{C}^A$-null class already is a $\mathcal{C}$-null class.

Since a set $Z \notin \mathcal{C}^A$ is in some $\mathcal{C}^A$-null class, this implies $\mathcal{C} \subseteq \mathcal{C}^A$, namely, A is low for $\mathcal{C}$. Being low for $\mathcal{C}$-null classes appears to be stronger than being low for $\mathcal{C}$: it could be that $\mathcal{C} \subseteq \mathcal{C}^A$ because each $Z \notin \mathcal{C}^A$ fails some $\mathcal{C}$-test that depends on Z. If there is a universal $\mathcal{C}$-test then the two are clearly equivalent. For deeper reasons, lowness for $\mathcal{C}$-null classes and lowness for $\mathcal{C}$ also coincide when $\mathcal{C}$ is SR, CR or W2R. For W2R this was shown already in Theorem 5.5.17. In this section we provide the proof for SR, and of course for CR it follows from the result that each set in Low(CR) is computable. No randomness notion $\mathcal{C}$ is known for which being low for $\mathcal{C}$-null classes is stronger than being low for $\mathcal{C}$.

The following uses computable measure machines (c.m.m.) via Theorem 8.2.23.

8.3.2 Proposition. *Each computably traceable set A is low for Schnorr null classes.*

Proof. As in Definition 3.2.6, for an oracle prefix-free machine M and an oracle X, we let $R_b^{M,X} = [\{x \in \{0,1\}^* \colon\ K_M^X(x) \le |x| - b\}]^{\prec}$. Let $(G_m)_{m \in \mathbb{N}}$ be a Schnorr test relative to A. By Lemma 3.5.18 relativized to A, there is a c.m.m. M relative to A such that $\bigcap_m G_m \subseteq \bigcap_d R_d^{M,A}$. By Theorem 8.2.23, there is a c.m.m. N and $b \in \mathbb{N}$ such that $\forall y\, [K_M^A(y) \ge K_N(y) - b]$. Then $(R_k^N)_{k \in \mathbb{N}}$ is a Schnorr test such that $\bigcap_m G_m \subseteq \bigcap_k R_k^N$. □

***The class* Low(MLR, SR)**

The main result of Terwijn and Zambella (2001) states that lowness for Schnorr null classes is equivalent to computable traceability. (The easier implication "$\Leftarrow$" has already been proved in Proposition 8.3.2.) This result represented an important advance, characterizing for the first time lowness for tests by a computability theoretic property.

Kjos-Hanssen, Nies and Stephan (2005) extended this, answering a question of Ambos-Spies and Kučera (2000): actually, lowness for Schnorr randomness is equivalent to computable traceability. To get around the problem that there is no universal Schnorr test, they proceeded indirectly.

Firstly, they proved that $A \in$ Low(MLR, SR) $\Leftrightarrow A$ is c.e. traceable, by adapting the measure-theoretic combinatorics of Terwijn and Zambella for their implication "$\Rightarrow$".

Secondly, they used a lemma of Bedregal and Nies (2003) that each set A in Low(CR, SR) is computably dominated. Each c.e. traceable computably dominated set is already computably traceable by Proposition 8.2.16. Therefore each set in Low(SR) is computably traceable.

8.3.3 Theorem. $A \in$ Low(MLR, SR) $\Leftrightarrow$ *A is c.e. traceable.*

Proof. Recall from page 71 that for a string v and a measurable class $\mathcal{C}$, we let $\mathcal{C} \mid v = \{X \colon vX \in \mathcal{C}\}$. Furthermore, $\lambda(\mathcal{C} \mid v) = 2^{|v|}\lambda(\mathcal{C} \cap [v])$ as in Definition 1.9.3. We will also use the notation $\lambda_v(\mathcal{C})$ as a short form of $\lambda(\mathcal{C} \mid v)$.

$\Leftarrow$: If Z is not Schnorr random in A, then by Theorem 3.5.19 there is a computable measure machine M relative to A such that $\forall b\, \exists n\ K_M^A(Z \upharpoonright_n) \le n - b$.

Since A is c.e. traceable, by Exercise 8.2.24 $K(y) \le^+ K_M^A(y)$ for each y. Hence Z is not ML-random.

$\Rightarrow$: Given $g \le_T A$, we will define a c.e. trace $(T_n)_{n\in\mathbb{N}}$ for g with bound $\lambda n.2^n$. We may assume that $g(n) > 0$ and $n \mid g(n)$ (that is, n divides $g(n)$) for each n. Otherwise, we replace g by the function $\overline{g} = \lambda n.ng(n) + n$, and a c.e. trace for $\overline{g}$ can be turned into a c.e. trace for g with the same bound.

The basic idea is to use a Schnorr test $(U_d^g)_{d\in\mathbb{N}}$ relative to A which contains a sufficient amount of information about g. For each d, the open set U_d^g consists of the sets Z such that for some $n > d$, there is a run of n zeros starting at position $g(n)$. By the hypothesis on A, we have $\bigcap_d U_d^g \subseteq$ Non-MLR, and in particular $\bigcap_d U_d^g \subseteq R$ where $R = \mathcal{R}_2 = [\{x \in \{0,1\}^*\colon\ K(x) \le |x| - 2\}]^{\prec}$.

Recall that $\lambda R \le 1/4$. We first need a lemma saying that, when we restrict ourselves to an appropriate basic open cylinder $[v]$, then $\lambda_v(R)$ is small but $\lambda_v(U_d^g - R) = 0$ for some d. The trace for g will be obtained from U_d^g and v. The lemma holds in more generality.

8.3.4 Lemma. *Let $\epsilon = 1/4$, and let $(U_d)_{d\in\mathbb{N}}$ be a sequence of open sets such that $\bigcap_d U_d \subseteq R$. Then there is a string v and $d \in \mathbb{N}$ such that $\lambda_v(R) \le \epsilon$ and $\lambda_v(U_d - R) = 0$.*

Subproof. Assume the Lemma fails. We define inductively a sequence of strings $(v_d)_{d\in\mathbb{N}}$ such that $v_0 \prec v_1 \prec \ldots$ and $\forall d\, \lambda(R \mid v_d) \le \epsilon$. Let v_0 be the empty string. Suppose v_d has been defined and $\lambda(R \mid v_d) \le \epsilon$. Then $\lambda((U_d - R) \mid v_d) > 0$ since the Lemma fails, so we can choose y such that $[y] \subseteq U_d$ and $\lambda([y] - R \mid v_d) > 0$; in particular, $y \succeq v_d$. By Theorem 1.9.4 we may choose $v_{d+1} \succ y$ such that $\lambda(R \mid v_{d+1}) \le \epsilon$. Now let $Z = \bigcup_d v_d$, then $Z \in \bigcap_d U_d$ and $Z \notin R$. ◇

Next we provide the details on how to code information about g into a Schnorr test relative to A. For each $n, k \in \mathbb{N}$, let

$$B_{n,k} = [\{x0^n\colon\ |x| = k\}]^{\prec}.$$

Thus, $B_{n,k}$ is the clopen class consisting of the sets that have a run of n consecutive zeros from position k on. Let

$$U_d^g = \bigcup_{n>d} B_{n,g(n)}.$$

Claim 1. *$(U_d^g)_{d\in\mathbb{N}}$ is a Schnorr test relative to A.*
Since $g \le_T A$ and $\lambda B_{n,k} = 2^{-n}$, $(U_d^g)_{d\in\mathbb{N}}$ is a ML-test relative to A. To show it is a Schnorr test relative to A, we have to verify that λU_d^g is computable in A uniformly in d. Let $q_{d,t} = \lambda \bigcup_{t\ge n>d} B_{n,g(n)}$, then $q_{d,t} \in \mathbb{Q}_2$, the function $\lambda t.\, q_{d,t}$ is uniformly computable in A, and $\lambda U_d^g - q_{d,t} \le \lambda \bigcup_{n>t} B_{n,g(n)} \le 2^{-t}$. ◇

We think of T_n as the set of k such that $B_{n,k} - R$ has small measure, in a sense to be specified (which depends on n). Since $B_{n,k} - R$ generally tends to have large measure, there can only be very few k for which $B_{n,k} - R$ has small measure. This leads to the bound on $\#T_n$.

What does this have to do with the function g? By the hypothesis, $\bigcap_d U_d^g \subseteq R$, so choose v, d for this test as in Lemma 8.3.4. We will actually carry out the idea sketched above within $[v]$. By Lemma 8.3.4, for $n > d$, $B_{n,g(n)} - R$ is null in $[v]$, so $g(n)$ is traced.

For the formal details, let $p = |v|$. Since $U_d^g \supseteq U_{d+1}^g$ for each d, we may assume that $d \geq p$, which implies that $g(n) \geq n \geq p$ for each $n \geq d$.

Let $\widetilde{R} = R \mid v$. For $n \leq d$, let $\widetilde{g}(n) = 0$ and $\widetilde{T}_n = \{0\}$; for $n > d$, let $\widetilde{g}(n) = g(n) - p$ and

$$\widetilde{T}_n = \{k\colon\ n|k+p \ \&\ \lambda(B_{n,k} - \widetilde{R}) < 2^{-(k+4)}\}.$$

In the remaining claims we show that $(\widetilde{T}_n)_{n\in\mathbb{N}}$ is a c.e. trace for $\widetilde{g}$ with bound $\lambda n.2^n$. Then $(T_n)_{n\in\mathbb{N}}$ is a c.e. trace for g with the same bound, where $T_n = \{g(n)\}$ for $n \leq d$, and $T_n = \{k + p\colon\ k \in \widetilde{T}_n\}$ for $n > d$.

Claim 2. $\forall n\, \widetilde{g}(n) \in \widetilde{T}_n$.
Firstly, by hypothesis $n \mid g(n) = \widetilde{g}(n) + p$. Secondly, since $B_{n,k} \mid v = B_{n,k-p}$ for each $k \geq p$,

$$U_d^g \mid v = \bigcup_{n>d} B_{n,g(n)} \mid v = \bigcup_{n>d} B_{n,\widetilde{g}(n)} = U_d^{\widetilde{g}}.$$

By the choice of v, this implies that $\lambda(U_d^{\widetilde{g}} - \widetilde{R}) = \lambda_v(U_d^g - R) = 0$, whence $\widetilde{g}(n) \in \widetilde{T}_n$ for each $n > d$. ◇

Claim 3. *$(\widetilde{T}_n)_{n\in\mathbb{N}}$ is a c.e. trace with bound $\lambda n.2^n$.*
For a Π_1^0 class P and $m \in \mathbb{N}$, $\lambda P < 2^{-m}$ is equivalent to $\exists s\, \lambda P_s < 2^{-m}$, so the relation $\{\langle n,k\rangle\colon\ \lambda(B_{n,k} - \widetilde{R}) < 2^{-(k+4)}\}$ is Σ_1^0. Let $a_n = \#\widetilde{T}_n$. We will show that $a_n < 2^n$. Note that

$$\lambda(\bigcup_{k\in\widetilde{T}_n} B_{n,k} - \widetilde{R}) \leq \sum_{k\in\widetilde{T}_n} \lambda(B_{n,k} - \widetilde{R}) \leq 1/8.$$

Since $\lambda\widetilde{R} \leq 1/4$, this implies that $\lambda \bigcup_{k\in\widetilde{T}_n} B_{n,k} \leq 3/8$.

Suppose $N \in \mathbb{N}\cup\{\infty\}$, and let $S_k \subseteq 2^{\mathbb{N}}$ be measurable for $k < N$. In probability theory, the events S_k are called *independent* if for each finite set $Y \subseteq \{k\colon\ k < N\}$, we have $\lambda \bigcap_{k\in Y} S_k = \prod_{k\in Y} \lambda S_k$. In this case, the events $2^{\mathbb{N}} - S_k$ are independent as well (see for instance Shiryayev 1984, Problem I.7.3). If $E \subseteq \mathbb{N}$ is a set such that $n \mid k+p$ for each $k \in E$, then the sets $B_{n,k}$ $(k \in E)$ are independent, and therefore

$$\lambda \bigcap_{k\in E} (2^{\mathbb{N}} - B_{n,k}) = (1 - 2^{-n})^{\#E}$$

(if E is infinite then the expression of the right is defined to be 0). For $E = \widetilde{T}_n$ we obtain

$$1/2 < \lambda(2^{\mathbb{N}} - \bigcup_{k\in\widetilde{T}_n} B_{n,k}) = \lambda \bigcap_{k\in\widetilde{T}_n} (2^{\mathbb{N}} - B_{n,k}) = (1 - 2^{-n})^{a_n} = (\tfrac{r}{r+1})^{a_n},$$

where $r = 2^n - 1$, and therefore $2 > \left(\frac{r+1}{r}\right)^{a_n}$. But $\left(\frac{r+1}{r}\right)^r \geq 2$ as $(r+1)^r \geq r^r + r^{r-1}\binom{r}{1} = 2r^r$. Therefore $a_n < 2^n$. □

Exercises.
8.3.5. (Kjos-Hanssen, Nies and Stephan, 2005) If $\Omega \in \mathsf{SR}^A$ then A is array computable (Definition 8.2.7). In particular, the class of array computable sets is conull.
8.3.6.[◇] Recall Definition 3.5.1, and let WR be the class of weakly random sets. Show that $A \in \mathrm{Low}(\mathsf{MLR}, \mathsf{WR})$ iff A does not have d.n.c. degree:

(i) (Kjos-Hanssen) (a) If A does not have d.n.c. degree then $A \in \mathrm{Low}(\mathsf{MLR}, \mathsf{WR})$.
(b) Conclude that if Z is ML-random and $D \subseteq Z$ is infinite then D has d.n.c. degree.
(ii) (Greenberg and Miller) If A has d.n.c. degree then A computes an infinite subset D of some ML-random set. In particular, $A \notin \mathrm{Low}(\mathsf{MLR}, \mathsf{WR})$.

Classes that coincide with Low(SR)

We give a number of characterizations for the class of sets that are low for Schnorr randomness. However, the main technical result of this subsection is on lowness for two randomness notions:

8.3.7 Theorem. $A \in \mathrm{Low}(\mathsf{CR}, \mathsf{SR}) \Leftrightarrow$ *A is computably traceable.*

Proof. $\Leftarrow$: A is low for Schnorr null classes by 8.3.2, so $A \in \mathrm{Low}(\mathsf{CR}, \mathsf{SR})$.
$\Rightarrow$: A is in $\mathrm{Low}(\mathsf{MLR}, \mathsf{SR})$, and hence c.e. traceable by Theorem 8.3.3. By Proposition 8.2.16, each c.e. traceable computably dominated set is computably traceable. So the following lemma of Bedregal and Nies (2003) suffices.

8.3.8 Lemma. *If $A \in \mathrm{Low}(\mathsf{CR}, \mathsf{SR})$ then A is computably dominated.*

To prove the lemma, suppose that A is not computably dominated. We will build a set Z that is computably random and not Schnorr random relative to A. For the former, we ensure that $\sup_n M(Z\upharpoonright_n) < \infty$ whenever M is a computable $\mathbb{Q}_2$-valued martingale. For the latter, we define a bounded request set relative to A with total weight (a real number) computable in A, in such a way that, if N is the associated c.m.m. relative to A, then $\forall b\, \exists x \prec Z\, \left[K_N^A(z) \leq z - b\right]$.

The uniform listing $(M_e)_{e \in \mathbb{N}}$ of the $\mathbb{Q}_2$-valued partial computable martingales was introduced on page 273. Recall that $\mathsf{TMG} = \{e\colon M_e \text{ total}\}$.

The basic strategy to make Z computably random but not Schnorr random relative to A is not unlike the one used in the proof of Theorem 7.4.8. We ensure that $M_e(Z) < \infty$ whenever M_e is total. In the following α, β, γ denote finite subsets of $\mathbb{N}$. Given α, in order to beat all the martingales M_e for $e \in \alpha$ together, we consider a linear combination M_α with positive rational coefficients of those martingales. For all $\alpha \subset \mathsf{TMG}$ (and inevitably some others) we will define strings x_α in such a way that $x_\alpha \prec x_{\alpha\cup\{e\}}$ when $e > \max(\alpha)$ and $x_{\alpha\cup\{e\}}$ is defined. We ensure that, for each total M_e, each γ containing e and each $x \prec x_\gamma$, $M_e(x)$ is bounded by a constant only depending on e. Then the set

$$Z = \bigcup\{x_{\mathsf{TMG}\cap\{0,\ldots,e\}}\colon e \in \mathbb{N}\}$$

is computably random.

Fix a function $g \leq_T A$ not dominated by any computable function. If $\alpha \subset \mathsf{TMG}$, there are infinitely many m such that, for all x of length at most m, $M_\alpha(x)$ has converged by stage $g(m)$. If $\alpha = \beta \cup \{e\}$ where $e > \max(\beta)$, we define x_α to be the leftmost non-ascending path of M_α of length m above x_β. The set A can recognize whether a string of length m is x_α. There are few x_α of each length, so a computable measure machine relative to A can assign short descriptions to each of them.

Let $\widetilde{M}_e = M_{e,0}$ where $M_{e,0}$ was defined in Fact 7.4.3; thus $\widetilde{M}_e$ is M_e scaled in such a way that the start capital is 1 (unless $M_e(\emptyset) = 0$). For each α let $n_\alpha = \sum_{e\in\alpha} 2^e$.

Inductive definition of partial computable martingales M_α and strings x_α. We maintain the condition that if x_α is defined, then

$$M_\alpha(x_\alpha) \text{ converges in } g(|x_\alpha|) \text{ steps and } M_\alpha(x_\alpha) < 2. \tag{8.2}$$

Let $x_\varnothing = \varnothing$, and let $M_\emptyset$ be the constant zero function. We may assume that $M_\emptyset(\varnothing)$ converges in $g(0)$ steps by increasing $g(0)$ if necessary, so (8.2) holds for $\alpha = \emptyset$. Now suppose $\alpha = \beta \cup \{e\}$ where $e > \max(\beta)$ if $\beta \neq \emptyset$, and inductively suppose that (8.2) holds for β. Let $M_\alpha = M_\beta + p_\alpha \widetilde{M}_e$ where the binary rational $p_\alpha > 0$ is chosen in such a way that $M_\alpha(x_\beta) < 2$: since $\widetilde{M}_e(x) \leq 2^{|x|}$ for each x, it is sufficient to let $p_\alpha = 2^{-|x_\beta|-1}(2 - M_\beta(x_\beta))$.

To define x_α, let $m > |x_\beta|$, $m \geq 2^{n_\alpha} + 20$ be least such that $M_j(z)$ converges in $g(m)$ steps for each $j \in \alpha$ and $|z| \leq m$. Choose $x_\alpha \succ x_\beta$ of length m in such a way that $M_\alpha(yr) \leq M_\alpha(y)$ for any string $y \succeq x_\beta$ and $r \in \{0,1\}$ such that $yr \preceq x_\alpha$. If m fails to exist then leave x_α undefined. *This completes the inductive definition.*

Claim 1. *If $\alpha \subset \mathsf{TMG}$ then x_α is defined.*
This is clear for $\alpha = \emptyset$. Suppose now the claim holds for β, and that $\alpha = \beta\cup\{e\} \subset \mathsf{TMG}$ where $e > \max(\beta)$ if $\beta \neq \emptyset$. Since the function f defined by

$$f(m) = \mu s. \forall e \in \alpha \forall x \left[\, |x| \leq m \to M_e(x)\downarrow [s]\,\right]$$

is computable, there is a least $m > |x_\beta|$, $m \geq 2^{n_\alpha} + 20$ such that $g(m) \geq f(m)$. Since M_α is a martingale, we have $M_\alpha(y0) \leq M_\alpha(y)$ or $M_\alpha(y1) \leq M_\alpha(y)$ for any y. So we can choose x_α as required. ◇

Claim 2. *Z is computably random.*
Suppose M_e is total, and let $\alpha = \mathsf{TMG} \cap \{0,\ldots,e\}$. We need to show that $\sup\{M_e(x)\colon x \prec Z\} < \infty$. Given x, let $k \in \mathsf{TMG}$ be largest such that $x_\gamma \preceq x$ where $\gamma = \mathsf{TMG} \cap \{0,\ldots,k\}$. Let $i > k$ be least such that $i \in \mathsf{TMG}$. Thus $x \prec x_{\gamma\cup\{i\}} \prec Z$. We may assume that $x_\alpha \preceq x_\gamma$. Then

$$p_\alpha \widetilde{M}_e(x) \leq M_\gamma(x) \leq M_\gamma(x_\gamma) < 2$$

by (8.2) for γ and the definition of $x_{\gamma\cup\{i\}}$. So $M_e(x)$ is bounded by a constant. ◇

Claim 3. *Z is not Schnorr random relative to A.*
We have $x_\alpha \prec Z$ for infinitely many α, so by 3.5.19 it is sufficient to show that

$$\{\langle |x_\alpha| - n_\alpha, x_\alpha\rangle\colon x_\alpha \text{ is defined}\}$$

is a bounded request set relative to A with total weight w computable in A.

Let $r_m = \sum_\alpha 2^{n_\alpha - m}[\![|x_\alpha| = m]\!]$, then $w = \sum_{m\geq 20} r_m$. Since $g \leq_T A$ we have $\{\langle y,\alpha\rangle\colon y = x_\alpha\} \leq_T A$. Thus the function $\lambda m.\, r_m$ is computable in A. Since

$|x_\alpha| \geq 2^{n_\alpha}$ for each α, the sum defining r_m contains at most m terms (far fewer, in fact). Then, since $m \geq 20$, and since $\log m \geq n_\alpha$ for each index α in the sum, $r_m \leq m2^{\log m - m} \leq m^2 2^{-m} \leq 2^{-m/2}$, so that $\sum_{m \geq i+1} r_m \leq 2^{-i/2}$ for each i. This shows that the real number w is computable in A. □

We summarize the characterizations of lowness for Schnorr randomness.

8.3.9 Theorem. *The following are equivalent for a set A.*

(i) A is computably traceable (Definition 8.2.15).
(ii) A is low for computable measure machines (8.2.22).
(iii) A is low for Schnorr null classes (8.3.1).
(iv) A is low for Schnorr randomness.
(v) $A \in$ Low(CR, SR). □

Note that (i)⇔(ii) is Theorem 8.2.23, (i)⇒(iii) is Proposition 8.3.2, the implications (iii)⇒(iv)⇒(v) are trivial, and (v)⇒(i) holds by Theorem 8.3.7.

By way of analogy, property (i) corresponds to K-triviality, (ii) to being low for K, (iv) to Low(MLR) and (v) to Low(W2R, MLR).

Low(MLR, CR) *coincides with being low for K*

The following theorem of Nies (2005*b*) concludes a series of preliminary results. The technique to wait till some clopen set is contained in a fixed c.e. open set R such that $\lambda R < 1$ was already present in the original proof that Low(MLR) $\subseteq \Delta^0_2$, which can be found in Nies (2005*a*), and could serve as a gentle introduction to the proof that follows here. Later, in Nies (2005*b*) the full result was obtained by combining this technique with the use of martingales. This was also the first proof of the fact that each set in Low(MLR) is low for K. Hirschfeldt, Nies and Stephan (2007) later obtained a simpler proof by considering bases for ML-randomness (see Theorem 5.1.22).

8.3.10 Theorem. $A \in$ Low(MLR, CR) ⇔ *A is low for K.*

8.3.11 Corollary. *If A is low for computable randomness then A is computable.*

Proof. A is in Low(CR, SR) and hence computably dominated by Lemma 8.3.8. On the other hand, A is in Low(MLR, CR), and hence low for K. The only Δ^0_2 computably dominated sets are the computable ones by 1.5.12. □

Nies (2005*a*) also gave a direct proof of Corollary 8.3.11, and extended his proof to show that the only sets in Low(PCR, CR) are the computable ones.

Recall from 7.6.19 that KLR denotes the class of KL-random sets. We proved that MLR ⊊ KLR. It is open whether an incomputable set in Low(KLR) exists.

8.3.12 Corollary. *Each set in* Low(KLR) *is low for K. The same holds for permutation randomness in place of* KLR. □

Proof of Theorem 8.3.10.

⇐: By Fact 5.1.8, each set that is low for K is in Low(MLR).

$\Rightarrow$: First we recall some notation and facts. For the remainder of this proof, an open set $S \subseteq 2^{\mathbb{N}}$ is identified with the set of strings y such that $[y] \subseteq S$. Thus we write $y \in S$ instead of $[y] \subseteq S$. For $R \subseteq \{0,1\}^*$ we write λR instead of $\lambda[R]^{\prec}$. We let $R = \mathcal{R}_1 = [\{x \in \{0,1\}^*\colon K(x) \le |x| - 1\}]^{\prec}$. There is a computable enumeration $(R_s)_{s \in \mathbb{N}}$ of R such that R_s contains only strings of a length up to s, and R_s is closed under extensions within those strings (see (1.16) on page 54). We may also assume that $x0, x1 \in R_s \to x \in R_s$ for each string x. For a string v and a measurable class $\mathcal{C}$, we use the notation $\lambda_v(\mathcal{C})$ instead of $\lambda(\mathcal{C} \mid v)$. By Exercise 3.3.4 there is a computable function g such that for each string v,

$$v \notin R \to \lambda_v(R) < 1 - 2^{-g(v)}. \tag{8.3}$$

8.3.13 Lemma. *Let N be a martingale such that $\mathsf{Succ}(N) \subseteq$ Non-MLR. Then there is $v \in \{0,1\}^*$ and $d \in \mathbb{N}$ such that $v \notin R$ and*

$$\forall x \succeq v\, \big[N(x) \ge 2^d \to x \in R\big]. \tag{8.4}$$

This is an analog of Lemma 8.3.4. Once again, a condition on subsets of $\mathbb{N}$ (the containment $\mathsf{Succ}(N) \subseteq$ Non-MLR) implies a condition on finite objects, which is easier to work with.

Subproof. Suppose the Lemma fails. We build a set $Z \notin R$ on which N succeeds. Define a sequence of strings $(v_d)_{d \in \mathbb{N}}$ as follows: let $v_0 = \emptyset$, and let v_{d+1} be some proper extension y of v_d such that $N(y) \ge 2^d$ but $y \notin R$. Then N succeeds on $Z = \bigcup_d v_d$. On the other hand $Z \notin R$, so $Z \in \mathsf{MLR}$. $\diamond$

We build a Turing functional L such that L^X is a (total) $\mathbb{Q}_2$-valued martingale for each oracle X. (We say that L is a *martingale functional.*) If A is in $\mathrm{Low}(\mathsf{MLR}, \mathsf{CR})$ then $\mathsf{Succ}(L^A) \subseteq$ Non-MLR, so we can apply the Lemma with $N = L^A$. At first we pretend to know a witness $\langle v, d\rangle$ as in the Lemma.

We will define an effective sequence $(T_s)_{s \in \mathbb{N}}$ of finite subtrees of $2^{<\omega}$. For each string γ, $T(\gamma) = \lim_s T_s(\gamma)$ exists. Further, A is a path of the tree T. Each path of T is low for K, for we enumerate a bounded request set W such that, for some constant c determined below, if $\gamma \in T$ and $K^\gamma(y) = r$ then $\langle r + c, y\rangle \in W$, which causes $K(y) \le^+ K^\gamma(y)$. The tree T_s is used to check whether the condition (8.4) looks correct at stage s for $N = L^\gamma$: if for some z we have defined $L^\gamma(z) \ge 2^d$ at a stage prior to s, then γ is only allowed to be on T_s if $z \in R_s$.

We define the martingale functional L in such a way that we avoid putting too much garbage into W. This ensures that W is a bounded request set. Garbage consists of requests $\langle r + c, y\rangle$ for strings γ such that $K^\gamma(y) = r$ but $\gamma \notin T$.

A *procedure* is a triple $\alpha = \langle \sigma, y, \gamma\rangle$, where $\sigma, y, \gamma \in 2^{<\omega}$, $|y| < |\gamma|$ and $|\sigma| \le |y| + 2\log|y| + c_K$ (here c_K is a constant such that $\forall y\, [K(y) \le |y| + 2\log|y| + c_K]$). We start α at a stage s that is least such that $\gamma \in T_s$ and $\mathbb{U}^\gamma_s(\sigma) = y$, and γ is the shortest among such strings. Now α wants to put $\langle r + c, y\rangle$ into W, where $r = |\sigma|$. It first causes a clopen set $\widetilde{C} \subseteq [v]$ such that $\lambda_v(\widetilde{C}) = 2^{-(r+c)}$ to go into R. The weight of the requests α puts into W is accounted against the measure of new enumeration into R (see 5.1.21). If the clopen sets belonging to different procedures are disjoint, then W is a bounded request set.

Roughly, the procedure α chooses a clopen set $\widetilde{C} = \widetilde{C}(\alpha)$ such that $\lambda_v(\widetilde{C}) = 2^{-(r+c)}$, and $\widetilde{C}$ is disjoint from R_s and the sets chosen by other procedures. Then α causes (in a way also to be specified) $L^X(z) \geq 2^d$ for each $X \succeq \gamma$ and each string $z \in \widetilde{C}$ of minimal length. If at a stage $t > s$ once again we have $\gamma \in T_t$, then $\widetilde{C} \subseteq R_t$, and α now has the permission to put $\langle r+c, y\rangle$ into W.

This approach has to be refined in order to guarantee the disjointness of clopen sets belonging to different procedures. Suppose $\beta \neq \alpha$ is a procedure that wants to choose its set $\widetilde{C}(\beta)$ at a stage $s' > s$. If $\gamma = (\alpha)_2$ is in some T_q with $s < q < s'$, then $\widetilde{C}(\alpha) \subseteq R_{s'}$, so there is no problem since β chooses its set disjoint from $R_{s'}$. However, if γ has not reappeared (it possibly never will), then α keeps away its set from assignment to other procedures, which may cause a conflict because $\widetilde{C}(\alpha)$ is relatively large. The solution to this problem is a "controlled risk strategy" similar to the ones used for the decanter constructions of Section 5.4. The procedure α builds up $\widetilde{C}(\alpha)$ in small portions $\widetilde{D}$ such that $\lambda_v\widetilde{D}$ is a fixed fraction of $2^{-(r+c)}$, and only assigns a new set $\widetilde{D}$ once the previous one is in R. If α always reappears on T_s at a stage s after assigning such a set, then eventually $\widetilde{C}(\alpha)$ reaches the required measure $2^{-(r+c)}$, in which case α is allowed to enumerate the request $\langle r+c, y\rangle$ into W. Otherwise, α reserves only one single set $\widetilde{D}$ not contained in R, which is so small that the union (over all procedures) of sets reserved has a measure of at most $2^{-(g(v)+2)}$. In the formal construction, $\widetilde{E}_t$ denotes the union of sets of strings appointed by some procedure at the end of stage t. Then $\lambda(\widetilde{E}_t - R_t) \leq 2^{-(u+2)}$ for any t.

The witness for Lemma 8.3.13 is not actually known, so we follow the plan outlined above for each potential witness. Fix an effective listing $(\delta_m)_{m\geq 1}$ of all pairs $\delta_m = \langle v, d\rangle$ where v is a string, and $d \in \mathbb{N}$. Uniformly in $m \geq 1$, we will build Turing functionals L_m almost as above; however, if $|x| \leq m$ then $L_m(x) = 2^{-m}$. Then $L = \sum_{m\geq 1} L_m$ is a martingale functional. Note that L is $\mathbb{Q}_2$-valued since the contributions of the L_m for $m > |w|$ add up to $2^{-|w|}$.

If $\delta_m = \langle v, d\rangle$ is fixed, we let

$$u = g(v) \text{ and } c = m + d + u + 3,$$

where g is the computable function in (8.3). Some δ_m represents a witness $\langle v, d\rangle$ in Lemma 8.3.13, and in that case we will be able to define a bounded request set W showing that A is low for K.

The procedure $\alpha = \langle \sigma, y, \gamma\rangle$ appoints certain strings z and ensures that $L_m^X(z) \geq 2^d$ for each $X \succeq \gamma$. Once activated, namely when $\mathbb{U}_s^\gamma(\sigma) = y$, the procedure α can claim the amount $\epsilon = 2^{-(r+m)}$ of the initial capital 2^{-m} of L_m^X, for any oracle $X \succeq \gamma$ (recall that $r = |\sigma|$). Therefore, given X, the total capital claimed by all activated procedures for the witness δ_m is $2^{-m}\Omega^X < 2^{-m}$. The procedure only appoints strings of the form $z = x0^{1+r+m+d}$. It "withdraws" its capital at x, increasing $L_m^X(x0)$ by ϵ for oracles $X \succeq \gamma$. To maintain the martingale property, it also decreases $L_m^X(x1)$ by ϵ. Now it doubles its capital along z, always betting all the capital on 0, and reaches an increase of 2^d at z. Every string extending x is called *used by* α.

The procedure α has to obey the following restrictions at a stage s.

1. Choose the extension z not in $[\widetilde{E}_{s-1}]^{\prec}$ where $\widetilde{E}_{s-1}$ is the set of strings previously appointed by other procedures β. (The open sets generated by the strings appointed by different procedures need to be disjoint.)
2. Let $C_t(\alpha)$ denote the set of strings x used by α up to stage t. Ensure that $x \notin [C_{s-1}(\alpha)]^{\prec}$ so that the capital of α is still available at x. Such a choice is possible for sufficiently many x, because for all t we have $\lambda_v[\widetilde{C}_t(\alpha)]^{\prec} \leq 2^{-(r+c)}$, so that $\lambda_v[C_t(\alpha)]^{\prec} \leq 2^{-(r+c)}2^{1+r+d+m} = 2^{-(u+2)}$.

There is no conflict between $\alpha = \langle \sigma, y, \gamma \rangle$ and any other procedure $\beta = \langle \sigma', y', \gamma' \rangle$ as far as the capital is concerned: if γ' is incomparable with γ then γ and γ' can only be extended by different oracles X. Otherwise, α and β own different parts of the initial capital of L_m^X for every set X extending their third components.

We now proceed to the formal definition of the martingale functionals L_m. The relevant objects are summarized in Table 8.2. For a procedure $\alpha = \langle \sigma, y, \gamma \rangle$, let $n_\alpha > \max(|\sigma| + m + d + 1, |\gamma|, |v|)$ be a natural number assigned to α in some effective one-one way. Each procedure α defines an auxiliary function $F_\alpha : 2^{<\omega} \to \mathbb{Q}_2$ (which is a generalized martingale, where negative values are allowed as in 7.1.10, with initial capital 0). The set $\widetilde{C}(\alpha)$ of appointed strings coincides with the set of minimal strings in $\{w\colon\ F_\alpha(w) \geq 2^d\}$. For each oracle set X let

$$L_m^X(w) = 2^{-m} + \sum_\alpha F_\alpha(w)\, [\![(\alpha)_2 \preceq X]\!]. \tag{8.5}$$

Given $\alpha = \langle \sigma, y, \gamma \rangle$, let $r = |\sigma|$. We ensure that

(B1) $F_\alpha(w) = 0$ if $|w| \leq |\gamma|$,

(B2) $F_\alpha(w) \geq -2^{-(r+m)}$, and $F_\alpha(w) = 0$ unless $\mathbb{U}_s^\gamma(\sigma) = y$, and

(B3) $\forall w\, \left[F_\alpha(w0) + F_\alpha(w1) = 2F_\alpha(w)\right]$.

Based on these properties, we verify that L_m is a martingale functional for each m. Firstly, $L_m^X(w) \in \mathbb{Q}_2$ for each set X and string w, since by (B1), only the finitely many procedures α such that $|(\alpha)_2| < |w|$ contribute to the sum in (8.5). Next, for $p = |w|$,

$$\begin{aligned} L_m^X(w0) + L_m^X(w1) &= 2^{-m+1} + \sum_\alpha F_\alpha(w0) + F_\alpha(w1)\, [\![(\alpha)_2 \preceq X \upharpoonright_{p+1}]\!] \\ &= 2(2^{-m} + \sum_\alpha F_\alpha(w)\, [\![(\alpha)_2 \preceq X \upharpoonright_{p+1}]\!]) \\ &= 2L_m^X(w). \end{aligned}$$

For the last equality we have used (B1). Finally, $L_m^X(w) \geq 0$ since $F_\alpha(w) \geq -2^{-(r+m)}$, and α contributes to the sum (8.5) only if the computation $\mathbb{U}^\gamma(\sigma) = y$ converges, where $r = |\sigma|$ and $\gamma \preceq X$. So $L_m^X(w) \geq 2^{-m}(1 - \Omega^X) \geq 0$ for each w.

TABLE 8.2. Summary of the objects occuring in the proof.

R	$\mathcal{R}_1 = [\{x \in \{0,1\}^*: K(x) \le \|x\| - 1\}]^{\prec}$
g	computable function as in (8.3)
δ_m	witness for Lemma 8.3.13, of the form $\langle v, d\rangle$
u	$g(v)$
c	$m + d + u + 3$
L_m	martingale functional for witness δ_m
α	procedure of form $\langle \sigma, y, \gamma\rangle$ where $\mathbb{U}^\gamma(\sigma) = y$; let $r = \|\sigma\|$
n_α	code number of α, chosen $> \max(\|\sigma\| + m + d + 1, \|\gamma\|, \|v\|)$
F_α	auxiliary function defined by α
$C_t(\alpha)$	set of strings x used by α up to (the end of) stage t
$\widetilde{C}_t(\alpha)$	set of strings appointed by α up to stage t, of form $x0^{r+m+d+1}$
T_s	tree for m at the end of stage s
W	bounded request set for m
$\widetilde{E}_t$	set of strings appointed by procedures up to stage t

Construction for parameter m. Let $\delta_m = \langle v, d\rangle$ and $u = g(v)$. The construction proceeds at stages which are powers of 2; the variables s and t denote such stages. At stage s, we define T_s and extend the functions $F_\alpha(w)$ to all w such that $s \le |w| < 2s$. For each w such that $s \le |w| < 2s$ and each string η (which may be shorter than w), by the end of stage s we can calculate

$$\overline{L}_m(\eta, w) = 2^{-m} + \sum_\alpha F_\alpha(w)\, [\![(\alpha)_2 \preceq \eta]\!]. \tag{8.6}$$

Stage 1. Let $T_1 = \{\varnothing\}$, and $F_\alpha(w) = 0$ for each α and each w, $|w| \le 1$. Let $\widetilde{E}_1 = \emptyset$.

Stage $s > 1$. Suppose T_t has been determined for $t < s$, and the functions $F_\alpha(w)$ have been defined for all w such that $|w| < s$. Let

$$T_s = \{\gamma\colon \forall w \succeq v\, [(|w| < s \;\&\; \overline{L}_m(\gamma, w) \ge 2^d) \to w \in R_s]\}.$$

(1) If $\lambda_v R_s \ge 1 - 2^{-u}$ goto (4). Note that if δ_m is a witness as in Lemma 8.3.13 then this case does not occur.

(2) For each $\alpha = \langle \sigma, y, \gamma\rangle$, $n_\alpha < s$, if $\mathbb{U}^\gamma_s(\sigma) = y$ and $\mathbb{U}^\gamma_{s/2}(\sigma)\uparrow$ and, for σ, y, the string γ is the shortest such string, then *start* the procedure α.

(3) Carry out the following for each procedure $\alpha = \langle \sigma, y, \gamma\rangle$ in the order of $n_\alpha < s$. Let $r = |\sigma|$.

(3a) If α has been started and $\gamma \in T_s$, first check whether the goal has been reached, namely $\lambda_v \widetilde{C}_{s/2}(\alpha) = 2^{-(r+c)}$. In that case put $\langle r+c, y\rangle$ into W. We say that α *ends*. Otherwise we say that α *acts*. Choose a set $D = D_\alpha \subseteq [v]$ of strings of length s such that $\lambda_v D = 2^{-(n_\alpha+u+2)}$ and

$$[D]^{\prec} \cap [R_s \cup \widetilde{E}_{s/2} \cup G \cup C_{s/2}(\alpha)]^{\prec} = \emptyset,$$

where $G = \bigcup\{D_\beta \colon \beta$ has so far acted at stage $s\}$. (In Claim 2 we will verify that D exists.) Let $\widetilde{D} = \{x0^{m+d+r+1} \colon x \in D\}$, put D into $C_s(\alpha)$, and put $\widetilde{D}$ into $\widetilde{C}_s(\alpha)$ and $\widetilde{E}_s$. (Note that $|w| < 2s$ for all strings $w \in \widetilde{D}$, since $m + d + r + 1 < n_\alpha < s$.)

(3b) For each $x \in D$, let $F_\alpha(x) = 0, F_\alpha(x1) = -\epsilon$ and $F_\alpha(x0) = \epsilon$, where $\epsilon = 2^{-(r+m)}$. Now double the capital along $x0^{r+d+m+1}$: for each string z such that $|z| \le r + m$, let $F_\alpha(x0z) = \epsilon 2^l$ if $z = 0^l$, and $F_\alpha(x0z) = 0$ otherwise. (This causes $\overline{L}_m(\gamma, w) \ge 2^d$ for each $w \in \widetilde{D}$.)

Process the next α.

(4) For each string w, $s \le |w| < 2s$, if $F_\alpha(w)$ is still undefined let $F_\alpha(w) = F_\alpha(w')$, where $w' \preceq w$ is longest such that $F_\alpha(w')$ is defined.

End of stage s.

Verification. We establish a series of claims. Let $\alpha = \langle \sigma, y, \gamma \rangle$ be a procedure.

Claim 1. *(B1)-(B3) are satisfied. Thus* L *is a martingale functional.*

Property (B1) holds because when we assign a nonzero value to $F_\alpha(w)$ at stage s, then $|w| \ge s > n_\alpha > |\gamma|$. (B2) and (B3) are satisfied since each x chosen in (3b) goes into $C(\alpha)$. So by the choice of D in (3a), no future definition of F_α on extensions of x is made except for by (4). ◇

Claim 2. Each procedure α *is able to choose a set* D_α *in (3a).*

- By the definition of T_s, for each $\beta = \langle \sigma', y', \gamma' \rangle$ and each $t \ge 2$, if $\gamma' \in T_t$ then $\widetilde{C}_{t/2}(\beta) \subseteq R_t$. Thus for each procedure β, $\lambda_v(\widetilde{C}_t(\beta) - R_t) \le 2^{-(n_\beta+u+2)}$ because $\widetilde{C}_t(\beta) - R_t$ consists of a single set $\widetilde{D}_\beta$. Then, letting $t = s/2$, we have $\lambda_v(\widetilde{E}_{s/2} - R_s) \le 2^{-(u+2)}$.
- Each set D_β chosen during stage s satisfies $\lambda_v(D_\beta) \le 2^{-(n_\beta+u+2)}$, hence $\lambda_v G$ never exceeds $2^{-(u+2)}$.
- For each s, $\lambda_v \widetilde{C}_s(\alpha) \le 2^{-(r+c)}$, and hence $\lambda_v C_s(\alpha) \le 2^{r+d+m+1} 2^{-(r+c)} = 2^{-(u+2)}$.

Since the query in (1) was answered negatively, $\lambda_v R_s < 1 - 2^{-u}$, so relative to $[v]$ a measure of $2^{-(u+2)}$ is available outside $[R_s \cup \widetilde{E}_{s/2} \cup G \cup C_{s/2}]$ for choosing D_α. All strings in $R_s \cup \widetilde{E}_{s/2} \cup G \cup C_{s/2}$ are shorter than s (for strings in $\widetilde{E}_{s/2}$ this holds by the comment at the end of (3a)), so the strings in D_α can be chosen of length s. ◇

Claim 3. *Each procedure α acts only finitely often.*

Each time α acts at s and $s' > s$ is least such that $\gamma \in T_{s'}$ we have increased $\lambda_v \widetilde{C}(\alpha)$ by the fixed amount of $2^{-(n_\alpha+c+r)}$. So either $\gamma \notin T_s$ for almost all s, or α ends. ◇

In (8.6) we defined $\overline{L}_m(\eta, w)$.

Claim 4. *For each string η, there is a stage s_η such that no procedure α with $(\alpha)_2 \preceq \eta$ acts at any stage $\ge s_\eta$. Further, for each $w \succeq v$, if $|w| \ge s_\eta$ then $\overline{L}_m(\eta, w) = \overline{L}_m(\eta, w')$ for some w' such that $v \preceq w' \preceq w$ and $|w'| < s_\eta$.*

This holds because there are only finitely many procedures α with $(\alpha)_2 \preceq \eta$. By Claim 3 there is a stage s_η by which these procedures have stopped acting, and further definitions $F_\alpha(w)$ are only made in (4). $\diamond$

Claim 5. *$T(\eta) = lim_s T_s(\eta)$ exists.*
Suppose $s \geq s_\eta$ is least such that $\eta \in T_s$. We show $\eta \in T_t$ for each $t \geq s$. Suppose $v \preceq w$, $|w| \leq t$ and $\overline{L}_m(\eta, t) \geq 2^d$. By Claim 4, $\overline{L}_m(\eta, w) = \overline{L}_m(\eta, w')$ for some $w' \preceq w$ of length $< s_\eta$. Since $\eta \in T_s$, we have $w' \in R_s$ and hence $w \in R_t$. $\diamond$

We now assume that δ_m is a witness for Lemma 8.3.13 where $N = L^A$.

Claim 6. *A is a path of T.*
Given l, let $\eta = A \restriction_l$. Suppose $|w'| < s_\eta$ and $\overline{L}_m(w', \eta) \geq 2^d$. Then $L^A(w') \geq 2^d$, since $L^A(w') \geq L^A_m(w') \geq \overline{L}_m(w, \eta)$. By (8.4) $w' \in R$. Let s be a stage so that all such w' are in R_s. Then by Claim 4 we have $\eta \in T_t$ for all $t \geq s$. $\diamond$

Claim 7. *W is a bounded request set.*
When $\alpha = \langle \sigma, y, \gamma \rangle$ ends at stage s and puts $\langle |\sigma| + c, y \rangle$ into W, then $\lambda_v \widetilde{C}_{s/2}(\alpha) = 2^{-(|\sigma|+c)}$. The sets $[\widetilde{C}(\alpha)]^{\prec}$ are contained in $[v]$, and pairwise disjoint by the choice of $D = D_\alpha$ in (3a). Thus the total weight of W is at most 1. $\diamond$

Claim 8. *Each path of T is low for K.*
Let M_e be a prefix machine for W according to Theorem 2.2.17. We claim that $K(y) \leq K^X(y) + c + e$ for each path X of T and each string y. For choose a shortest $\mathbb{U}^X$-description σ of y, and choose $\gamma \subseteq X$ shortest such that $|\gamma| > y$ and $\mathbb{U}^\gamma(\sigma) = y$. At some stage t, we start the procedure $\langle \sigma, y, \gamma \rangle$ since $\gamma \in T$. This procedure ends by Claim 3 and its proof, so we put $\langle |\sigma| + c, y \rangle$ into W, causing $K(y) \leq K^X(y) + c + e$. $\square$

There are similarities between Theorem 8.3.10 and results in earlier chapters. The theorem is closely related to Theorem 5.1.22 that each base for ML-randomness is low for K. In both cases we have procedures that rely on a computation $\mathbb{U}^\gamma(\sigma) = y$. The hungry sets $C^\gamma_{d,\sigma}$ correspond to the sets $\widetilde{C}(\alpha)$, $\alpha = \langle \sigma, y, \gamma \rangle$. In either case, the set has to reach a certain measure before the request can be enumerated. One of the reasons why Theorem 5.1.22 requires less effort to prove is that it is much easier to ensure that the hungry sets are disjoint.

Next, we will compare the proof of Theorem 8.3.10 with the proof of Theorem 5.4.1 that each K-trivial set is low for K. In 8.3.10 there are no levels of procedures. The procedure $\alpha = \langle \sigma, y, \gamma \rangle$ corresponds to a procedure $Q_{j,\sigma y w}$ at a *fixed* level j. Again, in each case a procedure is based on a computation, $\mathbb{U}^\gamma(\sigma) = y$ for 8.3.10, and $\mathbb{U}^A_s(\sigma) = y$ for 5.4.1. It becomes inactive (is cancelled, respectively) when its guess about A turns out to be wrong. It carries out its actions in small bits to avert too much damage in case this happens (a controlled risk strategy). Reserving only a small set D_α at a time corresponds to calling a procedure P_{j-1} with a small goal β.

Finally, there is some similarity between the foregoing proof and the proof of Theorem 1.7.20 on creative sets! The c.e. set F there corresponds to the c.e. open set R. There, we assign a number x that has just entered F as a value $p_{i,s}(e)$ to ensure that x does not coincide with a number used earlier. Here, we choose clopen sets disjoint from R (expecting they will enter R later) to make them disjoint from previously used sets that have already entered R.

Recall that W2R denotes the class of weakly 2-random sets. A modification of the proof of 8.3.10 yields a stronger result, which also entails Theorem 5.1.33 that Low(W2R, MLR) = Low(MLR).

8.3.14 Theorem. (Extends 8.3.10)
$A \in \text{Low}(\text{W2R}, \text{CR}) \Leftrightarrow$ *A is low for K.*

Proof. We only need to prove the implication "$\Rightarrow$". The construction of L_m is now based on potential witnesses of the form $\delta_m = \langle v, d, V, q, \epsilon \rangle$, where $v \in \{0,1\}^*$, $d \in \mathbb{N}$, V is a c.e. open set and $q, \epsilon \in \mathbb{Q}_2$, $0 < \epsilon, q$, and $q + \epsilon \leq 1$. Given such a potential witness δ_m, we let $R = [\{x \succeq v \colon \lambda_x(V) \geq q + \epsilon\}]^{\prec}$. Let $u \in \mathbb{N}$ be least such that $q/(q+\epsilon) < 1 - 2^{-u}$. In the construction of L_m, each procedure α is specified as before but with the new definitions of R and u. In particular, a procedure based on δ_m is active at stages s only as long as $\lambda_v(R_s) < 1 - 2^{-u}$. The function $x \to \lambda_x(V)$ is a martingale, so if $\lambda_v(V) \leq q$, then by Proposition 7.1.9 we have $\lambda_v(R) < 1 - 2^{-u}$.

Lemma 8.3.13 is modified: for an appropriate witness such that $\lambda_v(V) \leq q$, if a procedure α causes $L^A(x) \geq 2^d$ for some $x \succeq v$, then $\lambda_x(V) \geq q + \epsilon$.

8.3.15 Lemma. *Let N be a martingale such that* $\text{Succ}(N) \subseteq \text{non-W2R}$.
Then there is a $\delta_m = \langle v, d, V, q, \epsilon \rangle$ as above such that $\lambda_v(V) \leq q$ and

$$\forall x \succeq v \left[N(x) \geq 2^d \to \lambda_x(V) \geq q + \epsilon \right]. \tag{8.7}$$

Subproof. The argument is similar to the one in the proof of Theorem 5.1.33. (The negation of the lemma now plays the role of Claim 5.1.34.) As before, let $\{G_n^e\}_{e,n\in\omega}$ be a listing of all the generalized ML-tests (Definition 3.6.1) with no assumption on the uniformity in e. If the lemma fails we may inductively define a sequence $w_0 \prec w_1 \prec \ldots$ and numbers n_d ($d \in \mathbb{N}$) such that

$$N(w_d) \geq 2^d \ \& \ \lambda(V_d \mid w_d) \leq \gamma_d, \tag{8.8}$$

where $\gamma_d = 1 - 2^{-d}$ and $V_d = \bigcup_{i<d} G_{n_i}^i$. We may assume that $N(\varnothing) \geq 1$. We let $w_0 = \varnothing$. Then (8.8) holds for $d = 0$. In step $d \geq 0$ we choose n_d so large that $\lambda(G_{n_d}^d) \leq 2^{-|w_d|-d-2}$. Then $\lambda(G_{n_d}^d | w_d) \leq 2^{-(d+2)}$. Since $V_{d+1} = V_d \cup G_{n_d}^d$, we have $\lambda(V_{d+1} \mid w_d) \leq \gamma_d + 2^{-(d+2)} < \gamma_{d+1}$. Since the lemma fails for $\langle v, e, V, q, \epsilon \rangle$ where $v = w_d$, $e = d+1$, $V = V_{d+1}$, $q = \gamma_d + 2^{-(d+2)}$ and $\epsilon = 2^{-(d+2)}$, we may choose $w_{d+1} \succ w_d$ such that $N(w_{d+1}) \geq 2^{d+1}$ and $\lambda(V_{d+1} \mid w_{d+1}) \leq \gamma_{d+1}$. Then N succeeds on the weakly 2-random set $Z = \bigcup_d w_d$. ◇

Claims 1–8 are now verified as before. Hence A is low for K. □

We have characterized nine of the ten classes $\text{Low}(\mathcal{C}, \mathcal{D})$ where $\mathcal{C}, \mathcal{D} \in \{\text{W2R}, \text{MLR}, \text{CR}, \text{SR}\}$ and $\mathcal{C} \subseteq \mathcal{D}$. Five classes coincide with Low(MLR).

8.3.16.$^{\diamond}$ **Problem.** Show that the class Low(W2R, SR) coincides with Low(MLR, SR), and hence with being c.e. traceable (Bienvenu and Miller, 20xx).

8.4 Jump traceability

Our basic notion is c.e. traceability of a set, introduced in 8.2.2. In Definition 8.2.15 we strengthened this to computable traceability, where the traces are effective sequence of strong indices for finite sets. We can also stay with c.e. traces, but strengthen the basic notion by tracing functions partial computable in A. We show in Fact 8.4.2 that it suffices to require a c.e. trace for J^A.

8.4.1 Definition. Let g be an order function. The set A is *jump traceable with bound* g (Nies, 2002) if there is a c.e. trace $(T_e)_{e\in\mathbb{N}}$ with bound g such that

$$\forall e\left[J^A(e)\downarrow\rightarrow J^A(e)\in T_e\right].$$

We say that $(T_e)_{e\in\mathbb{N}}$ is a *jump trace* for A. The set A is called *jump traceable* if A is jump traceable with some bound.

For c.e. traceability and computable traceability, by the argument in Theorem 8.2.3, the growth rate of the bounds of the traces is irrelevant as long as they are order functions. The argument relies on the fact that only total functions are traced, so it is not surprising that the case of jump traceability is different: there is a proper hierarchy depending on the growth of the bounds by Theorem 8.5.2.

Later on in this section we will study sets that are jump traceable for every bound. We will see in Section 8.5 that this lowness property, called strong jump traceability, is much more restrictive than jump traceability. For instance, for c.e. sets, strong jump traceability strictly implies K-triviality, while jump traceability is equivalent to superlowness.

Just like the class $\mathrm{Low}(\Omega)$, the class of jump traceable sets is closed downward under $\leq_T$, and contained in GL_1. In fact, building a jump traceable set is a good way to obtain a set in GL_1. Friedberg (1957*a*) showed that for each set $C\geq_T\emptyset'$ there is a set A such that $C\equiv_T A\oplus\emptyset'\equiv_T A'$. We prove this in Corollary 8.4.5 via a jump traceable set A.

Unlike sets that are low for Ω, by Corollary 8.4.7 there is an incomputable jump traceable set that is computably dominated. No characterization via relative randomness is known for jump traceability. However, we will give one using C-complexity relative to an oracle.

Basics of jump traceability, and existence theorems

First we show that it suffices to require that the jump is traced.

8.4.2 Fact. *If A is jump traceable, there is a c.e. trace $(S_m)_{m\in\mathbb{N}}$ such that for each Turing functional Φ we have $\forall^\infty m\left[\Phi^A(m)\downarrow\rightarrow\Phi^A(m)\in S_m\right]$.*
In particular, A is c.e. traceable.

Proof. There is an effective listing $(p_i)_{i\in\mathbb{N}}$ of all the reduction functions in the sense of Fact 1.2.15. (This relies on the proof of the Parameter Theorem 1.1.2 which is applied to obtain Fact 1.2.15.) Suppose that A is jump traceable via the c.e. trace $(T_e)_{e\in\mathbb{N}}$ with bound g. Let $S_m=\bigcup_{i\leq m}T_{p_i(m)}$. Then S_m is c.e. uniformly in m, and $\lambda m.\sum_{i\leq m}g(p_i(m))$ is a bound for $(S_m)_{m\in\mathbb{N}}$.

Given a Turing functional Φ, choose a reduction function p_i for Φ. Then, for each $m \geq i$, $\Phi^A(m)\downarrow$ implies $\Phi^A(m) \in S_m$. □

If $B \leq_T A$ then J^B is partial computable in A. Therefore the class of jump traceable sets is closed downward under Turing reducibility.

8.4.3 Proposition. *Each jump traceable set A is in* GL_1. *Moreover, a reduction procedure for $A' \leq_T A \oplus \emptyset'$ can be obtained effectively from an index of a jump trace for A.*

Proof. Consider the Turing functional Ψ defined by $\Psi^X(n) \simeq \mu s.\, J^X(n)\downarrow [s]$. Choose a reduction function p for Ψ by 1.2.15. To determine whether $e \in A'$, first compute $t = \max T_{p(e)}$ using $\emptyset'$ as an oracle. Next, using A check whether $J^A(p(e))\downarrow$ in at most t steps. If so, answer "yes", otherwise answer "no". □

Function trees were introduced in the proof of Theorem 8.2.17 to build a perfect class of computably traceable sets. They can also be used to obtain a perfect class of jump traceable sets. Before, the purpose of the e-th function tree was to provide a trace for the Turing functional Φ_e. Now we only trace the jump. The s-th function tree is used to trace computations that converge by stage s. The result is due to Nies (2002).

8.4.4 Theorem. *There is a perfect Π^0_1 class P such that each set in P is jump traceable via a fixed c.e. trace $(T_e)_{e\in\mathbb{N}}$ with bound $\lambda e.2 \cdot 4^e$.*

Proof. We build a uniformly computable sequence of function trees $(F_s)_{s\in\mathbb{N}}$. We think of the $F_s(\sigma)$ as movable markers on binary strings. The active movement of a marker is to an extensions of its present value. Thus, for each s, σ, if $F_{s+1}(\sigma) \neq F_s(\sigma)$ then we have $F_{s+1}(\sigma) \succ F_s(\sigma)$ unless the change is caused by $F_{s+1}(\rho) \neq F_s(\rho)$ for some $\rho \prec \sigma$. We ensure that $F(\sigma) = \lim_s F_s(\sigma)$ exists for each σ, that is, the markers stabilize eventually. Then F is a function tree. Note that the class

$$P = \mathit{Paths}(F) = \{Z\colon \forall n\, \forall s\, \exists \eta\, [|\eta| \leq n \;\&\; Z\restriction_n \preceq F_s(\eta)]\}$$

is a perfect Π^0_1 class.

The idea is to move $F_s(\sigma)$, $|\sigma| = e$, in order to make a computation $J_s^{F(\sigma)}(e)$ converge whenever possible. Once achieved, we maintain this unless the movement of a marker $F_s(\rho)$ for $\rho \prec \sigma$ interferes. Note the similarity to meeting the lowness requirements in the proof of Theorem 1.6.4. The interference of a marker $F_s(\rho)$ for $\rho \prec \sigma$ corresponds to the injury of a lowness requirement L_e.

Construction of the function trees F_s. Let $F_0(\sigma) = \sigma$ for each σ.

Stage $s+1$. Look for the least $e < s$ such that for some σ of length e, chosen least, we have $J_s^{F_s(\sigma)}(e)\uparrow$ and

$$\exists \rho \succeq \sigma\, [|\rho| \leq s \;\&\; J_s^{F_s(\rho)}(e)\downarrow].$$

Let $F_{s+1}(\sigma\alpha) = F_s(\rho\alpha)$ for each string α (including $\alpha = \varnothing$). For all strings $\eta \not\succeq \sigma$, let $F_{s+1}(\eta) = F_s(\eta)$. (If e fails to exist, let $F_{s+1}(\eta) = F_s(\eta)$ for all η.)

Let us first verify merely that $Y' \leq_T Y \oplus \emptyset'$ for each $Y \in P$, as this was asked in Exercise 8.2.20. Given an input e, using the oracle $\emptyset'$, find the least stage t

such that $F_t(\sigma)$ has stabilized for all σ, $|\sigma| \le e+1$. Let σ be the string such that $F_t(\sigma) \prec Y$, then $J^Y(e) \simeq J_t^{F_t(\sigma)}(e)$.

For the full result, define a c.e. trace by $T_e = \{J_s^{F_s(\sigma)}(e)\colon s \in \mathbb{N}, |\sigma| = e\}$. For $|\sigma| = e$, a value $F_s(\sigma)$ can change at most $2^{e+1} - 1$ times because the marker $F_s(\sigma)$ moves at most once after $F_s(\rho)$ is stable for all $\rho \prec \sigma$. Therefore $\#T_e \le 2^e(2^{e+1} - 1) < 2 \cdot 4^e$. □

8.4.5 Corollary. *For each set $C \ge_T \emptyset'$ there is a jump traceable set A such that $C \equiv_T A \oplus \emptyset' \equiv_T A'$. In particular, there is a high jump traceable set.*

Proof. Let $A = \bigcup_{\sigma \prec C} F(\sigma)$ where F is the function tree introduced above. Since $F \le_T \emptyset'$, we have $C \le_T A \oplus F \le_T A \oplus \emptyset'$. Also $A \le_T C \oplus F \equiv_T C$. □

The construction in the proof of Theorem 8.4.4 can be modified in order to obtain a bound for the trace closer to $\lambda e.\, 2^e$. However, if the bound g grows slowly enough, then a set that is jump traceable for g is Δ^0_2 by Downey and Greenberg (2011).

By Cor. 8.4.5, a computably random set can be jump traceable (see Ex. 8.6.6).

8.4.6 Exercise. Extend Theorem 8.4.4 by also ensuring that P contains no computable sets (at the cost of a somewhat larger bound for the jump trace).

8.4.7 Corollary. *There is a perfect class of sets that are both computably traceable and jump traceable.*

Proof. Let $P \ne \emptyset$ be the Π^0_1 class of Exercise 8.4.6. By Theorem 1.8.44, there is a perfect subclass S of P such that each element of S is computably dominated. Each member of S is c.e. traceable and hence computably traceable. □

The foregoing result implies that a computably traceable set can be in GL_1. Not all computably traceable sets are in GL_1: Lerman (1983, Thm. V.3.12) constructed a set A of minimal Turing degree such that $A'' \equiv_T \emptyset''$ and $A \notin \mathrm{GL}_1$. Examining his construction reveals that A is computably traceable.

8.4.8 Exercise. Show in a direct way that if A is low for K, then A is jump traceable with a bound in $O(n \log^2 n)$. (See 5.5.12 for a proof via K-triviality.)

8.4.9.$^\diamond$ **Problem.** Characterize the sets that are low for Ω and jump traceable (each K-trivial set is). Characterize the sets that are computably traceable and jump traceable.

Jump traceability and descriptive string complexity

Figueira, Nies and Stephan (2008) proved that A is jump traceable iff $C^A(x)$ is not much smaller than $C(x)$ for each x. We need some notation to make this precise. For any function f, let

$$\widehat{f}(y) = y + f(y).$$

Let α be a reduction function for the plain optimal machine $\mathbb{V}$, namely, $\forall X\, \forall \sigma\, \mathbb{V}^X(\sigma) \simeq J^X(\alpha(\sigma))$. Let c be a constant such that $J^A(|x|)\downarrow$ implies $C^A(x, J^A(|x|)) \le |x| + c$ for each x. Such a c exists because given x we can compute $J^A(|x|)$ relative to A.

8.4.10 Theorem. *A is jump traceable $\Leftrightarrow$ $\forall x\, [C(x) \le^+ C^A(x) + h(C^A(x))]$ for some order function h. In more detail:*

(i) *Let A be jump traceable with bound g. Then* $\forall x\, [C(x) \le^+ \widehat{h}(C^A(x))]$, *where*
$$h(n) = \max\{2g(\alpha(\sigma)) + 1\colon\ |\sigma| = n\}.$$
(ii) *Suppose h is an order function such that* $\forall x\, [C(x) \le^+ \widehat{h}(C^A(x))]$. *Then A is jump traceable with bound* $\lambda e.\, 3^{h(e+c)}$.

Proof. (i) Let $(T_e)_{e\in\mathbb{N}}$ be a jump trace for A with bound g. We define a machine M by letting $M(0^{|r|}1r\sigma)$ be the r-th element in the enumeration of $T_{\alpha(\sigma)}$ if such an element exists, and leaving it undefined otherwise.

Suppose σ is a shortest $\mathbb{V}^A$-description of a string x. Then $x \in T_{\alpha(\sigma)}$. Since $2\#T_{\alpha(\sigma)} + 1 \le h(|\sigma|)$, we have $M(0^{|r|}1r\sigma) = x$ for some r such that $2r+1 \le h(|\sigma|)$. Thus $C(x) \le^+ \widehat{h}(|\sigma|)$ as required.

(ii) We need an auxiliary fact stating that, independently of x, we can bound the number of y such that $C(x,y)$ exceeds $C(x)$ by no more than $b \in \mathbb{N}$.

8.4.11 Lemma. $\#\{y\colon\ C(x,y) \le C(x) + b\} = O(2^{b+2\log b})$.

Subproof. Let M be the machine given by $M(\sigma) \simeq (\mathbb{V}(\sigma))_1$, namely, $M(\sigma)$ is the first component of $\mathbb{V}(\sigma)$ viewed as an ordered pair. By Lemma 5.2.21, $\#\{\sigma\colon\ M(\sigma) = x \ \&\ |\sigma| \le C(x) + b\} = O(2^{b+2\log b})$. If $C(x,y) \le C(x) + b$ then $\mathbb{V}(\sigma) = \langle x, y\rangle$ for some σ such that $|\sigma| \le C(x) + b$. Then $M(\sigma) = x$, so the required bound on $\#\{y\colon\ C(x,y) \le C(x) + b\}$ follows. $\diamond$

By the hypothesis of (ii) and the definition of c, for all e, if $J^A(e)\downarrow$ and $|x| = e$ then
$$C(x, J^A(|x|)) \le^+ \widehat{h}(C^A(x, J^A(|x|))) \le \widehat{h}(|x| + c).$$
Now let $T_e = \{y\colon\ \forall x\ [|x| = e \ \to\ C(x,y) \le \widehat{h}(e+c) + d]\}$, where d is the implicit constant in the inequality above. Thus, if $J^A(e)\downarrow$ then $J^A(e) \in T_e$. Clearly $(T_e)_{e\in\mathbb{N}}$ is uniformly c.e. For the bound on $\#T_e$, given e we choose x such that $|x| = e$ and $C(x) \ge e$. If $y \in T_e$ then $C(x,y) \le \widehat{h}(e+c) + d \le C(x) + c + h(e+c) + d$. Thus, for almost all e we have $\#T_e \le 3^{h(e+c)}$ by Lemma 8.4.11 where $b = c + h(e+c) + d$. Then a modification of $(T_e)_{e\in\mathbb{N}}$ at finitely many components is a jump trace for A with the required bound. $\square$

8.4.12 Exercise. Recall from 1.4.5 that V_e is the e-th ω-c.e. set.
Show that $\{e\colon\ V_e$ *is jump traceable*$\}$ is Σ^0_3-complete.
Similarly, show that $\{e\colon\ W_e$ *is jump traceable*$\}$ is Σ^0_3-complete.

The weak reducibility associated with jump traceability

In Section 5.6 we studied the weak reducibility $\le_{LR}$ associated with the lowness property Low(MLR). Simpson (2007) defined implicitly a weak reducibility $\le_{JT}$ associated with jump traceability. Also see Table 8.3 on page 363.

Note that a sequence of sets $(T_e)_{e\in\mathbb{N}}$ is a c.e. trace relative to B iff there is function $f \le_T B$ such that $T_e = W^B_{f(e)}$ for each e, and a bound $h \le_T B$ such that $\#T_e \le h(e)$ for each e. We may equivalently require that f is computable. For, let $f = \Gamma^B$, and let Θ be a Turing functional such that $\Theta^B(e,x)\downarrow$ iff

$x \in W^B_{\Gamma^B(e)}$. By the Parameter Theorem for oracles (mentioned before 1.2.10), there is a *computable* function r such that $\Phi^B_{r(e)}(x) \simeq \Theta^B(e,x)$ for each B, e, so $W^B_{f(e)} = W^B_{r(e)}$ for each e.

However, it is a restriction to require a computable bound h for the trace.

8.4.13 Definition. A is *jump traceable by* B, written $A \leq_{JT} B$, if there is a c.e. trace $(T_e)_{e\in\mathbb{N}}$ relative to B for J^A, and an order function h such that $\#T_e \leq h(e)$ for each e.

Being jump traceable *by* B is somewhat different from being jump traceable *relative* to B because we only require the existence of a c.e. trace for the function J^A, not for $J^{A\oplus B}$; on the other hand, the bound for this trace must be computable, not merely computable in B. See Barmpalias, Miller and Nies (2011) for more.

8.4.14 Fact. *The relation $\leq_{JT}$ is transitive.*

Proof. Suppose A is jump traceable by B via a trace $(S_n)_{n\in\mathbb{N}}$ with bound an order function g, and B is jump traceable by C via a trace $(T_i)_{i\in\mathbb{N}}$ with bound an order function h. There is a computable function β such that

$$J^B(\beta(\langle n,k\rangle)) \simeq \text{the } k\text{-th element enumerated into } S_n.$$

Let $V_n = \bigcup_{k<g(n)} T_{\beta(\langle n,k\rangle)}$, then $\#V_n \leq g(n)\cdot h(\beta(\langle n, g(n)\rangle))$ and A is jump traceable by C via the c.e. trace $(V_n)_{n\in\mathbb{N}}$. □

It is not hard to show that $\leq_{JT}$ is a Σ^0_3 relation on sets, that $A \leq_T B$ implies $A \leq_{JT} B$, and that $A' \not\leq_{JT} A$. Thus, $\leq_{JT}$ is indeed a weak reducibility in the sense of Section 5.6.

By Exercise 8.4.8 each set in Low(MLR) is jump traceable. This can be extended to the associated weak reducibilities by a result of Simpson (2007) relying on Kjos-Hanssen, Miller and Solomon (2011).

8.4.15 Theorem. *If $A \leq_{LR} B$ then $A \leq_{JT} B$ with trace bound $\lambda r.\, 2^r$.*

Proof. We apply Lemma 5.6.4 to the computable function f such that $f(\langle r,y\rangle) = r$ for each r, y, and the f-small set $I = \{\langle r,y\rangle \colon J^A(r) = y\}$. By the lemma there is a set $H \supseteq I$ such that H is c.e. in B, and f-small. Thus $\sum_{r,y} 2^{-r} [\![\langle r,y\rangle \in H]\!] < \infty$. Let $T_r = \{y \colon \langle r,y\rangle \in H\}$, then $(T_r)_{r\in\mathbb{N}}$ is uniformly c.e. in B and $\#T_r \leq 2^r$ for almost every r. Also, if $J^A(r) = y$ then $y \in T_r$. A modification of $(T_r)_{r\in\mathbb{N}}$ on finitely many indices r yields a trace as required. □

Nies (2002) proved that each c.e. jump traceable set is superlow (8.4.23 below). The following fact of Simpson (2007) strengthens this (letting $B = \emptyset$). It can be seen as a variant of Exercise 5.6.10 that $A \leq_T B'$ and $A \leq_{LR} B$ implies $A' \leq_T B'$ (we strengthen the first, but weaken the second hypothesis).

8.4.16 Theorem. *If A is c.e. in B and $A \leq_{JT} B$ then $A' \leq_{tt} B'$.*

Proof. Intuitively, because in the hypothesis the trace bound is computable, in the conclusion we obtain a plain truth-table reduction, not merely one relative to B. Let Ψ be the Turing functional given by

$$\Psi^X(e) \simeq \min\{\sigma \prec X\colon\ J^\sigma_{|\sigma|}(e)\downarrow\}.$$

Choose a reduction function p for Ψ. Thus, $\Psi^X(e) \simeq J^X(p(e))$ for each X and e. We write $\widehat{e}$ for $p(e)$.

Let $(A_s)_{s\in\mathbb{N}}$ be a B-computable enumeration of A. Suppose that $A \leq_{JT} B$ via a B-c.e. trace $(T_e)_{e\in\mathbb{N}}$ with computable bound h. We may assume that $J^\sigma_{|\sigma|}(\widehat{e})\downarrow$ for each $\sigma \in T_{\widehat{e}}$ by only admitting such strings into $T_{\widehat{e}}$. Let $\sigma_{e,k,s}$ be the k-th string enumerated into $T_{\widehat{e},s}$, if such a string exists, and undefined otherwise. Note that B can compute the value of $\sigma_{e,k,s}$, and if $\sigma_{e,k,s}$ is defined then $k < h(\widehat{e})$. Now $J^A(e)\downarrow \leftrightarrow \exists\sigma \in T_{\widehat{e}}\,[\sigma \prec A] \leftrightarrow \bigvee_{k<h(\widehat{e})}$

$$\exists s\,(\sigma_{e,k,s} \prec A_s)\ \&\ \neg\exists s\,\exists t\,(s < t\ \&\ \sigma_{e,k,s} \prec A_s\ \&\ \sigma_{e,k,s} \not\prec A_t). \tag{8.9}$$

The first statement in (8.9) is in $\Sigma^0_1(B)$ form, the second in $\Pi^0_1(B)$ form. These statements are obtained effectively from e and k, so $A' \leq_{tt} B'$. □

Recall from 6.3.13 that a set C is called superhigh if $\emptyset'' \leq_{tt} C'$. In Theorem 6.3.14 and the discussion preceding it we studied implications between highness properties. Simpson (2007) showed the nontrivial implication in (6.7) on page 256. This is now an immediate consequence of Theorem 8.4.15, together with Theorem 8.4.16 where $A = \emptyset'$ and $B = C$:

8.4.17 Corollary. *If $\emptyset' \leq_{LR} C$ then C is superhigh.* □

The weak reducibility $\leq_{JT}$ is conceptually close to $\leq_{LK}$ (see 5.6.1).

8.4.18 Corollary. $A \leq_{JT} B \Leftrightarrow$

$$\forall x\,\left[C^B(x) \leq^+ C^A(x) + h(C^A(x))\right] \text{ for some order function } h.$$

Proof. We adapt the proof of Theorem 8.4.10. We write $C^B(x)$ instead of $C(x)$, and $C^B(x,y)$ instead of $C(x,y)$. In (i), we are now given a B-c.e. trace $(T_e)_{e\in\mathbb{N}}$, while g still is a computable bound. The machine M uses B as an oracle. □

Exercises.

8.4.19. Prove the statements after Fact 8.4.14.

8.4.20. Show that the class $\mathsf{JTH} = \{C\colon\ C \geq_{JT} \emptyset'\}$ is Σ^0_3.

8.4.21. Define a weak reducibility $\leq_{CT}$ corresponding to computable traceability. Verify that $\leq_{CT}$ is transitive.

8.4.22.$^\diamond$ **Problem.** Decide whether $A \leq_{CT} B \Leftrightarrow \mathsf{SR}^B \subseteq \mathsf{SR}^A$ for each A, B.

Jump traceability and superlowness are equivalent for c.e. sets

Recall that a set A is superlow if $A' \leq_{tt} \emptyset'$ (Definition 1.5.3). Jump traceability and superlowness are in general quite different. While jump traceability expresses that the function J^A has a small possible *range*, superlowness restricts the computational complexity of its *domain* $A' = \{e\colon\ J^A(e)\downarrow\}$. There is a superlow ML-random set, but not a jump traceable one (see the remark after 8.4.24 below). There is a perfect class of jump traceable sets (see 8.4.4), while the class of superlow sets is countable. Nonetheless, for a c.e. set, the two properties are equivalent by a result of Nies (2002).

8.4.23 Theorem. *Let A be c.e. Then A is jump traceable $\Leftrightarrow$ A is superlow.*

Proof. $\Rightarrow$: (Range to domain) In Theorem 8.4.16 let $B = \emptyset$.

$\Leftarrow$: (Domain to range) We will write $j(X,e) \simeq \mathsf{use}\, J^X(e)$ and $j(X,e,s) \simeq \mathsf{use}\, J^X_s(e)$. By Proposition 1.4.4, A is superlow iff there is a binary $\{0,1\}$-valued computable function q and an order function g such that, for each e,

$$A'(e) = \lim_s q(e,s) \;\&\; g(e) \geq \#\{s > 0\colon\ q(e,s) \neq q(e,s-1)\}. \tag{8.10}$$

To obtain a jump trace $(T_e)_{e\in\mathbb{N}}$ for A, we will define an auxiliary Turing functional Ψ which copies computations of J with some delay. We assume a Turing functional $\widetilde{\Psi}$ is given. Let β be the reduction function effectively obtained from $\widetilde{\Psi}$ according to Fact 1.2.15. We build Ψ effectively from β. Then, by the Recursion Theorem we may assume that $\widetilde{\Psi} = \Psi$, so β is a reduction function for Ψ as well.

For each e let $\widehat{e} = \beta(e)$. At stage 0, Ψ is undefined for all inputs. At stage $s > 0$ we distinguish two cases.

(a) $q(\widehat{e},s) = 0$. If $\Psi^A(e)[s-1]\uparrow$ and $J^A(e)[s]\downarrow= v$, let $\Psi^A(e)[s] = v$ with use $j(A_s,e,s)$.

(b) $q(\widehat{e},s) = 1$. If $\Psi^A(e)[s]\downarrow$ then enumerate $y = J^A(e)[s]$ into T_e.

Note that Ψ merely copies computations of J at a later stage, so when a new computation $J^A(e)[s]$ appears, no computation $\Psi^A(e)[t]$ which was defined at $t < s$ persists to stage s.

Suppose $J^A(e) = z$, and let s be the least stage where this (final) computation appears. We show that $z \in T_e$. At a stage $t \geq s$, we may only define a new computation $\Psi^A(e)[t]$ in case $q(\widehat{e},t) = 0$. Since $\Psi^A(e)[t]$ remains undefined till this happens, by the definition of β, in fact there must be such a stage $t \geq s$. Then $\Psi^A(e)\downarrow$ since the use for $\Psi^A(e)[t]$ is $j(A_s,e,s)$ and $A_s\restriction_{j(A_s,e,s)}$ is stable. Hence $q(\widehat{e},r) = 1$ for some $r > t$, and at stage r we enumerate z into T_e.

Next, we show that $(T_e)_{e\in\mathbb{N}}$ is a c.e. trace with bound $h(e) = \lfloor\frac{1}{2}g(\beta(e))\rfloor$. Suppose that $v < r$ are stages at which distinct elements y,z are enumerated into T_e. Then $y = J^A(e)[v]$, $z = J^A(e)[r]$, and $q(\widehat{e},v) = q(\widehat{e},r) = 1$. Since $A_v\restriction_{j(A_v,e,v)} \neq A_r\restriction_{j(A_v,e,v)}$, no definition $\Psi^A(e)[v']$ issued at a stage $v' \leq v$ can be valid at stage r. (Here we have used the hypothesis that A is c.e.) So we must have made a new definition $\Psi^A(e)[t]$ at a stage t, $v < t < r$, whence $q(\widehat{e},t) = 0$. Since $q(\widehat{e},s)$ can change from 1 to 0 and back to 1 for at most $h(e)$ times, this proves that $\#T_e \leq h(e)$. □

The following consequence of Theorem 8.4.23 is not obvious from the definition of superlowness. To prove it we use the corresponding Fact 8.4.12 for jump traceability.

8.4.24 Corollary. *$\{e\colon W_e$ is superlow$\}$ is Σ^0_3-complete.* □

None of jump traceability and superlowness implies the other within the ω-c.e. sets. For let Z be superlow and ML-random by Theorem 1.8.38, then Z is of d.n.c degree and hence not even c.e. traceable. Furthermore, Nies (2002) built an ω-c.e. jump traceable set that is not superlow. In contrast, Ng has announced that the two properties coincide for the sets that are n-c.e. for some $n > 1$.

By Theorem 5.5.7, each K-trivial set A is Turing below a c.e. K-trivial set, which is superlow, hence jump traceable, and hence c.e. traceable. Thus A is c.e. traceable. By Proposition 8.2.29 this implies the following.

8.4.25 Corollary. *Each K trivial set is facile.* □

See the solution to Exercise 8.2.11 for an alternative proof that each K-trivial set is c.e. traceable.

8.4.26 Exercise. (Ishmukhametov, 1999) Recall Corollary 8.2.8. Show that each array computable c.e. set A is c.e. traceable.

More on weak reducibilities

The following extension of the implication from right to left in the foregoing Theorem 8.4.23 was proved by Simpson and Cole (20xx). It is a converse to Theorem 8.4.16 under the extra hypothesis that $B \leq_T A$.

8.4.27 Corollary. *Suppose A is c.e. in B and $B \leq_T A$. Then $A' \leq_{tt} B'$ implies $A \leq_{JT} B$. In particular, if C is Δ^0_2 and superhigh then $\emptyset' \leq_{JT} C$.*

Proof. At first we merely relativize "$\Leftarrow$" of Theorem 8.4.23 to B: if A is c.e. in B and $(A \oplus B)' \leq_{tt(B)} B'$ then there is a trace $(T_e)_{e\in\mathbb{N}}$ for $J^{A\oplus B}$ that is c.e. relative to B. Here $\leq_{tt(B)}$ denotes the reducibility where B can be used as an oracle to compute the truth table.

Since $B \leq_T A$ and $A' \leq_{tt} B'$, we actually have the stronger hypothesis that $(A \oplus B)' \leq_{tt} B'$. Thus, in (8.10) we can choose the binary function $q \leq_T B$, which now approximates $(A \oplus B)'$, in such a way that the number of its changes is bounded by a computable function g (not merely one computable in B).

When we view the proof of 8.4.23 relative to B, the reduction function β for the given functional $\widetilde{\Psi}$ is still computable. Note that $\#T_e \leq \lfloor \frac{1}{2} g(\beta(e)) \rfloor$ as before. This shows that $A \leq_{JT} B$. □

On the other hand, a superhigh set C can be jump traceable by Exercise 8.6.2. Thus $C \leq_{JT} \emptyset$ and hence $\emptyset' \not\leq_{JT} C$. Together with Theorem 6.3.14, this shows that all the highness properties in the first five rows of Table 8.3 on page 363 can be separated.

Strong jump traceability

Figueira, Nies and Stephan (2008) introduced the following lowness property.

8.4.28 Definition. A is *strongly jump traceable* if A is jump traceable with bound g (as defined in 8.4.1) for each order function g.

For such a set A, if Ψ is a Turing functional and h an order function, there is a c.e. trace for Ψ^A with bound h. To see this choose an increasing reduction function p, so that $\Psi^A(x) \simeq J^A(p(x))$ for each x. If $B \leq_T A$ then $J^B = \Psi^A$ for some Turing functional Ψ, so B is strongly jump traceable as well.

We will see in Section 8.5 that for the c.e. sets being strongly jump traceable implies most other lowness properties. For instance, by Corollary 8.5.5, a strongly jump traceable c.e. set is Turing below each ω-c.e. ML-random set, and hence low for K.

Figueira, Nies and Stephan (2008) built a promptly simple strongly jump traceable set. Their construction can be viewed as an adaptive cost function construction. Hence it does not qualify as being injury-free. In fact, in the proof we will modify the construction with injury of Theorem 1.7.10, which in itself extends Theorem 1.6.4 that there is a low simple set.

8.4.29 Theorem. *There is a promptly simple strongly jump traceable set A.*

Proof. As in Theorem 1.7.10 we meet the prompt simplicity requirements

$$PS_e\colon\ \#W_e = \infty \ \Rightarrow\ \exists s\, \exists x\, [x \in W_{e,\mathsf{at}\ s}\ \&\ x \in A_s]$$

(where $W_{e,\mathsf{at}\ s} = W_{e,s} - W_{e,s-1}$), and the lowness requirements L_k which attempt to stabilize a computation $J^A(k)$. However, now the priority ordering is dynamic. Recall from 2.1.22 that the function $\overline{C}$ given by $\overline{C}(x) = \min\{C(y)\colon\ y \geq x\}$ is dominated by each order function g. Note that $\overline{C}_s(x) = \min\{C_s(y)\colon\ y \geq x\}$ defines a nonincreasing computable approximation of $\overline{C}$. We stipulate that PS_e can only injure a requirement L_k at stage s if $e < \overline{C}_s(k)$. For almost all k we can compute a stage s such that $\overline{C}_s(k) \leq g(k)$. To obtain a jump trace $(T_k)_{k\in\mathbb{N}}$ for A with bound h, we enumerate all the values $J^A(k)$ into T_k which appear from that stage on. Each requirement PS_e acts at most once, so at most $g(k)$ values are enumerated.

Construction of A. Let $A_0 = \emptyset$.

Stage $s > 0$. For each $e < s$, if PS_e is not satisfied and there is $x \geq 2e$ such that $x \in W_{e,\mathsf{at}\ s}$ and

$$\forall k\, \big[(e \geq \overline{C}_s(k)\ \&\ J^A(k)[s-1]\downarrow)\ \rightarrow\ x > \mathsf{use}\ J^A(k)[s-1]\big], \qquad (8.11)$$

then put x into A_s and declare PS_e satisfied.

Verification. The first claim is immediate from the construction.

Claim 1. *Given a number k and a stage t such that $J^A(k)[t-1]\downarrow$, we have $J^A(k)[t] = J^A(k)[t-1]$ unless some PS_e such that $e < \overline{C}_t(k)$ acts at stage t.*

Claim 2. *Let g be an order function. Then A is jump traceable with bound g.*
By Proposition 2.1.22 there is n such that $\overline{C}(k) \leq g(k)$ for each $k \geq n$. Let f be the computable function such that $f(k) = 0$ for $k < n$, and $f(k) = \mu t. \overline{C}_t(k) \leq g(k)$ for $k \geq n$. Let $T_k = \{J^A(k)[t]\colon\ t \geq f(k)\}$. Since each requirement PS_e acts at most once, Claim 1 implies that $\#T_k \leq g(k)$ for each $k \geq n$. If $J^A(k)\downarrow$ for $k \geq n$ then $J^A(k) \in T_k$. ◇

Claim 3. *Each requirement PS_e is met. Thus A is promptly simple.*
Since each prompt simplicity requirement acts at most once, by Claim 1 there is a stage s_0 such that the computation $J^A(k)[s]$ is either undefined, or stable for each $s \geq s_0$, whenever $e \geq \overline{C}(k)$ (i.e., whenever L_k could restrain PS_e). If W_e is infinite then there are $x \geq 2e$ and $s \geq s_0$ such that $x \in W_{e,\mathsf{at}\ s}$ and

$x > \text{use } J^A(k)[s_0]$ whenever $e \geq \overline{C}(k)$ and $J^A(k)[s_0]$ is defined. We enumerate x into A at stage s if PS_e has not been met yet. □

To turn the proof into a cost function construction similar to the one in the proof of Theorem 5.3.35, one uses the adaptive cost function $c(x,s) = \max\{2^{-e} \colon \text{ (8.11) holds}\}$. In the following variant of the theorem we apply the Robinson guessing method already used in Theorem 5.3.22.

8.4.30 Theorem. *For each low c.e. set B, there is a strongly jump traceable c.e. set A such that $A \not\leq_T B$.*

Proof. We make the necessary changes to the proof of Theorem 5.3.22. In the construction, instead of (5.8) we now ask that $\Phi_e^B(x) = 0[s]$ and

$$\forall k \left[(\langle e, n\rangle \geq \overline{C}_s(k) \ \& \ J_{s-1}(\mathcal{A}_{s-1}; k)\downarrow) \ \rightarrow \ x > \text{use } J_{s-1}(\mathcal{A}_{s-1}; k) \right].$$

Informally, if P_e has acted n times before stage s, it has to obey the restraint of all L_k such that $\langle e, n\rangle \geq \overline{C}_s(k)$. The same condition, but with the given partial enumeration $(\widetilde{\mathcal{A}}_s)_{s\in\mathbb{N}}$ instead of $(\mathcal{A}_s)_{s\in\mathbb{N}}$, replaces condition (ii) in the $\Sigma_1^0(B)$ question for requirement P_e. One shows that $\mathcal{A}$ is total as in Claim 5.3.23 in the proof of Theorem 5.3.22. Claim 1 and Claim 2 in the proof of Theorem 8.4.29 are verified as before. Claim 3 is replaced by Claim 5.3.25, with the obvious changes. □

8.4.31 Remark. For each order function g there is a (much slower growing) order function h and a c.e. set A that is jump traceable with bound g but not with bound h. We give a new proof of this result of Ng (2008*a*) in Theorem 8.5.2. Ng (2008*a*) used similar methods to show that the class of strongly jump traceable c.e. sets has a Π_4^0-complete index set.

In Section 8.5 we will show that each strongly jump traceable c.e. set is low for K. Being strongly jump traceable is also related to relative C-complexity. A set that is low for C is computable by Exercise 5.2.22, but a weaker condition than being low for C yields a characterization of strongly jump traceability. This is an easy consequence of Theorem 8.4.10 due to Figueira *et al.* (2008):

8.4.32 Corollary. *The following are equivalent.*

(i) A is strongly jump traceable.

(ii) A is lowly for C, namely, for each order function h,

$$\forall x \left[C(x) \leq^+ C^A(x) + h(C^A(x)) \right]. \tag{8.12}$$

Proof. The function α and the constant c were introduced before Theorem 8.4.10.

(i) ⇒ (ii): Let h be an order function. There is an order function g such that $\widetilde{h}(n) = \max\{2g(\alpha(\sigma)) + 1 \colon |\sigma| = n\} \leq h(n)$ for almost all n. Since A is jump traceable with bound g, (i) of Theorem 8.4.10 implies (8.12).

(ii) ⇒ (i): Let g be an order function. There is an order function h such that $3^{h(e+c)} \leq g(e)$ for almost all e. Since (8.12) holds for h, (ii) of Theorem 8.4.10 implies that A is jump traceable with bound g. □

Downey and Greenberg (2011) have shown that each strongly jump traceable set is ω-c.e. (even with the number of changes dominated by any order function). Being in GL_1, this implies that each strongly jump traceable set is low.

Strong superlowness ⋆

This subsection mostly follows Figueira, Nies and Stephan (2008). We strengthen the property of superlowness by universally quantifying over all order functions, the same way we passed from jump traceability to strong jump traceability.

8.4.33 Definition. Let g be an order function. We say that A is *superlow with bound* g if there is a $\{0,1\}$-valued computable binary function q such that, for each x,

$$A'(x) = \lim_s q(x,s) \ \&\ \#\{s > 0\colon\ q(x,s) \neq q(x,s-1)\} \leq g(x). \tag{8.13}$$

A is *strongly superlow* if A is superlow with bound g for each order function g.

The construction in the proof of Theorem 8.4.29 yields a strongly superlow c.e. set (see Exercise 8.4.36). Actually, for c.e. sets, strong superlowness and strong jump traceability are equivalent by the techniques in the proof of Theorem 8.4.23. The implication from left to right holds for all sets. No direct proof of the latter result is known. Rather, we use Corollary 8.4.32.

8.4.34 Theorem. *Each strongly superlow set A is strongly jump traceable. In fact, for each order function g there is an order function b such that*

A is superlow with bound b $\Rightarrow$ A is jump traceable with bound g.

Proof idea. By Theorem 8.4.10(ii) it is sufficient to show that for each order function h, there is an order function b such that

$$A \text{ is superlow with bound } b \rightarrow \forall x\,[C(x) \leq^+ \widehat{h}(C^A(x))]$$

(where $\widehat{h}(y) = y + h(y)$ as before). For each pair of strings σ, x we have

$$C(x) \leq^+ |\sigma| + K(x \mid \sigma) \leq^+ |\sigma| + 2C(x \mid \sigma).$$

Now suppose that σ_x is a shortest $\mathbb{V}^A$-description of x. Since A is computationally weak, very little extra information is needed to obtain x from σ_x without the help of A, namely, $2C(x \mid \sigma_x) \leq^+ h(|\sigma_x|)$. To show this, for each n we code into A' the bits of the first τ of length n we find such that $x = \mathbb{V}^2(\tau, \sigma_x)$. If $n = n_x := C(x \mid \sigma_x)$, an appropriate approximation to A' changes a sufficiently small number of times on these bits (compared to h). Then a description of the corresponding τ from n_x, σ_x and this number of changes is used to show that $2C(x \mid \sigma_x) \leq^+ h(|\sigma_x|)$.

Proof details. Let Ψ^X be the Turing functional given by the following procedure (if it gets stuck on input y we leave $\Psi^X(y)$ undefined). On input $y = \langle i, n, \sigma \rangle$:

(1) Attempt to compute $x = \mathbb{V}^X(\sigma)$.
(2) Let s be least such that $\mathbb{V}^2_s(\tau, \sigma) = x$ for some τ of length n.
(3) If $i < n$ and $\tau(i) = 1$ then declare $\Psi^X(\langle i, n, \sigma \rangle)\downarrow$.

Let α be a reduction function for Ψ according to Fact 1.2.15. Thus, $J^X(\alpha(i,n,\sigma)) \simeq \Psi^X(\langle i, n, \sigma \rangle)$ for each X, i, n, σ, where we write $\alpha(i,n,\sigma)$ instead of $\alpha(\langle i, n, \sigma \rangle)$. Note that α is increasing in each argument. Let b be an order function such that $b(\alpha(n,n,\sigma)) \leq$

$nh(|\sigma|)$ for all $n > 0$ and all σ. To obtain b, for each k let

$$m_k = \max\{\alpha(n,n,\sigma)\colon\ n > 0\ \&\ nh(|\sigma|) \le k\}.$$

Note that m_k exists since h is an order function. Let $b(j) = 1$ for $j \le m_1$, and if $k > 1$ and $m_{k-1} < j \le m_k$ let $b(j) = k$.

Let σ_x be a shortest $\mathbb{V}^A$-description of x, and let $n_x = C(x|\sigma_x)$. Note that the procedure determining $\Psi^X(i, n_x, \sigma_x)$ for $i < n_x$ finds in (2) the first τ_x of length n_x such that $\mathbb{V}^2(\tau_x, \sigma_x) = x$, and declares $\Psi^A(i, n_x, \sigma_x)\downarrow$ iff $\tau_x(i) = 1$.

Let q be a binary computable function for the bound b as in (8.13). We may assume that for each $s > 0$ there is at most one x such that $q(x,s) \neq q(x, s-1)$. We can approximate τ_x at stage s by

$$\tau_{x,s} = q(\alpha(0, n_x, \sigma_x), s) \ldots q(\alpha(n_x - 1, n_x, \sigma_x), s),$$

and $d_x = \#\{s > 0\colon\ \tau_{x,s} \neq \tau_{x,s-1}\} \le n_x b(\alpha(n_x, n_x, \sigma_x)) \le n_x^2 h(|\sigma_x|)$.

There is a machine describing τ_x in terms of n_x, σ_x, and d_x for each x. Since $\mathbb{V}^2(\tau_x, \sigma_x) = x$, a further machine describes x in terms of n_x, σ_x, and d_x. Note that for $u, v \in \mathbb{N}$ we have $|\mathsf{bin}(u \cdot v)| =^+ |\mathsf{bin}(u)| + |\mathsf{bin}(v)|$, where $\mathsf{bin}(u) \in \{0,1\}^*$ is the binary representation of u. We temporarily write $|n|$ for the length of $\mathsf{bin}(n)$. Note that

$$n_x = C(x \mid \sigma_x) \le^+ 2|n_x| + |n_x^2 \cdot h(|\sigma_x|)| \le^+ 4|n_x| + |h(|\sigma_x|)|. \tag{8.14}$$

Claim. *We have $n_x \le^+ 2|h(|\sigma_x|)|$ for all x.*
There is a constant N such that $8|n| \le n$ for all $n \ge N$. Since h is an order function, $|h(|\sigma_x|)| \ge N$ for almost all x, hence for almost every x either $n_x \le |h(|\sigma_x|)|$ or $4|n_x| \le n_x/2$. In the latter case $n_x - 4|n_x| \ge n_x/2$ and by (8.14), $n_x/2 \le^+ |h(|\sigma_x|)|$. This proves the claim. To conclude the proof, note that

$$C(x) \le^+ |\sigma_x| + 2C(x \mid \sigma_x) \le^+ |\sigma_x| + 4|h(|\sigma_x|)| \le^+ |\sigma_x| + h(|\sigma_x|).$$

The rightmost expression equals $C^A(x) + h(C^A(x)) = \widehat{h}(C^A(x))$. □

8.4.35 Corollary. *Let A be computably enumerable. Then*
A is strongly jump traceable $\Leftrightarrow$ A is strongly superlow.

Proof. $\Leftarrow$: This implication holds for every set A.
$\Rightarrow$: We show that A is superlow with bound g for any order function g by looking at the proof of Theorem 8.4.16 for $B = \emptyset$. For each k there are two queries to $\emptyset'$ in (8.9). Let h be an order function such that $2h(\widehat{e}) \le g(e)$ for almost all e. Then the approximation to A' obtained by evaluating the truth-table reduction in (8.9) on $\emptyset'_s$ at stage s changes at most $g(e)$ times for almost all e. □

Exercises.

8.4.36. Show directly from the construction that the set A we build in the proof of Theorem 8.4.29 is strongly superlow.

8.4.37. Define weak reducibilities $\le_{SSL}$ and $\le_{SJT}$ corresponding to strong superlowness and strong jump traceability. Show that if $B \le_T A$, then $A \le_{SSL} B$ implies $A \le_{SJT} B$. (For work related to $\le_{SJT}$ see Ng 2010.)

8.4.38.$^\diamond$ **Problem.** Is each strongly jump traceable set strongly superlow?

8.5 Subclasses of the K-trivial sets

Note that the term "subclass" does not imply being proper.

By Theorem 8.3.9, lowness for Schnorr randomness can be characterized by the computability theoretic property of computable traceability. In Chapter 5 we showed that several properties coincide with lowness for Martin-Löf randomness, but none of them is purely computability theoretic. A characterization of this kind would be desirable for lowness for Martin-Löf randomness, for instance because it might lead to shorter proofs of some of those coincidences.

When the properties of strong jump traceability and strong superlowness were introduced in Figueira, Nies and Stephan (2008), the hope was that one of them is equivalent to Low(MLR), leading to such a characterization. Cholak, Downey and Greenberg (2008) showed that the strongly jump traceable c.e. sets form a subclass of the c.e. sets in Low(MLR), introducing the so-called box promotion method. However, they also proved that this subclass is proper, thereby destroying the hope that either property characterizes Low(MLR).

Following work of Greenberg and Nies, we give proofs using cost functions of these results in Cholak, Downey and Greenberg (2008). In 8.5.3 we will introduce *benign* cost functions, monotonic cost functions c with the property that the number of pairwise disjoint intervals $[x, s)$ such that $c(x, s) \geq 2^{-n}$ is bounded computably in n. The standard cost function is benign (Remark 5.3.14). If c is benign then by the box promotion method each strongly jump traceable c.e. set A has a computable enumeration obeying c. On the other hand, there is a c.e. set obeying c which is not strongly jump traceable.

We study further subclasses of the c.e. K-trivial sets that contain the strongly jump traceable c.e. sets, such as the class of c.e. sets that are Turing below each ML-random set $C \geq_{LR} \emptyset'$. None of these subclasses is known to be proper.

The results in this section are rather incomplete and leave plenty of opportunity for future research. For some updates see the end of this chapter.

Some K-trivial c.e. set is not strongly jump traceable

In Theorem 5.3.5 we built a promptly simple set obeying a given cost function with the limit condition. For the standard cost function $c_{\mathcal{K}}$, this construction is more flexible than one might expect: there is a K-trivial c.e. set A that is not strongly jump traceable. There are strategies R_e ($e \in \mathbb{N}$) to make A not strongly jump traceable. Each R_e is eventually able to enumerate $e + 1$ numbers into A at stages of its choice. The construction has to cope with a computably bounded number of failed attempts of R_e. Failure means that the sequence of enumerations is terminated before the $e + 1$-th number is reached.

In Exercise 8.5.8 the result is extended to an arbitrary benign cost function. An alternative proof of the result can be obtained from Corollary 8.5.5 and Exercise 8.5.25: each strongly jump traceable c.e. set, but not each K-trivial c.e. set, is below each ω-c.e. ML-random set.

8.5.1 Theorem. *There exists a c.e. K-trivial set A that is not strongly jump traceable. Indeed, for some Turing functional Ψ, there is no c.e. trace for Ψ^A with bound $\lambda z. \max(1, \lfloor 1/2 \log\log z \rfloor)$.*

Proof. We build a computable enumeration $(A_s)_{s\in\mathbb{N}}$ of a set A that obeys the standard cost function $c_{\mathcal{K}}$ (see 5.3.2 and 5.3.4 for the definitions). Then A is K-trivial.

Let $I_1, I_2, \ldots$ be the consecutive intervals in $\mathbb{N}$ such that $\#I_e = 2^{e2^e}$. Let h be the order function such that $h(z) = e$ for $z \in I_e$. Note that $1/2 \log\log z \le h(z)$ for each z because $\sum_{1\le i<e} 2^{i2^i} = \min I_e \le 2^{2^{2e}}$ for each $e > 0$. Let $(S^e_z)_{z\in\mathbb{N}}$ be the e-th c.e. trace with bound h in some uniformly c.e. listing of all such traces.

We build a Turing functional Ψ in such a way that for each $e > 0$ we have $\Psi^A(z) \notin S^e_z$ for some $z \in I_e$. Since $\#S^e_z \le e$ it is sufficient to have $e+1$ numbers available for enumeration into A at stages of our choice before defining $\Psi^A(z)$ for the first time with large use (say the stage number). From now on, whenever $y = \Psi^A(z)$ is defined and y has appeared in S^e_z, we enumerate the next among those numbers into A and redefine $\Psi^A(z)$ with a larger value than before and the stage number as the use.

The challenge is to implement this strategy while obeying the cost function $c_{\mathcal{K}}$. As a preparation, we will rephrase the construction of an incomputable K-trivial set in terms of procedures. We meet for all e the requirement that $A \neq \Phi_e$. The procedure $P^e(b)$ for this requirement is allowed to incur a cost of $1/b$. It wants to enumerate a number w into A at a stage s when $\Phi_{e,s}(w) = 0$ in order to ensure that $\Phi_e(w) = A(w)$ fails. It tries to find a number w such that $c_{\mathcal{K}}(w, s) \le 1/b$ at a stage s when it wants to enumerate w.

Procedure $P^e(b)$ $(e > 0, b \in \mathbb{N})$

(1) Let w be the current stage number.
(2) WAIT for a stage s such that $\Phi_{e,s}(w) = 0$. If s is found put w into A_s.
If $c_{\mathcal{K}}(w,t) > 1/b$ at a stage t during this waiting GOTO (1). We say that P^e is reset.

Since $\sum_n 2^{-K(n)} \le 1$ the procedure $P^e(b)$ is reset at most b times, so eventually w stabilizes. (This particular feature of $c_{\mathcal{K}}$ was already discussed in Remark 5.3.14.) At stage e of the construction we start $P^e(2^e)$. Since each procedure enumerates into A at most once, the total cost of all A-changes is finite.

Now suppose that for each e we have to meet the requirement that $(S^e_z)_{z\in\mathbb{N}}$ is not a trace for Ψ^A. The strategy for this requirement needs $e+1$ numbers available for enumeration into A, so it now involves $e+1$ levels of procedures P^e_j, $e \ge j \ge 0$. The strategy begins with P^e_e, which calls P^e_{e-1}, and so on down to P^e_0 which picks $z \in I_e$ and defines $\Psi^A(z)$ for the first time. Each time the current value $\Psi^A(z)$ appears in the trace set S^e_z, a procedure returns control to the procedure that called it, which now changes A in order to redefine $\Psi^A(z)$ to a larger value. The allowance of each run of P^e_j is the same amount $1/b_{e,j}$, where the numbers $b_{e,j} \in \mathbb{N}$ can be computed in advance.

It may now happen that for $j > 0$, a procedure P^e_j calls P^e_{j-1} but is reset (in the same way as above) before P^e_{j-1} returns. In this case P^e_{j-1} is cancelled. Typically it has incurred some cost already. To bound this type of garbage, P^e_j with the allowance of $1/b$ calls P^e_{j-1} with the smaller allowance of $1/b^2$. The run $P^e_j(b)$ is reset at most b times, and each time the cost at level $j-1$ is at most b^{-2}. The total cost at this level is therefore at most $1/b$. We will verify that we can afford this cost.

If a run of P^e_{j-1} is cancelled this also wastes inputs for Ψ^A, because the run may define computations $\Psi^A(z)$ when A is already stable up to the use. However, the size of I_e is at least the largest possible number of runs of P^e_0, so that each such run can choose a new number $z \in I_e$.

In the following let $e > 0$. In the *construction* of A we call $P^e_e(2^e)$ at stage e.

Procedure $P^e_0(b)$ (at level 0, the parameter b merely simplifies the notation later):
(1) At the current stage s, let z be the least unused number in I_e.
Define $y = \Psi^A(z)[s] = s$ with use s.
(2) WAIT for a stage $r > s$ such that $y \in S^e_{z,r}$. Meanwhile, if $A \upharpoonright_y$ changes, redefine $\Psi^A(z) = y$ with the same use y. If r is found RETURN z.

Procedure $P^e_j(b)$ $(b, j \in \mathbb{N},\ e > 0,\ j > 0)$.
(1) Let w be the current stage number. CALL $P^e_{j-1}(b^2)$.
(2) WAIT for a stage s at which this run of the procedure P^e_{j-1} returns a number z. If s is found put w into A_s. Since $w <$ use $\Psi^A(z)[s-1]$ this makes $\Psi^A(z)$ undefined. Let $y = s$ and redefine $\Psi^A(z)[s] = y$ with use y.

If $c_{\mathcal{K}}(w,t) > 1/b$ at a stage t during this waiting, then at the beginning of t cancel the run of P^e_{j-1} and all its subruns, and GOTO (1). We say that the run of P^e_j is *reset* (note that it is not cancelled, and has not incurred any cost yet).
(3) WAIT for a stage $r > s$ such that $y \in S^e_{z,r}$. Meanwhile, if $A_{t-1} \upharpoonright_y \neq A_t \upharpoonright_y$ for the current stage t, then redefine $\Psi^A(z)[t] = y$ with the same use y as before. If r is found RETURN z.

Claim 1. *A run $P^e_j(b)$ is reset fewer than b times.*
Suppose the run is started at w_0, and reset at stages $w_1 < \ldots < w_k$. Then $k/b < \sum_{0 \le i < k} c_{\mathcal{K}}(w_i, w_{i+1}) \le \Omega \le 1$, hence $k < b$. ◇

The numbers $b_{e,j}$ $(j \le e)$ are given by $b_{e,e} = 2^e$ and $b_{e,j-1} = b^2_{e,j}$ if $j > 0$; in other words, $b_{e,j} = 2^{e2^{e-j}}$. Note that we only call a run $P^e_j(b)$ for $b = b_{e,j}$.

Claim 2. *For each $j \le e$ the procedure P^e_j is called at most $b_{e,j}2^{-e}$ times. In particular, since $\#I_e = b_{e,0}$, a run of P^e_0 can always choose a new z in (1).*
We use reverse induction on $j \le e$. Procedure P^e_e is called once. Suppose the claim holds for $j > 0$. Since each run of P^e_j is called with parameter $b_{e,j}$, each run is reset fewer than $b_{e,j}$ times. Therefore P^e_{j-1} is called at most $b^2_{e,j}2^{-e} = b_{e,j-1}2^{-e}$ times. ◇

Claim 3. *There is $z \in I_e$ such that $\Psi^A(z)\downarrow \notin S^e_z$.*
Let $z \in I_e$ be the last number chosen by a run of P^e_0. No run of a procedure P^e_j, $j > 0$, is reset after the stage s when z is chosen, else a further number would be chosen later by a run of P^e_0. Suppose that $r \le e+1$ is largest such that for $j < r$, the run of P^e_j at stage s returns at a stage s_j. Let $y_j = \Psi^A(z)[s_j]$. Then

$y_0 < \ldots < y_{r-1}$ and $y_j \in S^e_z$, whence $r \le e$ since $\#S^e_z \le e$. By the actions at (3), or at (2) if $r = 0$, $y = \Psi^A(z)\downarrow$. Further, $y \notin S^e_z$ because the run of P^e_r does not return. $\diamond$

Claim 4. *A is K-trivial.*
It suffices to show that $(A_s)_{s\in\mathbb{N}}$ obeys $c_{\mathcal{K}}$. If $w \in A_s - A_{s-1}$ this is caused by (2) of a run $P^e_j(b_{e,j})$, $j > 0$. Then $c_{\mathcal{K}}(w, s) \le 1/b_{e,j}$, otherwise the run would be reset before w enters A. By Claim 2 P^e_j is called at most $b_{e,j}2^{-e}$ times. At most one number enters A per call, so the sum S in (5.6) on page 186 satisfies

$$S \le \sum_{0<e} \sum_{j=1}^{e} b_{e,j}2^{-e}/b_{e,j} \le \sum_{0<e} e2^{-e} < \infty. \qquad \square$$

We modify the construction to give a proof of the first statement in Remark 8.4.31, that the hierarchy of c.e. sets jump traceable for order functions g is downward proper.

8.5.2 Theorem. (Ng, 2008*a*) *For each order function g there is an order function h and a c.e. set A that is jump traceable with bound g but not with bound h.*

Proof. To ensure that A is jump traceable via g we meet the lowness requirements L_k as in Theorem 8.4.29. A requirement L_k can only be injured $g(k)$ times. As in the foregoing theorem, for $e > 0$ we have $e + 1$ levels of runs of P^e-type procedures that diagonalize against the e-th trace with a bound h defined shortly. To bound the number of times a run can be reset by a computable function, we will explicitly define finite subtrees V_e of $\{\alpha \in \mathbb{N}^* : |\alpha| \le e\}$. They consist of strings denoting runs, similar to the original proof of Theorem 8.5.1 in Cholak, Downey and Greenberg (2008). A run P^e_α, $\alpha \in V_e$, corresponds to a run $P^e_{e-|\alpha|}$ before.

To each pair $\langle e, \alpha\rangle$, $e > 0$, $\alpha \in \mathbb{N}^*$, we effectively assign a code number $n_{e,\alpha} \in \mathbb{N}$ in such a way that $\alpha \prec \beta$ implies $n_{e,\alpha} < n_{e,\beta}$. A run P^e_α is only allowed to injure L_k if $n_{e,\alpha} < g(k)$. Each run enumerates into A at most once, so A is jump traceable via g.

In the *construction*, the runs P^e_α for $|\alpha| = e$ act like P^e_0 before. P^e_α for $|\alpha| < e$ acts like $P^e_{e-|\alpha|}$ before, except that it now calls in (1) the run $P^e_{\alpha i}$ for the least i such that no such run has been called yet. Each time a computation $J^A(k)$ converges, all the runs P^e_α such that $g(k) \le n_{e,\alpha}$ are reset by cancelling their subruns and going to (1).

Let $\widehat{g}(n) = \max\{k : g(k) \le n\}$. To define the tree V_e we have to calculate an upper bound on how often P^e_α can be reset. Let $n = n_{e,\alpha}$. Each time some run with priority $m < n$ enumerates a number into A, all L_k, $\widehat{g}(m) < k \le \widehat{g}(n)$, may be injured, and hence may reset P^e_α when their computation $J^A(k)$ converges again. So P^e_α can be reset at most $r_{e,\alpha} = (n+1)(\widehat{g}(n)+1)$ times. Thus, each $\alpha \in V_e$ of length $< e$ has a successor αi on V_e for each $i \le r_{e,\alpha}$. Now let $(I_e)_{e\in\mathbb{N}^+}$ be the consecutive intervals with length the number of leaves of V_e, and let $h(z) = e$ for $z \in I_e$.

Claim 1 now states that P^e_α is reset at most $r_{e,\alpha}$ times, which is clear. Claim 2 is not needed, Claim 3 is as before, and Claim 4 (that A is jump traceable via g) is clear. $\square$

Strongly jump traceable c.e. sets and benign cost functions

Cost functions were defined in 5.3.1.

8.5.3 Definition. We say that a cost function c is *benign* if
$c(x + 1, s) \le c(x, s) \le c(x, s + 1)$ for each $x < s$ (monotonicity), and there is a computable function g such that

$$x_0 < x_1 < \ldots < x_k \;\&\; \forall i < k\,[c(x_i, x_{i+1}) \geq 2^{-n}] \text{ implies } k \leq g(n).$$

The standard cost function $c_{\mathcal{K}}$ is benign via $g(n) = 2^n$ (see 5.3.14). For a further example, suppose Y is an ω-c.e. set, and choose a computable approximation $(Y_s)_{s\in\mathbb{N}}$ of Y and a computable function g such that $Y_s(n)$ changes at most $g(n)$ times. Then the cost function c_Y defined in (5.7) on page 189 is benign via g. As in 5.3.21 one shows that each benign cost function satisfies the limit condition.

The following result of Greenberg and Nies is obtained by generalizing a proof of Cholak, Downey and Greenberg (2008), using the language of cost functions.

8.5.4 Theorem. *Let A be a strongly jump traceable c.e. set. If c is a benign cost function then A has a computable enumeration $(\widehat{A}_r)_{r\in\mathbb{N}}$ that obeys c.*

Item (i) of the following was proved by Cholak, Downey and Greenberg (2008).

8.5.5 Corollary. *Suppose A is a strongly jump traceable c.e. set.*
(i) A is K-trivial. (ii) If Y is a ML-random ω-c.e. set then $A \leq_{wtt} Y$ with a use function bounded by the identity.

Proof. For (i), some computable enumeration of A obeys $c_{\mathcal{K}}$; now we apply 5.3.10. For (ii), we use that the cost function c_Y introduced in (5.7) on page 189 is benign. Then by Fact 5.3.13 $A \leq_{wtt} Y$ with use bounded by the identity.

Note that (i) also follows from (ii), by choosing a ML-random set Y such that $Y <_T \emptyset'$ (say, a superlow ML-random set by 1.8.38). Then $Y \in \mathsf{MLR}^A$ by 3.4.13, hence A is a base for ML-randomness and thus K-trivial. □

Proof of Theorem 8.5.4. We begin with a lemma on order functions, which implies that a jump trace for A with a bound h that grows sufficiently slowly yields a c.e. trace with a desired bound h_d for Φ_d^A. Let α_d be the (strictly increasing) reduction function for Φ_d given by Fact 1.2.15; thus $J^X(\alpha_d(x)) \simeq \Phi_d^X(x)$ for each oracle X and each input x.

8.5.6 Lemma. *Let $(h_d)_{d\in\mathbb{N}^+}$ be an effective listing of order functions such that $h_d(0) \geq d$ for each d. Then there is an order function h such that $h(\alpha_d(z)) \leq h_d(z)$ for each $d > 0$ and $z \in \mathbb{N}$.*

Subproof. We define a computable sequence $0 = n_1 \leq n_2 \leq \ldots$ and let $h(y) = i$ for $y \in [n_i, n_{i+1})$. For $i > 1$ let $n_i = 1 + \max\{\alpha_d(z) \colon 0 < d < i \;\&\; h_d(z) \leq i\}$. Suppose $y = \alpha_d(z) \in [n_i, n_{i+1})$. If $d \geq i$ then $h_d(x) \geq i$ for each x, so $h(y) = i \leq h_d(z)$. If $0 < d < i$ then $h_d(z) > i$ by the definition of n_i, so again $h(y) \leq h_d(z)$. ◇

We will define in (8.15) a sequence of order functions $(h_d)_{d\in\mathbb{N}}$ as in Lemma 8.5.6, so let h be the order function obtained via 8.5.6. We show that some computable enumeration of A obeys c, assuming only that A is jump traceable with bound h. Let $(T_n)_{n\in\mathbb{N}}$ be a jump trace with bound h for A. In the construction we are given a parameter $d > 0$. We effectively obtain a Turing functional $\Phi = \Phi_{f(d)}$. Then, by the Recursion Theorem, there is $d > 0$ such that $\Phi_{f(d)} = \Phi_d$, so we can think of Φ as Φ_d. Let

$$T_z^d = T_{\alpha_d(z)}.$$

By the definition of α_d, $(T_z^d)_{z\in\mathbb{N}}$ is a c.e. trace for Φ^A with bound h_d.

We are given A with a computable enumeration, and define the new enumeration $(\widehat{A}_r)_{r\in\mathbb{N}}$. Roughly, to ensure that $(\widehat{A}_r)_{r\in\mathbb{N}}$ obeys c, the larger $c(y,r)$, the less frequent is the event that $y \in \widehat{A}_r - \widehat{A}_{r-1}$. To this end we also define $\Phi^A(z)$ with use greater than y for an appropriate z such that T_z^d is small. If $\Phi^A(z)$ enters T_z^d then Otto has spent one trace element. If y enters A after that, we can redefine $\Phi^A(z)$ with a larger value. We now say that z is *promoted.* This process takes place at most $\#T_z^d$ times.

A procedure R_n wants to ensure that there are at most $n+d$ stages r at which a number y enters $\widehat{A}_r$ such that $c(y,r) \geq 2^{-n}$. This suffices to make the total cost of changes bounded, for R_n actually only has to consider the y such that also $2^{-n+1} > c(y,r)$, else some R_m for $m < n$ does the job of R_n, and $\sum_n (n+d)2^{-n+1} < \infty$. The actions of R_n depend on the sequence $0 = x_0 < x_1 < \ldots$ where x_{i+1} is the least s such that $c(x_i, s) \geq 2^{-n}$. This sequence grows as R_n proceeds, but since c is benign via g its length is bounded by $g(n)$. By the monotonicity of c, if $c(y,r) \geq 2^{-n}$ then we have $y < x_m$, where x_m is the last in the sequence at stage r.

Once x_i is defined, R_n picks for x_i a number z such that $\#T_z^d \leq n+d$, and defines $\Phi^A(z)$ with value the stage number and with use x_i. It keeps $\Phi^A(z)$ defined with the same value and use (even if A changes) till this value appears in T_z^d. Now $A\restriction_{x_i}$ is called certified.

The new enumeration $(\widehat{A}_r)_{r\in\mathbb{N}}$ slows down $(A_s)_{s\in\mathbb{N}}$ in such a way that a number y with $c(y,r) \geq 2^{-n}$ can enter $\widehat{A}_r$ only after $A\restriction_{x_i}$ is certified, where i is least such that $x_i > y$. If y does enter $\widehat{A}_r$, Otto has to spend one of his numbers in T_z^d.

To keep its overall cost down, the strategy R_n wants to ensure that such a change of A after certification can happen at most $n+d$ times. The typical situation is that $j < i$, $A\restriction_{x_j}$ and $A\restriction_{x_i}$ are certified, first A changes in $[x_j, x_i)$, and later A changes below x_j. Now R_n has to be sure that the two changes can be used for promoting the same number z. To do so it promotes whole boxes of numbers.

R_n begins with an interval I_n, called its box, of inputs z for Φ^A. This box is split into $g(n)$ subboxes B_i of equal size. The same division process is applied to the subboxes and so on, for $n+d+1$ times. Since g is computable, we can determine in advance how large I_n has to be to allow for $n+d+1$ subdivisions. The trace bound is defined by $h_d(z) = n+d$ for $z \in I_n$.

When x_i appears, R_n defines $\Phi^A(z)$ as above for all $z \in B_i$. If i is least such that $A\restriction_{x_i}$ is certified and now A changes below x_i, then all the computations $\Phi^A(z)$ for $z \in B_i$ become undefined. So *all* $z \in B_i$ get promoted in the sense above. To ensure we can use this box of promoted numbers z in case later $A\restriction_{x_j}$ changes for $j < i$, R_n now uses B_i as its new box. It copies sufficiently many of the remaining computations $\Phi^A(y)$ for $y \in B_j$ onto inputs in the new box of R_n.

Proof details. In the following we fix $d > 0$. Let $(I_n)_{n\in\mathbb{N}^+}$ be the consecutive intervals of $\mathbb{N}$ such that $\#I_n = g(n)^{(n+d+1)}$. Given n, the variables k, l range over $\{1, \ldots, g(n)\}$. If B is an interval of $\mathbb{N}$ such that $g(n) \mid \#B$, then let B_k be the k-th subinterval of B of size $\#B/g(n)$.

The procedure R_n controls $\Phi^A(z)$ for each $z \in I_n$. We define the order function h_d by

$$h_d(z) = n + d \text{ if } z \in I_n. \tag{8.15}$$

To deal with the waiting for tracing, the construction (for parameter d) proceeds in *stages* $s(i)$, similar to the proof of Theorem 5.3.27. We will only consider stages of this sort, and use italics to indicate this. Let $s(0) = 0$. If $s(i)$ has been defined and the construction has been carried out up to *stage* $s(i)$, let

$$s(i+1) \simeq \mu s > s(i) \left[2 \mid s \,\&\, \forall z < s(i)\, (\Phi_s^A(z)\uparrow \ \vee \Phi_s^A(z) \in T^d_{z,s})\right].$$

If $\Phi = \Phi_d$ then $s(i+1)$ exists. Note that all the *stages* are even. We may assume that $A_{s+1} = A_s$ for each even s. If $t > 0$ is a *stage*, then $\overline{t}$ denotes the largest *stage* less than t. We say that $A_t \upharpoonright_x$ is *certified* if

$$A_t \upharpoonright_x = A_{\overline{t}} \upharpoonright_x.$$

Construction of the Turing functional Φ, given the parameter $d \in \mathbb{N}$. The procedures R_n act independently. We describe their actions at each *stage*. To ensure tracing of the relevant computations by the next *stage*, R_n is only active from *stage* $s_n^* = s(i+1)$ on, where i is least such that $s(i) > \max I_n$.

Action of R_n. At stage s_n^* let $B = I_n$, $m = 0$ and $x_0 = 0$.

Stage $s > s_n^*$.

(1) Check whether there is $k \le m$ such that

(◦) $\quad x = x_k$ is defined by the end of $\overline{s}$, $A_{\overline{s}} \upharpoonright_x$ is certified and $A_s \upharpoonright_x \neq A_{\overline{s}} \upharpoonright_x$.

If so, let k be the least such number. Note that $\Phi^A(z)[s]\uparrow$ for all $z \in B_k[s]$, because such a computation is only defined in (3) and has use x_k.

For each $y \in B[s]$ such that $g(n) \mid (y - \min B[s])$, let

$$y' = \min B_k[s] + (y - \min B[s])/g(n).$$

Then $y' \in B_k[s]$. If $\Phi^A(y)[s]\downarrow$, make a copy of this computation with input y': define $\Phi^A(y')[s+1] = \Phi^A(y)[s]$ with the same use.
Let $B[s+1] := B_k[s]$. We say that R_n *shrinks its box* at s.

(2) If $c(x_m, s) \ge 2^{-n}$, then increment m and let $x_m = s$.

(3) For each $l \le m$ and $z \in B_l[s+1]$, if $\Phi^A(z)[s]\uparrow$, define $\Phi^A(z)[s+1] = s+1$ with use x_l.

Verification. For the first two claims fix n.

Claim 1. *For each stage $s > s_n^*$, if $x = x_k$ is defined at $\overline{s}$ and $A_s \upharpoonright_x$ is certified, then $\Phi^A(z)[s]\downarrow \in T^d_{z,s}$ for each $z \in B_k[s+1]$.*

By the action in (3) at some *stage* $\le \overline{s}$ and the faithful copying, $w = \Phi^A(z)[\overline{s}+1]$ is defined with use x. Since $A_s \upharpoonright_x = A_{\overline{s}} \upharpoonright_x$, we have $w \in T^d_{x,s}$ by the definition of *stages* and since $s > s_n^*$. ◇

Claim 2. *The procedure R_n shrinks its box no more than $n + d$ times.*

Assume for a contradiction that $t_0 < \ldots < t_{n+d}$ are the first $n+d+1$ *stages* $s > s_n^*$ such that (∘) is satisfied, and let $r_i = \overline{t}_i$. Note that $B[t_{n+d}+1] \neq \emptyset$ as the set I_n was chosen large enough to accommodate $n+d+1$ shrinking processes. Fix $z \in B[t_{n+d}+1]$. Let k_i be the witness k for (∘) at *stage* t_i. For each $i \le n+d$ we have $z \in B_{k_i}[r_i+1] = B[t_i+1]$. Since $A_{r_i} \upharpoonright_{x_k}$ is certified, by Claim 1 we have $\Phi^A(z)[r_i] \downarrow \in T^d_{z,r_i}$. Since R_n shrinks its box at t_i, $\Phi^A(z)[t_i] \uparrow$. Therefore $\Phi^A(z)[r_i] < t_i \le \Phi^A(z)[r_{i+1}]$ for all $i < n+d$, whence $\#T^d_z > n+d$. But $z \in I_n$, so $h_d(z) = n+d$, contradiction. ◇

Now let $q(0) = 0$ and let $q(r+1)$ be the least *stage* $s > q(r)$ such that $A_s \upharpoonright_{q(r)}$ is certified. Similar to Theorem 5.5.2, let $\widehat{A}_r = A_{q(r+2)} \cap [0, r)$.

Claim 3. *$(\widehat{A}_r)_{r \in \mathbb{N}}$ obeys the cost function c.*

Suppose that $y < r$, $c(y,r) > 0$ and $\widehat{A}_r(y) \neq \widehat{A}_{r-1}(y)$. Let n be such that $2^{-n+1} > c(y,r) \ge 2^{-n}$ and consider the procedure R_n at stage $t = q(r)$, when $x_0, \ldots, x_m$ have been defined. By the monotonicity of c in the second argument we have $c(y,t) \ge 2^{-n}$. Also $c(x_m,t) < 2^{-n}$, so $y < x_m$ by the monotonicity in the first argument. There is a least *stage* s, $q(r+1) < s \le q(r+2)$, such that for some $i \le m$ (chosen least), where $x = x_i$ we have $A_s \upharpoonright_x \neq A_{\overline{s}} \upharpoonright_x$. Since $x < q(r)$, $A_{q(r+1)} \upharpoonright_{q(r)}$ is certified, and since s was chosen least, $A_{\overline{s}} \upharpoonright_x$ is certified. Thus (∘) holds at s and R_n shrinks its box. By Claim 2 we may conclude that

$$\sum_{y,r} c(y,r)[\![r > 0 \;\&\; y \text{ least s.t. } \widehat{A}_{r-1}(y) \neq \widehat{A}_r(y)]\!] \;\le\; \sum_n (n+d)2^{-n+1} \;<\; \infty.$$

□

Let c be the standard cost function $c_{\mathcal{K}}$. We study how slowly an order function h must grow if we require that each set jump traceable with bound h obeys c. The growth rate is given by the reduction functions α_d and therefore depends on our definition of the universal Turing functional J. If we use the universal functional given by $\Theta^Y(\langle d, x\rangle) \simeq \Phi^Y_d(x)$ then Φ_d has the reduction function $\alpha_d(x) = \langle d, x\rangle$, and we can provide a reasonable lower bound on the growth rate of some such h.

8.5.7 Corollary. *There is an order function h such that $\forall^\infty z \left[h(z) \ge \sqrt{(\log z)/3}\,\right]$, and, for any c.e. set A, if Θ^A has a c.e. trace with bound h, then A is K-trivial.*

Proof. The cost function $c_{\mathcal{K}}$ is benign via $g(n) = 2^n$. Given d, let $(I^d_n)_{n \in \mathbb{N}^+}$ be the consecutive intervals of $\mathbb{N}$ such that $\#I^d_n = 2^{n(n+d+1)}$. Note that $\max I^d_n < 2^{n(n+d+1)+1}$. As in (8.15) we define the order function h_d by $h_d(z) = n+d$ if $z \in I^d_n$. To prove the lower bound on h, we want to find an upper bound for the numbers n_i from the proof of Lemma 8.5.6.

If $\exists d < i\,[h_d(z) \le i]$ then $z \in I^d_n$ for some n such that $n + d \le i$. Therefore $z < 2^{n(n+d+1)+1} \le 2^{(n+d+1)^2} \le 2^{(i+1)^2}$. So, for almost all i,

$$n_i = 1 + \max\{\langle d, z\rangle \colon\ 0 < d < i \;\&\; h_d(z) \le i\} \le \langle i-1, 2^{(i+1)^2}\rangle \le 2^{3i^2}.$$

Hence $h(z) \ge \sqrt{(\log z)/3}$ for almost all z.

To show that some enumeration of A obeys $c_{\mathcal{K}}$, in the proof of Theorem 8.5.4 we let $(T_n)_{n \in \mathbb{N}}$ be a c.e. trace for Θ^A with bound h, and use the reduction functions $\alpha_d(z) = \langle d, z\rangle$. □

The following table shows similarities between the box promotion method and the decanter method, which was first used to show that each K-trivial set is Turing incomplete (page 202). However, the number of levels is fixed for the decanter method while in the case of box promotion, the requirement R_n needs $n+d+1$ levels.

Decanter method	Box promotion method
a number put into F_0	an input x for Φ^A
j-set	a box of numbers that have been promoted j times
Fact 5.4.2 about $\boldsymbol{k}$-sets	Claim 2
a garbage number	a number that is not promoted because it is in the wrong box when R_n shrinks its box

The golden run method (which builds on the decanter method) is less relevant, because in the proof of Theorem 8.5.4 we did not need a tree of runs, while parallel runs are necessary in the proof of Theorem 5.4.1. However, the present construction bears some resemblance to the calling of Q-type procedures in that proof. When $x = x_m$ is defined in (2) of the construction, one can think of calling such a procedure. It returns when $\Phi^A(z)[s+1]$ is traced. After that, a change of $A\restriction_{x_m}$ releases the procedure and causes R_n to shrink its box.

The proof of Theorem 8.5.1 can be viewed as a failed decanter type construction for each interval I_e. This is plausible since we want to ensure that A is not strongly jump traceable, so we are now on the side that Otto occupies in the proof of 8.5.4. If $(S^e_z)_{z\in\mathbb{N}}$ is a c.e. trace for Ψ^A with the indicated bound, then we can promote some number $z \in I_e$ for $e+1$ times. This means that the analog of Fact 5.4.2 fails.

Cholak, Downey and Greenberg (2008) proved that for any strongly jump traceable c.e. sets A_0, A_1, the set $A_0 \oplus A_1$ is strongly jump traceable. Thus, the degrees of the c.e. strongly jump traceable sets form a proper subideal of the ideal of c.e. K-trivial degrees. The proof uses the box promotion method as well.

Exercises.

8.5.8. Extend 8.5.1: if c is a benign cost function with bound g, then there exists a c.e. set A with an enumeration obeying c that is not strongly jump traceable.

8.5.9. Show that for each incomputable Δ^0_2 set A there is a monotonic cost function c with the limit condition such that no computable approximation of A obeys c.

8.5.10.$^\diamond$ Prove the converse of Theorem 8.5.4: if A is a c.e. set which for each benign cost function c has a computable enumeration obeying c, then A is strongly jump traceable.

The diamond operator

For a class $\mathcal{H} \subseteq 2^{\mathbb{N}}$ let $\mathcal{H}^\diamond = \{A\colon\ A \text{ is c.e. } \& \ \forall Y \in \mathcal{H} \cap \mathsf{MLR}\,[A \le_T Y]\}$.

Note that $\mathcal{H}^\diamond$ determines an ideal in the c.e. Turing degrees. By Theorem 5.1.12, the class $\{Y\colon\ A \le_T Y\}$ is null for each incomputable set A. Thus, if $\mathcal{H}^\diamond$ contains an incomputable set then $\mathcal{H}$ is null. On the other hand, for each null Σ^0_3 class $\mathcal{H}$ there is a promptly simple set in $\mathcal{H}^\diamond$ obtained via an injury-free construction by Theorem 5.3.15.

We will study several subclasses of the c.e. K-trivial sets that are of the form $\mathcal{H}^\diamond$. Often $\mathcal{H}$ is a highness property applying to some ML-random set $Z \not\ge_T \emptyset'$. The weak reducibilities $\le_{LR}$ and $\le_{JT}$ were defined in 5.6.1 and 8.4.13, respectively. The main highness properties relevant here are being LR-hard, being JT-hard, and being superhigh:

$$\begin{aligned} \mathsf{LRH} &= \{C\colon \emptyset' \leq_{LR} C\}, \\ \mathsf{JTH} &= \{C\colon \emptyset' \leq_{JT} C\}, \\ \mathsf{Shigh} &= \{C\colon \emptyset'' \leq_{tt} C'\} \end{aligned}$$

In Theorem 5.6.30 we proved that LRH coincides with the class of uniformly a.e. dominating sets. By 8.4.16 and 8.6.2, JTH is properly contained in the class of superhigh sets introduced in 6.3.13. The two latter classes coincide on the Δ^0_2 sets by 8.4.27. Hence, by Theorem 6.3.14, LRH is a proper subclass of JTH.

We provide a basic fact. For a stronger result see Exercise 8.5.21.

8.5.11 Proposition. *The sets that are* $high_1$ *form a null class.*

Proof. Otherwise $high_1$ is conull by the zero-one law 1.9.12, so $high_1 \cap \mathrm{GL}_1 = \{C\colon \emptyset'' \leq_T C \oplus \emptyset'\}$ is conull. By 5.1.12 relativized to $\emptyset'$, $\{C\colon A \leq_T C \oplus \emptyset'\}$ is null for each set $A \not\leq_T \emptyset'$. Letting $A = \emptyset''$ this yields a contradiction. □

8.5.12 Proposition. *(i) The classes* LRH *and* JTH *are null* Σ^0_3 *classes.*
(ii) (Simpson) Shigh *is contained in a null* Σ^0_3 *class.*

Proof. (i) Since $\mathsf{LRH} \subseteq \mathsf{JTH} \subseteq \mathsf{Shigh}$ the classes are null. By 8.4.20 JTH is Σ^0_3. To prove that LRH is Σ^0_3, recall from Theorem 5.6.5 that $\leq_{LR}$ is equivalent to $\leq_{LK}$, and note that $\emptyset' \leq_{LK} C$ is equivalent to $\exists b \, \forall y, \sigma, s \, \exists t \geq s$

$$\mathbb{U}^{\emptyset'}(\sigma) = y[s] \;\rightarrow\; (\mathbb{U}^{\emptyset'}(\sigma)[t]\uparrow \;\vee\; K_t^{C\upharpoonright_t}(y) \leq |\sigma| + b).$$

(ii) Note that a function f is d.n.c. relative to $\emptyset'$ if $\forall x \, \neg f(x) = J^{\emptyset'}(x)$. Let P be the $\Pi^0_1(\emptyset')$ class of $\{0,1\}$-valued functions that are d.n.c. relative to $\emptyset'$. By Exercise 5.1.15 relative to $\emptyset'$, the class $\{Z\colon \exists f \leq_T Z \oplus \emptyset' \, [f \in P]\}$ is null. Then, since GL_1 is conull, the class $\mathcal{H} = \{Z\colon \exists f \leq_{tt} Z' \, [f \in P]\}$ is also null. This class contains Shigh by 1.8.30 relative to $\emptyset'$ (note that q there remains computable).

To show that $\mathcal{H}$ is Σ^0_3, fix a Π^0_2 relation $R \subseteq \mathbb{N}^3$ such that a string σ is extended by a member of P iff $\forall u \, \exists v \, R(\sigma, u, v)$. Let $(\Psi_e)_{e \in \mathbb{N}}$ be an effective listing of truth-table reduction procedures. It suffices to show that $\{Z : \Psi_e(Z') \in P\}$ is a Π^0_2 class. To this end, note that, as use $\Psi_e^X(j)$ is independent of the oracle X,

$$\Psi_e(Z') \in P \;\leftrightarrow\; \forall x \, \forall t \, \forall u \, \exists s > t \, \exists v \, R(\Psi_e^{Z'} \upharpoonright_x [s], u, v). \quad \square$$

From Theorem 5.3.15 we conclude:

8.5.13 Corollary. *There is a promptly simple set in* $\mathsf{Shigh}^\diamond \subseteq \mathsf{JTH}^\diamond \subseteq \mathsf{LRH}^\diamond$. □

In Proposition 3.4.17 we showed that the 2-random set $\Omega^{\emptyset'}$ is high. This contrasts with the following fact, which is immediate from 8.5.12(ii).

8.5.14 Corollary. *No weakly 2-random set is superhigh.* □

A further results suggests that $high_1 \cap \mathsf{MLR}$ is much larger than $\mathsf{Shigh} \cap \mathsf{MLR}$. By unpublished work announced in Kučera (1990, Remark 8), for each incomputable Δ^0_2 set A there is a high Δ^0_2 ML-random set $Z \not\geq_T A$ (while $\mathsf{Shigh}^\diamond$ contains a promptly simple set A). In particular, if $A \in high_1^\diamond$ then A is computable. For the latter fact one can also argue that $\Omega^{\emptyset'}$ is high and 2-random (3.4.17), and therefore forms a minimal pair with the high ML-random set Ω.

We prove some inclusion relations among subclasses of the c.e. K-trivial sets.

8.5.15 Theorem. *Consider the following properties of a c.e. set A.*

(i) $A \in \mathsf{Shigh}^{\diamond}$; *(ii)* $A \in \mathsf{JTH}^{\diamond}$; *(iii)* $A \in \mathsf{LRH}^{\diamond}$;

(iv) A is ML-coverable, namely, there is a ML-random set $Z \geq_T A$ such that $\emptyset' \not\leq_T Z$ (see 5.1.23);

(v) for each ML-random set Z, if $\emptyset' \leq_T A \oplus Z$ then $\emptyset' \leq_T Z$;

(vi) A is K-trivial.

The following implications hold: (i)⇒(ii)⇒(iii); (iii)⇒(iv)⇒(vi); (iii)⇒(v)⇒(vi).

All these classes are closed downward under $\leq_T$ within the c.e. sets. The classes given by (i), (ii), (iii) and (vi) are even known to be ideals.

Proof. (i)⇒(ii)⇒(iii): Immediate since $\mathsf{Shigh} \supseteq \mathsf{JTH} \supseteq \mathsf{LRH}$.

(iii)⇒(iv): By Theorem 6.3.14 there is a ML-random set $C <_T \emptyset'$ such that $\emptyset' \leq_{LR} C$. If $A \in \mathsf{LRH}^{\diamond}$ then A is ML-coverable via C.

(iv)⇒(vi): See Corollary 5.1.23.

(iii)⇒(v): By the previous implications, A is K-trivial and hence low for ML-randomness. We show that for each set A that is low for ML-randomness, if Z is ML-random and $\emptyset' \leq_T A \oplus Z$ then $\emptyset' \leq_{LR} Z$. (Intuitively, since A is computably weak, Z must be computationally strong.) The proof, due to Hirschfeldt, involves relativizing the van Lambalgen Theorem 3.4.6 to a set D: for any sets Z and R such that $Z \in \mathsf{MLR}^D$, we have $R \oplus Z \in \mathsf{MLR}^D$ iff $R \in \mathsf{MLR}^{Z\oplus D}$. Then, for each set R, we have $R \in \mathsf{MLR}^Z \to R \oplus Z \in \mathsf{MLR} \to R \oplus Z \in \mathsf{MLR}^A \to R \in \mathsf{MLR}^{Z\oplus A} \to R \in \mathsf{MLR}^{\emptyset'}$. Thus $\emptyset' \leq_{LR} Z$.

(v)⇒(vi): If A is not K-trivial then A is not a base for ML-randomness by Theorem 5.1.22. Hence $A \not\leq_T \Omega^A$. Thus $\emptyset' \not\leq_T \Omega^A$ while $\emptyset' \leq_T A \oplus \Omega^A$ by Fact 3.4.16. □

8.5.16 Remark. A is called *ML-cuppable* if $A \oplus Z \equiv_T \emptyset'$ for some ML-random set $Z <_T \emptyset'$ (such a Z is called a cupping partner for A). The class of c.e. *non*-ML-cuppable sets clearly contains the class given by (v); we show that it is still contained in (vi). Suppose that A is not K-trivial. First assume that A is low, then $\Omega^A <_T \emptyset'$ is a cupping partner. If A is not low, by 1.6.10 let $A \equiv_{wtt} A_0 \oplus A_1$ for low c.e. sets A_0, A_1. One of the sets A_0, A_1 is not K-trivial, say A_0. Then Ω^{A_0} is a cupping partner for A_0 and hence for A.

Nies (2007) proved by a direct construction that there is a promptly simple ML-noncuppable set. An easier proof is obtained by noting that the class LRH is Σ^0_3, and hence $\mathsf{LRH}^{\diamond}$ contains a promptly simple set by Theorem 5.3.15.

It is unknown whether $\mathsf{LRH}^{\diamond}$ coincides with the c.e. K-trivial sets, that is, whether each c.e. K-trivial set is Turing below each LR-hard ML-random set. If so, then all the classes (iii)–(vi) in 8.5.15 would coincide. The same applies to $\mathsf{JTH}^{\diamond}$. Kjos-Hanssen and Nies (2009) have shown that $\mathsf{Shigh}^{\diamond}$ is a proper subclass of the c.e. K-trivial sets.

The following strengthens 8.5.13.

8.5.17 Theorem. *Each strongly jump traceable c.e. set A is in $\mathsf{JTH}^{\diamond}$.*

Proof. Our plan is to define a class $\mathcal{H} \supseteq \mathit{JTH}$ and a benign cost function c such that $\mathcal{H}^\diamond$ contains every Δ^0_2 set with a computable approximation obeying c (this is similar to Fact 5.3.13). Then we apply Theorem 8.5.4: some computable enumeration of A obeys c, so $A \in \mathcal{H}^\diamond \subseteq \mathit{JTH}^\diamond$. This is like the proof of 8.5.5(ii).

For the present proof we develop some theory of interest by itself. We actually define first a class $\mathcal{G} \supseteq \mathit{JTH}$, and then in a second step the class $\mathcal{H} \supseteq \mathcal{G}$. For an order function h, we consider uniform sequences of c.e. operators $(T_n)_{n\in\mathbb{N}}$ such that $(T^Y_n)_{n\in\mathbb{N}}$ is a Y-c.e. trace with bound h, for each oracle Y. Let

$$\begin{aligned}\mathcal{G} = \{Y\colon\ & \exists \text{ order function } h\, \forall f \leq_{wtt} \emptyset' \\ & \quad \exists \text{ uniform sequence of c.e. operators } (T_n)_{n\in\mathbb{N}} \\ & \quad\quad (T^Y_n)_{n\in\mathbb{N}} \text{ is a } Y\text{-c.e. trace with bound } h \text{ for } f\}.\end{aligned}$$

To see that $\mathit{JTH} \subseteq \mathcal{G}$, note that, by definition, $Y \in \mathit{JTH}$ iff $\emptyset'$ is jump traceable by Y, which implies that $\emptyset'$ is c.e. traceable by Y (defined similar to 8.4.13), and hence $Y \in \mathcal{G}$.

By the usual argument in Theorem 8.2.3 we may fix an order function h (say the identity) in the definition of $\mathcal{G}$. Then we may as well let the sequence of trace operators be the universal sequence $V = (V^Y_e)_{e\in\mathbb{N}}$ for this order function, obtained as in the proof of Corollary 8.2.4 but for c.e. operators. This yields the simpler expression $\mathcal{G} = \{Y\colon \forall f \leq_{wtt} \emptyset' \, \forall^\infty n\, f(n) \in V^Y_n\}$.

For the sake of simplicity we formulate the next two lemmas for this universal sequence $V = (V^Y_e)_{e\in\mathbb{N}}$ with this fixed bound h, even if they would actually work for any uniform sequence of c.e. operators $(T_n)_{n\in\mathbb{N}}$ with bound h as above. The first lemma is based on an argument due to Hirschfeldt.

8.5.18 Lemma. *There is a function $f \leq_{wtt} \emptyset'$ such that $2^{-n} \geq \lambda\{Y\colon f(n) \in V^Y_n\}$ for each n.*

Subproof. We define a computable approximation for f in the sense of Definition 1.4.6. Let $f_0(n) = 1$ for each n. Let $s > 0$. If $2^{-n} < \lambda\{Y\colon f_{s-1}(n) \in V^Y_{n,s}\}$ then we let $f_s(n) = f_{s-1}(n) + 1$, otherwise $f_s(n) = f_{s-1}(n)$.

Claim. *For each n, s we have $f_s(n) \leq 2^n h(n)$.*
We apply Exercise 1.9.15 for $\epsilon = 2^{-n}$. Suppose $f_s(n)$ is incremented $N > 2^n h(n)$ times. Let $C_i = \{Y\colon i \in V^Y_e\}$. For each $i \leq N$, since i is not the final value of $f_s(n)$, we have $\lambda C_i \geq \epsilon$. Also $N\epsilon > h(n)$, so by 1.9.15 there is $F \subseteq \{1, \ldots, N\}$ such that $\#F = h(n) + 1$ and $\bigcap_{i\in F} C_i \neq \emptyset$. If $Y \in \bigcap_{i\in F} C_i$ then $F \subseteq V^Y_n$, contradiction. $\diamond$

For a function f let $\mathcal{H}(f) = \{Y\colon \forall^\infty n\, f(n) \in V^Y_n\}$. The class $\mathcal{H} \supseteq \mathcal{G}$ will be of the form $\mathcal{H}(f)$ for f as in the foregoing lemma.

8.5.19 Lemma. *Suppose f is a function as in Lemma 8.5.18. Then there is a benign cost function c_f such that, if $(B_s)_{s\in\mathbb{N}}$ is a computable approximation of a set B obeying c_f, then $B \leq_T Y$ for each ML-random set $Y \in \mathcal{H}(f)$.*

Subproof. We modify the argument in Fact 5.3.13. Let $\lambda n, s.\, f_s(n)$ be a computable approximation of f. We may suppose that $2^{-n} \geq \lambda\{Y\colon f_s(n) \in V^Y_{n,s}\}$ for each n, s. The cost function c_f is defined as in (5.7) on page 189: let $c_f(x, s) = 2^{-x}$ for each $x \geq s$. If $x < s$ and $n < x$ is least such that $f_{s-1}(n) \neq f_s(n)$, let $c_f(x, s) = \max(c_f(x, s-1), 2^{-n})$. We now follow the proof of 5.3.13. As before, we build an interval Solovay test G: when $B_{s-1}(x) \neq B_s(x)$ and $c_f(x, s) = 2^{-n}$ we list in G all the minimal strings σ such that $f_s(n) \in V^{\sigma}_{n,s}$. Then G is an interval Solovay test since the computable approximation of B obeys c_f.

Choose $n_0 > 0$ such that $f(n) \in V_n^Y$ for each $n \geq n_0$, and choose $s_0 > n_0$ such that $f(m)$ is stable at s_0 for $m \leq n_0$ and $\sigma \not\prec Y$ for each σ enumerated into G after stage s_0. To show $B \leq_T Y$, given an input $x \geq s_0$, using Y as an oracle compute $t > x$ such that $\forall i\, [n_0 \leq i < x \rightarrow f_t(i) \in T^Y_{i,t}]$. We claim that $B(x) = B_t(x)$. Otherwise $B_s(x) \neq B_{s-1}(x)$ for some $s > t$. Let $n \leq x$ be the largest number such that $f_r(i) = f_t(i)$ for all i, $n_0 \leq i < n$ and all r, $t < r \leq s$. Then $n > n_0$ by choice of s_0. If $n = x$ then $c_f(x, s) \geq 2^{-x}$. If $n < x$ then $f(n)$ changes in that interval of stages, so $c_f(x, s) \geq 2^{-n}$. Hence, by the choice of t, we list an initial segment σ of Y in G at stage $s \geq s_0$, contradiction.

So far we only have used the weaker hypothesis that $f \leq_T \emptyset'$. If $f \leq_{wtt} \emptyset'$, choose a computable approximation of f with a computably bounded number of changes as in 1.4.6. Then c_f is benign. $\diamond$

Choose $f \leq_{wtt} \emptyset'$ as in Lemma 8.5.18 and note that $\mathcal{H}(f) \supseteq \mathcal{G}$. By Theorem 8.5.4 the strongly jump traceable c.e. set A has a computable enumeration that obeys c_f. Hence $A \in \mathcal{H}(f)^{\diamond} \subseteq \mathcal{G}^{\diamond} \subseteq \mathit{JTH}^{\diamond}$. $\square$

By Exercise 8.5.8 there exists a c.e. set A with an enumeration obeying the benign cost function c_f that is not strongly jump traceable. We conclude:

8.5.20 Proposition. *Some c.e. set in $\mathit{JTH}^{\diamond}$ is not strongly jump traceable.*

Exercises.

8.5.21. We say that Z is weakly 3-random if Z is in no null Π^0_3 class. Show that no weakly 3-random set is high.

8.5.22. Show that the Δ^0_2 sets form a Σ^0_4 class that is not Σ^0_3.

Let $\mathcal{H}$ be the class of ω-c.e. sets. The following exercises show that $\mathcal{H}^{\diamond}$ is a *proper* subclass of the c.e. K-trivial sets which contains a promptly simple set. The same reasoning works for the class $\widetilde{\mathcal{H}} \subset \mathcal{H}$ of superlow sets.

8.5.23. Show that $\mathcal{H}$ and $\widetilde{\mathcal{H}}$ are Σ^0_3.

Thus, merely from Theorem 5.3.15 we may conclude that there is a promptly simple set A in $\mathcal{H}^{\diamond}$ and hence in $\widetilde{\mathcal{H}}^{\diamond}$. (As a consequence, one cannot achieve $Y \leq_{wtt} \emptyset'$ in Theorem 1.8.39 when $P = 2^{\mathbb{N}} - \mathcal{R}_1$ and $B = A$, so there is a low but not ω-c.e. ML-random set.) Actually each strongly jump traceable c.e. set is in $\mathcal{H}^{\diamond}$ by Corollary 8.5.5.

8.5.24. Suppose $A \in \widetilde{\mathcal{H}}^{\diamond}$. Show that if B is superlow then $A \oplus B$ is superlow as well. In particular, if A is c.e., then A is not superlow cuppable, so by Theorem 5.3.15 we have a new proof of the result of Diamondstone (2009) that some promptly simple set fails to be superlow cuppable (see page 247).
Hint. Use Exercise 1.8.41.

8.5.25.$^\diamond$ Prove that for each nonempty Π^0_1 class P there is a superlow set $Y \in P$ and a c.e. K-trivial set B such that $B \not\leq_T Y$. Conclude that there is a c.e. K trivial set $B \notin \mathcal{H}^\diamond$, and in fact $B \notin \widetilde{\mathcal{H}}^\diamond$. (Actually, for each benign cost function c there is such a Y and a B obeying c.)

8.5.26.$^\diamond$ **Problem.** Is the class *Shigh* of superhigh sets Σ^0_3?

8.6 Summary and discussion

We summarize the results on the absolute computational complexity of sets. We also discuss the interplay between computational complexity and the degree of randomness and interpret some results from this point of view.

A diagram of downward closed properties

Figure 8.1 displays most of our properties closed downward under $\leq_T$. It contains the lowness properties and the complements of highness properties. The properties occurring frequently are framed in boldface.

The lines indicate implications from properties lower down in the diagram to properties higher up in the diagram. Up to a few exceptions, we know that no

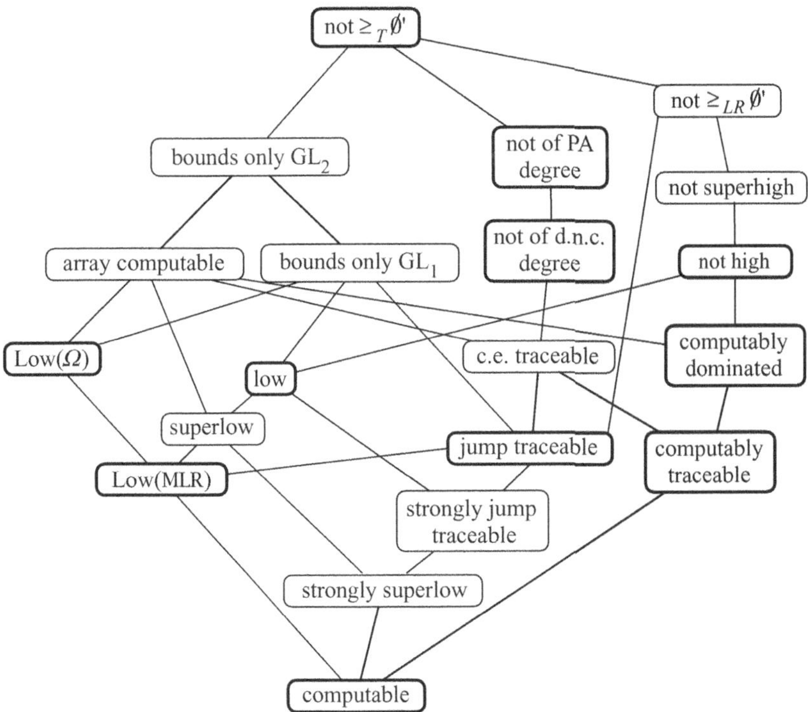

FIG. 8.1. Downward closed properties.

other implications hold. Open questions include whether strong jump traceability is equivalent to strong superlowness, or at least whether it implies superlowness. For further questions regarding possible implications see Problem 8.6.4.

There are interesting intersection relations:

Low(Ω) $\cap$ computably dominated $\Leftrightarrow$ computable (8.1.18), and
c.e. traceable $\cap$ computably dominated $\Leftrightarrow$ computably traceable (8.2.16).

The original, much larger diagram is due to Kjos-Hanssen (2004). It includes for instance the properties of computing no ML-random set, or of being computed by a 2-random set; see Exercise 8.6.1 below.

For the c.e. sets, several classes coincide, as Fig. 8.2 shows. No further implications hold. The coincidences, starting from the bottom, are proved in 8.4.35, 5.1.27, 8.4.23, and 8.4.26.

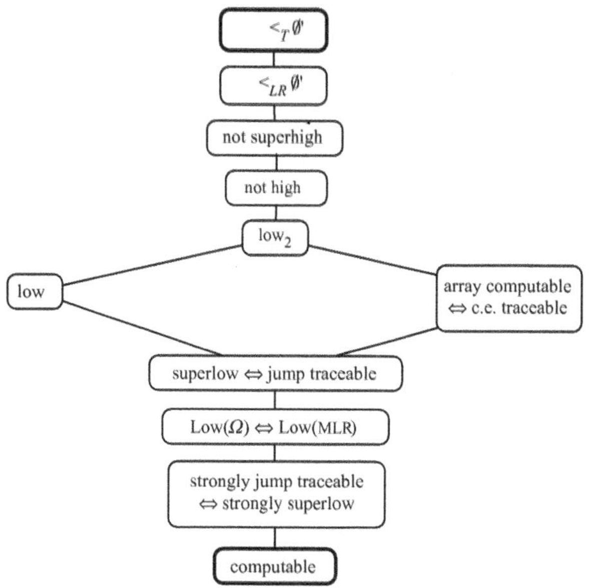

FIG. 8.2. Downward closed properties for c.e. sets.

Weak reducibilities, defined in Section 5.6, provide a general framework for lowness properties and highness properties. A weak reducibility $\leq_W$ determines the lowness property $C \leq_W \emptyset$ and the dual highness property $C \geq_W \emptyset'$. Table 8.3 gives an overview. For related results see Simpson and Cole (2007).

Exercises.

8.6.1. Insert the classes (i) $BN1R = \{A\colon \forall Z \leq_T A\,[Z \notin \mathsf{MLR}]\}$ and (ii) $BB2R = \{A\colon \exists Z \geq_T A\,[Z \text{ is 2-random}]\}$ into Fig. 8.1.
(iii) Identify the conull classes in Fig. 8.1.

8.6.2.$^{\diamond}$ (Kjos-Hanssen and Nies, 20xx) Note that a jump traceable set is not JT-hard and hence not LR-hard. However, show that there is a jump traceable superhigh set. *Hint.* Use the proof of the jump inversion theorem for truth-table reducibility due to Mohrherr (1984): for every set $A \geq_{tt} \emptyset'$ there is a set B such that $B' \equiv_{tt} A$.

TABLE 8.3. Some weak reducibilities.

Weak reducibility	Lowness property	Highness prop.	Defined in
$\leq_T$	computable	$\geq_T \emptyset'$	1.2.6
$\leq_{LR}$	Low(MLR)	u.a.e.d	5.6.1
$\leq_{JT}$	jump traceable	$\geq_{JT} \emptyset'$	8.4.13
$A' \leq_{tt} B'$	superlow	superhigh	1.2.20
$A' \leq_T B'$	low	high	pg. 25
$\leq_{CT}$	comp. traceable	$\geq_T \emptyset'$	8.4.21
$\leq_{\mathrm{cdom}}$	comp. dominated	$\geq_T \emptyset'$	5.6.8

8.6.3.$^\diamond$ For as many classes $\mathcal{C}$ in Fig. 8.1 as possible, show that $\mathcal{C}$ is only contained in the indicated classes. (Most counterexamples are somewhere in this book.)

8.6.4.$^\diamond$ **Problem.** We ask whether no further implications hold in Fig. 8.1.
(i) Can a c.e. traceable set be LR-hard? (ii) Can a set that bounds only GL_1 sets be LR-hard? (iii) Can a set that is low for Ω be superhigh?

8.6.5.$^\diamond$ **Problem.** Decide whether the sets that are computably dominated and in GL_1 are closed downward under $\leq_T$.

8.6.6. Show that some computably random set is jump traceable.

Note that by Theorem 8.2.29, each jump traceable set is facile, so Ex. 8.6.6 yields a new, indirect proof that some computably random set Z is facile (Theorem 7.4.11). However, the methods there can be extended to make Z a Δ^0_2 set, which is not possible for the set in 8.6.6 by Prop. 3.5.13 and since a jump traceable set is GL_1.

Theorem 7.5.2 states that for each high set C there is a computably random set $Z \leq_T C$. A partial computably random set is not facile by Theorem 7.6.7, so Exercise 8.6.6 also shows that Theorem 7.5.2 cannot be strengthened in general to make Z partial computably random. There is even a superhigh set C without a partial computably random below, by Exercise 8.6.2. How about LRH ?

Computational complexity versus randomness

Many results relate the computational complexity of a set with its degree of randomness. As much as we would like a paradigm such as "to be more random means to be less complex", in fact the relationship has no overall direction. A particular computational complexity class only exposes one aspect of the information that can be extracted from the set, and the direction of the relationship depends on this aspect. It also depends on further properties of the sets under discussion, such as being in a class of descriptive complexity. For instance, within the left-c.e. sets, more random means *more* complex: Schnorr randomness implies being high, and ML-randomness even implies being weak truth-table complete (3.2.31). If we widen the scope to the ω-c.e. sets, the direction of the relationship between complexity and randomness is already less clear because there are superlow ML-random sets. There still is a promptly simple set Turing below all the ω-c.e. ML random sets (8.5.23). On the other hand, a set that is Turing below all the Δ^0_2 ML-random sets is computable (1.8.39).

We now look at ML-random sets in general, without imposing extra restrictions on their descriptive complexity. Numerous results suggest now that more random means computationally *less* complex. A ML-random set either computes $\emptyset'$, or is not even of PA-degree (4.3.8). Randomness notions stronger than ML-randomness are incompatible with computing $\emptyset'$. In fact, if Z is ML-random, then Z is weakly 2-random if and only if Z and $\emptyset'$ form a minimal pair (page 135). In particular, a weakly 2-random set is not of PA degree. We review further results in this direction.

1. For $Z \in \mathsf{MLR}$ we have Z is 2-random $\Leftrightarrow$ Z is low for Ω (3.6.19).
2. For any set C, MLR^C is closed downward under $\leq_T$ within MLR (5.1.16).
3. If A and B are ML-random then $A \leq_{vL} B \Leftrightarrow A \geq_{LR} B$ (5.6.2).

For some particular aspects of the information extracted from the set, the opposite can happen.

1. Each 2-random set is of hyperimmune degree (3.6.15), while a ML-random set can be computably dominated.
2. For $n > 1$, each n-random set computes an n-fixed point free function (4.3.17), while an $(n-1)$-random set may fail to do so.

Some updates

In recent work two classes have been shown to coincide with the strongly jump traceable c.e. sets: (1) the class $(\omega\text{-c.e.})^\diamond$ of Exercise 8.5.23 and (2) the class $\mathsf{Shigh}^\diamond$ of Theorem 8.5.15. Each of these results implies that one can obtain a promptly simple strongly jump traceable set A via an injury-free construction. Namely, one builds a promptly simple set A that obeys the non-adaptive cost function for being in $\mathcal{H}^\diamond$ for the appropriate Σ^0_3 class $\mathcal{H}$ (Theorem 5.3.15). The results also yield new proofs of the theorem of Cholak, Downey and Greenberg (2008) that the strongly jump traceable c.e. sets are closed under $\oplus$.

(1) For the first result, after Corollary 8.5.5 due to Greenberg and Nies (20xx) it remains to show the implication that each c.e. set in $(\omega\text{-c.e.})^\diamond$ is strongly jump traceable. This has been proved by Greenberg, Hirschfeldt and Nies (20xx). Given an order function h, they build a ML-random ω-c.e. set Z such that if A is a superlow c.e. set, then $A \leq_T Z$ implies that A is jump traceable with bound h. Roughly, they make Z a "bad set" in that it tries to not compute A by exploiting the changes of A. This forces A to change little.

(2) The second result is due to Nies (20xx). Using appropriate benign cost functions he shows that each strongly jump traceable c.e. set A is Turing below each superhigh ML-random set. For the converse implication, he combines the techniques used for (1) with a version of Kucera coding introduced in Kjos-Hanssen and Nies (20xx) to make the bad set Z superhigh.

The bound in Theorem 8.5.1 can be improved to $\lambda z. \max(1, \lfloor\sqrt{\log z}\rfloor)$. One modifies the construction by letting $P^e_j(b)$ call $P^e_{j-1}(b2^e)$ in (1). In Claim 1 one now proves that $P^e_j(b)$ for $j < e$ is reset at most 2^e times before the run P^e_{j+1} that called it also reset. This uses that $c_{\mathcal{K}}(x, y) + c_{\mathcal{K}}(y, z) \leq c_{\mathcal{K}}(x, z)$ for $x < y < z$.

9

HIGHER COMPUTABILITY AND RANDOMNESS

Recall that a randomness notion is introduced by specifying a test concept. Such a test concept describes a particular kind of null class, and a set is random in that sense if it avoids each null class of that kind. (For instance, in 3.6.1 we introduced weak 2-randomness by defining generalized ML-tests, which describe null Π^0_2 classes. So a set is weakly 2-random iff it is in no null Π^0_2 class.)

Up to now, all the null classes given by tests were arithmetical. In this chapter we introduce tests beyond the arithmetical level. We give mathematical definitions of randomness notions using tools from higher computability theory. The main tools are the following. A relation $\mathcal{B} \subseteq \mathbb{N}^k \times (2^{\mathbb{N}})^r$ is Π^1_1 if it is obtained from an arithmetical relation by a universal quantification over sets. The relation $\mathcal{B}$ is Δ^1_1 if both $\mathcal{B}$ and its complement are Π^1_1. There is an equivalent representation of Π^1_1 relations where the members are enumerated at stages that are countable ordinals. For Π^1_1 sets (of natural numbers) these stages are in fact computable ordinals, i.e., the order types of computable well-orders.

Effective descriptive set theory (see Ch. 25 of Jech 2003) studies effective versions of notion from descriptive set theory. For instance, the Δ^1_1 classes are effective versions of Borel classes. Higher computability theory (see Sacks 1990) studies analogs of concepts from computability theory at a higher level. For instance, the Π^1_1 sets form a higher analog of the c.e. sets, and, to some extent, the Δ^1_1 sets can be seen as an analog of the computable sets.

A main difference between effective descriptive set theory and higher computability is that the latter also studies sets of countable ordinals. However, we still focus on sets of natural numbers: only the stages of enumerations are ordinals. Thus, for us there is no need to distinguish between the two.

While analogs of many notions from the computability setting exist in the setting of higher computability, the results about them often turn out different. The reason is that there are two new closure properties.

(C1) The Π^1_1 and Δ^1_1 relations are closed under number quantification.
(C2) If a function f maps each number n in a certain effective way to a computable ordinal, then the range of f is bounded by a computable ordinal. This is the Bounding Principle 9.1.22.

The theory of Martin-Löf randomness and K-complexity of strings is ultimately based on c.e. sets, so it can be viewed within the new setting. In Section 9.2, we transfer several results to the setting of Π^1_1 sets, for instance the Machine Existence Theorem 2.2.17. However, using the closure properties above we also prove that the analog of Low(MLR) coincides with the Δ^1_1 sets, while the analog of the K-trivial sets does not. These results are due to Hjorth and Nies (2007).

In Section 9.3 we study Δ^1_1-randomness and Π^1_1-randomness. The tests are simply the null Δ^1_1 classes and the null Π^1_1 classes, respectively. The implications are

$$\Pi^1_1\text{-randomness} \Rightarrow \Pi^1_1\text{-ML-randomness} \Rightarrow \Delta^1_1\text{-randomness.}$$

The converse implications fail.

Martin-Löf (1970) was the first to study randomness in the setting of higher computability theory. He suggested Δ^1_1-randomness as the appropriate mathematical concept of randomness. His main result was that the union of all Δ^1_1 null classes is a Π^1_1 class that is not Δ^1_1 (see 9.3.11). Later it turned out that Δ^1_1-randomness is the higher analog of both Schnorr and computable randomness.

The strongest notion we consider is Π^1_1-randomness, which has no analog in the setting of computability theory. This notion was first mentioned in Exercise 2.5.IV of Sacks (1990), but called Σ^1_1-randomness there. Interestingly, there is a universal test, namely, a largest Π^1_1 null class (9.3.6).

In Section 9.4 we study lowness properties in the setting of higher computability. For instance, we consider sets that are Δ^1_1 dominated, the higher analog of being computably dominated. Using the closure properties (C1) and (C2), we show that each Π^1_1-random set is Δ^1_1 dominated. We also prove that lowness for Δ^1_1-randomness is equivalent to a higher analog of computable traceability. Lowness for Π^1_1-randomness implies lowness for Δ^1_1-randomness. It is unknown whether some such set is not Δ^1_1.

A citation such as "Sacks 5.2.I" refers to Sacks (1990), our standard reference for background on higher computability. References to the original results can be found there. Some of the results we only quote rely on elaborate technical devices such as ranked formulas.

9.1 Preliminaries on higher computability theory

Π^1_1 *and other relations*

For sets $X_1, \ldots, X_m$, we let $X_1 \oplus \ldots \oplus X_m = \{mz + i \colon i < m \ \&\ z \in X_i\}$. Extending 1.8.55, we say that a relation $\mathcal{A} \subseteq \mathbb{N}^k \times (2^{\mathbb{N}})^m$ is Σ^0_n if the relation $\{\langle e_1, \ldots, e_k, X_1 \oplus \ldots \oplus X_m\rangle \colon \langle e_1, \ldots, e_k, X_1, \ldots, X_m\rangle \in \mathcal{A}\}$ is Σ^0_n, and similarly for Π^0_n relations. As before, a relation is *arithmetical* if it is Σ^0_n for some n.

9.1.1 Definition. Let $k, r \geq 0$ and $\mathcal{B} \subseteq \mathbb{N}^k \times (2^{\mathbb{N}})^r$. We say that $\mathcal{B}$ is Π^1_1 if there is an arithmetical relation $\mathcal{A} \subseteq \mathbb{N}^k \times (2^{\mathbb{N}})^{r+1}$ such that

$$\langle e_1, \ldots, e_k, X_1, \ldots, X_r\rangle \in \mathcal{B} \leftrightarrow \forall Y\, \langle e_1, \ldots, e_k, X_1, \ldots, X_r, Y\rangle \in \mathcal{A}.$$

For $k = 1$ and $r = 0$, $\mathcal{B}$ is called a Π^1_1 *set* (of numbers). For $k = 0$ and $r = 1$, $\mathcal{B}$ is called a Π^1_1 *class* (of sets). $\mathcal{B}$ is called Σ^1_1 if its complement is Π^1_1, and $\mathcal{B}$ is Δ^1_1 if it is both Π^1_1 and Σ^1_1. A Δ^1_1 set is also called *hyperarithmetical*.

In defining Π^1_1 relations, one can equivalently require that $\mathcal{A}$ be Σ^0_2 (Sacks 1.5.I).

The Π^1_1 relations are closed under the application of number quantifiers. So are the Σ^1_1 and Δ^1_1 relations. For instance, consider a Σ^1_1 relation $\exists Y\, R(e, n, Y)$.

For a set Z let $p_n(Z) = \{x\colon \langle x, n\rangle \in Z\}$. Let $\widehat{R} = \{\langle e, n, Z\rangle\colon R(e, n, p_n(Z))\}$. Then $\forall n\, \exists Y\, R(e, n, Y) \Leftrightarrow \exists Z\, \forall n\, \widehat{R}(e, n, Z)$: to show the implication from left to right, suppose that for each n there is Y_n such that $R(e, n, Y_n)$. Using the countable axiom of choice one can combine these witnesses into a single witness $Z = \bigcup_n Y_n \times \{n\}$, and $p_n(Z) = Y_n$ for each n. Then $\widehat{R}(e, n, Z)$.

9.1.2 Exercise. If $R \subseteq 2^{\mathbb{N}}$ is open then in a uniform way we have the following equivalence: R is a Π^1_1 class $\Leftrightarrow A_R = \{x\colon [x] \subseteq R\}$ is a Π^1_1 set.

Well-orders and computable ordinals

In the following we will consider binary relations $W \subseteq \mathbb{N} \times \mathbb{N}$ with domain an initial segment of $\mathbb{N}$. They can be encoded by sets $R \subseteq \mathbb{N}$ via the usual pairing function. We identify the relation with its code. Linear orders will be irreflexive. Note that a c.e. linear order is computable. A linear order R is a *well-order* if each nonempty subset of its domain has a least element. The class of well-orders is Π^1_1. Furthermore, the index set $\{e\colon W_e \text{ is a well-order}\}$ is Π^1_1.

Given a well-order R and an ordinal α, we let $|R|$ denote the *order type* of R, namely, the ordinal α such that $(\alpha, \in)$ is isomorphic to R. We say that α is computable if $\alpha = |R|$ for a computable well-order R. Each initial segment of a computable well-order is also a computable well-order, so the computable ordinals are closed downwards. We let ω_1^Y denote the least ordinal that is not computable in Y. The least incomputable ordinal is ω_1^{ck} (which equals $\omega_1^{\emptyset}$).

An important example of a Π^1_1 class is

$$\mathcal{C} = \{Y\colon \omega_1^Y > \omega_1^{\mathrm{ck}}\}. \tag{9.1}$$

To see that this class is Π^1_1, note that

$$Y \in \mathcal{C} \leftrightarrow \exists e\, \left[\Phi_e^Y \text{ is well-order } \&\ \forall i\, [W_i \text{ is computable relation} \to \Phi_e^Y \not\cong W_i]\right].$$

This can be put into Π^1_1 form because the Π^1_1 relations are closed under number quantification. If $\omega_1^Y = \omega_1^{\mathrm{ck}}$ we say that Y is *low for* ω_1^{ck}.

Representing Π^1_1 *relations by well-orders*

A c.e. set can be either described by a Σ_1 formula in arithmetic or by a computable enumeration. A Σ^0_1 class, of the form $\{X\colon \exists y\, R(X\restriction_y)\}$ for computable R, can be thought of as being enumerated at stages $y \in \mathbb{N}$.

In Definition 9.1.1 we introduced Π^1_1 classes by formulas in second order arithmetic. Equivalently, they can be described by a generalized type of enumeration where the stages are countable ordinals. It becomes possible to look at a whole set X in the course of a generalized computation (this is the global view of Section 1.3). While one could introduce a formal model of infinite computations, we will be content with viewing the order type of a well-order computed from X via a Turing functional as the stage when X enters the class.

9.1.3 Theorem. *Let $k, r \geq 0$. Given a Π^1_1 relation $\mathcal{B} \subseteq \mathbb{N}^k \times (2^{\mathbb{N}})^r$, one can effectively obtain a computable function $p\colon \mathbb{N}^k \to \mathbb{N}$ such that*

$$\langle e_1, \ldots, e_k, X_1 \oplus \ldots \oplus X_r \rangle \in \mathcal{B} \leftrightarrow \Phi^{X_1 \oplus \ldots \oplus X_r}_{p(e_1,\ldots,e_k)} \text{ is a well-order.} \tag{9.2}$$

Moreover, the expression on the right hand side is always a linear order. If $r = 0$ the oracle is the empty set by convention. If $k = 0$ then p is a constant.

The order type of $\Phi^{X_1 \oplus \ldots \oplus X_r}_{p(e_1,\ldots,e_k)}$ is the stage at which the element enters $\mathcal{B}$, so for a countable ordinal α, we let

$$\mathcal{B}_\alpha = \{\langle e_1, \ldots, e_k, X_1 \oplus \ldots \oplus X_r \rangle \colon |\Phi^{X_1 \oplus \ldots \oplus X_r}_{p(e_1,\ldots,e_k)}| < \alpha\}. \tag{9.3}$$

Thus, $\mathcal{B}_\alpha$ contains the elements that enter $\mathcal{B}$ before stage α.

The right hand side in (9.2) can be expressed by a universal set quantification of an arithmetical relation. So, conversely, each relation given by (9.2) is Π^1_1.

The idea to prove Theorem 9.1.3 is as follows, say in the case of a Π^1_1 class $\mathcal{B}$.
(1) One expresses $\mathcal{B}$ by universal quantification over functions, rather than sets: $X \in \mathcal{B} \leftrightarrow \forall f \, \exists n \, R(\overline{f}(n), X)$, where $\overline{f}(n)$ denotes the tuple $(f(0), \ldots, f(n-1))$ as in Sacks 5.2.I. The advantage of taking functions is that one can choose R computable, rather than merely Σ^0_2.
(2) This yields a Turing functional $\Psi_\mathcal{B}$ such that, for each X, $\Psi_\mathcal{B}(X)$ is a set of codes for tuples in $\mathbb{N}^*$ (the sequence numbers as defined before 1.8.69), and $X \in \mathcal{B}$ iff $\Psi_\mathcal{B}(X)$ is well-founded under the reverse prefix relation $\succ$ on sequence numbers (Sacks 5.4.II).
(3) Using the length-lexicographical (also called Kleene–Brouwer) ordering, one can effectively "linearize" $\Psi_\mathcal{B}(X)$ (Sacks 3.5.III), and thus obtain a Turing functional Φ_p such that $X \in \mathcal{B} \leftrightarrow \Phi^X_p$ is a well-order. Also see Jech (2003, Thm. 25.3, 25.12).

Indices

An index for p as a computable function (or the constant p itself in the case of Π^1_1 classes) serves as an index for $\mathcal{B}$ in (9.2). Via (9.3) such an index yields an approximation of $\mathcal{B}$ at stages which are countable ordinals. We are mostly interested in the cases of Π^1_1 sets ($k = 1$ and $r = 0$) and Π^1_1 classes ($k = 0$ and $r = 1$). In the former case, by the convention that the oracle is $\emptyset$, we have

$$e \in \mathcal{B} \leftrightarrow R_e \text{ is a well-order,}$$

where $R_e = \Phi^\emptyset_{p(e)}$. We may replace R_e by a linear order of type $\omega R_e + e + 1$, so we will assume that at each stage at most one element is enumerated, and none are enumerated at a limit stage. By the above,

$$O = \{e \colon W_e \text{ is a well-order}\}$$

is a Π^1_1*-complete set.* That is, O is Π^1_1 and $S \le_m O$ for each Π^1_1 set S. The set of ordinal notations is Π^1_1-complete as well (Sacks 2.1.I). In most cases it is inessential which Π^1_1-complete set one chooses; we will stay with the one given above.

For $p \in \mathbb{N}$, we let $\mathcal{Q}_p$ denote the Π^1_1 class with index p. Thus,

$$\mathcal{Q}_p = \{X \colon \Phi^X_p \text{ is a well-order}\}. \tag{9.4}$$

Note that $\mathcal{Q}_{p,\alpha} = \{X \colon |\Phi^X_p| < \alpha\}$, so that $X \in \mathcal{Q}_p$ implies $X \in \mathcal{Q}_{p,|\Phi^X_p|+1}$.

An *index for a* Δ^1_1 *set* $S \subseteq \mathbb{N}$ is a pair of Π^1_1 indices for S and $\mathbb{N} - S$. An *index for a* Δ^1_1 *class* $\mathcal{S} \subseteq 2^{\mathbb{N}}$ is a pair of Π^1_1 indices for the classes $\mathcal{S}$ and $2^{\mathbb{N}} - \mathcal{S}$.

Relativization

The notions introduced above can be relativized to a set A. It suffices to include A as a further set variable in (9.2). For instance, $S \subseteq \mathbb{N}$ is a $\Pi^1_1(A)$ set if $S = \{e\colon \langle e, A\rangle \in \mathcal{B}\}$ for a Π^1_1 relation $\mathcal{B} \subseteq \mathbb{N} \times 2^{\mathbb{N}}$, and $\mathcal{C} \subseteq 2^{\mathbb{N}}$ is a $\Pi^1_1(A)$ class if $\mathcal{C} = \{X\colon \langle X, A\rangle \in \mathcal{B}\}$ for a Π^1_1 relation $\mathcal{B} \subseteq 2^{\mathbb{N}} \times 2^{\mathbb{N}}$.

The following set is $\Pi^1_1(A)$-complete:

$$O^A = \{e\colon W^A_e \text{ is a well-order}\}.$$

We frequently apply that a Π^1_1 object can be approximated by Δ^1_1 objects.

9.1.4 Approximation Lemma. *(i) For each* Π^1_1 *set* S *and each* $\alpha < \omega_1^{\mathrm{ck}}$, *the set* S_α *is* Δ^1_1. *(ii) For each* Π^1_1 *class* $\mathcal{B}$ *and each countable ordinal* α, *the class* $\mathcal{B}_\alpha$ *is* $\Delta^1_1(R)$, *for every well-order* R *such that* $|R| = \alpha$.

Proof. We prove (ii) and leave (i) as an exercise. Let $p \in \mathbb{N}$ be an index for $\mathcal{B}$. Then $X \in \mathcal{B}_\alpha \leftrightarrow \exists g\,[g$ is an isomorphism of Φ^X_p and a proper initial segment of $R]$, which is $\Sigma^1_1(R)$. Moreover, $X \notin \mathcal{B}_\alpha \leftrightarrow \Phi^X_p$ is not a well-order $\vee \exists g\,[g$ is an isomorphism of R and an initial segment of $\Phi^X_p]$, which is also $\Sigma^1_1(R)$. □

9.1.5 Remark. Borel classes were introduced on page 66. One can show that $\mathcal{C} \subseteq 2^{\mathbb{N}}$ is Borel $\Leftrightarrow \mathcal{C}$ is $\Delta^1_1(A)$ for some A. For the implication from left to right one lets A be a Borel code for $\mathcal{C}$, namely, a set that codes how $\mathcal{C}$ is built up from open sets using the operations of complementation and countable intersection. The implication from right to left follows from the Lusin separation theorem (Kechris, 1995, 14.7).

Exercises.

9.1.6. (Reduction Principle for Π^1_1 classes) Given Π^1_1 classes $\mathcal{P}$ and $\mathcal{Q}$, one can effectively determine disjoint Π^1_1 classes $\widetilde{\mathcal{P}} \subseteq \mathcal{P}$ and $\widetilde{\mathcal{Q}} \subseteq \mathcal{Q}$ such that $\mathcal{P} \cup \mathcal{Q} = \widetilde{\mathcal{P}} \cup \widetilde{\mathcal{Q}}$.

9.1.7. Improve the indexing of the Δ^1_1 classes introduced after (9.4): there are computable functions g and h such that $(\mathcal{Q}_{g(n)})_{n\in\mathbb{N}}$ ranges over all the Δ^1_1 classes and, for each n, we have $\mathcal{Q}_{g(n)} \cap \mathcal{Q}_{h(n)} = \emptyset$, and $\mathcal{Q}_{g(n)} \cup \mathcal{Q}_{h(n)} = 2^{\mathbb{N}}$ unless $\mathcal{Q}_{g(n)} = \emptyset$.

9.1.8. Show that $\omega_1^{\mathrm{ck}} < \omega_1^O$.

Π^1_1 *classes and the uniform measure*

9.1.9 Theorem. (Lusin; see Sacks 6.2.II) *Each* Π^1_1 *class is measurable.*

Proof. $\aleph_1$ denotes the least uncountable ordinal. Recall from (9.4) that $\mathcal{Q}_p$ is the Π^1_1 class with index p. For each $\alpha < \aleph_1$, $\mathcal{Q}_{p,\alpha}$ is Borel by 9.1.4 and hence measurable. Since $\mathbb{R}$ has a countable dense subset, for each j there is a countable ordinal β_j such that $\lambda\mathcal{Q}_{j,\alpha} = \lambda\mathcal{Q}_{j,\beta_j}$ for each $\alpha \geq \beta_j$. Let $\beta = \sup_j \beta_j$. It suffices to show that $\mathcal{Q}_p - \mathcal{Q}_{p,\beta}$ is null for each p. If $Y \in \mathcal{Q}_p - \mathcal{Q}_{p,\beta}$ then the well-order Φ^Y_p has an initial segment of order type β, which can be computed by Y. Therefore $|\Phi^Y_j| = \beta$ for some j. This shows that $\mathcal{Q}_p - \mathcal{Q}_{p,\beta} \subseteq \bigcup_{j\in\mathbb{N}}\{Y\colon |\Phi^Y_j| = \beta\}$, which is null by the choice of β. □

For the following frequently used result see Sacks 1.11.IV. It states that the measure of a class has the same descriptive complexity as the class itself. Note that (ii) follows from (i).

9.1.10 Measure Lemma. *(i) For each Π^1_1 class $\mathcal{B}$, $\lambda\mathcal{B}$ is left-Π^1_1. A Π^1_1 index for $\{q \in \mathbb{Q}_2 \colon q < \lambda\mathcal{B}\}$ can be computed from an index for $\mathcal{B}$ as a Π^1_1 class.*

(ii) If $\mathcal{S}$ is a Δ^1_1 class then $\lambda\mathcal{S}$ is left-Δ^1_1. A Δ^1_1 index for $\{q \in \mathbb{Q}_2 \colon q < \lambda\mathcal{S}\}$ can be computed from an index for $\mathcal{S}$ as a Δ^1_1 class. □

For the following basis theorem of Sacks and Tanaka see Sacks 2.2.IV.

9.1.11 Theorem. *A Π^1_1 class that is not null has a hyperarithmetical member.*

Reducibilities

Turing reducibility has two analogs in the new setting. Intuitively, as the stages are now countable ordinals, it is possible to look at the whole oracle set during a "computation". If full access to the oracle set is granted we obtain *hyperarithmetical reducibility*: $X \le_h A$ iff $X \in \Delta^1_1(A)$. If only a finite initial segment of the oracle can be used we have the restricted version $\le_{\mathit{fin}\text{-}h}$. Instead of introducing a formal oracle machine model for infinite computations, we rely on Π^1_1 relations. This is similar to considering Turing functionals as particular types of c.e. relations in Fact 6.1.1. The graph of a function ψ is the set $\{\langle n, y\rangle \colon \psi(n) = y\}$.

9.1.12 Definition. For a function $f \colon \mathbb{N} \to \mathbb{N}$ and a set A, we write $f \le_h A$ if the graph of f is $\Pi^1_1(A)$. A *reduction procedure for $f \le_h A$* is a Π^1_1 relation $\mathcal{S} \subseteq \mathbb{N} \times \mathbb{N} \times 2^{\mathbb{N}}$ such that $\{\langle n, y\rangle \colon \langle n, y, A\rangle \in \mathcal{S}\}$ is the graph of f.

It is equivalent to require that the graph of f is $\Delta^1_1(A)$. (We say that f is a $\Delta^1_1(A)$ function.) We may assume that $\{\langle n, y\rangle \colon \langle n, y, X\rangle \in \mathcal{S}\}$ is the graph of a partial function for each X by Exercise 9.1.18. The following are used frequently.

9.1.13 Theorem. (Sacks, 1969) *$A \notin \Delta^1_1 \Leftrightarrow \{X \colon X \ge_h A\}$ is null.*

This is an analog of 5.1.12. For a proof see Sacks 2.4.IV.

Next, we reconsider the class (9.1) of sets that are not low for ω_1^{ck}.

9.1.14 Theorem. (Spector, 1955) *$O \le_h X \Leftrightarrow \omega_1^{\mathrm{ck}} < \omega_1^X$.*

Proof. $\Rightarrow$: By Exercise 9.1.8 we have $\omega_1^{\mathrm{ck}} < \omega_1^O$. By Exercise 9.1.24 below $\omega_1^X = \{|R| \colon R \le_h X \;\&\; R \text{ is a well-order}\}$, hence $\omega_1^O \le \omega_1^X$.

$\Leftarrow$: If $\omega_1^{\mathrm{ck}} < \omega_1^X$ then by the Approximation Lemma 9.1.4(i) relative to X, $O = \{e \colon W_e \text{ is a well-order}\}$ is a $\Delta^1_1(X)$ set. □

9.1.15 Corollary. *The Π^1_1 class $\{Y \colon \omega_1^Y > \omega_1^{\mathrm{ck}}\}$ in (9.1) is null.* □

The following result, known as the Gandy Basis Theorem, is an analog of the Low Basis Theorem 1.8.37. The proof differs from the proof of 1.8.37 because Σ^1_1 classes are not closed in general.

9.1.16 Theorem. *Let $\mathcal{S} \subseteq 2^{\mathbb{N}}$ be a nonempty Σ^1_1 class. Then there is $A \in \mathcal{S}$ such that $A \le_T O$ and $O^A \le_h O$ (whence $A <_h O$).*

To prove it, firstly, by Sacks 1.4.III and its proof there is a set $A \le_T O$, $A <_h O$ such that $A \in \mathcal{S}$. Then $\omega_1^A = \omega_1^{ck}$. Secondly, by Sacks 7.6.II, $\omega_1^A < \omega_1^B$ and $A \le_h B$ imply $O^A \le_h B$. This is applied for $B = O$ to obtain $O^A \le_h O$.

Spector (1955) proved that if A is a Π^1_1 set then either $A \in \Delta^1_1$ or $O \le_h A$ (Sacks 7.2.II). Recall that we view the Π^1_1 sets as analogs of the c.e. sets. The reducibility $\le_h$ on the Π^1_1 sets certainly behaves differently from $\le_T$ on the c.e. sets. Finite hyperarithmetical reducibility $\le_{\mathit{fin}\text{-}h}$ is more like $\le_T$.

9.1.17 Definition. A *fin-h reduction procedure* is a partial function $\Phi\colon \{0,1\}^* \to \{0,1\}^*$ with Π^1_1 graph such that $\mathrm{dom}(\Phi)$ is closed under prefixes and, if $\Phi(x)\downarrow$ and $y \preceq x$, then $\Phi(y) \preceq \Phi(x)$. We write $A = \Phi^Z$ if $\forall n \exists m\, \Phi(Z\upharpoonright_m) \succeq A\upharpoonright_n$, and $A \le_{\mathit{fin}\text{-}h} Z$ if $A = \Phi^Z$ for some *fin-h* reduction procedure Φ.

If A is hyperarithmetical then $\Phi = \{\langle x, A\upharpoonright_{|x|}\rangle \colon x \in \{0,1\}^*\}$ is Π^1_1, so $A \le_{\mathit{fin}\text{-}h} Z$ via Φ for any Z.

Many constructions of c.e. sets can be adapted to the setting of Π^1_1 sets and $\le_{\mathit{fin}\text{-}h}$. For instance, one can transfer the proof of Theorem 1.6.8 to the new setting to show that there are Π^1_1 sets A and B such that $A \,|_{\mathit{fin}\text{-}h}\, B$. Note that, even though the use of a *fin-h* reduction procedure is finite, the stages are now computable ordinals.

Exercises.

9.1.18. Prove the statements after Definition 9.1.12.

9.1.19. Define a binary function $g \le_T (O)'$ such that for each Δ^1_1 function h there is e such that $h(x) = g(e,x)$ for each x. Conclude that O is not Δ^1_1.

A set theoretical view

Reasoning about Π^1_1 sets and Π^1_1 classes can often be simplified by emphasizing a set theoretical view. From now on we will use the symbols $\mathbb{N}$ and ω interchangeably. We assume familiarity with the constructible hierarchy, defined by recursion on ordinals: for a set S we let $L(0,S)$ be the transitive closure of $\{S\} \cup S$. $L(\alpha+1, S)$ contains the sets that are first-order definable with parameters in $(L(\alpha,S), \in)$, and $L(\eta, S) = \bigcup_{\alpha<\eta} L(\alpha, S)$ for a limit ordinal η. We write $L(\alpha)$ for $L(\alpha, \emptyset)$. For details see Sacks 1.4.VII or Kunen (1980, pg. 180).

A Δ_0 formula is a first-order formula in the language of set theory which involves only bounded quantification, namely, quantification of the form $\exists z \in y$ and $\forall z \in y$. A Σ_1 formula has the form $\exists x_1 \exists x_2 ... \exists x_n\, \varphi_0$ where φ_0 is Δ_0.

9.1.20 Remark. The structures $L_A := L(\omega_1^A, A)$ for $A \subseteq \mathbb{N}$ are of particular interest because one can show they satisfy the axiom schemes of Δ_0 separation and Σ_1 bounding. These schemes are versions of the separation and replacement schemes from set theory. The relevant formulas are restricted to Δ_0 and Σ_1 formulas, respectively (Sacks 1.2.VII). Note that the formulas may contain parameters.

L_A can be introduced without referring to first-order definability in set theory: one codes hereditarily countable sets by certain subsets of $\mathbb{N}$, and L_A consists of the sets with a code in $\Delta^1_1(A)$ (Sacks 1.11.VII). In particular, the subsets of $\mathbb{N}$ that are in L_A are precisely the $\Delta^1_1(A)$ sets.

The principle of *transfinite recursion in L_A* (Sacks 1.6.VII) states that for each Σ_1 over L_A function $I\colon L_A \to \omega_1^A$ there is a Σ_1 over L_A function $f\colon \omega_1^A \to \omega_1^A$ such that $f(\alpha) = I(f\upharpoonright_\alpha)$ for each $\alpha < \omega_1^A$. (Here $f\upharpoonright_\alpha$ denotes the restriction of f to $\alpha = \{\beta\colon \beta < \alpha\}$.)

By Theorem 9.1.3 we can view Π^1_1 sets as being enumerated at stages that are computable ordinals. The following important theorem provides a further view of this existential aspect of Π^1_1 sets.

9.1.21 Theorem. *$S \subseteq \mathbb{N}$ is Π^1_1 iff there is a Σ_1-formula $\varphi(y)$ such that*

$$S = \{y \in \omega\colon (L(\omega_1^{ck}), \in) \models \varphi(y)\}.$$

To see that each Π^1_1 set S is of this form, suppose $y \in S \leftrightarrow R_y$ is well-ordered where R_y is a computable relation effectively determined from y. The formula $\varphi(y)$ expresses that R_y is isomorphic to an ordinal: $\exists\alpha\,\exists g\,[g\colon R_y \cong (\alpha, \in)]$. Such an isomorphism g is in $L(\omega_1^{ck})$ by transfinite induction. For the converse see for instance Sacks 1.3.VII. Note that the proof of the converse works even when φ contains parameters from $L(\omega_1^{ck})$.

We say that $D \subseteq (L_A)^k$ is Σ_1 *over* L_A if there is a Σ_1 formula φ such that $D = \{\langle x_1, \ldots, x_k\rangle \in (L_A)^k\colon (L_A, \in) \models \varphi(x_1, \ldots, x_k)\}$. Thus, by 9.1.21, $S \subseteq \mathbb{N}$ is Π^1_1 iff S is Σ_1 over $L(\omega_1^{ck})$.

We often consider partial functions from L_A to L_A with a graph defined by a Σ_1 formula with parameters. We say the function is Σ_1 over L_A. Such functions are an analog of functions partial computable in A. We provide a useful tool.

9.1.22 Bounding Principle. *Suppose $f\colon \omega \to \omega_1^A$ is Σ_1 over L_A. Then there is an ordinal $\alpha < \omega_1^A$ such that $f(n) < \alpha$ for each n.*

Proof. Since L_A satisfies Σ_1 replacement by 9.1.20, there is a set $y \in L_A$ such that $\mathrm{ran}(f) \subseteq y$. Since L_A satisfies Δ_0 separation, the set s of ordinals in y is in L_A. Hence $\beta = \bigcup \mathrm{ran}(f) \subseteq \bigcup s$ is in L_A. Now let $\alpha = \beta + 1$. □

9.1.23 Remark. One can use Theorem 9.1.21 to build Π^1_1 sets S. An enumeration of S over $L(\omega_1^{ck})$ is a Σ_1 function $f\colon \omega_1^{ck} \to \omega \cup \{\mathrm{nil}\}$ such that $S = \mathrm{ran} f \cap \omega$ (here nil is a further element, say ω). A construction C of S is given by a Σ_1 function telling us what to enumerate at stage α given the enumeration up to α. Formally, C is a Σ_1 over $L(\omega_1^{ck})$ function mapping $f\upharpoonright_\alpha$ to the number to be enumerated at α, or to nil if no number is enumerated. By transfinite recursion in $L(\omega_1^{ck})$, for each construction C a unique enumeration f exists.

9.1.24 Exercise. Use the Bounding Principle to show that each Δ^1_1 well-order R is isomorphic to a computable well-order.

9.2 Analogs of Martin-Löf randomness and K-triviality

We develop an analog of the theory of ML-randomness and K-triviality based on Π^1_1 sets. To do so we view some important questions of Chapters 2, 3, and 5 in the higher setting. The results of this section were obtained in Hjorth and Nies (2007).

Π^1_1 Machines and prefix-free complexity

9.2.1 Definition. A Π^1_1-*machine* is a possibly partial function $M\colon \{0,1\}^* \to \{0,1\}^*$ with a Π^1_1 graph. For $\alpha \le \omega_1^{\mathrm{ck}}$ we let $M_\alpha(\sigma) = y$ if $\langle \sigma, y\rangle \in M_\alpha$. We say that M is *prefix-free* if $\mathrm{dom}(M)$ is prefix-free.

9.2.2 Proposition. *There is an effective listing $(M_d)_{d\in\mathbb{N}}$ of all the prefix-free Π^1_1-machines.*

Proof. Let $(S_d)_{d\in\mathbb{N}}$ be an effective listing of the Π^1_1 subsets of $\{0,1\}^* \times \{0,1\}^*$. Thus $\langle \sigma, y\rangle \in S_d \leftrightarrow R^d_{\sigma,y}$ is a well-order, where $(R^d_{\sigma,y})$ is a uniformly c.e. sequence of linear orders as after Theorem 9.1.3. Now let $\langle \sigma, y\rangle \in M_d \leftrightarrow$
$R^d_{\sigma,y}$ is a well-order &
$\forall \langle \rho, z\rangle \, \forall g \, [(\rho \prec \sigma \ \vee \ (\rho = \sigma \ \& \ z \neq y)) \to$
g is not an isomorphism of $R^d_{\rho,z}$ with an initial segment of $R^d_{\sigma,y}]$.
(Informally, the machine has not halted on a proper prefix of σ, and has not produced an output other than y on input σ, before the stage given by the order type of $R^e_{\sigma,y}$.) Clearly this is a Π^1_1 condition uniformly in d. If S_d is a prefix-free Π^1_1-machine then $M_d = S_d$. □

As a consequence, there is an optimal prefix-free Π^1_1-machine.

9.2.3 Definition. The prefix-free Π^1_1-machine $\underline{\mathbb{U}}$ is given by $\underline{\mathbb{U}}(0^d 1 \sigma) \simeq M_d(\sigma)$. Let $\underline{K}(y) = \min\{|\sigma| \colon \underline{\mathbb{U}}(\sigma) = y\}$. For $\alpha \le \omega_1^{\mathrm{ck}}$ let $\underline{K}_\alpha(y) = \min\{|\sigma| \colon \underline{\mathbb{U}}_\alpha(\sigma) = y\}$.

Since $\underline{\mathbb{U}}$ has Π^1_1 graph, the relation "$\underline{K}(y) \le u$" is Π^1_1 and, by 9.1.4, for $\alpha < \omega_1^{\mathrm{ck}}$ the relation "$\underline{K}_\alpha(y) \le u$" is Δ^1_1. Moreover $\underline{K} \le_T O$. (Here $(\rho)_0$ is the first component r of the pair $\rho = \langle r, y\rangle$.)

Similar to 2.2.15, a Π^1_1 set $W \subseteq \mathbb{N} \times \{0,1\}^*$ is called a Π^1_1 *bounded request set* if $\sum_\rho 2^{-(\rho)_0} [\![\rho \in W]\!] \le 1$. (Here $(\rho)_0$ is the first component r of the pair $\rho = \langle r, y\rangle$.)

9.2.4 Theorem. *From a Π^1_1 bounded request set W one can effectively obtain a prefix-free Π^1_1-machine M such that $\forall \langle r, y\rangle \in W \, \exists w \, [|w| = r \ \& \ M(w) = y]$.*

Proof. By Theorem 9.1.3 let g be a computable function such that $x \in W \leftrightarrow S_x$ is a well-order, where $S_x = \Phi^{\emptyset}_{g(x)}$ is a computable linear order. Recall that for $x \in W$, we view the order type $\alpha = |S_x|$ as the stage when x is enumerated into W. As noted after 9.1.3 we may assume that at each stage at most one element is enumerated, and none at a limit stage. We turn this enumeration into a construction (in the sense of Remark 9.1.23) of a prefix-free Π^1_1-machine M. We let $R_0 = \{\varnothing\}$. At each stage $\gamma \ge 0$ we define an antichain $R_\gamma \in L(\omega_1^{\mathrm{ck}})$ of strings. The set of extensions of strings in R_γ is our reservoir of future w-values. Strings in this set are called *unused*. We are given R_α for $\alpha < \gamma$. As in Theorem 2.2.17, with each string x we associate the half-open interval $I(x) \subseteq [0,1)$ of real numbers with binary representation extending x.

Construction of M and of finite sets $R_\gamma \subseteq \{0,1\}^*$ ($\gamma < \omega_1^{\mathrm{ck}}$).
At a successor stage $\alpha = \beta + 1$, if a request $\langle r, y\rangle$ is enumerated into W, we will find a string w of length r and put $\langle w, y\rangle$ into M. Let z be the longest string in R_β of length $\le r$. Choose $w = w_\alpha$ so that $I(w)$ is the leftmost subinterval of $I(z)$ of length 2^{-r}, i.e., let $w = z0^{r-|z|}$. To obtain R_α, first remove z from R_β. If $w \neq z$ then also add the strings $z0^i1$, $0 \le i < r - |z|$ to R_α.

For a limit ordinal η, we let $R_\eta = \{x\colon \exists\gamma < \eta \forall\alpha\, [\gamma < \alpha < \eta \to x \in R_\alpha]\}$.

Verification. We will see that a string x can appear in R_α at most once, so that actually $R_\eta(x) = \lim_{\gamma\to\eta} R_\gamma(x)$ for a limit ordinal η. In the claim below we verify a number of properties in order to show that for each request $\langle r, y\rangle$, z as above exists, and therefore we can assign a string w of length r to the request. Let

$$E_\alpha = \bigcup\{I(x)\colon\ x \in R_\alpha\}$$

be the set of real numbers corresponding to R_α. At a limit stage η, let

$$G_\eta = \bigcap\nolimits_{\beta<\eta} E_\beta.$$

The measure of the unused strings at stage η is $\lambda(G_\eta)$. To be able to get beyond this limit stage we want to replace G_η by E_η. The main statement, (i) below, says that this replacement is allowed because $E_\eta \subseteq G_\eta$ and $G_\eta - E_\eta$ is null.

We first illustrate the construction with an example showing that this null class may be nonempty. Suppose at stage $i < \omega$ the request $\langle 2i+1, y_i\rangle$ is enumerated. Then $G_\omega - E_\omega = \{1/3\}$. For $R_0 = \{\emptyset\}$, $z_0 = \emptyset$, $w_0 = 00$; $R_1 = \{01, 1\}$, $z_1 = 01$, $w_1 = 0100$; $R_2 = \{0101, 011, 1\}$, $z_2 = 0101$, $w_2 = 010100$ etc. Then $R_\omega = \{(01)^i 1\colon\ i \in \omega\}$. 1/3 has the binary representation $0.010101\ldots$, so that $1/3 \in E_i$ for each i, but $1/3 \notin E_\omega$.

Claim. (i) *$E_{\alpha+1} \subseteq E_\alpha$ for each stage α. If $\alpha = \eta$ is a limit ordinal then $E_\eta \subseteq G_\eta$ and $G_\eta - E_\eta$ is null.*
(ii) *If a request is enumerated at stage $\alpha = \beta+1$, one can choose z and hence $w = w_\alpha$.*
(iii) *The strings in R_α have different lengths and form an antichain. (In fact, for $x, y \in R_\alpha$ we have $|x| < |y| \leftrightarrow x <_L y$.)*
(iv) *The intervals $\{I(z)\colon\ z \in R_\alpha\} \cup \{I(w_\beta)\colon\ \beta \le \alpha$ & w_β defined$\}$ form a partition of a conull subset P_α* of $[0, 1)$.

Inductively assume (i)–(iv) for all ordinals $\gamma < \alpha$.

(i) Clearly $E_{\alpha+1} \subseteq E_\alpha$. If $\alpha = \eta$ is a limit ordinal, to show $E_\eta \subseteq G_\eta$, let $\beta < \eta$. If $r \in E_\eta$, then $r \in I(x)$ for some $x \in R_\eta$, so there is γ, $\beta < \gamma < \eta$, such that $x \in R_\gamma$. Inductively $E_\gamma \subseteq E_\beta$. Thus $r \in E_\beta$.

We verify $\lambda E_\eta \ge \lambda G_\eta$ by showing that $\lambda E_\eta \ge \lambda G_\eta - 2^{-k+1}$ for each $k \in \omega$. Write λG_η in binary form: $\lambda(G_\eta) = \sum_{d\in A} 2^{-d}$ where $A \subseteq \omega$ is co-infinite. Since $(\lambda E_\gamma)_{\gamma<\eta}$ is non-increasing and converges to λG_η, there is $\gamma < \eta$ such that

$$2^{-k+1} + \sum\nolimits_{d\in A\cap k} 2^{-d} \ge \lambda E_\gamma.$$

Let $A \cap k = \{d_1, d_2, \ldots, d_N\}$ where the d_i are pairwise distinct. For each α, $\gamma < \alpha < \eta$, let z_i^α $(1 \le i \le N)$ be the elements of R_α such that $|z_i^\alpha| = d_i$ (such strings exist by the inductive hypothesis (iii) for α). If $z \in R_\beta - R_{\beta+1}$ for some $\beta < \eta$ then $z \preceq w_\beta$, so $z \notin R_\delta$ for each δ, $\beta < \delta < \eta$ by inductive hypothesis (iv) for δ (in brief, z cannot reappear after disappearing). Since there are only 2^{d_i} possibilities for z_i^α, we eventually settle on some strings z_i of length d_i, hence $z_i \in R_\eta$. Thus,
$\lambda(E_\eta) \ge \sum_{1\le i\le N} 2^{-|z_i|} \ge \lambda(E_\gamma) - 2^{-k+1} \ge \lambda(G_\eta) - 2^{-k+1}$, as required.

(ii) Suppose the request $\langle r, y\rangle$ is enumerated at stage $\alpha = \beta+1$. If z_α fails to exist, then r is less than the length of each string in R_β. By (iii) for β, $\lambda E_\beta = \sum_z 2^{-|z|}\, [\![z \in R_\beta]\!]$, so by (iv) for β, $1 < 2^{-r} + \sum_m 2^{-m}\, [\![$a request $\langle m, z\rangle$ is enumerated at a stage $\le \beta]\!]$, contrary to the assumption that W is a bounded request set.

(iii) This is clear for successor stages α, because the intervals $I(w_\gamma)$, $\gamma \le \alpha$ and w_γ defined, are disjoint. The property persists to limit stages by the definition of R_η.

(iv) Again, this is clear for successor stages $\alpha = \beta + 1$, in which case we may define $P_\alpha = P_\beta$. If $\alpha = \eta$ is a limit stage, let P_η be the intersection of the sets P_γ and the complements of the null classes $G_\gamma - E_\gamma$ from (i) for $\gamma \leq \eta$. Then for each $\beta < \eta$, the set P_η is partitioned by E_β and $I(w_\gamma)$, $\gamma \leq \beta$, w_γ defined. So P_η is partitioned into G_η and the sets $I(w_\gamma)$ for $\gamma < \eta$. Since G_η is partitioned on P_η into the intervals $I(z)$, $z \in R_\eta$, we have shown (iv) for η.

By (iv) the Π^1_1-machine M is prefix-free. □

One can use the foregoing Theorem 9.2.4 to characterize $\underline{K}$ by its minimality, similar to Theorem 2.2.19. We skip this, but we do adapt the Coding Theorem 2.2.25 by applying 9.2.4. The probability that a prefix-free Π^1_1-machine D outputs a string x is $P_D(x) = \lambda[\{\sigma\colon D(\sigma) = x\}]^\prec$. For $\alpha \leq \omega_1^{\mathrm{ck}}$, let $P_{D,\alpha}(x) = \lambda[\{\sigma\colon D_\alpha(\sigma) = x\}]^\prec$. Clearly $2^{-\underline{K}(x)} \leq P_{\underline{\mathbb{U}}}(x)$. We show $\forall x\ 2^{\mathbf{c}}2^{-\underline{K}(x)} > P_D(x)$ for some constant $\mathbf{c}$. This also holds at certain ordinal stages. For $g\colon \omega_1^{\mathrm{ck}} \to \omega_1^{\mathrm{ck}}$, we say that a limit ordinal $\eta \leq \omega_1^{\mathrm{ck}}$ is *g-closed* if $\forall \alpha < \eta\,[g(\alpha) < \eta]$.

9.2.5 Theorem. *For each prefix-free Π^1_1-machine D there is a Σ_1 over $L(\omega_1^{\mathrm{ck}})$ function $g_D\colon \omega_1^{\mathrm{ck}} \to \omega_1^{\mathrm{ck}}$ and a constant $\mathbf{c}$ such that for each g_D-closed $\eta \leq \omega_1^{\mathrm{ck}}$ we have $\forall x\ 2^{\mathbf{c}}2^{-\underline{K}_\eta(x)} > P_{D,\eta}(x)$.*

Proof. We build a Π^1_1 bounded request set W, accounting the enumeration of requests $\langle r, x\rangle$ against the open sets generated by the D-descriptions of x. At stage α, if x is a string, $r \in \mathbb{N}$ is least such that $P_{D,\alpha}(x) \geq 2^{-r+1}$, and the request $\langle r, x\rangle$ is not in W yet, then we put $\langle r, x\rangle$ into W.

For a string x, let α_x be the greatest stage at which a request $\langle r, x\rangle$ is put into W. Then $P_{D,\alpha_x}(x) \geq 2^{-r+1}$. Hence all such requests together contribute at most 1/2. The total weight of all requests $\langle r', x\rangle$ enumerated at previous stages is at most 2^{-r}, since $r' > r$ for such a request, and there is at most one for each length r'. Thus W is a bounded request set.

Let c_W be the coding constant for W given by Theorem 9.2.4. The function g_D is the delay it takes the optimal Π^1_1-machine to react to an enumeration of a request into W. Thus $g_D(\alpha) = \mu\beta.\forall \langle r, x\rangle \in W_\alpha\,[\underline{K}_\beta(x) \leq r + c_W]$ for $\alpha < \omega_1^{\mathrm{ck}}$, which is Σ_1 over $L(\omega_1^{\mathrm{ck}})$. If r is least such that $P_{D,\eta}(x) > 2^{-r+1}$, then at the least stage $\alpha < \eta$ where $P_{D,\alpha}(x) \geq 2^{-r+1}$, we enumerate $\langle r, x\rangle$ and cause $\underline{K}_\eta(x) \leq \underline{K}_{g_D(\alpha)}(x) \leq r + c_W$, by the hypothesis that η is g_D-closed. Since r was chosen least, we have $2^{-r+2} \geq P_{D,\eta}(x)$, hence $2^{c_W+3}2^{-\underline{K}_\eta(x)} > 2^{-r+2} \geq P_{D,\eta}(x)$. Thus $\mathbf{c} = c_W + 3$ is as required. □

As in Theorem 2.2.26, one can apply Theorem 9.2.5 to bound the number of strings of a given length with a low $\underline{K}$-complexity.

9.2.6 Theorem. *There is a constant $\mathbf{c} \in \mathbb{N}$ and a Σ_1 over $L(\omega_1^{\mathrm{ck}})$ function $g\colon \omega_1^{\mathrm{ck}} \to \omega_1^{\mathrm{ck}}$ such that, for each g-closed $\eta \leq \omega_1^{\mathrm{ck}}$,*

$$\forall b\,\forall n\,\#\{x\colon |x| = n\ \&\ \underline{K}_\eta(x) \leq \underline{K}_\eta(n) + b\} \leq 2^{\mathbf{c}}2^b.$$

Proof. Let D be the prefix-free machine given by $D(\sigma) = |\underline{\mathbb{U}}(\sigma)|$. Let g be the function g_D obtained in the foregoing theorem for D, and let $\mathbf{c}$ be the constant such that $2^{\mathbf{c}}2^{-\underline{K}_\eta(n)} > P_{D,\eta}(n)$ for each n and each g-closed η. Now we conclude the argument as in the proof of Theorem 2.2.26. □

9.2.7$^\diamond$ Carry out the Π^1_1 version of 2.2.24(ii) for a simple proof of a variant of 9.2.4.

A version of Martin-Löf randomness based on Π^1_1 sets

A Π^1_1-*ML-test* is a sequence $(G_m)_{m\in\mathbb{N}}$ of open sets such that $\forall m \in \mathbb{N}\, \lambda G_m \le 2^{-m}$ and the relation $\{\langle m,\sigma\rangle \colon [\sigma] \subseteq G_m\}$ is Π^1_1 (by Exercise 9.1.2 it is equivalent to require that the classes G_m be uniformly Π^1_1). A set Z is Π^1_1-*ML-random* if $Z \notin \bigcap_m G_m$ for each Π^1_1-ML-test $(G_m)_{m\in\mathbb{N}}$. Let $\underline{\mathsf{MLR}}$ denote the class of Π^1_1-ML-random sets. For $b \in \mathbb{N}$ let $\underline{\mathcal{R}}_b = [\{x \in \{0,1\}^* \colon \underline{K}(x) \le |x| - b\}]^{\prec}$.

9.2.8 Proposition. *$(\underline{\mathcal{R}}_b)_{b\in\mathbb{N}}$ is a Π^1_1-ML-test.*

Proof. By the remark after 9.2.3 the relation $\{\langle b,\sigma\rangle \colon [\sigma] \subseteq \underline{\mathcal{R}}_b\}$ is Π^1_1. One shows $\lambda\underline{\mathcal{R}}_b \le 2^{-b}$ as in the proof of Proposition 3.2.7. □

Hjorth and Nies (2007) proved the higher analog of Schnorr's Theorem 3.2.9. Thus Z is Π^1_1-ML-random iff $Z \in 2^{\mathbb{N}} - \underline{\mathcal{R}}_b$ for some b. (The proof is harder in the new setting because of the limit stages.) Since $\bigcap_b \underline{\mathcal{R}}_b$ is Π^1_1, this implies that the class of Π^1_1-ML-random sets is Σ^1_1.

We provide two examples of Π^1_1-ML-random sets.

1. By the Gandy Basis Theorem 9.1.16 there is a Π^1_1-ML-random set $Z \le_T O$ such that $O^Z \le_h O$.
2. Let $\underline{\Omega} = \lambda[\operatorname{dom}\underline{\mathbb{U}}]^{\prec} = \sum_\sigma 2^{-|\sigma|}\, [\![\underline{\mathbb{U}}(\sigma)\downarrow]\!]$. Note that $\underline{\Omega}$ is left-Π^1_1. The proof of Theorem 3.2.11 shows that $\underline{\Omega}$ is Π^1_1-ML-random (the stage variable t now denotes a computable ordinal).

The reducibility $\le_{\mathit{fin}\text{-}h}$ was introduced in 9.1.17. We say that $A \le_{\mathit{wtt}\text{-}h} Z$ if $A = \Phi^Z$ for some *fin-h* reduction such that the use function is computably bounded. The following is analogous to Theorem 3.3.2.

9.2.9 Theorem. *Let Q be a closed Σ^1_1 class of Π^1_1-ML-random sets such that $\lambda Q \ge 1/2$ (say $Q = 2^{\mathbb{N}} - \underline{\mathcal{R}}_1$). For each set A there is $Z \in Q$ such that $A \le_{\mathit{wtt}\text{-}h} Z$.*

Proof. Let f be the function from the proof of Theorem 3.3.2. Let $\widehat{Q} = \mathit{Paths}(T)$ where T is the tree defined by (3.4). Then $\widehat{Q}$ is a closed Σ^1_1 class. Using the purely measure-theoretic Lemma 3.3.1, we may define Z exactly as in that proof, and $Z \in \widehat{Q}$ because $\widehat{Q}$ is closed. For a Σ^1_1 class $\mathcal{S}$ given by an index, emptiness is a Π^1_1 property. Hence it becomes apparent at an ordinal stage $\alpha < \omega_1^{\mathrm{ck}}$.

We describe a reduction for $A \le_{\mathit{wtt}\text{-}h} Z$, where $f(n+1)$ is the computable bound on the use for input n. To determine $A(r)$, let $x = Z\restriction_{f(r)}$, and let $y = Z\restriction_{f(r+1)}$. Find a stage $\alpha < \omega_1^{\mathrm{ck}}$ such that $\widehat{Q}_\alpha \cap [\{v \colon x \preceq v \;\&\; |v| = |y| \;\&\; v <_L y\}]^{\prec} = \emptyset$ or $\widehat{Q}_\alpha \cap [\{v : x \preceq v \;\&\; |v| = |y| \;\&\; v >_L y\}]^{\prec} = \emptyset$. In the first case output 0, in the second case output 1. □

9.2.10 Exercise. Show $\underline{\Omega} \equiv_{\mathit{wtt}} O$. *Hint.* Modify the proof of Proposition 3.2.30.

An analog of K-triviality

Recall from 9.2.3 that $\underline{K}(x)$ denotes the prefix-free complexity of x in the setting of Π^1_1 sets. We study sets that are $\underline{K}$-trivial. For technical reasons we also consider $\underline{K}$-triviality at certain limit stages.

9.2.11 Definition. A is *$\underline{K}$-trivial* if $\exists b \forall n\, \underline{K}(A\restriction_n) \le \underline{K}(n) + b$. Given a limit ordinal $\eta \le \omega_1^{\mathrm{ck}}$, we say that A is *$\underline{K}$-trivial at η* if $\exists b \forall n\, \underline{K}_\eta(A\restriction_n) \le \underline{K}_\eta(n) + b$. (Thus, being $\underline{K}$-trivial is the same as being $\underline{K}$-trivial at ω_1^{ck}.)

One can modify the proof of Proposition 5.3.11 in order to show:

9.2.12 Proposition. *There is a $\underline{K}$-trivial Π^1_1 set A that is not Δ^1_1.* □

The proof is by a construction in the sense of Remark 9.1.23. It uses the Machine Existence Theorem in the version 9.2.4.

Fix b and $\eta \le \omega_1^{\mathrm{ck}}$. The sets that are $\underline{K}$-trivial via b at η are the paths through the tree $T_{\eta,b} = \{z\colon \forall u \le |z|\, \underline{K}_\eta(z\restriction_u) \le \underline{K}_\eta(u) + b\}$. If $\eta < \omega_1^{\mathrm{ck}}$ then $T_{\eta,b}$ is Δ^1_1 because $\underline{\mathbb{U}}_\eta$ is a Δ^1_1 set by 9.1.4(i) and Δ^1_1 sets are closed under number quantification. Let g_D be the function obtained in 9.2.5 where $D(x) \simeq |\underline{\mathbb{U}}(x)|$.

9.2.13 Proposition. *Let $\eta \le \omega_1^{\mathrm{ck}}$ be g_D-closed.*

(i) There is $\mathbf{c} \in \mathbb{N}$ such that the following holds: for each b, at most $2^{\mathbf{c}+b}$ sets are $\underline{K}$-trivial at η with constant b.

(ii) If $\eta < \omega_1^{\mathrm{ck}}$ and A is $\underline{K}$-trivial at η then A is hyperarithmetical.

(iii) Each $\underline{K}$-trivial set is computable in O.

Proof. Let $\mathbf{c}$ be as in Theorem 9.2.6. The size of each level of $T_{\eta,b}$ is at most $2^{\mathbf{c}+b}$, which shows (i).

Note that each path A of $T_{\eta,b}$ is isolated, and hence computable in $T_{\eta,b}$. For (ii), if $\eta < \omega_1^{\mathrm{ck}}$ this shows that A is hyperarithmetical. For (iii), note that since $\underline{K} \le_T O$, the tree $T_{\omega_1^{\mathrm{ck}},b}$ is computable in O. Now we argue as in (ii). □

9.2.14 Lemma. *If A is $\underline{K}$-trivial and $\omega_1^A = \omega_1^{\mathrm{ck}}$ then A is hyperarithmetical.*

Proof. Suppose A is $\underline{K}$-trivial via b. We show that A is $\underline{K}$-trivial at η via b for some g_D-closed $\eta < \omega_1^{\mathrm{ck}}$, where g_D is the function from the proof of 9.2.13. We define by transfinite recursion (see before 9.1.23) a function $h\colon \omega \to \omega_1^{\mathrm{ck}}$ which is Σ_1 over L_A: let $h(0) = 0$ and

$$h(n+1) = \mu\beta > g_D(h(n)).\, \forall m \le n\, \underline{K}_\beta(A\restriction_m) \le \underline{K}_\beta(m) + b.$$

Since A is $\underline{K}$-trivial, $h(n)$ is defined for each $n \in \omega$. Let $\eta = \sup(\operatorname{ran} h)$, then $\eta < \omega_1^A = \omega_1^{\mathrm{ck}}$ by the Bounding Principle 9.1.22, so η is as required. □

Adapting the relevant proofs from Section 5.2, one can show that the class $\underline{\mathcal{K}}$ of $\underline{K}$-trivial sets is closed under $\oplus$ and closed downward under $\le_{\textit{wtt-h}}$. In particular, O is not $\underline{K}$-trivial, as $\underline{\Omega} \equiv_{wtt} O$ by 9.2.10. The golden run method of Section 5.4 can be adapted to show that $\underline{\mathcal{K}}$ is even closed downward under $\le_{\textit{fin-h}}$. However, $\underline{\mathcal{K}}$ is not closed downward under $\le_h$, since $O \le_h A$ for the Π^1_1 set $A \in \underline{\mathcal{K}}$ from Proposition 9.2.12.

Lowness for Π^1_1-ML-randomness

$\underline{\mathsf{MLR}}^A$ denotes the class of sets which are Π^1_1-ML-random relative to A, and A is *low for Π^1_1-ML-randomness* if $\underline{\mathsf{MLR}}^A = \underline{\mathsf{MLR}}$. A is called a *base for Π^1_1-ML-randomness* if $A \le_{\textit{fin-h}} Z$ for some $Z \in \underline{\mathsf{MLR}}^A$ (see Definition 9.1.17 for $\le_{\textit{fin-h}}$).

By Theorem 9.2.9, a set that is low for Π^1_1-ML-randomness is a base for Π^1_1-ML-randomness. The following contrasts with Theorem 5.1.19:

9.2.15 Theorem. *A is a base for* Π^1_1*-ML-randomness* $\Leftrightarrow$ *A is* Δ^1_1.

Proof. $\Leftarrow$: $A \le_{\mathit{fin}\text{-}h} Z$ for each Z, so A is a base for Π^1_1-ML-randomness.
$\Rightarrow$: Suppose $A = \Phi^Z$ for some *fin-h* reduction procedure Φ (see 9.1.17) and $Z \in \underline{\mathsf{MLR}}^A$. Assume for a contradiction that A is not Δ^1_1.
1. We show that $\omega_1^A = \omega_1^{\mathrm{ck}}$. Note that the class $\{Y\colon A = \Phi^Y\}$ is null by Theorem 9.1.13 (or by directly adapting Theorem 5.1.12 to the setting of $\le_{\mathit{fin}\text{-}h}$). Similar to Remark 5.1.13, for each n let

$$V_n = S^{\Phi}_{A,n} = [\{\rho\colon A\upharpoonright_n \preceq \Phi^\rho\}]^{\prec} = [\{\rho\colon \exists \alpha < \omega_1^{\mathrm{ck}}\, A\upharpoonright_n \preceq \Phi^\rho_\alpha\}]^{\prec}.$$

If $\omega_1^{\mathrm{ck}} < \omega_1^A$ then by the Approximation Lemma 9.1.4(i) relative to A, the set $\{\langle n, \rho\rangle\colon \exists \alpha < \omega_1^{\mathrm{ck}}\, A\upharpoonright_n \preceq \Phi^\rho_\alpha\}$ is $\Delta^1_1(A)$. So by the Measure Lemma 9.1.10(ii) the function $h(n) = \mu n.\, \lambda V_n \le 2^{-n}$ is $\Delta^1_1(A)$. Then $(V_{h(n)})_{n\in\mathbb{N}}$ is a Π^1_1-ML-test relative to A which succeeds on Z, contrary to the hypothesis that $Z \in \underline{\mathsf{MLR}}^A$.
2. We show that A is $\underline{K}$-trivial, which together with $\omega_1^A = \omega_1^{\mathrm{ck}}$ implies that A is Δ^1_1 by Lemma 9.2.14. We want to adapt the proof of Theorem 5.1.22, which (in the setting of computability theory) states that each base for ML-randomness A is low for K. First we observe that (still in that setting) the proof can be modified to show directly that A is K-trivial: we replace $\mathbb{U}$ by the oracle prefix-free machine M given by $M^X(\sigma) \simeq X\upharpoonright_{\mathbb{U}(\sigma)}$. The modified proof now shows that $\forall y\, K(y) \le^+ K^A_M(y)$. For $y = A\upharpoonright_n$, we obtain $K(A\upharpoonright_n) \le^+ K^A_M(A\upharpoonright_n) \le^+ K(n)$.

Now we adapt this (already modified) proof to the present setting, where Φ is a *fin-h* reduction procedure. Fix a parameter $d \in \mathbb{N}$. If $\underline{\mathbb{U}}^\tau(\sigma)$ converges at a stage $\alpha < \omega_1^{\mathrm{ck}}$, we start feeding open sets $C^\tau_{d,\sigma}$, and continue to do so as long as $\lambda C^\tau_{d,\sigma} < 2^{-|\sigma|-d}$. We restrict the enumeration into $C^\tau_{d,\sigma}$ to successor stages. For limit stages η we define $C^\tau_{d,\sigma}[\eta] = \bigcup_{\alpha<\eta} C^\tau_{d,\sigma}[\alpha]$. The verification that A is $\underline{K}$-trivial is as before, now using the Machine Existence Theorem in the version 9.2.4. □

From Theorems 9.2.9 and 9.2.15 we obtain:

9.2.16 Corollary. *A is low for* Π^1_1*-ML-randomness* $\Leftrightarrow$ *A is* Δ^1_1. □

9.2.17$^\diamond$ **Problem.** A *generalized* Π^1_1*-ML-test* is a sequence $(G_m)_{m\in\mathbb{N}}$ of uniformly Π^1_1 open sets such that $\bigcap_m G_m$ is a null class. Z is Π^1_1*-weakly 2-random* if Z passes each generalized Π^1_1-ML-test. Is this stronger than Π^1_1-ML-randomness?

9.3 Δ^1_1-randomness and Π^1_1-randomness

We study two randomness concepts, the first weaker and the second stronger than Π^1_1-ML-randomness. Background was given at the beginning of this chapter.

9.3.1 Definition. Z is Δ^1_1*-random* if Z avoids each null Δ^1_1 class.

We will show that Δ^1_1-randomness coincides with the higher analogs of both Schnorr randomness and computable randomness. In fact the higher versions of

Schnorr tests and hyperarithmetical martingales are all equivalent in strength to null Δ^1_1 classes (9.3.3). Moreover, Chong, Nies and Yu (2008) proved that each null Σ^1_1 class is contained in a null Δ^1_1 class, so Σ^1_1-randomness also coincides with Δ^1_1-randomness. However, adapting the methods of Section 7.4, there is a Δ^1_1-random set of slowly growing initial segment complexity, which therefore is not Π^1_1-ML-random (Theorem 9.3.4).

9.3.2 Definition. Z is Π^1_1-*random* if Z avoids each null Π^1_1 class.

Each Π^1_1-random set Z satisfies $\omega_1^Z = \omega_1^{\mathrm{ck}}$ because the Π^1_1 class $\{A\colon \omega_1^A > \omega_1^{\mathrm{ck}}\}$ is null by 9.1.15. Thus, the Π^1_1-ML-random set $\underline{\Omega}$ is not Π^1_1-random, as $\underline{\Omega} \equiv_{wtt} O$ by 9.2.10. In fact we show that Z is Π^1_1-random iff Z is Δ^1_1-random and $\omega_1^Z = \omega_1^{\mathrm{ck}}$. Therefore, within the Δ^1_1-random sets, the Π^1_1-random sets are characterized by a lowness property (for the new setting, one that is closed downward under $\leq_h$). This is similar to the characterizations of the weakly 2-random and the 2-random sets within the ML-random sets in Section 3.6.

Notions that coincide with Δ^1_1-randomness

A Π^1_1-*Schnorr test* is a Π^1_1-ML-test $(G_m)_{m\in\mathbb{N}}$ such that λG_m is left-Δ^1_1 uniformly in m. A supermartingale $M\colon \{0,1\}^* \to \mathbb{R}^+ \cup \{0\}$ is *hyperarithmetical* if its undergraph $\{\langle x, q\rangle : q \in \mathbb{Q}_2 \;\&\; M(x) > q\}$ is Δ^1_1. Recall that the success class $\mathsf{Succ}(M)$, defined in 7.1.1, is null by 7.1.15. By the following, Δ^1_1-randomness coincides with the higher analogs of both Schnorr and computable randomness. This difference to the computability theoretic case stems from the closure properties (C1) and (C2) in the chapter introduction.

9.3.3 Theorem. *(i) Let $\mathcal{A}$ be a null Δ^1_1 class. Then $\mathcal{A} \subseteq \bigcap G_m$ for some Π^1_1-Schnorr test $\{G_m\}_{m\in\mathbb{N}}$ such that $\lambda G_m = 2^{-m}$ for each m. In fact, the relation $\{\langle m, \sigma\rangle\colon [\sigma] \subseteq G_m\}$ is Δ^1_1.*

(ii) If $(G_m)_{m\in\mathbb{N}}$ is a Π^1_1-Schnorr test then $\bigcap_m G_m \subseteq \mathsf{Succ}(M)$ for a hyperarithmetical martingale M.

(iii) $\mathsf{Succ}(M)$ is a null Δ^1_1 class for each hyperarithmetical supermartingale M.

Proof. (i) By Sacks 1.8.IV, for each m we may effectively obtain a Δ^1_1 open set G_m such that $\mathcal{A} \subseteq G_m$ and $\lambda(G_m - \mathcal{A}) \leq 2^{-m}$. It now suffices to show that from a Δ^1_1 open set S and a rational $q \geq \lambda S$ we may in an effective way obtain a Δ^1_1 open set $\widetilde{S}$ such that $S \subseteq \widetilde{S}$ and $\lambda\widetilde{S} = q$. One can easily adapt the proof of Lemma 1.9.19 to the new setting. As before, we define a function $f\colon [0,1)_{\mathbb{R}} \to [0,1)_{\mathbb{R}}$ by $f(r) = \lambda(S \cup [0,r))$. For $x \in \mathbb{Q}_2$, $f(x)$ is a uniformly left-Δ^1_1 real by 9.1.10. So, by the closure under number quantification, the least t such that $f(t) = q$ is Δ^1_1 effectively in q. Now let $\widetilde{S} = S \cup [0,t)$.

(ii) Let $f(m,k)$ be the least α such that $\lambda G_m - \lambda G_{m,\alpha} \leq 2^{-k}$. By 9.1.15 $f(m,k) < \omega_1^{\mathrm{ck}}$ for each m,k. By 9.1.10 and 9.1.21, f is Σ_1 over $L(\omega_1^{\mathrm{ck}})$, so by the Bounding Principle 9.1.22 there is $\beta < \omega_1^{\mathrm{ck}}$ such that $\forall m,k\, f(m,k) < \beta$.

By 7.1.15, $\bigcap_m G_m \subseteq \mathsf{Succ}(M)$ for the martingale $M = \sum_m M_{G_m}$. Since $\lambda G_m = \lambda G_{m,\beta}$ we have $M = \sum_m M_{G_{m,\beta}}$, so M is hyperarithmetical by 9.1.10 and 9.1.4.

(iii) We have $Z \in \mathsf{Succ}(M) \leftrightarrow \forall d \exists n\, M(Z\upharpoonright_n) \geq d$, so $\mathsf{Succ}(M)$ is Δ^1_1 by the closure of Δ^1_1 relations under number quantification. $\square$

The foregoing characterization of Δ^1_1-randomness via hyperarithmetical martingales can be used to separate it from Π^1_1-ML-randomness.

9.3.4 Theorem. *For every unbounded nondecreasing hyperarithmetical function h there is a Δ^1_1-random set Z such that $\forall^\infty n\, \underline{K}(Z\upharpoonright_n \mid n) \leq h(n)$.*

Proof. We adapt the proof of 7.4.8. Most of the basics developed for computable martingales carry over to the case of hyperarithmetical martingales, for instance Proposition 7.3.8 that one can assume the values are in $\mathbb{Q}_2$. The partial computable martingales of Definition 7.4.1 now become partial martingales with a Π^1_1 graph. As before, they can be listed effectively. The partial computable function F in Lemma 7.4.10 is now a function with a Π^1_1 graph. $\square$

The higher analog of Schnorr's Theorem 3.2.9 in Hjorth and Nies (2007) implies:

9.3.5 Corollary. *There is a Δ^1_1-random set that is not Π^1_1-ML-random.* $\square$

By Theorem 9.1.11 the class of Δ^1_1-random sets is not Π^1_1. In particular, there is no largest null Δ^1_1 class. However, the class of Δ^1_1-random sets is Σ^1_1 by 9.3.11.

More on Π^1_1-randomness

There are only countably many Π^1_1 classes, so the union of all null Π^1_1 classes is null. We show that it is also a Π^1_1 class. In particular, there is a universal test for Π^1_1-randomness. Theorem 1A-2 in Kechris (1975) is more general: it not only implies this result, but also shows that there is a largest countable Π^1_1 class, and a largest thin Π^1_1 class (a class is thin if it has no perfect subclass). Also see Moschovakis (1980, Thm 4F.4).

9.3.6 Theorem. *There is a null Π^1_1 class $\mathcal{Q}$ such that $\mathcal{S} \subseteq \mathcal{Q}$ for each null Π^1_1 class $\mathcal{S}$.*

Proof. We claim that one may effectively determine from a Π^1_1 class $\mathcal{S}$ a null Π^1_1 class $\widehat{\mathcal{S}} \subseteq \mathcal{S}$ such that $\widehat{\mathcal{S}} = \mathcal{S}$ in case $\mathcal{S}$ is already null. Assuming this, let $\mathcal{Q}_p$ be the Π^1_1 class with index p as defined in (9.4). Then $\mathcal{Q} = \bigcup_p \widehat{\mathcal{Q}}_p$ is Π^1_1, so $\mathcal{Q}$ is as required.

To prove the claim, if $\mathcal{S} = \mathcal{Q}_p$ let $\widehat{\mathcal{S}} = \{X \in \mathcal{S} \colon \mathcal{S}_{|\Phi^X_p|+1} \text{ is null}\}$. By the Approximation Lemma 9.1.4(ii) (and its proof) we can compute from p (independently of X) a number that is an index for $\mathcal{S}_{|\Phi^X_p|+1}$ as a $\Delta^1_1(X)$ class if $R = \Phi^X_p$ is well-ordered.

The proof of (ii) of the Measure Lemma 9.1.10 can be extended to show that there is a computable function taking us (still independently of X) from an index for a $\Delta^1_1(X)$ class to an index for its measure as a left-$\Delta^1_1(X)$ real number. To express that this real number equals 0 merely involves quantification over the rationals. Therefore $\widehat{\mathcal{S}}$ is a Π^1_1 class, and its index can be computed from an index for $\mathcal{S}$. Since $\{X \colon \omega^X_1 > \omega^{\mathrm{ck}}_1\}$ is null by 9.1.15, $\widehat{\mathcal{S}}$ is contained in the union of a null class and all the $\mathcal{S}_\alpha$, $\alpha < \omega_1^{\mathrm{ck}}$ that are null, hence $\widehat{\mathcal{S}}$ is null. When $\mathcal{S}$ is null every $\mathcal{S}_\alpha$ for $\alpha < \omega_1$ is null, and therefore $\widehat{\mathcal{S}} = \mathcal{S}$. □

Applying the Gandy Basis Theorem 9.1.16 to the Σ_1^1 class $\mathcal{S} = 2^{\mathbb{N}} - \mathcal{Q}$ yields:

9.3.7 Corollary. *There is a Π_1^1-random set $Z \leq_T O$ such that $O^Z \leq_h O$.* □

This contrasts with the fact that in the computability setting already a weakly 2-random set forms a minimal pair with $\emptyset'$ (page 135) that $\{Y : \omega_1^Y > \omega_1^{\mathrm{ck}}\}$ is a null Π_1^1 class:

For each Π_1^1 class $\mathcal{S}$ we have $\mathcal{S} \subseteq \{Y : \omega_1^Y > \omega_1^{\mathrm{ck}}\} \cup \bigcup_{\alpha<\omega_1^{\mathrm{ck}}} \mathcal{S}_\alpha$, because $Y \in \mathcal{S}$ implies $Y \in \mathcal{S}_\alpha$ for some $\alpha < \omega_1^Y$. For the largest null Π_1^1 class $\mathcal{Q}$, equality holds by 9.1.15 that $\{Y : \omega_1^Y > \omega_1^{\mathrm{ck}}\}$ is a null Π_1^1 class:

9.3.8 Fact. $\mathcal{Q} = \{Y : \omega_1^Y > \omega_1^{\mathrm{ck}}\} \cup \bigcup_{\alpha<\omega_1^{\mathrm{ck}}} \mathcal{Q}_\alpha$. □

For $\alpha < \omega_1^{\mathrm{ck}}$ the null class $\mathcal{Q}_\alpha$ is Δ_1^1 by the Approximation Lemma 9.1.4(ii). So the foregoing fact yields a characterization of the Π_1^1-random sets within the Δ_1^1-random sets by a lowness property in the higher setting, analogous to 3.5.13.

9.3.9 Theorem. *Z is Π_1^1-random $\Leftrightarrow$ $\omega_1^Z = \omega_1^{\mathrm{ck}}$ & Z is Δ_1^1-random.* □

Exercises.

9.3.10. Let $\mathcal{Q}$ be as in 9.3.6. Show that $\mathcal{Q} \cap \mathcal{S} \neq \emptyset$ for each nonempty Π_1^1 class $\mathcal{S}$.

9.3.11. (Martin-Löf, 1970) Let $\mathcal{G}$ be the union of all the null Δ_1^1 classes. Show that $\mathcal{G}$ is Π_1^1. (Thus, all the three higher randomness notions we study are Σ_1^1.)

9.3.12. (Hjorth and Nies, 2007) Let $\mathcal{R}^Y = 2^{\mathbb{N}} - \mathcal{Q}^Y$ be the class of Π_1^1-random sets relative to Y. Show an analog of the van Lambalgen Theorem 3.4.6: for every pair of sets X and Y we have $X \oplus Y \in \mathcal{R} \Leftrightarrow X \in \mathcal{R}^Y$ & $Y \in \mathcal{R}$.

9.3.13. Suppose $X \leq_K Y$ (see 5.6.1). Show the following. (i) If X is Δ_1^1-random then Y is Δ_1^1-random. (ii) If X is Π_1^1-random then Y is Π_1^1-random.
Hint. In (i) use that Z is Δ_1^1-random $\Leftrightarrow$ Z is ML-random in each hyperarithmetical set S. (This follows from results in Sacks (1990) about the sets H_n for ordinal notations n, together with 9.3.3(iii).)

9.4 Lowness properties in higher computability theory

We study some properties of sets that are closed downward under $\leq_h$, and relate them to higher randomness notions. The results in this section are due to Chong, Nies and Yu (2008).

Hyp-dominated sets

We consider an analog of being computably dominated. We will see that both Π_1^1-random sets and sets that are low for Π_1^1-randomness satisfy this property.

9.4.1 Definition. We say that A is *hyp-dominated* if each function $f \leq_h A$ is dominated by a hyperarithmetical function.

9.4.2 Fact. *A is hyp-dominated $\Rightarrow$ $\omega_1^A = \omega_1^{\mathrm{ck}}$.*

Proof. Let $g \le_T (O)'$ be as in Exercise 9.1.19. If $\omega_1^A > \omega_1^{ck}$ then $O \le_h A$ by Spector's Theorem 9.1.14, and hence $g \le_h A$. Let $f = \lambda n.\, g(n,n) + 1$, then $f \le_h A$, and $\exists n\, f(n) > h(n)$ for each hyperarithmetical function h. Therefore A is not hyp-dominated. □

Chong, Nies and Yu (2008) proved that the converse implication fails: if $G \subseteq \mathbb{N}$ is a member of each dense Δ_1^1 open set (this property is a higher analog of weak 1-genericity) then, adapting 1.8.48, p_G is not dominated by any hyperarithmetical function. A stronger genericity condition (corresponding to 1-genericity) implies that $\omega_1^G = \omega_1^{ck}$.

9.4.3 Theorem. *Each Π_1^1-random set A is hyp-dominated.*

Proof. Suppose the Π_1^1 relation $\mathcal{S}$ is a reduction procedure for $f \le_h A$ as in 9.1.12 such that $\{\langle n,y\rangle : \langle n,y,X\rangle \in \mathcal{S}\}$ is the graph of a partial function for each X. The function $n \mapsto \alpha_n = \min\{\alpha : \exists y \langle n,y,A\rangle \in \mathcal{S}_\alpha\}$ is Σ_1 over L_A, so by the Bounding Principle 9.1.22 relative to A and since $\omega_1^A = \omega_1^{ck}$, there is $\beta < \omega_1^{ck}$ such that $\alpha_n < \beta$ for each n.

Let $r(n)$ be the least k such that $2^{-n} \ge \lambda\{X : \exists i \ge k\, \langle n,i,X\rangle \in \mathcal{S}_\beta\}$. Note that $r(n)$ exists by our functionality assumption on $\mathcal{S}^X$, and that the function r is hyperarithmetical because S_β is Δ_1^1 and by the Measure Lemma 9.1.10(ii). Then $\{X : \exists^\infty n\, \exists i \ge r(n) [\langle n,i,X\rangle \in \mathcal{S}_\beta]\}$ is a null Δ_1^1 class. Since A avoids this class, f is dominated by r. □

As a consequence, by Corollary 9.3.7 there is a hyp-dominated set $Z \notin \Delta_1^1$ such that $Z \le_T O$ and $O^Z \le_h O$.

Exercises.

9.4.4. An analog of highness is being *hyp-high*, namely, some $f \le_h A$ dominates each hyperarithmetical function. Show that if A is not hyp-high, then $\omega_1^A = \omega_1^{ck}$. (This strengthens 9.4.2).

9.4.5. Show that there is a hyp-high set A such that $A \le_T O$ and $O^A \le_h O$ (and hence $\omega_1^A = \omega_1^{ck}$).

9.4.6. (Kjos-Hanssen, Nies, Stephan and Yu 20xx) We say that Z is weakly Δ_1^1-random if Z is in no *closed* null Δ_1^1 class. Show that Z is Π_1^1-random $\Leftrightarrow$ Z is hyp-dominated and weakly Δ_1^1-random.

Traceability

Traceability in the computability setting was studied in Section 8.2. In contrast to the computability setting, the analogs of c.e. and of computable traceability coincide, again because of the Bounding Principle. Moreover, this class characterizes lowness for Δ_1^1-randomness.

9.4.7 Definition. (i) Let h be a nondecreasing Δ_1^1 function. A Π_1^1 *trace* (Δ_1^1 *trace*) with *bound* h is a uniformly Π_1^1 (uniformly Δ_1^1) sequence of sets $(T_n)_{n\in\omega}$ such that $\forall n\, \#T_n \le h(n)$. $(T_n)_{n\in\omega}$ is a trace for the function f if $f(n) \in T_n$ for each n.
(ii) A is Π_1^1 *traceable* (Δ_1^1 traceable) if there is an unbounded nondecreasing hyperarithmetical function h such that each function $f \le_h A$ has a Π_1^1 trace (Δ_1^1 trace) with bound h.

The argument in Theorem 8.2.3 shows that the particular choice of the bound h does not matter in (ii). In the higher setting the two notions coincide:

9.4.8 Proposition. *If A is Π^1_1 traceable then A is Δ^1_1 traceable.*

Proof. First we show that $\omega_1^A = \omega_1^{\mathrm{ck}}$. Otherwise $A \geq_h O$, so it is sufficient to show that O is not Π^1_1-traceable. Since each Π^1_1 set is many-one reducible to O, there is a uniformly O-computable list $(T^e_n)_{e,n\in\mathbb{N}}$ of all the Π^1_1 traces for the bound $h(e) = e+1$. Define a function $g \leq_T O$ by $g(e) = \mu n.\, n \notin T^e_e$, then g does not have a Π^1_1 trace with bound h.

Now let $f \leq_h A$ and choose a reduction procedure $\mathcal{S}$ for f as in 9.1.12. Choose a Π^1_1-trace $(T_n)_{n\in\omega}$ for f with bound h. Let

$$\alpha_n = \min\{\alpha < \omega_1^A \colon \exists m \in T_{n,\alpha}\, [\langle n, m, A\rangle \in \mathcal{S}_\alpha]\}$$

Then by the Bounding Principle 9.1.22 relative to A there is $\beta < \omega_1^A = \omega_1^{\mathrm{ck}}$ such that $\alpha_n < \beta$ for each n. For a fixed $\beta < \omega_1^{\mathrm{ck}}$, the Approximation Lemma 9.1.4(i) provides an index for the Δ^1_1 set R_β uniformly in an index for a Π^1_1 set R. Therefore $(T_{n,\beta})_{n\in\mathbb{N}}$ is a Δ^1_1 trace as required. □

Chong, Nies and Yu (2008) showed that there are $2^{\aleph_0}$ Δ^1_1 traceable sets (an analog of 8.2.17). In fact each Sacks generic set for forcing with perfect Δ^1_1 trees (Sacks 4.5.IV) is Δ^1_1 traceable. There is a perfect class of generic sets in that sense. Also by Sacks 4.10.IV there a Sacks generic set $Z \leq_h O$. Then Z is Δ^1_1 traceable and $Z \notin \Delta^1_1$.

Δ^1_1 traceability characterizes lowness for Δ^1_1-randomness. The result is analogous to Theorem 8.3.3, and also to some of the equivalences in Theorem 8.3.9.

9.4.9 Theorem. *The following are equivalent for a set A.*

(i) A is Δ^1_1-traceable (or equivalently, Π^1_1 traceable).

(ii) Each null $\Delta^1_1(A)$ class is contained in a null Δ^1_1 class.

(iii) A is low for Δ^1_1-randomness.

(iv) Each Π^1_1-ML-random set is $\Delta^1_1(A)$-random.

Proof. (i) ⇒ (ii): This implication corresponds to Proposition 8.3.2. While the proof there relied on computable measure machines, here we give a direct proof.

Let $\mathcal{S}$ be a null $\Delta^1_1(A)$ class. By Theorem 9.3.3(i) relativized to A, $\mathcal{S} \subseteq \bigcap G_m$ for a Π^1_1-Schnorr test $\{G_m\}_{m\in\mathbb{N}}$ relative to A such that $\{\langle m, \sigma\rangle \colon [\sigma] \subseteq G_m\}$ is $\Delta^1_1(A)$ and $\lambda G_m = 2^{-m}$ for each m. We view the k-th finite set D_k as a subset of $\{0,1\}^*$. Define a function $f \leq_h A$ as follows: let $f(\langle m, s\rangle)$ be the least k such that $[D_k]^\prec \subseteq G_m$, $D_{f(\langle m,s-1\rangle)} \subseteq D_k$ if $s > 0$, $[\sigma] \subseteq G_m$ for $|\sigma| \leq s$ implies $\sigma \in D_k$, and $\lambda[D_k]^\prec > 2^{-m}(1 - 2^{-s})$. Let $G_{m,s} = [D_{f(\langle m,s\rangle)}]^\prec$, then $G_{m,s} \subseteq G_{m,s+1}$ and $G_m = \bigcup_{s\in\mathbb{N}} G_{m,s}$.

Let $T = (T_e)_{e\in\mathbb{N}}$ be a Δ^1_1 trace for f such that $\#T_e \leq e+1$ for each e. We define a Δ^1_1 trace $(\widehat{T}_e)_{e\in\mathbb{N}}$. By recursion on s, define $\widehat{T}_{\langle m,s\rangle}$ to be the set of $k \in T_{\langle m,s\rangle}$ such that $2^{-m}(1 - 2^{-s}) \leq \lambda[D_k]^\prec \leq 2^{-m}$ and $D_k \supseteq D_l$ for some $l \in \widehat{T}_{\langle m,s-1\rangle}$ (where $\widehat{T}_{\langle m,-1\rangle} = \mathbb{N}$). Let

$$V_m = \bigcup\{[D_k]^\prec \colon k \in \widehat{T}_{\langle m,s\rangle},\ s \in \mathbb{N}\}.$$

Then $\lambda V_m \le 2^{-m}\#\widehat{T}_{\langle m,0\rangle} + \sum_{s\in\mathbb{N}} 2^{-s-m}\#\widehat{T}_{\langle m,s\rangle}$. Since $\#\widehat{T}_{\langle m,s\rangle} \le \#T_{\langle m,s\rangle} \le \langle m,s\rangle + 1 \le (m+s+1)^2+1$, it is clear that $\lim_n \sum_{s\in\mathbb{N}} 2^{-s-m}\#\widehat{T}_{\langle m,s\rangle} = 0$, and hence $\lim_m \lambda V_m = 0$. Since $\forall m \forall s\, f(\langle m,s\rangle) \in \widehat{T}_{\langle m,s\rangle}$, we have $\forall m\, G_m \subseteq V_m$. So $\bigcap_m G_m$ is contained in the null Δ^1_1 class $\bigcap_m V_m$.
(ii) ⇒ (iii) and (iii) ⇒ (iv) are immediate.
(iv) ⇒ (i) is a straightforward adaptation of the implication "⇒" in Theorem 8.3.3. We are now given a function $g \le_h A$, and want to define a Π^1_1 trace $(T_n)_{n\in\mathbb{N}}$ for g with bound $\lambda n.2^n$. Now let $R = \underline{\mathcal{R}}_2$ as defined before 9.2.8. The measure-theoretic Lemma 8.3.4 can be used as before. We define a Π^1_1-Schnorr test $(U^g_d)_{d\in\mathbb{N}}$ relative to A. The definitions of $(\widetilde{T}_n)_{n\in\mathbb{N}}$ and $(T_n)_{n\in\mathbb{N}}$ are as before. The relation $\{\langle n,k\rangle\colon\ \lambda(B_{n,k} - \widetilde{R}) < 2^{-(k+4)}\}$ is now Π^1_1. □

For each set A there is a largest null $\Pi^1_1(A)$ class $\mathcal{Q}(A)$ by relativizing 9.3.6. Clearly $\mathcal{Q} \subseteq \mathcal{Q}(A)$; we say that A is low for Π^1_1-randomness if they are equal.

9.4.10 Lemma. *If A is low for Π^1_1-randomness then $\omega_1^A = \omega_1^{\mathrm{ck}}$.*

Proof. Otherwise, $A \ge_h O$ by Theorem 9.1.14. By Corollary 9.3.7 there is a Π^1_1-random set $Z \le_h O$, and Z is not even $\Delta^1_1(A)$ random. □

9.4.11 Open question. *Is each set that is low for Π^1_1-randomness in Δ^1_1?*

By the following result each such set is low for Δ^1_1-randomness. We say that A is *Π^1_1-random cuppable* if $A \oplus Y \ge_h O$ for some Π^1_1-random set Y.

9.4.12 Theorem. *A is low for Π^1_1-randomness ⇔*
(a) *A is not Π^1_1-random cuppable* & (b) *A is low for Δ^1_1-randomness.*

Proof. ⇒: (a) By 9.4.10 $A \not\ge_h O$. Therefore the $\Pi^1_1(A)$ class $\{Y\colon\ Y \oplus A \ge_h O\}$ is null, by relativizing Corollary 9.1.15 to A. Thus A is not Π^1_1-random cuppable.
(b) Suppose for a contradiction that Y is Δ^1_1-random but $Y \in \mathcal{C}$ for a null $\Delta^1_1(A)$ class $\mathcal{C}$. The union $\mathcal{D}$ of all null Δ^1_1 classes is Π^1_1 by 9.3.11. Thus Y is in the $\Sigma^1_1(A)$ class $\mathcal{C} - \mathcal{D}$. By the Gandy Basis Theorem 9.1.16 relative to A there is $Z \in \mathcal{C} - \mathcal{D}$ such that $\omega_1^{Z\oplus A} = \omega_1^A = \omega_1^{\mathrm{ck}}$. Then Z is Δ^1_1-random but not $\Delta^1_1(A)$-random, so by Theorem 9.3.9 and its relativization to A, Z is Π^1_1-random but not $\Pi^1_1(A)$-random, a contradiction.
⇐: By Fact 9.3.8 relative to A we have

$$\mathcal{Q}(A) = \{Y\colon\ \omega_1^{Y\oplus A} > \omega_1^A\} \cup \bigcup_{\alpha<\omega_1^A} \mathcal{Q}(A)_\alpha.$$

By hypothesis (a) $O \not\le_h A$ and hence $\omega_1^A = \omega_1^{\mathrm{ck}}$, so $\omega_1^{Y\oplus A} > \omega_1^A$ is equivalent to $O \le_h Y \oplus A$. If Y is Π^1_1-random then firstly $O \not\le_h Y \oplus A$ by (a), and secondly $Y \notin \mathcal{Q}(A)_\alpha$ for any $\alpha < \omega_1^A$ by hypothesis (b). Therefore $Y \notin \mathcal{Q}(A)$ and Y is $\Pi^1_1(A)$-random. □

SOLUTIONS TO THE EXERCISES

Solutions to Chapter 1

Answers to many of the exercises in this chapter can be found in Soare (1987) or Odifreddi (1989). Most of the solutions we provide are for exercises that are exceptions to this.

Section 1.1

1.1.12 Use an effective forth-and-back argument.

Section 1.2

1.2.25 $\Rightarrow$: By 1.2.22.

$\Leftarrow$: Let Z be the graph of f. The Turing functional Φ on input n with oracle set Y looks for the least $w \le h(n)$ such that $\langle n, w\rangle \in Y$. If found it outputs w, otherwise 0.

Section 1.4

1.4.7 (i) is like the proof of the Limit Lemma, substituting g for Z and $g_s(x)$ for $Z_s(x)$. (ii) $\Leftarrow$: Similar to the proof of the corresponding implication in 1.4.4.

$\Rightarrow$: If g is ω-c.e. we obtain a weak truth-table reduction of g to the change set by adapting the proof of the Limit Lemma. However, we cannot define a Boolean expression as in 1.4.4, because the outputs are now natural numbers.

1.4.8 Suppose $Z = \Phi^E$ for a Turing functional Φ with use function bounded by g. Given $i \in \mathbb{N}$ and a string σ of length $g(i)$, "$\Phi^\sigma(i) = 1$" is a Σ^0_1 property uniformly, so there is a computable binary function h such that $\Phi^\sigma(i) = 1 \leftrightarrow E$ satisfies the Boolean expression $h(i,\sigma)$. Then $i \in Z \leftrightarrow E$ satisfies the Boolean expression which is a disjunction over all σ of length $g(i)$ of expressions stating that σ is an initial segment of the oracle and the oracle satisfies $h(i,\sigma)$.

1.4.19 By the Recursion Theorem there is an e such that $W_e = \{e\}$. By the Padding Lemma 1.1.3 there is $i \neq e$ such that $W_i = W_e$; then $i \notin \emptyset'$.

1.4.20 For (i), (iii), and (iv) see Soare (1987). (ii) Choose a computable sequence $(Z_s)_{s\in\mathbb{N}}$ as in Proposition 1.4.17. If $n \notin Z_s$ then put $[0, s)$ into X_n.

1.4.22 By 1.4.20(ii) there is a uniformly c.e. double sequence $(X_{e,n})_{e,n\in\mathbb{N}}$ such that each $X_{e,n}$ is an initial segment of $\mathbb{N}$ and $e \in S \leftrightarrow \exists n\, X_{e,n} = \mathbb{N}$. Now let $A_{\langle e,n\rangle} = W_e \cap X_{e,n}$ [let $A_{\langle e,n\rangle} = V_e \cap X_{e,n}$], then $(A_k)_{k\in\mathbb{N}}$ is the required uniform listing of S.

1.4.23 Let U, V be disjoint Σ^0_3 sets that are effectively inseparable relative to $\emptyset''$ (see Soare 1987, pg. 44). Note that if $\widetilde{U}$ is Σ^0_3 and $U \subseteq \widetilde{U}$, $\widetilde{U} \cap V = \emptyset$ then $\widetilde{U}$ is Σ^0_3 complete. Fix a c.e. set $G \notin S$. Uniformly in e we will enumerate a c.e. set $A_e \subseteq G$ such that $e \in U \rightarrow A_e$ is computable, and $e \in V \rightarrow A_e =^* G$. There is a computable triple sequence $(x_{e,i,s})_{e,i,s\in\mathbb{N}}$, nondecreasing in s, such that $e \in U \leftrightarrow \exists n \lim_s x_{e,2n,s} = \infty$, and $e \in V \leftrightarrow \exists n \lim_s x_{e,2n+1,s} = \infty$. At stage $s > 0$, if i is least such that $x_{e.i,s} > x_{e,i,s-1}$ do the following. If i is even *restrain* A_e up to s with priority i. If i is odd put into A_e all $x \in G_s$ not restrained with priority $< i$.

Let $\widetilde{U} = \{e\colon A_e \in S\}$, then $\widetilde{U}$ is Σ^0_3 complete and m-reducible to the index set of S.

Section 1.5

1.5.5 Let $C' = \Theta(\emptyset')$ for a truth-table reduction Θ with use bound a computable function q. The set $B = \{\langle e, x\rangle\colon \Phi^C_e(x) = 1\}$ is c.e. in C, hence $B \le_m C'$ via some computable function r. If $Y = \Phi_e^C$ then $Y \le_{tt} \emptyset'$ because $x \in Y \leftrightarrow \Theta(\emptyset'; r(\langle e, x\rangle)) = 1$. A use bound is $h(x) = \max_{k \le x} q(r(\langle k, x\rangle))$ for all $x \ge e$, which suffices.

1.5.6 $\Rightarrow$: Let $f(x) = \max\{\Phi_e^A(x) \colon e \le x \ \&\ \Phi_e^A(x)\downarrow\}$. Since $f \le_T A' \equiv_T A \oplus \emptyset'$, the function f is as required. $\Leftarrow$: Since $\psi(x) \simeq \mu s\, J_s^A(x)\downarrow$ is partial computable in A, we have $J^A(x)\downarrow \leftrightarrow J^A_{f(x)}(x)\downarrow$ for almost all x. Thus $A' \le_T A \oplus \emptyset'$.

1.5.13 (a) By the Limit Lemma 1.4.2 relative to C, there is a C-computable approximation $(A_s)_{s\in\mathbb{N}}$ of A. If g is as in 1.5.12 then $g \le_T A$ since $C \le_T A$. Now argue as before in order to show that $A \le_T C$.

(b) If $A' \equiv_T C'$ then $C <_T A <_T C'$, so A is not computably dominated.

1.5.14 By 1.4.17 there is a computable sequence of strong indices $(A_s)_{s\in\mathbb{N}}$ such that $x \in A \leftrightarrow \forall^\infty x\,[x \in A_s]$. Define $g \le_T A$ by

$$g(\langle x, t\rangle) = \begin{cases} \mu s > t.\, x \notin A_s & \text{if } x \notin A, \\ t & \text{if } x \in A. \end{cases}$$

There is a computable function f such that $f(\langle x, t\rangle) \ge g(\langle x, t\rangle)$ for all x, t. Then $x \in A \leftrightarrow \exists t \forall u \in [t, f(\langle x, t\rangle))\, [x \in A_u]$ so A is c.e. and hence computable.

1.5.15 By 1.4.20 relative to A, the set $\mathsf{Tot}^A = \{e \colon \mathrm{dom}(\Phi_e^A) = \mathbb{N}\}$ is $\Pi^0_2(A)$-complete. Hence $A'' \equiv_T \mathsf{Tot}^A$ and Tot^A is $\Pi^0_1(A')$. Further,

$$e \in \mathsf{Tot}^A \leftrightarrow \exists i\, [\Phi_i \text{ total } \&\ \forall x\, \Phi^A_{e,\Phi_i(x)}(x)\downarrow].$$

Therefore Tot^A is $\Sigma^0_1(A' \oplus \emptyset'')$. This implies $A'' \equiv_T \mathsf{Tot}^A \le_T A' \oplus \emptyset''$.

1.5.16 Suppose the function $g \le_T A$ is not dominated by any computable function. We build a set $Y \le_T A$ such that $Y \not\le_{tt} A$. Suppose $Y \restriction_m$ has been defined. To define $Y(m)$, look for the least $e \le m$ such that $\Phi^A_{e,g(m)} \restriction_m = Y \restriction_m$ and $r = \Phi^A_{e,g(m)}(m)\downarrow$, and define $Y(m)$ to be not equal to r. If there is no such e define $Y(m) = 0$.
Clearly each e is considered at most once. If $Y \le_{tt} A$, then by Proposition 1.2.22 $Y = \Phi_e^A$ for some e such that the number of steps to compute $\Phi_e^A(m)$ is bounded by $t(m)$, for a strictly increasing computable function t. Since there are infinitely many m such that $g(m) \ge t(m)$ we cause $Y \ne \Phi_e^A$, contradiction.

1.5.17 (J. Miller) Clearly a computable set is of that kind. Now assume A is uniformly computably dominated via r. By 1.5.12 it suffices to show that $A \le_T \emptyset'$. In fact, $A \le_T \emptyset'$ follows already from the weaker hypothesis that $1 + \Phi_{r(e)}$ is not dominated by Φ_e^A for each e. Let p be a computable binary function such that for each e, n we have

$$\Phi^A_{p(e,n)}(x) \simeq \begin{cases} 1 + \Phi_{r(e)}(x) & \text{if } n \in A \\ 0 & \text{else.} \end{cases}$$

Let q be the function obtained when we require "$n \notin A$" in the first line instead. By the Recursion Theorem with Parameters there are computable functions h, l such that $\Phi^A_{p(h(n),n)} = \Phi^A_{h(n)}$ and $\Phi^A_{q(l(n),n)} = \Phi^A_{l(n)}$ for each n. If $n \in A$ then $\alpha_n = \Phi_{r(h(n))}$ is partial, otherwise $\Phi^A_{h(n)}$ would be total and dominate $1 + \alpha_n$. If $n \notin A$ then α_n is total. Similarly, $\beta_n = \Phi_{r(l(n))}$ is partial iff $n \notin A$. Using $\emptyset'$ as an oracle we can decide which of α_n, β_n is partial.

1.5.21 Relativize Theorem 1.5.19 to A.

Section 1.6

1.6.7 The strategy for the lowness requirements L_e is as before. To ensure that A' is not ω-c.e. we have to rule out each potential approximation $\Phi_i^2(x, t)$, with the computable bound $\Phi_j^1(x)$ on the number of changes, of A'. Let $U_{\langle i,j\rangle}$ be the corresponding

requirement. We build a Turing functional $\Gamma^Y(x)$ (see Section 6.1), and by Fact 1.2.15 and the Recursion Theorem for functionals we are *given* a reduction function p for Γ, namely $\forall Y \, \forall x \, \Gamma^Y(x) \simeq J^Y(p(x))$.
The strategy for $U_{\langle i,j\rangle}$ is as follows. Choose x greater than the last stage when $U_{\langle i,j\rangle}$ was initialized. Let $z = p(x)$. Wait for $n = \Phi_j^1(z)\downarrow$. Let $t^* = 0$. For at most $n+1$ times, do the following:
(a) whenever t is greater than t^* and $\Phi_i^2(z, t-1) = 1$, $\Phi_i^2(z,t) = 0$ then declare $\Gamma^A(x)\downarrow$ with large use u (in particular $u > \langle n+1, \langle i,j\rangle\rangle$). Wait for a stage v such that $J^A(p(x))\downarrow [v]$ and let $t^* = v$;
(b) whenever t is greater than t^* and $\Phi_i^2(z,t-1) = 0$, $\Phi_i^2(z,t) = 1$ then make $\Gamma^A(x)$ undefined by enumerating the greatest number $m \in \mathbb{N}^{[\langle i,j\rangle]}$ such that $m < u$ into A. Wait for a stage v such that $J^A(z)\uparrow [v]$ and let $t^* = v$.
If the strategy for $U_{\langle i,j\rangle}$ is no more initialized, it ensures that Φ_i^2 does not approximate A' with bound Φ_j^1. It only acts finitely often, so eventually it stops injuring the lowness requirements of weaker priority. Thus A is low.

1.6.10 See the main text.

Section 1.7

1.7.7 Given a set Z, let $\widehat{Z} = \{Z\restriction_n : n \in \mathbb{N}\}$, then $Z \equiv_{tt} \widehat{Z}$ and $\widehat{Z}$ is introreducible.

1.7.11 (Also see Soare 1987, pg. 283.)
$\Rightarrow$: Define a uniformly c.e. sequence $(G_e)_{e\in\mathbb{N}}$ as follows. When $x \in W_{e,\text{at } s}$ then put x into $G_{e,s}$ unless $x \in A_s$. By Corollary 1.1.11 there is a computable function q such that $G_e = W_{q(e)}$ for each e. Then q is as desired: the first two condition are immediate, and (2) follows from (1.11).
$\Leftarrow$: Fix some computable enumeration $(\widetilde{A}_s)_{s\in\mathbb{N}}$. We will speed up this enumeration in order to show that A is promptly simple. Let $A_0 = \emptyset$. Given $s > 0$, if there is x and e such that $x \in W_{e,\text{at } s}$, let t be the least stage such that $x \in \widetilde{A}_t$ or $x \in W_{q(e),t}$. This stage exists since $W_e - A = W_{q(e)} - A$. If the first alternative applies let $A_s = A_{s-1} \cup \widetilde{A}_t$, otherwise let $A_s = A_{s-1} \cup \widetilde{A}_s$. If W_e is infinite then for some x, s we have $x \in W_{e,\text{at } s}$ and $x \notin W_{q(e)}$. Then $x \in A_s$.

1.7.12 See Soare (1987, pg. 283).

Section 1.8

1.8.9 "$\Leftarrow$" is the use principle. For "$\Rightarrow$" let $A = \{\langle \sigma, x\rangle : \forall Z \succeq \sigma \, L(Z) \succeq x\}$.

1.8.12 For each pair of sets X, Y we have $\text{abs}(F(X) - F(Y)) \le d(X,Y)$ (see 1.8.7 for the definition of the distance function d), so $\widetilde{F}$ is continuous. To show that $\widetilde{F}^{-1}$ is continuous, note that for any cylinder $[\sigma]$ we have

$$\widetilde{F}([\sigma] \cap \mathcal{X}) = (0.\sigma, 0.\sigma + 2^{-|\sigma|}) - \mathbb{Q}_2,$$

which is open in $[0,1)_\mathbb{R} - \mathbb{Q}_2$.

1.8.13 The homeomorphism is given by $1^{n_0}01^{n_1}01^{n_2}0\ldots \to (n_0, n_1, n_2, \ldots)$.

1.8.18 Given $k \in \mathbb{N}$, for each $i \le k$ we may compute $q_i \in \mathbb{Q}_2$ such that $0 \le r_i - q_i < 2^{-k-i-1}$. Then $0 \le r - \sum_{i \le k} q_i < 2^{-k+1}$.

1.8.40 At stages $2e+2$, $e = nk + i$, $0 \le i < k$, in (a) work towards $B_i \not\le_T Y$. .

1.8.41 For a string τ, let

$$Q_\tau = \{Y \in P : \forall e < |\tau| \, [\tau(e) = 0 \;\rightarrow\; J^{Y\oplus B}(e)\uparrow]\}.$$

Note that Q_τ is a $\Pi_1^0(B)$ class uniformly in τ. Emptyness of such a class is a $\Sigma_1^0(B)$ condition, so there is a computable function g such that $Q_\tau = \emptyset \;\leftrightarrow\; g(\tau) \in B'$. Thus, there is a Turing functional Ψ such that Ψ^X is total for each oracle X and $\Psi(B';e) = \tau_e$, where τ_e is the leftmost string τ of length $e+1$ such that $Q_\tau \neq \emptyset$. Let $Y \in \bigcap_e Q_{\tau_e}$. Then $e \in (Y \oplus B)' \leftrightarrow \tau_e(e) = 1$, so $(Y \oplus B)' \leq_{tt} B'$.

1.8.46 *Construction relative to* $\emptyset'$ *of* Π^0_1 *classes* $(P^i)_{i\in\mathbb{N}}$. Let $P^0 = P$.

Stage $2i+1$. If $P^{2i} \cap \{X\colon J^X(i)\uparrow\} \neq \emptyset$, then let P^{2i+1} be this class. Otherwise, let $P^{2i+1} = P^{2i}$.

Stage $2i+2$. See whether there is $e \leq i$ which has not been active so far such that for some $m \leq c_B(i)$ we have $Q^i_{e,m} := P^{2i+1} \cap \{X\colon \Phi^X_e(m)\uparrow\} \neq \emptyset$. If so let e be least, let m be least for e, and let $P^{2i+2} = Q^i_{e,m}$. Say that e is *active*. Otherwise, let $P^{2i+2} = P^{2i+1}$.

Verification. Let $Y \in \bigcap_r P^r$. Since B can determine an index for each P^r, we have $Y' \leq B$ by the usual argument of the Low Basis Theorem. Each e is active at most once, and if so then Φ^Y_e is partial. Suppose now that Φ^Y_e is total. We claim that there is r such that Φ^Z_e is total for each $Y \in P^r$, and therefore Φ^Z_e is computably dominated by the argument in the proof of 1.8.42. If the claim fails then $B \leq_T \emptyset'$, as follows. Let s_0 be a stage such that no $j < e$ is active from s_0 on. Given $i \geq s_0$, using the oracle $\emptyset'$ find the least m such that $Q^i_{e,m} \neq \emptyset$. Then $c_B(i) \leq m$ (otherwise we would now ensure $\Phi^Y_e(m)$ is undefined), so that $B_m \upharpoonright_i = B \upharpoonright_i$.

1.8.54 Given a string σ and $n \in \mathbb{N}$, $\emptyset'$ can decide whether $J^Y(n)\uparrow$ for each $Y \succeq \sigma$ (by 1.8.28). To decide whether $n \in G'$ using $G \oplus \emptyset'$ as an oracle, search for the least string $\sigma \prec G$ such that either $J^\sigma_{|\sigma|}(n)\downarrow$ or $J^Y(n)\uparrow$ for each $Y \succeq \sigma$. In the first case output "yes", in the second case "no".

1.8.63 Y is c.e. iff $\exists e \forall n, s\, \exists t > s\, W_{e,t}(n) = Y(n)$. For the computable sets apply this and 1.1.9.

1.8.64 Suppose the class of c.e. sets is contained in a Π^0_2 class $S = \bigcap_n R_n$ where the R_n are uniformly Σ^0_1. Then each R_n is dense, so S contains every weakly 1-generic set. The same argument works for the computable sets. Alternatively, if the class of computable sets is Π^0_2, then the index set $\{e\colon W_e \text{ is computable}\}$ is in $\Sigma^0_3 \cap \Pi^0_2(\emptyset')$, and hence in Δ^0_3. This contradicts the fact that this index set is Σ^0_3 complete, Exercise 1.4.20(iv).

1.8.65 Let $\mathcal{A} = \bigcap_y \mathcal{B}_y$ where the class $\mathcal{B}_y$ is Σ^0_1 uniformly in y. Then $\mathcal{B}^*_y = \{Z\colon \exists m\, [\Psi^{Z\upharpoonright_m}] \subseteq \mathcal{B}_y\}$ is Σ^0_1 uniformly in y, so $\mathcal{C} = \{Z\colon \Psi^Z \text{ is total}\} \cap \bigcap_y \mathcal{B}^*_y$ is Π^0_2.

1.8.66 Let $\mathcal{B} = \bigcup_n \mathcal{A}_n$ for an effective sequence of Π^0_2 classes. For each n, e let $\mathcal{C}_{n,e}$ be the class uniformly obtained via 1.8.65 from $\mathcal{A}_n$ and $\Psi = \Phi_e$. Then $\mathcal{G} = \bigcup_{n,e} \mathcal{C}_{n,e}$ is Σ^0_3.

1.8.67 (i) Suppose $\mathcal{C} = \{Z\colon \forall n\, Z\upharpoonright_n \in R\}$ where $R \leq_T \emptyset'$. By the Limit Lemma choose a computable approximation $(R_t)_{t\in\mathbb{N}}$ for R. Then $\mathcal{C} = \{Z\colon \forall n \forall s\, \exists t > s\, Z\upharpoonright_n \in R_t\}$, so $\mathcal{C}$ is Π^0_2. (ii) Every $\Pi^0_1(\emptyset')$ singleton is Δ^0_2 by Fact 1.8.33.

1.8.71 $\Rightarrow$: If there is σ such that σu is on T_P for infinitely many u, then $\{f\colon \sigma \not\prec f\} \cup \bigcup_u [\sigma u]$ is an open covering of P without a finite subcovering.
$\Leftarrow$: If $T_P \subseteq \{\sigma \in \mathbb{N}^*\colon \forall i < |\sigma|\, [\sigma(i) < g(i)]\}$ for a function g, then P is a closed subset of the space $\prod_i \{0, \ldots, g(i)\}$. This space is compact by Tychonoff's Theorem.

1.8.72 Adapt the proof of 1.8.37. By the hypothesis on T_P the set $\emptyset'$ can decide whether a Π^0_1 subclass Q of P is empty. Thus we can define a $\emptyset'$-computable sequence $(P^e)_{e\in\mathbb{N}}$ as before. Its intersection is nonempty since $P^0 = P$ is compact.

Section 1.9

1.9.5 $\mu_r(\emptyset) = 0$ since $\emptyset$ is open. Monotonicity is straightforward. Countable subadditivity is first verified for families of open sets. In the general case, given a family

$(\mathcal{C}_i)_{i\in\mathbb{N}}$ and $\epsilon > 0$, chose open sets $\mathcal{A}_i \supseteq \mathcal{C}_i$ such that $\mu_r(\mathcal{A}_i) \le \mu_r(\mathcal{C}_i) + \epsilon 2^{-i}$. Then $\mu_r(\bigcup_i \mathcal{C}_i) \le \mu_r \bigcup_i(\mathcal{A}_i) \le \sum_i \mu_r(\mathcal{A}_i) \le 2\epsilon + \sum_i \mu_r(\mathcal{C}_i)$.

1.9.13 If $\epsilon > 0$ and $2^{-n}\#(S \cap \{0,1\}^n) \ge \epsilon$ then the strings of length n contribute at least ϵ to $\lambda[S]^{\prec}$. So there are at most $1/\epsilon$ such n.

1.9.14 Assume for a contradiction that $\mathcal{U}$ is measurable. Since $X \in \mathcal{U} \leftrightarrow \mathbb{N} - X \notin \mathcal{U}$ and the complementation operation on subsets of $\mathbb{N}$ preserves λ, we have $\lambda\mathcal{U} = 1/2$. But $\mathcal{U}$ is closed under finite variants whence $\lambda\mathcal{U} \in \{0,1\}$ by the zero-one law.

1.9.15 (J. Miller) For a class $\mathcal{C}$ let $I_{\mathcal{C}}$ denote the function given by $I_{\mathcal{C}}(Z) = 1$ if $Z \in \mathcal{C}$ and $I_{\mathcal{C}}(Z) = 0$ otherwise. Consider the function $g\colon 2^{\mathbb{N}} \to \{0,\dots,N\}$ defined by $g(Z) = \#\{i\colon Z \in \mathcal{C}_i\} = \sum_{i=1}^{N} I_{\mathcal{C}_i}(Z)$. Since $\int_{2^{\mathbb{N}}} g(Z)\lambda(dZ) \ge \sum_{i=1}^{N} \epsilon > k$, the class $\{Z\colon g(Z) > k\}$ is not a null class.

1.9.20 Let $(q_n)_{n\in\mathbb{N}}$ be a nondescending computable sequence of rationals such that $\lim_n q_n = r$, Fact 1.8.15. Introducing some repetitions if necessary, we may assume that each q_n is of the form $k2^{-n}$ for some $k \in \mathbb{N}$. We define an effective sequence $(C_n)_{n\in\mathbb{N}}$ of clopen sets such that $\lambda C_n = q_n$ and $C_n \subseteq C_{n+1}$ for each n, and $R = \bigcup_n C_n$ is the required c.e. open set. Each set C_n is generated by strings of length k_n, where $k_n < k_{n+1}$ and the sequence $(k_n)_{n\in\mathbb{N}}$ is computable. If $|\sigma| = k_n$ then $[\sigma] \subseteq C_n \leftrightarrow [\sigma] \subseteq R$, so that A_R is computable. Let $k_0 = 0$ and $C_0 = \emptyset$.

Step $n > 0$. Let $b \in \mathbb{N}^+$ be least such that $q_n - q_{n-1} \le (1 - 2^{-b})(1 - q_{n-1})$, and let $k_n = k_{n-1} + b$. Pick a clopen set D such that $\lambda D = q_n - q_{n-1}$, $D \cap C_{n-1} = \emptyset$ and D is given by strings of length k_n none of which are of the form $\sigma 1^b$ for $|\sigma| = k_{n-1}$. Let $C_n = D \cup C_{n-1}$.

1.9.21 Define a computable $Z \in P$, $Z = \bigcup_n \sigma_n$, as follows. Let $\sigma_0 = \varnothing$. If σ_n is defined such that $\lambda(P \mid \sigma_n) > 0$, then let σ_{n+1} be the leftmost extension σ of σ_n of length $n+1$ such that $\lambda(P \mid \sigma) > 0$. Use 1.9.18 to show that we can compute the sequence $(\sigma_n)_{n\in\mathbb{N}}$.

1.9.22 For $n = 1$, this is Fact 1.9.16. Now let $n > 1$, and inductively assume the result holds for Π^0_{n-1} classes. A Σ^0_n class $\mathcal{C}$ is an effective union of Π^0_{n-1} classes $\mathcal{D}_i$ where $\mathcal{D}_i \subseteq \mathcal{D}_{i+1}$ for each i. Each $\lambda\mathcal{D}_i$ is left-Π^0_{n-1} and hence left-Σ^0_n, uniformly in i. Thus $\lambda\mathcal{C} = \lim_i \lambda\mathcal{D}_i$ is left-Σ^0_n. The case of Π^0_n classes is similar.

Solutions to Chapter 2

Section 2.1

2.1.3 Consider the machine M given by $M(\tau) \simeq \mathbb{V}(\mathbb{V}(\tau))$. Each $\mathbb{V}$-description of σ is an M-description of x, hence $C_M(x) \le C(\sigma)$. If $M = \Phi_b$, $b > 0$, then $C(x) \le C_M(x) + b$, so that $|\sigma| = C(x) \le C(\sigma) + b$.

2.1.4 When the computation $\mathbb{V}(\sigma) = x$ converges at a stage, declare $R(\sigma) = x$ unless x already has an R-description of the same length at a previous stage, or there is a $\mathbb{V}$-description $\rho <_L \sigma$, $|\rho| = |\sigma|$, such that $\mathbb{V}(\rho) = x$ converges at the same stage.

2.1.5 If $\mathbb{V}(\sigma)\downarrow$ and $|\sigma| \equiv i \mod d$ for $0 \le i < d$ then let $R(0^{d-i-1}1\sigma) = \mathbb{V}(\sigma)$. Clearly $C_R(x) \le C(x) + d$ for each x.

2.1.6 Let $b = 2 + d$ where d is the coding constant for the copying machine Φ_1 relative to R. For each y of length $n+1$, since $\Phi_1(y) = y$, there is an R-description σ_y of y such that $|\sigma_y| < n + b$. Fewer than 2^n of the σ_y's have length less than n. The remaining ones show that $2^n \le s_{n,b}$. The inequality $s_{n,b} < 2^{n+b}$ is trivial.

2.1.7 Use the machine M that on input $a\sigma$, $a \in \{0,1\}$, first simulates $\mathbb{V}(\sigma)$. If $y = \mathbb{V}(\sigma)$ where $|y| = n$, it outputs $0y_0 0y_1 \dots 0y_{n-1}$ in case $a = 0$, and $0y_0 0y_1 \dots 0y_{n-1}0$ in case $a = 1$.

2.1.17 By Fact 2.1.13 and since $d \geq 2$, we have $2^n \geq r_n > 2^n - 2^{n-d+1} \geq 2^{n-d+1}$. Thus $n \geq \log r_n \geq n - d + 1$, whence we can obtain n from r_n and a constant number of bits.
Let w_n be the leftmost d-incompressible$_C$ string of length n. Since the set of d-compressible$_C$ strings is c.e., if we know r_n and n, we can compute all the d-compressible$_C$ strings of length n, and therefore determine w_n. Thus $n - d < C(w_n) \leq^+ C(r_n)$. (The cases $d = 0, 1$ would be interesting to settle in view of Ex. 2.1.19(ii).)
2.1.18 Let $D(x) = 2m$ for each x of length $2m$, and also $D(0^{2m+1}) = 2m$. Then D satisfies the counting condition. Given a machine M, for each m there is x of length $2m$ or $2m+1$ such that $D(x) = 2m$ but $C_M(x) = 2m$ fails. (C_M satisfies the stronger variant of the counting condition $\forall i\, \#\{x\colon C_M(x) = i\} \leq 2^i$.)
2.1.19 (i) Since R is optimal, we can choose $d \in \mathbb{N}$ such that $C_R(x) < |x| + d$ for each x, and $C_R(zz) < C_R(z) + d$ for each z. Let $P(y)$ be the leftmost among the shortest R-descriptions of a string y.
Now assume for a contradiction that $n > d$ and $C_R(x) \leq |x|$ for all x such that $n \leq |x| < n + d$. Then $|x| < n + d$ implies $C_R(x) < n + d$, so the 1-1 function P is onto when restricted to the set of strings of length $< n + d$. On the other hand, if z is any string of length n and $y = zz$, then $|P(y)| = C_R(y) < n + d \leq |y|$, contrary to the fact that P is 1-1.
(ii) (Cai) Suppose that n is a number such that $C_R(x) \leq |x|$ for each string x of length n. By assumption we also have $C_R(y) \leq |y| + 1$ for each string y, so $C_R(x) \leq n$ for each string x of length *at most* n. This is only possible if $R(\sigma)$ is defined for each string σ of length at most n.
If there are infinitely many n like that, then the machine R is total, and hence C_R is computable. This contradicts the optimality of R by Cor. 2.1.29.
(iii) Let S be the machine defined by $S(0^i 1\sigma) \simeq \mathbb{V}(\sigma)$ if $i < b$ and b divides $|\sigma| + i + 1$.
2.1.23 Let d be a constant such that $\forall z\, C(z+1) \leq C(z) + d$. If $C(z_{n+d}) < n$ we have $C(z_{n+d} + 1) < n + d$, contrary to the fact that z_{n+d} is the largest such number.
2.1.24 Since $2^n \leq s_{n,b} < 2^{n+b}$, we have $n \leq \log s_{n,b} \leq n + b$. So we can describe n by $s_{n,b}$ and a constant number of bits. Since R is partial computable, if we know $s_{n,b}$ (and hence n), we can evaluate all converging computations $R(\sigma)$ for $n \leq |\sigma| < n + b$. Since $C(z_{n+b}) \geq n$ by the previous exercise, z_{n+b} is the largest value obtained this way. Hence $n \leq C(z_{n+b}) \leq^+ C(s_{n,b})$.
2.1.25 (Stephan) Let L be the machine that outputs the number of steps $\mathbb{V}$ needs to stop. That is, $L(\sigma) \simeq \mu s\, \mathbb{V}_s(\sigma)\downarrow$. (Of course, we let the computation $L(\sigma)$ converge at this same stage s.) By convention, s is greater than the output of $\mathbb{V}$ on input σ.
The machine R is the "join" of $\mathbb{V}$ and L, namely, $R(0\sigma) \simeq \mathbb{V}(\sigma)$ and $R(1\sigma) \simeq L(\sigma)$. Note that $C_L(x)$ can only decrease at stage x, and only from ∞ to a finite value. We use this to show that

$$x \in D_R \;\Leftrightarrow\; \exists y > x\, \exists s \geq y\, C_{R,s}(y) \leq C_{R,s}(x),$$

whence D_R is c.e. The implication from left to right is clear. For the converse implication, if $C_{R,s}(x)$ is not the final value, let $t > s$ be greatest such that $C_R(x)$ decreases at stage t. This is because $C_{\mathbb{V}}(x)$ decreases. Hence $C_R(t) \leq C_L(t)+1 \leq C_{\mathbb{V}}(x)+1 \leq C_R(x)$. Thus, t is the required witness showing that $x \in D_R$.

2.1.26 Let y_n be as in the proof of Proposition 2.1.22 for C_R, and note that $p_{\overline{D_R}}(n) \geq y_n$. If $n \to r_n$ is an increasing computable function, then the argument in that proof shows that $y_n > r_n$ for almost all n. So $p_{\overline{D_R}}$ dominates the function $n \to r_n$.

2.1.30 A is c.e. via the computable enumeration given by $A_s = \{\langle x, n\rangle : C_s(x) \leq n\}$, and $B \leq_m A$ for the weak truth-table complete set B from 2.1.28.

2.1.31 Otherwise $C(x) = C_{g(|x|)}(x)$ for almost all x, so C is computable. There are infinitely many x such that $C(x) = C^g(x) = |x| + 1$, so this cannot be improved to $\forall^\infty x\, C(x) < C^g(x)$.

Section 2.2

2.2.3 There are $2^n(n+1)$ pairs $\langle x, y\rangle$ such that $|xy| = n$, so one of them must satisfy $C(\langle x, y\rangle) \geq^+ n + \log n$. Clearly $n + \log n \geq^+ C(x) + C(y) + \log n$.

2.2.10 It suffices to carry out the obvious modifications of the solutions for C.

2.2.11 Let $n = |x|$. Then $K(n) \leq^+ |n| + 2\log|n|$. So

$$K(x) \leq K(n) + |x| + 1 \leq^+ 2\log\log n + \log n + |x|.$$

2.2.12 (i) $\mathcal{C}$ is the set of paths of the Π^0_1 tree

$$B = \{x \colon \forall\sigma\,\forall\rho \neq \sigma\,[x(\sigma) = x(\rho) = 1 \to \sigma \mid \rho] \;\&\; \forall\sigma\,[U(\sigma)\downarrow \to x(\sigma) = 1]\}.$$

Now use the Low Basis Theorem.

(ii) For $n < m \leq \infty$, let $S^{n,m} = \{x \in S : n \leq |x| < m\}$. Note that $\lim_n \lambda[S^{n,\infty}]^{\prec} = 0$. We build a prefix-free machine M. A number d is given which, by the Recursion Theorem (and since the standard universal prefix-free machine is simulated by U), we think of as a coding constant for M relative to U (see Remark 2.2.21). *Compute n such that* $\lambda[S^{n,n+d+1}]^{\prec} \leq 2^{-d-2}$ (this uses that S is Π^0_1). *Declare $M(y) = y$ for all y of length n.* If d actually is a coding constant for M then, for each y of length n, there is a string σ_y of length at most $n + d$ such that $U(\sigma_y) = y$. Clearly, at most half of the σ_y's have length $< n$. The remaining ones contribute a measure of at least $2^{n-1}2^{-d-n} = 2^{-d-1}$ to $\lambda[S^{n,n+d+1}]^{\prec}$, contrary to the choice of n.

2.2.22 We may assume $\forall n\, f(n) \leq n$. If the sum is finite, for large enough c the function $D(x) = K(x) - \log f(K(x)) + c$ satisfies the weight condition and is computably approximable from above. Now 2.2.19 yields a contradiction.

2.2.24

(i) Let k be least such that $2^{-k+1} \leq \delta$. Let i be least such that $i2^{-k} \geq p$. Then $(i+1)2^{-k} < p + \delta$. Thus, if w is the string of length k such that $0.w = i2^{-k}$ then $I(w) \subseteq [i2^{-k}, (i+1)2^{-k}) \subseteq [p, p+\delta)$. Also $\delta < 2^{-k+2}$ by choice of k.

(ii) As before let the sequence $\langle r_n, y_n\rangle_{n<N}$ list the bounded request set W. To define the prefix-free machine M, at stage n let $p = \sum_{i<n} 2^{-r_i}$, $\delta = 2^{-r_n}$ and compute $w = w_n$ as in (i). Let $M(w_n) = y_n$.

2.3.4 "$\leq^+$" is trivial. For "$\geq^+$", consider the machine given by $M(\sigma) \simeq \langle \mathbb{V}(\sigma), |\sigma|\rangle$.

2.3.5 The binary machine M on inputs τ, z searches for a decomposition $\tau = \sigma\rho$ such that $x = \mathbb{U}^2(\sigma, \mathbb{U}^2(\rho, z))$ converges. If one is found it outputs x. Verify that M is prefix-free when fixing the second component.

2.5.10 For each n and ϵ let $B_{n,\epsilon} = \{x \in \{0,1\}^n \colon \mathrm{abs}(S_n(x)/n - 1/2) \geq \epsilon\}$. Then $B_{n,\epsilon}$ is the event that the number of ones differs by at least $n\epsilon$ from the expected value $n/2$. By the Chernoff bounds (2.17) already used in the proof of Theorem 2.5.8, we have $P(B_{n,\epsilon}) \leq 2e^{-2n\epsilon^2}$. We are interested in the case that $n\epsilon = \sqrt{n \ln n}$, so let $\epsilon_n = \sqrt{\ln n / n}$. Then $2e^{-2n\epsilon_n^2} \leq 2n^{-2}$. Define the computable set G by letting

$G_n = B_{n,\epsilon_n}$. Then $\#G_n \le 2^{n+1}/n^2$. Let $W = \{\langle n - \log n, x\rangle : n \ge c \ \&\ x \in G_n\}$. For each k, there are at most two n such that $n - \log n = k$. Thus, if c is chosen appropriately, for each k there are at most 2^k requests in W with first component k. Now, by Proposition 2.1.14, $C(x) \le^+ n - \log n$ for each $x \in G_n$. Also see Li and Vitányi (1997, Lemma 2.2 on pg. 128).

Solutions to Chapter 3

Section 3.2

3.2.25 Once again, split off a prefix-free descriptions (see the beginning of Section 2.4). On input τ, the machine N first searches for a decomposition $\tau = \sigma z$ such that $n = \mathbb{U}(\sigma)\downarrow$ and $|z| = n$. Once it finds such a decomposition, N prints $z0^{|\sigma|-r}$. If n is the starting position of a block of zeros of length $K(n) - r$, then for $m = n + K(n) - r$, $\sigma = n^*$, and $z = Z\restriction_n$ we have $N(\sigma z) = Z\restriction_m$, so $K_N(Z\restriction_m) \le^+ m$.

3.2.26 Otherwise, there is d such that $Z \in [A_{m,d}]^{\prec}$ for infinitely many m, where $A_{m,d}$ is the set of strings of length m defined in (2.16) on page 101. But $([A_{m,d}]^{\prec})_{m\in\mathbb{N}}$ is a Solovay test by (2.17) and since the geometric series $\sum_m e^{-2m/d^2}$ converges.

3.2.33 See Downey, Hirschfeldt and Nies (2002).

Section 3.3

3.3.3 Given a number $d > 1$, we build a bounded request set L (we think of d as a coding constant for L provided by the Machine Existence Theorem 2.2.17). Let $c = d + 1$. By (1.17) on page 55, each Π^0_1 class P (given by an index for Π^0_1 classes) has an effective approximation $(P_s)_{s\in\mathbb{N}}$ by clopen sets P_s, which are given by a strong index for a set of strings of length s. The hypothesis "$\lambda(P^e \cap Q) \le 2^{-K(e)-c}$" is a Σ^0_1-property of e, because $\lambda(P^e_t \cap Q_t)$ is nonincreasing in t and converges to $\lambda(P^e \cap Q)$, while $2^{-K_t(e)-c}$ is nondecreasing and converges to $2^{-K(e)-c}$.

Construction of L. If t is least such that $\lambda(P^e_t \cap Q_t) \le 2^{-K_t(e)-c}$, then for each $y \in P^e_t \cap Q_t$ enumerate a request $\langle |y| - c, y\rangle$ into L at stage t.

The enumeration for P^e adds a weight of at most $2^{-K(e)}$ to L. Then, for *any* given d, L is a bounded request set since $\sum_e 2^{-K(e)} \le 1$. Now suppose that, by Remark 2.2.21, $d > 1$ is a coding constant for L (note that this fixed point, and hence c, can be found effectively). If $\lambda(P^e \cap Q) \le 2^{-K(e)-c}$ then at some stage t the enumeration into L causes $K(Z\restriction_t) \le t - 1$ for each $Z \in P^e \cap Q$. This contradicts $Q = 2^{\mathbb{N}} - \mathcal{R}_1$, hence $P^e \cap Q = \emptyset$.

3.3.4 Apply 3.3.3 to the particular effective listing of Π^0_1 classes given by $P^x = [x]$.

Section 3.4

3.4.3 (i) Use the binary machine $M(\sigma, \alpha) \simeq \mathbb{V}^\alpha(\sigma)$.
(ii) Let $\alpha = 0^n$, then $C^\alpha(n) =^+ C(n)$.

3.4.4 Suppose that $A = \Phi(B)$ for a Turing functional Φ. Define a prefix-free oracle machine M by $M^X(\sigma) \simeq \mathbb{U}^{\Phi(X)}(\sigma)$. Namely, on input σ, M looks for $\alpha \prec \Phi(X)$ such that $\mathbb{U}^\alpha(\sigma) = y$ and, if it is found, outputs y. Then $\forall y\, K^B(y) \le^+ K_{M^B}(y) =^+ K^A(y)$.

3.4.9 For "$\Leftarrow$" it suffices to adapt the proof of this implication in 3.4.6. $(V_d)_{d\in\mathbb{N}}$ is now a ML-test relative to $\emptyset^{(k-1)}$, and $(S_d)_{d\in\mathbb{N}}$ is a Solovay test relative to $\emptyset^{(k-1)}$. The open sets H_d are uniformly c.e. in $B \oplus \emptyset^{(k-1)}$. If $A \in H_d$ for each $d \ge d_0$, then A is not ML-random relative to $B^{(k-1)}$.

The proof of "$\Rightarrow$" needs some additional idea; see van Lambalgen (1987).

3.4.14 If Z is ω-c.e., then f is ω-c.e. in the sense of Exercise 1.4.7, and hence $f \leq_{wtt} \emptyset'$. Let

$$\widehat{G}_e = \begin{cases} [Z_s \upharpoonright_{e+1}] & \text{if } s \text{ is the stage such that } J_s^{A_s}(e) \text{ converges } A\text{-correctly} \\ \emptyset & \text{if } J^A(e)\uparrow . \end{cases}$$

Then for almost all e, $J^A(e)\downarrow \leftrightarrow J^A(e)[f(e)]\downarrow$, which shows that $A' \leq_{wtt} \emptyset'$.

3.4.15 (Greenberg and Stephan) The set $S = \{e\colon\ J(e)\downarrow\ \&\ J(e) \in X\}$ is $\Sigma_1^0(X)$, so there is a computable increasing function f such that $e \in S \leftrightarrow f(e) \in X'$. If X is superlow we can fix a Turing functional Γ with use bound an increasing computable function g such that $X' = \Gamma(Z)$. Since $J(e) \simeq \Phi_e(e)$, we can pick a computable sequence of numbers $a_0 < a_1 < \ldots$ such that

$$g(f(a_n)) < J(a_n) < a_{n+1}$$

for each n: we let a_n be an index e such that $\Phi_e(x) = g(f(x)) + 1$ for each x and, if $n > 0$, then $e > J(a_{n-1})$ by the usual "padding".
Now fix n and let $r = J(a_n)$. If $\sigma \prec Z$ is a string such that $|\sigma| = 2r$, then $Z(2r) = X(J(a_n)) = X'(f(a_n)) = \Gamma^Z(f(a_n))$, where the use bound is $g(f(a_n)) < r$. Thus $Z(2r)$ can be computed from $Z\upharpoonright_r$. Since $J(a_0) < J(a_1) < \ldots$ is a computable set, this implies that Z is not ML-random, similar to Proposition 3.3.6 (that a ML-random set is not autoreducible). Note that in fact, Z is not even partial computable random as defined in 7.4.2.

3.4.21. If $0^n1 \prec X$, let $M^X(\sigma)\downarrow$ for all strings σ of length n such that $\sigma \neq 1^n$. Thus $\Omega_M^X = 1 - 2^{-n}$. We have $r_1 = 1$, and this supremum is not assumed by Ω_M.

3.4.22 By Fact 3.4.19, for every $\delta > 0$, $\{X\colon \operatorname{abs}(\Omega_M^A - \Omega_M^X) < \delta\} = \{X\colon \Omega_M^X > \Omega_M^A - \delta\}$ is open.

3.4.23 If the real r is right-c.e. and r is left-c.e. relative to A, then $r \leq_T A$, so r is not ML-random relative to A. Therefore, if $r = 1 - \Omega = \Omega^A$ then r is not ML-random in A, contradiction.

Section 3.5

3.5.6 Clear when B is finite. Otherwise, consider the Π_1^0 class $P = \{Z\colon B \subseteq Z\}$. For each finite set F, we have $\lambda\{Z\colon F \subseteq Z\} \leq 2^{-\#F}$. Thus P is a null class containing B.

3.5.7 By Theorem 3.3.7, if $Z \in P$ is not autoreducible then the Π_1^0 class P is uncountable.

3.5.16 Let L_0 be the bounded request set $\{\langle|\sigma|, x\rangle\colon M(\sigma) = x\}$, and let L_1 be a bounded request set with total weight $1 - \Omega_M$. Then the total weight of $L = L_0 \cup L_1$ is 1, so the prefix-free machine for L is as required.

3.5.17 Compute a stage s such that $\Omega_M - \Omega_{M,s} < 2^{-r}$. Then $\{y\colon K_M(y) \leq r\} = \{y\colon K_{M,s}(y) \leq r\}$ as in the proof of Proposition 3.5.15.

Section 3.6

3.6.6 Relativize Proposition 3.5.5 to $\emptyset'$.

3.6.7 Proceed similar to the solution for 3.5.6, but relative to $\emptyset'$.

3.6.9 Barmpalias, Downey and Ng (20xx) show that there is a weakly 2-random set $A \oplus B$ such that A is not weakly 2-random relative to B (and B is not weakly 2-random relative to A). Thus, one implication of van Lambalgens theorem fails for weak 2-randomness. However, they also show that the other implication holds: if B is weakly 2-random and A is weakly 2-random relative to B, then $A \oplus B$ is weakly 2-random.

3.6.16 The class of compression functions for K forms a bounded Π^0_1 class described by the condition $\forall n\, [F\restriction_n \in S]$, where S is the computable set $\{\alpha \in \mathbb{N}^*:$ $\forall i, j < |\alpha|\, [i \neq j \rightarrow \alpha(i) \neq \alpha(j)]\ \&\ \sum_{i<|\alpha|} 2^{-|\alpha(i)|} \leq 1\ \&\ \forall i < |\alpha|\, [|\alpha(i)| \leq K_{|\alpha|}(i)]\}$. This class is nonempty because it contains the function $F(x) = x^*$.

3.6.21 Since $A \leq_T \emptyset'$ we have $\forall n\, K^{\emptyset'}(Z\restriction_n) =^+ K^{\emptyset'}((Z \triangle A)\restriction_n)$. Now use Schnorr's Theorem 3.2.9 relativized to $\emptyset'$.

3.6.22 Let A be weakly 2-random but not 2-random (for instance, let A be a computably dominated ML-random set). A forms a minimal pair with $\emptyset'$, so the class $\mathcal{C} = \{X\colon\ A \oplus X$ forms a minimal pair with $\emptyset'\}$ is conull. Then there is a set X in $\mathcal{C}$ that is 2-random relative to A, and hence of hyperimmune degree. By the van Lambalgen Theorem 3.4.6, $R = A \oplus X$ is ML-random. Since $R \in \mathcal{C}$, R is weakly 2-random. If a set $D \equiv_T R$ is 2-random, then the ML-random set $A \leq_T D$ is 2-random as well.

3.6.27 Choose a ML-random set Z that is superlow, then Z is ω-c.e. and hence not Demuth random.

3.6.28 (i) Let $(G_m)_{m\in\mathbb{N}}$ be a generalized ML-test. Since $\emptyset'$ can compute λG_m, there is $h \leq_T \emptyset'$ such that $\lambda G_{h(m)} \leq 2^{-m}$ for each m. Then $(G_{h(m)})_{m\in\mathbb{N}}$ is a Schnorr test relative to $\emptyset'$.
(ii) Let $(S_m)_{m\in\mathbb{N}}$ be a Demuth test. Then $\emptyset'$ can compute a c.e. index for S_m, so $G_n = \bigcup_{m>n} S_m$ is a $\Sigma^0_1(\emptyset')$ class uniformly in n. Also $\emptyset'$ can compute λG_n uniformly in n. (Given k, $\emptyset'$ finds a stage t such that $\lambda S_i - \lambda S_{i,t} \leq 2^{-k-i-1}$ for all i, $n < i \leq k$. Then $\lambda G_n - \lambda \bigcup_{n<i\leq k} S_{i,t} \leq 2^{-k+1}$.) Thus $(G_n)_{n\in\mathbb{N}}$ is a Schnorr test relative to $\emptyset'$. If Z fails the Demuth test $(S_m)_{m\in\mathbb{N}}$ then Z fails $(G_n)_{n\in\mathbb{N}}$.
(iii) Choose Y Schnorr random relative to $\emptyset'$ but not 2-random.

3.6.31 By the low basis theorem relative to $\emptyset'$, there is a 2-random set Z such that $(Z \oplus \emptyset')' \leq_T \emptyset''$. By Corollary 3.6.20(ii) Z is GL_1. Thus Z is Low_2.

Solutions to Chapter 4

Section 4.1

4.1.5. Let p be a reduction function (Fact 1.2.15) such that $\psi(e) \simeq J(p(e))$ for each e, and let $\widetilde{f} = f \circ p$. Then $\psi(e)\downarrow$ implies that $\psi(e) = J(p(e)) \neq f(p(e)) = \widetilde{f}(e)$.

4.1.6 Suppose Φ^G is total. We prove that the function Φ^G is not d.n.c. It suffices to show that $S = \{\rho\colon\ \exists e\, \Phi^\rho(e) = J(e)\}$ is dense along G. We define an auxiliary partial computable function ψ. We are given its reduction function p in 1.2.15 by the Recursion Theorem. On input σ, search for $\rho \succeq \sigma$ such that $y = \Phi^\rho(p(\sigma))\downarrow$, and give the first such y found (if any) as an output. If $\sigma \prec G$ then $\psi(\sigma)$ is defined via some $\rho \succeq \sigma$ because Φ^G is total. Then $J(p(\sigma)) = \psi(\sigma) = \Phi^\rho(p(\sigma))$. Hence $\rho \in S$.

4.1.7 By the remark after Fact 1.2.15, let π be a computable permutation of $\mathbb{N}$ such that $\widehat{J}(x) \simeq J(\pi(x))$ for each x. Let $\widehat{f} = f \circ \pi$. Then $\widehat{f} \equiv_T f$, and for each x, $\widehat{J}(x) = \widehat{f}(x) \leftrightarrow J(\pi(x)) = f(\pi(x))$.

4.1.8 By Ex. 1.1.12, let π be a computable permutation of $\mathbb{N}$ such that $\forall x\, \widehat{W}_x = W_{\pi(x)}$. Let $\widehat{g} = \pi^{-1} \circ g \circ \pi$. Then $\widehat{g} \equiv_T g$, and for each x, $\widehat{W}_{\widehat{g}(x)} = \widehat{W}_x \leftrightarrow W_{g(\pi(x))} = W_{\pi(x)}$.

4.1.14 $\emptyset'$ is of d.n.c. degree relative to A. Hence $\emptyset'$ is Turing complete relative to A, that is, A is in GL_1.

4.1.15 See Odifreddi (1989, III.8.16).

4.1.16 Let h be a computable function such that $\forall n\, p_{\overline{A}}(n) \le h(n)$. To compute $f(e)$ in the proof of Proposition 4.1.13, we merely need to query the oracle A on numbers k less than $h(g(e)) + 1$. Thus $f \le_{wtt} A$, and A is wtt-complete by Theorem 4.1.11 and the proof of 4.1.4.

Section 4.2

4.2.7 In the following let $i \in \{0, 1\}$. Suppose that Y_i is of d.n.c. degree via $f^i \le_T Y_i$. We define partial computable functions α_i. By the Double Recursion Theorem we are given reduction functions p^i such that $\forall e\, [\alpha^i(e) \simeq J(\alpha_i(e))]$. We replace (4.1) by

$$x \in W_{e,s} - W_{e,s-1} \;\&\; \forall i \in \{0,1\} \forall t_{x<t<s}\; f^i_t(p^i(e)) = f^i_s(p^i(e)).$$

When we put x into A for the sake of PS_e, we define $\alpha^i(e) = f^i_s(p(e))$.
One verifies that $A \le f^i$ as before.

Section 4.3

4.3.3

(i) $\Rightarrow$ (ii): Let $g(x) = \psi(x)$ if $\psi(x)\downarrow$, and $g(x) = 0$ otherwise. Then $g \le_T D$.
(ii) $\Rightarrow$ (i): Let $g \le_T D$ extend the function $\psi(x) \simeq \mu s.\, x \in \emptyset'_s$. Then $\forall x\, [x \in \emptyset' \leftrightarrow x \in \emptyset'_{g(x)}]$, whence $\emptyset' \le_T D$.

4.3.4 By the Low Basis Theorem relative to Z, there is a set S of PA degree relative to Z such that $Y' \equiv_T Z' \equiv_T \emptyset'$ where $Y = S \oplus Z$. Then Y is of PA-degree, so by Theorem 4.2.1 there is a promptly simple set $A \le_T Y$, and $Z \oplus A \le_T Y$ is low.

4.3.5. It is sufficient to show that for each n, if a d.n.c. function f does not exceed $2^{n+1} - 1$, then there is a d.n.c. function $g \le_T f$ that does not exceed $2^n - 1$. By Fact 1.2.15 we can fix a binary computable function α such that $J(\alpha(x, y)) \simeq J(x) + J(y) + 1$ for each x, y. Let $f_0, f_1 \le_T f$ be functions that do not exceed $2^n - 1$ such that $f = f_0 + f_1 + 1$. If there is x such that $f_1(\alpha(x, y)) \ne J(y)$ for each y then let $g = \lambda y. f_1(\alpha(x, y))$. Otherwise, given x, using f_1 as an oracle we can compute the first s such that $f_1(\alpha(x, y_x)) = J_s(y_x)$ for some $y_x < s$ (chosen least for stage s). Then $f_0(\alpha(x, y_x)) \ne J(y)$ for each x. So let $g = \lambda x. f_0(\alpha(x, y_x))$.

4.3.6 $\Rightarrow$: The set of completions of PA can be seen as a Π^0_1 class P by 1.8.32. The theory represented by a set $Z \in P$ is $\{\alpha_n \colon Z(n) = 1\} \cup \{\neg\alpha_n \colon Z(n) = 0\}$. Now use Theorem 4.3.2(iii).

$\Leftarrow$: Given a partial computable $\{0, 1\}$-valued function ψ, define a function $g \le_T B$ as follows: given a number n, if there is $k \in \{0, 1\}$ such that $B \vdash \exists t\, \psi_t(\dot{n}) = k$, then $g(n) = k$, otherwise $g(n) = 0$. If $\psi(n) = k$ then $\psi_s(n) = k$ for some $s \in \mathbb{N}$, so $B \vdash \psi_{\dot{s}}(\dot{n}) = \dot{k}$ since PA decides all the Δ^0_1 sentences. Thus g extends ψ.

4.3.7 After 4.3.6, it remains to prove the implication "$\Rightarrow$". For $r \in \mathbb{N}, k \in \{0, 1\}$, the pair $\langle r, k\rangle$ is identified with the sentence α_r if $k = 1$, and with the sentence $\neg\alpha_r$ if $k = 0$. We define a permutation $r_0, r_1, \ldots$ of $\mathbb{N}$. At step i we determine r_i and $B(r_i)$. Inductively we are given $F_i = (\langle r_0, B(r_0)\rangle, \ldots, \langle r_{i-1}, B(r_{i-1})\rangle)$. We let the variable F range over sequences of pairs of this kind, and write $\mathrm{dom}(F)$ for the set $\{r_0, \ldots, r_{i-1}\}$. (Note that F is a finite assignment in the sense of Section 1.3.)

Step i for even i. We code $D(n)$ into B. By the Gödel incompleteness theorem, from any F we may effectively determine a sentence that cannot be decided on the base of PA $\cup F$: we find a number $c(F) \in \mathbb{N}$, $c(F) > \max \mathrm{dom}(F)$ such that, if PA $\cup F$ is consistent, then so are PA $\cup F \cup \{\alpha_{c(F)}\}$ and PA $\cup F \cup \{\neg\alpha_{c(F)}\}$. Let $r_i = c(F)$ and $B(r_i) = D(n)$.

Step i for odd i. We ensure that the next sentence left open is decided by B. This is done in a way similar to (ii)⇒(iii) in the proof of Theorem 4.3.2. Consider the partial computable binary function ψ given by $\psi(F,k) = 1$ if we first find a proof of a contradiction from $\mathsf{PA} \cup F \cup \{\neg\alpha_k\}$, and $\psi(F,k) = 0$ if we first find a proof of a contradiction from $\mathsf{PA} \cup F \cup \{\alpha_k\}$. Let $g \leq_T D$ be a total $\{0,1\}$-valued extension of ψ. Let r_i be the least number not in $\mathrm{dom}(F_i)$, and let $B(r_i) = g(F_i, r_i)$.
Clearly this construction determines a set $B \leq_T D$ that is a completion of PA. Also $D \leq_T B$ since each number r_i is determined effectively from F_i, and $D(n) = B(r_{2n})$.
4.3.10 (Kučera) One can extend 1.8.40 to infinite sequences $(B_i)_{i\in\mathbb{N}}$ of incomputable sets. Letting $(B_i)_{i\in\mathbb{N}}$ list the incomputable Δ^0_2 sets, for each nonempty Π^0_1 class P we obtain a set $A \in P$ such that $A, \emptyset'$ form a minimal pair. Now let P be the class of $\{0,1\}$-valued d.n.c. functions. If $Z \geq_T A$ is ML-random then $Z \geq_T \emptyset'$ by 4.3.8, so Z is not weakly 2-random.
4.3.18 Let $B >_T \emptyset'$ be a Σ^0_2 set such that $B' \equiv_T \emptyset''$. By 1.8.46 there is a computably dominated ML-random set Y such that $Y \leq_T B$. Thus Y is weakly 2-random. If $g \leq_T Y$ is 2-f.p.f then there is 2-d.n.c. function $f \leq_T Y$, whence $\emptyset'' \leq_T B \oplus \emptyset'$ by Theorem 4.1.11 relative to $\emptyset'$, contradiction.

Solutions to Chapter 5

Section 5.1

5.1.4 $K(Z\upharpoonright_n) =^+ K^A(Z\upharpoonright_n) =^+ K^A((Z\triangle A)\upharpoonright_n) =^+ K((Z\triangle A)\upharpoonright_n)$.
5.1.5 Y is 2-random by Exercise 3.6.21, so A forms a minimal pair with both Y and Z (page 135). Since $A \leq_T Y \oplus Z$, this implies $Y \mid_T Z$.
5.1.6 Suppose A is low for K, and consider Φ^A_e. For each n, if $m = \Phi^A_e(n)$, then

$$K(m) =^+ K^A(\Phi^A_e(n)) \leq^+ K^A(n) \leq^+ 2\log n.$$

Let $F(n) = 2\max\{m \colon K(m) \leq 3\log n\}$. Then F dominates Φ^A_e for each A, e as above. Also, F is ω-c.e., and hence $F \leq_{wtt} \emptyset'$ by Exercise 1.4.7(ii).
5.1.11 $W_e \in \mathrm{Low}(\mathsf{MLR}) \leftrightarrow \exists R$ c.e. open
$\exists k\, \forall s\, \lambda[R_s]^{\prec} \leq 1 - 2^{-k}\ \&\ \forall z\, \forall s\, \exists t \geq s\, (K^{W_e}(z)[t] \geq |z| \ \vee\ [z] \subseteq [R_t]^{\prec})$.
5.1.15 Otherwise, the class is conull by the 0-1 law, and therefore its intersection with the class of ML-random sets is conull. However, by Theorem 4.3.8, each ML-random set of PA degree is Turing above $\emptyset'$, so by Theorem 5.1.12 this intersection is a null class.
It is possible to only apply a technique used in the proof of Theorem 4.3.8 for a simple direct solution. If the class is conull then by the Lebesgue Density Theorem 1.9.4 there is a Turing functional Φ such that $\Phi^X(w) \in \{0,1\}$ if defined, and $\{Z \colon \Phi^Z$ is total and d.n.c. $\}$ has measure at least $3/4$. We define a partial computable α and are given a reduction function p for α:
If s is least such that $1/4 < \lambda\{Z \colon \Phi^Z_s(p(0)) = b\}$, *define* $\alpha(0) = b$. Then $\Phi^Z(p(0)) = J(p(0))$ for more than $1/4$ of all sets, contradiction.
5.1.16 Suppose that $A = \Phi^Y$ for a Turing functional Φ, and let c be the constant of Proposition 5.1.14.
(i) and (ii): Given a c.e. open set R, we will effectively obtain a c.e. open set $\widehat{R}$ such that $\lambda\widehat{R} \leq 2^c\lambda R$. If A fails a test $(G_n)_{n\in\mathbb{N}}$ for the randomness notion in question, then Y fails the test $(\widehat{G}_{n+c})_{n\in\mathbb{N}}$.

For $x \in \{0,1\}^*$, let S_x be the effectively given c.e. set which follows the canonical computable enumeration of $\{\sigma\colon x \preceq \Phi^\sigma\}$ as long as the measure of the open set generated does not exceed $2^{c-|x|}$. From a c.e. open set R we can effectively obtain a (finite or infinite) c.e. antichain $\{x_0, x_1, \ldots\}$ such that $R = \bigcup_i [x_i]$. Let $\widehat{R} = \bigcup_i [S_{x_i}]^\prec$. Since $[S_{x_i}]^\prec \cap [S_{x_j}]^\prec = \emptyset$ for $i \neq j$, we have $\lambda\widehat{R} = \sum_i \lambda S_{x_i} \leq 2^c \lambda R$. Moreover, $A \in R$ implies $x_i \prec A$ for some i and hence $Y \in \widehat{R}$ by the hypothesis on c. Clearly $(\widehat{G}_{n+c})_{n\in\mathbb{N}}$ is a test for the same randomness notion that succeeds on Y.
(iii): argue in a similar way but with R and $\widehat{R}$ being $\Sigma^0_1(C)$ classes.

5.1.27 Since $\Omega \equiv_T \emptyset'$, such an A is a base for ML-randomness and hence low for K.

5.1.28 Let A be a superlow ML-random set and let $Z = A$.

5.1.29 Relativizing the Low Basis Theorem to A, one obtains a $\{0,1\}$-valued function f such that $\forall e\, \neg f(e) = J^A(e)$ and $(f \oplus A)' \equiv_T A'$. Let $D = f \oplus A$.

5.1.30 Z is ML-random relative to Y by Theorem 3.4.6, and hence ML-random relative to A. Thus A is a base for ML-randomness.

5.1.31 Suppose that A is c.e. and not low for K, $\Phi^Z = A$, and $\emptyset' \not\leq_T Z$. We will show that Z is not ML-random by building a Solovay test $(E_d)_{d\in\mathbb{N}}$ for Z.
Define $C^\eta_{d,\sigma}$ as in the proof of Theorem 5.1.22. For each d, the bounded request set L_d fails to show that A is low for K, so we must have $\lambda C^\eta_{d,\sigma} < 2^{-|\sigma|-d}$ for some $\eta \prec A$ and σ such that $\mathbb{U}^\eta(\sigma)$ converges. $C^\eta_{d,\sigma}$ keeps asking for new strings, thus, for each d, there are n and σ such that $Z \upharpoonright_n \in C^\eta_{d,\sigma}$ for some $\eta \prec A$. Now let $g(d)$ be the least stage s such that $Z \upharpoonright_n$ is in $C^\eta_{d,\sigma,s}$ for some $n, \sigma < s$ and $\eta \prec A_s$. Then $g \leq_T Z$, so if we let $m(d) \simeq \mu s.\ d \in \emptyset'_s$, as in the proof of Lemma 3.4.13, then since $\emptyset' \not\leq_T Z$, there are infinitely many d such that $m(d) > g(d)$. Now define $E_d = \emptyset$ if $d \notin \emptyset'$, and otherwise

$$E_d = \bigcup\{C^\eta_{d,\sigma,m(d)}\colon \eta \prec A_{m(d)} \;\&\; \sigma \in \{0,1\}^*\}.$$

Then the sets E_d are uniformly c.e. and $\lambda E_d \leq 2^{-d}$, so $([E_d]^\prec)_{d\in\mathbb{N}}$ is a Solovay test. Furthermore, if $m(d) > g(d)$, there are n, σ such that $Z \upharpoonright_n$ is in $C^\eta_{d,\sigma}$ for some $\eta \prec A_{m(d)}$, so $Z \upharpoonright_n$ is in E_d. Thus Z is not ML-random.

Section 5.2

5.2.7
$\Rightarrow$: If A is low for K then for each n we have $K(A \upharpoonright_n) =^+ K^A(A \upharpoonright_n) =^+ K^A(n)$.
$\Leftarrow$: For each n we have $K(n) \leq^+ K(A \upharpoonright_n) \leq^+ K^A(n)$, so A is low for K.

5.2.8 (Stephan) We deal with the case of K; for the case of C one uses the plain optimal machine $\mathbb{V}$ instead of $\mathbb{U}$.
The prefix-free machine M operates as follows on input σ.
(1) Wait for $\mathbb{U}(\sigma)\downarrow = n$. (2) If t is least such that $n \in \emptyset'_t$, output $A_t \upharpoonright_n$.
Since A is wtt-incomplete, there are infinitely many $n \in \emptyset'$ such that $A_t \upharpoonright_n = A \upharpoonright_n$, where n enters $\emptyset'$ at stage t. Thus $K_M(A \upharpoonright_n) \leq |\sigma|$ for each $\mathbb{U}$-description σ of such an n, and hence $K(A \upharpoonright_n) \leq^+ K(n)$.

5.2.9 (i) For each partial computable one-one f, $\forall n\, K(f(n)) =^+ K(n)$; see Fact 2.3.1. Also, from $A \upharpoonright_{r_n}$ one can compute $A \upharpoonright_n$.
(ii) We have $K(A \upharpoonright_n) \leq^+ K(A \upharpoonright_{r_n}) \leq^+ K(r_n) \leq^+ K(n)$. The increase of the constant that is hidden in the inequality $K(A \upharpoonright_n) \leq^+ K(n)$ does not depend on b because it is due to the first and third inequalities.

5.2.10 To make A not K-trivial, meet the requirements

$$R_b\colon \exists n\, K(A \upharpoonright_n) > K(n) + b.$$

The strategy is as follows. Start by picking a sufficiently large candidate n, and ensure that $K(n)$ is small by enumerating a request into a bounded request set W. From now on, change $A\restriction_n$ whenever $K(A\restriction_n) \le K(n) + b$, but only for up to 2^{c+b} times for each approximation to $K(n)$. If the approximation to $K(n)$ is final, then this number of A-changes will suffice by (i) of Theorem 5.2.4. One needs to know in advance how many A-changes are needed necessary so that n can be chosen large enough.
To ensure that A is low, meet the usual lowness requirements L_e from Theorem 1.6.4: initialize the requirements R_b, $b > e$, whenever $J^A(e)$ newly converges.
Here are the details for the R_b-strategy. Let d be the coding constant for the bounded request set W given in advance via the Recursion Theorem.
Phase 1. If R_b has been initialized for k times so far, let $r = k + b + d + 2$. Choose $n = n_0 + 2^{c+b}r$, where n_0 is the largest number mentioned so far in the construction. Put the request $\langle k+b+1, n\rangle$ into W and wait for $K(n) < r$ for s such that $K_S(n) < r$.
Phase 2. Up to 2^{c+b} times, whenever $K(A\restriction_n) \le K(n) + b$ for the current stage s, put the greatest number in $[0, n) - A_s$ into A and initialize the requirements L_e, $e \ge b$. If $K(n)$ decreases, start counting from 0 again.
Since $K_S(n) < r$, there are enough numbers in $[n_0, n)$ available for enumeration in Phase 2. It is easy to check that W is a bounded request set. Also, there is a computable bound on the number of injuries to requirement L_e, so A is superlow.

5.2.14 (a) For each $y \prec A$ we have $K(xy) \le K(x) + K(y) \le K(x) + K(|y|) + c \le^+ 2K(x) + K(|xy|) + c$. The last inequality holds because $|y|$ can be computed from $|xy|$ and x. (b) For $x' \preceq x$ we have $K(x') \le^+ K(x) + K(|x'|)$.
The maximum r_0 of the two constants implicit in (a) and (b) is as required.

5.2.15 Assume h is such a function, then $h(b) = \lim_s h_s(b)$ for an effective approximation $(h_s)_{s\in\mathbb{N}}$. As before d is given. Let $r \ge d$ be a constant such that $\emptyset$ is K-trivial via r. At each stage s, let $b_s \ge r$ be least such that $h_s(b_s) < 2^{-d}$. Whenever $s > 0$ is such that $b_s \ne b_{s-1}$, then start a new attempt at building 2^{b_s-d} K-trivial sets, namely the sets $A_x = 0^s1x0^\infty$ where $|x| = b_s - d$. As long as $b_t = b_s$ at stages $t > s$, we put requests $\langle K_t(n) + b_s - d, A_x\restriction_n\rangle$ into L, but we only do this for $s < n \le t$, as the case $n \le s$ is covered by the choice of r. It is now easy to verify that L is a bounded request set. Eventually b_s settles, and we obtain a contradiction as before.

5.2.19 We may assume A and B are co-infinite. Given n, let $a \in \mathbb{N}$ be the number with binary representation $1A\restriction_n$. Likewise define b, c from $B\restriction_n, C\restriction_n$. Then $c = a + b$ or $c = a + b + 1$. Now argue as in 5.2.17.

5.2.22 Prove that A is C-trivial as in 5.2.3, but with $\mathbb{V}$ and C instead of $\mathbb{U}$ and K.

5.2.23 An essential feature near the end of the proof is the existence of a computable upper bound for $C(n)$ that is achieved frequently: for each k some $u \in [k, f(k)]$ has the highest possible C complexity $1 + \log(u + 1)$, where f is a computable function (in fact $f(k) = 2k$). This condition fails for K.

5.2.24 For the implication (i)$\Rightarrow$(ii), let the binary machine M be given by $M(\varnothing, n) = A\restriction_n$. If d is the coding constant for M then $K(A\restriction_n | n) \le d$ for each n. The implication (ii)$\Rightarrow$(iii) is trivial. For the implication (iii)$\Rightarrow$(i), note that $C(A\restriction_n | n) \le b$ implies $C(A\restriction_n) \le C(n) + 2b + O(1)$, and apply Theorem 5.2.20(ii).

Section 5.3

5.3.7 Since $K(2^j) \le^+ 2\log j$, there is an increasing computable function f and a number j_0 such that $\forall j \ge j_0\, K_{f(j)}(2^j) \le j - 1$. Enumerate the set $A = \mathbb{N}$ in order, but so slowly that for each $j \ge j_0$ the elements of $(2^{j-1}, 2^j]$ are enumerated only after stage $f(j)$, one by one. Each such enumeration costs at least $2^{-(j-1)}$, so the cost for each interval $(2^{j-1}, 2^j]$ is 1.

5.3.8 Suppose the limit condition fails for e. Choose s_0 such that

$$\sum_{s\geq s_0}\sum_{x<s} c(x,s)\,[\![x \text{ is least s.t. } A_{s-1}(x)\neq A_s(x)]\!] \leq 2^{-e}.$$

To compute A, on input n find $s > s_0, n$ such that $c(n,s) > 2^{-e}$. Then $A_s(n) = A(n)$.

5.3.9 The condition that Φ_i is total is Π^0_2. Given that Φ_i is total, the condition that the sum S in (5.6) is finite is Σ^0_2. In that case $\lim_s D_{\Phi_i(s)}(x)\downarrow$, so the condition that Φ_i is an approximation of V_e is Π^0_2.

5.3.17 $\{\Omega^{\emptyset'}\}$ is a $\Pi^0_2(\emptyset')$ class, and hence Π^0_3. Being 2-random, $\Omega^{\emptyset'}$ forms a minimal pair with $\emptyset'$. So each c.e. set $A \leq_T \Omega^{\emptyset'}$ is computable.

5.3.18 Let $x < s$. Let e be least such that $Y_t(e) \neq Y_{t-1}(e)$ for some t, $x < t \leq s$. Then $e \leq x$, so $c_Y(x,s) = 2^{-e}$. Also $V_{x,s} = [Y_t\restriction_e]$, so $c(x,s) = \lambda V_{x,s} = 2^{-e}$.

5.3.19 In general, V_x contains strings of a length that exceeds x by far. (For instance, $\mathcal{C}$ could be not contained in any closed null class, in which case we cannot expect that all the V_x are clopen.) When computing A from a ML-random set $Y \in \mathcal{C}$, given an input $x \geq s_0$, using Y as an oracle, we need to search for $t > x$ such that $[Y\restriction_t] \subseteq V_{x,t}$. We cannot compute a bound on t from x.

5.3.20 (i) Y and $\emptyset'$ form a minimal pair by Theorem 5.3.16, so Y and Z form a minimal pair by Exercise 5.1.30 and the fact that each set that is low for K is Δ^0_2.
(ii) By 1.8.43 let $Y \oplus Z$ be a *low*$_2$ computably dominated set in $2^{\mathbb{N}} - \mathcal{R}_1$. Then Y and Z are weakly 2-random.

5.3.26 In (i) of the $\Sigma^0_1(B)$ question and in (5.8), add the condition that E permit the enumeration of x into A, namely $E_s\restriction_x \neq E_{s-1}\restriction_x$. Claim 5.3.25 requires more work now.

5.3.30 Given i, j, compute k such that $V_k = B_i \oplus B_j$. Let $c = 3\max(d_i, d_j) + O(1)$, where the implicit constant is as in Theorem 5.2.17. Let $f(i,j) = \langle k, c\rangle$. Since V_k is K-trivial via c, we have $B_{f(i,j)} = V_k$.

5.3.31 Suppose the effective sequence $(e_n, b_n)_{n\in\mathbb{N}}$ shows that the index set of the computable sets is uniformly Σ^0_3. Let $S = \{e_n \colon e_n \in W_{b_n}\}$, then S is computable but $S \neq W_{e_n}$ for each n.

5.3.32 $\Rightarrow$: The definition gives us directly that $i \in C \leftrightarrow \exists n\, W_i = W_{e_n}$.
$\Leftarrow$: The relation defined by $P(e,b) \leftrightarrow W_e = W_{e_b}$ is Π^0_2. Let $b_n = n$. Then $(e_n, b_n)_{n\in\mathbb{N}}$ is a sequence as required.

5.3.33 Assume for a contradiction that $i \in C \leftrightarrow \exists n\, W_i = W_{e_n}$ for a computable sequence of indices $(e_n)_{n\in\mathbb{N}}$. One can define a co-infinite computable set S such that $S \cap W_{e_n} \neq \emptyset$ for each n. Then $\mathbb{N} - S = W_{e_r}$ for some r, contradiction.

5.3.38 By Theorem 5.1.9 it suffices to show that there is a c.e. open S such that $\lambda S < 1$ and $\mathcal{R}^A_1 \subseteq S$. Let B^A be an antichain that is c.e. in A such that $[B^A]^\prec = \mathcal{R}^A_1$. Use the adaptive cost function $c_A(x,s) = \lambda[\{y \colon y \in B^A[s-1] \textit{ with use} > x\}]^\prec$ and proceed in a way similar to the proof of 5.3.35.

Section 5.4

5.4.5 A web search (2007) revealed that they are still available. (One bottle will set you back around US\$1500.) Recall that your bottle of wine corresponds to a weight of 1. In each of the constructions we show that L is a bounded request set. This means that we never need to pour more than one bottle into the topmost decanter(s).

Section 5.5

5.5.9 Suppose not, then for each superlow set A there is a c.e. superlow set $C \geq_T A$ by Proposition 5.3.6. This is clearly not the case: for instance A could be ML-random, and hence of d.n.c. degree, so $C \equiv_T \emptyset'$.

5.5.10 For (i) use Exercise 1.8.53. For (ii), if a Schnorr random set is low then it is already ML-random by 3.5.13.

5.5.13 For some c, the set $L = \{\langle \log h(m) + c, m\rangle : m \in \mathbb{N}\}$ is a bounded request set. It shows that each m has a $\mathbb{U}$-description σ_m such that $|\sigma_m| \leq^+ \log h(m)$. Now argue as before.

5.5.16 If A is K-trivial then Ω_U^A is left-c.e.; since Ω_U^A is ML-random, this implies that $A \leq_T \Omega_U^A$. If $A \leq_T \Omega_U^A$ then A is a base for ML-randomness and hence K-trivial.

5.5.19 Downey and Ng (2009) have shown that lowness for Demuth randomness implies being computably dominated. Using this, Bienvenu, Downey, Greenberg, Nies and Turetsky (20xx) have given a full characterization via a concept called BLR-traceability, which implies jump traceability defined in 8.4.1. They build a perfect Π^0_1 class of sets that are low for Demuth randomness.

Section 5.6

5.6.6 $\Rightarrow$: $K^B(A\upharpoonright_n) \leq^+ K^A(A\upharpoonright_n) =^+ K^A(n)$.
$\Leftarrow$: For each n, $K^B(n) \leq^+ K^B(A\upharpoonright_n) \leq^+ K^A(n)$. Thus $A \leq_{LK} B$.

5.6.7 Argue as in the proof of the implication "$\Leftarrow$" of Theorem 5.6.5.

5.6.8 Suppose $\emptyset' \leq_{\text{cdom}} C$. Then there is $g \leq_T C$ such that $g(n) \geq \mu s. \emptyset'(n) = \emptyset'_s(n)$ for each n. Thus $\emptyset' \leq_T C$.

5.6.10 Modify the proof of the implication (iii)$\Rightarrow$(i) of Theorem 5.6.9.
The class $U_1 = \{Z : Z \leq_L A'\}$ is $\Pi^0_1(A')$, and hence a $\Pi^0_2(A)$ class of measure $0.A'$. The class $U_2 = \{Z : Z \leq_L \mathbb{N} - A'\}$ is a $\Pi^0_2(A)$ class of measure $1 - 0.A'$. Now argue as before that both $0.A'$ and $1 - 0.A'$ are right-c.e. relative to B', whence $A' \leq_T B'$.

5.6.22 Note that $\emptyset' \leq_T A \oplus Z$, for otherwise $Z \in \mathsf{MLR}^A$ by Proposition 3.4.13, which implies $Z \not\leq_{LR} A$. By Lemma 5.6.20, $A \in \mathcal{K}(Z)$, and hence $\emptyset' \leq_T A \oplus Z \in \mathcal{K}(Z)$. Thus $\emptyset' \in \mathcal{K}(Z)$ by Theorem 5.6.17(iii).

5.6.23 (i) Let $S >_T \emptyset'$ be a Σ^0_2 set in $\mathcal{K}(\emptyset')$. By the Sacks jump theorem (page 160) there is a c.e. set A such that $A' \equiv_T S$. Then $A' \equiv_{LR} \emptyset'$. (ii) This follows from 5.6.21.

5.6.25 We show that $A \in \text{Low}(\mathsf{MLR}^X)$. If $R \in \mathsf{MLR}^X$ then $R \oplus X$ is ML-random, and hence ML-random relative to A. By van Lambalgen's Theorem 3.4.6 relative to A, this implies that R is ML-random relative to $A \oplus X$.

5.6.32 Let $G = \bigcap U_n$ where the U_n are uniformly Σ^0_2. By the uniformity of the implication "$\Rightarrow$" in Lemma 5.6.29, each U_n is covered by a class V_n of the same measure, where the V_n are uniformly $\Pi^0_2(C)$. Then $\bigcap_n V_n$ is a $\Pi^0_2(C)$ class as required.

5.6.33 Let $H = \bigcap_n S_n$ as in Lemma 5.6.29. Similar to 5.6.28, given e, n, the oracle $\emptyset'$ can determine clopen classes $C_{n,e} \subseteq S_n$ such that $\lambda(S_n - C_{n,e}) \leq 2^{-n-e}$. Let $L_e = \bigcap_n C_{n,e}$. Then L_e is uniformly $\Pi^0_1(\emptyset')$, $L_e \subseteq H$, and $\lambda(H - L_e) \leq \lambda \bigcup_n (S_n - C_{e,n}) \leq 2^{-e+1}$. Now $L = \bigcup_e L_e$ is as required.

Solutions to Chapter 6

6.1.2 Consider a prefix-free machine M such that $M(0^n 1)$ converges when n enters A.

6.3.7 Let us first show how to make C incomputable. Run the construction almost as before, with the only difference that now the marker $\gamma(2i)$ is used for coding $\emptyset'(i)$, while

$\gamma(2i+1)$ is used to satisfy the requirement $C \neq \Phi_i$ (i.e., C is not computed by Φ_i). At stage s, in step (1) find the least i such that $i \in \emptyset'_s - \emptyset'_{s-1}$ or $\Phi_i(\gamma_{s-1}(2i+1)) = 0$. In the first case put $\gamma_{s-1}(2i) - 1$ into C, in the second put $\gamma_{s-1}(2i+1)$ into C.
To ensure C is of promptly simple degree, meet the requirement PSD_i from (6.4) via $\gamma(2i+1)$. When a number $x > \gamma_{s-1}(2i+1)$ enters W_i at stage s then attempt to put $\gamma_{s-1}(2i+1)$ into C.
6.3.16 Use 6.3.15(iv) with $A = \emptyset'$ and $B = S$.

Solutions to Chapter 7

7.1.10 (i) is clear. (ii) Suppose $C = \alpha E_x + \beta E_y = 0$ for $\alpha, \beta \in \mathbb{R}$. Say $|x| \le |y|$, and choose $r \in \{0,1\}$ such that $xr \not\preceq y$. Then $C(xr) = \alpha 2^{|x|} = 0$, so $\alpha = 0$ and hence $\beta = 0$. (iii) B equals the convex combination $\sum_{|x|=n} B(x)2^{-n}E_x$.

7.1.11 Let $S(0^k) = 2^k$ and $S(x) = 0$ for $x \notin \{0\}^*$. For each b of the form 2^k we have $\lambda\{Z\colon \exists n\, S(Z\restriction_n) \ge b\} = 2^{-k} = 1/b$.

7.1.14 We may assume that $S(x) > 0$ for each $x \in \{0,1\}^*$, and $S(\varnothing) < 1$. We let $R = G + E$, where $G(x) \in \mathbb{N}$ is the current savings and E is a supermartingale bounded by 2. Whenever $a \in \{0,1\}$ and $E(xa) > 1$ then define $G(xa) = G(x) + 1$, otherwise $G(xa) = G(x)$. In the former case, for strings $y \succeq xa$, E begins with capital $E(xa) - 1$, using the same betting factors $S(yb)/S(y)$ as S for $b \in \{0,1\}$. If $y \succeq x$ then $R(y) - R(x) \ge E(y) - E(x) \ge -2$. If $Z \in \mathsf{Succ}(S)$ then $\lim_n G(Z\restriction_n) = \infty$, whence $\lim_n R(Z\restriction_n) = \infty$. We have $R \le_T S$ because the S-computable real numbers form a field.

7.2.5 View W_e as a subset of $\{0,1\}^* \times \mathbb{Q}_2$ and let $U_e = \{\langle x, q\rangle \in W_e \colon x = \varnothing \rightarrow q \le 1\}$. The construction in Fact 7.2.4 applied to $U = U_e$ yields a supermartingale approximation of a c.e. supermartingale S_e. If U_e is the undergraph of a c.e. supermartingale S such that $S(\varnothing) \le 1$ then $S_e = S$.

7.2.7 (i) is immediate by 7.1.7(ii), and (ii) is (almost) trivial. For (iii), let S be the martingale such that $S(0^k) = 2^k$ for each k (see the solution to 7.1.11). We show that $\lim_k \mathsf{FS}(0^k)2^{-k} = 0$, whence FS does not multiplicatively dominate S. Note that

$$\mathsf{FS}(0^k)2^{-k} = \textstyle\sum_{i<k} 2^{-K(0^i)+i-k} + \sum_y 2^{-K(y)}\, [\![0^k \preceq y]\!].$$

Given $r \in \mathbb{N}$ choose $n \in \mathbb{N}$ so large that $\sum_y 2^{-K(y)}\, [\![0^n \preceq y]\!] \le 2^{-r}$. Then for each $k > n$ we have $\sum_{n \le i < k} 2^{-K(0^i)+i-k} \le 2^{-r}$. So $\mathsf{FS}(0^k)2^{-k} \le 2^{-r+2}$ for large k.

7.2.9 (i) Let n_k be the least n be least such that $\sum_{n \ge k} 2^{-f(n)} \le 2^{-2k}$. We may assume that $n_0 = 0$. If $n \in [n_k, n_{k+1})$ let $g(n) = f(n) - k$.
(ii) Let g be as in (i) for $f = f_Z$. There is a ML-random set Z such that $\forall n\, f_Z(n) \le^+ g(n)$. Then $\lim_n(f_Y(n) - f_Z(n)) = \infty$ whence $Z <_K Y$.

7.3.5 For a string z, the real number $r_z = \sum_v 2^{-K_M(v)}\, [\![z \preceq v]\!]$ is computable uniformly in z. Now use that $\mathsf{FS}_M(x) = \sum_z 2^{|z|} r_z\, [\![z \preceq x]\!]$.
If Z is not Schnorr random then $Z \in \bigcap_b R_b^M$ for some computable measure machine M by Theorem 3.5.19. Hence $Z \in \mathsf{Succ}(\mathsf{FS}_M)$.

7.3.6 In (2.16) on page 101 we defined the sets $A_{m,d}$. If Z fails the law of large numbers then after possibly replacing Z by $\mathbb{N} - Z$ we have $\exists^\infty m\, Z\restriction_m \in A_{m,d}$ for some d. Let $R = \bigcup_m A_{m,d}$, then $\sum_{x \in R} 2^{-|x|} = \sum_m 2^{-m} \# A_{m,d}$ is finite and computable:

let $q = e^{-2/d^2} < 1$, then $2^{-m}\#A_{m,d} \leq 2q^m$ by (2.17), hence the tail sum taken from r on is bounded by $2\sum_{m\geq r} q^m = 2q^r/(1-q)$ which effectively tends to 0 with r. Hence Z is not Schnorr random by 7.3.4.

7.3.11 We may assume that S is $\mathbb{Q}_2$ valued. By 7.3.10 the leftmost non-ascending path for S is computable.

7.4.5 Suppose $Y =^* Z$ and B_k succeeds on Y. Choose n such that $Y(m) = Z(m)$ for all $m \geq n$, then $B_{k,n}$ succeeds on Z.

7.4.6 If $x = E\restriction_n$ has been defined already wait for $B(x0)$ to converge. If $B(x0) \leq B(x1)$ let $E(n) = 0$, else $E(n) = 1$. If $B(x0)$ diverges for some x define $E = x0^\infty$.

7.5.11 Some set $Z \equiv_T C$ is weakly 1-generic by 1.8.50, and hence weakly random. Such a set is not Schnorr random because it fails the law of large numbers.

7.5.12 We meet the usual minimal pair requirements N_e, and requirements

$$T_{(k,r)}\colon\ \exists n\, B_k(Z_r\restriction_n)\uparrow\ \vee\ B_k(Z_r) < \infty\ (k \in \mathbb{N}, r \in \{0,1\}),$$

where (k,r) denotes $2k+r$. We associate strategies α of length $|2e|$ with N_e, and strategies β of length $2i+1$ with T_i. For such a β, if $i = (k,r)$, the outcome 0 means that $B_k(Z_r\restriction_n)$ is defined for arbitrarily large n. The $T_{(k,r)}$ strategy β is as follows: at stage s let $s_\beta < s$ be the greatest stage t when β was initialized (namely $t = 0$ or $\delta_t <_L \beta$). *If $n \geq s_\beta$ is least such that $B_k(Z_r\restriction_n 0) > B_k(Z_r\restriction_n 1)$ and $Z_r(n) = 0$ (β requires attention) then leave $Z_r\restriction_n$ unchanged, let $Z_r(n) = 1$ and $Z_r(m) = 0$ for* $n < m < s$. At stage s let the shortest $\beta \subseteq \delta_s$ that requires attention act.
For $|\alpha| = 2e$, a computation $\Phi_e^{Z_r}(x)[s]$ is *α-correct* at stage s if for each $i < e$, $i = (k,r)$, if $\alpha(2i+1) = 0$ then we have $B_k(Z_r\restriction_{n+1}) \leq B_k(Z_r\restriction_n)$ for $s_\beta \leq n < u$, where $\beta = \alpha\restriction_{2i+1}$ and u is the use of $\Phi_e^{Z_r}(x)[s]$. Define α-expansionary stages in terms of the maximal length of agreement $\Phi_e^{Z_0}(x) = \Phi_e^{Z_1}(x)$ between α-correct computations.

7.6.25 Suppose $\rho = (S,B)$ succeeds on $Z = Z_0 \oplus Z_1$. We may suppose $B(x) > 0$ for each x. Define KL betting strategies $\rho_r = (S_r, B_r)$, $r \in \{0,1\}$ with total scan rules and total B_r, one of which succeeds on Z. The strategy ρ_r simulates ρ on the positions of Z_r. It merely reads positions in Z_{1-r} without betting while waiting for converging computations giving the next position in Z_r, and what to bet on them.
Suppose so far it has simulated ρ on $\alpha \in \mathsf{FinA}$, $\alpha = (\langle d_0, r_0\rangle, \ldots, \langle d_{n-1}, r_{n-1}\rangle)$. Let $x = r_0 \ldots r_{n-1}$.

(1) Wait for computations $k = S(\alpha)$ and $q = B(x0)$ to converge. Meanwhile, read new positions of Z_{1-r} in ascending order without betting.
(2) If k is a position of Z_r, bet with factor $q/B(x)$ that the next value is 0, otherwise do not bet. Now request the value v at position k, let $\alpha := \alpha\langle k, v\rangle$ and goto (1).

7.6.26 (i) Define A and an auxiliary (possibly finite) sequence of finite assignments $\alpha_0 \prec \alpha_1 \prec \ldots$ such that $\alpha_i \lhd A$ for each i. Let $A_0 = \emptyset$ and $\alpha_0 = \emptyset$. If α_n has been defined and $S(\alpha_n)\uparrow$ then $S(A)$ is not a set. Otherwise let $\alpha_{n+1} = \alpha_n\langle S(\alpha_n), E(n)\rangle$. If $E(n) = 1$ then put $S(\alpha_n)$ into A.
(ii) By 7.4.6 there is a computable set E such that $\exists n\, B(E\restriction_n)\uparrow$ or $B(E) < \infty$ for some n. Hence (S,B) does not succeed on A.

7.6.27 $\Leftarrow$: This is clear since each binary string x can be seen as a finite assigment α_x of the same length, and $K(x) =^+ K(\alpha_x)$.
$\Rightarrow$: It suffices to show that for each finite assignment α and string z,

$$K(\alpha) \leq |\alpha| - b\ \&\ \alpha \lhd z\ \&\ |z| = \max(\mathrm{dom}(\alpha)) + 1\ \rightarrow\ K(z) \leq^+ |z| - b.$$

(In z the positions left open by α have been filled in, up to the largest position in $\mathrm{dom}(\alpha)$.) The prefix-free machine M on input τ waits for $\mathbb{U}(\sigma) = \alpha$ where $\tau = \sigma y$. Then it checks whether $|\alpha| + |y| = \max(\mathrm{dom}(\alpha)) + 1$; if so it prints the string z where the bits of y in ascending order are used to fill in those positions. For instance, if $\alpha = (\langle 4, 0\rangle, \langle 0, 1\rangle)$ and $y = 110$ then $z = 11100$.

7.6.28 Suppose this probability is 1. In the proof of 7.6.24, replace (7.20) by

$$n_{i+1} = \mu n > n_i . 2^{-n} \#\{y\colon\ |y| = n \ \&\ \exists r\, [0, n_i) \subseteq \mathrm{dom}\, \Theta^y_S(r)\} \geq 1 - 2^{-i}.$$

Let C_i be the clopen set generated by the y that are left out, then $\lambda C_i \leq 2^{-i}$. If S does not scan all places of the set Y then $Y \in C_i$ for almost every i. Since $(\bigcup_{i>m} C_i)_{m\in\mathbb{N}}$ is a Schnorr test, Y is not Schnorr random.
If Y passes this Schnorr test then a suitable variant of D succeeds on Y. Suppose $Y \not\in C_i$ for $i \geq n_{2r}$. D only bets from strings of length n_{2r} on. For $k \geq r$, one only considers strings u of length m such that $[xu] \cap C_{2k+1} = \emptyset$ in (7.21) and the subsequent discussion, because r_u and w_u are only defined for such strings.

Solutions to Chapter 8

Section 8.1

8.1.7 If $P \subseteq \mathsf{MLR}$, each set Z in $P \cap \mathrm{Low}(\Omega)$ is 2-random, and hence not Σ^0_2.

8.1.8 Let $r = \lim_k r_k$ where $(r_k)_{k\in\mathbb{N}}$ is an effective sequence of rationals. Since $\lim_k \Omega_k^{X\upharpoonright k} = \Omega^X$, we have $X \in G_r \leftrightarrow \forall n\, \exists k \geq n\, [\mathrm{abs}(\Omega_k^{X\upharpoonright k} - r_k) \leq 1/n]$.

8.1.11 A is low for Ω, hence a base for ML-randomness, and therefore low for K.

8.1.12 Suppose $C \leq_{LR} X, Y$. We show that C is a base for ML-randomness.
Note that X and Y are 2-random relative to each other by 3.4.9. They are low for Ω, and hence in GL_1. Since Y is 2-random in X, Y is ML-random in $X \oplus \Omega$, so by the van Lambalgen Theorem relative to X, $Y \oplus \Omega$ is ML-random in X, and hence ML-random in C. Finally $C \leq_T Y' \equiv_T Y \oplus \Omega$ by Corollary 8.1.10, since Y is low for Ω.

8.1.13 See Barmpalias and Lewis (20xx).

8.1.17 (i) "$\leq^+$": Modify the request operator L in 8.1.16 as follows. For each m, i, if $K_s(X\upharpoonright_{\langle m,i\rangle}) < K_{s-1}(X\upharpoonright_{\langle m,i\rangle})$, put the request $\langle K_s(X\upharpoonright_{\langle m,i\rangle}) - \langle m, i\rangle + c + 1, m\rangle$ into L_s^X, as long as the total weight of L^X does not exceed 1.
"$\geq^+$": Given m, for some i we want an M-description σ of length $\leq^+ K^X(m) + \langle m, i\rangle$ for $X\upharpoonright_{\langle m,i\rangle}$. The machine M simulates computations $\mathbb{U}^X(\rho)$: when a new bit of either ρ or of the oracle is required, it takes the next bit of its input. If the bit is an oracle bit it is also printed. If $\mathbb{U}^X(\rho) = m$ and $\langle m, i\rangle = \mathsf{use}\ \mathbb{U}^X(\rho)$ then for an appropriate σ of length $|\rho| + \langle m, i\rangle$ we have $M(\sigma) = X\upharpoonright_{\langle m,i\rangle}$. One can verify that M is prefix-free. (With the modified optimal oracle machine $\mathbb{U}$ associate a partial computable scan rule that works monotonically on both the input and on the oracle side, but switches between them. Now apply Remark 7.6.18.)
Clearly (i) implies (ii). For (iii), if $X \geq_{LR} Y$ where $Y \in \mathsf{MLR}$ and X is n-random, then let $S \equiv_T \emptyset^{(n-1)}$ be ML-random. By the van Lambalgen Theorem 3.4.6 we have $S \in \mathsf{MLR}^X$, so $S \in \mathsf{MLR}^Y$ and hence $Y \in \mathsf{MLR}^S$. Thus Y is n-random as well.
If $X \leq_K Y$ and X is n-random then $Y \in \mathsf{MLR}$ by Schnorr's Theorem, hence $X \geq_{LR} Y$ and Y is n-random.

8.1.20 $\mathrm{GL}_1 \cap \mathcal{C}$ contains a perfect class by 8.2.20 below, and $\mathrm{GL}_1 \cap \mathcal{D}$ contains the perfect class $2^{\mathbb{N}} - \mathcal{R}_1^{\emptyset'}$ of 2-random sets. For $\mathcal{C} \cap \mathcal{D}$, by 1.8.44 choose a perfect subclass of $2^{\mathbb{N}} - \mathcal{R}_1$ consisting of computably dominated sets.

Section 8.2

8.2.9 It suffices to show that some function $g \le_{wtt} \emptyset'$ dominates each computable function. To this end, let $g(n) = \max\{\Phi_e(n) \colon \Phi_e(n)\downarrow \ \& \ e \le n\}$.

8.2.10 As in Exercise 1.5.5, let h be a computable function such that $Y \le_T C$ implies $Y \le_{tt} \emptyset'$ with use bound h for each Y. We claim that the ω-c.e. function $g(m) = \mu s. \emptyset'_s \upharpoonright_{h(m)} = \emptyset' \upharpoonright_{h(m)}$ dominates each function computed by C.
Assume for a contradiction that the function $f \le_T C$ is not dominated by g. We build a set $Y \le_T C$ such that Y is not truth-table reducible to $\emptyset'$ with use bound h. Let $(\Psi_e)_{e \in \mathbb{N}}$ be an effective listing of all (possibly partial) truth-table reduction procedures with use bound h defined on initial segments of $\mathbb{N}$, similar to the listing introduced before 1.4.5. We define $Y(m)$ inductively. Look for the least $e \le m$ such that $\Psi_{e,f(m)}(\emptyset'_{f(m)}) \upharpoonright_m = Y \upharpoonright_m$ and $r = \Psi_{e,f(m)}(\emptyset'_{f(m)}; m)\downarrow$, and define $Y(m)$ to be not equal to r. If there is no such e define $Y(m) = 0$.
Each e eventually has strongest priority at a stage m such that $f(m) \ge g(m)$. Since $\emptyset'_{f(m)} \upharpoonright_{h(m)}$ is stable, we cause $Y \ne \Psi_e(\emptyset')$.

8.2.11 A is low for K. If $f \le_T A$ then $K(f(x)) \le^+ K^A(f(x)) \le^+ K^A(x) \le^+ 2|x|$. To obtain a c.e. trace that works for all f on almost all inputs, let $T_x = \{y \colon K(y) \le 3|x|\}$ if this set is nonempty and $\{0\}$ otherwise. Note that $\#T_x = O(x^3 + 1)$.

8.2.12 We meet the lowness requirements L_e from the proof of Theorem 1.6.4. We build a functional Γ such that $f = \Gamma^A$ is total, and the requirements $R_k \colon \exists n > k\, f(n) \notin V_n$ are met, where $(V_n)_{n \in \mathbb{N}}$ is a universal c.e. trace as in Corollary 8.2.4. The priority ordering is $R_0 > L_0 > R_1 > \ldots$; if L_e $(e < k)$ acts then R_k is initialized. The strategy for R_k, after being last initialized at stage t, chooses $n = \langle t, k \rangle$ as a candidate and keeps redefining $\Gamma^A(n)$ by changing A, until $\Gamma^A(n) \notin V_n$.
The set A is not superlow because the final candidate of R_k depends on when one of the L_e for $e < k$ acts for the last time. This determines $\#V_n$, and hence the number of injuries to L_k.

8.2.13 Let $g(n, s) = \max V_{n,s}$ where $(V_n)_{n \in \mathbb{N}}$ is a universal c.e. trace as in Corollary 8.2.4. Then $e \in \mathrm{Tot}^A \Leftrightarrow \exists x \exists s \forall t \ge s$

$$\forall n < x \,\mathsf{use}\, \Phi_e^A(n)[t] \le s \ \& \ \forall n \ge x\, \exists v \ge t \left[\mathsf{use}\, \Phi_e^A(n)[v] \le g(n, v)\right].$$

The implication "$\Rightarrow$" holds since $\lambda e.\, \mathsf{use}\, \Phi^A(e)$ is an A–computable function. The converse implication holds because the right hand side implies that for each n there are only finitely many possibilities for $\Phi_e^A(n)[v]$.
The Σ^0_3 index for Tot^A is obtained effectively from the computable approximation of A.

8.2.14 See Barmpalias, Downey and Ng (20xx).

8.2.19 Clearly $\operatorname{ran} F \le_T F$ because $|F(\sigma)| \ge |\sigma|$ for each σ. To show $F \le_T \operatorname{ran} F$, suppose that $y = F(\sigma)$ has already been determined. Using $\operatorname{ran} F$ as an oracle, find the immediate successors $y_0 <_L y_1$ of $F(\sigma)$ in $\operatorname{ran} F$. Then $y_a = F(\sigma a)$ for $a \in \{0, 1\}$.

8.2.20 For the first statement, see the proof of a stronger result, Theorem 8.4.4.
For the second statement, apply the basis theorem for computably dominated sets 1.8.42.

8.2.21 For an interval $I = [k, l)$ we write $f \upharpoonright I$ for the tuple $(f(k), \ldots, f(l-1))$. Let $n(x) = n$ if $x \in I_n$.
$\Rightarrow$: Given $f \le_T A$ let $(T_n)_{n \in \mathbb{N}}$ be a c.e. [computable] trace with bound h for the function $\lambda n. f \upharpoonright I_n$. We may assume T_n only contains tuples of length $h(n)$. [We may also assume that $\#T_n = h(n)$.] Suppose $x \in I_n$, $x = \min I_n + i$. Let $\psi(x) \simeq$ the i-th

entry of the $i+1$st elemented enumerated in T_n. If $f \restriction I_n$ is the $i+1$st tuple in T_n then $\psi(x) = f(x)$.
$\Leftarrow$: Let $\widetilde{f}(x) = f \restriction I_{n(x)}$. Pick a [total] function ψ for $\widetilde{f}$. We may assume that $\psi(y)$ is a tuple of length $h(n(y))$ for each y. Let S_x be the union of the sets of entries occurring in $\psi(y)$ for some $y \in I_{n(x)}$. Then $(S_x)_{x\in\mathbb{N}}$ is a c.e. [computable] trace for f. Note that $\#S_x \le h(n(x))^2$.
8.2.24 Modify the proof of Theorem 8.2.23.
$\Rightarrow$: There is a c.e. trace $(W_{g(n)})_{n\in\mathbb{N}}$ with bound $\lambda n.2^n$ for f. Define L as before but using $W_{g(n)}$ instead of $D_{g(n)}$. Then L is a bounded request set, and the corresponding prefix-free machine N satisfies $K^A_M(y) \ge K_N(y) - 1$ for every y.
$\Leftarrow$: Given $f \le_T A$, define M as before. Then $\forall y\, K(y) \le K^A_M(y) + b$ for some b. Let g be a computable function such that $W_{g(x)} = \{y\colon K(y) \le 3\log x\}$, then $\#W_{g(x)} \le 2x^3+1$ for each x, and $f(x) \in W_{g(x)}$ for almost all x.

8.2.25 Adapt the proof of 5.2.3. Let $\widetilde{M}$ be the computable measure machine relative to A such that $\widetilde{M}(\sigma) \simeq A\restriction_{M(\sigma)}$ for each σ. Choose a c.m.m. N for $\widetilde{M}$ as in 8.2.22, then for each n we have $K_N(A\restriction_n) \le^+ K_{\widetilde{M}}(A\restriction_n) \le^+ K_M(n)$.
8.2.31 $\forall n\, K(Z\restriction_n \mid n) \le^+ \log n$, so $K(Z\restriction_n) \le^+ K(Z\restriction_n \mid n) + K(n) = O(\log n)$.
8.2.32 Let $\widetilde{h}(n) = \lfloor h(n)/3 \rfloor$. There are infinitely many n such that $K(n) < \widetilde{h}(n)$ and $K(Z\restriction_n \mid n) \le \widetilde{h}(n)$. For every such n, we have

$$K(Z\restriction_n) \le^+ K(Z\restriction_n \mid n) + K(n) \le^+ \widetilde{h}(n) + K(n) \le^+ 2\widetilde{h}(n).$$

Section 8.3
8.3.5 We show that the function $g(n) = \mu s.\, \Omega_s\restriction_{2n} = \Omega\restriction_{2n}$ dominates each function $f \le_T A$. Since $g \le_{wtt} \Omega \le_{wtt} \emptyset'$, this implies that A is array computable.
The set $L = \{\langle n+1, \Omega_{f(n)}\restriction_{2n}\rangle\colon n \in \mathbb{N}\}$ is a bounded request set relative to A of total weight 1. Let M be the corresponding computable measure machine relative A. If $g(n) \le f(n)$ then $K^A_M(\Omega\restriction_{2n}) \le n+1$. If there are infinitely many such n, then Ω is not Schnorr random relative to A.

8.3.6 (i.a) Let C_n be the n-th clopen set in some effective listing of the clopen sets. Suppose $(G_m)_{m\in\mathbb{N}}$ is a Kurtz test relative to A (Definition 3.5.1). Then there is $f \le_T A$ such that $\forall m\, C_{f(m)} = G_m$. Let $S_m = C_{J(m)}$ if $J(m)\downarrow$ and $\lambda C_{J(m)} \le 2^{-m}$; otherwise let $S_m = \emptyset$. Since A does not have d.n.c. degree there are infinitely many m such that $G_m = S_m$. So, whenever Z fails $(G_m)_{m\in\mathbb{N}}$ then Z fails the Solovay test $(S_m)_{m\in\mathbb{N}}$.
(i.b) Clearly $Z \notin \mathsf{WR}^D$. Thus $D \notin \mathrm{Low}(\mathsf{MLR}, \mathsf{WR})$, hence D has d.n.c. degree by (i).
(ii) Let $f \le_T A$ be a d.n.c. function. Let $Q = 2^{\mathbb{N}} - \mathcal{R}_1$ be a nonempty Π^0_1 class of ML-random sets. Using f compute a sequence $d_0 < d_1 < \ldots$ such that for each n the Π^0_1 class $\{Z \in Q\colon d_0, \ldots, d_{n-1} \in Z\}$ is nonempty. Let $D = \{d_0, d_1, d_2, \ldots\}$. By compactness $\{Z \in Q\colon D \subseteq Z\}$ is nonempty.
Suppose we have determined $d_0 < \ldots < d_{n-1}$ such that $P = \{Z \in Q\colon d_0, \ldots, d_{n-1} \in Z\}$ is nonempty. The set $G = \{m\colon \forall Z \in P\, [Z(m) = 0]\}$ is c.e. uniformly in an index for P. We will determine $d_n \notin G$.
Since $P \subseteq \mathsf{MLR}$ we can by 3.3.3 compute k such that $2^{-k} < \lambda P$ and hence $\#G \le k$.
Identify $\mathbb{N}^*$ with $\mathbb{N}$ via some computable bijection. Let $(S_y)_{y\in\mathbb{N}^*}$ be a uniformly computable sequence of sets such that $S_\varnothing = \mathbb{N}$ and for each y, $(S_{y\frown i})_{i\in\mathbb{N}}$ is a partition of S_y. Let $\Psi(y) = i$ if i is the first number such that some element of $S_{y\frown i}$ is enumerated in G. Let α be the effectively obtained reduction function for Ψ. Thus $J(\alpha(y)) \simeq \Psi(y)$ and hence $\neg f(\alpha(y)) = \Psi(y)$ for each y.

Now let $y_0 = \varnothing$ and $y_{i+1} = y_i\widehat{\ }f(\alpha(y_i))$ for $i < k$. Clearly $G \cap S_{y_k} = \emptyset$ since for each $i < k$ some element of G is in some $S_{y_i\widehat{\ }r}$ for $r \neq f(\alpha(y_i))$ (unless already $G \cap S_{y_i} = \emptyset$). Choose $d_n > d_{n-1}$ in S_{y_k}.

Section 8.4

8.4.6 The markers $F_s(\sigma)$ for $|\sigma| = 2e$ are now used to make $J_s^{F_s(\sigma)}(e)$ convergent whenever possible, by the same strategy as before. The markers $F_s(\sigma)$, $|\sigma| = 2e+1$, ensure that no path of the function tree F is computable. For each such σ, we satisfy the requirement

$$R_e\colon\ \exists k < |F(\sigma)|\left[\neg\Phi_e(k) = F(\sigma)(k)\right].$$

At stage $s+1$, if R_e is not satisfied along σ and $\Phi_{e,s}(k) = 0$ where $k = |F_s(\sigma)|$, then let $F_{s+1}(\sigma\alpha) = F_s(\sigma 1\alpha)$ for each string α, so that $F_{s+1}(\sigma)(k) = 1$. As before, for each i, a value $F_s(\sigma)$, $|\sigma| = i$, can change at most $2^{i+1} - 1$ times. Let $T_e = \{J_s^{F_s(\sigma)}(e)\colon s \in \mathbb{N}, |\sigma| = 2e\}$, then $(T_e)_{e\in\mathbb{N}}$ is a c.e. trace with bound $\lambda e.2^{2e}(2^{2e+1} - 1)$.

8.4.8 For each σ, X we have $K^X(J^X(\mathbb{U}(\sigma))) \leq^+ |\sigma|$ if $J^X(\mathbb{U}(\sigma))$ is defined. Hence $K(J^A(n)) \leq^+ K(n) \leq^+ \log n + 2\log\log n$ for each n. Thus for sufficiently large b the c.e. trace given by $T_n = \{y\colon K(y) \leq b + \log n + 2\log\log n\}$ is as required.

8.4.12 V_e is jump-traceable $\Leftrightarrow$ there is a u.c.e. sequence $(T_n)_{n\in\mathbb{N}}$ and a computable h such that $\forall n\, \#T_n \leq h(n)$ & $\forall n \forall s \exists t \geq s\left[J^{V_e}(n)[t]\uparrow \ \vee\ J^{V_e}(n) \in T_{n,t}\right]$). (The implication "$\Rightarrow$" is clear. For the converse implication, note that if $J^{V_e}(n)\downarrow$ then the condition implies $J^{V_e}(n) \in T_n$.)
For each $e \in \mathbb{N}$ we can effectively obtain $\widehat{e}$ such that $W_e = V_{\widehat{e}}$. This proves that the second index set is Σ^0_3. For Σ^0_3-hardness it suffices to consider the c.e. case. Apply Exercise 1.4.23.

8.4.19 The first two are routine. To show $A' \not\leq_{JT} A$, suppose that $(T_n)_{n\in\mathbb{N}}$ is a uniform sequence of c.e. operators. Let Ψ be a Turing functional such that for each X and n we have $\Psi(X'; n) \simeq \min(\mathbb{N} - T^X_{p(n)})$, where p is a reduction function for Ψ by the Recursion Theorem. If $T^A_{p(n)}$ is finite then $J(A'; p(n)) = \Psi(A'; n) \notin T^A_{p(n)})$. Hence $(T^A_n)_{n\in\mathbb{N}}$ is not a jump trace for A' relative to A.

8.4.20 C is JT-hard $\leftrightarrow$ $\exists$ *computable functions* g, r

$$\forall i\, \forall s\, \#W^C_{r(i),s} \leq g(i) \ \&\ \forall i\, \forall s\, \exists t > s\, [J^{\emptyset'}(i)[t]\uparrow \ \vee\ J^{\emptyset'}(i)[t] \in W^C_{r(i),t})].$$

8.4.21 We say that $A \leq_{CT} B$ if there is an order function h as follows: for each $f \leq_T A$ there is $p \leq_T B$ such that $\forall n\, f(n) \in D_{p(n)}$ and $\forall n\, \#D_{p(n)} \leq h(n)$. We may assume that $\#D_{p(n)} = h(n)$.
Suppose $A \leq_{CT} B$. If $f \leq_T A$, pick p and h as in the definition for $A \leq_{CT} B$. There is a binary function $q \leq_T B$ such that $q(\langle n, k\rangle)$ is the k-th element in $D_{p(n)}$ in order of magnitude for $k < h(n)$, and $q(\langle n, k\rangle) = 0$ otherwise.
Suppose further $B \leq_{CT} C$ via bound $\overline{h}$, and pick for q a C-computable trace $(T_i)_{i\in\mathbb{N}}$ with bound $\overline{h}$. Let $V_n = \bigcup_{k<h(n)} T_{\langle n,k\rangle}$. Then f is traced by $(V_n)_{n\in\mathbb{N}}$. The common bound on the traces for all such f is $\lambda n.\, h(n) \cdot \overline{h}(\langle n, h(n)\rangle)$.

8.4.26 This is somewhat similar to the implication "$\Leftarrow$" in 8.4.23 but simpler. Suppose the strictly increasing function $h \leq_{wtt} \emptyset'$ dominates each function computable in A. Let $h(e) = \lim_s h(e,s)$ where $h(e, s-1) \leq h(e,s)$ for each $s > 0$, and $h(e,s)$ changes at most $g(e)$ times for some computable function g. Suppose now that Φ^A is total, and let $\Psi^A(e) = \mu s.\, [\Phi^A(e)\downarrow [s]$ with A stable below the use]. We define a c.e. trace with bound g for Φ^A. At stage s, if $y = \Phi^A(e)[s]\downarrow$ and $\Psi^A(e)[s] \leq h(e,s)$ then enumerate y into T_e. If for a stage $t > s$ we have $\Phi^A(e)[t]\uparrow$, then the next value $\Psi^A(e)$ is greater than t

(here we used that A is c.e., and thus cannot turn back to a previous configuration). Therefore $\#T_e \le g(e)$. A finite variant of $(T_e)_{e\in\mathbb{N}}$ is a trace for Φ^A.

8.4.36 Given an order function g, let n and f be as in the proof of Claim 2. Let $q(k,t) = A'(k)$ for $k < n$. For $k \ge n$, if $t < f(k)$ let $q(k,t) = 0$. If $t \ge f(k)$, let $q(k,t) = 1$ in the case that $J^A(k)[t]\downarrow$, and $q(k,t) = 0$ otherwise.

8.4.37 $A \le_{SSL} B$ if for each order function b we have $A' \le_{tt} B'$ with the number of queries bounded by b. $A \le_{SJT} B$ if $A \le_{JT} B$ with bound h for each order function h. For the implication $\le_{SSL} \Rightarrow \le_{SJT}$ under the given hypotheses, we extend the proof of the second statement in Theorem 8.4.34 in the same manner as we did in the proof of Corollary 8.4.18. We write $C^B(x)$ instead of $C(x)$, and $C^B(x \mid y)$ instead of $C(x \mid y)$. Instead of $\mathbb{V}^2$ we use the oracle machine $\mathbb{V}^{B,2}$. We fix a Turing functional Γ such that $B = \Gamma(A)$. To define Ψ, in (2) we ask that $\mathbb{V}^{\Gamma(X),2}(\tau,\sigma)[s] = x$. This yields a machine with oracle B describing x in terms of n_x, σ_x, and d_x. The rest is as before.

Section 8.5

8.5.8 Modify the proof of 8.5.1. Let $r(b) = g(1+\log b)$. Then $x_0 < x_1 < \ldots < x_k$ and $\forall i < k\,[c(x_i,x_{i+1}) \ge 1/b]$ implies $k \le r(b)$. For $j > 0$, P^e_j with an allowance of $1/b$ now calls P^e_{j-1} with an allowance of $1/(br(b))$. Each run $P^e_j(b)$ is reset fewer than $r(b)$ times. Let $b_{e,e} = 2^e$ and $b_{e,j-1} = b_{e,j}r(b_{e,j})$ if $j > 0$. Let $\#I_e = b_{e,0}$. P^e_j is called at most $b_{e,j}2^{-e}$ times, and the argument in Claim 4 shows that the enumeration of A obeys c.

8.5.9 By 1.8.39 let Y be a ML-random Δ^0_2 set such that $Y \not\ge_T A$. Let c be $ the cost function c_Y defined in (5.7) on page 189.

8.5.10 The set A obeys $c_{\mathcal{K}}$. Hence A is K-trivial, and therefore jump traceable via a c.e. trace $(S_n)_{n\in\mathbb{N}}$ with bound a computable function b. Let Ψ be the Turing functional such that $\Psi^X(n) \simeq \mu s.\, J^X_s(n)\downarrow$ and pick a reduction function q for Ψ. Recall that $\Psi^A(x)[s]$ denotes $\Psi^{A_s}_s(x)$, and similarly for J. We write $J^A(n)\downdownarrows[s]$ if $J^A(n)\downarrow[s]$ and $\Psi^A(n)[s] \in S_{q(n),s}$. We think of $J^A(n)[s]$ as certified; note that for at most $b(q(n))$ times $J^A(n)$ can diverge after being certified. We say that $J^A(n)\downdownarrows[s]$ *newly* if $J^A(n)\downdownarrows[s]$ but not $J^A(n)\downdownarrows[s-1]$.

Given an order function h, we will define a jump trace $(T_n)_{n\in\mathbb{N}}$ with bound $O(h)$ for A, which suffices as we can replace h a given h by $\lfloor\sqrt{h}\rfloor$. Define a cost function c similar to (5.7) on page 189: let $c(x,s) = 0$ for each $x \ge s$. If $x < s$ and $n < x$ is least such that $J^A(n)\downdownarrows[s]$ newly, then let $c(x,s) = \max(c(x,s-1), 1/h(n))$. It is easy to see that c is benign, so choose a computable enumeration $(\widehat{A}_r)_{r\in\mathbb{N}}$ of A obeying c.

Let $s(0) = 0$, and $s(i+1) = \mu t > s(i).\, A_t\upharpoonright_{s(i)} = \widehat{A}_t\upharpoonright_{s(i)}$. We call the $s(i)$ *stages*. For a *stage* $s > 0$ we denote by $\overline{s}$ the preceding *stage*. We define the jump trace $(T_n)_{n\in\mathbb{N}}$ as follows: if $y = J^A(n)\downdownarrows[\overline{s}]$ with use u and $A_s\upharpoonright_u = A_{\overline{s}}\upharpoonright_u$ then put y into T_n at *stage* s. Clearly $J^A(n) \in T_n$ if it is defined.

We claim that $\#T_n = O(h(n))$. Suppose distinct numbers enter T_n at *stages* $t_1 < t_2 < \ldots$, then for each $j > 0$ we have $J^A(n)\downdownarrows$ newly at some r (with use $\le r$) such that $t_{j-1} < r \le \overline{t}_j$ (here $t_0 = 0$). Hence $c(x,s) \ge 1/h(n)$ for all x,s such that $x < r \le s$. Since $r \le \overline{t}_j$, by the definition of *stages* we have $\widehat{A}_{t_{j+1}}\upharpoonright_r \neq \widehat{A}_{t_j}\upharpoonright_r$, which incurs a cost of at least $1/h(n)$. Hence the length of the sequence (t_j) is bounded by $O(h(n))$.

8.5.21 Let $(B_t)_{t\in\mathbb{N}}$ be an enumeration of $\emptyset''$ relative to $\emptyset'$. For each Turing functional Φ let $C_\Phi = \{Z : \emptyset'' = \Phi(Z \oplus \emptyset')\}$. Then C_Φ is $\Pi^0_2(\emptyset')$, because $Z \in C_\Phi \leftrightarrow \forall n, s\, \exists t \ge s\, B_t\upharpoonright_n = \Phi_t(Z \oplus \emptyset')\upharpoonright_n$. Hence C_Φ is Π^0_3. Also, C_Φ is null by Sacks' Theorem 5.1.12 relativized to $\emptyset'$. Each weakly 3-random set Z is 2-random, and hence in GL_1. If Z is also high then Z is in some class C_Φ, contradiction.

8.5.22 The Δ_2^0 sets form a Σ_4^0 class since "$Y \le_T \emptyset'$" amounts to the Σ_4^0 condition $\exists e\, \forall n\, \exists s\, \forall t \ge s\, [Y(n) = \Phi_e^{\emptyset'}(n)[t]]$. If this class would be Σ_3^0 then by Theorem 5.3.15 there would be a low incomputable c.e. set B Turing below each ML-random Δ_2^0 set, contrary to Theorem 1.8.39.

8.5.23 By 1.4.4 $Y \in \mathcal{H}$ iff there is a truth-table reduction procedure Φ such that $\forall n, s\, \exists t \ge s\, [\Phi(\emptyset'; n)[t] = Y(n)]$.
For $\widetilde{\mathcal{H}}$, use instead the expression $\forall n, s\, \exists t \ge s\, [\Phi(\emptyset'; n)[t] = Y'_t(n)]$.

8.5.24 By 1.8.41 with $P = 2^{\mathbb{N}} - \mathcal{R}_1$, there is a ML-random set Y such that $(Y \oplus B)' \le_{tt} B' \le_{tt} \emptyset'$. Then $Y \in \widetilde{\mathcal{H}}$, so $A \le_T Y$. Therefore $(A \oplus B)' \le_m (Y \oplus B)' \le_{tt} \emptyset'$.

8.5.25 The second assertion follows from the first by letting $P = 2^{\mathbb{N}} - \mathcal{R}_1$. For the first assertion, we combine the framework given by the proof of Theorem 1.8.39 with a cost function construction of a c.e. K-trivial set B. Instead of working relative to $\emptyset'$, we provide at each stage s approximations of the objects to be constructed. For each $i < s$ we have an index $P^{i,s}$ for a Π_1^0 class. This index eventually stabilizes. We let $Y \in \bigcap_i P^i$, where P^i is the (class given by the) final index. To ensure that Y is superlow we meet the requirements

$$L_e\colon\ Y'(e) = \lim_s f(e, s),$$

where f is a computable binary "guessing" function defined during the construction such that $\lim_s f(e, s)$ exists and the number of changes is computably bounded. The strategy for L_e controls P^{2e+1}. Further, we meet the requirements

$$S_e\colon\ B \neq \Phi_e^Y$$

by a strategy that controls P^{2e+2} and also maintains a candidate x_e targeted for B. If m_e is the number of times S_e has been initialized so far, then the enumeration of x_e is allowed to incur a cost of at most 2^{-e-m_e}. Whenever the potential cost of enumerating x_e exceeds this quantity, the strategy picks a new large candidate. Such a change of candidates can occur up to 2^{m_e+e} times. The priority ordering is $S_0 > L_0 > S_1 > L_1 > \ldots$, and when a strategy changes its Π_1^0 class it initializes the strategies of weaker priority, because the environment they work in has changed.

Construction of B and Y. For a Π_1^0 class G we let $G[s]$ be the usual approximation at stage s by a clopen class given by (1.17). Let $P^{0,s} = P$ and $f(e, s) = 0$ for each e such that $2e + 1 \ge s$.

Carry out substages i for $1 \le i < s$.

Substage $i = 2e+1$. If $G[s] \neq \emptyset$ where $G = P^{2e,s} \cap \{Y\colon\ \Phi_e^Y(e)[s]\uparrow\}$, then let $P^{2e+1,s} = G$ and define $f(e, s) = 0$; otherwise let $P^{2e+1,s} = P^{2e,s}$ and $f(e, s) = 1$. In this case, if $f(e, s-1) = 0$, initialize the strategies of weaker priority (we say that L_e acts).

Substage $i = 2e+2$. (a) If S_e is not satisfied, and x_e is undefined or $c_{\mathcal{K}}(x_e, s) > 2^{-e-m_e}$, then let $x_e = \langle s, e\rangle$ and initialize the strategies of weaker priority. We say that S_e acts.
(b) If $G[s] \neq \emptyset$ where $G = P^{2e+1,s} \cap \{Y\colon\ \neg\Phi_e^Y(x_e)[s] = 0\}$, let $P^{2e+2,s} = G$; otherwise let $P^{2e+2,s} = P^{2e+1,s}$, put x_e into B and declare S_e satisfied. We say that S_e acts.

Verification. Let R_i be the i-th requirement in the priority list. We show that there is a computable function g such that R_i acts no more than $g(i)$ times. Suppose we have defined $g(j)$ for $1 \le j < i$. Let $k = \sum_{j<i} g(j)$. If $R_i = L_e$ then it acts no more than $g(i) := k + 1$ times. Now suppose that $R_i = S_e$. After each initialization, by the benignity feature of the standard cost function $c_{\mathcal{K}}$, the candidate x_e changes at most 2^{e+k} times. Thus R_e acts no more than $g(i) := k(2^{e+k} + 1)$ times.

If $f(e,s) \neq f(e,s-1)$ then some R_i acts at stage s where $i \leq 2e+1$, so the number of changes of $f(e)$ is computably bounded. Clearly $\lim_s f(e,s) = Y'(e)$, so Y is superlow. Let x be the final candidate of S_e. If $\neg\Phi_e^Y(x) = 0$ then $x \notin B$. If $\Phi_e^Y(x) = 0$ then eventually the alternative (b) in S_e applies at substage $2e+2$, so $x \in B$. In either case S_e is met. Finally, B is K-trivial because its enumeration obeys $c_{\mathcal{K}}$.

Section 8.6

8.6.1 (i) The implications are: "not of d.n.c. degree" $\Rightarrow$ *BN1R* $\Rightarrow$ "not of PA degree". To argue in terms of highness properties, each set of PA degree bounds a ML-random set, but a low ML-random set is not of PA degree. Each ML-random set is of d.n.c. degree, but there is a set of d.n.c. degree that does not compute a ML-random set by Ambos-Spies, Kjos-Hanssen, Lempp and Slaman (2004). (ii) The implications are: computable $\Rightarrow$ *BB2R* $\Rightarrow$ Low(Ω). The second implication is strict because an incomputable set in Low(MLR) is not computed by a 2-random set. (iii) The minimal conull classes in the diagram are Low(Ω), "not of PA degree" and "not high".

8.6.3 We discuss two classes. (1) The sets that are not of d.n.c. degree: by 6.3.14 let C be a Turing incomplete LR-hard c.e. set. Then C is not of d.n.c. degree. Also C is high and hence not in GL_2. (2) The sets that are computably dominated: such a set can be of PA degree by 1.8.42. It can fail to be in GL_1 by the discussion after 8.4.7.

8.6.4 (i) has been solved by Barmpalias (2011). He shows that in fact no array computable set can be LR-hard. This yields a new implication in in Fig. 8.1.

8.6.6 Combine Corollary 8.4.5 with Theorem 7.5.2.

Solutions to Chapter 9

9.1.2 $\Rightarrow$: $A_R = \{x\colon \forall Y \succ x\,[Y \in R]\}$.
$\Leftarrow$: The relation $\mathcal{B} = \{\langle n, Y\rangle\colon Y\upharpoonright_n \in A_R\}$ is Π^1_1, and $R = \{Y\colon \exists n\, \langle n, Y\rangle \in \mathcal{B}\}$.

9.1.6 We are given indices e, i for the Π^1_1 classes $\mathcal{P}$ and $\mathcal{Q}$. Replacing Φ_e^X by a linear order of type $|\Phi_e^X|+\omega+1$, and Φ_i^X by a linear order of type $|\Phi_i^X|+\omega+2$, we may assume that sets enter $\mathcal{P}$ and $\mathcal{Q}$ at different stages. Let $\widetilde{\mathcal{P}} = \{X \in \mathcal{P}\colon \Phi_i^X$ is not isomorphic to an initial segment of $\Phi_e^X\}$ (X enters $\mathcal{P}$ at a stage when it has not entered $\mathcal{Q}$). Symmetrically, let $\widetilde{\mathcal{Q}} = \{X \in \mathcal{Q}\colon \Phi_e^X$ is not isomorphic to an initial segment of $\Phi_i^X\}$.

9.1.7 For each pair i, j let $n = \langle i, j\rangle$, and let $\mathcal{Q}_{\overline{g}(n)}$ be the Π^1_1 class $\{X \in \mathcal{Q}_i\colon \forall Z\,[Z \in \mathcal{Q}_i \vee Z \in \mathcal{Q}_j]\}$. Now apply 9.1.6 to the classes $\mathcal{Q}_{\overline{g}(n)}$ and $\mathcal{Q}_j$ in order to obtain $\mathcal{Q}_{g(n)}$ and $\mathcal{Q}_{h(n)}$.

9.1.8 Let $\widetilde{W}_e$ be the relation (with domain $\mathbb{N}$) isomorphic to $\omega + W_e$. Let R be the well-order isomorphic to the ω-sum of all the $\widetilde{W}_e$ that are well-orders. Then $|R| = \omega_1^{\mathrm{ck}}$ as each proper initial segment of R is a computable well-order. Since $R \leq_T O$ we have $|R| < \omega_1^O$.

9.1.18 The first statement is clear because $\langle n, y\rangle$ is not in the graph of f iff $\langle n, z\rangle$ is in the graph for some $z \neq y$. For the second, let g be a computable function such that $\langle n, y, X\rangle \in \mathcal{S} \leftrightarrow \Phi^X_{g(n,y)}$well-ordered. Replace $\mathcal{S}$ by the reduction procedure $\widetilde{\mathcal{S}} = \{\langle n, y, X\rangle \in \mathcal{S}\colon \forall z \neq y\,[\Phi^X_{g(n,z)}$ is not a proper initial segment of $\Phi^X_{g(n,y)}$ & $[\Phi^X_{g(n,z)} \cong \Phi^X_{g(n,y)} \to y < z]]\}$. Then $f \leq_h A$ via $\widetilde{\mathcal{S}}$, and $\widetilde{\mathcal{S}}$ is as required.

9.1.19 Let $\psi_e = \Phi_e^\emptyset$ be the e-th partial computable function. Each Π^1_1 set is of the form $\psi_e^{-1}(\mathcal{O})$ for some total ψ_e. A total function f is Δ^1_1 iff its graph is Π^1_1. Let $g(e,x) = \min\{y\colon \psi_e(\langle x, y\rangle) \in \mathcal{O}\}$ if there is such a y, and $g(e,x) = 0$ otherwise. Then g is as required. If O is Δ^1_1 then $\lambda e.\, g(e,e)+1$ is Δ^1_1, contradiction.

9.1.24 We may assume the domain of R is $\mathbb{N}$. Note that $R \in L(\omega_1^{ck})$ by 9.1.21. Let $\gamma = |R|$. By transfinite recursion the isomorphism $g\colon \gamma \to R$ is Σ_1 over $L(\omega_1^{ck})$ (since $g(\alpha) \simeq$ the least element with respect to R in $\omega - \text{ran}(g \restriction_\alpha)$.) Let $f(n)$ be the ordinal isomorphic to the initial segment of R given by n. Then f is Σ_1 over $L(\omega_1^{ck})$. Hence by the Bounding Principle the range of f is bounded, whence γ is a computable ordinal.
9.2.7 By transfinite induction over $L(\omega_1^{ck})$, for each Π_1^1 set S the function $\alpha \to S_\alpha$ is Σ_1 over $L(\omega_1^{ck})$.
We extend the solution to 2.2.24. We may assume that at most one request enters W at each stage α. Define the prefix-free Π_1^1-machine M as follows. If $\langle r, y\rangle$ enters W at stage α, let $k = r+1$. Let i_α be the least i such that for $X = W_\alpha$

$$\forall F \subseteq X \text{ finite } \left[i2^{-k} \geq \textstyle\sum_{l,z} 2^{-l} [\![\langle l, z\rangle \in F]\!]\right].$$

Note that the foregoing expression in variables X, i, k is Δ_0 over $L(\omega_1^{ck})$. Now as before let w_α be the string w of length k such that $0.w = i2^{-k}$. The function taking α to w_α (and to $\{\text{nil}\}$ if no element is enumerated at α) is Σ_1 over $L(\omega_1^{ck})$. Define $M(w_\alpha) = y$.
9.3.10 If $\mathcal{S}$ is null then $\mathcal{S} \subseteq \mathcal{Q}$. Otherwise, by the Sacks–Tanaka Theorem 9.1.11, $\mathcal{S}$ has a hyperarithmetic member X, so $\{X\}$ is a Π_1^1 null class, whence $X \in \mathcal{Q}$.
9.3.11 We claim that $\mathcal{G} = \bigcup_n \widehat{\mathcal{Q}}_{g(n)}$, where g is as in Exercise 9.1.7 and $\widehat{\mathcal{S}}$ for a Π_1^1 class $\mathcal{S}$ is defined in the proof of 9.3.6. Then $\mathcal{G}$ is Π_1^1 because we uniformly in n have an index for $\widehat{\mathcal{Q}}_{g(n)}$ as a Π_1^1 class.
If $X \in \mathcal{G}$ then $X \in \mathcal{S}$ for some null Δ_1^1 class $\mathcal{S}$. Then $\mathcal{S} = \mathcal{Q}_{g(n)}$ for some n and $\widehat{\mathcal{S}} = \mathcal{S}$. Now suppose that $X \in \widehat{\mathcal{Q}}_{g(n)}$. If $\mathcal{Q}_{g(n)}$ is null we are done. Otherwise let γ be the least ordinal such that $\mathcal{Q}_{g(n),\gamma}$ is not null. Then $\gamma < \omega_1^{ck}$ by 9.1.15, so $X \in \widehat{\mathcal{Q}}_{g(n),\alpha+1}$ where $\alpha + 1 < \gamma$. Therefore X is in a null Δ_1^1 class by the Approximation Lemma 9.1.4(ii).
9.3.12 $\Rightarrow$: The class $\mathcal{L} = \{X \oplus Y\colon X \in \mathcal{R}^Y \ \&\ Y \in \mathcal{R}\}$ is Σ_1^1. Since $\mathcal{R}^Y$ is co-null for each Y, by Fubini's Theorem $\mathcal{L}$ is conull. Hence $\mathcal{R} \subseteq \mathcal{L}$.
$\Leftarrow$: Let $\mathcal{R}[Y] = \{X\colon X \oplus Y \in \mathcal{R}\}$. Then the class $\{Y\colon \mathcal{R}[Y] \text{ is conull}\}$ is Σ_1^1 by the Measure Lemma 9.1.10, and conull by Fubini's Theorem: otherwise there are rationals $\epsilon > 0$ and $q < 1$ such that $\lambda\{Y\colon \lambda\mathcal{R}[Y] \leq q\} \geq \epsilon$, so that $\lambda\mathcal{R} = \int_Y (\lambda\mathcal{R}[Y]) d\lambda \leq \epsilon q + (1-\epsilon) < 1$. Thus, if $Y \in \mathcal{R}$ then $\mathcal{R}[Y]$ is conull. Since $\mathcal{R}[Y]$ is $\Sigma_1^1(Y)$, $X \in \mathcal{R}^Y$ implies $X \in \mathcal{R}[Y]$, that is, $X \oplus Y \in \mathcal{R}$.
9.3.13 (i) Since X is ML-random, by Remark 5.6.2 we have $X \leq_{vL} Y$. Each set $S \geq_T \emptyset'$ is Turing equivalent to a ML-random set. Thus, by the hint, Y is Δ_1^1-random.
(ii) Suppose X is 2-random. Then X is low for Ω. If $X \leq_K Y$ then $Y \leq_{LR} X$ by 8.1.17. Thus, by 8.1.10, we have $Y \leq_T X'$ and hence $Y \leq_h X$. Now we use Theorem 9.3.9: if X is Π_1^1-random then $\omega_1^X = \omega_1^{ck}$, so $\omega_1^Y = \omega_1^{ck}$. As Y is Δ_1^1-random by (i), Y is Π_1^1-random.
9.4.4 Modify the proof of Fact 9.4.2. Let $f = \lambda n.\, \max_{e \leq n} g(e, n) + 1$, then f dominates each hyperarithmetical function. If $\omega_1^{ck} < \omega_1^A$ then $f \leq_h O \leq_h A$.
9.4.5 (L. Yu) Using the Spector–Gandy Theorem 9.1.21, the graphs of functions dominating each hyperarithmetical function form a nonempty Σ_1^1 class. Now apply 9.1.16.
9.4.6 By 9.4.3 it remains to prove the implication "$\Leftarrow$". Since $\omega_1^Z = \omega_1^{ck}$ it suffices to show that Z is Π_1^1-ML-random. Otherwise $Z \in \bigcap_m G_m$ for a Π_1^1-ML-test $(G_m)_{m \in \mathbb{N}}$. By the bounding principle 9.1.22 there is $\alpha < \omega_1^Z$ such that $Z \in \bigcap_m G_{m,\alpha}$. Let $f(m) = \min\{r\colon [Z \restriction_r] \subseteq G_{m,\alpha}\}$, then $f \leq_h Z$. So choose a hyperarithmetical function g such that $g(m) \geq f(m)$ for each m. Let $P = \bigcap_m [\{x\colon [x] \subseteq G_{m,\alpha} \ \&\ |x| \leq g(m)\}]^{\prec}$, then P is a closed null Δ_1^1 class such that $Z \in P$.

REFERENCES

Ambos-Spies, K., Jockusch, Jr., Carl G., Shore, Richard A., and Soare, Robert I. (1984). An algebraic decomposition of the recursively enumerable degrees and the coincidence of several degree classes with the promptly simple degrees. *Trans. Amer. Math. Soc.*, **281**, 109–128.

Ambos-Spies, K., Kjos-Hanssen, B., Lempp, S., and Slaman, T. (2004). Comparing DNR and WWKL. *J. Symbolic Logic*, **69**(4), 1089–1104.

Ambos-Spies, K. and Kučera, A. (2000). Randomness in computability theory. In *Computability Theory and Its Applications: Current Trends and Open Problems* (ed. P. Cholak, S. Lempp, M. Lerman, and R. Shore). American Mathematical Society.

Ambos-Spies, K., Mayordomo, E., Wang, Y., and Zheng, X. (1996). Resource-bounded balanced genericity, stochasticity and weak randomness. In *STACS 96 (Grenoble, 1996)*, Volume 1046 of *Lecture Notes in Comput. Sci.*, pp. 63–74. Springer, Berlin.

Ambos-Spies, K., Weihrauch, K., and Zheng, X. (2000). Weakly computable real numbers. *J. Complexity*, **16**(4), 676–690.

Arslanov, M. M. (1981). Some generalizations of a fixed-point theorem. *Izv. Vyssh. Uchebn. Zaved. Mat.* (5), 9–16.

Barmpalias, G. (2010). Relative randomness and cardinality. *Notre Dame J. Form. Log.*, **51**(2), 195–205.

Barmpalias, G. (2011). Tracing and domination in the Turing degrees. *Ann. Pure Appl. Logic*. To appear.

Barmpalias, G., Downey, R., and Greenberg, N. (2009). K-trivial degrees and the jump-traceability hierarchy. *Proc. Amer. Math. Soc.*, **137**(6), 2099–2109.

Barmpalias, G., Downey, R., and Ng, S. (20xx). Jump inversions inside effectively closed sets and applications to randomness. *J. Symbolic Logic*. To appear.

Barmpalias, G. and Lewis, A. (20xx). Chaitin's halting probability and the compression of strings by oracles. To appear.

Barmpalias, G., Lewis, A., and Soskova, M. (2008). Randomness, Lowness and Degrees. *J. Symbolic Logic*, **73**(2), 559–577.

Barmpalias, G., Lewis, A., and Stephan, F. (2008). Π^0_1 classes, LR degrees and Turing degrees. *Ann. Pure Appl. Logic*, **156**(1), 21–38.

Barmpalias, G., Miller, J., and Nies, A. (2011). Randomness notions and partial relativization. *Israel J. Math.* To appear.

Barmpalias, G. and Nies, A. (2011). Upper bounds for ideals in the computably enumerable degrees. *Ann. Pure Appl. Logic*. To appear.

Barmpalias, G. and Sterkenburg, T. (20xx). On the number of infinite sequences with trivial initial segment complexity. To appear.

Bedregal, B. and Nies, A. (2003). Lowness properties of reals and hyper-immunity. In *Wollic 2003*, Volume 84 of *Electronic Notes in Theoretical Computer Science*. Elsevier.

Bickford, M. and Mills, F. (1982). Lowness properties of r.e. sets. Manuscript, UW Madison, 1982.

Bienvenu, L. and Downey, R. (2009). Kolmogorov complexity and Solovay functions. In *Symposium on Theoretical Aspects of Computer Science (STACS 2009)*, Volume 09001 of *Dagstuhl Seminar Proceedings*, http://drops.dagstuhl.de/opus/volltexte/2009/1810, pp. 147–158.

Bienvenu, L., Downey, R., Greenberg, N., Nies, A., and Turetsky, D. (20xx). Lowness for Demuth randomness. In preparation.

Bienvenu, L. and Miller, J. (20xx). Randomness and lowness notions via open covers. *Ann. Pure Appl. Logic*. To appear.

Binns, S., Kjos-Hanssen, B., Lerman, M., and Solomon, R. (2006). On a conjecture of Dobrinen and Simpson concerning almost everywhere domination. *J. Symbolic Logic*, **71**(1), 119–136.

Calhoun, W. (1993). Incomparable prime ideals of the recursively enumerable degrees. *Ann. Pure Appl. Logic*, **63**, 36–56.

Calude, C., Hertling, P., Khoussainov, B., and Wang, Y. (2001). Recursively enumerable reals and Chaitin Ω numbers. *Theoret. Comput. Sci.*, **255**(1-2), 125–149.

Calude, Cristian S., Nies, André, Staiger, Ludwig, and Stephan, Frank (2008). Universal recursively enumerable sets of strings. In *Developments in language theory*, Volume 5257 of *Lecture Notes in Comput. Sci.*, pp. 170–182. Springer, Berlin.

Carathéodory, C. (1968). *Vorlesungen über reelle Funktionen*. Third (corrected) edition. Chelsea Publishing Co., New York.

Chaitin, G. (1975). A theory of program size formally identical to information theory. *J. Assoc. Comput. Mach.*, **22**, 329–340.

Chaitin, G. (1976). Information-theoretical characterizations of recursive infinite strings. *Theoretical Computer Science*, **2**, 45–48.

Cholak, P., Downey, R., and Greenberg, N. (2008). Strongly jump-traceability I: the computably enumerable case. *Adv. in Math.*, **217**, 2045–2074.

Cholak, P., Greenberg, N., and Miller, J. (2006). Uniform almost everywhere domination. *J. Symbolic Logic*, **71**(3), 1057–1072.

Chong, C., Nies, A., and Yu, L. (2008). The theory of higher randomness. To appear in Israel J. Math.

Chong, C. T. and Yang, Yue (2000). Computability theory in arithmetic: provability, structure and techniques. In *Computability theory and its applications (Boulder, CO, 1999)*, Volume 257 of *Contemp. Math.*, pp. 73–81. Amer. Math. Soc., Providence, R. I.

Church, A. (1940). On the concept of a random sequence. *Bull. Amer. Math. Soc*, **46**, 130–135.

Cooper, B., Harrington, L., Lachlan, A.., Lempp, S., and Soare, R. (1991). The d.r.e. degrees are not dense. *Ann. Pure Appl. Logic*, **55**, 125–151.

Csima, B. and Montalbán, A. (2005). A minimal pair of K-degress. *Proceedings of the Amer. Math. Soc.*, **134**, 1499–1502.

de Leeuw, K., Moore, E., Shannon, C., and Shapiro, N. (1956). Computability by probabilistic machines. In *Automata studies*, Annals of mathematics studies, no. 34, pp. 183–212. Princeton University Press, Princeton, N. J.

Demuth, O. (1988). Remarks on the structure of tt-degrees based on constructive measure theory. *Comment. Math. Univ. Carolin.*, **29**(2), 233–247.

Diamondstone, D. (2009). Promptness does not imply superlow cuppability. *J. Symbolic Logic*, **74**(4), 1264–1272.

Ding, D., Downey, R., and Yu, L. (2004). The Kolmogorov complexity of random reals. *Ann. Pure Appl. Logic*, **129**(1-3), 163–180.

Dobrinen, N. and Simpson, S. (2004). Almost everywhere domination. *J. Symbolic Logic*, **69**(3), 914–922.

Doob, J. L. (1994). *Measure theory*, Volume 143 of *Graduate Texts in Mathematics*. Springer-Verlag, New York.

Downey, R. and Greenberg, N. (2011). Strong jump traceability II: the general case. *Israel J. Math.*. In press.

Downey, R., Greenberg, N., Mikhailovich, N., and Nies, A. (2008). Lowness for computable machines. In *Computational prospects of infinity II*, Volume 15 of *IMS Lecture Notes Series*, pp. 79–86. World Scientific.

Downey, R. and Griffiths, E. (2004). Schnorr randomness. *J. Symbolic Logic*, **69**(2), 533–554.

Downey, R., Griffiths, E., and LaForte, G. (2004). On Schnorr and computable randomness, martingales, and machines. *MLQ Math. Log. Q.*, **50**(6), 613–627.

Downey, R. and Hirschfeldt, D. (2010). *Algorithmic randomness and complexity*. Springer-Verlag, Berlin. 855 pages.

Downey, R., Hirschfeldt, D., Miller, J., and Nies, A. (2005). Relativizing Chaitin's halting probability. *J. Math. Log.*, **5**(2), 167–192.

Downey, R., Hirschfeldt, D., and Nies, A. (2002). Randomness, computability and density. *SIAM J. Computing*, **31**, 1169–1183.

Downey, R., Hirschfeldt, D., Nies, A., and Stephan, F. (2003). Trivial reals. In *Proceedings of the 7th and 8th Asian Logic Conferences*, Singapore, pp. 103–131. Singapore University Press.

Downey, R., Hirschfeldt, D., Nies, A., and Terwijn, S. (2006). Calibrating randomness. *Bull. Symbolic Logic*, **12**(3), 411–491.

Downey, R. and Miller, J. (2006). A basis theorem for Π^0_1 classes of positive measure and jump inversion for random reals. *Proc. Amer. Math. Soc.*, **134**(1), 283–288.

Downey, R. and Ng, S. (2009). Lowness for Demuth randomness. In *Mathematical Theory and Computational Practice*, Volume 5635 of *Lecture Notes in Computer Science*, pp. 154–166. Springer Berlin / Heidelberg. Fifth Conference on Computability in Europe, CiE 2009, Heidelberg, Germany, July 19-July 24.

Downey, R., Nies, A., Weber, R., and Yu, L. (2006). Lowness and Π^0_2 nullsets. *J. Symbolic Logic*, **71**(3), 1044–1052.

Downey, R. and Shore, R. (1997). There is no degree-invariant half-jump. *Proc. Amer. Math. Soc.*, **125**, 3033–3037.

Downey, R. G., Jockusch, Jr., Carl G., and Stob, M. (1990). Array nonrecursive sets and multiple permitting arguments. In *Recursion Theory Week, Oberwolfach 1989* (ed. K. Ambos-Spies, G. H. Muller, and G. E. Sacks), Volume 1432 of *Lecture Notes in Mathematics*, Heidelberg, pp. 141–174. Springer–Verlag.

Figueira, Santiago, Miller, Joseph S., and Nies, André (2009). Indifferent sets. *J. Logic Comput.*, **19**(2), 425–443.

Figueira, S., Nies, A., and Stephan, F. (2008). Lowness properties and approximations of the jump. *Ann. Pure Appl. Logic*, **152**, 51–66.

Figueira, S., Stephan, F., and Wu, G. (2006). Randomness and universal machines. *J. Complexity*, **22**(6), 738–751.

Franklin, J. and Stephan, F. (2010). Schnorr trivial sets and truth-table reducibility. *J. Symbolic Logic*, **75**(2), 501–521.

Friedberg, R. M. (1957*a*). A criterion for completeness of degrees of unsolvability. *J. Symbolic Logic*, **22**, 159–160.

Friedberg, R. M. (1957*b*). Two recursively enumerable sets of incomparable degrees of unsolvability (solution to post's problem, 1944). *Proc. Nat. Acad. Sci. U.S.A.*, **43**, 236–238.

Friedberg, R. M. and Rogers, Jr., H. (1959). Reducibility and completeness for sets of integers. *Z. Math. Logik Grundlag. Math.*, **5**, 117–125.

Gács, P. (1974). The symmetry of algorithmic information. *Dokl. Akad. Nauk SSSR*, **218**, 1265–1267.

Gács, P. (1986). Every sequence is reducible to a random one. *Inform. and Control*, **70**, 186–192.

Gödel, K. (1931). Über formal unentscheidbare Sätze der Principia Mathematica und verwandter Systeme I. *Monatsh. Math. Phys.*, **38**, 349–360.

Hájek, P. and Kučera, A. (1989). On recursion theory in $I\Sigma_1$. *J. Symbolic Logic*, **54**(2), 576–589.

Harrington, L. and Soare, R. (1991). Post's program and incomplete recursively enumerable sets. *Proc. Nat. Acad. Sci. U.S.A.*, **88**, 10242–10246.

Hirschfeldt, D., Nies, A., and Stephan, F. (2007). Using random sets as oracles. *J. Lond. Math. Soc. (2)*, **75**(3), 610–622.

Hjorth, G. and Nies, A. (2007). Randomness via effective descriptive set theory. *J. London Math. Soc.*, **75**(2), 495–508.

Hölzl, R, Kräling, T., Stephan, S., and Wu, G. (20xx). Initial segment complexities of randomness notions. To appear.

Ishmukhametov, S. (1999). Weak recursive degrees and a problem of Spector. In *Recursion theory and complexity (Kazan, 1997)*, Volume 2 of *de Gruyter Ser. Log. Appl.*, pp. 81–87. de Gruyter, Berlin.

Jech, T. (2003). *Set theory.* Springer Monographs in Mathematics. Springer-Verlag, Berlin. The third millennium edition, revised and expanded.

Jockusch, Jr., C. (1977). Simple proofs of some theorems on high degrees of unsolvability. *Canad. J. Math.*, **29**(5), 1072–1080.

Jockusch, Jr., C. (1989). Degrees of functions with no fixed points. In *Logic, methodology and philosophy of science, VIII (Moscow, 1987)*, Volume 126 of *Stud. Logic Found. Math.*, pp. 191–201. North-Holland, Amsterdam.

Jockusch, Jr., C., Lerman, M., Soare, R., and Solovay, R. (1989). Recursively enumerable sets modulo iterated jumps and extensions of Arslanov's completeness criterion. *J. Symbolic Logic*, **54**(4), 1288–1323.

Jockusch, Jr., C. (1969). Relationships between reducibilities. *Trans. Amer. Math. Soc.*, **142**, 229–237.

Jockusch, Jr., C. (1980). Degrees of generic sets. In *Recursion Theory: Its Generalizations and Applications, Proceedings of Logic Colloquium '79, Leeds, August 1979* (ed. F. R. Drake and S. S. Wainer), Cambridge, U. K., pp. 110–139. Cambridge University Press.

Jockusch, Jr., C. and Shore, R. (1984). Pseudo-jump operators II: Transfinite iterations, hierarchies, and minimal covers. *J. Symbolic Logic*, **49**, 1205–1236.

Jockusch, Jr., C. and Soare, R. (1972*a*). Degrees of members of Π^0_1 classes. *Pacific J. Math.*, **40**, 605–616.

Jockusch, Jr., C. and Soare, R. (1972*b*). Π^0_1 classes and degrees of theories. *Trans. Amer. Math. Soc.*, **173**, 33–56.

Kastermans, B. and Lempp, S. (2010). Comparing notions of randomness. *Theoretical Computer Science*, **411**(3), 602 – 616.

Kaye, R. (1991). *Models of Peano arithmetic*, Volume 15 of *Oxford Logic Guides*. Oxford University Press, New York.

Kechris, A. (1975). The theory of countable analytical sets. *Trans. Amer. Math. Soc.*, **202**, 259–297.

Kechris, Alexander S. (1995). *Classical Descriptive Set Theory*, Volume 156 of *Graduate Texts in Mathematics*. Springer–Verlag, Heidelberg.

Kjos-Hanssen, B. (2004). Classes of computational complexity: a diagram.

Kjos-Hanssen, B. (2007). Low for random reals and positive-measure domination. *Proc. Amer. Math. Soc.*, **135**(11), 3703–3709.

Kjos-Hanssen, B., Merkle, W., and Stephan, F. (2006). Kolmogorov complexity and the Recursion Theorem. In *STACS 2006*, Volume 3884 of *Lecture Notes in Comput. Sci.*, pp. 149–161. Springer, Berlin.

Kjos-Hanssen, B., Merkle, W., and Stephan, F. (2011). Kolmogorov complexity and the Recursion Theorem. To appear in Trans. of the AMS.

Kjos-Hanssen, B., Miller, J., and Solomon, R. (2011). Lowness notions, measure, and domination. *J. London Math. Soc. (2)*, **84**.

Kjos-Hanssen, B. and Nies, A. (2009). Superhighness. *Notre Dame J. Form. Log.*, **50**(4), 445–452 (2010).

Kjos-Hanssen, B., Nies, A. and Stephan, F. (2005). Lowness for the class of Schnorr random sets. *SIAM J. Computing*, **35**(3), 647–657.

Kleene, S. C. (1938). On notations for ordinal numbers. *J. Symbolic Logic*, **3**, 150–155.

Kleene, S. and Post, E. (1954). The upper semi-lattice of degrees of recursive unsolvability. *Ann. of Math. (2)*, **59**, 379–407.

Knuth, D. (1992). Two notes on notation. *Amer. Math. Monthly*, **99**(5), 403–422.

Kolmogorov, A. N. (1963). On tables of random numbers. *Sankhyā Ser. A*, **25**, 369–376. Reprinted in Theoret. Comput. Sci. 207(2) (1998) 387-395.

Kolmogorov, A. N. (1965). Three approaches to the definition of the concept "quantity of information". *Problemy Peredači Informacii*, **1**(vyp. 1), 3–11.

Kraft, L. (1949). *A device for quantizing grouping and coding amplitude modulated pulses*. MS Thesis, MIT, Cambridge, Mass.

Kučera, A. (1985). Measure, Π^0_1-classes and complete extensions of PA. In *Recursion theory week (Oberwolfach, 1984)*, Volume 1141 of *Lecture Notes in Math.*, pp. 245–259. Springer, Berlin.

Kučera, A. (1986). An alternative, priority-free, solution to Post's problem. In *Mathematical foundations of computer science, 1986 (Bratislava, 1986)*, Volume 233 of *Lecture Notes in Comput. Sci.*, pp. 493–500. Springer, Berlin.

Kučera, A. (1988). On the role of $\mathbf{0}'$ in recursion theory. In *Logic colloquium '86 (Hull, 1986)*, Volume 124 of *Stud. Logic Found. Math.*, pp. 133–141. North-Holland, Amsterdam.

Kučera, A. (1989). On the use of diagonally nonrecursive functions. In *Logic Colloquium '87 (Granada, 1987)*, Volume 129 of *Stud. Logic Found. Math.*, pp. 219–239. North-Holland, Amsterdam.

Kučera, A. (1990). Randomness and generalizations of fixed point free functions. In *Recursion theory week (Oberwolfach, 1989)*, Volume 1432 of *Lecture Notes in Math.*, pp. 245–254. Springer, Berlin.

Kučera, A. (1993). On relative randomness. *Ann. Pure Appl. Logic*, **63**, 61–67.

Kučera, A. and Slaman, T. (2001). Randomness and recursive enumerability. *SIAM J. Comput.*, **31**(1), 199–211.

Kučera, A. and Slaman, T. (2007). Turing incomparability in Scott sets. *Proc. Amer. Math. Soc.*, **135**(11), 3723–3731.

Kučera, A. and Slaman, T. (20xx). Low upper bounds of ideals. To appear.

Kučera, A. and Terwijn, S. (1999). Lowness for the class of random sets. *J. Symbolic Logic*, **64**, 1396–1402.

Kummer, M. (1996). On the complexity of random strings (extended abstract). In *STACS '96: Proceedings of the 13th Annual Symposium on Theoretical Aspects of Computer Science*, London, UK, pp. 25–36. Springer-Verlag.

Kunen, K. (1980). *Set theory: An introduction to independence proofs*, Volume 102 of *Studies in Logic and the Foundations of Mathematics*. North-Holland Publishing Co., Amsterdam.

Kurtz, S. (1981). *Randomness and genericity in the degrees of unsolvability*. Ph.D. Dissertation, University of Illinois, Urbana.

Kurtz, S. (1983, September). Notions of weak genericity. *J. Symbolic Logic*, **48**, 764–770.

Lachlan, Alistair H. (1966). Lower bounds for pairs of recursively enumerable degrees. *Proc. London Math. Soc. (3)*, **16**, 537–569.

Lachlan, A. H. (1968). Degrees of recursively enumerable sets which have no maximal supersets. *J. Symbolic Logic*, **33**, 431–443.

Lerman, M. (1983). *Degrees of Unsolvability*. Perspectives in Mathematical Logic. Springer–Verlag, Heidelberg. 307 pages.

Levin, L. A. (1973). The concept of a random sequence. *Dokl. Akad. Nauk SSSR*, **212**, 548–550.

Levin, L. A. (1976). The various measures of the complexity of finite objects (an axiomatic description). *Dokl. Akad. Nauk SSSR*, **227**(4), 804–807.

Levin, L. A. and Zvonkin, A. K. (1970). The complexity of finite objects and the basing of the concepts of information and randomness on the theory of algorithms. *Uspehi Mat. Nauk*, **25**(6:156), 85–127.

Lewis, A., Montalban, A., and Nies, A. (2007). A weakly 2-random set that is not generalized low. In *CiE 2007*, pp. 474–477.

Li, M. and Vitányi, P. (1997). *An introduction to Kolmogorov complexity and its applications* (Second edn). Graduate Texts in Computer Science. Springer-Verlag, New York.

Loveland, D. (1966). A new interpretation of the von Mises' concept of random sequence. *Z. Math. Logik Grundlagen Math.*, **12**, 279–294.

Loveland, D. (1969). A variant of the Kolmogorov concept of complexity. *Information and Control*, **15**, 510–526.

Maass, W. (1982). Recursively enumerable generic sets. *J. Symbolic Logic*, **47**, 809–823.

Marchenkov, S. S. (1976). A class of incomplete sets. *Math. Z.*, **20**, 473–487.

Martin, D. A. (1966*a*). Classes of recursively enumerable sets and degrees of unsolvability. *Z. Math. Logik Grundlag. Math.*, **12**, 295–310.

Martin, Donald A. (1966*b*). Completeness, the recursion theorem, and effectively simple sets. *Proc. Amer. Math. Soc.*, **17**, 838–842.

Martin, D. A. and Miller, W. (1968). The degrees of hyperimmune sets. *Z. Math. Logik Grundlag. Math.*, **14**, 159–166.

Martin-Löf, P. (1966). The definition of random sequences. *Inform. and Control*, **9**, 602–619.

Martin-Löf, P. (1970). On the notion of randomness. In *Intuitionism and Proof Theory (Proc. Conf., Buffalo, N.Y., 1968)*, pp. 73–78. North-Holland, Amsterdam.

Merkle, W. (2003). The complexity of stochastic sequences. In *Conference on Computational Complexity 2003*, Volume 64, pp. 230–235. IEEE Computer Society Press.

Merkle, W., Miller, J., Nies, A., Reimann, J., and Stephan, F. (2006). Kolmogorov-Loveland randomness and stochasticity. *Ann. Pure Appl. Logic*, **138**(1-3), 183–210.

Merkle, W. and Stephan, F. (2007). On C-degrees, H-degrees, and T-degrees. In *Conference on Computational Complexity 2007*, pp. 60–69. IEEE Computer Society Press.

Miller, D. (1981). High recursively enumerable degrees and the anticupping property. In *Logic Year 1979–80 (Proc. Seminars and Conf. Math. Logic, Univ. Connecticut, Storrs, Conn., 1979/80)*, Volume 859 of *Lecture Notes in Math.*, pp. 230–245. Springer, Berlin.

Miller, J. (2004). Every 2-random real is Kolmogorov random. *J. Symbolic Logic*, **69**, 907–913.

Miller, J. (2008). Contrasting plain and prefix-free Kolmogorov complexity. Preprint.

Miller, J. (2009). The K-degrees, low for K-degrees, and weakly low for K sets. *Notre Dame J. Form. Log.*, **50**(4), 381–391 (2010).

Miller, J. and Nies, A. (2006). Randomness and computability: Open questions. *Bull. Symbolic Logic*, **12**(3), 390–410.

Miller, J. and Yu, L. (2008). On initial segment complexity and degrees of randomness. *Trans. Amer. Math. Soc.*, **360**, 3193–3210.

Miller, J. and Yu, Liang (20xx). Oscillation in the initial segment complexity of random reals. To appear in Advances in Mathematics.

Mohrherr, J. (1984). Density of a final segment of the truth-table degrees. *Pacific J. Math.*, **115**(2), 409–419.

Mohrherr, J. (1986). A refinement of low_n and high_n for the r.e. degrees. *Z. Math. Logik Grundlag. Math.*, **32**(1), 5–12.

Moschovakis, Y. (1980). *Descriptive set theory*, Volume 100 of *Studies in Logic and the Foundations of Mathematics*. North-Holland Publishing Co., Amsterdam.

Muchnik, A. A. (1956). On the unsolvability of the problem of reducibility in the theory of algorithms. *Dokl. Akad. Nauk SSSR*, **N. S. 108**, 194–197.

Muchnik, Andrei A., Semenov, A., and Uspensky, V. (1998). Mathematical metaphysics of randomness. *Theoret. Comput. Sci.*, **207**(2), 263–317.

Myhill, John (1955). Creative sets. *Z. Math. Logik Grundlagen Math.*, **1**, 97–108.

Mytilinaios, M. E. (1989). Finite injury and Σ_1-induction. *J. Symbolic Logic*, **54**, 38–49.

Ng, K.M. (2008*a*). On strongly jump traceable reals. *Ann. Pure Appl. Logic*, **154**, 51–69.

Ng, K.M. (2008*b*). On very high degrees. *J. Symbolic Logic*, **73**(1), 309–342.

Ng, K.M. (2010). Beyond strong jump traceability. *Proceedings of the London Mathematical Society*. To appear.

Nies, A. (2002). Reals which compute little. In *Logic Colloquium '02*, Lecture Notes in Logic, pp. 260–274. Springer–Verlag.

Nies, A. (2005*a*). Low for random sets: the story. Preprint, available at `http://www.cs.auckland.ac.nz/nies/papers/`.

Nies, A. (2005*b*). Lowness properties and randomness. *Adv. in Math.*, **197**, 274–305.

Nies, A. (2007). Non-cupping and randomness. *Proc. Amer. Math. Soc.*, **135**(3), 837–844.

Nies, A. (2011). Studying randomness through computation. In: H.Zenil, editor, Randomness through computation, World Scientific, pp. 207-223.

Nies, A., Shore, R., and Slaman, T. (1998). Interpretability and definability in the recursively enumerable Turing-degrees. *Proc. Lond. Math. Soc.*, **3**(77), 241–291.

Nies, A., Stephan, F., and Terwijn, S. (2005). Randomness, relativization and Turing degrees. *J. Symbolic Logic*, **70**(2), 515–535.

Odifreddi, Piergiorgio (1989). *Classical Recursion Theory (Volume I)*. North–Holland Publishing Co., Amsterdam.

Odifreddi, P. G. (1999). *Classical recursion theory. Vol. II*. North-Holland Publishing Co., Amsterdam.

Post, Emil L. (1944). Recursively enumerable sets of positive integers and their decision problems. *Bull. Amer. Math. Soc.*, **50**, 284–316.

Raichev, A. (2005). Relative randomness and real closed fields. *J. Symbolic Logic*, **70**(1), 319–330.

Robinson, R. W. (1971). Interpolation and embedding in the recursively enumerable degrees. *Ann. of Math. (2)*, **93**, 285–314.

Sacks, G.E. (1963*a*). Recursive enumerability and the jump operator. *Trans. Amer. Math. Soc.*, **108**, 223–239.

Sacks, G. E. (1990). *Higher Recursion Theory*. Perspectives in Mathematical Logic. Springer–Verlag, Heidelberg.

Sacks, G. E. (1963*b*). *Degrees of Unsolvability*, Volume 55 of *Annals of Mathematical Studies*. Princeton University Press.

Sacks, G. E. (1963*c*). On the degrees less than $\mathbf{0}'$. *Ann. of Math. (2)*, **77**, 211–231.

Sacks, G. E. (1964). The recursively enumerable degrees are dense. *Ann. of Math. (2)*, **80**, 300–312.

Sacks, G. E. (1969). Measure-theoretic uniformity in recursion theory and set theory. *Trans. Amer. Math. Soc.*, **42**, 381–420.

Schnorr, C.P. (1971). *Zufälligkeit und Wahrscheinlichkeit. Eine algorithmische Begründung der Wahrscheinlichkeitstheorie*. Springer-Verlag, Berlin. Lecture Notes in Mathematics, Vol. 218.

Schnorr, C.P. (1973). Process complexity and effective random tests. *J. Comput. System Sci.*, **7**, 376–388. Fourth Annual ACM Symposium on the Theory of Computing (Denver, Colo., 1972).

Shiryayev, A. N. (1984). *Probability*, Volume 95 of *Graduate Texts in Mathematics*. Springer-Verlag, New York. Translated from the Russian by R. P. Boas.

Shoenfield, Joseph R. (1959). On degrees of unsolvability. *Ann. of Math. (2)*, **69**, 644–653.

Shore, R. (2007). Local definitions in degree structures: the Turing jump, hyperdegrees and beyond. *Bull. Symbolic Logic*, **13**(2), 226–239.

Shore, R. and Slaman, T. (1999). Defining the Turing jump. *Math. Res. Lett.*, **6**(5-6), 711–722.

Shore, R. (1981). The theory of the degrees below $0'$. *J. London Math. Soc.*, **24**, 1–14.

Simpson, S. (2007). Almost everywhere domination and superhighness. *Math. Log. Quart.*, **53**(4-5), 462–482.

Simpson, S. and Cole, J (20xx). Mass problems and hyperarithmeticity. To appear in J. Math. Logic.

Slaman, T. (2005). Aspects of the Turing jump. In *Logic Colloquium 2000*, Volume 19 of *Lect. Notes Log.*, pp. 365–382. Assoc. Symbol. Logic, Urbana, IL.

Soare, Robert I. (1987). *Recursively Enumerable Sets and Degrees.* Perspectives in Mathematical Logic, Omega Series. Springer–Verlag, Heidelberg.

Solomonoff, R. J. (1964). A formal theory of inductive inference. I. *Information and Control*, **7**, 1–22.

Solovay, R. (1975). Handwritten manuscript related to Chaitin's work. IBM Thomas J. Watson Research Center, Yorktown Heights, NY, 215 pages.

Spector, C. (1955). Recursive well-orderings. *J. Symb. Logic*, **20**, 151–163.

Stephan, F. (2006). Marin-Löf random and PA-complete sets. In *Logic Colloquium '02*, Volume 27 of *Lect. Notes Log.*, pp. 342–348. Assoc. Symbol. Logic, La Jolla, CA.

Stephan, F. and Yu, L. (2006). Lowness for weakly 1-generic and Kurtz-random. In *Theory and applications of models of computation*, Volume 3959 of *Lecture Notes in Comput. Sci.*, pp. 756–764. Springer, Berlin.

Terwijn, S. and Zambella, D. (2001). Algorithmic randomness and lowness. *J. Symbolic Logic*, **66**, 1199–1205.

Trahtenbrot, B. A. (1970). Autoreducibility. *Dokl. Akad. Nauk SSSR*, **192**, 1224–1227.

Turing, A. M. (1936). On computable numbers with an application to the Entscheidungsproblem. *Proc. London Math. Soc. (3)*, **42**, 230–265. A correction, 43:544–546.

van Lambalgen, M. (1987). *Random Sequences.* University of Amsterdam.

von Mises, R. (1919). Grundlagen der Wahrscheinlichkeitsrechnung. *Math. Zeitschrift*, **5**, 52–99.

Yates, C. E. M. (1966). A minimal pair of recursively enumerable degrees. *J. Symbolic Logic*, **31**, 159–168.

Zambella, D. (1990). On sequences with simple initial segments. ILLC technical report ML 1990-05, Univ. Amsterdam.

NOTATION INDEX

General

The absolute value of a number $r \in \mathbb{R}$ is denoted by $\mathrm{abs}(r)$.
The cardinality of a set X is denoted by $\#X$.
See the preface, page viii, for conventions on variables.

Notation	Meaning
$\forall^\infty$	for almost every x
$\exists^\infty$	for infinitely many x
$\alpha \simeq \beta$	both expressions α, β are undefined, or both are defined with the same value 4
$\varnothing$	the empty string
$\emptyset$	the empty set
$\#X$	cardinality of the set X
$X \triangle Y$	$(X-Y)\cup(Y-X)$, the elements on which sets X and Y disagree 70
$X =^* Y$	$(X-Y)\cup(Y-X)$ is finite 70
$\mathrm{dom}\,\psi$ or $\mathrm{dom}(\psi)$	domain of a function ψ
$\mathrm{ran}\,\psi$ or $\mathrm{ran}(\psi)$	range of ψ
$\mathbb{N}$	set of natural numbers $0, 1, 2 \ldots$
$\mathbb{R}$	set of real numbers
$\mathbb{R}_0^+$	set of non-negative real numbers 68
$\mathrm{abs}(r)$	absolute value of $r \in \mathbb{R}$
$\mathbb{Q}_2$	$\{z2^{-n} \colon z \in \mathbb{Z}, n \in \mathbb{N}\}$, set of dyadic rationals 50
$\langle m, n\rangle$	$m + (m+n)(m+n+1)/2$, number coding the ordered pair of m and n
$(\langle m, n\rangle)_0$	m, the first component of such a pair
$(\langle m, n\rangle)_1$	n, the second component of such a pair
$\langle n_0, \ldots, n_k\rangle$	$\langle \ldots \langle n_0, n_1\rangle, n_2\rangle, \ldots, n_k\rangle$
$X \times Y$	$\{\langle m, n\rangle \colon m \in X \;\&\; n \in Y\}$, cartesian product of $X, Y \subseteq \mathbb{N}$
$\mathbb{N}^{[i]}$	$\mathbb{N} \times \{i\} = \{\langle m, i\rangle \colon m \in \mathbb{N}\}$
$A^{[i]}$	$A \cap \mathbb{N}^{[i]}$
$\sum_n \tau(n)\,[\![\theta(n)]\!]$	sum of terms $\tau(n)$ over all n such that condition $\theta(n)$ holds, as for instance in $\infty = \sum_n 1/n\,[\![n \text{ is prime}]\!]$. Note that the value of $\tau(n)$ can be repeated, which is the advantage of this notation over $\sum\{\tau(n) \colon \theta(n)\}$.

Chapter 1

Notation	Meaning
P_e^k	e-th Turing program with k inputs 3
Φ_e^k	partial computable function or functional given by P_e^k 3, 10
Φ_e	short for Φ_e^1 3
W_e	$\mathrm{dom}(\Phi_e)$, the e-th c.e. set 6
$\emptyset'$	$\{e \colon e \in W_e\}$, the halting problem 6
$\Phi_{e,s}(x) = y$	P_e on input x yields y in at most s steps 7
$W_{e,s}$	$\mathrm{dom}(\Phi_{e,s})$ 7

Chapter 2

Chapter 3

Chapter 4

Chapter 5

Chapter 6

$\gamma^Y(x)$	use $\Gamma^Y(x)$, for a Turing functional Γ 240
use $(e \in W^C)$	use $\Phi^C(e)$, where the functional Φ defines the c.e. operator W 248

Chapter 7

$S(Z)$	$\sup_n S(Z\upharpoonright_n)$, for a supermartingale S 260, 262
$\mathsf{Succ}(S)$	$\{Z\colon S(Z)=\infty\}$, the success class of S. 260, 262
$B_{\mathcal{C}}(x)$	$\lambda(\mathcal{C} \mid x)$, the martingale associated with a measurable class $\mathcal{C}$ 261
FS	$\sum_y 2^{-K(y)} E_y$, a universal c.e. martingale 266
CR	class of computably random sets 268
PCR	class of partial computably random sets 273
B_k	a martingale copying Φ_k as long as possible 273
TMG	$\{k\colon B_k \text{ total}\}$ 274
$B_{k,n}$	a supermartingale copying a scaled version of B_k from length n on 274
G_n	the martingale such that $G_n(x)=1$ for $\lvert x\rvert<n$ and $G_n(x)=0$ else 274
I_k	$[n_k, n_{k+1})$ in the context of the sequence $n_0 < n_1 < \ldots$ 275
$\mathsf{Int}(z)$	the number k such that $\lvert z\rvert \in I_k$ 275
B_k^*	$B_{k,n_{k+1}}$ 275
$y^L_{x,k+1,0}, y^L_{x,k+1,1}$	extensions of string x of length $\lvert x\rvert+k+1$ defined in (7.12) 282
$Z\upharpoonright I$	the string $Z(n)\ldots Z(m-1)$, where $I=[n,m)$ 290
FinA	set of finite assigments $\alpha = (\langle d_0, r_0\rangle, \ldots, \langle d_{n-1}, r_{n-1}\rangle)$ 297
$\alpha \lhd Y$	$Y(d_i)=r_i$ for each $i<n$ 297
$\lambda_v\mathcal{C}$	short form of $\lambda(\mathcal{C} \mid v)$ defined in Chapter 1 323
$\mathcal{H}^\diamond$	$\{A\colon A \text{ c.e. } \& \ \forall Y \in \mathcal{H} \cap \mathsf{MLR}\,[A \le_T Y]\}$ for a class $\mathcal{H}$ 356

Chapter 9

$X_1 \oplus \ldots \oplus X_m$	$\{mz+i\colon i<m \ \& \ z \in X_i\}$ 366
$\lvert R\rvert$	ordinal α such that $(\alpha,<)$ is isomorphic to R 367
ω_1^Y	least ordinal not computable in a set Y 367
ω_1^{ck}	least noncomputable ordinal, $\omega_1^\emptyset$ 367
$\mathcal{B}_\alpha$	the elements of a Π^1_1 relation $\mathcal{B}$ that enter before stage α 368
O	$\{e\colon W_e \text{ encodes a well-order}\}$ 369
$\mathcal{Q}_p$	the Π^1_1 class with index p 368
$f \le_h A$	$f \in \Delta^1_1(A)$, f is hyperarithmetical in A 370
$\le_{\mathit{fin\text{-}h}}$	finite hyperarithmetical reducibility 370
$L(\alpha)$	level α in the constructible hierarchy 371
L_A	$L(\omega_1^A, A)$ 371
$\underline{\mathbb{U}}$	optimal prefix-free Π^1_1-machine 373
$\underline{K}(y)$	$\min\{\lvert\sigma\rvert\colon \underline{\mathbb{U}}(\sigma)=y\}$ 373
$\underline{K}_\alpha(y)$	$\min\{\lvert\sigma\rvert\colon \underline{\mathbb{U}}_\alpha(\sigma)=y\}$ 373
$\underline{\mathsf{MLR}}$	class of Π^1_1-ML-random sets 376
$\underline{\Omega}$	$\sum_\sigma 2^{-\lvert\sigma\rvert} [\![\underline{\mathbb{U}}(\sigma)\downarrow]\!]$, the halting probability of $\underline{\mathbb{U}}$ 376
$\le_{\mathit{wtt\text{-}h}}$	$\le_{\mathit{fin\text{-}h}}$ with the use function computably bounded 376

INDEX

The manufacturer's authorised representative in the EU for product safety is Oxford University Press España S.A. of El Parque Empresarial San Fernando de Henares, Avenida de Castilla, 2 - 28830 Madrid (www.oup.es/en or product.safety@oup.com). OUP España S.A. also acts as importer into Spain of products made by the manufacturer.

Printed and bound by CPI Group (UK) Ltd, Croydon, CR0 4YY

22/04/2026

02094914-0014